Mechanical
Engineering Design

McGraw-Hill Series in Mechanical Engineering

Jack P. Holman, Southern Methodist University Consulting Editor

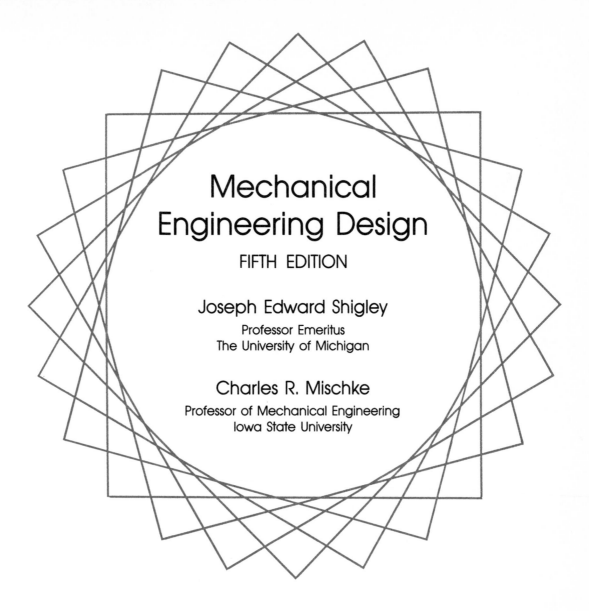

Mechanical Engineering Design

FIFTH EDITION

Joseph Edward Shigley
Professor Emeritus
The University of Michigan

Charles R. Mischke
Professor of Mechanical Engineering
Iowa State University

McGraw-Hill, Inc.
New York St. Louis San Francisco Auckland Bogotá
Caracas Lisbon London Madrid Mexico City Milan
Montreal New Delhi San Juan Singapore
Sydney Tokyo Toronto

Mechanical Engineering Design

90 AGM AGM 9987654

ISBN 0-07-056899-5

This book was set in Times Roman by York Graphic Services, Inc.
The editors were John J. Corrigan and James W. Bradley; the designer was
Elliot Epstein; the production supervisor was Leroy A. Young.
Drawings were done by J & R Services, Inc.
Arcata Graphics/Martinsburg was printer and binder.

LIBRARY OF CONGRESS
Library of Congress Cataloging-in-Publication Data

Shigley, Joseph Edward.
   Mechanical engineering design/Joseph Edward Shigley, Charles R.
Mischke. —5th ed.
      p. cm. —(McGraw-Hill series in mechanical engineering)
   Bibliography: p.
   Includes index.
   ISBN 0-07-056899-5
   1. Machinery—Design.   I. Mischke, Charles R.   II. Title.
III. Series.
TJ230.S5 1989   88-12698
621.8'15—dc19

This book is printed on acid-free paper.

About the Authors

Charles R. Mischke has held positions on the faculty of the University of Kansas and was professor and chairman of Mechanical Engineering at Pratt Institute in New York. He is currently a professor at Iowa State University, involved in research activities and he works as a consultant to industry. He received the Ralph Teeter Award of the Society of Automotive Engineers in 1977 and Iowa State's Outstanding Teacher Award in 1980. He is the author of "Elements of Mechanical Analysis," "Introduction to Computer-Aided Design," "Mathematical Model Building," as well as many technical papers. He serves on the Reliability, Stress Analysis, and Failure Prevention Executive Committee of the American Society of Mechanical Engineers. Mischke is also a Fellow of the society. He is co-editor of the new Standard Handbook of Machine Design (McGraw-Hill). His B.S.M.E. and M.M.E. are from Cornell University, and the Ph.D. is from the University of Wisconsin.

Joseph E. Shigley is the well-known co-editor of the *Standard Handbook of Machine Design*. Shigley, Professor Emeritus, the University of Michigan, is a Fellow of the American Society of Mechanical Engineers. He received the ASME Mechanisms Committee Award in 1974, the Worcester Reed Warner Medal in 1977, and the Machine Design Award in 1985. He is the author or coauthor of eight books, including "Mechanical Engineering Design," "Theory of Machines and Mechanisms" (with J. J. Uicker, Jr.), and "Applied Mechanics of Materials" (all McGraw-Hill). He received his B.S.E.E. and B.S.M.E. from Purdue University and his M.S. from the University of Michigan.

To
Opal
and
Margaret

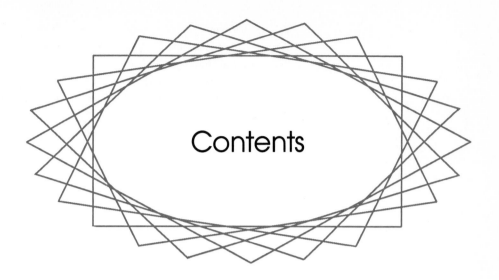

Contents

*Italic section numbers denote optional-reading sections.

3
Deflection and Stiffness
91

4
Statistical Considerations
145

PART TWO
FAILURE PREVENTION

5
Materials
185

6
Steady Loading
223

7
Variable Loading
269

PART THREE
DESIGN OF MECHANICAL ELEMENTS

8
The Design of Screws, Fasteners, and Connections
325

9
Welded, Brazed, and Bonded Joints
383

10
Mechanical Springs
413

11
Rolling Contact Bearings
451

12
Lubrication and Journal Bearings
479

17
Flexible Mechanical Elements
665

18
Shafts, Axles, and Spindles
697

Appendix
725

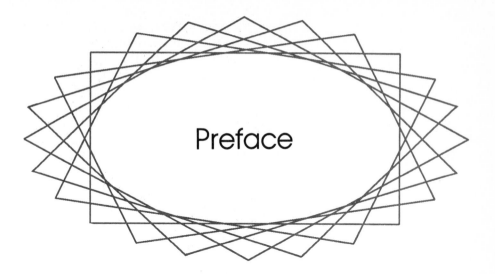

Preface

This book has been written for engineering students who are beginning a course of study in mechanical engineering design. Such students will have acquired a set of engineering tools consisting, essentially, of mathematics, computer languages, and the ability to use the English language to express themselves in the spoken and written forms. Mechanical design involves a great deal of geometry, too; therefore, another useful tool is the ability to sketch and draw the various configurations which arise. Students will also have studied a number of basic engineering sciences, including physics, engineering mechanics, materials and processes, and the thermal-fluid sciences. These, the tools and sciences, constitute the foundation for the practice of engineering, and so, at this stage of undergraduate education, it is appropriate to introduce some professional aspects of engineering. These professional studies should integrate and use the tools and the sciences in the accomplishment of an engineering objective. The pressures upon the undergraduate curricula today require that we do this in the most efficient manner. Most engineering educators are agreed that mechanical design integrates and utilizes a greater number of the tools and the sciences than any other professional study. Mechanical design is also the very core of other professional and design types of studies in mechanical engineering. Thus studies in mechanical design seem to be the most effective method of starting the student in the practice of mechanical engineering.

One of the reasons for writing a new edition now is the recent increased emphasis on the creative aspects of design in so many colleges and universities. In the early 1950s a committee on evaluation of engineering education of the American Society for Engineering Education stated:

> Training for the creative and practical phases of economic design, involving analysis, synthesis, development and engineering research, is the most distinctive feature of professional engineering education.

> The technical goal of engineering education is preparation for performance of the functions of analysis and design or of the functions of construction, production or operation with full knowledge of analysis and design of the structure, machine, or process involved.

Though these goals were stated over a generation ago, they are valid today. Ways must be found to involve the engineering student in genuine design experiences.

The approach in this book is to suggest and to present short design problems or situations to illustrate the decision-making processes of design without demanding an inordinate amount of the student's precious time. A proposal to include a chapter or two devoted to realistic design projects was abandoned. Investigation revealed that such projects were rarely used more than once by instructors and hence the space could not be justified. Good, short design projects are certainly needed in the professional design studies. These are most effective when they are created from the instructor's own professional background and presented with the enthusiasm and thoroughness which this background allows. With such an approach new and updated projects can always be devised to meet current trends and ideas.

There are many additional reasons for publishing a new edition. The major factors for this one are:

* Product liability

* Reliability

* Quality control

* Personal computers

* New problems

Designing to a reliability specification, or to exceed specified minimums, and to a quality-control goal, will help to assure a design free from liability problems. At the request of some users, we have taken a dual approach to dimensional design. This combines the classical *factor-of-safety approach* and an optional *reliability* or *stochastic method*. These methods are presented in parallel with each other, but the stochastic method has been clearly designated for optional reading (headings for these sections are printed in black rather than color). The instructional time available, or a lack of student preparation, could require that only the factor-of-safety method be presented. Nevertheless the material used in the reliability approach will give the engineer much of the background needed to apply quality-control procedures in professional practice.

In developing the dual approach to design, we found it necessary to define the factor-of-safety method much more carefully and exactly than ever before. It is now a very sound well-documented approach. For this reason the factor-of-safety method, as presented here, also contributes to the first three goals detailed above.

Access to a personal computer or to a programmable calculator is now very important in mechanical-design studies. Their use makes it possible to solve many problems using numerical methods and eases much repetitive computation. Specific programs are not included in this book because of user's requests, and because most people prefer to create their own. The analyses presented together with programming suggestions should facilitate the use of computers and programmable calculators.

Many of the users of past editions of this book have expressed a need for new problems for student homework. Most of the problems in this edition are new; only a few were kept from previous editions and most of those have been revised.

The book now is in three parts. Part 1 is basic, and includes the introduction as well as definitions, stress analysis, deflection and stiffness determination, and the statistical tools needed for reliability and quality-control analyses. At some schools the students will be somewhat prepared in some of these topics, but only at some schools. But this basic material must still be presented in a comprehensive manner if only to serve basic reference purposes and to organize the terminology and symbolism.

Part 2 is on failure prevention, and it makes use of and integrates the fundamentals of Part 1 toward the goal of analyzing and designing mechanical elements to achieve satisfactory levels of safety, quality, reliability, and life.

Part 3 deals with the design and analysis of specific mechanical elements such as gears, clutches, springs, bearings, and the like. The new material in Parts 1 and 2 has greatly enhanced the development and presentation of the analyses presented in Part 3 and gives the instructor wider opportunities to augment, enrich, and address particular goals outside the scope of this book.

Chapter 1 contains an important introduction to the dual approach and a new section on preferred units.

In Chapter 2, new material on principal shear stresses and octahedral shear stresses has been added. Users have also requested more on equilibrium and free-body diagramming; these topics as well as all new problems have been added. The subject of stress concentration has been relocated and is now in Chapter 2. A new section on stresses in rings has been included.

In Chapter 3, on deflection and stiffness, additions and changes include (1) computer methods using numerical integration; (2) shock and impact and development of formulas using piecewise differential equations; (3) more on Castigliano's theorem, including indeterminate problems, rings, and other curved members; and (4) complete revision and rearrangement of the material on columns with new material on short compression members.

Chapter 4, on statistics, is completely new and replaces the old Chapter 5. The chapter now contains the background material needed to apply quality-control requirements and for the reliability analyses used in many of the chapters that follow. This chapter has been developed very carefully so as to contain the minimum amount of material necessary to develop these objectives, with particular attention paid to how the material is used in the chapters that follow. Some of the sections have been marked for optional reading for those who elect not to study the entire chapter.

Chapter 5, on materials, is a revision of the old Chapter 4. The changes include:

* New material on strength, cold work, and hardness, including stochastic properties

* Expanded sections on metal processing

* New sections on notch sensitivity and fracture toughness

* Completely new problems for home solution

Chapter 6 is the old Chapter 6 expanded to contain the dual factor-of-safety and reliability approach. The changes include a rewriting of the failure theories and the addition of other theories. The second part of the chapter is all new and contains the stochastic approach. This includes general interference theory, design and analysis, tolerance setting, and a numerical approach for mixed statistical distributions.

Chapter 7, on variable loading, or fatigue, has been substantially revised, though it is based on the previous outline. Because of the inclusion of the reliability method, it was necessary to research the endurance-limit modifying factors quite thoroughly. As a result these factors have been re-evaluated and, in our opinion, now express the available knowledge much more fairly than in the past. This is important whether the deterministic (factor-of-safety) method or the stochastic (reliability) approach is used. In accomplishing this objective some juggling of the various factors was done. The result has cleared some past confusion and produced a more logical and useful approach. The new sections dealing with fatigue strength as a random variable are presented as optional reading.

In addition to the complete revision of the Marin k factors, other new material in Chapter 7 includes:

* Fatigue strength treated as a random variable
* Reliability analyses
* More on load lines versus factor of safety
* Torsional fatigue under pulsating stresses
* Combined loading modes
* The fracture-mechanics approach to fatigue

Chapter 8 begins Part 3 with a presentation on screws, fasteners, and connections. The presentation of bolted joints in the previous edition provoked much discussion and analysis. Advantage was taken of the many user's comments, and a completely new presentation of the subject is a feature of this edition. All the methods of analyzing factor of safety of a preloaded bolted joint are presented together with their significance. The material on joint compression has been expanded and new material added on the use of cap screws, set screws, and keys.

The use of statistics in Chapters 9 and 10 make it possible, for example, to determine how tolerancing affects the performance of the element or design under study. The inclusion of such material in this edition adds substantially to the treatment of springs, for example, where it is shown how tolerancing of wire diameter, coil diameter, and free length of a helical spring affects the spring rate and deflection of the springs produced. The use of statistics in spring design is thus a valuable quality-control tool.

An important feature of this edition is the new arrangement of the material on gearing. Chapter 13 is now devoted to the basics of gearing. This includes fundamental concepts, descriptions and terminology of gears in general, gear trains, and force analysis. Chapter 14 deals with spur and helical gears and explains and uses the AGMA methods of analysis and design. This chapter is important because it explains the theory and basic thinking upon which the AGMA approach is established. Chapter 15 continues with AGMA methods applied to bevel gears. As a change of pace, however, worm and crossed-helical gears are treated using the British standards.

Chapter 17, dealing with belts, roller chain, and wire rope, has been substantially revised and, we think, improved.

Chapter 18, on the design and analysis of shafting, is now the final chapter. It occupies this important position in order to make available all the material of the previous chapters in its development. This is a completely new chapter, and it treats the design of shafting from the initial concept to the final detail drawing. The subject of reliability of shafting is treated thoroughly together with a detailed illustrative example. This chapter also contains all new home problems together with some interesting short design projects.

Improvements in the appendix include new tables needed for stochastic analysis, which include a new table of values for the normal distribution and one on the gamma function. The subject of limits and fits is now treated using the more logical International System. And the former tables on threaded fasteners are replaced with shorter and more useful displays.

This edition has been significantly influenced by the publisher's reviewers. Their comments and suggestions were sound and worthwhile and, indeed, kept us from straying too far from the central theme of the book. They made known the need for certain additional topics and alternate presentations; these ideas were of considerable value to us and most have been included. These reviewers were:

Richard M. Alexander, Texas A & M University

Eugene J. Fisher, University of Santa Clara

Geza Kardos, Carleton University

Alois van Eyken, Queen's University

John L. Mathieson, State University of New York, Maritime College

Ralph S. Blanchard

J. D. Pfeiffer, McGill University

We are very grateful to these people for their attention and sound advice.

We have also received much help and many valuable suggestions from the users of the previous edition. We thank them very much and have listed their names under the acknowledgements.

We are both grateful to Iowa State University for granting a sabbatical leave to Professor Mischke which permitted extended visitation with the user-community both on campus and in industry. This helped frame the scope and treatment of the material in this edition. We acknowledge the contribution that this opportunity made to the usefulness of this edition.

Our editors were John J. Corrigan, successor to Anne C. Duffy, and James W. Bradley. The copy reader was Richard K. Mickey. There is a great deal more involved in the publication of a book than that of printing a typed manuscript and rendering the drawings executed by the writers. The editors and their backup people take the raw manuscript and subject it to a very detailed procedure, leading, eventually, to the finished product. We think they have done a great job with our book. We hope you do, too.

ACKNOWLEDGEMENTS

We are grateful to the following people for their suggestions and ideas:
George G. Adams, Northeastern University, Boston, Massachusetts; Robert W. Adamson, California Polytechnic State University, San Luis Obispo, California; James M. Allman, Northrop University, Inglewood, California; Lola Boyce, University of Texas, San Antonio, Texas; Parviz Dadras, Wright State University, Dayton, Ohio;

Joseph Datsko, The University of Michigan, Ann Arbor, Michigan; Kenneth S. Edwards, University of Texas, El Paso, Texas; A. F. Abdel Azim El-Sayed, Zagazig University, Cairo, Egypt; A. van Eyken, Queen's University, Kingston, Ontario, Canada; Eugene J. Fisher, University of Santa Clara, Santa Clara, California; J. Darrell Gibson, Rose-Hulman Institute, Terre Haute, Indiana; Vladimir Glozman, California Polytechnic State University, San Luis Obispo, California; Hans J. Goettler, North Dakota State University, Fargo, North Dakota; Itzhak Green, Georgia Institute of Technology, Atlanta, Georgia; Louis T. Hayes, Widner University, Chester, Pennsylvania; Peter C. Hills, Royal Military College of Science, United Kingdom; Harold L. Johnson, Georgia Institute of Technology, Atlanta, Georgia; S. F. Johnston, New South Wales Institute of Technology, Australia; William D. Jordan, University of Alabama, University, Alabama; Raymond W. Kaupila, Michigan Technological University, Houghton, Michigan; Boris L. Krayterman, University of Maryland, College Park, Maryland; Shankar Lal, Naval Postgraduate School, Monterey, California; F. C. Liu, University of Alabama, Huntsville, Alabama; Leo R. Maier, Jr., Ohio Northern University, Ada, Ohio; Kurt Marshek, University of Texas, Austin, Texas; Larry D. Mitchell, Virginia Polytechnic Institute and State University, Blacksburg, Virginia; Joseph Motherway, University of Massachusetts, Amherst, Massachusetts; Mohammad Naji, Wichita State University, Wichita, Kansas; J. David Pfeiffer, McGill University, Montreal, Quebec, Canada; George Piatrowski, University of Florida, Gainesville, Florida; Fred Rimrott, University of Toronto, Toronto, Ontario, Canada; Cyril Samónov, Freiburg, West Germany; John Schober, Macmillan B Ltd, Sturgeon Falls, Ontario, Canada; G. D. Sellwood, British Aerospace, Bedfordshire, United Kingdom; Raj S. Sodhi, Wichita State University, Wichita, Kansas; Ajit K. Srivastava, Michigan State University, East Lansing, Michigan; John Steffin, Valparaiso University, Valparaiso, Indiana; J. Wolak, University of Washington, Seattle, Washington; Chia Hsang Wu, Taipei Institute of Technology, Taipei, Taiwan; and Y. C. Yong, California Polytechnic University, San Luis Obispo, California.

Joseph Edward Shigley
Charles R. Mischke

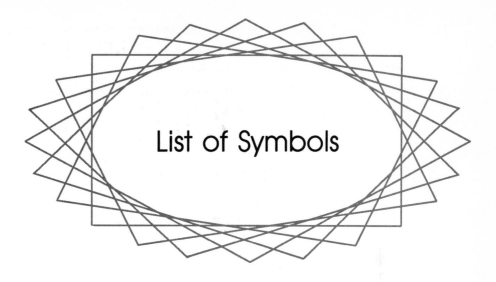
List of Symbols

See Table 14-1, page 586, for gearing symbols

A Area, coefficient

$\mathbf{A}$ Area variate, vector

a Distance

B Coefficient, life

$\mathbf{B}$ Vector

b Distance, Weibull shape parameter

C Basic load rating, bolted-joint constant, center distance, coefficient of variation, column end condition, constant, correction factor, heating coefficient, specific heat, spring index

c Distance

D Helix diameter

E Energy, error quantity, modulus of elasticity

e Distance, eccentricity, efficiency

F Force

$\mathbf{F}$ Force variate, force vector

f Coefficient of friction, frequency

G Modulus of rigidity

g Acceleration due to gravity

H Heat, power

H_B Brinell hardness

h distance, film thickness

I	Integral, mass moment of inertia, second moment of area
$\mathbf{I}$	Variate of I
$\mathbf{i}$	Unit vector in x direction
J	Mechanical equivalent of heat, polar second moment of area
$\mathbf{j}$	Unit vector in y direction
K	Stress-concentration factor, stress-correction factor, torque coefficient
k	Endurance-limit modifying factor
$\mathbf{k}$	Unit vector in z direction
L	Length, life
l	Length
M	Moment
$\mathbf{M}$	Moment vector
m	Mass, slope, strain-strengthening exponent
N	Normal force, number, rotational speed
n	Factor of safety, load factor, rotational speed
P	Force, unit bearing load
p	pitch, pressure, probability
Q	First moment of area, imaginary force, volume
q	Distributed load, notch sensitivity
R	Radius, reaction force, reliability, Rockwell hardness, stress ratio
$\mathbf{R}$	Vector reaction force
r	Correlation coefficient, radius
$\mathbf{r}$	Length vector
S	Sommerfeld number, strength
$\mathbf{S}$	Variate of S
s	Distance, sample standard deviation
T	Temperature, tolerance, torque
$\mathbf{T}$	Torque vector
t	Distance, time
U	Strain energy
u	Unit strain energy
V	Linear velocity, shear force
v	Linear velocity
W	Cold-work factor, load, weight
w	Distance, unit load
X	Coordinate
$\mathbf{x}$	Variate of x
x	Coordinate, Weibull guaranteed parameter

Y	Coordinate
$\mathbf{y}$	Variate of y
y	Coordinate, deflection
Z	Coordinate, section modulus, viscosity
$\mathbf{z}$	Variate of z
z	Coordinate, unit standard deviation
α	Coefficient, coefficient of thermal expansion, end-condition constant (for springs), thread angle
β	Bearing angle, coefficient
δ	Deviation, elongation
ϵ	Eccentricity ratio, engineering unit strain
ε	True or logarithmic strain
Γ	Gamma function
γ	Pitch angle, shear strain, unit weight
λ	Slenderness ratio (for springs)
μ	Absolute viscosity, coefficient of friction, population mean
ν	Poisson's ratio
ω	Angular velocity, circular frequency
ϕ	Angle, wavelength
ψ	Slope integral
ρ	Radius of curvature
σ	Normal stress
$\boldsymbol{\sigma}$	Normal stress variate
$\hat{\sigma}$	Population standard deviation
τ	Shear stress
θ	Angle, Weibull characteristic parameter

Mechanical
Engineering Design

PART ONE

BASICS

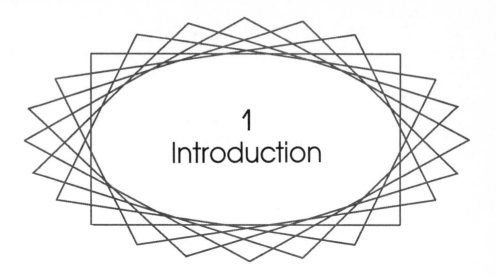

1
Introduction

This book is a study of the decision-making processes with which mechanical engineers formulate plans for the physical realization of machines, devices, and systems. These processes are common to all disciplines in the field of engineering design—not just to mechanical engineering design. But since our subject *is* mechanical engineering design, we will use mechanical engineering as the vehicle for understanding these decision-making processes, and for applying them to practical situations where we can see them pay off.

This book consists of three parts. Part 1 begins by explaining the difference between *design* and *analysis* and introducing some fundamental notions and approaches to design. It then proceeds with a review of stress analysis and an introduction to stiffness and deflection analysis, both of which will be drawn on extensively in the chapters that follow. Part 1 also includes a chapter on statistical methods (Chap. 4). A knowledge of statistical analysis is not really required in order to study the balance of the book. However, you will find some useful ideas in Chap. 4, as well as methods to aid your understanding and to perfect your designs.

Part 2, on failure prevention, consists of one chapter on materials and two chapters on the prevention of failure of mechanical parts. The chapter on materials—Chap. 5—is included so that definitions and analyses relating to mechanical properties used in design can be presented in one place and at the same time be related to a variety of materials. The subject of why machine parts fail and how they can be designed to prevent failure is a long one, and so we take two chapters for the purpose, one on the prevention of failure due to static loads and the other on preventing failure due to fatigue and dynamic loads.

In Part 3 the material of Parts 1 and 2 is applied to the design of specific mechanical elements such as shafts, springs, gears, and brakes, and to many typical design situations which arise in the design or the selection and application of these elements and others.

1-1 THE MEANING OF DESIGN

To design is to formulate a plan for the satisfaction of a human need. The particular need to be satisfied may be quite well defined from the beginning. Here are two examples in which needs are well defined:

1 How can we obtain large quantities of power cleanly, safely, and economically without using fossil fuels and without damaging the surface of the earth?

2 This gearshaft is giving trouble; there have been eight failures in the last six weeks. Do something about it.

On the other hand, the statement of a particular need to be satisfied may be so nebulous and ill defined that a considerable amount of thought and effort is necessary in order to state it clearly as a problem requiring a solution. Here are two examples:

1 Lots of people are killed in airplane accidents.

2 In big cities there are too many automobiles on the streets and highways.

This second type of design situation is characterized by the fact that neither the need nor the problem to be solved has been identified. Note, too, that the situation may contain not one problem but many.

We can classify design, too. For instance, we speak of:

1 Clothing design	7 Bridge design
2 Interior design	8 Computer-aided design
3 Highway design	9 Heating system design
4 Landscape design	10 Machine design
5 Building design	11 Engineering design
6 Ship design	12 Process design

In fact, there are an endless number, since we can classify design according to the particular article or product or according to the professional field.

In contrast to scientific or mathematical problems, design problems have no unique answers; it is absurd, for example, to request the "correct answer" to a design problem, because there is none. In fact, a "good" answer today may well turn out to be a "poor" answer tomorrow, if there is a growth of knowledge during the period or if there are other structural or societal changes.

Almost everyone is involved with design in one way or another, even in daily living, because problems are posed and situations arise which must be solved. Consider the design of a family vacation. There may be seven different places to go, all at different distances from home. The costs of transportation are different for each, and some of the options require overnight stops on the way. The children would like to go to a lake or seashore resort. The wife would prefer to go to a large city with department store shopping, theaters, and nightclubs. The husband prefers a resort with a golf course and perhaps a nearby mountain trout stream. When these needs and desires are related to time and money, various solutions may be found. Of these, there may or may not be one or more solutions. But the solution chosen will include the travel route, the stops, the mode of transportation, and the names and locations of resorts, motels, camping sites, or other away-from-home facilities. It is not hard to see that there are really a rather large group of interrelated complex factors involved in arriving at one of the solutions to the vacation design problem.

A design is always subject to certain problem-solving constraints. For example, two of the constraints on the vacation design problem are the time and money available for the vacation. Note, too, that there are also constraints on the solution. In the case above, some of those constraints are the desires and needs of each of the family members. Finally, the design solution chosen might well be optimal. In this case an optimal solution is obtained when each and every family member can say that he or she has had a good time.

A design problem is *not* a hypothetical problem at all. Design has an authentic purpose—*the creation of an end result by taking definite action, or the creation of something having physical reality*. In engineering, the word *design* conveys different meanings to different persons. Some think of a designer as one who employs the drawing board to draft the details of a gear, clutch, or other machine member. Others think of design as the creation of a complex system, such as a communications network. In some areas of engineering the word *design* has been replaced by other terms such as *systems engineering* or *applied decision theory*. But no matter what words are used to describe the design function, in engineering it is still the process in which scientific principles and the tools of engineering—mathematics, computers, graphics, and English—are used to produce a plan which, when carried out, will satisfy a human need.

1-2 MECHANICAL ENGINEERING DESIGN

Mechanical design means the design of things and systems of a mechanical nature—machines, products, structures, devices, and instruments. For the most part, mechanical design utilizes mathematics, the materials sciences, and the engineering-mechanics sciences.

Mechanical engineering design includes all mechanical design, but it is a broader study, because it includes all the disciplines of mechanical engineering, such as the thermal and fluids sciences, too. Aside from the fundamental sciences that are required, the first studies in mechanical engineering design are in mechanical design, and hence this is the approach taken in this book.

1-3 THE PHASES OF DESIGN

The total design process is of interest to us in this chapter. How does it begin? Does the engineer simply sit down at his or her desk with a blank sheet of paper and jot down some ideas? What happens next? What factors influence or control the decisions which have to be made? Finally, how does this design process end?

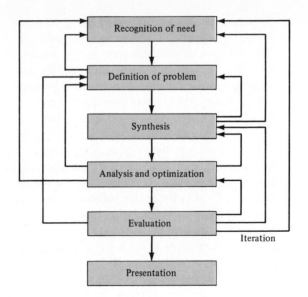

FIGURE 1-1

The phases of design.

The complete process, from start to finish, is often outlined as in Fig. 1-1. The process begins with a recognition of a need and a decision to do something about it. After many iterations, the process ends with the presentation of the plans for satisfying the need. In the next several sections, we shall examine these steps in the design process in detail.

1-4   RECOGNITION AND IDENTIFICATION

Sometimes, but not always, design begins when an engineer recognizes a need and decides to do something about it. *Recognition of the need* and phrasing the need often constitute a highly creative act, because the need may be only a vague discontent, a feeling of uneasiness, or a sensing that something is not right. The need is often not evident at all; recognition is usually triggered by a particular adverse circumstance or a set of random circumstances which arise almost simultaneously. For example, the need to do something about a food-packaging machine may be indicated by the noise level, by the variation in package weight, and by slight but perceptible variations in the quality of the packaging or wrap.

It is evident that a sensitive person, one who is easily disturbed by things, is more likely to recognize a need—and also more likely to do something about it. And for this reason sensitive people are more creative. A need is easily recognized after someone else has stated it. Thus the need in many countries for cleaner air and water, for more parking facilities in the cities, for better public transportation systems, and for faster traffic flow has become quite evident.

There is a distinct difference between the statement of the need and the identification of the problem which follows this statement (Fig. 1-1). The problem is more specific. If the need is for cleaner air, the problem might be that of reducing the dust discharge from power-plant stacks, of reducing the quantity of irritants from automotive exhausts, or of quickly extinguishing forest fires.

Definition of the problem must include all the specifications for the thing that is to be designed. The specifications are the input and output quantities, the characteristics and dimensions of the space the thing must occupy, and all the limitations on these quantities. We can regard the thing to be designed as something in a black box. In this case we must specify the inputs and outputs of the box, together with their characteristics and limitations. The specifications define the cost, the number to be manufactured, the expected life, the range, the operating temperature, and the reliability. Obvious items in the specifications are the speeds, feeds, temperature limitations, maximum range, expected variations in the variables, and dimensional and weight limitations.

There are many implied specifications which result either from the designer's particular environment or from the nature of the problem itself. The manufacturing processes which are available, together with the facilities of a certain plant, constitute restrictions on a designer's freedom, and hence are a part of the implied specifications. It may be that a small plant, for instance, does not own cold-working machinery. Knowing this, the designer selects other metal-processing methods which can be performed in the plant. The labor skills available and the competitive situation also constitute implied constraints. Anything which limits the designer's freedom of choice is a constraint. Many materials and sizes are listed in supplier's catalogs, for instance, but these are not all easily available and shortages frequently occur. Furthermore, inventory economics requires that a manufacturer stock a minimum number of materials and sizes.

After the problem has been defined and a set of written and implied specifications has been obtained, the next step in design, as shown in Fig. 1-1, is the *synthesis* of the optimum solution. Now synthesis cannot take place without both *analysis* and *optimization,* because the system under design must be analyzed to determine whether the performance complies with the specifications. The analysis may reveal that the system is not an optimal one. If the design fails either or both of these tests, the synthesis procedure must begin again.

We have noted, and we shall do so again and again, that design is an iterative process in which we proceed through several steps, evaluate the results, and then return to an earlier phase of the procedure. Thus we may synthesize several components of a system, analyze and optimize them, and return to synthesis to see what effect this has on the remaining parts of the system. Both analysis and optimization require that we construct or devise abstract models of the system which will admit some form of mathematical analysis. We call these models *mathematical models.* In creating them it is our hope that we can find one which will simulate the real physical system very well.

1-5   EVALUATION AND PRESENTATION

As indicated in Fig. 1-1, *evaluation* is a significant phase of the total design process. Evaluation is the final proof of a successful design and usually involves the testing of a prototype in the laboratory. Here we wish to discover if the design really satisfies the need or needs. Is it reliable? Will it compete successfully with similar products? Is it economical to manufacture and to use? Is it easily maintained and adjusted? Can a profit be made from its sale or use? How likely is it to result in product-liability lawsuits? And is insurance easily and cheaply obtained? Is it likely that recalls will be needed to replace defective parts or systems?

Communicating the design to others is the final, vital step in the design process. Undoubtedly, many great designs, inventions, and creative works have been lost to posterity simply because the originators were unable or unwilling to explain their accomplishments to others. Presentation is a selling job. The engineer, when presenting a new solution to administrative, management, or supervisory persons, is attempting to sell or to prove to them that this solution is a better one. Unless this can be done successfully, the time and effort spent on obtaining the solution have been largely wasted. When designers sell a new idea, they also sell themselves. If they are repeatedly successful in selling ideas, designs, and new solutions to management, they begin to receive salary increases and promotions; in fact, this is how anyone succeeds in his or her profession.

Basically, there are only three means of communication. These are the *written*, the *oral*, and the *graphical* forms. Therefore the successful engineer will be technically competent and versatile in *all three forms* of communication. An otherwise competent person who lacks ability in any one of these forms is severely handicapped. If ability in all three forms is lacking, no one will ever know how competent that person is! The three forms of communication—writing, speaking, and drawing—are skills, that is, abilities which can be developed or acquired by any reasonably intelligent person. Skills are acquired only by practice—hour after monotonous hour of it. Musicians, athletes, surgeons, typists, writers, dancers, aerialists, and artists, for example, are skillful because of the number of hours, days, weeks, months, and years they have practiced. Nothing worthwhile in life can be achieved without work, often tedious, dull, and monotonous, and lots of it; and engineering is no exception.

Ability in writing can be acquired by writing letters, reports, memos, papers, and articles. It does not matter whether or not the articles are published—the practice is the important thing. Ability in speaking can be obtained by participating in fraternal, civic, church, and professional activities. This participation provides abundant opportunities for practice in speaking. To acquire drawing ability, pencil sketching should be employed to illustrate every idea possible. The written or spoken word often requires study for comprehension, but pictures are readily understood and should be used freely.

The competent engineer should not be afraid of the possibility of not succeeding in a presentation. In fact, occasional failure should be expected, because failure or criticism seems to accompany every really creative idea. There is a great deal to be learned from a failure, and the greatest gains are obtained by those willing to risk defeat. In the final analysis, the real failure would lie in deciding not to make the presentation at all.

The purpose of this section is to note the importance of *presentation* as the final step in the design process. No matter whether you are planning a presentation to your teacher or to your employer, you should communicate thoroughly and clearly, for this is the payoff. Helpful information on report writing, public speaking, and sketching or drafting is available from countless sources, and you should take advantage of these aids.

1-6 DESIGN CONSIDERATIONS

Sometimes the strength required of an element in a system is an important factor in the determination of the geometry and the dimensions of the element. In such a situation we say that *strength* is an important design consideration. When we use the expression *design consideration,* we are referring to some characteristic which influences the design of the element or, perhaps, the entire system. Usually quite a number of such characteristics must be considered in a given design situation. Many of the important ones are as follows:

1 Strength	13 Noise
2 Reliability	14 Styling
3 Thermal properties	15 Shape
4 Corrosion	16 Size
5 Wear	17 Flexibility
6 Friction	18 Control
7 Processing	19 Stiffness
8 Utility	20 Surface finish
9 Cost	21 Lubrication
10 Safety	22 Maintenance
11 Weight	23 Volume
12 Life	24 Liability

Some of these have to do directly with the dimensions, the material, the processing, and the joining of the elements of the system. Other considerations affect the configuration of the total system. We shall be giving our attention to these factors and other considerations throughout the book.

In this book you will be faced with a great many design situations in which engineering fundamentals must be applied, usually in a mathematical approach, to resolve the problem or problems. This is completely correct and appropriate in an academic atmosphere, where the need is actually to utilize these fundamentals in the resolution of professional problems. To keep the correct perspective, however, it should be observed that in many design situations the important design considerations are such that no calculations or experiments are necessary in order to define an element or system. Students, especially, are often confounded when they run into situations in which it is virtually impossible to make a single calculation and yet an important design decision must be made. These are not extraordinary occurrences at all; they happen every day. Suppose that it is desirable from a sales standpoint—for example, in medical laboratory machinery—to create an impression of great strength and durability. Thicker parts assembled with larger-than-usual oversize bolts can be used to create a rugged-looking machine. Sometimes machines and their parts are designed purely from the standpoint of styling and nothing else. These points are made here so that you will not be misled into believing that there is a rational mathematical approach to every design decision.

1-7   CODES AND STANDARDS

Once upon a time there were no standards for bolts, nuts, and screw threads. One manufacturer would produce, say, $\frac{1}{2}$-in bolts with 9 threads per inch; another used 12. Some fasteners had left-handed threads and sometimes the thread profiles differed. It wasn't unusual in the early days of the automobiles to see a mechanic lay out the fasteners in a line as they were disassembled in order to avoid mixing them during reassembly. This lack of standards and uniformity was costly and inefficient for a great variety of reasons. It is no wonder that a person, disgusted with his or her inability to find a replacement for a damaged fastener, sometimes used baling wire to fasten parts together.

A *standard* is a set of specifications for parts, materials, or processes intended to achieve uniformity, efficiency, and a specified quality. One of the important purposes of a standard is to place a limit on the number of items in the specifications so as to provide a reasonable inventory of tooling, sizes, shapes, and varieties.

A *code* is a set of specifications for the analysis, design, manufacture, and construction of something. The purpose of a code is to achieve a specified degree of safety, efficiency, and performance or quality. It is important to observe that safety codes *do not* imply *absolute safety*. In fact, absolute safety is impossible to obtain. Sometimes the unexpected event really does happen. Designing a building to withstand a 120 mi/h wind does not mean that the designer thinks a 140 mi/h wind is impossible; it simply means that he or she thinks it is highly improbable.

All of the organizations and societies listed below have established specifications for standards and safety or design codes. The name of the organization provides a clue to the nature of the standard or code. Some of the standards and codes, as well as addresses, can be obtained in most technical libraries. The organizations of interest to mechanical engineers are

Aluminum Association (AA)

American Gear Manufacturers Association (AGMA)

American Institute of Steel Construction (AISC)

American Iron and Steel Institute (AISI)

American National Standards Institute (ANSI)*

American Society for Metals (ASM)

American Society of Mechanical Engineers (ASME)

American Society of Testing and Materials (ASTM)

American Welding Society (AWS)

Anti-Friction Bearing Manufacturers Association (AFBMA)

British Standards Institution (BSI)

---

*In 1966 the American Standards Association (ASA) changed its name to the United States of America Standards Institute (USAS). Then, in 1969, the name was again changed, to American National Standards Institute, as shown above and as it is today. This means that you may occasionally find ANSI standards designated as ASA or USAS.

Industrial Fasteners Institute (IFI)

Institution of Mechanical Engineers (I. Mech. E.)

International Bureau of Weights and Measures (BIPM)

International Standards Organization (ISO)

National Bureau of Standards (NBS)

Society of Automotive Engineers (SAE)

1-8   STRESS AND STRENGTH CONSIDERATIONS

Strength is a *property* of a material or of a mechanical element. The strength of an element depends upon the choice, the treatment, and the processing of the material. Consider, for example, a shipment of 1000 springs. We can associate a strength S_i with the ith spring. When this spring is incorporated into a machine, external forces are applied that result in stresses in the spring, the magnitudes of which depend upon its geometry and are independent of the material and its processing. If the spring is removed from the machine unharmed, the stress due to the external forces will drop to zero, its value before assembly. But the strength S_i remains as one of the properties of the spring. Remember, then, that *strength is an inherent property of a part,* a property built into the part because of the use of a particular material and process.

Various metalworking and heat-treating processes, such as forging, rolling, and cold forming, cause variations in the strength from point to point throughout a part. The spring cited above is quite likely to have a strength on the outside of the coils different from its strength on the inside because the spring has been formed by a cold winding process and the two sides may not have been deformed by the same amount. Remember, too, therefore, that a strength value given for a part may apply to only a particular point or set of points on the part.

In this book we shall use the capital letter S to denote strength, as above, with appropriate markings and subscripts to denote the kind of strength. Thus, S_s is a shear strength, S_y a yield strength, S_u an ultimate strength, and $\overline{S}$ a mean strength obtained by sampling test data.

In accordance with accepted practice, we shall employ the Greek letters σ (sigma) and τ (tau) to designate normal and shear stresses, respectively. Again, various markings and subscripts will indicate some special characteristic. For example, σ_1 is a principal stress, σ_y a stress component in the y direction, and σ_r a stress component in the radial direction.

One of the basic problems in dealing with stress and strength is how to relate the two in order to develop a safe, economical, and efficient design. In studying this problem it will be useful to examine this relationship as it applies in a particular branch of engineering design. The American Institute of Steel Construction (AISC), founded in 1921, is a nonprofit society whose objectives are to improve and advance the use of fabricated structural steel. To accomplish these objectives, the AISC publishes manuals, textbooks, specifications, and technical booklets. The best known and most widely used is the *Manual of Steel Construction,* * which holds a highly respected position in

---

*Now in its eighth edition; new editions are published periodically.

TABLE 1-1

Specified Minimum Strengths
of Certain ASTM Steels

STEEL TYPE	ASTM NO.	S_y, kpsi	S_u, kpsi	SIZE, in, UP TO
Carbon	A36	36	58	8
Carbon	A529	42	60	$\frac{1}{2}$
Low alloy	A572	42	60	6
Low alloy	A572	50	65	2
Stainless	A588	50	70	4
Alloy Q&T	A514	100	110	$2\frac{1}{2}$

engineering literature. Important standards included in this manual are the Specification for the Design, Fabrication, and Erection of Structural Steel for Buildings and the Code of Standard Practice for Steel Buildings and Bridges.

Table 1-1 lists minimum yield strengths S_y and minimum tensile strengths S_u for certain ASTM steels as given in the AISC specifications. In this table, S_y represents either the minimum yield point, if the material has one, or the minimum yield strength.

The ANSI-ASTM standard B483-78 defines *minimum strength* as follows:

> **A2.1** Standard mechanical property limits for the respective size ranges are based on an analysis of data from standard production material and are established at a level [at] which at least 99 percent of the population of values obtained from all standard material in the size range meets the established value.

Many readers may prefer to qualify *minimum* by always using the adjective *ASTM,* as in *ASTM minimum strength*. The unqualified word *minimum* may be misleading, since there is a chance that up to 1 percent of the materials involved may have a strength smaller than the ASTM minimum.

To ensure that materials having these minimum strengths are actually used in the construction, the AISC specification states that:

> Certified mill test reports or certified reports of tests made by the fabricator or a testing laboratory in accordance with ASTM A6 or A568, as applicable, and the governing specification shall constitute sufficient evidence of conformity with one of the ASTM standards. Additionally, the fabricator shall, if requested, provide an affidavit stating that the structural steel furnished meets the requirements of the grade specified.

The language of designers often includes the terms *stress allowables, allowable stresses,* or simply *allowables*. These terms refer to reduced values of strengths that are used in design to determine the geometrical dimensions of parts sized according to strength. Finding these reduced values constitutes the first part of the AISC procedure. Let us designate allowable normal stress as σ_{all} and allowable shear stress as τ_{all}. Then the relationship between allowable stresses and specified minimum strengths using the AISC code is specified as:

TENSION $\quad 0.45S_y \leq \sigma_{all} \leq 0.60S_y$

SHEAR $\quad \tau_{all} = 0.40S_y$

BENDING $\quad 0.60S_y \leq \sigma_{all} \leq 0.75S_y$ $\qquad$ (1-1)

BEARING $\quad \sigma_{all} = 0.90S_y$

TABLE 1-2		
AISC Service Factors *K* for Use in Eq. (1-2)	For supports of elevators	$K = 2$
	For cab-operated traveling-crane support girders and their connections	$K = 1.25$
	For pendant-operated traveling-crane support girders and their connections	$K = 1.10$
	For supports of light machinery, shaft- or motor-driven	$K \geq 1.20$
	For supports of reciprocating machinery or power-driven units	$K \geq 1.50$
	For hangers supporting floors and balconies	$K = 1.33$

The next part of the AISC code deals with determining the loads or forces that are used to obtain the stresses. The procedure can be presented succinctly by the equation

$$F = \sum W_d + \sum W_l + \sum KF_l + F_w + \sum F_{\text{misc}} \qquad (1\text{-}2)$$

where F is the force to be used in the appropriate stress equation; see Chap. 2, for example. The components of this force are defined as follows.

The term ΣW_d is the sum of the *dead loads*. These consist of the weight of the steelwork, the materials fastened to it, and the parts supported by it.

The term ΣW_l is the sum of all the stationary or static *live loads*. This includes the weight of equipment, occupants, fixtures, and the snow load if specified by an applicable code.

The force or the resultant of forces due to equipment that may cause impact or dynamic loading is also considered to be a live load and is represented by the term F_l. This factor is to be multiplied by a *service factor K* obtained from Table 1-2.

The term F_w in Eq. (1-2) is the wind load on the structure. Appropriate guidelines for this may be specified by local or regional codes.

The term ΣF_{misc} must be included in some localities to account for the effects of earthquakes, hurricanes, or other extraordinary regional conditions.

The final step in the AISC procedure is to select dimensions of the member to be sized such that the design stress computed from the force F does not exceed the allowable stress as given by Eq. (1-1); in other words, to select dimensions or geometry such that

$$\sigma \leq \sigma_{\text{all}} \qquad \text{or} \qquad \tau \leq \tau_{\text{all}} \qquad (1\text{-}3)$$

where σ and τ may be called the design values of the normal and shear stresses, respectively.

In summary, Eq. (1-1) defines allowable stresses as the specified minimum strengths reduced by multiplication factors varying from 40 to 90 percent to ensure safety. Equation (1-2) accounts for all possible loads, and the service factors in Table 1-2 provide an additional degree of safety for dynamic loading.

1-9   FACTOR OF SAFETY

The method of relating stress and strength shown in the previous section is also used in some other specialized design areas. However, it is not a general approach, since it addresses specific materials and loadings.

A general approach to the stress strength problem is the *factor-of-safety method,* a method as old as engineering design itself, and hence often called the *classical method of design.* A *design factor of safety n_d,* or *n,* sometimes called simply *design factor,* is defined by the relation

$$n_d = \frac{\text{strength}}{\text{stress}} \tag{1-4}$$

In this equation, the "strength" can be anything the designer chooses it to be. We shall often use such strengths as minimum, mean, yield, tensile, fatigue, and shear, as well as others. Of course, the stress used must correspond in type and units to the strength. If the strength is a shear strength in pounds-force per square inch (psi), then the stress must be a shear stress in psi too. Also, both the strength and the stress must apply to the same point or set of points on the member being designed.

Equation (1-1) in the preceding section is needed to account for any uncertainties regarding the actual strength of a member. And Eq. (1-2) must account for the actual loads, including their uncertainties. But the design factor in Eq. (1-4) is used to account for *both* sets of uncertainties—one involving strength and the other involving the loading.

Sometimes it is useful to express the allowances for these uncertainties separately. By selecting the symbols n_S and n_L for design factors used to account for uncertainties of strength and load, respectively, we have

$$n_d = n_S n_L \tag{1-5}$$

Thus, if the service factors of Table 1-2 are to be used, then we would select n_L equal to K at the very least. Circumstances might require an even larger value for n_L.

Equation (1-4) applies only when stress is linearly proportional to load. Cases frequently arise in which this is not true. Then Eq. (1-4) must be cast in the form

$$n_d = \frac{\text{strength in force units}}{\text{applied force or load}} \tag{1-6}$$

In specifying the sizes of machine parts, it is nearly always true that stock sizes must be used. Thus, in a particular application, it may turn out that use of 40-mm tubing is subject to a normal stress greater than the allowable stress whereas 50-mm tubing yields a value of the normal stress that is less than the allowable stress. If there are no stock sizes available between 40 and 50 mm, then the 50-mm size must be selected. For reasons similar to and including this one it is desirable to define *realized factor of safety n_r,* or *n,* as the ratio of the strength to the actual or computed stress. Then the realized factor of safety is defined by either of the equations

$$n_r = \frac{S}{\sigma} \qquad n_r = \frac{S_s}{\tau} \tag{1-7}$$

where now σ and τ are the stresses computed using the final size selection. Thus *design* factor of safety represents our intention at the beginning of design, while *realized* factor of safety tells us what has actually been obtained in the design.

1-10 RELIABILITY*

In these days of greatly increasing numbers of liability lawsuits and the need to conform to regulations issued by governmental agencies such as EPA and OSHA, it is very important for the designer and the manufacturer to know the reliability of their product. The *reliability method* of design is one in which we learn or determine the distribution of stresses and the distribution of strengths and then relate these two in order to achieve an acceptable success rate.

The statistical measure of the probability that a mechanical element will not fail in use is called the *reliability* of that element. The reliability R can be expressed by a number having the range

$$0 \le R < 1 \tag{1-8}$$

A reliability of $R = 0.90$ means that there is a 90 percent chance that the part will perform its proper function without failure. The failure of 6 parts out of every 1000 manufactured might be considered an acceptable failure rate for a certain class of products. This represents a reliability of

$$R = 1 - \frac{6}{1000} = 0.994$$

or 99.4 percent.

In the *reliability method of design,* the designer's task is to make a judicious selection of materials, processes, and geometry (size) so as to achieve a reliability goal. Thus, if the objective reliability is to be 99.4 percent, as above, what combination of materials, processing, and dimensions are needed to meet this goal?

Analyses which lead to an assessment of reliability address uncertainties, or their estimates, in parameters which describe the situation. Stochastic variables such as stress, strength, load, or size are described in terms of their means, standard deviations, and distributions. If bearing balls are produced by a manufacturing process in which a diameter distribution is created, we can say upon choosing a ball that there is uncertainty as to size. If we wish to consider weight or second moment of area in rolling, this size uncertainty can be considered to be *propagated* to our knowledge of weight or inertia. There are ways of estimating the statistical parameters describing weight and inertia from those describing size and density. These methods are variously called *propagation of error, propagation of uncertainty,* or *propagation of dispersion.* These methods are integral parts of analysis or synthesis tasks when probability of failure is involved.

The *factor-of-safety method of design* is a reliable, time-proven method. When properly used, sound and safe designs are obtained using it.

The *statistical* or *reliability approach to design* is relatively new. It can be used only when all the needed stochastic data are available. There are many situations where such data are not available in sufficient quantities to make a sound analysis. It would be dangerous, indeed, to design to a reliability specification when the quality of the data is suspect. For these reasons, you may decide not to study the reliability approach at all.

---

*Headings that are printed black indicate optional reading material, dealing with statistical methods.

This is perfectly acceptable, and, as previously noted, headings for material involving statistical considerations are printed in black type rather than color to call your attention to the nature of the content.

Of course, the reliability method is more expensive, because such a large quantity of the data must be obtained by testing. Quite naturally, the resulting product is priced higher to pay for the laboratory testing (as well as the increased liability insurance that made reliability testing necessary in the first place).

1-11   ECONOMICS

The consideration of cost plays such an important role in the design decision process that we could easily spend as much time in studying the cost factor as in the study of the entire subject of design. Here we introduce only a few general approaches and simple rules.

First, observe that nothing can be said in an absolute sense concerning costs. Materials and labor usually show an increasing cost from year to year. But the costs of processing the materials can be expected to exhibit a decreasing trend because of the use of automated machine tools and robots. The cost of manufacturing a single product will vary from city to city and from one plant to another because of overhead, labor, taxes, and freight differentials and the inevitable slight manufacturing variations.

Standard Sizes

The use of standard or stock sizes is a first principle of cost reduction. An engineer who specifies an AISI 1020 bar of hot-rolled steel 53 mm square, called a hot-rolled square, has added cost to the product, provided a bar 50 or 60 mm square, both of which are preferred sizes, would do equally well. The 53-mm size can be obtained by special order or by rolling or machining a 60-mm square, but these approaches add cost to the product. To ensure that standard or preferred sizes are specified, the designer must have access to stock lists of the materials he or she employs. These are available in libraries or can be obtained directly from the suppliers.

A further word of caution regarding the selection of preferred sizes is necessary. Although a great many sizes are usually listed in catalogs, they are not all readily available. Some sizes are used so infrequently that they are not stocked. A rush order for such sizes may mean more expense and delay. Thus you should also have access to a list such as those in Table A-17 for preferred inch and millimeter sizes.

There are many purchased parts, such as motors, pumps, bearings, and fasteners, which are specified by designers. In the case of these, too, you should make a special effort to specify parts that are readily available. Parts that are made and sold in large quantities usually cost somewhat less than the odd sizes. The cost of rolling bearings, for example, depends more upon the quantity of production by the bearing manufacturer than upon the size of the bearing.

Large Tolerances

Among the effects of design specifications on costs, those of tolerances are perhaps most significant. Tolerances in design influence the producibility of the end product in

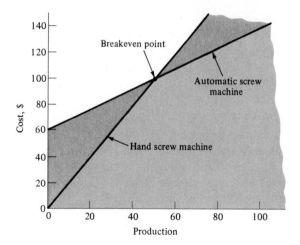

FIGURE 1-2

many ways; close tolerances may necessitate additional steps in processing or even render a part completely impractical to produce economically. Tolerances cover dimensional variation and surface-roughness range and also the variation in mechanical properties resulting from heat treatment and other processing operations.

Since parts having large tolerances can often be produced by machines with higher production rates, labor costs will be smaller than if skilled workers were required. Also, fewer such parts will be rejected in the inspection process, and they are usually easier to assemble.

Breakeven Points

Sometimes it happens that, when two or more design approaches are compared for cost, the choice between the two depends upon a set of conditions such as the quantity of production, the speed of the assembly lines, or some other condition. There then occurs a point corresponding to equal cost which is called the *breakeven point*.

As an example, consider a situation in which a certain part can be manufactured at the rate of 25 parts per hour on an automatic screw machine or 10 parts per hour on a hand screw machine. Let us suppose, too, that the setup time for the automatic is 3 h and that the labor cost for either machine is $20 per hour, including overhead. Figure 1-2 is a graph of cost versus production by the two methods. The breakeven point corresponds to 50 parts. If the desired production is greater than 50 parts, the automatic machine should be used.

Cost Estimates

There are many ways of obtaining relative cost figures so that two or more designs can be roughly compared. A certain amount of judgment may be required in some instances. For example, we can compare the relative value of two automobiles by comparing the dollar cost per pound of weight. Another way to compare the cost of one design with another is simply to count the number of parts. The design having the smaller number of parts is likely to cost less. Many other cost estimators can be used, depending upon the application, such as area, volume, horsepower, torque, capacity, speed, and various performance ratios.

1-12   SAFETY AND PRODUCT LIABILITY

The *strict liability* concept of product liability generally prevails in the United States. This concept states that the manufacturer of an article is liable for any damage or harm that results because of a defect. And it doesn't matter whether the manufacturer knew about the defect, or even could have known about it. For example, suppose an article was manufactured, say, 10 years ago. And suppose at that time the article could not have been considered defective on the basis of all technological knowledge then available. Ten years later, according to the concept of strict liability, the manufacturer is still liable. Thus, under this concept, the plaintiff only needs to prove that the article was defective and that the defect caused some damage or harm. Negligence of the manufacturer need not be proved.

One of the touchy subjects that sometimes comes up in the practice of engineering is what to do if you detect something that you consider poor engineering. If possible, of course, you should attempt to correct it or run sufficient tests to prove that your fears are groundless. If neither of these approaches is feasible, then another approach is to place a memo in the design file and to keep a copy of the memo at home in case the original is ''lost.'' While this approach may protect you, it may also result in your being passed over for promotion, or even in your being discharged. In the long run, if you really feel strongly about the fact that the engineering is poor, and you cannot live with it, then you should transfer or seek a new position.

The best approaches to the prevention of product liability are good engineering in both analysis and design, quality control, and comprehensive testing procedures. Advertising managers often make glowing promises in the warranties and sales literature for a product. These statements should be reviewed carefully by the engineering staff to eliminate excessive promises and to insert adequate warnings and instructions for use.

1-13   UNITS

In the symbolic equation of Newton's second law

$$F = MLT^{-2} \tag{1-9}$$

F stands for force, M for mass, L for length, and T for time. Units chosen for any three of these quantities are called *base units*. The first three having been chosen, the fourth unit is called a *derived unit*. When force, length, and time are chosen as base units, the mass is the derived unit and the system that results is called a *gravitational system of units*. When mass, length, and time are chosen as base units, force is the derived unit and the system that results is called an *absolute system of units*.

In the English-speaking countries, the *U.S. customary foot-pound-second system* (fps) and the *inch-pound-second system* (ips) are the two standard gravitational systems most used by engineers. In the fps system the unit of mass is

$$M = \frac{FT^2}{L} = \frac{(\text{pound-force})(\text{second})^2}{\text{foot}} = \text{lbf} \cdot \text{s}^2/\text{ft} = \text{slug} \tag{1-10}$$

Thus, length, time, and force are the three base units in the fps gravitational system.

The unit of time in the fps system is the second, abbreviated s. The unit of force in the fps system is the pound, more properly the *pound-force*. We shall seldom abbreviate this unit as lbf; the abbreviation lb is permissible, since we shall be dealing only with the U.S. customary gravitational system. In some branches of engineering it is useful to represent 1000 lb as a kilopound and to abbreviate it as kip. Many writers add the letter ''s'' to kip to obtain the plural, but to be consistent with the practice of using only singular forms for units we shall not do so here. Thus, 1 kip and 3 kip are used to designate 1000 and 3000 lb, respectively. Finally, we note in Eq. (1-10) that the derived unit of mass in the fps gravitational system is the lb · s²/ft, called a *slug*; there is no abbreviation for slug.

The unit of mass in the ips gravitational system is

$$M = \frac{FT^2}{L} = \frac{(\text{pound-force})(\text{second})^2}{\text{inch}} = \text{lb} \cdot \text{s}^2/\text{in} \tag{1-11}$$

This unit of mass has *not* been given a special name.

The International System of Units (SI) is an absolute system. The base units are the meter, the kilogram (for mass), and the second. The unit of force is derived using Newton's second law and is called the *newton* to distinguish it from the kilogram, which, as indicated, is the unit of mass. The units constituting the newton (N) are

$$F = \frac{ML}{T^2} = \frac{(\text{kilogram})(\text{meter})}{(\text{second})^2} = \text{kg} \cdot \text{m/s}^2 = \text{N} \tag{1-12}$$

The weight of an object is the force exerted upon it by gravity. Designating the weight as W and the acceleration due to gravity as g, we have

$$W = mg \tag{1-13}$$

In the fps system, standard gravity is $g = 32.1740$ ft/s². For most cases this is rounded off to 32.2. Thus the weight of a mass of 1 slug in the fps system is

$$W = mg = (1 \text{ slug})(32.2 \text{ ft/s}^2) = 32.2 \text{ lb}$$

In the ips system, standard gravity is 386.088 or about 386 in/s². Thus, in this system, a unit mass weighs

$$W = (1 \text{ lb} \cdot \text{s}^2/\text{in})(386 \text{ in/s}^2) = 386 \text{ lb}$$

With SI units, standard gravity is 9.806 or about 9.80 m/s². Thus, the weight of a 1-kg mass is

$$W = (1 \text{ kg})(9.80 \text{ m/s}^2) = 9.80 \text{ N}$$

In view of the fact that weight is the force of gravity acting upon a mass, the following quotation is pertinent:

The greater advantage of SI units is that there is one, and only one unit for each physical quantity—the meter for length, the kilogram for mass, the newton for force, the second for time, etc. To be consistent with this unique feature, it follows that a given unit or word should not be used as an accepted technical name for two physical quantities. However, for generations the term ''weight'' has been used in both technical and nontechnical fields to mean either the force of gravity acting upon a body or the mass of the body itself. The reason for

<table>
<tr><td>TABLE 1-3</td></tr>
<tr><td>SI Base Units</td></tr>
</table>

QUANTITY	NAME	SYMBOL
Length	meter	m
Mass	kilogram	kg
Time	second	s
Electric current	ampere	A
Thermodynamic temperature	kelvin	K
Amount of matter	mole	mol
Luminous intensity	candela	cd

this double use of the term ''weight'' for two different physical quantities—force and mass—is attributed to the dual use of the pound units in our present customary gravitational system in which we often use weight to mean both force and mass.*

The seven SI base units, with their symbols, are shown in Table 1-3. These are dimensionally independent. Lowercase letters are used for the symbols unless they are derived from a proper name; then a capital is used for the first letter of the symbol. Note that the unit of mass uses the prefix kilo; this is the only base unit having a prefix.

Table 1-3 shows that the SI unit of temperature is the kelvin. The Celsius temperature scale (once called centigrade) is not a part of SI, but a difference of one degree on the Celsius scale equals one kelvin.

*From ''S.I., The Weight/Mass Controversy,'' *Mech. Eng,* vol. 99, no. 9, September 1977, p. 40, and vol. 101, no. 3, March 1979, p. 42.

TABLE 1-4

Examples of SI Derived Units*

QUANTITY	UNIT	SI SYMBOL	FORMULA
Acceleration	meter per second squared		$m \cdot s^{-2}$
Angular acceleration	radian per second squared		$rad \cdot s^{-2}$
Angular velocity	radian per second		$rad \cdot s^{-1}$
Area	square meter		m^2
Circular frequency	radian per second	ω	$rad \cdot s^{-1}$
Density	kilogram per cubic meter		$kg \cdot m^{-3}$
Energy	joule	J	$N \cdot m$
Force	newton	N	$kg \cdot m \cdot s^{-2}$
Force couple	newton-meter		$N \cdot m$
Frequency	hertz	Hz	s^{-1}
Power	watt	W	$J \cdot s^{-1}$
Pressure	pascal	Pa	$N \cdot m^{-2}$
Quantity of heat	joule	J	$N \cdot m$
Speed (rotary)	revolution per second		s^{-1}
Stress	pascal	Pa	$N \cdot m^{-2}$
Torque	newton-meter		$N \cdot m$
Velocity	meter per second		$m \cdot s^{-1}$
Volume	cubic meter		m^3
Work	joule	J	$N \cdot m$

*In this book, negative exponents are seldom used; thus, circular frequency, for example, would be expressed in rad/s.

A second class of SI units comprises the derived units, many of which have special names. Table 1-4 is a list of those we shall find most useful in our work in this book.

The radian (symbol rad) is a supplemental unit in SI for a plane angle.

A series of names and symbols to form multiples and submultiples of SI units has been established to provide an alternative to the writing of powers of 10. Table A-1 includes these prefixes and symbols.

1-14   RULES FOR USE OF SI UNITS

The International Bureau of Weights and Measures (BIPM), the international standardizing agency for SI, has established certain rules and recommendations for the use of SI. These are intended to eliminate differences which occur among various countries of the world in scientific and technical practices.

Number Groups

Numbers having four or more digits are placed in groups of three and separated by a space instead of a comma. However, the space may be omitted for the special case of numbers having four digits. A period is used as a decimal point. These recommendations avoid the confusion caused by certain European countries in which a comma is used as a decimal point, and by the English use of a centered period. Examples of correct and incorrect usage are as follows:

> 1924 or 1 924 but not 1,924
>
> 0.1924 or 0.192 4 but not 0.192,4
>
> 192 423.618 50 but not 192,423.61850

The decimal point should always be preceded by a zero for numbers less than unity.

Use of Prefixes

The multiple and submultiple prefixes in steps of 1000 only are recommended (Table A-1). This means that length can be expressed in mm, m, or km, but not in cm, unless a valid reason exists.

When SI units are raised to a power, the prefixes are also raised to the same power. This means that the km^2 is defined thus:

$$1 \ km^2 = (1000 \ m)^2 = (1000)^2 \ m^2 = 10^6 \ m^2$$

Similarly, the mm^2 is defined as follows:

$$(0.001 \ m)^2 = (0.001)^2 \ m^2 = 10^{-6} \ m^2$$

When raising prefixed units to a power, it is permissible, though not often convenient, to use the nonpreferred prefixes such as cm^2 or dm^3.

Except for the kilogram, which is a base unit, prefixes should not be used in the denominators of derived units. Thus the meganewton per square meter, MN/m^2, is satisfactory, but the newton per square millimeter, N/mm², is not to be used. Note that this recommendation avoids a proliferation of derived units.

Double prefixes should not be used. Thus, instead of millimillimeters (mmm), use micrometers (μm).

1-15   PREFERRED UNITS

As a general rule, it is both convenient and good practice to select prefixes such that the number strings will contain no more than four digits to the left of the decimal point. By applying this rule, we find that some of the preferred units are kilopounds per square inch (kpsi) and megapascals (MPa) for stress, pounds per square inch (psi) and kilopascals (kPa) for air or hydraulic pressure, pounds (lb) and kilonewtons (kN) for force, and inches to the fourth power (in⁴) and centimeters to the fourth power (cm⁴) for second moment of area.

Table A-1 is of particular value when you are using units with mixed prefixes in an equation. Suppose we wish to solve the deflection equation (see Chap. 4)

$$y = \frac{64Fl^3}{3\pi d^4 E}$$

where $F = 1.30$ kN, $l = 300$ mm, $d = 2.5$ cm, and $E = 207$ GPa. It is convenient to show the solution in two parts, the first containing the numbers, and the second containing the prefixes. Thus

$$y = \frac{64(1.30)(300)^3}{3\pi(2.5)^4(207)} \frac{(kilo)(milli)^3}{(centi)^4(giga)}$$

Now compute the numerical value of the first part and substitute the prefix values in the second. This gives

$$y = [29.48(10)^3]\left[\frac{10^3(10^{-3})^3}{(10^{-2})^4(10^9)}\right] = 29.48(10)^{-4} \text{ m}$$

$$= 2.948 \text{ mm}$$

In many cases, similar equations are used so often that we remember the prefixes to use in order to obtain a result having a convenient prefix. Thus, in the above equation, if we use F in kN, l and d in mm, and E in GPa, the result y is in mm. Try it. Tables A-3 and A-4 will also be of help to you.

PROBLEMS

1-1   Many times in this book it will be necessary to convert from rectangular to polar coordinates. Make sure now that your computer has this routine, or create one of your own that will be at hand when needed. Angles should be in degrees. Be sure to record how your routine treats the sign of the angle θ for each of the four quadrants.

1-2   Sometimes you will want to interpolate to obtain values from tabular data. Suppose $Y = f(X)$ and neighboring points are close enough to assume linearity. Write a computer program that will give you the value of Y_2 when X_1, X_2, X_3, Y_1, and Y_3 are given. Save it for future use.

1-3   Convert the following to appropriate SI units:
(a) A stress of 20 000 psi
(b) A force of 350 lb
(c) A moment of 1200 lb · in
(d) An area of 2.4 in^2
(e) A second moment of area of 17.4 in^4
(f) An area of 3.6 mi^2
(g) A modulus of elasticity of 21 Mpsi
(h) A speed of 45 mi/h
(i) A volume of 60 in^3

1-4   Convert the following to appropriate SI units:
(a) A length of 60 in
(b) A stress of 90 kpsi
(c) A pressure of 160 psi
(d) A section modulus of 11.2 in^3
(e) A unit weight of 2.61 lb/ft
(f) A deflection of 0.002 in
(g) A velocity of 1200 ft/min
(h) A unit strain of 0.0021 in/in
(i) A volume of 8 gal (U.S.)

1-5   Generally, design results should be rounded off or fixed to three significant digits because the given data cannot justify a greater display. In addition, prefixes should be selected so as to limit number strings to no more than four digits to the left of the decimal point. Using these rules, as well as those for the choice of prefixes, solve the following relations:
(a) $\sigma = M/Z$, where $M = 200$ N · m and $Z = 15.3$ cm^3.
(b) $\sigma = F/A$, where $F = 42$ kN and $A = 6$ cm^2.
(c) $y = Fl^3/3EI$, where $F = 1200$ N, $l = 800$ mm, $E = 207$ GPa, and $I = 6.4$ cm^4.
(d) $\theta = Tl/GJ$, where $J = \pi d^4/32$, $T = 1100$ N · m, $l = 250$ mm, $G = 79.3$ GPa, and $d = 25$ mm. Convert results to degrees of angle.

1-6   Repeat Prob. 1-5 for the following:
(a) $\sigma = F/wt$, where $F = 600$ N, $w = 20$ mm, and $t = 6$ mm
(b) $I = bh^3/12$, where $b = 8$ mm and $h = 24$ mm
(c) $I = \pi d^4/64$, where $d = 32$ mm
(d) $\tau = 16T/\pi d^3$, where $T = 16$ N · m and $d = 25$ mm

1-7   Repeat Prob. 1-5 for:
(a) $\tau = F/A$, where $A = \pi d^2/4$, $F = 120$ kN, and $d = 20$ mm
(b) $\sigma = 32Fa/\pi d^3$, where $F = 800$ N, $a = 800$ mm, and $d = 32$ mm
(c) $Z = (\pi/32d)(d^4 - d_i^4)$ for $d = 36$ mm and $d_i = 26$ mm
(d) $k = d^4G/8D^3N$, where $d = 1.6$ mm, $G = 79.3$ GPa, $D = 19.2$ mm, and $N = 32$ (a dimensionless number)

1-8   The size of a certain hollow shaft is found to be governed by the relation $e = 0.5 - [d/(d^4 - d_i^4)]$, where d and d_i are the outside (OD) and inside (ID) diameters, respectively, and e is an error number. Find suitable values for d and d_i to the nearest $\frac{1}{8}$ in by requiring that e be a small positive number.

1-9   Numerical integration is often forced upon the designer. Simpson's rule is of the form

$$\int_a^b f(x)\ dx = \frac{h}{3}\,(y_0 + 4y_1 + 2y_2 + 4y_3 + \cdots + 4y_{n-1} + y_n)$$

where the y's are evenly spaced function evaluations of $f(x)$ in the interval $[a, b]$ and the constant h is the interval between function evaluations. The number of intervals n should be an even number, and there are advantages to making it an even number divisible by 4. If this is done, the error in the integration can be estimated by integrating again using every other ordinate, calling this integration $I_{n/2}$ and estimating the numerical integration error to be

$$E = \frac{I_n - I_{n/2}}{15}$$

We express the integral as $I_n + E$ in applying Richardson's correction to improve Simpson's value. Apply Simpson's rule to the integral

$$I = \int_0^1 x^5\ dx$$

using but four intervals. Apply Richardson's correction and show that the result is exact.

1-10   Use your experience with Prob. 1-9 and others as the basis for a computer program to perform a Simpson's rule integration. Such a program will be useful to you in the future.

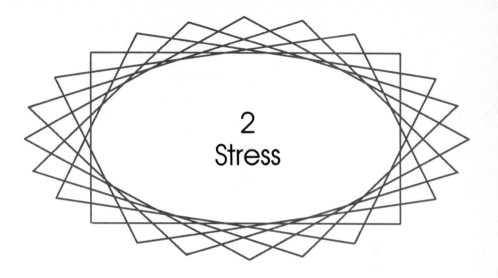

2
Stress

One of the first problems encountered by the designer is to ensure that the strength of the part being designed always exceeds the stress due to any loads that may be imposed upon it. Thus this chapter deals with stress. Later chapters will deal with strength and how to relate stress and strength to obtain a safe and reliable design.

2-1 STRESS COMPONENTS

Figure 2-1a is a general three-dimensional stress element, showing three normal stresses σ_x, σ_y, and σ_z, all positive; and six shear stresses τ_{xy}, τ_{yx}, τ_{yz}, τ_{zy}, τ_{zx}, and τ_{xz}, also all positive. The element is in static equilibrium and hence

$$\tau_{xy} = \tau_{yx} \qquad \tau_{yz} = \tau_{zy} \qquad \tau_{zx} = \tau_{xz} \tag{2-1}$$

Outwardly directed normal stresses are called tension or tensile stresses and are considered positive. Shear stresses on a positive face of an element are positive if they act in the positive direction of a reference axis; such is the case in Fig. 2-1a. The first subscript of a shear-stress component is the coordinate normal to the element face. The shear-stress component is parallel to the axis of the second subscript. Since the element shown is in static equilibrium, the negative faces will have shear stresses acting in the opposite direction; these are also considered positive.

What we have just described is the classical sign convention for stress. In this book we shall use a sign convention different from the classical, in order to obtain agreement in direction in the measurement of angles and shear stresses. Figure 2-1b illustrates a state of *biaxial*, or *plane, stress*. The two normal stresses are shown in the positive direction. Shear stresses will be taken as positive when they are in the clockwise (cw) direction. Thus, in Fig. 2-1b, τ_{yx} is cw and positive; τ_{xy} is ccw (counterclockwise) and negative.

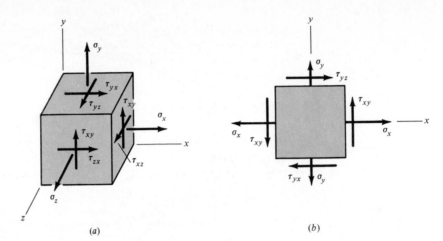

FIGURE 2-1   (a)   (b)

2-2   MOHR'S CIRCLE

Suppose the element of Fig. 2-1b is cut by an oblique plane at angle ϕ to the x axis as shown in Fig. 2-2. This section is concerned with the stresses σ and τ which act upon this oblique plane. By summing the forces caused by all the stress components to zero, the stresses σ and τ are found to be

$$\sigma = \frac{\sigma_x + \sigma_y}{2} + \frac{\sigma_x - \sigma_y}{2} \cos 2\phi + \tau_{xy} \sin 2\phi \tag{2-2}$$

$$\tau = -\frac{\sigma_x - \sigma_y}{2} \sin 2\phi + \tau_{xy} \cos 2\phi \tag{2-3}$$

Differentiating the first equation with respect to ϕ and setting the result equal to zero gives

$$\tan 2\phi = \frac{2\tau_{xy}}{\sigma_x - \sigma_y} \tag{2-4}$$

Equation (2-4) defines two particular values for the angle 2ϕ, one of which defines the maximum normal stress σ_1 and the other, the minimum normal stress σ_2. These two

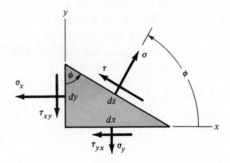

FIGURE 2-2

stresses are called the *principal stresses,* and their corresponding directions, the *principal directions.* The angle ϕ between the principal directions is 90°.

In a similar manner, we differentiate Eq. (2-3), set the result equal to zero, and obtain

$$\tan 2\phi = -\frac{\sigma_x - \sigma_y}{2\tau_{xy}} \tag{2-5}$$

Equation (2-5) defines the two values of 2ϕ at which the shear stress τ reaches an extreme value.

It is interesting to note that Eq. (2-4) can be written in the form

$$2\tau_{xy} \cos 2\phi = (\sigma_x - \sigma_y) \sin 2\phi$$

or

$$\sin 2\phi = \frac{2\tau_{xy} \cos 2\phi}{\sigma_x - \sigma_y} \tag{a}$$

Now substitute Eq. (*a*) for $\sin 2\phi$ in Eq. (2-3). We obtain

$$\tau = -\frac{\sigma_x - \sigma_y}{2} \frac{2\tau_{xy} \cos 2\phi}{\sigma_x - \sigma_y} + \tau_{xy} \cos 2\phi = 0 \tag{2-6}$$

Equation (2-6) states that the shear stress associated with both principal directions is zero.

Solving Eq. (2-5) for $\sin 2\phi$, in a similar manner, and substituting the result in Eq. (2-2) yields

$$\sigma = \frac{\sigma_x + \sigma_y}{2} \tag{2-7}$$

Equation (2-7) tells us that the two normal stresses associated with the directions of the two maximum shear stresses are equal.

Formulas for the two principal stresses can be obtained by substituting the angle 2ϕ from Eq. (2-4) in Eq. (2-2). The result is

$$\sigma_1, \sigma_2 = \frac{\sigma_x + \sigma_y}{2} \pm \sqrt{\left(\frac{\sigma_x - \sigma_y}{2}\right)^2 + \tau_{xy}^2} \tag{2-8}$$

In a similar manner the two extreme-value shear stresses are found to be

$$\tau_1, \tau_2 = \pm\sqrt{\left(\frac{\sigma_x - \sigma_y}{2}\right)^2 + \tau_{xy}^2} \tag{2-9}$$

Your particular attention is called to the fact that an extreme value of the shear stress may not be the same as the maximum value. See Sec. 2-3.

A graphical method for expressing the relations developed in this section, called a *Mohr's circle diagram,* is a very effective means of visualizing the stress state at a point and keeping track of the directions of the various components associated with plane stress. In Fig. 2-3 we create a coordinate system with normal stresses plotted along the abscissa and shear stresses plotted as the ordinates. On the abscissa, tensile (positive) normal stresses are plotted to the right of the origin O and compressive

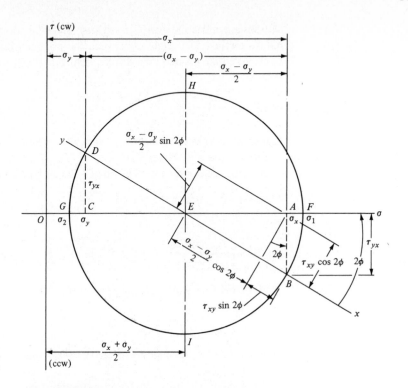

FIGURE 2-3

Mohr's circle diagram.

(negative) normal stresses to the left. On the ordinate, positive (cw) shear stresses are plotted up; counterclockwise (ccw) shear stresses are negative and are plotted down.

Using the stress state of Fig. 2-1b, we plot the Mohr's circle diagram (Fig. 2-3) by laying off σ_x as OA, τ_{xy} as AB, σ_y as OC, and τ_{yx} as CD. The line DEB is the diameter of the Mohr's circle with center at E on the σ axis. Point B represents the stress coordinates σ_x, τ_{xy} on the x faces, and point D the stress coordinates σ_y, τ_{yx} on the y faces. Thus EB corresponds to the x axis and ED to the y axis. The angle 2ϕ, measured counterclockwise from EB to ED, is $180°$, which corresponds to $\phi = 90°$, measured counterclockwise from x to y, on the stress element of Fig. 2-1b. The maximum principal normal stress σ_1 occurs at F, and the minimum principal normal stress σ_2 at G. The two extreme-value shear stresses, one clockwise and one counterclockwise, occur at I and H, respectively. You should demonstrate for yourself that the geometry of Fig. 2-3 satisfies all the relations developed in this section.

Though Eqs. (2-4) and (2-8) can be solved directly for the principal stresses and directions, a semigraphical approach is easier and quicker and offers fewer opportunities for error. This method is illustrated by the following example.

EXAMPLE 2-1   A stress element has $\sigma_x = 80$ MPa and $\tau_{xy} = 50$ MPa cw.* Find the principal stresses and directions and show these on a stress element correctly aligned with respect to the xy system. Draw another stress element to show τ_1 and τ_2, find the corresponding normal stresses, and label the drawing completely.

*Any stress components such as σ_y, τ_{zx}, etc., that are not given in a problem are always taken as zero.

Solution We shall construct a Mohr's circle corresponding to the given data and then read the results directly from the diagram. You can construct this diagram with compass and scales and find the required information with the aid of scales and protractor, if you choose to do so. Such a complete graphical approach is perfectly satisfactory and sufficiently accurate for most purposes.

In the semigraphical approach used here, we first make an approximate freehand sketch of the Mohr's circle and then use the geometry of the figure to obtain the desired information.

Draw the σ and τ axes first (Fig. 2-4a) and locate $\sigma_x = 80$ MPa along the σ axis. Then, from σ_x, locate $\tau_{xy} = 50$ MPa in the cw direction of the τ axis to establish point A. Corresponding to $\sigma_y = 0$, locate $\tau_{yx} = 50$ MPa in the ccw direction along the τ axis to obtain point D. The line AD forms the diameter of the required circle, which can now be drawn. The intersection of the circle with the σ axis defines σ_1 and σ_2 as shown. The x axis is the line CA; the y axis is the line CD. Now, noting the triangle ABC, indicate on the sketch the length of the legs AB and BC as 50 and 40 MPa, respectively. The length of the hypotenuse AC is

$$\tau_1 = \sqrt{(50)^2 + (40)^2} = 64.0 \text{ MPa}$$

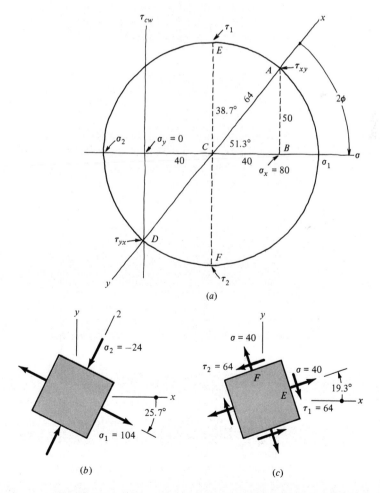

(a)

(b)

(c)

FIGURE 2-4

All stresses in MPa.

and this should be labeled on the sketch too. Since intersection C is 40 MPa from the origin, the principal stresses are now found to be

$$\sigma_1 = 40 + 64 = 104 \text{ MPa} \qquad \sigma_2 = 40 - 64 = -24 \text{ MPa}$$

The angle 2ϕ from the x axis cw to σ_1 is

$$2\phi = \tan^{-1} \tfrac{50}{40} = 51.3°$$

If your calculator has rectangular-to-polar conversion, you should always use it to obtain τ_1 and 2ϕ. But see Prob. 1-1.

To draw the principal stress element (Fig. 2-4b), sketch the x and y axes parallel to the original axes. The angle ϕ on the stress element must be measured in the *same* direction as is the angle 2ϕ on the Mohr's circle. Thus, from x measure 25.7° (half of 51.3°) clockwise to locate the σ_1 axis. The stress element can now be completed and labeled as shown.

The two maximum shear stresses occur at points E and F in Fig. 2-4a. The two normal stresses corresponding to these shear stresses are each 40 MPa, as indicated. Point E is 38.7° ccw from point A on the Mohr's circle. Therefore, in Fig. 2-4c, draw a stress element oriented 19.3° (half of 38.7°) ccw from x. The element should then be labeled with magnitudes and directions as shown.

In constructing these stress elements it is important to indicate the x and y directions of the original reference system. This completes the link between the original machine element and the orientation of its principal stresses.

2-3   TRIAXIAL STRESS

As in the case of plane or biaxial stress, a particular orientation of the stress element occurs in space for which all shear-stress components are zero. When an element has this particular orientation, the normals to the faces correspond to the principal directions, and the normal stresses associated with these faces are the principal stresses. Since there are six faces, there are three principal directions and three principal stresses σ_1, σ_2, and σ_3.

In our studies of plane stress we were able to specify any stress state σ_x, σ_y, and τ_{xy} and find the principal stresses and principal directions. But six components of stress are required to specify a general state of stress in three dimensions, and the problem of determining the principal stresses and directions is much more difficult. It turns out that it is rarely necessary in design, and so we shall not investigate the problem in this book. The process involves finding the three roots to the cubic equation

$$\sigma^3 - (\sigma_x + \sigma_y + \sigma_z)\sigma^2 + (\sigma_x\sigma_y + \sigma_x\sigma_z + \sigma_y\sigma_z - \tau_{xy}^2 - \tau_{yz}^2 - \tau_{zx}^2)\sigma$$
$$- (\sigma_x\sigma_y\sigma_z + 2\tau_{xy}\tau_{yz}\tau_{zx} - \sigma_x\tau_{yz}^2 - \sigma_y\tau_{zx}^2 - \sigma_z\tau_{xy}^2) = 0 \qquad (2\text{-}10)$$

In plotting Mohr's circles for triaxial stress, the principal normal stresses are ordered so that $\sigma_1 > \sigma_2 > \sigma_3$. Then the result appears as in Fig. 2-5a. The stress coordinates σ_N, τ_N for any arbitrarily located plane will always lie within the shaded area.

Figure 2-5a also shows the three *principal shear stresses* $\tau_{1/2}$, $\tau_{2/3}$, and $\tau_{1/3}$.* Each

---

*Note the difference between this notation and that for a shear stress, say, τ_{xy}. The use of the shilling is not accepted practice, but is used here to emphasize the distinction.

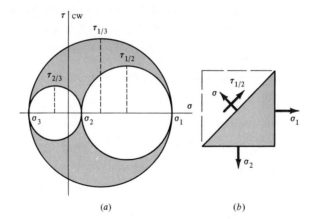

FIGURE 2-5

Mohr's circles for triaxial stress.

(a)   (b)

of these occurs on the two planes, one of which is shown in Fig. 2-5*b*. The figure shows that the principal shear stresses are given by the equations

$$\tau_{1/2} = \frac{\sigma_1 - \sigma_2}{2} \qquad \tau_{2/3} = \frac{\sigma_2 - \sigma_3}{2} \qquad \tau_{1/3} = \frac{\sigma_1 - \sigma_3}{2} \qquad (2\text{-}11)$$

Of course, $\tau_{max} = \tau_{1/3}$ when the normal principal stresses are ordered ($\sigma_1 > \sigma_2 > \sigma_3$).

Octahedral Stresses

Now visualize a principal stress element having the stresses σ_1, σ_2, and σ_3, as shown in Fig. 2-6. Cut the stress element by a plane that forms equal angles with each of the three principal stresses as shown by plane *ABC*. This is called an *octahedral plane*. Note that the solid cut from the stress element retains one of the original corners. Since there are eight of these corners in all, there are a total of eight such planes.

Figure 2-6 may be regarded as a free-body diagram when each of the stress components shown is multiplied by the area over which it acts to obtain the corresponding force. It is possible, then, to sum these forces to zero in each of the three coordinate

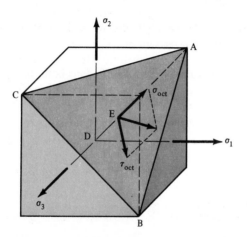

FIGURE 2-6

The octahedral plane.

directions. When this is done, it is found that a force, called the *octahedral force*, exists on plane *ABC*. When this force is divided by the area of face *ABC*, it can be resolved into two components. One of these is normal to plane *ABC* and so is called the *octahedral normal stress*. The other is in plane *ABC* and is called the *octahedral shear stress*.* The magnitudes of these stress components are

$$\tau_{oct} = \tfrac{2}{3}(\tau_{1/2}^2 + \tau_{2/3}^2 + \tau_{1/3}^2)^{1/2}$$

$$= \tfrac{1}{3}[(\sigma_1 - \sigma_2)^2 + (\sigma_2 - \sigma_3)^2 + (\sigma_3 - \sigma_1)^2]^{1/2}$$

$$= \tfrac{1}{3}[(\sigma_x - \sigma_y)^2 + (\sigma_y - \sigma_z)^2 + (\sigma_z - \sigma_x)^2 + 6(\tau_{xy}^2 + \tau_{xz}^2 + \tau_{yz}^2)]^{1/2} \tag{2-12}$$

$$\sigma_{oct} = \tfrac{1}{3}(\sigma_1 + \sigma_2 + \sigma_3) = \tfrac{1}{3}(\sigma_x + \sigma_y + \sigma_z) \tag{2-13}$$

The importance of these equations is that any three-dimensional state of stress can be represented by a single pair of stress components. In later chapters we shall find this to be a very powerful tool in design analysis.

2-4   UNIFORMLY DISTRIBUTED STRESSES

The assumption of a uniform distribution of stress is frequently made in design. The result is then often called *pure tension, pure compression*, or *pure shear*, depending upon how the external load is applied to the body under study. The word *simple* is sometimes used instead of *pure* to indicate that there are no other complicating effects. The tension rod is typical. Here a tension load F is applied through pins at the ends of the bar. The assumption of uniform stress means that if we cut the bar at a section remote from the ends and remove one piece, we can replace its effect by applying a uniformly distributed force of magnitude σA to the cut end. So the stress σ is said to be uniformly distributed. It is calculated from the equation

$$\sigma = \frac{F}{A} \tag{2-14}$$

This assumption of uniform stress distribution requires that:

• The bar be straight and of a homogeneous material

• The line of action of the force contains the centroid of the section

• The section be taken remote from the ends and from any discontinuity or abrupt change in cross section

The same equation and assumptions hold for simple compression. A slender bar in compression may, however, fail by buckling, and this possibility must be eliminated from consideration before Eq. (2-14) is used.†

Use of the equation

$$\tau = \frac{F}{A} \tag{2-15}$$

---

*For a complete derivation see J. O. Smith and O. M. Sidebottom, *Elementary Mechanics of Deformable Bodies*, Macmillan, New York, 1969, p. 131.

†See Sec. 3-12.

for a body, say a bolt, in shear assumes a uniform stress distribution too. It is very difficult in practice to obtain a uniform distribution of shear stress; the equation is included because occasions do arise in which this assumption is made.

2-5   ELASTIC STRAIN

When a straight bar is subjected to a tensile load, the bar becomes longer. The amount of stretch, or elongation, is called *strain*. The elongation per unit length of the bar is called *unit strain*. In spite of these definitions, however, it is customary to use the word *strain* to mean *unit strain* and the expression *total strain* to mean total elongation, or deformation, of a member. Using this custom here, the expression for strain is

$$\epsilon = \frac{\delta}{l} \qquad (2\text{-}16)$$

where δ is the total elongation (total strain) of the bar within the length l.

Shear strain γ is the change in a right angle of a stress element subjected to pure shear.

Elasticity is that property of a material which enables it to regain its original shape and dimensions when the load is removed. Hooke's law states that, within certain limits, the stress in a material is proportional to the strain which produced it. An elastic material does not necessarily obey Hooke's law, since it is possible for some materials to regain their original shape without the limiting condition that stress be proportional to strain. On the other hand, a material which obeys Hooke's law is elastic. For the condition in which stress is proportional to strain, we can write the relations

$$\sigma = E\epsilon \qquad \tau = G\gamma \qquad (2\text{-}17)$$

where E and G are the constants of proportionality. Since the strains are dimensionless numbers, the units of E and G are the same as the units of stress. The constant E is called the *modulus of elasticity*. The constant G is called the *shear modulus of elasticity,* or sometimes, the *modulus of rigidity.* Both E and G, however, are numbers which are indicative of the stiffness or rigidity of the materials. These two constants represent fundamental properties.

By substituting $\sigma = F/A$ and $\epsilon = \delta/l$ into Eq. (2-17) and rearranging, we obtain the equation for the total deformation of a bar loaded in axial tension or compression:

$$\delta = \frac{Fl}{AE} \qquad (a)$$

Experiments demonstrate that when a material is placed in tension, there exists not only an axial strain, but also a lateral strain. Poisson demonstrated that these two strains were proportional to each other within the range of Hooke's law. This constant is expressed as

$$\nu = -\frac{\text{lateral strain}}{\text{axial strain}} \qquad (2\text{-}18)$$

and is known as *Poisson's ratio*. These same relations apply for compression, except that a lateral expansion takes place instead.

The three elastic constants are related to each other as follows:

$$E = 2G(1 + \nu) \tag{2-19}$$

2-6 STRESS-STRAIN RELATIONS

There are many experimental techniques which can be used to measure strain. Thus, if the relationship between stress and strain is known, the stress state at a point can be calculated after the state of strain has been measured. We define the *principal strains* as the strains in the direction of the principal stresses. It is true that the shear strains are zero, just as the shear stresses are zero, on the faces of an element aligned in the principal directions. Table 2-1 contains the relations for all three types of stress. See Table A-5 for values of Poisson's ratio ν.

2-7 EQUILIBRIUM

The law of particle motion states that *any force F acting on a particle of mass will produce an acceleration of the particle.* If we assume that all members to be studied are motionless or, at most, have a constant velocity, then every particle has zero acceleration. Applying the law of particle motion gives

$$\mathbf{F}_1 + \mathbf{F}_2 + \mathbf{F}_3 + \cdots + \mathbf{F}_i = \sum \mathbf{F} = \mathbf{0} \tag{a}$$

where $\Sigma \mathbf{F}$ is the vector sum of all the forces acting on the particle.

TABLE 2-1

Elastic Stress-Strain Relations

TYPE OF STRESS	PRINCIPAL STRAINS	PRINCIPAL STRESSES
Uniaxial	$\epsilon_1 = \dfrac{\sigma_1}{E}$	$\sigma_1 = E\epsilon_1$
	$\epsilon_2 = -\nu\epsilon_1$	$\sigma_2 = 0$
	$\epsilon_3 = -\nu\epsilon_1$	$\sigma_3 = 0$
Biaxial	$\epsilon_1 = \dfrac{\sigma_1}{E} - \dfrac{\nu\sigma_2}{E}$	$\sigma_1 = \dfrac{E(\epsilon_1 + \nu\epsilon_2)}{1 - \nu^2}$
	$\epsilon_2 = \dfrac{\sigma_2}{E} - \dfrac{\nu\sigma_1}{E}$	$\sigma_2 = \dfrac{E(\epsilon_2 + \nu\epsilon_1)}{1 - \nu^2}$
	$\epsilon_3 = -\dfrac{\nu\sigma_1}{E} - \dfrac{\nu\sigma_2}{E}$	$\sigma_3 = 0$
Triaxial	$\epsilon_1 = \dfrac{\sigma_1}{E} - \dfrac{\nu\sigma_2}{E} - \dfrac{\nu\sigma_3}{E}$	$\sigma_1 = \dfrac{E\epsilon_1(1 - \nu) + \nu E(\epsilon_2 + \epsilon_3)}{1 - \nu - 2\nu^2}$
	$\epsilon_2 = \dfrac{\sigma_2}{E} - \dfrac{\nu\sigma_1}{E} - \dfrac{\nu\sigma_3}{E}$	$\sigma_2 = \dfrac{E\epsilon_2(1 - \nu) + \nu E(\epsilon_1 + \epsilon_3)}{1 - \nu - 2\nu^2}$
	$\epsilon_3 = \dfrac{\sigma_3}{E} - \dfrac{\nu\sigma_1}{E} - \dfrac{\nu\sigma_2}{E}$	$\sigma_3 = \dfrac{E\epsilon_3(1 - \nu) + \nu E(\epsilon_1 + \epsilon_2)}{1 - \nu - 2\nu^2}$

Whenever Eq. (*a*) holds, the forces acting on the particle are said to be *balanced* and the particle is said to be in *equilibrium*. The phrase *static equilibrium* is also used to imply that the particle is *at rest*.

The word *system* will be used to denote any part or portion of a machine or structure, including all of it if desired, that we may wish to study. A system, under this definition, may consist of a particle, several particles, a part of a rigid body or an entire rigid body, or even several rigid bodies.

An *internal force* or *internal moment* is an action of one part of a system on another part of the same system. An *external force* or an *external moment* is an action that is applied to the system from the outside.

Equation (*a*) expresses the conditions for the equilibrium of a single particle. But what about the equilibrium of a system containing many particles? We can apply Eq. (*a*) to each particle in the system. We shall select one of these, say the *j*th one. Let $\mathbf{F}_e$ be the sum of the external forces and $\mathbf{F}_i$ be the sum of the internal forces. Then for the *j*th particle, Eq. (*a*) becomes

$$\sum \mathbf{F}_j = \mathbf{F}_e + \mathbf{F}_i = \mathbf{0} \qquad\qquad (b)$$

If there are *n* particles in the system and if we add them all together, we get

$$\sum_1^n \mathbf{F}_e + \sum_1^n \mathbf{F}_i = \mathbf{0} \qquad\qquad (c)$$

Now Newton's third law, called the law of action and reaction, states that *when two particles react, a pair of interacting forces come into existence, that these forces have the same magnitude, have opposite senses, and act along the straight line common to the two particles*. This means that, in the system under study, all the internal forces acting between particles occur in pairs. Each force of the pair, according to the third law, is equal in magnitude and opposite in direction to its mate. Therefore the second term in Eq. (*c*) is zero, and we have left that

$$\sum_1^n \mathbf{F}_e = \mathbf{0} \qquad\qquad (2\text{-}20)$$

which states that *the sum of all the external force vectors acting upon a system in equilibrium is zero*.

A similar procedure can be used to demonstrate that *the sum of all the external moment vectors acting upon a system in equilibrium is zero*. Or, in mathematical form,

$$\sum_1^n \mathbf{M}_e = \mathbf{0} \qquad\qquad (2\text{-}21)$$

The statements accompanying Eqs. (2-20) and (2-21) can be reversed; thus it is also true that *if these two equations are simultaneously satisfied,* then the system is in *static equilibrium*.

Free-Body Diagrams

In examining the problem of analyzing the behavior, performance, or efficiency of a complex structure or device, such as a bridge, typewriter, or tractor, the beginner is faced with a bewildering array of complicated parts and geometries. Fortunately, the problem is not as difficult as it seems. One of the most powerful analytical techniques of mechanics is that of isolating or freeing a portion of a system in our imagination in order to study the behavior of one of its segments. When the segment is isolated, the original effect of the system on the segment is replaced by the interacting forces and moments. Figure 2-7 is a symbolic illustration of the process. Let Fig. 2-7a be a total system, such as a bridge. Then we might decide to analyze just one part or segment of the bridge, such as a beam, or even several members joined together. We remove this segment, as in Fig. 2-7b, and replace the effect of the whole system on the segment by various forces and moments that would necessarily act at the interfaces of the segment and the system. Although these forces may be internal effects upon the whole system, they are external effects when applied to the segment. These interface forces are represented symbolically by the force vectors $\mathbf{F}_1$, $\mathbf{F}_2$, $\mathbf{F}_3$, and $\mathbf{F}_i$ in Fig. 2-7b. The isolated subsystem that results, together with all forces and moments due to any external effects and the reactions with the main system, is called a *free-body diagram*.

We can greatly simplify the analysis of a very complex structure or machine by successively isolating each element and studying and analyzing it by the use of free-body diagrams. When all the members have been treated in this manner, the knowledge can be assembled to yield information concerning the behavior of the total system. Thus, free-body diagramming is essentially a means of breaking a complicated problem into manageable segments, analyzing these simple problems, and then, usually, putting the information together again.

Using free-body diagrams for force analysis serves the following important purposes:

- The diagram establishes the directions of reference axes, provides a place to record the dimensions of the subsystem and the magnitudes and directions of the known forces, and helps in assuming the directions of unknown forces.

- The diagram simplifies your thinking because it provides a place to store one thought while proceeding to the next.

- The diagram provides a means of communicating your thoughts clearly and unambiguously to other people.

- Careful and complete construction of the diagram clarifies fuzzy thinking by bringing out various points that are not always apparent in the statement or in the geometry of the total problem. Thus, the diagram aids in understanding all facets of the problem.

- The diagram helps in the planning of a logical attack on the problem and in setting up the mathematical relations.

- The diagram helps in recording progress in the solution and in illustrating the methods used.

- The diagram allows others to follow your reasoning.

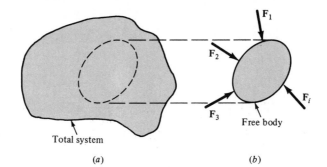

FIGURE 2-7

The isolation of a subsystem.

Total system

(a)

Free body

(b)

2-8 SHEAR AND MOMENT

Figure 2-8a *shows a beam supported by* reactions R_1 and R_2 and loaded by the concentrated forces F_1, F_2, and F_3. The direction chosen for the y axis is the clue to the sign convention for the forces. F_1, F_2, and F_3 are negative because they act in the negative y direction; R_1 and R_2 are positive.

If the beam is cut at some section located at $x = x_1$ and the left-hand portion is removed as a free body, an *internal shear force V* and *bending moment M* must act on the cut surface to ensure equilibrium. The shear force is obtained by summing the forces to the left of the cut section. The bending moment is the sum of the moments of the forces to the left of the section taken about an axis through the section. Shear force and bending moment are related by the equation

$$V = \frac{dM}{dx} \tag{2-22}$$

Sometimes the bending is caused by a distributed load. Then, the relation between shear force and bending moment may be written

$$\frac{dV}{dx} = \frac{d^2M}{dx^2} = -w \tag{2-23}$$

where w is a downward-acting load of w units of force per unit length.

The sign conventions used for bending moment and shear force in this book are shown in Fig. 2-9.

The loading w of Eq. (2-23) is uniformly distributed. A more general distribution

FIGURE 2-8

Free-body diagram of simply supported beam with V and M shown in positive directions.

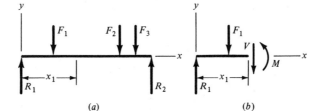

(a)

(b)

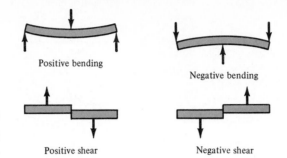

FIGURE 2-9

Sign conventions for bending and shear.

can be defined by the equation

$$q = \lim_{\Delta x \to 0} \frac{\Delta F}{\Delta x}$$

where q is called the *load intensity;* thus $q = -w$.

Equations (2-22) and (2-23) reveal additional relations if they are integrated. Thus, if we integrate between, say, x_A and x_B, we obtain

$$\int_{V_A}^{V_B} dV = \int_{x_A}^{x_B} q\ dx = V_B - V_A \tag{2-24}$$

which states that *the change in shear force from A to B is equal to the area of the loading diagram between x_A and x_B.*

In a similar manner,

$$\int_{M_A}^{M_B} dM = \int_{x_A}^{x_B} V\ dx = M_B - M_A \tag{2-25}$$

which states that *the change in moment from A to B is equal to the area of the shear-force diagram between x_A and x_B.*

2-9   SINGULARITY FUNCTIONS

The five singularity functions defined in Table 2-2 constitute a useful and easy means of integrating across discontinuities. By their use, general expressions for shear force and bending moment in beams can be written when the beam is loaded by concentrated moments or forces. As shown in the table, the concentrated moment and force functions are zero for all values of x not equal to a. The functions are undefined for values of $x = a$. Note that the unit step, ramp, and parabolic functions are zero only for values of x that are less than a. The integration properties shown in the table constitute a part of the mathematical definition too. The examples which follow show how these functions are used.

TABLE 2-2

Singularity and Macaulay Functions

FUNCTION	GRAPH OF $f_n(x)$	MEANING
Concentrated moment (unit doublet)	$\langle x - a \rangle^{-2}$	$\langle x - a \rangle^{-2} = 0 \qquad x \neq a$ $\displaystyle\int_{-\infty}^{x} \langle x - a \rangle^{-2}\, dx = \langle x - a \rangle^{-1}$ $\langle x - a \rangle^{-2} = \pm\infty \qquad x = a$
Concentrated force (unit impulse)	$\langle x - a \rangle^{-1}$	$\langle x - a \rangle^{-1} = 0 \qquad x \neq a$ $\displaystyle\int_{-\infty}^{x} \langle x - a \rangle^{-1}\, dx = \langle x - a \rangle^{0}$ $\langle x - a \rangle^{-1} = +\infty \qquad x = a$
Unit step	$\langle x - a \rangle^{0}$	$\langle x - a \rangle^{0} = \begin{cases} 0 & x < a \\ 1 & x \geq a \end{cases}$ $\displaystyle\int_{-\infty}^{x} \langle x - a \rangle^{0}\, dx = \langle x - a \rangle^{1}$
Ramp	$\langle x - a \rangle^{1}$	$\langle x - a \rangle^{1} = \begin{cases} 0 & x < a \\ x - a & x \geq a \end{cases}$ $\displaystyle\int_{-\infty}^{x} \langle x - a \rangle^{1}\, dx = \dfrac{\langle x - a \rangle^{2}}{2}$
Parabolic	$\langle x - a \rangle^{2}$	$\langle x - a \rangle^{2} = \begin{cases} 0 & x < a \\ (x - a)^{2} & x \geq a \end{cases}$ $\displaystyle\int_{-\infty}^{x} \langle x - a \rangle^{2}\, dx = \dfrac{\langle x - a \rangle^{3}}{3}$

EXAMPLE 2-2

Derive expressions for the loading, shear-force, and bending-moment diagrams for the beam of Fig. 2-10.

Solution

Using Table 2-2 and $q(x)$ for the loading function, we find

$$q = R_1 \langle x \rangle^{-1} - F_1 \langle x - a_1 \rangle^{-1} - F_2 \langle x - a_2 \rangle^{-1} + R_2 \langle x - l \rangle^{-1} \tag{1}$$

Next, we use Eq. (2-24) to get the shear force. Note that $V = 0$ at $x = -\infty$.

$$V = \int_{-\infty}^{x} q\, dx = R_1 \langle x \rangle^{0} - F_1 \langle x - a_1 \rangle^{0} - F_2 \langle x - a_2 \rangle^{0} + R_2 \langle x - l \rangle^{0} \tag{2}$$

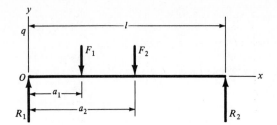

FIGURE 2-10

A second integration, in accordance with Eq. (2-25), yields

$$M = \int_{-\infty}^{x} V\,dx = R_1\langle x\rangle^1 - F_1\langle x - a_1\rangle^1 - F_2\langle x - a_2\rangle^1 + R_2\langle x - l\rangle^1 \tag{3}$$

The reactions R_1 and R_2 can be found by taking a summation of moments and forces as usual, or they can be found by noting that the shear force and bending moment must be zero everywhere except in the region $0 \le x \le l$. This means that Eq. (2) should give $V = 0$ at x slightly larger than l. Thus

$$R_1 - F_1 - F_2 + R_2 = 0 \tag{4}$$

Since the bending moment should also be zero in the same region, we have, from Eq. (3),

$$R_1 l - F_1(l - a_1) - F_2(l - a_2) = 0$$

Equations (4) and (5) can now be solved for the reactions R_1 and R_2.

EXAMPLE 2-3

Figure 2-11a shows the loading diagram for a beam cantilevered at O and having a uniform load w acting on the portion $a \le x \le l$. Derive the shear-force and moment relations. M_1 and R_1 are the support reactions.

Solution

Following the procedure of Example 2-2, we find the loading function to be

$$q = -M_1\langle x\rangle^{-2} + R_1\langle x\rangle^{-1} - w\langle x - a\rangle^0 \tag{1}$$

Then integrating successively gives

$$V = \int_{-\infty}^{x} q\,dx = -M_1\langle x\rangle^{-1} + R_1\langle x\rangle^0 - w\langle x - a\rangle^1 \tag{2}$$

$$M = \int_{-\infty}^{x} V\,dx = -M_1\langle x\rangle^0 + R_1\langle x\rangle^1 - \frac{w}{2}\langle x - a\rangle^2 \tag{3}$$

The reactions are found by making x slightly larger than l because both V and M are zero in this region. Equation (2) will then give

$$-M_1(0) + R_1 - w(l - a) = 0 \tag{4}$$

which can be solved for R_1. From Eq. (3) we get

$$-M_1 + R_1 l - \frac{w}{2}(l - a)^2 = 0 \tag{5}$$

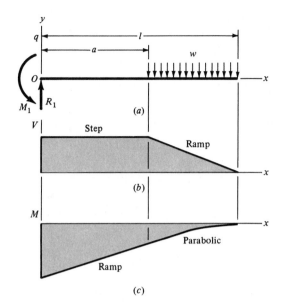

FIGURE 2-11

(a) Loading diagram for a beam cantilevered at O; (b) shear-force diagram; (c) bending-moment diagram.

which can be solved for M_1. Figure 2-11b and c shows the shear-force and bending-moment diagrams.

2-10   NORMAL STRESSES IN FLEXURE

In deriving the relations for the normal bending stresses in beams, we make the following idealizations:

1 The beam is subjected to pure bending; this means that the shear force is zero, and that no torsion or axial loads are present.

2 The material is isotropic and homogeneous.

3 The material obeys Hooke's law.

4 The beam is initially straight with a cross section that is constant throughout the beam length.

5 The beam has an axis of symmetry in the plane of bending.

6 The proportions of the beam are such that it would fail by bending rather than by crushing, wrinkling, or sidewise buckling.

7 Cross sections of the beam remain plane during bending.

In Fig. 2-12a we visualize a portion of a beam acted upon by the positive bending moment M. The y axis is the axis of symmetry. The x axis is coincident with the *neutral axis* of the section, and the xz plane, which contains the neutral axes of all cross sections, is called the *neutral plane*. Elements of the beam coincident with this plane

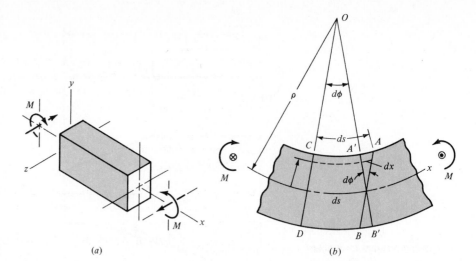

FIGURE 2-12   (a)   (b)

have zero strain. The location of the neutral axis with respect to the cross section has not yet been defined.

Application of the positive moment will cause the upper surface of the beam to bend downward, and the neutral axis will then be curved, as shown in Fig. 2-12b. Because of the curvature, a section AB originally parallel to CD, since the beam was straight, will rotate through the angle $d\phi$ to $A'B'$. Since AB and $A'B'$ are both straight lines, we have utilized the assumption that plane sections remain plane during bending. If we now specify the radius of curvature of the neutral axis as ρ, the length of a differential element of the neutral axis as ds, and the angle subtended by the two adjacent sides CD and $A'B'$ as $d\phi$, then, from the definition of curvature, we have

$$\frac{1}{\rho} = \frac{d\phi}{ds} \tag{a}$$

As shown in Fig. 2-12b, the deformation of a "fiber" at distance y from the neutral axis is

$$dx = y\,d\phi \tag{b}$$

The strain is the deformation divided by the original length, or

$$\epsilon = -\frac{dx}{ds} \tag{c}$$

where the negative sign indicates compression. Solving Eqs. (a), (b), and (c) simultaneously gives

$$\epsilon = -\frac{y}{\rho} \tag{d}$$

Thus the strain is proportional to the distance y from the neutral axis. Now, since $\sigma = E\epsilon$, we have for the stress

$$\sigma = -\frac{Ey}{\rho} \qquad\qquad (e)$$

We are now dealing with pure bending, which means that there are no axial forces acting on the beam. We can state this in mathematical form by summing all the horizontal forces acting on the cross section and equating this sum to zero. The force acting on an element of area dA is $\sigma\, dA$; therefore

$$\int \sigma\, dA = -\frac{E}{\rho} \int y\, dA = 0 \qquad\qquad (f)$$

Equation (f) defines the location of the neutral axis. The moment of the area about the neutral axis is zero, and hence the neutral axis passes through the centroid of the cross-sectional area.

Next we observe that equilibrium requires that the internal bending moment created by the stress σ be the same as the external moment M. In other words,

$$M = \int y\sigma\, dA = \frac{E}{\rho} \int y^2\, dA \qquad\qquad (g)$$

The second integral in Eq. (g) is the *second moment of area* about the z axis. This is,

$$I = \int y^2\, dA \qquad\qquad (2\text{-}26)$$

If we next solve Eqs. (g) and $(2\text{-}26)$ and rearrange them, we have

$$\frac{1}{\rho} = \frac{M}{EI} \qquad\qquad (2\text{-}27)$$

This is an important equation in the determination of the deflection of beams, and we shall employ it in Chap. 3. Finally, we eliminate ρ from Eqs. (e) and $(2\text{-}27)$ and obtain

$$\sigma = -\frac{My}{I} \qquad\qquad (2\text{-}28)$$

Equation $(2\text{-}28)$ states that the bending stress σ is directly proportional to the distance y from the neutral axis and the bending moment M, as shown in Fig. 2-13. It is custom-

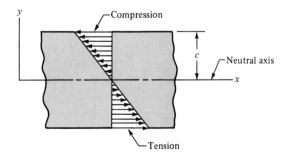

FIGURE 2-13

ary to designate $c = y_{max}$, to omit the negative sign, and to write

$$\sigma = \frac{Mc}{I} \tag{2-29}$$

where it is understood that Eq. (2-29) gives the maximum stress. Then tensile or compressive maximum stresses are determined by inspection when the sense of the moment is known.

Equation (2-29) is often written in the two alternative forms

$$\sigma = \frac{M}{I/c} \qquad \sigma = \frac{M}{Z} \tag{2-30}$$

where $Z = I/c$ is called the *section modulus*.

EXAMPLE 2-4   A beam having a T section with the dimensions shown in Fig. 2-14 is subjected to a bending moment of 1600 N · m. Locate the neutral axis and find the maximum tensile and compressive bending stresses. The sign of the moment produces tension at the top surface.

Solution   The area of the composite section is $A = 1956$ mm². Now divide the T section into two rectangles, numbered 1 and 2, and sum the moments of these areas about the top edge. We then have

$$1956c_1 = 12(75)(6) + 12(88)(56)$$

and hence $c_1 = 33.0$ mm. Therefore $c_2 = 100 - 33.0 = 67.0$ mm.

Next we calculate the second moment of area of each rectangle about its own centroidal axis. Using Table A-18, we find for the top rectangle

$$I_1 = \frac{bh^3}{12} = \frac{7.5(1.2)^3}{12} = 1.08 \text{ cm}^4$$

As indicated in Sec. 1-14, it is permissible to use nonpreferred prefixes when raising prefixed units to a power. In this case the use of cm in this equation instead of mm or

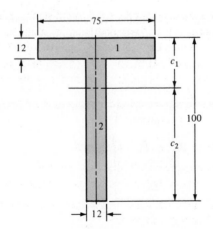

FIGURE 2-14

Dimensions in millimeters.

m yields the shortest number string and, hence, is preferred. For the bottom rectangle, we have

$$I_2 = \frac{bh^3}{12} = \frac{1.2(8.8)^3}{12} = 68.15 \text{ cm}^4$$

We now employ the parallel-axis theorem to obtain the second moment of area of the composite figure about its own centroidal axis. This theorem states

$$I_x = I_{cg} + Ad^2$$

where I_{cg} is the second moment of area about its own centroidal axis and I_x is the second moment of area about any parallel axis a distance d removed. For the top rectangle, the distance is

$$d_1 = 33.0 - 6 = 27.0 \text{ mm}$$

and for the bottom rectangle,

$$d_2 = 67.0 - 44 = 23.0 \text{ mm}$$

Using the parallel-axis theorem twice, we now find that

$$I = [1.08 + 9(2.70)^2] + [68.15 + 10.56(2.30)^2]$$
$$= 190.7 \text{ cm}^4$$

Finally, then, the maximum tensile stress, which occurs at the top surface, is found to be

$$\sigma = \frac{Mc_1}{I} = \frac{1600(3.30)}{190.7} = 27.69 \text{ MPa}$$

Note that we have used Table A-3 in solving this equation. Similarly, the maximum compressive stress at the lower surface is found to be

Answer
$$\sigma = -\frac{Mc_2}{I} = -\frac{1600(6.70)}{190.7} = -56.21 \text{ MPa}$$

EXAMPLE 2-5

Determine the diameter for the solid round shaft, 18 in long, shown in Fig. 2-15a. The shaft is supported by self-aligning bearings at the ends. Mounted upon the shaft are a V-belt sheave, which contributes a radial load of 400 lb to the shaft, and a gear, which contributes a radial load of 150 lb. The two loads are in the same plane and have the same directions. The bending stress is not to exceed 10 kpsi.

Solution

As indicated in Chap. 1, certain assumptions are necessary. In this problem we decide on the following conditions:

- The weight of the shaft is neglected.

- Since the bearings are self-aligning, the shaft is assumed to be simply supported and the loads and bearing reactions to be concentrated.

- The normal bending stress is assumed to govern the design at the location of the maximum bending moment.

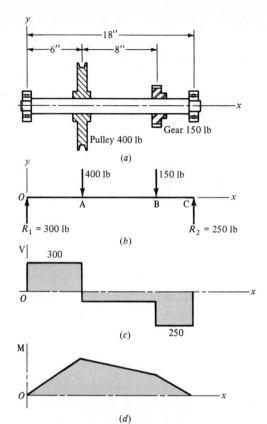

FIGURE 2-15

(a) Shaft drawing; dimensions in inches; (b) loading diagram; (c) shear-force diagram; (d) bending-moment diagram.

The assumed loading diagram is shown in Fig. 2-15b. Using Eq. (2-21) by taking moments about C gives

$$\Sigma M_C = -18R_1 + 12(400) + 4(150) = 0$$
$$R_1 = 300 \text{ lb}$$

To get R_2, we can sum forces using Eq. (2-20), or we can use Eq. (2-21) again for point O. With either method, we find

$$R_2 = 250 \text{ lb}$$

This statics problem can also be solved by vectors. We define the unit vector triad **ijk** in association with the x, y, and z axes, respectively. Then position vectors from the origin O to loading points A, B, and C are

$$\mathbf{r}_A = 6\mathbf{i} \qquad \mathbf{r}_B = 14\mathbf{i} \qquad \mathbf{r}_C = 18\mathbf{i}$$

Also, the force vectors are

$$\mathbf{R}_1 = R_1\mathbf{j} \qquad \mathbf{F}_A = -400\mathbf{j} \qquad \mathbf{F}_B = -150\mathbf{j} \qquad \mathbf{R}_2 = R_2\mathbf{j}$$

Then taking moments about O yields the vector equation

$$\mathbf{r}_A \times \mathbf{F}_A + \mathbf{r}_B \times \mathbf{F}_B + \mathbf{r}_C \times \mathbf{R}_2 = \mathbf{0} \tag{1}$$

The terms in Eq. (1) are called *vector cross products*. The cross product of the two vectors

$$\mathbf{A} = x_A\mathbf{i} + y_A\mathbf{j} + z_A\mathbf{k} \qquad \mathbf{B} = x_B\mathbf{i} + y_B\mathbf{j} + z_B\mathbf{k}$$

is

$$\mathbf{A} \times \mathbf{B} = (y_A z_B - z_A y_B)\mathbf{i} + (z_A x_B - x_A z_B)\mathbf{j} + (x_A y_B - y_A x_B)\mathbf{k} \qquad (2\text{-}31)$$

But this relation is more conveniently viewed as the determinant

$$\mathbf{A} \times \mathbf{B} = \begin{vmatrix} \mathbf{i} & \mathbf{j} & \mathbf{k} \\ x_A & y_A & z_A \\ x_B & y_B & z_B \end{vmatrix} \qquad (2\text{-}32)$$

Thus we compute the terms in Eq. (1) as follows:

$$\mathbf{r}_A \times \mathbf{F}_A = \begin{vmatrix} \mathbf{i} & \mathbf{j} & \mathbf{k} \\ 6 & 0 & 0 \\ 0 & -400 & 0 \end{vmatrix} = -2400\mathbf{k}$$

$$\mathbf{r}_B \times \mathbf{F}_B = \begin{vmatrix} \mathbf{i} & \mathbf{j} & \mathbf{k} \\ 14 & 0 & 0 \\ 0 & -150 & 0 \end{vmatrix} = -2100\mathbf{k}$$

$$\mathbf{r}_C \times \mathbf{R}_2 = \begin{vmatrix} \mathbf{i} & \mathbf{j} & \mathbf{k} \\ 18 & 0 & 0 \\ 0 & R_2 & 0 \end{vmatrix} = 18R_2\mathbf{k}$$

Substituting these three terms into Eq. (1) and solving the resulting algebraic equation gives $R_2 = 250$ lb. Hence $\mathbf{R}_2 = 250\mathbf{j}$. Next, we write

$$\mathbf{R}_1 = -\mathbf{F}_A - \mathbf{F}_B - \mathbf{R}_2 = -(-400\mathbf{j}) - (-150\mathbf{j}) - 250\mathbf{j} = 300\mathbf{j} \text{ lb}$$

Though the vector approach does seem more laborious, it is easy to program and is especially useful for three-dimensional problems.

The next step is to draw the shear-force and bending-moment diagrams (Fig. 2-15c and d). The maximum bending moment is found to be

$$M = 300(6) = 1800 \text{ lb} \cdot \text{in}$$

The section modulus is

$$\frac{I}{c} = \frac{\pi d^3}{32} = 0.0982d^3 \qquad (2)$$

Then, using Eq. (2-30),

$$\sigma = \frac{M}{I/c} = \frac{1800}{0.0982d^3}$$

Substituting $\sigma = 10\ 000$ psi and solving for d yields

$$d = \sqrt[3]{\frac{1800}{0.0982(10\ 000)}} = 1.22 \text{ in}$$

Therefore we select $d = 1\frac{1}{4}$ in for the shaft diameter.

2-11 BEAMS WITH ASYMMETRICAL SECTIONS

The relations developed in Sec. 2-10 can also be applied to beams having asymmetrical sections, provided that the plane of bending coincides with one of the two principal axes of the section. We have found that the stress at a distance y from the neutral axis is

$$\sigma = -\frac{Ey}{\rho} \qquad (a)$$

Therefore, the force on the element of area dA in Fig. 2-16 is

$$dF = \sigma\, dA = -\frac{Ey}{\rho}\, dA$$

Taking moments of this force about the y axis and integrating across the section gives

$$M_y = \int z\, dF = \int \sigma z\, dA = -\frac{E}{\rho} \int yz\, dA \qquad (b)$$

We recognize that the last integral in Eq. (b) is the product of inertia I_{yz}. If the bending moment on the beam is in the plane of one of the principal axes, then

$$I_{yz} = \int yz\, dA = 0 \qquad (c)$$

With this restriction, the relations developed in Sec. 2-10 hold for any cross-sectional shape. Of course, this means that the designer has a special responsibility to ensure that the bending loads do, in fact, come onto the beam in a principal plane.

2-12 SHEAR STRESSES IN BEAMS

Most beams have *both* shear forces and bending moments present. It is only occasionally that we encounter beams subjected to pure bending, that is to say, beams having

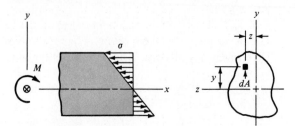

FIGURE 2-16

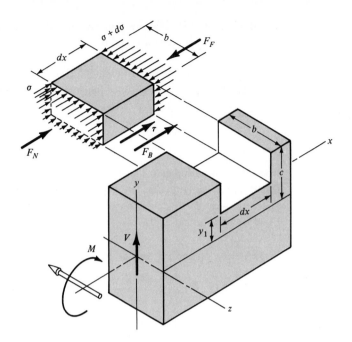

FIGURE 2-17

zero shear force. And yet, the flexure formula was developed using the assumption of pure bending. As a matter of fact, the reason for assuming pure bending was simply to eliminate the complicating effects of shear force in the development. For engineering purposes, the flexure formula is valid no matter whether a shear force is present or not. For this reason, we shall utilize the same normal bending-stress distribution [Eqs. (2-28) and (2-29)] when shear forces are present too.

In Fig. 2-17, we show a beam of constant cross section subjected to a shear force V and a bending moment M. The direction of the bending moment is easier to visualize by associating the hollow vector with your right hand. The hollow vector points in the negative z direction. If you will place the thumb of your right hand in the negative z direction, then your fingers, when bent, will indicate the direction of the moment M. By Eq. (2-22), the relationship of V to M is

$$V = \frac{dM}{dx} \tag{a}$$

At some point along the beam, we cut a transverse section of length dx down to a distance y_1 above the neutral axis, as illustrated. We remove this section to study the forces which act. Because a shear force is present, the bending moment is changing as we move along the x axis. Thus, we can designate the bending moment as M on the near side of the section and as $M + dM$ on the far side. The moment M produces a normal stress σ, and the moment $M + dM$, a normal stress $\sigma + d\sigma$, as shown. These normal stresses produce normal forces on the vertical faces of the element, the compressive force on the far side being greater than on the near side. The resultant of these two would cause the section to tend to slide in the $-x$ direction, and so this resultant must be balanced by a shear force acting in the $+x$ direction on the bottom of the section. This shear force results in a shear stress τ, as shown. Thus, there are three resultant forces acting on the element: F_N, due to σ, acts on the near face; F_F, due to $\sigma + d\sigma$, acts on the far face; and F_B, due to τ, acts on the bottom face. Let us evaluate these forces.

For the near face, select an element of area dA. The stress acting on this area is σ, and so the force is the stress times the area, or $\sigma\, dA$. The force acting on the entire near face is the sum of all the $\sigma\, dA$'s, or

$$F_N = \int_{y_1}^{c} \sigma\, dA \tag{b}$$

where the limits indicate that we integrate from the bottom $y = y_1$ to the top $y = c$. Using $\sigma = My/I$ [Eq. (2-28)], Eq. (b) becomes

$$F_N = \frac{M}{I} \int_{y_1}^{c} y\, dA \tag{c}$$

The force on the far face is found in a similar manner. It is

$$F_F = \int_{y_1}^{c} (\sigma + d\sigma)\, dA = \frac{M + dM}{I} \int_{y_1}^{c} y\, dA \tag{d}$$

The force on the bottom face is the shear stress τ times the area of the bottom face. Since this area is $b\, dx$, we have

$$F_B = \tau b\, dx \tag{e}$$

Summing these three forces in the x direction gives

$$\Sigma F_x = +F_N - F_F + F_B = 0 \tag{f}$$

If we substitute Eqs. (c) and (d) for F_N and F_F and solve the result for F_B, we get

$$F_B = F_F - F_N = \frac{M + dM}{I} \int_{y_1}^{c} y\, dA - \frac{M}{I} \int_{y_1}^{c} y\, dA = \frac{dM}{I} \int_{y_1}^{c} y\, dA \tag{g}$$

Next, using Eq. (e) for F_B and solving for the shear stress gives

$$\tau = \frac{dM}{dx} \frac{1}{Ib} \int_{y_1}^{c} y\, dA \tag{h}$$

Then, by the use of Eq. (a), we finally get the shear-stress formula as

$$\tau = \frac{V}{Ib} \int_{y_1}^{c} y\, dA \tag{2-33}$$

In this equation, the integral is the first moment of the area of the vertical face about the neutral axis. This moment is usually designated Q. Thus,

$$Q = \int_{y_1}^{c} y\, dA \tag{2-34}$$

With this final simplification, Eq. (2-33) may be written as

$$\tau = \frac{VQ}{Ib} \tag{2-35}$$

In using this equation, note that b is the width of the section at the particular distance y_1 from the neutral axis. Also, I is the second moment of area of the entire section about the neutral axis.

2-13   SHEAR STRESSES IN RECTANGULAR-SECTION BEAMS

The purpose of this section is to show how the equations of the preceding section are used to find the shear-stress distribution in a beam having a rectangular cross section. Figure 2-18 shows a portion of a beam subjected to a shear force V and a bending moment M. As a result of the bending moment, a normal stress σ is developed on a cross section such as A-A, which is in compression above the neutral axis and in tension below. To investigate the shear stress at a distance y_1 above the neutral axis, we select an element of area dA at a distance y above the neutral axis. Then, $dA = b\,dy$, and so Eq. (2-34) becomes

$$Q = \int_{y_1}^{c} y\, dA = b \int_{y_1}^{c} y\, dy = \left.\frac{by^2}{2}\right|_{y_1}^{c} = \frac{b}{2}(c^2 - y_1^2) \tag{a}$$

Substituting this value for Q into Eq. (2-35) gives

$$\tau = \frac{V}{2I}(c^2 - y_1^2) \tag{2-36}$$

This is the general equation for shear stress in a rectangular beam. To learn something about it, let us make some substitutions. From Table A-18, we learn that the second

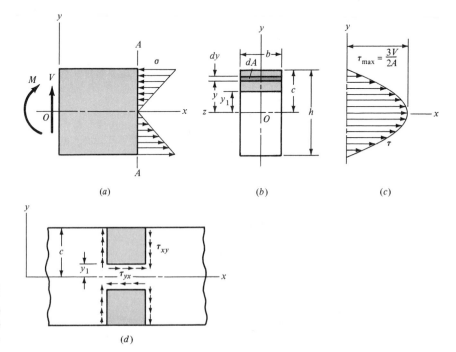

(a)   (b)   (c)

(d)

FIGURE 2-18

Note in part d that τ_{xy} denotes the vertical shear and τ_{yx} denotes the horizontal shear.

TABLE 2-3

Variation of Shear Stress
$\tau = C(V/A)$

DISTANCE y_1	0	0.2c	0.4c	0.6c	0.8c	c
Factor C	1.50	1.44	1.26	0.96	0.54	0

moment of area for a rectangular section is $I = bh^3/12$; substituting $h = 2c$ and $A = bh = 2bc$ gives

$$I = \frac{Ac^2}{3} \qquad (b)$$

If we now use this value of I for Eq. (2-36) and rearrange, we get

$$\tau = \frac{3V}{2A}\left(1 - \frac{y_1^2}{c^2}\right) \qquad (2\text{-}37)$$

Now let us substitute various values of y_1, beginning with $y_1 = 0$ and ending with $y_1 = c$. The results are displayed in Table 2-3. We note that the maximum shear stress exists with $y_1 = 0$, which is at the neutral axis. Thus

$$\tau_{max} = \frac{3V}{2A} \qquad (2\text{-}38)$$

TABLE 2-4

Formulas for Maximum Shear
Stress Due to Bending

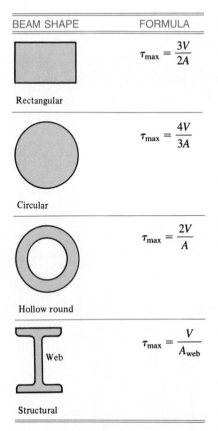

BEAM SHAPE	FORMULA
Rectangular	$\tau_{max} = \dfrac{3V}{2A}$
Circular	$\tau_{max} = \dfrac{4V}{3A}$
Hollow round	$\tau_{max} = \dfrac{2V}{A}$
Structural (Web)	$\tau_{max} = \dfrac{V}{A_{web}}$

for a rectangular section. As we move away from the neutral axis, the shear stress decreases until it is zero at the outer surface where $y_1 = c$. This is a parabolic distribution and is shown in Fig. 2-18c. It is particularly interesting and significant here to observe that the shear stress is maximum at the neutral axis, where the normal stress due to bending is zero, and that the shear stress is zero at the outer surfaces, where the bending stress is a maximum. Horizontal shear stress is always accompanied by vertical shear stress of the same magnitude, and so the distribution can be diagrammed as shown in Fig. 2-18d. Figure 2-18c shows that the shear τ_{xy} on the vertical surfaces varies with y. We are almost always interested in the horizontal shear, τ_{yx} in Fig. 2-18d, which is nearly uniform with constant y. The maximum horizontal shear occurs where the vertical shear is largest. This is usually at the neutral axis but may not be if the width b is smaller somewhere else. Furthermore, if the section is such that b can be minimized on a plane not horizontal, then the horizontal shear stress occurs on an inclined plane. For example, with tubing, the horizontal shear stress occurs on a radial plane and the corresponding "vertical shear" is not vertical, but tangential.

Formulas for the maximum flexural shear stress for the most commonly used shapes are listed in Table 2-4.

2-14   TORSION

Any moment vector that is collinear with an axis of a mechanical element is called a *torque vector,* because the moment causes the element to be twisted about that axis. A bar subjected to such a moment is also said to be in *torsion.*

As shown in Fig. 2-19, the torque T applied to a bar can be designated by drawing arrows on the surface of the bar to indicate direction or by drawing torque-vector arrows along the axes of twist of the bar. Torque vectors are the hollow arrows shown on the x axis in Fig. 2-19. Note that they conform to the right-hand rule for vectors.

The angle of twist for a solid round bar is

$$\theta = \frac{Tl}{GJ} \tag{2-39}$$

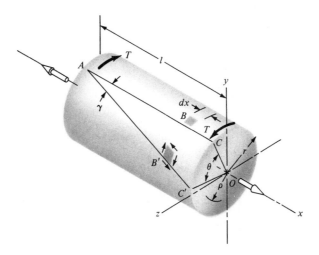

FIGURE 2-19

where T = torque

$\quad l$ = length

$\quad G$ = modulus of rigidity

$\quad J$ = polar second moment of area

For a solid round bar, the shear stress is zero at the center and maximum at the surface. The distribution is proportional to the radius ρ and is

$$\tau = \frac{T\rho}{J} \tag{2-40}$$

Designating r as the radius to the outer surface, we have

$$\tau_{max} = \frac{Tr}{J} \tag{2-41}$$

The assumptions used in the analysis are:

- The bar is acted upon by a pure torque, and the sections under consideration are remote from the point of application of the load and from a change in diameter.
- Adjacent cross sections originally plane and parallel remain plane and parallel after twisting, and any radial line remains straight.
- The material obeys Hooke's law.

Equation (2-41) applies only to circular sections. For a solid round section,

$$J = \frac{\pi d^4}{32} \tag{2-42}$$

where d is the diameter of the bar. For a hollow round section,

$$J = \frac{\pi}{32} (d_o^4 - d_i^4) \tag{2-43}$$

where the subscripts o and i refer to the outside and inside diameters, respectively (OD and ID).

In using Eq. (2-41) it is often necessary to obtain the torque T from a consideration of the power and speed of a rotating shaft. For convenience, three forms of this relation are

$$H = \frac{2\pi Tn}{33\ 000(12)} = \frac{FV}{33\ 000} = \frac{Tn}{63\ 000} \tag{2-44}$$

where H = power, hp

$\quad T$ = torque, lb · in

$\quad n$ = shaft speed, rev/min

$\quad F$ = force, lb

$\quad V$ = velocity, ft/min

When SI units are used, the equation is

$$H = T\omega \tag{2-45}$$

where H = power, W

T = torque, N · m

ω = angular velocity, rad/s

The torque T corresponding to the power in watts is given approximately by

$$T = 9.55 \frac{H}{n} \tag{2-46}$$

where n is in revolutions per minute.

Determination of the torsional stresses in noncircular members is a difficult problem, generally handled experimentally using a soap-film or membrane analogy, or analytically using finite-element techniques. The following approximate formula is useful for estimating the maximum torsional stress in a rectangular section:

$$\tau_{max} = \frac{T}{wt^2}\left(3 + 1.8\,\frac{t}{w}\right) \tag{2-47}$$

In this equation, w and t are the width and thickness of the bar, respectively; they cannot be interchanged, since t must be the shortest dimension. For thin plates in torsion, t/w is small, and the second term may be neglected. The equation is also approximately valid for equal-sided angles; these can be considered as two rectangles, each of which is capable of carrying half the torque.

EXAMPLE 2-6

Figure 2-20 shows a crank loaded by a force $F = 300$ lb which causes twisting and bending of a $\frac{3}{4}$-in-diameter shaft fixed to a support at the origin of the reference system. In actuality, the support may be an inertia which we wish to rotate, but for the purposes of a stress analysis we can consider this as a statics problem.

(a) Draw separate free-body diagrams of the shaft AB and the arm BC, and compute the values of all forces, moments, and torques which act. Label the directions of the coordinate axes on these diagrams.

(b) Compute the maxima of the torsional stress and the bending stress in the arm BC and indicate where these act.

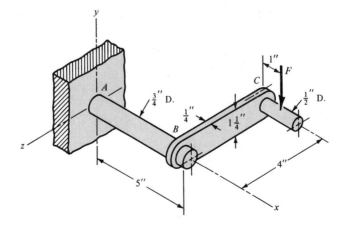

FIGURE 2-20

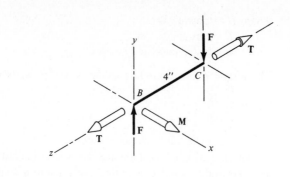

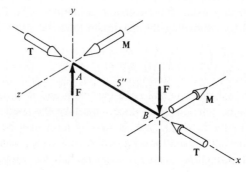

FIGURE 2-21

(c) Locate a stress element on the top surface of the shaft at A, and calculate all the stress components which act upon this element.

Solution   (a) The two free-body diagrams are shown in Fig. 2-21. The results are

AT C: $\mathbf{F} = -300\mathbf{j}$ lb, $\mathbf{T} = -300\mathbf{k}$ lb · in

AT END B OF ARM BC: $\mathbf{F} = 300\mathbf{j}$ lb, $\mathbf{M} = 1200\mathbf{i}$ lb · in, $\mathbf{T} = 300\mathbf{k}$ lb · in

AT END B OF SHAFT AB: $\mathbf{F} = -300\mathbf{j}$ lb, $\mathbf{T} = -1200\mathbf{i}$ lb · in, $\mathbf{M} = -300\mathbf{k}$ lb · in

AT A: $\mathbf{F} = 300\mathbf{j}$ lb, $\mathbf{M} = 1800\mathbf{k}$ lb · in, $\mathbf{T} = 1200\mathbf{i}$ lb · in

(b) For arm *BC*, the bending stress will reach a maximum near the shaft at *B*. If we assume this is 1200 lb · in, then the bending stress for a rectangular section will be

Answer $$\sigma = \frac{M}{I/c} = \frac{6M}{bh^2} = \frac{6(1200)}{0.25(1.25)^2} = 18\,400 \text{ psi}$$

Of course, this is not exactly correct, because at *B* the moment is actually being transferred into the shaft, probably through a weldment.

For the torsional stress, use Eq. (2-47). Thus

Answer $$\tau_{\max} = \frac{T}{wt^2}\left(3 + 1.8\,\frac{t}{w}\right) = \frac{300}{1.25(0.25)^2}\left(3 + 1.8\,\frac{0.25}{1.25}\right) = 12\,900 \text{ psi}$$

This stress occurs at the middle of the $1\frac{1}{4}$-in side.

(c) For a stress element at A, the bending stress is

Answer $\sigma_x = \dfrac{M}{I/c} = \dfrac{32M}{\pi d^3} = \dfrac{32(1800)}{\pi(0.75)^3} = 43\ 400$ psi

The torsional stress is

Answer $\tau_{xz} = \dfrac{T}{J/c} = \dfrac{16T}{\pi d^3} = \dfrac{16(1200)}{\pi(0.75)^3} = 14\ 500$ psi

2-15   STRESS CONCENTRATION

In the development of the basic stress equations for tension, compression, bending, and torsion, it was assumed that no irregularities occurred in the member under consideration. But it is quite difficult to design a machine without permitting some changes in the cross sections of the members. Rotating shafts must have shoulders designed on them so that the bearings can be properly seated and so that they will take thrust loads; and the shafts must have key slots machined into them for securing pulleys and gears. A bolt has a head on one end and screw threads on the other end, both of which account for abrupt changes in the cross section. Other parts require holes, oil grooves, and notches of various kinds. Any discontinuity in a machine part alters the stress distribution in the neighborhood of the discontinuity so that the elementary stress equations no longer describe the state of stress in the part. Such discontinuities are called *stress raisers,* and the regions in which they occur are called areas of *stress concentration.*

To help in understanding this effect, examine Fig. 2-22. Note that the stress trajectories are uniform everywhere except in the vicinity of the hole. But at the hole these lines of force must bend to get around. Stress concentration is a highly localized effect. The stress on the tension plate is highest at the edge of the hole on plane A-A; this stress

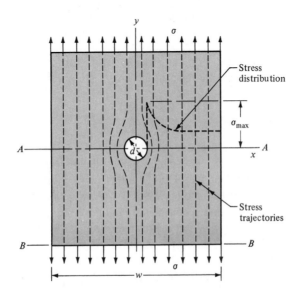

FIGURE 2-22

Stress distribution near a hole in a plate loaded in tension. The tensile stress on a section at B-B, remote from the hole, is $\sigma = F/A$, where $A = wt$ and t is the plate thickness. On a section at A-A, through the hole, the area is $A_0 = (w - d)t$ and the nominal stress is $\sigma_0 = F/A_0$. Note the difference between the nominal stress and the stress at a section remote from the discontinuity.

drops rapidly as points are examined farther from the hole edge and soon becomes uniform again.

A *theoretical*, or *geometric, stress-concentration factor* K_t or K_{ts} is used to relate the actual maximum stress at the discontinuity to the nominal stress. The factors are defined by the equations

$$K_t = \frac{\sigma_{\max}}{\sigma_0} \qquad K_{ts} = \frac{\tau_{\max}}{\tau_0} \tag{2-48}$$

where K_t is used for normal stresses and K_{ts} for shear stresses. The nominal stress σ_0 or τ_0 is more difficult to define. Generally, it is the stress calculated by using the elementary stress equations and the net area, or net cross section. But sometimes the gross cross section is used instead, and so it is always wise to check before calculating the maximum stress.

The subscript t in K_t means that this stress-concentration factor depends for its value only on the *geometry* of the part. That is, the particular material used has no effect on the value of K_t. This is why it is called a *theoretical* stress-concentration factor.

The analysis of geometric shapes to determine stress-concentration factors is a difficult problem, and not many solutions can be found. One such solution is that of an infinite plate containing an elliptical hole loaded in uniform tension. The result is

$$K_t = 1 + \frac{2b}{a} \tag{2-49}$$

where, after replacing the hole in Fig. 2-22 with an ellipse, b is the half-width, a is the half-height, and $w = \infty$. Thus, for a circular hole, $b = a$ and $K_t = 3$.

Note that Eq. (2-49) can be applied to a transverse crack, where $b \gg a$, or to a longitudinal crack ($b \ll a$).

Most stress-concentration factors are found by using experimental techniques.* Though the finite-element method has been used, the fact that the elements are indeed finite prevents finding the true maximum stress. Experimental approaches generally used include photoelasticity, grid methods, brittle-coating methods, and electrical strain-gauge methods. Of course, the grid and strain-gauge methods both suffer from the same drawback as the finite-element method.

Stress-concentration factors for a variety of geometries may be found in Tables A-15 and A-16.

2-16   STRESSES IN CYLINDERS

Cylindrical pressure vessels, hydraulic cylinders, gun barrels, and pipes carrying fluids at high pressures develop both radial and tangential stresses with values that are dependent upon the radius of the element under consideration. In determining the radial stress σ_r and the tangential stress σ_t, we make use of the assumption that the longitudinal elongation is constant around the circumference of the cylinder. In other words, a right section of the cylinder remains plane after stressing.

---

*The best source book is R. E. Peterson, *Stress Concentration Factors*, Wiley, New York, 1974.

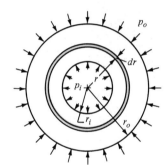

FIGURE 2-23

A cylinder subjected to both internal and external pressure.

Referring to Fig. 2-23, we designate the inside radius of the cylinder by r_i, the outside radius by r_o, the internal pressure by p_i, and the external pressure by p_o. Then it can be shown that tangential and radial stresses exist whose magnitudes are

$$\sigma_t = \frac{p_i r_i^2 - p_o r_o^2 - r_i^2 r_o^2 (p_o - p_i)/r^2}{r_o^2 - r_i^2}$$

$$\sigma_r = \frac{p_i r_i^2 - p_o r_o^2 + r_i^2 r_o^2 (p_o - p_i)/r^2}{r_o^2 - r_i^2}$$

(2-50)

As usual, positive values indicate tension and negative values, compression.

The special case of $p_o = 0$ gives

$$\sigma_t = \frac{r_i^2 p_i}{r_o^2 - r_i^2} \left(1 + \frac{r_o^2}{r^2}\right)$$

$$\sigma_r = \frac{r_i^2 p_i}{r_o^2 - r_i^2} \left(1 - \frac{r_o^2}{r^2}\right)$$

(2-51)

The equations of set (2-51) are plotted in Fig. 2-24 to show the distribution of stresses over the wall thickness.

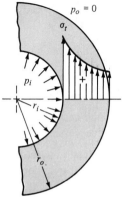

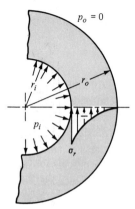

FIGURE 2-24

Distribution of stresses in a thick-walled cylinder subjected to internal pressure.

(a) Tangential stress distribution

(b) Radial stress distribution

It should be realized that longitudinal stresses exist when the end reactions to the internal pressure are taken by the pressure vessel itself. This stress is found to be

$$\sigma_l = \frac{p_i r_i^2}{r_o^2 - r_i^2} \tag{2-52}$$

We further note that Eqs. (2-50), (2-51), and (2-52) apply only to sections taken a significant distance from the ends and away from any areas of stress concentration.

Thin-Walled Vessels

When the wall thickness of a cylindrical pressure vessel is about one-twentieth, or less, of its radius, the radial stress which results from pressurizing the vessel is quite small compared with the tangential stress. Under these conditions the tangential stress can be obtained as follows: Let an internal pressure p be exerted on the wall of a cylinder of thickness t and inside diameter d_i. The force tending to separate two halves of a unit length of the cylinder is pd_i. This force is resisted by the tangential stress, also called the *hoop stress,* acting uniformly over the stressed area. We then have $pd_i = 2t\sigma_t$, or

$$\sigma_{t,\text{av}} = \frac{pd_i}{2t} \tag{2-53}$$

This equation gives the *average* tangential stress and is valid regardless of the wall thickness. For a thin-walled vessel an approximation to the maximum tangential stress is

$$\sigma_{t,\text{max}} = \frac{p(d_i + t)}{2t} \tag{2-54}$$

where $d_i + t$ is the average diameter.

In a closed cylinder, the longitudinal stress σ_l exists because of the pressure upon the ends of the vessel. If we assume this stress is also distributed uniformly over the wall thickness, we can easily find it to be

$$\sigma_l = \frac{pd_i}{4t} \tag{2-55}$$

EXAMPLE 2-7

An aluminum-alloy pressure vessel is made of tubing having an outside diameter of 8 in and a wall thickness of $\frac{1}{4}$ in.

(a) What pressure can the cylinder carry if the permissible tangential stress is 12 kpsi and the theory for thin-walled vessels is assumed to apply?

(b) On the basis of the pressure found in part (a), compute all of the stress components using the theory for thick-walled cylinders.

Solution

(a) Here $d_i = 8 - 2(0.25) = 7.5$ in, $r_i = 7.5/2 = 3.75$ in, and $r_o = 8/2 = 4$ in. Then $t/r_i = 0.25/3.75 = 0.067$. Since this ratio is greater than $\frac{1}{20}$, the theory for thin-walled vessels may not yield safe results.

We first solve Eq. (2-54) to obtain the allowable pressure. This gives

Answer $$p = \frac{2t\sigma_{t,\max}}{d_i + t} = \frac{2(0.25)(12)(10)^3}{7.5 + 0.25} = 774 \text{ psi}$$

Then, from Eq. (2-55), we find the average longitudinal stress to be

$$\sigma_l = \frac{pd_i}{4t} = \frac{774(7.5)}{4(0.25)} = 5800 \text{ psi}$$

(b) The maximum tangential stress will occur at the inside radius, and so we use $r = r_i$ in the first equation of the pair (2-51). This gives

Answer $$\sigma_t = p_i \frac{r_o^2 + r_i^2}{r_o^2 - r_i^2} = 774 \frac{4^2 + 3.75^2}{4^2 - 3.75^2} = 12\ 000 \text{ psi}$$

Similarly, the maximum radial stress is found, from the second equation of the pair (2-51), to be

Answer $$\sigma_r = -p_i = -774 \text{ psi}$$

Equation (2-52) gives the longitudinal stress as

Answer $$\sigma_l = \frac{p_i r_i^2}{r_o^2 - r_i^2} = \frac{774(3.75)^2}{4^2 - 3.75^2} = 5620 \text{ psi}$$

These three stresses σ_t, σ_r, and σ_l are principal stresses, since there is no shear. Note that there is no significant difference in the tangential stresses in parts (a) and (b), and so the thin-wall theory can be considered satisfactory.

2-17   ROTATING RINGS

Many rotating elements, such as flywheels and blowers, can be simplified to a rotating ring to determine the stresses. When this is done it is found that the same tangential and radial stresses exist as in the theory for thick-walled cylinders except that they are caused by inertial forces acting on all the particles of the ring. The tangential and radial stresses so found are subject to the following restrictions:

- The outside radius of the ring, or disk, is large compared with the thickness $r_o \geq 10t$.
- The thickness of the ring or disk is constant.
- The stresses are constant over the thickness.

The stresses are

$$\sigma_t = \rho\omega^2 \left(\frac{3 + \nu}{8}\right) \left(r_i^2 + r_o^2 + \frac{r_i^2 r_o^2}{r^2} - \frac{1 + 3\nu}{3 + \nu} r^2\right)$$

$$\sigma_r = \rho\omega^2 \left(\frac{3 + \nu}{8}\right) \left(r_i^2 + r_o^2 - \frac{r_i^2 r_o^2}{r^2} - r^2\right)$$

(2-56)

where r is the radius to the stress element under consideration, ρ is the mass density, and ω is the angular velocity of the ring in radians per second. For a rotating disk, use $r_i = 0$ in these equations.

2-18   PRESS AND SHRINK FITS

When two cylindrical parts are assembled by shrinking or press-fitting one part upon another, a contact pressure is created between the two parts. The stresses resulting from this pressure may easily be determined with the equations of the preceding sections.

Figure 2-25b shows two cylindrical members which have been assembled with a shrink fit. A contact pressure p exists between the members at the transition radius R, causing radial stresses $\sigma_r = -p$ in each member at the contacting surfaces. From Sec. 2-16, we find the tangential stress at the transition radius of the inner member to be

$$\sigma_{it} \text{ (at } R) = -p\, \frac{R^2 + r_i^2}{R^2 - r_i^2} \tag{2-57}$$

In the same manner, the tangential stress at the inner surface of the outer member is found to be

$$\sigma_{ot} \text{ (at } R) = p\, \frac{r_o^2 + R^2}{r_o^2 - R^2} \tag{2-58}$$

These equations cannot be solved until the contact pressure is known. In obtaining a shrink fit, the radius of the male member is made larger than the radius of the female member. The difference in these dimensions is called the *radial interference* and is the radial deformation which the two members must experience. Since these dimensions are usually known, the deformation should be introduced in order to evaluate the stresses. As shown in Fig. 2-25a, δ_i and δ_o symbolize the changes in the radii of the inner and outer members, respectively. The total radial interference is, therefore,

$$\delta = |\delta_i| - |\delta_o| \tag{a}$$

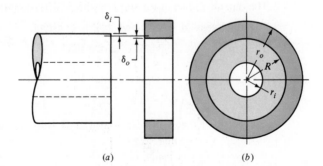

FIGURE 2-25

Notation for press and shrink fits.
(a) Unassembled parts; (b) after assembly.

(a)   (b)

The tangential strain at the transition radius of the outer cylinder is measured by the change in circumference, and is

$$\epsilon_{ot} = \frac{2\pi(R + \delta_o) - 2\pi R}{2\pi R} = \frac{\delta_o}{R} \qquad (b)$$

and so $\delta_o = R\epsilon_{ot}$. But, since

$$\epsilon_{ot} = \frac{\sigma_{ot}}{E_o} - \frac{\nu_o \sigma_{or}}{E_o} \qquad (c)$$

then, from Eqs. (2-57) and (2-58), we have

$$\delta_o = \frac{pR}{E_o} \left(\frac{r_o^2 + R^2}{r_o^2 - R^2} + \nu_o \right) \qquad (d)$$

This is the change in radius of the outer member. In a similar manner, the change in radius of the inner member is found to be

$$\delta_i = -\frac{pR}{E_i} \left(\frac{R^2 + r_i^2}{R^2 - r_i^2} - \nu_i \right) \qquad (e)$$

Then, from Eq. (a), we have for the total deformation

$$\delta = \frac{pR}{E_o} \left(\frac{r_o^2 + R^2}{r_o^2 - R^2} + \nu_o \right) + \frac{pR}{E_i} \left(\frac{R^2 + r_i^2}{R^2 - r_i^2} - \nu_i \right) \qquad (2\text{-}59)$$

This equation can be solved for the pressure p when the radial interference δ is given. If the two members are of the same material, $E_o = E_i = E$, $\nu_o = \nu_i$, and the relation simplifies to

$$p = \frac{E\delta}{R} \left[\frac{(r_o^2 - R^2)(R^2 - r_i^2)}{2R^2(r_o^2 - r_i^2)} \right] \qquad (2\text{-}60)$$

The value of the interface pressure p from either Eq. (2-59) or Eq. (2-60) can now be used to obtain the stress state at the specified radius in either cylinder.

Assumptions

In addition to the assumptions both stated and implied by the development, it is necessary to assume that both members have the same length. In the case of a hub which has been press-fitted to a shaft, this assumption would not be true, and there would be an increased pressure at each end of the hub. It is customary to allow for this condition by the employment of a stress-concentration factor. The value of this factor depends upon the contact pressure and the design of the female member, but its theoretical value is seldom greater than 2.

2-19 TEMPERATURE EFFECTS

When the temperature of an unrestrained body is uniformly increased, the body expands, and the normal strain is

$$\epsilon_x = \epsilon_y = \epsilon_z = \alpha(\Delta T) \qquad (2\text{-}61)$$

TABLE 2-5

Coefficients of Thermal Expansion (Linear Mean Coefficients for the Temperature Range 0–100°C)

MATERIAL	CELSIUS SCALE	FAHRENHEIT SCALE
Aluminum	$23.9(10)^{-6}$	$13.3(10)^{-6}$
Brass, cast	$18.7(10)^{-6}$	$10.4(10)^{-6}$
Carbon steel	$10.8(10)^{-6}$	$6.0(10)^{-6}$
Cast iron	$10.6(10)^{-6}$	$5.9(10)^{-6}$
Magnesium	$25.2(10)^{-6}$	$14.0(10)^{-6}$
Nickel steel	$13.1(10)^{-6}$	$7.3(10)^{-6}$
Stainless steel	$17.3(10)^{-6}$	$9.6(10)^{-6}$
Tungsten	$4.3(10)^{-6}$	$2.4(10)^{-6}$

where α is the coefficient of thermal expansion and ΔT is the temperature change, in degrees. In this action the body experiences a simple volume increase with the components of shear strain all zero.

If a straight bar is restrained at the ends so as to prevent lengthwise expansion and then is subjected to a uniform increase in temperature, a compressive stress will develop because of the axial constraint. The stress is

$$\sigma = \epsilon E = \alpha(\Delta T)E \tag{2-62}$$

In a similar manner, if a uniform flat plate is restrained at the edges and also subjected to a uniform temperature rise, the compressive stress developed is given by the equation

$$\sigma = \frac{\alpha(\Delta T)E}{1 - \nu} \tag{2-63}$$

The stresses expressed by Eqs. (2-62) and (2-63) are called *temperature stresses*. They arise because of a temperature change in a clamped or restrained member. Such stresses, for example, occur during welding, since parts to be welded must be clamped before welding. Table 2-5 lists approximate values of the coefficients of thermal expansion.

A *thermal stress* is one which arises because of the existence of a *temperature gradient* in a member. Figure 2-26 is an example. Shown are the stress distributions within a slab of infinite dimensions during heating and cooling. During cooling, the maximum stress is the surface tension. During heating, the external surfaces are hot and tend to expand but are restrained by the cooler center. This causes compression in the surface and tension in the center as shown.

FIGURE 2-26

Thermal stresses in an infinite slab during heating and cooling.

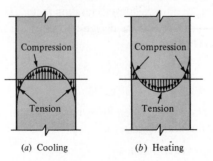

(a) Cooling   (b) Heating

2-20   CURVED MEMBERS IN FLEXURE

The distribution of stress in a curved flexural member is determined by using the following assumptions:

- The cross section has an axis of symmetry in a plane along the length of the beam.
- Plane cross sections remain plane after bending.
- The modulus of elasticity is the same in tension as in compression.

We shall find that the neutral axis and the centroidal axis of a curved beam, unlike a straight beam, are not coincident and also that the stress does not vary linearly from the neutral axis. The notation shown in Fig. 2-27 is defined as follows:

r_o = radius of outer fiber
r_i = radius of inner fiber
h = depth of section
c_o = distance from neutral axis to outer fiber
c_i = distance from neutral axis to inner fiber
r_n = radius of neutral axis
R = radius of centroidal axis
e = distance from centroidal axis to neutral axis

Figure 2-27 shows that the neutral and centroidal axes are not coincident.* It turns out that the location of the neutral axis with respect to the center of curvature O is given by the equation

$$r_n = \frac{A}{\displaystyle\int \frac{dA}{r}} \tag{2-64}$$

---

*For a complete development of the relations in this section, see Joseph E. Shigley, *Mechanical Engineering Design,* First Metric Edition, McGraw-Hill, New York, 1986, pp. 72–75.

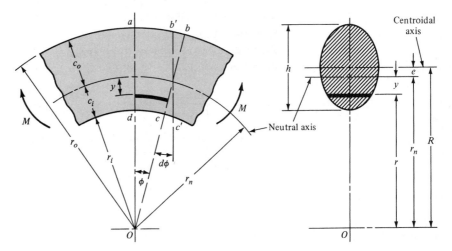

FIGURE 2-27

Note that y is positive in the direction toward point O.

The stress distribution can be found by balancing the external applied moment against the internal resisting moment. The result is found to be

$$\sigma = \frac{My}{Ae(r_n - y)} \tag{2-65}$$

where M is positive in the direction shown in Fig. 2-27. Equation (2-65) shows that the stress distribution is hyperbolic. The critical stresses occur at the inner and outer surfaces and are

$$\sigma_i = \frac{Mc_i}{Aer_i} \qquad \sigma_o = -\frac{Mc_o}{Aer_o} \tag{2-66}$$

These equations are valid for pure bending. In the usual and more general case, such as a crane hook, the U frame of a press, or the frame of a clamp, the bending moment is due to forces acting to one side of the cross section under consideration. In this case the bending moment is computed about the *centroidal axis,* not the neutral axis. Also, an additional axial tensile or compressive stress must be added to the bending stresses given by Eqs. (2-65) and (2-66) to obtain the resultant stresses acting on the section.

EXAMPLE 2-8

Plot the distribution of stresses across section *A-A* of the crane hook shown in Fig. 2-28a. The cross section is rectangular, with $b = 0.75$ in and $h = 4$ in, and the load is $F = 5000$ lb.

Solution

Since $A = bh$, we have $dA = b\,dr$ and, from Eq. (2-64),

$$r_n = \frac{A}{\displaystyle\int \frac{dA}{r}} = \frac{bh}{\displaystyle\int_{r_i}^{r_o} \frac{b}{r}\,dr} = \frac{h}{\ln \dfrac{r_o}{r_i}} \tag{1}$$

From Fig. 2-28b, we see that $r_i = 2$ in, $r_o = 6$ in, $R = 4$ in, and $A = 3$ in². Thus, from Eq. (1),

$$r_n = \frac{h}{\ln (r_o/r_i)} = \frac{4}{\ln \frac{6}{2}} = 3.641 \text{ in}$$

and so the eccentricity is $e = R - r_n = 4 - 3.641 = 0.359$ in. The moment M is positive and is $M = FR = 5000(4) = 20\,000$ lb · in. Adding the axial component of stress to Eq. (2-65) gives

$$\sigma = \frac{F}{A} + \frac{My}{Ae(r_n - y)} = \frac{5000}{3} + \frac{(20\,000)(3.641 - r)}{3(0.359)r} \tag{2}$$

Substituting values of r from 2 to 6 in results in the stress distribution shown in Fig. 2-28c. The stresses at the inner and outer radii are found to be 16.9 and −5.6 kpsi, respectively, as shown.

Sections most frequently encountered in the stress analysis of curved beams are shown in Fig. 2-29. Formulas for the rectangular section were developed in Example

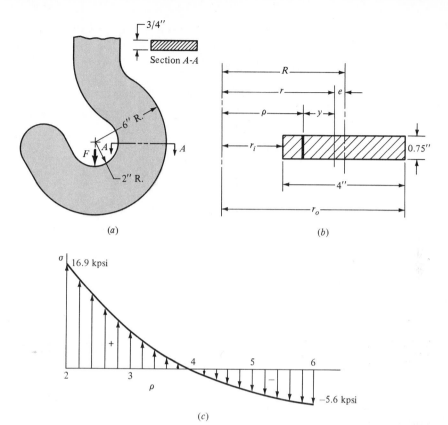

FIGURE 2-28

(a) Front view of crane hook;
(b) cross section and notation;
(c) resulting stress distribution.

2-8, but they are repeated here for convenience:

$$R = r_i + \frac{h}{2} \tag{2-67}$$

$$r_n = \frac{h}{\ln (r_o/r_i)} \tag{2-68}$$

For the trapezoidal section in Fig. 2-29b, the formulas are

$$R = r_i + \frac{h}{3} \frac{b_i + 2b_o}{b_i + b_o} \tag{2-69}$$

$$r_n = \frac{A}{b_o - b_i + [(b_i r_o - b_o r_i)/h] \ln (r_o/r_i)} \tag{2-70}$$

For the T section in Fig. 2-29c, we have

$$R = r_i + \frac{b_i c_1^2 + 2b_o c_1 c_2 + b_o c_2^2}{2(b_o c_2 + b_i c_1)} \tag{2-71}$$

$$r_n = \frac{b_i c_1 + b_o c_2}{b_i \ln [(r_i + c_1)/r_i)] + b_o \ln [r_o/(r_i + c_1)]} \tag{2-72}$$

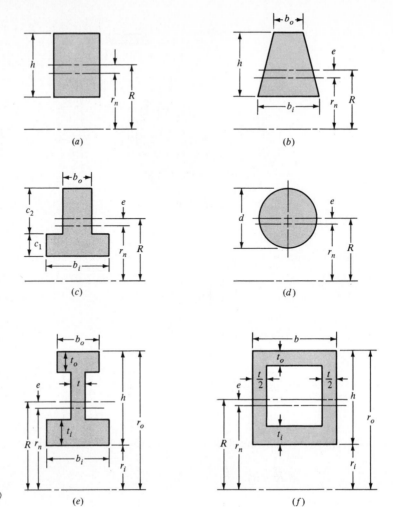

FIGURE 2-29

The equations for the solid round section of Fig. 2-29 are

$$R = r_i + \frac{d}{2} \tag{2-73}$$

$$r_n = \frac{d^2}{4(2R - \sqrt{4R^2 - d^2})} \tag{2-74}$$

For the I shape in Fig. 2-29e, we have

$$R = r_i + \frac{\frac{1}{2}h^2 t + \frac{1}{2}t_i^2(b_i - t) + t_o(b_o - t)(h - t_o/2)}{t_i(b_i - t) + t_o(b_o - t) + ht} \tag{2-75}$$

$$r_n = \frac{t_i(b_i - t) + t_o(b_o - t) + ht_o}{b_i \ln \dfrac{r_i + t}{r_i} + t \ln \dfrac{r_o - t_o}{r_i + t_i} + b_o \ln \dfrac{r_o}{r_o - t_o}} \tag{2-76}$$

Finally, for the rectangular tubing in Fig. 2-29f, the results are

$$R = r_i + \frac{\frac{1}{2}h^2 t + \frac{1}{2}t_i^2(b - t) + t_o(b - t)(h - t_o/2)}{ht + (b - t)(t_i + t_o)} \tag{2-77}$$

$$r_n = \frac{(b - t)(t_i + t_o) + ht}{b\left(\ln \dfrac{r_i + t_i}{r_i} + \ln \dfrac{r_o}{r_o - t_o}\right) + t \ln \dfrac{r_o - t_o}{r_i + t_i}} \tag{2-78}$$

Formulas for other sections can be obtained by performing the integration indicated by Eq. (2-64).

Many cases arise in which numerical integration must be used. These may occur because

* A digital computer is being used.

* It is not possible to integrate the function by any other means.

* The function to be integrated is described only by data.

A method of integration by Simpson's rule consists of defining equally spaced ordinates in the integration interval. Then parabolic curves are assumed to pass through each contiguous set of three ordinates. Using the notation of Fig. 2-30, the area under the curve AB, by Simpson's rule, is

$$I = \frac{H}{3}(Y_0 + 4Y_1 + 2Y_2 + 4Y_3 + 2Y_4 + \cdots + 4Y_{N-1} + Y_N)$$

$$= \frac{H}{3}(Y_0 + Y_N + 4\,\Sigma Y_{\text{odd}} + 2\,\Sigma Y_{\text{even}}) \tag{2-79}$$

where H is the width of the interval and is

$$H = \frac{X_N - X_0}{N} \tag{2-80}$$

The terms ΣY_{odd} and ΣY_{even} are the sums, respectively, of the odd-numbered and

FIGURE 2-30

Notation for integration by Simpson's rule. Note that N is an even number.

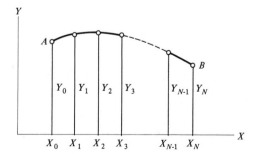

even-numbered subscripted ordinates. Equation (2-79) then gives Simpson's approximation to the equation

$$I = \int_{X_0}^{X_N} F(X)\, dX \tag{2-81}$$

Unfortunately, Eq. (2-79) does not always give good accuracy. The result can be checked using *Richardson's error estimate*.* This is obtained by performing the integration twice, once with all the ordinates, and again with every other ordinate. Designating the first integration by I_1 and the second by I_2, Richardson's error is

$$E = \frac{I_1 - I_2}{15} \tag{2-82}$$

The sign of the result is significant. Once E has been obtained from Eq. (2-82), a better estimate of the integral is

$$I = I_1 + E \tag{2-83}$$

For the analysis of a curved beam of any arbitrary cross section, divide the cross section into an even number of strips of thickness Δr and length b_I, where b_I is the length of the Ith strip. Then the equations to be solved are

$$A = \int_{r_i}^{r_o} b\, dr \tag{2-84}$$

$$R = \int_{r_i}^{r_o} \frac{br\, dr}{A} \tag{2-85}$$

$$r_n = \frac{A}{\displaystyle\int_{r_i}^{r_o} \frac{b\, dr}{r}} \tag{2-86}$$

$$e = R - r_n \tag{2-87}$$

Numerical integration methods are easy to program; see Fig. 2-31 for a simplified flow diagram.

---

*See B. Carnahan, H. A. Luther, and J. O Wilkes, *Applied Numerical Analysis*, Wiley, New York, 1969, p. 79.

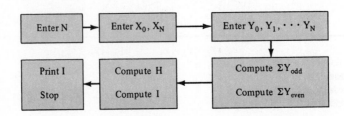

FIGURE 2-31

Flow diagram for computer solution of Simpson's rule for integration.

2-21 CONTACT STRESSES

When two bodies having curved surfaces are pressed together, point or line contact changes to area contact, and the stresses developed in the two bodies are three-dimensional. Contact-stress problems arise in the contact of a wheel and a rail, in automotive valve cams and tappets, in mating gear teeth, and in the action of rolling bearings. Typical failures are seen as cracks, pits, or flaking in the surface material.

The most general case of contact stress occurs when each contacting body has a double radius of curvature; that is, when the radius in the plane of rolling is different from the radius in a perpendicular plane, both planes taken through the axis of the contacting force. Here we shall consider only the two special cases of contacting spheres and contacting cylinders.* The results presented here are due to Hertz and so are frequently known as *Hertzian stresses*.

When two solid spheres of diameters d_1 and d_2 are pressed together with a force F, a circular area of contact of radius a is obtained. Specifying E_1, ν_1 and E_2, ν_2 as the respective elastic constants of the two spheres, the radius a is given by the equation

$$a = \sqrt[3]{\frac{3F}{8} \frac{(1 - \nu_1^2)/E_1 + (1 - \nu_2^2)/E_2}{1/d_1 + 1/d_2}} \tag{2-88}$$

The pressure within each sphere has a semielliptical distribution, as shown in Fig. 2-32. The maximum pressure occurs at the center of the contact area and is

---

*A good explanation of the general case may be found in Arthur P. Boresi, Omar M. Sidebottom, Fred B. Seely, and James O. Smith, *Advanced Mechanics of Materials,* 3d ed., Wiley, New York, 1978, pp. 581–627.

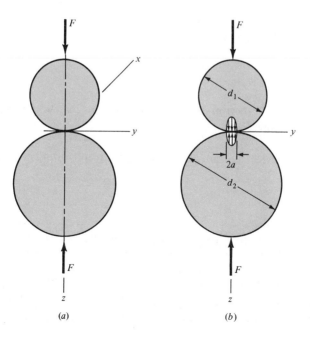

FIGURE 2-32

(a) Two spheres held in contact by force *F*. (b) Contact stress has an elliptical distribution at face of contact of width 2*a*.

$$p_{max} = \frac{3F}{2\pi a^2} \tag{2-89}$$

Equations (2-88) and (2-89) are perfectly general and also apply to the contact of a sphere and a plane surface or of a sphere and an internal spherical surface. For a plane surface, use $d = \infty$. For an internal surface, the diameter is expressed as a negative quantity.

The maximum stresses occur on the z axis, and these are principal stresses. Their values are

$$\sigma_x = \sigma_y = -p_{max}\left[\left(1 - \frac{z}{a}\tan^{-1}\frac{1}{\frac{z}{a}}\right)(1 + \mu) - \frac{1}{2\left(1 + \frac{z^2}{a^2}\right)}\right] \tag{2-90}$$

$$\sigma_z = \frac{-p_{max}}{1 + \frac{z^2}{a^2}} \tag{2-91}$$

These equations are valid for either sphere, but the value used for Poisson's ratio must correspond with the sphere under consideration. The equations are even more complicated when stress states off the z axis are to be determined, because here the x and y coordinates must also be included. But these are not required for design purposes, because the maxima occur on the z axis.

The Mohr's circles for the stress state described by Eqs. (2-90) and (2-91) are a point and two coincident circles. Since $\sigma_x = \sigma_y$, we have $\tau_{xy} = 0$ and

$$\tau_{xz} = \tau_{yz} = \frac{\sigma_x - \sigma_z}{2} = \frac{\sigma_y - \sigma_z}{2} \tag{2-92}$$

Figure 2-33 is a plot of Eqs. (2-90) and (2-91) for a distance of $3a$ below the surface.

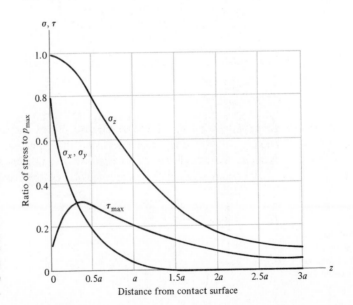

FIGURE 2-33

Magnitude of the stress components below the surface as a function of the maximum pressure for contacting spheres. Note that the maximum shear stress is slightly below the surface and is approximately $0.3p_{max}$. The chart is based on a Poisson's ratio of 0.30. Note that the normal stresses are all compressive stresses.

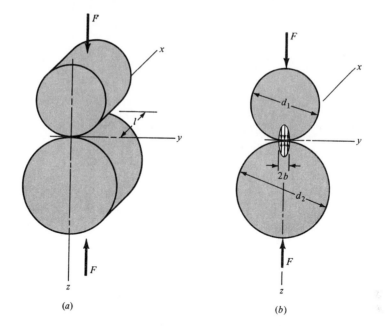

FIGURE 2-34

(a) Two cylinders held in contact by force F uniformly distributed along cylinder length l. (b) Contact stress has an elliptical distribution at face of contact of width 2b.

(a)

(b)

Note that the shear stress reaches a maximum value slightly below the surface. It is the opinion of many authorities that this maximum shear stress is responsible for the surface fatigue failure of contacting elements. The explanation is that a crack originates at the point of maximum shear stress below the surface and progresses to the surface and that the pressure of the lubricant wedges the chip loose.

Figure 2-34 illustrates a similar situation in which the contacting elements are two cylinders of length l and diameters d_1 and d_2. As shown in Fig. 2-34b, the area of contact is a narrow rectangle of width $2b$ and length l, and the pressure distribution is elliptical. The half-width b is given by the equation

$$b = \sqrt{\frac{2F}{\pi l} \frac{(1 - \nu_1^2)/E_1 + (1 - \nu_2^2)/E_2}{1/d_1 + 1/d_2}} \tag{2-93}$$

The maximum pressure is

$$p_{max} = \frac{2F}{\pi b l} \tag{2-94}$$

Equations (2-93) and (2-94) apply to a cylinder and a plane surface, such as a rail, by making $d = \infty$ for the plane surface. The equations also apply to the contact of a cylinder and an internal cylindrical surface; in this case d is made negative.

The stress state on the z axis is given by the equations

$$\sigma_x = -2\nu p_{max} \left(\sqrt{1 + \frac{z^2}{b^2}} - \frac{z}{b} \right) \tag{2-95}$$

$$\sigma_y = -p_{max} \left[\left(2 - \frac{1}{1 + \frac{z^2}{b^2}} \right) \sqrt{1 + \frac{z^2}{b^2}} - 2\frac{z}{b} \right] \tag{2-96}$$

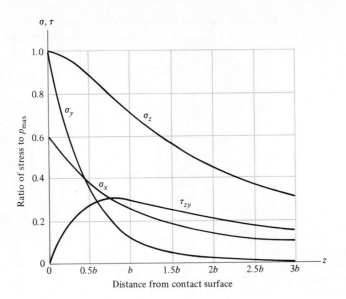

FIGURE 2-35

Magnitude of the stress components below the surface as a function of the maximum pressure for contacting cylinders. τ_{zy} becomes the largest of the three shear stresses at about $z/b = 0.75$; its maximum value is $0.30 p_{max}$. The chart is based on a Poisson's ratio of 0.30. Can you tell which two principal stresses are used to determine τ_{max} when $z/b < 0.75$? Note that all normal stresses are compressive stresses.

$$\sigma_z = \frac{-p_{max}}{\sqrt{1 + z^2/b^2}} \qquad (2\text{-}97)$$

These three equations are plotted in Fig. 2-35 up to a distance of $3b$ below the surface. Though τ_{zy} is not the largest of the three shear stresses for all values of z/b, it is a maximum at about $z/b = 0.75$ and is larger at that point than either of the other two shear stresses for any value of z/b.

Figure 2-36 shows the variation of the octahedral shear stress below the surface for contacting spheres and contacting cylinders. This stress is sometimes used, instead of the maximum shear stress, to define failure. The octahedral shear stress and the maximum shear stress reach their highest values on the curves at the same depth z, but the location of this point is very sensitive to the value of Poisson's ratio.

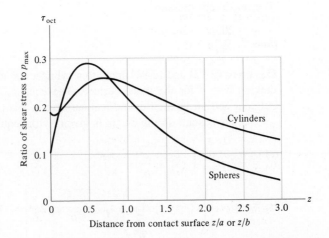

FIGURE 2-36

Plot of the octahedral shear stress for distances z/a and z/b below the surface for contacting spheres and cylinders. Based on $\nu = 0.30$.

PROBLEMS

2-1†   For each of the stress states listed below, draw a Mohr's circle diagram properly labeled, find the principal normal and shear stresses, and determine the angle from the x axis to σ_1. Draw a stress element as in Fig. 2-5b and label all details.

(a) $\sigma_x = 12$, $\sigma_y = 6$, $\tau_{xy} = 4$ cw
(b) $\sigma_x = 16$, $\sigma_y = 9$, $\tau_{xy} = 5$ ccw
(c) $\sigma_x = 10$, $\sigma_y = 24$, $\tau_{xy} = 6$ ccw
(d) $\sigma_x = 9$, $\sigma_y = 19$, $\tau_{xy} = 8$ cw

2-2   Repeat Prob. 2-1 for:

(a) $\sigma_x = -4$, $\sigma_y = 12$, $\tau_{xy} = 7$ ccw
(b) $\sigma_x = 6$, $\sigma_y = -5$, $\tau_{xy} = 8$ ccw
(c) $\sigma_x = -8$, $\sigma_y = 7$, $\tau_{xy} = 6$ cw
(d) $\sigma_x = 9$, $\sigma_y = -6$, $\tau_{xy} = 3$ cw

2-3   Repeat Prob. 2-1 for:

(a) $\sigma_x = -42$, $\sigma_y = -81$, $\tau_{xy} = 30$ cw
(b) $\sigma_x = 24$, $\sigma_y = 14$, $\tau_{xy} = 19$ cw
(c) $\sigma_x = -30$, $\sigma_y = -60$, $\tau_{xy} = 25$ ccw
(d) $\sigma_x = 120$, $\sigma_y = -40$, $\tau_{xy} = 50$ ccw

2-4   Repeat Prob. 2-1 for:

(a) $\sigma_x = 20$, $\sigma_y = -10$, $\tau_{xy} = 8$ cw
(b) $\sigma_x = 30$, $\sigma_y = -10$, $\tau_{xy} = 10$ ccw
(c) $\sigma_x = -10$, $\sigma_y = 18$, $\tau_{xy} = 9$ cw
(d) $\sigma_x = -12$, $\sigma_y = 22$, $\tau_{xy} = 12$ cw

2-5†   For each of the stress states listed below, find all three principal normal and shear stresses, and the octahedral shear stress. Draw a complete Mohr's three-circle diagram and label all points of interest.

(a) $\sigma_x = 10$, $\sigma_y = -4$
(b) $\sigma_x = 10$, $\tau_{xy} = 4$ ccw
(c) $\sigma_x = -2$, $\sigma_y = -8$, $\tau_{xy} = 4$ cw
(d) $\sigma_x = 10$, $\sigma_y = -30$, $\tau_{xy} = 10$ ccw

2-6   Repeat Prob. 2-5 for:

(a) $\sigma_x = 20$, $\sigma_y = 20$, $\sigma_z = 20$
(b) $\sigma_x = 20$, $\sigma_z = 20$
(c) $\sigma_y = -80$, $\sigma_z = 20$
(d) $\sigma_x = 100$

2-7   Repeat Prob. 2-5 for:

(a) $\sigma_x = -80$, $\sigma_y = -30$, $\tau_{xy} = 20$ cw
(b) $\sigma_x = 30$, $\sigma_y = -60$, $\tau_{xy} = 30$ cw
(c) $\sigma_x = 40$, $\sigma_z = -30$, $\tau_{xy} = 20$ ccw
(d) $\sigma_x = 50$, $\sigma_z = -20$, $\tau_{xy} = 30$ cw

---

†Stress components are always assumed to be zero unless given. Thus, in this example, $\sigma_z = 0$ for all four cases.

2-8 Repeat Prob. 2-5 for:
(a) $\tau_{xy} = 30$
(b) $\sigma_x = 60$
(c) $\sigma_x = 60$, $\sigma_y = 60$
(d) $\sigma_y = -80$
(e) $\sigma_x = 60$, $\sigma_y = 60$, $\sigma_z = 60$

2-9 A $\frac{1}{2}$-in-diameter steel tension rod is 72 in long and carries a load of 2000 lb. Find the tensile stress, the total deformation, the unit strains, and the change in the rod diameter.

2-10 A dime is squeezed by a force of 800 lb. Compute the change in diameter and thickness. Did you make any assumptions?

2-11 Twin diagonal aluminum alloy tension rods 15 mm in diameter are used in a rectangular frame to prevent collapse. The rods can safely support a tensile stress of 135 MPa. If the rods are initially 3 m in length, how much must they be stretched to develop this stress?

2-12 A brass member is subjected to stresses $\sigma_x = 20$ MPa and $\tau_{xy} = 15$ MPa. Find the principal stresses and strains.

2-13 Electrical strain gauges were applied to a notched specimen to determine the stresses in the notch. The results were $\epsilon_x = 0.0021$ and $\epsilon_y = -0.000\,67$. Find σ_x, σ_y, and the octahedral shear stress if the material is carbon steel.

2-14 The symbol W is used in the various figure parts to specify the weight of an element. If not given, assume the parts are weightless. For each figure part sketch a free-body diagram of each element including the frame. Try to get the forces in the proper directions, but do not compute magnitudes.

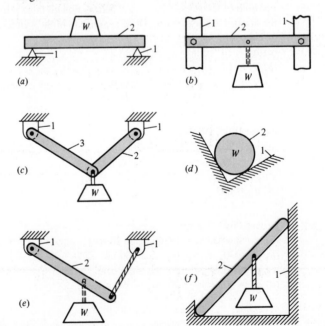

PROBLEM 2-14

2-15   Using the figure part selected by your instructor, sketch a free-body diagram of each element in the figure. Compute the magnitude and direction of each using an algebraic or vector method, as specified.

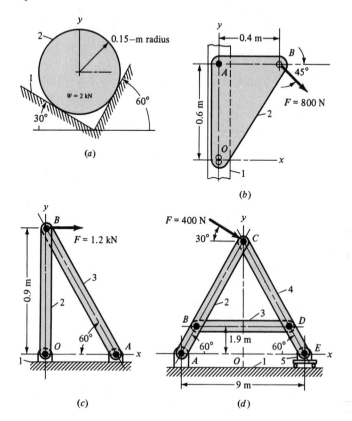

PROBLEM 2-15   (c)   (d)

2-16   Find the reactions at the supports and plot the shear-force and bending-moment diagrams for each of the beams shown in the figure. Label the diagrams properly.

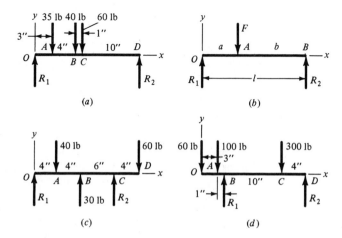

PROBLEM 2-16
*(continued on
next page)*

(c)   (d)

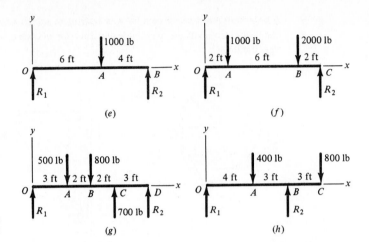

(e)

(f)

(g)

(h)

PROBLEM 2-16
(concluded)

2-17   Select a beam from Table A-9 and find general expressions for the loading, shear-force, bending-moment, and support reactions. Use the method specified by your instructor.

2-18*   A beam carrying a uniform load is simply supported with the supports set back a distance a from the ends as shown in the figure. The bending moment at x can be found from summing moments to zero at section x:

$$\Sigma M = M_x + \tfrac{1}{2}w(a + x)^2 - \tfrac{1}{2}wlx = 0$$

or

$$M_x = \frac{w}{2}[lx - (a + x)^2]$$

where w is the loading intensity in lb/in. The designer wishes to minimize the necessary weight of the supporting beam by choosing a setback a resulting in the smallest possible maximum bending stress.

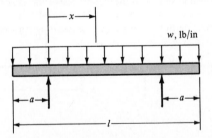

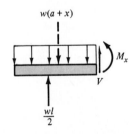

PROBLEM 2-18

(a) If the beam is configured with $a = 2.25$ in, $l = 10$ in, and $w = 100$ lb/in, find the magnitude of the severest bending stress in the beam.
(b) Since the configuration in part (a) is not optimal, find the optimal setback which will result in the lightest-weight beam.

*Throughout the remainder of the problems for Chap. 2, the asterisk denotes a design-type problem.

2-19*   An artist wishes to construct a mobile using pendants, string, and span wire with eyelets as shown in the figure.

(a) At what positions w, x, y, and z should the suspension strings be attached to the span wires?

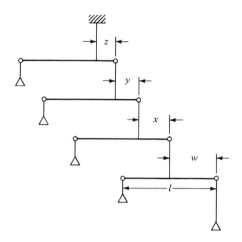

PROBLEM 2-19

(b) Is the mobile stable? If so, justify; if not, suggest a remedy.

2-20   Devise a set of programs for the following subroutines:
(a) Given $R \underline{/\theta}$, find $x\mathbf{i} + y\mathbf{j}$.
(b) Given $x\mathbf{i} + y\mathbf{j}$; find $R \underline{/\theta}$.
(c) Given θ; find $\hat{\mathbf{R}} = \bar{x}\mathbf{i} + \bar{y}\mathbf{j}$, where $\hat{\mathbf{R}}$ is a unit vector and $\bar{x}$ and $\bar{y}$ are the direction cosines.
(d) Given $\mathbf{F}_1, \mathbf{F}_2, \mathbf{F}_3, \cdots$ in x, y, and z components, find $\Sigma\mathbf{F}$.
(e) Given $\mathbf{C}$ and $\mathbf{C}'$ in x, y, and z components; find $\mathbf{C} \times \mathbf{C}'$.

2-21   Using the results of Example 2-6, find the stress components on two additional stress elements at A, one on the front of the shaft along the z axis and the other on the bottom of the shaft along the y axis.

2-22   For each section illustrated, find the second moment of area, the location of the neutral axis, and
to   the distances from the neutral axis to the top and bottom surfaces. Suppose a positive bending
2-24   moment of 10 kip · in is applied; find the resulting stresses at the top and bottom surfaces and at every abrupt change in cross section.

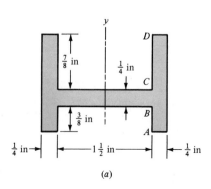

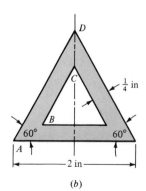

PROBLEM 2-22   (a)   (b)

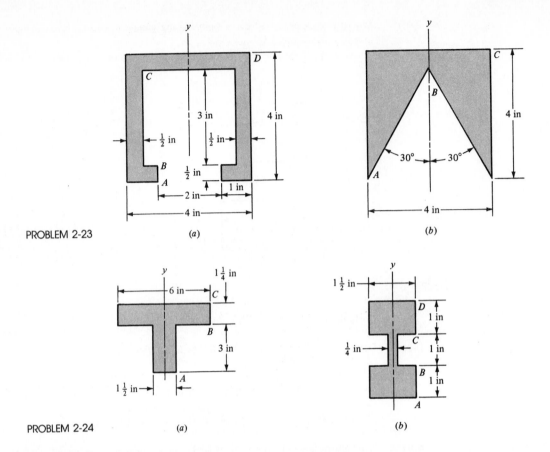

PROBLEM 2-23   (a)   (b)

PROBLEM 2-24   (a)   (b)

2-25   Find the x and y coordinates of the center of curvature corresponding to the place where the beam is bent the most, for each beam shown in the figure. The beams are both made of Douglas fir (see Table A-5) and have rectangular sections.

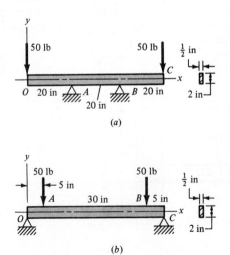

(a)

(b)

PROBLEM 2-25   (b)

2-26   For each beam illustrated, find the locations and magnitudes of the maximum tensile bending stress and the maximum shear stress.

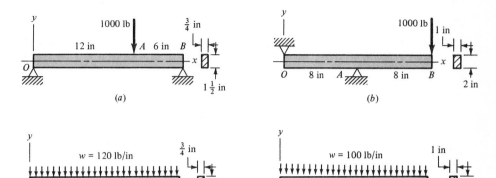

(a)

(b)

(c)

(d)

PROBLEM 2-26

2-27   The figure illustrates a number of beam sections. Use an allowable stress of 1.2 kpsi for wood and 12 kpsi for steel and find the maximum safe uniformly distributed load that each beam can carry if the given lengths are between simple supports.

(a) Wood joist $1\frac{1}{2}$ by $9\frac{1}{2}$ in and 12 ft long
(b) Steel tube, 2 in OD by $\frac{3}{8}$-in wall thickness, 48 in long
(c) Hollow steel tube 3 by 2 in, outside dimensions, formed from $\frac{3}{16}$-in material and welded, 48 in long
(d) Steel angles $3 \times 3 \times \frac{1}{4}$ in and 72 in long
(e) A 5.4-lb, 4-in steel channel, 72 in long
(f) A 4-in $\times$ 1-in steel bar, 72 in long

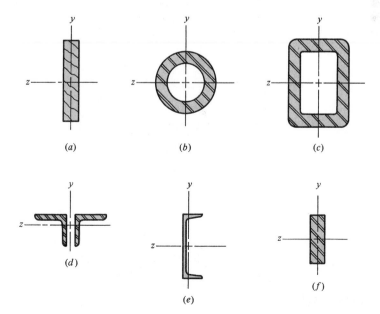

(a)

(b)

(c)

(d)

(e)

(f)

PROBLEM 2-27

2-28   Using a maximum allowable shear stress of 8000 psi, find the shaft diameter needed to transmit 50 hp when
(*a*) The shaft speed is 2000 rev/min.
(*b*) The shaft speed is 200 rev/min.

2-29   A 15-mm-diameter steel bar is to be used as a torsion spring. If the torsional stress in the bar is not to exceed 110 MPa when one end is twisted through an angle of 30°, what must be the length of the bar?

2-30   A 70-mm-diameter solid steel shaft, used as a torque transmitter, is replaced with a 70-mm hollow shaft having a 6-mm wall thickness. If both materials have the same strength, what is the percentage reduction in torque transmission? What is the percent reduction in shaft weight?

2-31   A hollow steel shaft is to transmit 5400 N · m of torque and is to be sized so that the torsional stress does not exceed 150 MPa.
(*a*) If the inside diameter is three-fourths of the outside diameter, what size shaft should be used? Use preferred sizes.
(*b*) What is the stress on the inside of the shaft when full torque is applied?

2-32   A pin in a knuckle joint carrying a tensile load *F* deflects somewhat on account of this loading, making the distribution of reaction and load as shown in part *b* of the figure. The usual designer's assumption of loading is shown in part *c*; others sometimes choose the loading shown in part *d*. If *a* = 0.5 in, *b* = 0.75 in, *d* = 0.5 in, and F = 1000 lb, estimate the maximum bending stress and the maximum shear stress for each approximation.

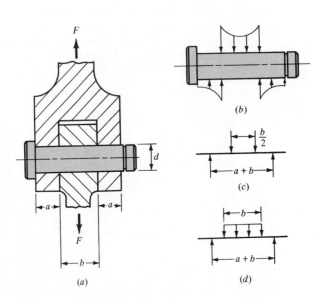

PROBLEM 2-32   (*a*)   (*d*)

2-33*   The figure illustrates two gears which are to be mounted on a shaft on the *x* axis. The gear loads are $F_1 = 2$ kip and $F_2 = 1.1$ kip. If one bearing is located at *A*, should the second bearing be located between the gears, or outboard, say, near *B*? Possible solutions will involve designing for minimum bending moment, or for equal bearing reactions.

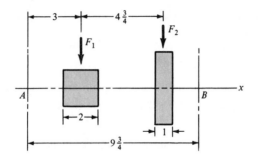

PROBLEM 2-33
Dimensions in inches.

2-34*   The figure illustrates a pin tightly fitted into a hole of a substantial member. A usual analysis is that of assuming concentrated reactions R and M at distance l from F. Suppose the reaction is distributed along distance a. Is the resulting moment reaction larger or smaller than the concentrated reaction? What is the loading intensity q? What do you think of using the usual assumption?

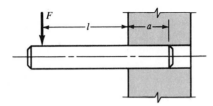

PROBLEM 2-34

2-35*   The figure shows an endless-belt conveyor drive roll. The roll has a diameter of 6 in and is driven at 5 rev/min by a geared-motor source rated at 1 hp. Determine a suitable shaft diameter d_C for an allowable torsional stress of 14 kpsi.

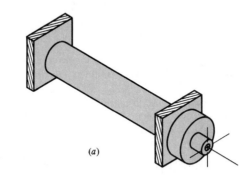

(a)

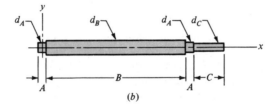

PROBLEM 2-35   (b)

(a) What would be the stress in the shaft you have sized if the motor starting torque is twice the running torque?

(b) Is bending stress likely to be a problem? What is the effect of different roll lengths B on bending?

2-36   The conveyer drive roll in the figure for Prob. 2-35 is 150 mm in diameter and is driven at 8 rev/min by a geared-motor source rated at 1 kW. Find a suitable shaft diameter d_C based on an allowable torsional stress of 75 MPa.

2-37   If the tension-loaded plate of Fig. 2-22 is infinitely wide, then the stress state anywhere in the plate can be described using polar coordinates as

$$\sigma_r = \frac{\sigma}{2}\left(1 - \frac{d^2}{4r^2}\right) - \frac{\sigma}{2}\left(1 - \frac{d^2}{4r^2}\right)\left(1 - \frac{3d^2}{4r^2}\right)\cos 2\theta$$

$$\sigma_\theta = \frac{\sigma}{2}\left(1 + \frac{d^2}{4r^2}\right) + \frac{\sigma}{2}\left(1 + \frac{3d^4}{16r^4}\right)\cos 2\theta$$

$$\tau_{r\theta} = \frac{\sigma}{2}\left(1 - \frac{d^2}{4r^2}\right)\left(1 + \frac{3d^2}{4r^2}\right)\sin 2\theta$$

for the radial, tangential, and shear components, respectively. Here r is the distance to the point of interest and θ is measured positive from the x axis.

(a) Find the stress components at the top and side of the hole for $r = d/2$.

(b) Plot a graph of the stress distribution, similar to that of Fig. 2-22, out to $r = 20$ mm for $d = 10$ mm.

2-38   Develop the formulas for the maximum radial and tangential stresses in a thick-walled cylinder due to internal pressure only.

2-39   Repeat Prob. 2-38 where the cylinder is subject to external pressure only. At what radii do the maximum stresses occur?

2-40   Develop the stress relations for a thin-walled spherical pressure vessel.

2-41   A pressure cylinder has an outside diameter of 6 in and has a $\frac{1}{4}$-in wall thickness. What pressure can this vessel carry if the maximum shear stress is not to exceed 4000 psi?

2-42   A pressure vessel has an outside diameter of 240 mm and a wall thickness of 10 mm. If the internal pressure is 2400 kPa, what is the maximum shear stress in the vessel walls?

2-43   An AISI 1020 cold-drawn steel tube has an ID of $1\frac{1}{4}$ in and an OD of $1\frac{3}{4}$ in. What maximum external pressure can this tube take if the largest principal normal stress is not to exceed 80 percent of the minimum yield strength of the material?

2-44   An AISI 1020 cold-drawn steel tube has an ID of $1\frac{3}{4}$ in and an OD of $2\frac{3}{8}$ in. What maximum internal pressure can this tube take if the largest principal normal stress is not to exceed 80 percent of the minimum yield strength of the material?

2-45   Find the maximum shear stress in a 10-in circular saw if it runs idle at 7200 rev/min. The saw is 14-gauge (0.0747 in) and is used on a $\frac{3}{4}$-in arbor. The thickness is uniform. What is the maximum radial component of stress?

2-46 The maximum recommended speed for a 300-mm-diameter abrasive grinding wheel is 2069 rev/min. Assume that the material is isotropic; use a bore of 25 mm, $\nu = 0.24$, and a mass density of 3320 kg/m³; and find the maximum tensile stress at this speed.

2-47 An abrasive cutoff wheel has a diameter of 6 in, is $\frac{1}{16}$ in thick, and has a 1-in bore. It weighs 6 oz and is designed to run at 10 000 rev/min. If the material is isotropic and $\nu = 0.20$, find the octahedral shear stress at the design speed.

2-48 to 2-53 The table below lists the maximum and minimum hole and shaft dimensions for a variety of standard press and shrink fits. The materials are both hot-rolled steel. Find the maximum and minimum values of the radial interference and the interface pressure. Use a collar diameter of 80 mm for the metric sizes and 3 in for those in inch units.

PROBLEM NUMBER	FIT DESIGNATION*	BASIC SIZE	HOLE		SHAFT	
			D_{max}	D_{min}	d_{max}	d_{min}
2-48	40H7/p6	40 mm	40.025	40.000	40.042	40.026
2-49	(1.5 in)H7/p6	1.5 in	1.5010	1.5000	1.5016	1.5010
2-50	40H7/s6	40 mm	40.025	40.000	40.059	40.043
2-51	(1.5 in)H7/s6	1.5 in	1.5010	1.5000	1.5023	1.5017
2-52	40H7/u6	40 mm	40.025	40.000	40.076	40.060
2-53	(1.5 in)H7/u6	1.5 in	1.5010	1.5000	1.5030	1.5024

*See Table 4–5 for description of fits.

2-54* An engineer wishes to determine the shearing strength of a certain epoxy cement. The problem is to devise a test specimen such that the joint is subject to pure shear. The joint shown in the figure, in which two bars are offset at an angle θ so as to keep the loading force F centroidal with the straight shanks, seems to accomplish this purpose. Using the contact area A and designating S_{su} as the *ultimate shearing strength*, the engineer obtains

$$S_{su} = \frac{F}{A} \cos \theta$$

The engineer's supervisor, in reviewing the test results, says the expression should be

$$S_{su} = \frac{F}{A} (1 + \tfrac{1}{4} \tan^2 \theta)^{1/2} \cos \theta$$

Resolve the discrepancy. What is your position?

PROBLEM 2-54

2-55* Using the results of Sec. 2-19, design a compact thermostat for use at room temperatures.

2-56* For a given cross section, a uniformly loaded beam of length l is simply supported as shown in the figure. If the supports are set back a short distance, the largest transverse bending moment is reduced. If the largest transverse bending moment can be minimized, the smallest beam section can be used. Write a computer program to discover the optimal setback a. Submit a listing of the

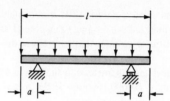

PROBLEM 2-56

programs you have written, the input and output of the production run, and the analysis on which your programming is based.

2-57* A rotary lawn-mower blade rotates at 3000 rev/min. The blade has a uniform cross section $\frac{1}{4}$ in thick by $1\frac{1}{4}$ in wide, and has a $\frac{1}{2}$-in-diameter hole in the center as shown in the figure. Estimate the nominal tensile stress at the central section due to rotation.

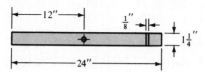

PROBLEM 2-57

2-58 A utility hook was formed from a 1-in-diameter round rod into the geometry shown in the figure. What are the stresses at the inner and outer surfaces at section *A-A* if the load *F* is 1000 lb?

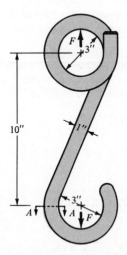

PROBLEM 2-58

2-59 The steel eyebolt shown in the figure is loaded with a force *F* of 100 lb. The bolt is formed of $\frac{1}{4}$-in-diameter wire to a $\frac{3}{8}$-in radius in the eye and at the shank. Estimate the stresses at the inner and outer surfaces at sections *A-A* and *B-B*.

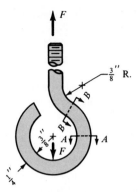

PROBLEM 2-59

2-60 Shown in the figure is a 12-gauge (0.1094-in) by $\frac{3}{4}$-in latching spring which supports a load of $F = 3$ lb. The inside radius of the bend is $\frac{1}{8}$ in. Estimate the stresses at the inner and outer surfaces at the critical section.

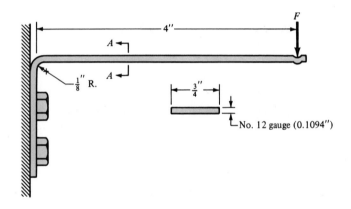

PROBLEM 2-60

2-61 A cast-iron bell-crank lever which is depicted in the figure is acted upon by forces F_1 of 250 lb and F_2 of 333 lb. The section A-A at the central pivot has a curved inner surface with a radius of $r_i = 1$ in. Estimate the stresses at the inner and outer surfaces of the curved portion of the lever.

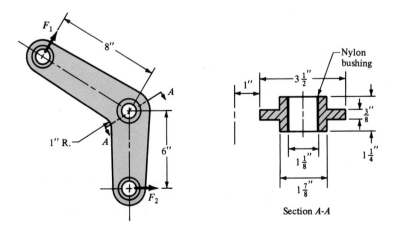

Section A-A

PROBLEM 2-61

2-62   Find the stress at the inner and outer surfaces at section *A-A* of the crane hook shown in Fig. 2-28 and Example 2-8, using Simpson's rule and numerical integration with ordinates at $r = 2, 3, 4, 5,$ and 6 in. What is the error in percent in the distance between the centroidal radius R and the neutral surface radius r_n? What can be done in the integration to reduce this error?

2-63   The crane hook depicted in Fig. 2-28 has a 1-in-diameter hole in the center of the critical section. For a load of 5 kip, estimate the bending stresses at the inner and outer surfaces at the critical section.

2-64   A 20-kip load is carried by the crane hook shown in the figure. The cross section of the hook uses two concave flanks. The width of the cross section is given by $b = 2/r$, where r is the radius from the center. The inside radius r_i is 2 in, and the outside radius $r_o = 6$ in. Find the stresses at the inner and outer surfaces at the critical section by (*a*) exact integration and (*b*) Simpson's rule, using ordinates placed at $r = 2, 3, 4, 5,$ and 6 in.

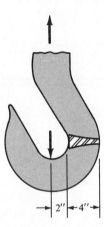

PROBLEM 2-64

2-65   An offset tensile link is shaped to clear an obstruction with a geometry as shown in the figure. The cross section at the critical location is elliptical, with a major axis of 4 in and a minor axis of 2 in. For a load of 20 kip, estimate the stresses at the inner and outer surfaces of the critical section.

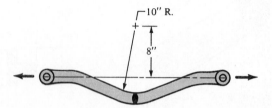

PROBLEM 2-65

2-66   A cast-steel C frame as shown in the figure has a rectangular cross section of 1 in by 1.6 in, with a 0.4-in-radius semicircular notch on both sides which forms midflank fluting as shown. Estimate A, R, r_n, and e, and for a load of 3000 lb, estimate the inner and outer surface stresses at the throat C.

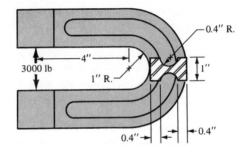

PROBLEM 2-66

2-67 Two carbon steel balls, each 25 mm in diameter, are pressed together by a force F. In terms of the force F, find the maximum values of the principal stress, the shear stress, and the octahedral shear stress, in MPa.

2-68 One of the balls in Prob. 2-67 is replaced by a flat carbon steel plate. If $F = 18$ N, at what depth does the maximum octahedral shear stress occur?

2-69 An aluminum alloy roller with diameter 1 in and length 2 in rolls on the inside of a cast-iron ring

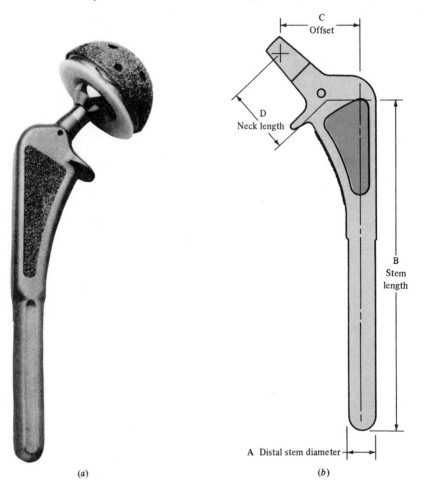

PROBLEM 2-70
Porous hip prosthesis. (*Photo-graph and drawing courtesy of Zimmer, Inc., Warsaw, Indiana.*)

(*a*)   (*b*)

having an inside radius of 4 in, which is 2 in thick. Find the maximum contact force F that can be used if the octahedral shear stress is not to exceed 4000 psi.

2-70*   The figure shows a hip prosthesis consisting of a stem which is cemented into a reamed cavity in the femur. The cup is cemented and fastened to the hip with bone screws. Shown are porous layers of titanium into which bone tissue will grow to form a longer-lasting bond than that afforded by cement alone. The bearing surfaces are a plastic cup and a titanium femoral head. The lip shown in the figures bears against the cutoff end of the femur to transfer the load to the leg from the hip. Walking will induce several million stress fluctuations per year for an average person, so there is danger that the prosthesis will loosen the cement bonds or that metal cracks may occur because of the many repetitions of stress. Prostheses like this are made in many different sizes. Typical dimensions are ball diameter 50 mm, stem diameter 15 mm, stem length 155 mm, offset 38 mm, and neck length 39 mm. Develop an outline to follow in making a complete stress analysis of this prosthesis. Describe the material properties needed, the equations required, and how the loading is to be defined.

2-71   Based on torsion what is the relative increase in cost of material when choosing a square shaft over a round shaft of the same material for a high-production part?

ANSWERS

2-1 (a) $\sigma_1 = 14$, $\sigma_2 = 4$, $\sigma_3 = 0$, $\tau_{max} = 7$, $\phi = -26.56°$

2-7 (d) $\tau_{1/2} = 39.05$, $\tau_{2/3} = 2.98$, $\tau_{1/3} = 40.0$, $\tau_{oct} = 37.3$

2-11 $\delta = 5.7$ mm

2-15 (a) 1.732 kN, 1 kN; (b) $F_A = 1100$ N; (c) $F_A = 2.4$ kN, $F_O = 2.06$ kN; (d) pin C, $F_{2C} = 204$ N, $F_{4C} = 503$ N

2-18 (a) 253 lb · in; (b) $a/l = (\sqrt{2} - 1)/2$

2-22 (b) $A = 1.175$ in^2, $\bar{y} = 0.577$ in, $I = 0.259$ in^4, 22.3 kpsi, 12.6 kpsi, -25.3 kpsi, -44.6 kpsi

2-26 (d) Maximum tensile stress on top at A; maximum shear stress on neutral axis at A

2-29 2.83 m

2-36 45 mm

2-39 $\sigma_{t,max} = -2p_o r_o^2/(r_o^2 - r_i^2)$ at $r = r_i$; $\sigma_{r,max} = -p_o$ at $r = r_o$

2-42 $\tau_{max} = 15.0$ MPa

2-45 $\tau_{max} = 4280$ psi at $r = r_i$

2-48 $p_{min} = 0$, $p_{max} = 50.5$ MPa

2-51 $p_{min} = 750$ psi, $p_{max} = 12.75$ kpsi

2-58 26.3 kpsi, -15.8 kpsi

2-61 3880 psi, -1550 psi

2-63 17.1 kpsi, -5.7 kpsi

2-65 By numerical integration, 34.1 kpsi, -20.7 kpsi

2-68 $\sigma_z = -251(F)^{1/3}$ MPa, τ_{oct} (max) $= 73.8(F)^{1/3}$ MPa, τ_{max} (max) $= 78.3(F)^{1/3}$ MPa at $z = 0.056$ mm

3
Deflection and Stiffness

A structure or mechanical element is said to be *rigid* when it does not bend, deflect, or twist too much when an external force, moment, or torque is applied. But if movement due to an external disturbance is large, the member is said to be *flexible*. The words *rigidity* and *flexibility* are qualitative terms which depend upon the situation. Thus if the floor of a building bends only 0.1 in under the weight of a machine placed upon it, it will be considered very rigid if the machine is heavy. But a surface plate which bends 0.01 in because of its own weight will be considered too flexible.

Deflection analysis enters into design situations in many ways. A snap ring, or retaining ring, must be flexible enough to be bent without permanent deformation and assembled with other parts; and then it must be rigid enough to hold the assembled parts together. In a transmission, the gears must be supported by a rigid shaft. If the shaft bends too much, that is, if it is too flexible, the teeth will not mesh properly, and the result will be excessive impact, noise, wear, and early failure. In rolling sheet or strip steel to prescribed thicknesses, the rolls must be crowned, that is, curved, so that the finished product will be of uniform thickness. Thus, to design the rolls it is necessary to know exactly how much they will bend when a sheet of steel is rolled between them. Sometimes mechanical elements must be designed to have a particular force-deflection characteristic. The suspension system of an automobile, for example, must be designed within a very narrow range to achieve an optimum bouncing frequency for all conditions of vehicle loading, because the human body is comfortable only within a limited range of frequencies.

3-1 SPRING RATES

Elasticity is that property of a material which enables it to regain its original configuration after having been deformed. A *spring* is a mechanical element which exerts a force when deformed. Figure 3-1*a* shows a straight beam of length *l* simply supported at the ends and loaded by the transverse force *F*. The deflection *y* is linearly related to the

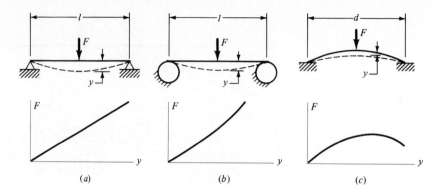

FIGURE 3-1

(a) A linear spring; (b) a stiffening spring; (c) a softening spring.

force, as long as the elastic limit of the material is not exceeded, as indicated by the graph. This beam can be described as a *linear spring*.

In Fig. 3-1b a straight beam is supported on two cylinders such that the length between supports decreases as the beam is deflected by the force F. A larger force is required to deflect a short beam than a long one, and hence the more this beam is deflected, the stiffer it becomes. Also, the force is not linearly related to the deflection, and hence this beam can be described as a *nonlinear stiffening spring*.

Figure 3-1c is a dish-shaped round disk. The force necessary to flatten the disk increases at first and then decreases as the disk approaches a flat configuration, as shown by the graph. Any mechanical element having such a characteristic is called a *nonlinear softening spring*.

If we designate the general relationship between force and deflection by the equation

$$F = F(y) \qquad (a)$$

then *spring rate* is defined as

$$k(y) = \lim_{\Delta y \to 0} \frac{\Delta F}{\Delta y} = \frac{dF}{dy} \qquad (3\text{-}1)$$

where y must be measured in the direction of F and at the point of application of F. Most of the force-deflection problems encountered in this book are linear, as in Fig. 3-1a. For these, k is a constant, also called the *spring constant;* consequently Eq. (3-1) is written

$$k = \frac{F}{y} \qquad (3\text{-}2)$$

We might note that Eqs. (3-1) and (3-2) are quite general and apply equally well for torques and moments, provided angular measurements are used for y. For linear displacements, the units of k are often pounds per inch or newtons per meter, and for angular displacements, pound-inches per radian or newton-meters per radian.

3-2   TENSION, COMPRESSION, AND TORSION

The relation for the total extension or deformation of a uniform bar has already been developed in Sec. 2-5, Eq. (*a*). It is repeated here for convenience:

$$\delta = \frac{Fl}{AE} \tag{3-3}$$

This equation does not apply to a long bar loaded in compression if there is a possibility of buckling. Using Eqs. (3-2) and (3-3), we see that the spring constant of an axially loaded bar is

$$k = \frac{AE}{l} \tag{3-4}$$

The angular deflection of a uniform round bar subjected to a twisting moment T was given in Eq. (2-39), and is

$$\theta = \frac{Tl}{GJ} \tag{3-5}$$

where θ is in radians. If we multiply Eq. (3-5) by $180/\pi$ and substitute $J = \pi d^4/32$ for a solid round bar, we obtain

$$\theta = \frac{583.6Tl}{Gd^4} \tag{3-6}$$

where θ is in degrees.

Equation (3-5) can be rearranged to give the torsional spring rate as

$$k = \frac{T}{\theta} = \frac{GJ}{l} \tag{3-7}$$

When the word *simple* is used to describe the loading, the meaning is that no other load is present and that no geometric complexities are present. Thus, a bar loaded in *simple tension* is a uniform bar acted upon by a tensile load directed along the centroidal axis, and acted upon by no other loads.

3-3   FLEXURE

Beams deflect a great deal more than axially loaded members, and the problem of bending probably occurs more often than any other loading problem in design. Shafts, axles, cranks, levers, springs, brackets, and wheels, as well as many other elements, must often be treated as beams in the design and analysis of mechanical structures and systems. The subject of bending, however, is one which you should have studied as preparation for reading this book. It is for this reason that we include here only a brief review to establish the nomenclature and conventions to be used throughout this book.

In Sec. 2-10 we developed the relation for the curvature of a beam subjected to a bending moment M [Eq. (2-27)]. The relation is

$$\frac{1}{\rho} = \frac{M}{EI} \tag{3-8}$$

where ρ is the radius of curvature. From studies in mathematics we also learn that the curvature of a plane curve is given by the equation

$$\frac{1}{\rho} = \frac{d^2y/dx^2}{[1 + (dy/dx)^2]^{3/2}} \tag{3-9}$$

where the interpretation here is that y is the deflection of the beam at any point x along its length. The slope of the beam at any point x is

$$\theta = \frac{dy}{dx} \tag{a}$$

For many problems in bending, the slope is very small, and for these the denominator of Eq. (3-9) can be taken as unity. Equation (3-8) can then be written

$$\frac{M}{EI} = \frac{d^2y}{dx^2} \tag{b}$$

Noting Eqs. (2-22) and (2-23) and successively differentiating Eq. (b) yields

$$\frac{V}{EI} = \frac{d^3y}{dx^3} \tag{c}$$

$$\frac{q}{EI} = \frac{d^4y}{dx^4} \tag{d}$$

It is convenient to display these relations in a group as follows:

$$\frac{q}{EI} = \frac{d^4y}{dx^4} \tag{3-10}$$

$$\frac{V}{EI} = \frac{d^3y}{dx^3} \tag{3-11}$$

$$\frac{M}{EI} = \frac{d^2y}{dx^2} \tag{3-12}$$

$$\theta = \frac{dy}{dx} \tag{3-13}$$

$$y = f(x) \tag{3-14}$$

The nomenclature and conventions are illustrated by the beam of Fig. 3-2. Here, a beam of length $l = 20$ in is loaded by the uniform load $w = 80$ lb per inch of beam length. The x axis is positive to the right, and the y axis positive upward. All quantities—loading, shear, moment, slope, and deflection—have the same sense as y; they are positive if upward, negative if downward.

The values of the quantities at the ends of the beam, where $x = 0$ and $x = l$, are

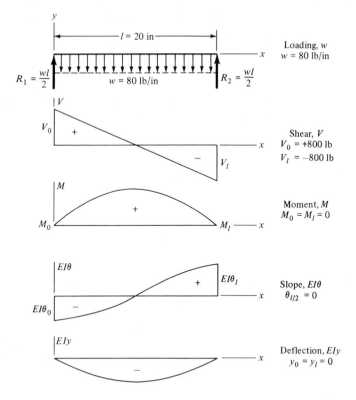

FIGURE 3-2

called their *boundary values*. For this reason the beam problem is often called a *boundary-value problem*. The reactions $R_1 = R_2 = +800$ lb and the shear forces $V_0 = +800$ lb and $V_l = -800$ lb are easily computed using the methods of Chap. 2. The bending moment is zero at each end because the beam is simply supported. Note that the beam-deflection curve must have a negative slope at the left boundary and a positive slope at the right boundary. This is easily seen by examining the deflection of Fig. 3-2. The magnitude of the slope at the boundaries is, as yet, unknown; because of symmetry, however, the slope is known to be zero at the center of the beam. We note, finally, that the deflection is zero at each end.

3-4   THE AREA-MOMENT METHOD

In Sec. 2-8, we learned methods for deriving the moment diagram directly from the loading diagram by employing the principles of statics. We saw in Sec. 3-3 that the moment diagram can also be obtained from the loading diagram by integrating twice. But the deflection diagram can be obtained from the moment diagram by integrating twice also. Thus it should be possible to derive the deflection diagram from the moment diagram by applying the principles of statics.

This method of deflection determination is called the *area-moment method*. It can be stated as follows: *The vertical distance between any point A on a deflection curve*

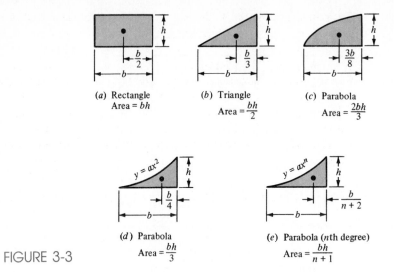

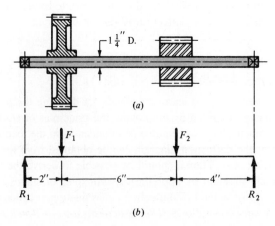

(a) Rectangle
Area = bh

(b) Triangle
Area = $\dfrac{bh}{2}$

(c) Parabola
Area = $\dfrac{2bh}{3}$

(d) Parabola
Area = $\dfrac{bh}{3}$

(e) Parabola (nth degree)
Area = $\dfrac{bh}{n+1}$

FIGURE 3-3

and a tangent through point B on the curve is the moment with respect to A of the area of the moment diagram between A and B divided by the stiffness EI. To use this statement, first find the area of the parts of the moment diagram or, if preferred, the M/EI diagram. Then, second, multiply the areas by their centroidal distances from the axis of moments. Areas and centroidal distances for typical portions of moment diagrams are shown in Fig. 3-3.

EXAMPLE 3-1   Figure 3-4a shows a $1\frac{1}{4}$-in-diameter steel shaft upon which are mounted two gears. If the shaft bends excessively, the gears will mesh improperly and an early failure can be expected. In this example the gear forces, shown in Fig. 3-4b, are assumed to be in the same plane and of magnitude $F_1 = 120$ lb and $F_2 = 90$ lb. Find the shaft deflection at each gear. The loading diagram indicates that the bearings are self-aligning.

FIGURE 3-4

Solution   This problem can be solved by employing the area-moment method only once, but we shall apply it twice, to illustrate the method of superposition.

The second moment of area is $I = \pi d^4/64 = \pi(1.25)^4/64 = 0.120$ in^4. Therefore $EI = 30(10)^6(0.120) = 3.6(10)^6$ lb $\cdot$ in^2. Our first step is to calculate the deflections due only to the action of F_1. In Fig. 3-5 the loading and moment diagrams have been constructed. By statics, the two reactions are found, and then the moments at A and B. An exaggerated deflection curve is drawn on the loading diagram, and a tangent is constructed at the left-hand reaction. Then, by the area-moment method, the distance $C'C$ is equal to the moment of the area of the moment diagram about C, divided by EI. Thus

$$EI(C'C) = \overset{\text{area}}{[(200/2)(2)]}\ \overset{\text{arm}}{(10\tfrac{2}{3})} + \overset{\text{area}}{[(200/2)(10)]}\ \overset{\text{arm}}{(6\tfrac{2}{3})} = 8800$$

Thus

$$C'C = \frac{8800}{3.6(10)^6} = 2.44(10)^{-3}\text{ in}$$

In this approach, the areas may be either positive or negative, but the moment arms are *always* positive. The distance $C'C$ is a positive quantity, measured from C' to C. However, it is easier to keep track of the signs in this approach using carefully made sketches of the moment and deflection diagrams. By similar triangles we also find

$$A'A = 4.074(10)^{-4}\text{ in} \qquad B'B = 1.630(10)^{-3}\text{ in}$$

Next, we find the distance $A'A''$ by taking moments about A:

$$EI(A'A'') = \overset{\text{area}}{[(200/2)(2)]}\ \overset{\text{arm}}{(\tfrac{2}{3})} = 133$$

$$A'A'' = \frac{133}{3.6(10)^6} = 3.704(10)^{-5}\text{ in}$$

The distance $B'B''$ is found by taking the moments of two triangular areas and one

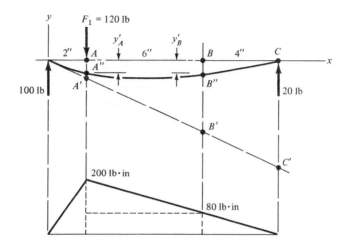

FIGURE 3-5

rectangular area about B:

$$EI(B'B'') = \underset{\text{area}}{[(200/2)(2)]} \; \underset{\text{arm}}{(6\tfrac{2}{3})} + \underset{\text{area}}{[(80)(6)]} \; \underset{\text{arm}}{(3)} + \underset{\text{area}}{[(120/2)(6)]} \; \underset{\text{arm}}{(4)} = 4213$$

$$B'B'' = \frac{4213}{3.6(10)^6} = 1.17(10)^{-3} \text{ in}$$

The deflection at A due only to F_1 is now found to be

$$y_A' = A'A'' - A'A = 3.704(10)^{-4} \text{ in}$$

In a similar manner,

$$y_B' = B'B'' - B'B = -4.60(10)^{-4} \text{ in}$$

The next step is to calculate the deflections at A and B due only to F_2. This is done in a similar manner, and the results are

$$y_A'' = -3.45(10)^{-4} \text{ in} \qquad y_B'' = -7.11(10)^{-4} \text{ in}$$

By superposition, the total deflection is the sum of the deflections caused by each load acting separately. Hence

$$y_A = -7.15(10)^{-4} \text{ in} \qquad y_B = -11.71(10)^{-4} \text{ in}$$

3-5 FINDING DEFLECTIONS BY USE OF SINGULARITY FUNCTIONS

The following examples are illustrations of the use of singularity functions.

EXAMPLE 3-2   As a first example, we choose the beam of Table A-9-6, which is a simply supported beam having a concentrated load F not in the center. Writing Eq. (3-10) for this loading gives

$$EI\frac{d^4y}{dx^4} = q = -F\langle x - a \rangle^{-1} \qquad 0 < x < l \tag{1}$$

Note that the reactions R_1 and R_2 do not appear in this equation, as they did in Chap. 2, because of the range chosen for x. If we now integrate from 0 to x—not $-\infty$ to x—according to Eq. (3-11), we get

$$EI\frac{d^3y}{dx^3} = V = -F\langle x - a \rangle^0 + C_1 \tag{2}$$

Using Eq. (3-12) this time, we integrate again and obtain

$$EI\frac{d^2y}{dx^2} = M = -F\langle x - a \rangle^1 + C_1x + C_2 \tag{3}$$

At $x = 0$, we have $M = 0$, and Eq. (3) gives $C_2 = 0$. At $x = l$, we again have $M = 0$, and Eq. (3) gives

$$C_1 = \frac{F(l - a)}{l} = \frac{Fb}{l}$$

Substituting these terms for C_1 and C_2 in Eq. (3) gives

$$EI\frac{d^2y}{dx^2} = M = \frac{Fbx}{l} - F\langle x - a \rangle^1 \tag{4}$$

Note that we could have obtained this equation by summing moments about a section a distance x from the origin. Next, integrate Eq. (4) twice in accordance with Eqs. (3-13) and (3-14). This yields

$$EI\frac{dy}{dx} = EI\theta = \frac{Fbx^2}{2l} - \frac{F\langle x - a \rangle^2}{2} + C_3 \tag{5}$$

$$EIy = \frac{Fbx^3}{6l} - \frac{F\langle x - a \rangle^3}{6} + C_3x + C_4 \tag{6}$$

The constants of integration C_3 and C_4 are evaluated using the two boundary conditions $y = 0$ at $x = 0$ and $y = 0$ at $x = l$. The first condition, substituted in Eq. (6), gives $C_4 = 0$. The second condition, substituted in Eq. (6), yields

$$0 = \frac{Fbl^2}{6} - \frac{Fb^3}{6} + C_3l$$

whence

$$C_3 = -\frac{Fb}{6l}(l^2 - b^2)$$

Upon substituting these results for C_3 and C_4 in Eq. (6), we obtain the deflection relation as

$$y = \frac{F}{6EIl}[bx(x^2 + b^2 - l^2) - l\langle x - a \rangle^3] \tag{7}$$

Compare Eq. (7) with the two deflection equations in Table A-9-6, and note that the use of singularity functions enables us to express the entire relation with a single equation.

EXAMPLE 3-3   Find the deflection relation, the maximum deflection, and the reactions for the statically indeterminate beam shown in Fig. 3-6a using singularity functions.

Solution   The loading diagram and approximate deflection curve are shown in Fig. 3-6b. Based upon the range $0 < x < l$, the loading equation is

$$q = R_2\langle x - a \rangle^{-1} - w\langle x - a \rangle^0 \tag{1}$$

We now integrate this equation four times in accordance with Eqs. (3-10) to (3-14).

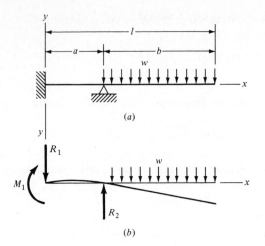

FIGURE 3-6

The results are

$$V = R_2\langle x - a\rangle^0 - w\langle x - a\rangle^1 + C_1 \tag{2}$$

$$M = R_2\langle x - a\rangle^1 - \frac{w}{2}\langle x - a\rangle^2 + C_1 x + C_2 \tag{3}$$

$$EI\theta = \frac{R_2}{2}\langle x - a\rangle^2 - \frac{w}{6}\langle x - a\rangle^3 + \frac{C_1}{2}x^2 + C_2 x + C_3 \tag{4}$$

$$EIy = \frac{R_2}{6}\langle x - a\rangle^3 - \frac{w}{24}\langle x - a\rangle^4 + \frac{C_1}{6}x^3 + \frac{C_2}{2}x^2 + C_3 x + C_4 \tag{5}$$

As in the previous example, the constants are evaluated by applying appropriate boundary conditions. First we note from Fig. 3-6b that both $EI\theta = 0$ and $EIy = 0$ at $x = 0$. This gives $C_3 = 0$ and $C_4 = 0$. Next we observe that at $x = 0$ the shear force is the same as the reaction. So Eq. (2) gives $V(0) = R_1 = C_1$. Since the deflection must be zero at the reaction R_2, where $x = a$, we have from Eq. (5) that

$$\frac{C_1}{6}a^3 + \frac{C_2}{2}a^2 = 0 \quad \text{or} \quad C_1\frac{a}{3} + C_2 = 0 \tag{6}$$

Also the moment must be zero at the overhanging end where $x = l$; Eq. (3) then gives

$$R_2(l - a) - \frac{w}{2}(l - a)^2 + C_1 l + C_2 = 0$$

Simplifying and noting that $l - a = b$ gives

$$C_1 a + C_2 = -\frac{wb^2}{2} \tag{7}$$

Equations (6) and (7) are now solved simultaneously for C_1 and C_2. The results are

$$C_1 = R_1 = -\frac{3wb^2}{4a} \qquad C_2 = \frac{wb^2}{4}$$

With R_1 known, we can sum the forces in the y direction to zero to get R_2. Thus

$$R_2 = -R_1 + wb = \frac{3wb^2}{4a} + wb = \frac{wb}{4a}(4a + 3b)$$

The moment reaction M_1 is obtained from Eq. (3) with $x = 0$. This gives

$$M(0) = M_1 = C_2 = \frac{wb^2}{4}$$

The complete equation of the deflection curve is obtained by substituting the known terms for R_2 and for the constants C_i in Eq. (5). The result is

$$EIy = \frac{wb}{24a}(4a + 3b)\langle x - a\rangle^3 - \frac{w}{24}\langle x - a\rangle^4 - \frac{wb^2x^3}{8a} + \frac{wb^2x^2}{8} \tag{8}$$

The maximum deflection occurs at the free end of the beam, at $x = l$. By making this substitution in Eq. (8) and manipulating the resulting expression, one finally obtains

$$y_{max} = \frac{wb^3l}{8EI} \tag{9}$$

Examination of the deflection curve of Fig. 3-6b reveals that the curve will have a zero slope at some point between R_1 and R_2. By replacing the constants in Eq. (4), setting $\theta = 0$, and solving the result for x we readily find that the slope is zero at $x = 2a/3$. The corresponding deflection at this point can be obtained by using the value of x in Eq. (8). The result is

$$y(x = 2a/3) = \frac{wa^2b^2}{54EI} \tag{10}$$

3-6   FINDING DEFLECTIONS BY NUMERICAL INTEGRATION

In this section we will develop an exact method of slope and deflection analysis of any simply supported beam having concentrated loads with any number of step changes in cross section. The objective here is to demonstrate the approach in sufficient detail to permit computer programming.

Trapezoidal integration can be explained by reference to Fig. 3-7. The integral of a function $y = f(x)$ is the area between the function $f(x)$ and the x axis. Thus, in Fig. 3-7a, the areas of the first two segments are, respectively,

$$\frac{y_0 + y_1}{2}\Delta x \qquad \text{and} \qquad \frac{y_1 + y_2}{2}\Delta x$$

But if Δx differs from each pair of x values, then the areas can be expressed as

$$\frac{y_0 + y_1}{2}(x_1 - x_0) \qquad \text{and} \qquad \frac{y_1 + y_2}{2}(x_2 - x_1)$$

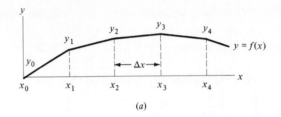

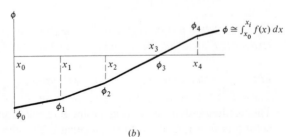

FIGURE 3-7

(a) Function to be integrated.
(b) Integral; note that ϕ_0 must
have a known value, and is here
considered to have a negative
starting value.

Thus, for each segment, the area is the mean value of the two ordinates times the spacing Δx. If we designate the integral as ϕ, then for the first segment, we have

$$\phi_1 = \int_{x_0}^{x_1} f(x)\, dx = \frac{y_0 + y_1}{2}(x_1 - x_0) + \phi_0 \qquad (a)$$

where ϕ_0 is the value of the integral up to x_0. The integral at the end of the second segment is

$$\phi_2 = \int_{x_0}^{x_2} f(x)\, dx = \frac{y_1 + y_2}{2}(x_2 - x_1) + \phi_1 \qquad (b)$$

Thus, in general,

$$\phi_i = \frac{y_{i-1} + y_i}{2}(x_i - x_{i-1}) + \phi_{i-1} \qquad (3\text{-}15)$$

Equation (3-15) gives exact results when the function is piecewise linear, as in Fig. 3-7a.

A stepped shaft with simple supports and concentrated loads, such as the one shown in Fig. 3-8, can be analyzed using the methods of Chap. 2, to develop an M/EI diagram. Such a diagram will be piecewise linear for each segment AB, BC, etc. We can, therefore, integrate Eq. (3-12) using Eq. (3-15) to obtain the slope. Thus, using

$$\phi = \int_0^x \frac{M}{EI}\, dx \qquad (c)$$

we find the slope to be

$$\theta = \frac{dy}{dx} = \int_0^x \frac{M}{EI}\, dx + C_1 = \phi + C_1 \qquad (d)$$

where C_1 is the slope at $x = 0$.

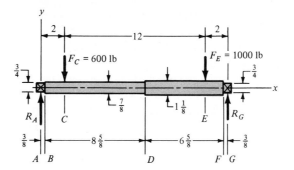

FIGURE 3-8

Simply supported stepped shaft loaded by forces F_C and F_E and supported by bearing reactions R_A and R_G. All dimensions in inches. $E = 30$ Mpsi.

Unfortunately, the slope is a piecewise quadratic function* and the use of the trapezoidal rule for a second integration to get the deflection is inexact. We can overcome this problem by using Simpson's rule, which is based on fitting a parabola to three of the points of the function to be integrated. Thus trapezoidal integration is carried out over one step, but Simpson's rule utilizes two steps.

Designate the integral of the slope as

$$\psi = \int_0^x \phi \, dx \tag{e}$$

Using Eq. (d), we then get the deflection as

$$y = \psi + C_1 + C_2 \tag{f}$$

It is convenient to write Eqs. (d) and (f) as

$$\theta = K(\phi + C_1)$$
$$y = K(\psi + C_1 x + C_2) \tag{3-17}$$

where K depends upon the units used. These equations are called the *prediction equations*.

Locating supports at $x = a$ and $x = b$ and specifying zero deflection at these supports provides the two conditions for finding C_1 and C_2. The results are

$$C_1 = \frac{\psi_b - \psi_a}{a - b} \qquad C_2 = \frac{b\psi_a - a\psi_b}{a - b} \tag{3-18}$$

Now, solving Eq. (c) using the trapezoidal rule gives

$$\phi_{i+2} = \phi_i + \frac{1}{2}\left[\left(\frac{M}{EI}\right)_{i+1} + \left(\frac{M}{EI}\right)_i\right](x_{i+2} - x_i) \tag{3-19}$$

Applying Simpson's rule [Eq. (2-79)] to the next integration gives

$$\psi_{i+4} = \psi_i + \tfrac{1}{6}(\phi_{i+4} + 4\phi_{i+2} + \phi_i)(x_{i+4} - x_i) \tag{3-20}$$

*See Charles R. Mischke, ''An Exact Numerical Method for Determining the Bending Deflection and Slope of Stepped Shafts,'' in *Advances in Reliability and Stress Analysis,* Proceedings of the Winter Annual Meeting of ASME, San Francisco, December 1978, pp. 101–115.

Equations (3-19) and (3-20) are used in a marching manner. Thus, using Eq. (3-19) we successively compute ϕ_1, ϕ_3, ϕ_5, . . . beginning at x_1 and ending at x_N, where N is the number of M/EI values. Similarly, Eq. (3-20) is solved successively to yield ψ_1, ψ_5, ψ_9, . . . , ψ_N.

After these two integrations have been performed, the constants C_1 and C_2 can be found from equation pair (3-18) and then Eqs. (3-16) and (3-17) can be solved for the slope and deflection. These terms will have the same indices as the integral ψ.

The details of the method are best explained by an example. The shaft of Fig. 3-8 has all points of interest designated by the station letters $A, B, C, ...$ These stations must include the locations of

- All supports and concentrated loads
- Cross-sectional changes
- Points at which the deflection and slope are desired

Refer now to Table 3-1, and note that coordinates x tabulated in column 2 correspond to each station. Note also the presence of additional x coordinates; these are selected as halfway stations and are needed to satisfy the need of Simpson's rule for two intervals.

Column 4 of Table 3-1 shows that two M/EI values must be computed for each coordinate x. These are needed to account for the fact that M/EI changes abruptly at some of the stations because of a change in cross section.

The approach described here can be used to approximate the slope and deflection for a beam having uniformly distributed loads. Simply divide the uniformly loaded interval into segments and use an equivalent concentrated force at the middle of each segment.

3-7   SHOCK AND IMPACT

Impact refers to the collision of two masses with initial relative velocity. In some cases it is desirable to achieve a known impact in design; for example, this is the case in the design of coining, stamping, and forming presses. In other cases, impact occurs because of excessive deflections, or because of clearances between parts, and in these cases it is desirable to minimize the effects. The rattling of mating gear teeth in their tooth spaces is an impact problem caused by shaft deflection and the clearance between the teeth. This impact causes gear noise and fatigue failure of the tooth surfaces. The clearance space between a cam and follower or between a journal and its bearing may result in crossover impact and also cause excessive noise and rapid fatigue failure.

Shock is a more general term which is used to describe any suddenly applied force or disturbance. Thus the study of shock includes impact as a special case. There are two general approaches to the study of shock, depending upon whether only statics is used in the analysis or both statics and dynamics are used. Lifetimes have been spent investigating shock and impact phenomena. For reasons of space, the material presented here is intended only to indicate the scope of the subject and to give a basic understanding of what is involved.

TABLE 3-1
Summary of Beam Computations*

1 STATION	2 x, in	3 N	4 M/EI, $(in^{-1})(10^{-6})$	5 ϕ, (10^{-6})	6 ψ, $(in)(10^{-6})$	7 y, in	8 θ, rad
A	0	0 1	0 0	0	0	0	$-1.028E-02$
	0.188	2 3	261.6 261.6	24.59			
B	0.375	4 5	523.2 282.4	98.0	12.27	$-3.8444E-03$	$-1.019E-02$
	1.188	6 7	894.6 894.6	576.4			
C	2	8 9	1506 1506	1 551	1 083	-0.0195	$-8.733E-03$
	5.5	10 11	1708.7 1708.7	7 177			
D	9	12 13	1911.4 699.5	13 512	52 149	$-4.0408E-02$	$3.228E-03$
	11.5	14 15	752.5 752.5	15 327			
E	14	16 17	805.5 805.5	17 274	128 894	$-1.5084E-02$	$6.990E-03$
	14.813	18 19	478.1 478.1	17 796			
F	15.625	20 21	151.0 764.7	18 052	157 741	$-2.9488E-03$	$7.768E-03$
	15.813	22 23	382.3 382.3	18 159			
G	16	24 25	0 0	18 195	164 546	0	$7.911E-03$

*The values of the constants used are

$$C_1 = \frac{\psi_b - \psi_a}{a - b} = \frac{164\ 546(10^{-6}) - 0}{0 - 16} = -0.010\ 284 \text{ rad}$$

$$C_2 = \frac{b\psi_a - a\psi_b}{a - b} = \frac{16(0) - 0(164\ 546)(10^{-6})}{0 - 16} = 0 \text{ in}$$

Thus

$\theta = \phi + C_1 = \phi - 0.010\ 284 \text{ rad}$
$y = \psi + C_1 x + C_2 = \psi - 0.010\ 284x \text{ in}$

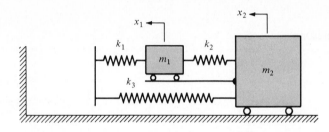

FIGURE 3-9

Two-degree-of-freedom mathematical model of an automobile in collision with a rigid obstruction.

Figure 3-9 represents a highly simplified mathematical model of an automobile in collision with a rigid obstruction. Here m_1 is the lumped mass of the engine. The displacement, velocity, and acceleration are described by the coordinate x_1 and its time derivatives. The lumped mass of the vehicle less the engine is denoted by m_2, and its motion by the coordinate x_2 and its derivatives. Springs k_1, k_2, and k_3 represent the linear and nonlinear stiffnesses of the various structural elements which compose the vehicle. Friction can and should be included, but is not shown in this model. The determination of the spring rates for such a complex structure will almost certainly have to be performed experimentally. Once these values—the k's, m's, and frictional coefficients—are obtained, a set of nonlinear differential equations can be written and a computer solution obtained for any impact velocity.

Figure 3-10 is another impact model. Here mass m_1 has an initial velocity v and is just coming into contact with spring k_1. The part or structure to be analyzed is represented by mass m_2 and spring k_2. The problem facing the designer is to find the maximum deflection of m_2 and the maximum force exerted by k_2 against m_2. In the analysis it doesn't matter whether k_1 is fastened to m_1 or to m_2, since we are interested only in a solution up to the point in time for which x_2 reaches a maximum. That is, the solution for the rebound isn't needed. The differential equations are not difficult to derive. They are

$$m_1\ddot{x}_1 + k_1(x_1 - x_2) = 0$$
$$m_2\ddot{x}_2 + k_2 x_2 - k_1(x_1 - x_2) = 0$$

(3-21)

Equation pair (3-21) is rather awkward to use in obtaining a general analytical solution. But, if the values of the m's and k's are specified, you can probably find an integration routine in your computer-program library for use in obtaining a solution.

3-8 ANALYSIS OF IMPACT

A simple case of impact is illustrated in Fig. 3-11a. Here a weight W, moving at a constant velocity v on a frictionless surface, strikes a cantilever of stiffness EI and

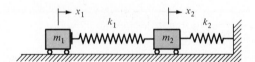

FIGURE 3-10

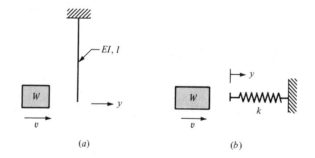

FIGURE 3-11

(a)   (b)

length l. We want to find the maximum deflection and the maximum bending moment in the beam due to the impact.

Figure 3-11b shows an abstract model of the system. Using Table A-9-1, we find the spring rate to be $k = F/y = 3EI/l^3$. We choose to count the time at the instant the weight strikes the spring. Thus at $t = 0$ the motion is started with $y = 0$ and $\dot{y} = v$. The differential equation is

$$\frac{W}{g}\ddot{y} = -ky \tag{a}$$

where the spring force ky is negative because it is opposite to the deflection y. The solution to this equation is well known, and is

$$y = A \cos\left(\frac{kg}{W}\right)^{1/2} t + B \sin\left(\frac{kg}{W}\right)^{1/2} t \tag{b}$$

The velocity is

$$\dot{y} = -A\left(\frac{kg}{W}\right)^{1/2} \sin\left(\frac{kg}{W}\right)^{1/2} t + B\left(\frac{kg}{W}\right)^{1/2} \cos\left(\frac{kg}{W}\right)^{1/2} t \tag{c}$$

The constants A and B are evaluated using the starting conditions in Eqs. (b) and (c). These are

$$A = 0 \qquad B = \frac{v}{(kg/W)^{1/2}}$$

Substituting these back in Eq. (b) gives the solution as

$$y = \frac{v}{(kg/W)^{1/2}} \sin\left(\frac{kg}{W}\right)^{1/2} t \tag{3-22}$$

Of course, this solution is valid only as long as the weight remains in contact with the beam. The maximum deflection is

$$y_{max} = \frac{v}{(kg/W)^{1/2}} = v\left(\frac{Wl^3}{3EIg}\right)^{1/2} \tag{3-23}$$

Also, the maximum bending moment is the product of the spring force and the beam length. The result is

$$M_{max} = kly_{max} = y\left(\frac{3EIW}{gl}\right)^{1/2} \tag{3-24}$$

As a second example, consider the weight W, in Fig. 3-12a, falling a distance h and impacting some structure or member whose spring rate is k. We choose the origin of the coordinate y corresponding to the position of the weight when time t is zero, as before. Two free-body diagrams, shown in Fig. 3-12b and c, are necessary—one when $y \le h$, and another when $y > h$—to account for the spring force. For each of these free-body diagrams we can write Newton's law by stating that the inertia force $(W/g)\ddot{y}$ is equal to the sum of the external forces acting on the weight. We then have

$$\frac{W}{g}\ddot{y} = W \qquad\qquad y \le h$$
$$\frac{W}{g}\ddot{y} = -k(y - h) + W \qquad y > h \tag{d}$$

We must also include in the mathematical statement of the problem the knowledge that the weight is released with zero initial velocity. Equation pair (d) constitutes a set of *piecewise differential equations*. Each equation is linear, but each applies only for a certain range of y. The solution to the set is valid for all values of t, but, as before, we are interested in values of y only up until the time that the spring or structure reaches its maximum deflection.

The solution to the first equation in the set is

$$y = \frac{gt^2}{2} \qquad y \le h \tag{3-25}$$

and you can verify this by direct substitution. Equation (3-25) is no longer valid after $y = h$; call this time t_1. Then

$$t_1 = \left(\frac{2h}{g}\right)^{1/2} \tag{e}$$

Differentiating Eq. (3-25) to get the velocity gives

$$\dot{y} = gt \qquad y \le h \tag{f}$$

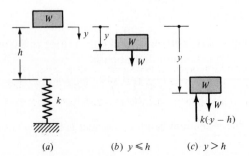

FIGURE 3-12   (a)   (b) $y \le h$ (c) $y > h$

and so the velocity of the weight at $t = t_1$ is

$$\dot{y}_1 = g t_1 = (2gh)^{1/2} \tag{g}$$

Having moved from $y = 0$ to $y = h$, we then need to solve the second equation of the set (d). It is convenient to define a new time $t' = t - t_1$. Thus $t' = 0$ at the instant the weight strikes the spring. Applying your knowledge of differential equations, you should find the solution to be

$$y = A \cos \left(\frac{kg}{W}\right)^{1/2} t' + B \sin \left(\frac{kg}{W}\right)^{1/2} t' + h + \frac{W}{k} \qquad y > h \tag{h}$$

If you cannot derive Eq. (h), you should at least prove to yourself that it is a solution by substituting it and its second derivative in the second equation of set (d) to show that the equality in Eq. (h) is satisfied identically.

The constants A and B are evaluated as shown in the previous solution. When these are substituted back, the result can be transformed to

$$y = \left[\left(\frac{W}{k}\right)^2 + \frac{2Wh}{k}\right]^{1/2} \cos \left[\left(\frac{kg}{W}\right)t' - \phi\right] + h + \frac{W}{k} \qquad y > h \tag{3-26}$$

where ϕ, though of no interest here, is given by

$$\phi = \frac{\pi}{2} + \tan^{-1} \left(\frac{W}{2kh}\right)^{1/2} \tag{i}$$

The maximum deflection of the spring (structure) occurs when the cosine term in Eq. (3-26) is unity. We designate this as δ and, after rearranging, find it to be

$$\delta = y - h = \frac{W}{k} + \frac{W}{k}\left[1 + \left(\frac{2hk}{W}\right)\right]^{1/2} \tag{3-27}$$

The maximum force acting on the spring or structure is now found to be

$$F = k\delta = W + W\left[1 + \left(\frac{2hk}{W}\right)\right]^{1/2} \tag{3-28}$$

Note, in this equation, that if $h = 0$, then $F = 2W$. This says that when the weight is released while in contact with the spring but is not exerting any force on the spring, the largest force is double the weight.

Most systems are not as ideal as those explored here, so be wary about using these relations for nonideal systems.

3-9   STRAIN ENERGY

The external work done on an elastic member in deforming it is transformed into *strain*, or *potential*, *energy*. If the member is deformed a distance y, this energy is equal to the product of the average force and the deflection, or

$$U = \frac{F}{2} y = \frac{F^2}{2k} \tag{a}$$

This equation is general in the sense that the force F also means torque, or moment, provided, of course, that consistent units are used for k. By substituting appropriate expressions for k, strain-energy formulas for various simple loadings may be obtained. For tension and compression and for torsion, for example, we employ Eqs. (3-4) and (3-7) and obtain

$$U = \frac{F^2 l}{2AE} \qquad \text{tension and compression} \tag{3-29}$$

$$U = \frac{T^2 l}{2GJ} \qquad \text{torsion} \tag{3-30}$$

To obtain an expression for the strain energy due to direct shear, consider the element with one side fixed in Fig. 3-13a. The force F places the element in pure shear, and the work done is $U = F\delta/2$. Since the shear strain is $\gamma = \delta/l = \tau/G = F/AG$, we have

$$U = \frac{F^2 l}{2AG} \qquad \text{direct shear} \tag{3-31}$$

The strain energy stored in a beam or lever by bending may be obtained by referring to Fig. 3-13b. Here AB is a section of the elastic curve of length ds having a radius of curvature ρ. The strain energy stored in this element of the beam is $dU = (M/2)\,d\theta$. Since $\rho\,d\theta = ds$, we have

$$dU = \frac{M\,ds}{2\rho} \tag{b}$$

We can eliminate ρ by using Eq. (3-8). Thus

$$dU = \frac{M^2\,ds}{2EI} \tag{c}$$

For small deflections, $ds \approx dx$. Then, for the entire beam

$$U = \int \frac{M^2\,dx}{2EI} \qquad \text{bending} \tag{3-32}$$

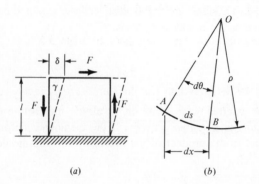

FIGURE 3-13   (a)   (b)

TABLE 3-2

Strain-Energy Correction
Factors for Shear*

BEAM CROSS-SECTIONAL SHAPE	FACTOR C
Rectangular	1.50
Circular	1.33
Tubular, round	2.00
Box sections†	1.00
Structural sections†	1.00

*Source: Arthur P. Boresi, Omar M. Sidebottom, Fred B. Seely, and
James O. Smith, *Advanced Mechanics of Materials,* 3d ed., Wiley,
New York, 1978, p. 173.

†Use area of web only.

Sometimes the strain energy stored in a unit volume is a useful quantity. By dividing Eqs. (3-29) to (3-31) by the total volume lA, we obtain

$$u = \begin{cases} \dfrac{\sigma^2}{2E} & \text{tension and compression} \\[2ex] \dfrac{\tau^2}{2G} & \text{direct shear} \\[2ex] \dfrac{\tau_{max}^2}{4G} & \text{torsion} \end{cases} \qquad (3\text{-}33)$$

It is interesting to note, from Eq. (3-33), that the development of a high stress in a material with a low modulus of elasticity, or rigidity, will result in the greatest amount of energy storage.

Equation (3-32) is exact only when a beam is subject to pure bending. Even when shear is present, Eq. (3-32) continues to give quite good results, except for very short beams. The strain energy due to shear loading of a beam is a complicated problem. An approximate solution can be obtained by using Eq. (3-31) with a correction factor whose value depends upon the shape of the cross section. If we use C for the correction factor and V for the shear force, then the strain energy due to shear in bending is the integral of Eq. (3-31), or

$$U = \int \frac{CV^2 \, dx}{2AG} \qquad \text{bending shear} \qquad (3\text{-}34)$$

Values of the factor C are listed in Table 3-2.

EXAMPLE 3-4   Find the strain energy due to shear in a rectangular cross-section beam, simply supported, and having a uniformly distributed load.

Solution   Using Appendix Table A-9-7, we find the shear force to be

$$V = \frac{wl}{2} - wx$$

Substituting into Eq. (3-34), with $C = 1.5$, gives

Answer $$U = \frac{1.5}{2AG} \int_0^l \left(\frac{wl}{2} - wx \right)^2 dx = \frac{3w^2l^3}{48AG}$$

EXAMPLE 3-5   A cantilever has a concentrated load F at the end, as shown in Fig. 3-14. Find the strain energy in the beam by neglecting shear.

Solution   At any point x along the beam, the moment is $M = -Fx$. Substituting this value of M into Eq. (3-32), we find

Answer $$U = \int_0^l \frac{F^2x^2 \, dx}{2EI} = \frac{F^2l^3}{6EI}$$

3-10   CASTIGLIANO'S THEOREM

A most unusual, powerful, and often surprisingly simple approach to deflection analysis is afforded by an energy method called Castigliano's theorem. It is a unique way of analyzing deflections and is even useful for finding the reactions of indeterminate structures. Castigliano's theorem states that *when forces act on elastic systems subject to small displacements, the displacement corresponding to any force, collinear with the force, is equal to the partial derivative of the total strain energy with respect to that force*. The terms *force* and *displacement* in this statement are broadly interpreted to apply equally to moments and angular displacements. Mathematically, the theorem of Castigliano is

$$\delta_i = \frac{\partial U}{\partial F_i} \tag{3-35}$$

where δ_i is the displacement of the point of application of the force F_i in the direction of F_i.

As an example, apply Castigliano's theorem using Eqs. (3-29) and (3-30) to get the axial and torsional deflections. The results are

$$\delta = \frac{\partial}{\partial F} \left(\frac{F^2l}{2AE} \right) = \frac{Fl}{AE} \tag{a}$$

$$\theta = \frac{\partial}{\partial T} \left(\frac{T^2l}{2GJ} \right) = \frac{Tl}{GJ} \tag{b}$$

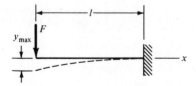

FIGURE 3-14

Compare Eqs. (*a*) and (*b*) with Eqs. (3-3) and (3-5). In Example 3-5, the strain energy for a cantilever having a concentrated end load was found. According to Castigliano's theorem, the deflection at the end of the beam is

$$y = \frac{\partial U}{\partial F} = \frac{\partial}{\partial F}\left(\frac{F^2 l^3}{6EI}\right) = \frac{F l^3}{3EI}$$

which checks with Table A-9-1.

Castigliano's theorem can be used to find the deflection at a point even though no force or moment acts there. The procedure is:

1 Set up the equation for the total strain energy U by including the energy due to a fictitious force or moment Q_i acting at the point whose deflection is to be found.

2 Find an expression for the desired deflection δ_i by taking the derivative of the total strain energy with respect to Q_i as follows:

$$\delta_i = \frac{\partial U}{\partial Q_i} \tag{3-36}$$

3 Since Q_i is a fictitious force, solve the expression obtained in step 2 by setting Q_i equal to zero.

EXAMPLE 3-6

The cantilever of Example 3-5 is a carbon steel bar 10 in long with a 1-in diameter and is loaded by a force $F = 100$ lb.

(*a*) Find the maximum deflection using Castigliano's theorem, including that due to shear.

(*b*) What error is introduced if shear is neglected?

Solution

(*a*) From Eq. (3-34) and Example 3-4 data, the total strain energy is

$$U = \frac{F^2 l^3}{6EI} + \int_0^l \frac{CV^2 \, dx}{2AG} \tag{1}$$

For the cantilever, $V = F$. Also, $C = 1.33$, from Table 3-2. Performing the integration and substituting these values in Eq. (1) gives, for the total strain energy,

$$U = \frac{F^2 l^3}{6EI} + \frac{1.33 F^2 l}{2AG} \tag{2}$$

Then, according to the theorem, the deflection of the end is

$$y = \frac{\partial U}{\partial F} = \frac{F l^3}{3EI} + \frac{1.33 F l}{AG} \tag{3}$$

We also find that

$$I = \frac{\pi d^4}{64} = \frac{\pi (1)^4}{64} = 0.0491 \text{ in}^4$$

$$A = \frac{\pi d^2}{4} = \frac{\pi (1)^2}{4} = 0.7854 \text{ in}^2$$

Substituting these values, together with $F = 100$ lb, $l = 10$ in, $E = 30$ Mpsi, and $G = 11.5$ Mpsi, in Eq. (3) gives

Answer $y = 0.022\ 629 + 0.000\ 147 = 0.022\ 776$ in

Note that the result is positive because it is in the *same* direction as the force F.

(*b*) The error in neglecting shear is found to be about 0.65 percent.

EXAMPLE 3-7   Find the maximum deflection of a cantilever of length l loaded by a concentrated force F in the middle. Neglect shear.

Solution   The problem is illustrated in Fig. 3-15, where the force F acts in the center and a fictitious force Q has been placed at the end where the deflection is desired.
The moments are

$$M_{AB} = F\left(x - \frac{l}{2}\right) + Q(x - l) \tag{1}$$

$$M_{BC} = Q(x - l) \tag{2}$$

The total strain energy is

$$U = \int_0^{l/2} \frac{M_{AB}{}^2\ dx}{2EI} + \int_{l/2}^l \frac{M_{BC}{}^2\ dx}{2EI} \tag{3}$$

Then, by Castigliano's theorem, the deflection is

$$y = \frac{\partial U}{\partial Q} = \frac{1}{2EI}\left[\int_0^{l/2} 2M_{AB}\left(\frac{\partial M_{AB}}{\partial Q}\right) dx + \int_{l/2}^l 2M_{BC}\left(\frac{\partial M_{BC}}{\partial Q}\right) dx\right] \tag{4}$$

Next, we have

$$\frac{\partial M_{AB}}{\partial Q} = \frac{\partial M_{BC}}{\partial Q} = x - l \tag{5}$$

Making the appropriate substitutions in Eq. (4) yields

$$y = \frac{1}{EI}\left\{\int_0^{l/2}\left[F\left(x - \frac{l}{2}\right) + Q(x - l)\right](x - l)\ dx + \int_{l/2}^l [Q(x - l)](x - l)\ dx\right\} \tag{6}$$

Since Q is fictitious, we substitute $Q = 0$ in Eq. (6) to get

Answer $y = \dfrac{F}{EI}\displaystyle\int_0^{l/2}\left(x - \frac{l}{2}\right)(x - l)\ dx = \dfrac{5Fl^3}{48EI}$

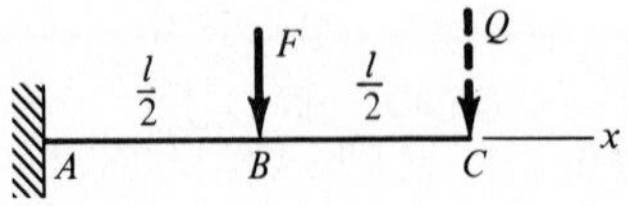

FIGURE 3-15

3-11   STATICALLY INDETERMINATE PROBLEMS

A system in which the laws of statics are not sufficient to determine all the unknown forces or moments is said to be *statically indeterminate*. Problems of which this is true are solved by writing the appropriate equations of static equilibrium and additional equations pertaining to the deformation of the part. In all, the number of equations must equal the number of unknowns.

A simple example of a statically indeterminate problem is furnished by the nested helical spring in Fig. 3-16*a*. When this assembly is loaded by the compressive force F, it deforms through the distance δ. What is the compressive force in each spring?

Only one equation of static equilibrium can be written. It is

$$\sum F = F - F_1 - F_2 = 0 \qquad (a)$$

which simply says that the total force F is resisted by a force F_1 in spring 1 plus the force F_2 in spring 2. Since there are two unknowns and only one equation, the system is statically indeterminate.

To write another equation, note the deformation relation in Fig. 3-16*b*. The two springs have the same deformation. Thus, we obtain the second equation as

$$\delta_1 = \delta_2 = \delta \qquad (b)$$

If we now substitute Eq. (3-2) in Eq. (*b*), we have

$$\frac{F_1}{k_1} = \frac{F_2}{k_2} \qquad (c)$$

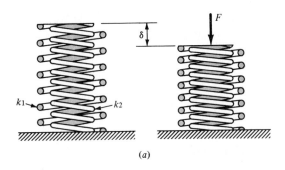

(*a*)

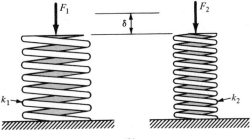

FIGURE 3-16   (*b*)

Now we solve Eq. (*c*) for F_1 and substitute the result in Eq. (*a*). This gives

$$F - \frac{k_1}{k_2}F_2 - F_2 = 0 \qquad \text{or} \qquad F_2 = \frac{k_2 F}{k_1 + k_2} \qquad (d)$$

This completes the solution, because with F_2 known, F_1 can be found from Eq. (*c*).

A more elegant approach is afforded by Castigliano's theorem. The method is based on the fact that the deflection at the point of and in the direction of any reaction is zero. Thus the derivative of the total strain energy with respect to the reaction should also be zero. The procedure is:

1 Choose the redundant reaction; this is usually a force or a moment.

2 Write and solve the equations of static equilibrium for the remaining reactions in terms of the redundant reaction.

3 Write the equation for the total strain energy U.

4 Find an expression for the redundant reaction by taking the derivative of the total strain energy with respect to the reaction, say R, as follows:

$$\frac{\partial U}{\partial R} = 0 \qquad (3\text{-}37)$$

5 Solve the resulting expression for the reaction R.

EXAMPLE 3-8   The indeterminate beam of Appendix Table A-9-11 is reproduced in Fig. 3-17. Determine the reactions using Castigliano's theorem to resolve the indeterminacy.

Solution   The reactions are shown in Fig. 3-17*b*. We choose M_1 as the redundant one. Applying the rule of static equilibrium yields the two equations

$$\Sigma F = R_1 - F + R_2 = 0$$

$$\Sigma M_O = M_1 - \frac{Fl}{2} + R_2 l = 0 \qquad (1)$$

Since M_1 is selected as the indeterminate reaction, we next solve equation set (1) for R_1 and R_2 in terms of M_1. Thus

$$R_1 = \frac{F}{2} + \frac{M_1}{l} \qquad R_2 = \frac{F}{2} - \frac{M_1}{l} \qquad (2)$$

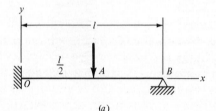
(a)

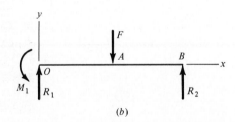
(b)

FIGURE 3-17

Neglecting shear, the total strain energy, from Eq. (3-32), is

$$U = \int_0^{l/2} \frac{M_{OA}^2 \, dx}{2EI} + \int_{l/2}^l \frac{M_{AB}^2 \, dx}{2EI} \tag{3}$$

The moments from O to A and from A to B are

$$M_{OA} = -M_1 + R_1 x = M_1 \left(\frac{x}{l} - 1\right) + \frac{Fx}{2}$$

$$M_{AB} = -M_1 + R_1 x - F\left(x - \frac{l}{2}\right) = M_1\left(\frac{x}{l} - 1\right) + \frac{F}{2}(l - x) \tag{4}$$

The reaction M_1 is now found by taking the derivative of Eq. (3) and setting the result equal to zero. Thus

$$\frac{\partial U}{\partial M_1} = \frac{1}{2EI}\left[\int_0^{l/2} 2M_{OA}\left(\frac{\partial M_{OA}}{\partial M_1}\right) dx + \int_{l/2}^l 2M_{AB}\left(\frac{\partial M_{AB}}{\partial M_1}\right) dx\right] \tag{5}$$

Taking the derivatives of the moments of equation set (4) yields

$$\frac{\partial M_{OA}}{\partial M_1} = \frac{\partial M_{AB}}{\partial M_1} = \frac{x}{l} - 1 \tag{6}$$

We next substitute Eqs. (4) and (6) in Eq. (5), perform the indicated multiplication, and cancel terms. We then have

$$\frac{\partial U}{\partial M_1} = \int_0^{l/2}\left[M_1\left(\frac{x}{l} - 1\right)^2 + \frac{Fx}{2}\left(\frac{x}{l} - 1\right)\right] dx$$

$$+ \int_{l/2}^l \left[M_1\left(\frac{x}{l} - 1\right)^2 + \frac{F}{2}(l - x)\left(\frac{x}{l} - 1\right)\right] dx = 0 \tag{7}$$

After Eq. (7) is integrated and the limits substituted, we obtain

$$\frac{\partial U}{\partial M_1} = \frac{7M_1 l}{24} - \frac{Fl^2}{24} + \frac{M_1 l}{24} - \frac{Fl^2}{48} = 0 \tag{8}$$

Answer   The redundant reaction is now found to be $M_1 = 3Fl/16$. This value of M_1 can now be substituted in Eq. (2) to obtain R_1 and R_2 in terms of the force F.

3-12   DEFLECTION OF CURVED MEMBERS

Machine frames, springs, clips, fasteners, and the like frequently occur as curved shapes. The determination of stresses in curved members has already been illustrated in Sec. 2-18. Castigliano's theorem is particularly useful for the analysis of deflections in curved parts too. Consider, for example, the curved frame of Fig. 3-18a. We are interested in finding the deflection of the frame due to F and in the direction of F. The total strain energy consists of four terms, and we shall consider each separately. The

FIGURE 3-18

(a) Curved bar loaded by force F. R = radius to centroidal axis of section; h = section thickness. (b) Diagram showing forces acting on section taken at angle θ. $F_r = V$ = shear component of F; F_θ is component of F normal to section; M is moment caused by force F.

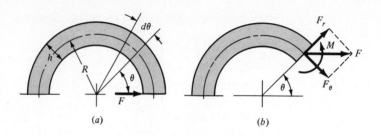

(a)   (b)

first is due to the bending moment and is

$$U_1 = \int \frac{M^2\, d\theta}{2AeE} \tag{3-38}$$

In this equation, the eccentricity e is

$$e = R - r_n \tag{3-39}$$

where r_n is the radius of the neutral axis as defined in Sec. 2-20 and shown in Fig. 2-27.

An approximate result can be obtained using the equation

$$U_1 \approx \int \frac{M^2 R\, d\theta}{2EI} \qquad \frac{R}{h} > 10 \tag{3-40}$$

which is obtained directly from Eq. (3-32). Note the limitation on the use of Eq. (3-40).

The strain energy component due to the normal force F_θ consists of two parts, one of which is axial and analogous to Eq. (3-29). This part is

$$U_2 = \int \frac{F_\theta^2 R\, d\theta}{2AE} \tag{3-41}$$

The force F_θ also produces a moment, which opposes the moment M in Fig. 3-18b. The resulting strain energy will be subtractive and is

$$U_3 = -\int \frac{MF_\theta\, d\theta}{AE} \tag{3-42}$$

The negative sign of Eq. (3-42) can be appreciated by referring to both parts of Fig. 3-18. Note that the moment M tends to decrease the angle $d\theta$. On the other hand, the moment due to F_θ tends to increase $d\theta$. Thus U_3 is negative. If F_θ had been acting in the opposite direction, then both M and F_θ would tend to decrease the angle $d\theta$.

The fourth and last term is the shear energy due to F_r. Adapting Eq. (3-34) gives

$$U_4 = \int \frac{CF_r^2 R\, d\theta}{2AG} \tag{3-43}$$

where C is the correction factor of Table 3-2.

Combining the four terms gives for the total strain energy

$$U = \int \frac{M^2\, d\theta}{2AeE} + \int \frac{F_\theta^2 R\, d\theta}{2AE} - \int \frac{MF_\theta\, d\theta}{AE} + \int \frac{CF_r^2 R\, d\theta}{2AG} \tag{a}$$

The deflection produced by the force F can now be found. It is

$$\delta = \frac{\partial U}{\partial F} = \int_0^\pi \frac{M}{AeE}\left(\frac{\partial M}{\partial F}\right) d\theta + \int_0^\pi \frac{F_\theta R}{AE}\left(\frac{\partial F_\theta}{\partial F}\right) d\theta$$

$$- \int_0^\pi \frac{1}{AE}\frac{\partial(MF_\theta)}{\partial F} d\theta + \int_0^\pi \frac{CF_r R}{AG}\left(\frac{\partial F_r}{\partial F}\right) d\theta \quad (b)$$

Using Fig. 3-18b, we find

$$M = FR \sin \theta \qquad \frac{\partial M}{\partial F} = R \sin \theta$$

$$F_\theta = F \sin \theta \qquad \frac{\partial F_\theta}{\partial F} = \sin \theta$$

$$MF_\theta = F^2 R \sin^2 \theta \qquad \frac{\partial MF_\theta}{\partial F} = 2FR \sin^2 \theta$$

$$F_r = F \cos \theta \qquad \frac{\partial F_r}{\partial F} = \cos \theta$$

Substituting all these into Eq. (b) and factoring yields

$$\delta = \frac{FR^2}{AeE}\int_0^\pi \sin^2 \theta \, d\theta + \frac{FR}{AE}\int_0^\pi \sin^2 \theta \, d\theta - \frac{2FR}{AE}\int_0^\pi \sin^2 \theta \, d\theta$$

$$+ \frac{CFR}{AG}\int_0^\pi \cos^2 \theta \, d\theta$$

$$= \frac{\pi FR^2}{2AeE} + \frac{\pi FR}{2AE} - \frac{\pi FR}{AE} + \frac{\pi CFR}{2AG} = \frac{\pi FR^2}{2AeE} - \frac{\pi FR}{2AE} + \frac{\pi CFR}{2AG} \quad (3\text{-}44)$$

Because the first term contains the square of the radius, the second two terms will be small if the frame has a large radius. Also, if $R/h > 10$, Eq. (3-40) can be used. An approximate result then turns out to be

$$\delta \approx \frac{\pi FR^3}{2EI} \quad (3\text{-}45)$$

The determination of the deflection of a curved member loaded by forces at right angles to the plane of the member is more difficult, but the method is the same.* We shall include here only one of the more useful solutions to such a problem, though the methods for all are similar. Figure 3-19 shows a cantilevered ring segment having a span angle ϕ. The strain energy is obtained from the equation

$$U = \int_0^\phi \frac{M^2 R \, d\theta}{2EI} + \int_0^\phi \frac{T^2 R \, d\theta}{2GJ} \quad (3\text{-}46)$$

The moments and torques acting on a section at B, due to the force F, are

$$M = FR \sin \theta \qquad T = FR(1 - \cos \theta)$$

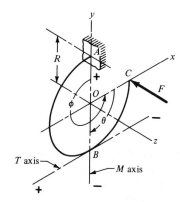

FIGURE 3-19

Ring ABC in xy plane subject to force F parallel to z axis. Corresponding to a ring segment CB at angle θ from point of application of F, the moment axis is a line BO and the torque axis is a line in the xy plane tangent to the ring at B. Note the positive directions of the T and M axes.

---

*For more solutions than are included here, see Joseph E. Shigley, "Curved Beams and Rings," chap. 16 in Joseph E. Shigley and Charles R. Mischke (eds.), *Standard Handbook of Machine Design*, McGraw-Hill, New York, 1986.

The deflection δ of the ring segment at F and in the direction of F is then found to be

$$\delta = \frac{\partial U}{\partial F} = \frac{FR^3}{2} \left(\frac{\alpha}{EI} + \frac{\beta}{GJ} \right) \tag{3-47}$$

where the coefficients α and β are dependent on the span angle ϕ and are defined as follows:

$$\begin{aligned} \alpha &= \phi - \sin \phi \cos \phi \\ \beta &= 3\phi - 4 \sin \phi + \sin \phi \cos \phi \end{aligned} \tag{3-48}$$

3-13   COMPRESSION MEMBERS—GENERAL

The analysis and design of compression members differs significantly from that of members loaded in tension or in torsion. If you were to take a long rod or pole, such as a meterstick, and apply gradually increasing forces at each end, nothing would happen at first, but then the stick would bend (buckle), and finally bend so much as to fracture. Try it. The other extreme would occur if you were to saw off, say, a 5-mm length of the meterstick and perform the same experiment on the short piece. You would then observe that the failure exhibits itself as a mashing of the specimen, that is, a simple compressive failure. For these reasons it is convenient to classify compression members according to their length and according to whether the loading is central or eccentric. The term *column* is applied to all such members except those in which failure would be by simple or pure compression. Columns can be categorized then as:

1 Long columns with central loading

2 Intermediate-length columns with central loading

3 Columns with eccentric loading

4 Struts or short columns with eccentric loading

Classifying columns as above makes it possible to develop methods of analysis and design specific to each category. Furthermore, these methods will also reveal whether or not you have selected the category appropriate to your particular problem. The four sections which follow correspond, respectively, to the four categories of columns listed above.

3-14   LONG COLUMNS WITH CENTRAL LOADING

The relationship between the critical load and the column material and geometry is developed with reference to Fig. 3-20a. We assume a bar of length l loaded by a force P acting along the centroidal axis on rounded or pinned ends. The figure shows that the bar is bent in the positive y direction. This requires a negative moment, and hence

$$M = -Py \tag{a}$$

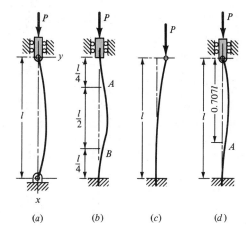

FIGURE 3-20

(a) Both ends rounded or pivoted;
(b) both ends fixed; (c) one end
free, one end fixed; (d) one end
rounded and guided and one end
fixed.

If the bar should happen to bend in the negative y direction, a positive moment would result, and so $M = -Py$, as before. Using Eq. (3-12), we write

$$\frac{d^2y}{dx^2} = -\frac{P}{EI} y \qquad (b)$$

or

$$\frac{d^2y}{dx^2} + \frac{P}{EI} y = 0 \qquad (3\text{-}49)$$

This resembles the well-known differential equation for simple harmonic motion. The solution is

$$y = A \sin \sqrt{\frac{P}{EI}} x + B \cos \sqrt{\frac{P}{EI}} x \qquad (c)$$

where A and B are constants of integration and must be determined from the boundary conditions of the problem. We evaluate them using the conditions that $y = 0$ at $x = 0$ and at $x = l$. This gives $B = 0$, and

$$0 = A \sin \sqrt{\frac{P}{EI}} l \qquad (d)$$

The trivial solution of no buckling occurs with $A = 0$. However, if $A \neq 0$, then

$$\sin \sqrt{\frac{P}{EI}} l = 0 \qquad (e)$$

Equation (e) is satisfied by $\sqrt{P/EI}\, l = n\pi$, where $n = 1, 2, 3, \ldots$. Solving for P when $n = 1$ gives the first critical load

$$P_{cr} = \frac{\pi^2 EI}{l^2} \qquad (3\text{-}50)$$

which is called the *Euler column formula;* it applies only to rounded-end columns. If we substitute these results back in Eq. (c), we get the equation of the deflection curve as

$$y = A \sin \frac{\pi x}{l} \qquad (f)$$

which indicates that the deflection curve is a half-wave sine. We are interested only in the minimum critical load, which occurs with $n = 1$. However, though it is not of any importance here, values of n greater than 1 result in deflection curves which cross the axis at points of inflection and are multiples of half-wave sines.

Using the relation $I = Ak^2$, where A is the area and k the radius of gyration, enables us to rearrange Eq. (3-50) into the more convenient form

$$\frac{P_{cr}}{A} = \frac{\pi^2 E}{(l/k)^2} \qquad (3\text{-}51)$$

where l/k is called the *slenderness ratio*. This ratio, rather than the actual column length, will be used in classifying columns according to length categories.

The quantity P_{cr}/A in Eq. (3-51) is the *critical unit load*. It is the load per unit area necessary to place the column in a condition of *unstable equilibrium*. In this state any small crookedness of the member, or slight movement of the support or load, will cause the column to collapse. The unit load has the same units as strength, but this is the strength of a specific column, not of the column material. Doubling the length of a member, for example, will have a drastic effect on the value of P_{cr}/A but no effect at all on, say, the yield strength S_y of the column material itself.

Equation (3-51) shows that the critical unit load depends only upon the modulus of elasticity and the slenderness ratio. Thus a column obeying the Euler formula made of high-strength alloy steel is no stronger than one made of low-carbon steel, since E is the same for both.

The critical loads for columns with different end conditions can be obtained by solving the differential equation or by comparison. Figure 3-20b shows a column with both ends fixed. The inflection points are at A and B, a distance $l/4$ from the ends. The distance AB is the same curve as a rounded-end column. Substituting the length $l/2$ for l in Eq. (3-50), we obtain

$$P_{cr} = \frac{\pi^2 EI}{(l/2)^2} = \frac{4\pi^2 EI}{l^2} \qquad (3\text{-}52)$$

In Fig. 3-20c is shown a column with one end free and one end fixed. This curve is equivalent to half the curve for columns with rounded ends, so that if a length of $2l$ is substituted in Eq. (3-50), the critical load becomes

$$P_{cr} = \frac{\pi^2 EI}{(2l)^2} = \frac{\pi^2 EI}{4l^2} \qquad (3\text{-}53)$$

A column with one end fixed and one end rounded, as in Fig. 3-20d, occurs frequently. The inflection point is at A, a distance of $0.707l$ from the rounded end. Therefore

$$P_{cr} = \frac{\pi^2 EI}{(0.707l)^2} = \frac{2\pi^2 EI}{l^2} \qquad (3\text{-}54)$$

We can account for these various end conditions by writing the Euler equation in the two following forms:

$$P_{cr} = \frac{C\pi^2 EI}{l^2} \qquad \frac{P_{cr}}{A} = \frac{C\pi^2 E}{(l/k)^2} \tag{3-55}$$

Here, the factor C is called the *end-condition constant*, and it may have any one of the theoretical values $\frac{1}{4}$, 1, 2, and 4, depending upon the manner in which the load is applied. In practice it is difficult, if not impossible, to fix the column ends so that the factor $C = 2$ or $C = 4$ would apply. Even if the ends are welded, some deflection will occur. Because of this, some designers never use a value of C greater than unity. However, if liberal factors of safety are employed, and if the column load is accurately known, then a value of C not exceeding 1.2 for both ends fixed, or for one end rounded and one end fixed, is not unreasonable, since it supposes only partial fixation. Of course, the value $C = \frac{1}{4}$ must always be used for a column having one end fixed and one end free. These recommendations are summarized in Table 3-3.

When Eq. (3-55) is solved for various values of the unit load P_{cr}/A in terms of the slenderness ratio l/k, we obtain the curve PQR shown in Fig. 3-21. Since the yield strength of the material has the same units as the unit load, the horizontal line through S_y and Q has been added to the figure. This would appear to make the figure cover the entire range of compression problems from the shortest to the longest compression member. Thus it would appear that any compression member having an l/k value less than $(l/k)_Q$ should be treated as a pure compression member while all others are to be treated as Euler columns. Unfortunately, this is not true.

In the actual design of a member which functions as a column, the designer will be aware of the end conditions shown in Fig. 3-20. He or she will endeavor to configure the ends, using bolts, welds, or pins, for example, so as to achieve the required ideal end condition. In spite of these precautions, the result, following manufacture, is likely to contain defects such as initial crookedness or load eccentricities. The existence of such defects and the methods of accounting for them will usually involve a factor-of-safety approach or a stochastic analysis. These methods work well for long columns and for simple compression members. However, tests show numerous failures for columns with slenderness ratios below and in the vicinity of point Q, as shown in the shaded area in Fig. 3-21. These have been reported as occurring even when near-perfect geometric specimens were used in the testing procedure.

A column failure is always sudden, total, and unexpected, and hence dangerous. There is no advance warning. A beam will bend and give visual warning that it is

TABLE 3-3

End-Condition Constants for Euler Columns [to Be Used with Eq. (3-55)]

COLUMN END CONDITIONS	END-CONDITION CONSTANT C		
	THEORETICAL VALUE	CONSERVATIVE VALUE	RECOMMENDED VALUE*
Fixed-free	$\frac{1}{4}$	$\frac{1}{4}$	$\frac{1}{4}$
Rounded-rounded	1	1	1
Fixed-rounded	2	1	1.2
Fixed-fixed	4	1	1.2

*To be used only with liberal factors of safety when the column load is accurately known.

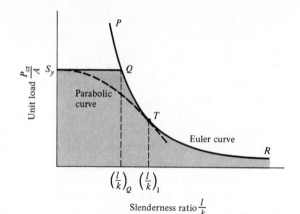

FIGURE 3-21

Euler's curve plotted using Eq. (3-55) with $C = 1$.

overloaded; but not so for a column. For this reason neither simple compression methods nor the Euler column equation should be used when the slenderness ratio is near $(l/k)_Q$. Then what should we do? The usual approach is to choose some point T on the Euler curve of Fig. 3-21. If the slenderness ratio is specified as $(l/k)_1$ corresponding to point T, then use the Euler equation only when the actual slenderness ratio is greater than $(l/k)_1$. Otherwise, use one of the methods in the sections which follow. See Examples 3-9 and 3-10.

Most designers select point T such that $P_{cr}/A = S_y/2$. Using Eq. (3-55), we find the corresponding value of $(l/k)_1$ to be

$$\left(\frac{l}{k}\right)_1 = \left(\frac{2\pi^2 CE}{S_y}\right)^{1/2} \tag{3-56}$$

3-15   INTERMEDIATE-LENGTH COLUMNS WITH CENTRAL LOADING

Over the years there have been a number of column formulas proposed and used for the range of l/k values for which the Euler formula is not suitable. Many of these are based on the use of a single material; others, on a so-called safe unit load rather than the critical value. Most of these formulas are based on the use of a linear relationship between the slenderness ratio and the unit load. The *parabolic* or *J. B. Johnson, formula* now seems to be the preferred one among designers in the machine, automotive, aircraft, and structural-steel construction fields.

The general form of the parabolic formula is

$$\frac{P_{cr}}{A} = a - b\left(\frac{l}{k}\right)^2 \tag{3-57}$$

where a and b are constants that are evaluated by fitting a parabola to the Euler curve of Fig. 3-21 as shown by the dashed line ending at T. If the parabola is begun at S_y, then $a = S_y$. If point T is selected as previously noted, then Eq. (3-56) gives the value of

$(l/k)_1$ and the constant b is found to be

$$b = \left(\frac{S_y}{2\pi}\right)^2 \frac{1}{CE} \qquad (a)$$

Upon substituting the known values of a and b into Eq. (3-57), we obtain, for the parabolic equation,

$$\frac{P_{cr}}{A} = S_y - \left(\frac{S_y}{2\pi}\frac{l}{k}\right)^2 \frac{1}{CE} \qquad \frac{l}{k} \le \left(\frac{l}{k}\right)_1 \qquad (3\text{-}58)$$

3-16   COLUMNS WITH ECCENTRIC LOADING

We have noted before that deviations from an ideal column, such as load eccentricities or crookedness, are likely to occur during manufacture and assembly. Though these deviations are often quite small, it is still convenient to have a method of dealing with them. Frequently, too, problems occur in which load eccentricities are unavoidable.

Figure 3-22a shows a column in which the line of action of the column forces is separated from the centroidal axis of the column by the eccentricity e. This problem is developed using the free-body diagram of Fig. 3-22b and summing the moments to zero about the origin O. This gives

$$\Sigma M_O = M + Pe + Py = 0 \qquad (a)$$

Using Eq. (3-12) and rearranging produces the differential equation

$$\frac{d^2y}{dx^2} + \frac{P}{EI}y = -\frac{Pe}{EI} \qquad (b)$$

Equation (b) is solved in a manner quite similar to others we have solved in this chapter. The boundary conditions are found to be

$$x = 0 \qquad y = 0$$
$$x = \frac{l}{2} \qquad \frac{dy}{dx} = 0$$

By substituting $x = l/2$ in the resulting solution, we find the deflection at midspan and the maximum bending moment to be

$$\delta = e\left[\sec\left(\frac{l}{2}\sqrt{\frac{P}{EI}}\right) - 1\right] \qquad (3\text{-}59)$$

$$M_{max} = -P(e + \delta) = -Pe \sec\left(\frac{l}{2}\sqrt{\frac{P}{EI}}\right) \qquad (3\text{-}60)$$

The maximum compressive stress at midspan is found by superposing the axial component and the bending component. This gives

$$\sigma_c = \frac{P}{A} - \frac{Mc}{I} = \frac{P}{A} - \frac{Mc}{Ak^2} \qquad (c)$$

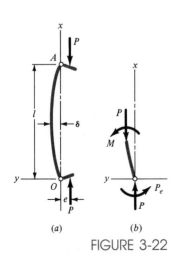

(a)   (b)

FIGURE 3-22

Notation for an eccentric column.

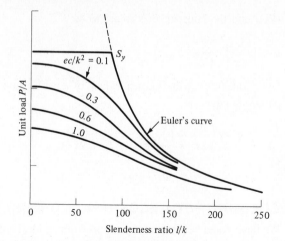

FIGURE 3-23

Comparison of the secant and
Euler formulas.

Substituting $M_{\max}$ from Eq. (3-60) yields

$$\sigma_c = \frac{P}{A}\left[1 + \frac{ec}{k^2}\sec\left(\frac{l}{2k}\sqrt{\frac{P}{EA}}\right)\right]\tag{3-61}$$

Note that the length l occurs only in the secant function. Since the secant of an acute angle is a number greater than unity, the length amplifies the eccentricity through the secant term. By imposing the yield strength S_{yc} as the maximum value of σ_c, we can write Eq. (3-61) in the form

$$\frac{P}{A} = \frac{S_{yc}}{1 + (ec/k^2)\sec\left[(l/2k)\sqrt{P/AE}\right]}\tag{3-62}$$

This is called the *secant column formula*. The term ec/k^2 is called the *eccentricity ratio*. Figure 3-23 is a plot of Eq. (3-62) for a steel having a compressive (and tensile) yield strength of 40 kpsi. Note how the P/A contours asymptotically approach the Euler curve as l/k increases.

Equation (3-62) cannot be solved explicitly for the load P. Design charts, in the fashion of Fig. 3-23, can be prepared for a single material if much column design is to be done. Otherwise, a root-finding technique using numerical methods must be used. With these, successive substitution with assured convergence can be applied.*

EXAMPLE 3-9   Develop specific Euler equations for the sizes of columns having

(*a*) Round cross sections

(*b*) Rectangular cross sections

---

*See Charles R. Mischke, "Computational Considerations in Design," Chap 5 in Joseph E. Shigley and Charles R. Mischke (eds.), *Standard Handbook of Machine Design*, McGraw-Hill, New York, 1986, p. 513.

Solution   (a) Using $A = \pi d^2/4$ and $k = d/4$ with Eq. (3-55) gives

Answer
$$d = \left(\frac{64P_{cr}l^2}{\pi^3 CE}\right)^{1/4} \tag{3-63}$$

(b) For the rectangular column, we specify a height h and a width b with the restriction that $h \le b$. If the end conditions are the *same* for buckling in both directions, then buckling will occur in the direction of the least thickness. Therefore

$$I = \frac{bh^3}{12} \qquad A = bh \qquad k^2 = \frac{h^2}{12}$$

Substituting these in Eq. (3-55) gives

Answer
$$b = \frac{12P_{cr}l^2}{\pi^2 CEh^3} \tag{3-64}$$

Note very particularly, though, that rectangular columns do not generally have the same end conditions in both directions.

EXAMPLE 3-10   Specify the diameter of a round column 1.5 m long which is to carry a maximum load estimated to be 22 kN. Use a design factor $n_d = 4$ and consider the ends as pinned (rounded). The column material selected has a minimum yield strength of 500 MPa and a modulus of elasticity of 207 GPa.

Solution   We shall design the column for a critical load of

$$P_{cr} = n_d P = 4(22) = 88 \text{ kN}$$

Then, using Eq. (3-63) with $C = 1$ gives

$$d = \left(\frac{64P_{cr}l^2}{\pi^3 CE}\right)^{1/4} = \left[\frac{64(88)(1.5)^3}{\pi^3(1)(207)}\right]^{1/4}\left(\frac{10^3}{10^9}\right)^{1/4}(10^3) = 37.48 \text{ mm}$$

Table A-17 shows that the preferred size is 40 mm. The slenderness ratio for this size is

$$\frac{l}{k} = \frac{l}{d/4} = \frac{1.5(10^3)}{40/4} = 150$$

To be sure that this is an Euler column, we use Eq. (3-56) and obtain

$$\left(\frac{l}{k}\right)_1 = \left(\frac{2\pi^2 CE}{S_y}\right)^{1/2} = \left[\frac{2\pi^2(1)(207)}{500}\right]^{1/2}\left(\frac{10^9}{10^6}\right)^{1/2} = 90.4$$

which indicates that it is indeed an Euler column. So select

Answer $d = 40$ mm

EXAMPLE 3-11   Repeat Example 3-9 for J. B. Johnson columns.

Solution   (a) For round columns,

Answer
$$d = 2\left(\frac{P_{cr}}{\pi S_y} + \frac{S_y l^2}{\pi^2 CE}\right)^{1/2} \tag{3-65}$$

(b) For a rectangular section with dimensions $h \leq b$, we find

$$b = \frac{P_{cr}}{hS_y\left(1 - \dfrac{3l^2 S_y}{\pi^2 CEh^2}\right)} \qquad h \leq b \tag{3-66}$$

EXAMPLE 3-12

Choose a set of dimensions for a rectangular link which is to carry a maximum compressive load of 5000 lb. The material selected has a minimum yield strength of 75 kpsi and a modulus of elasticity $E = 30$ Mpsi. Use a design factor of 4 and an end condition constant $C = 1$ for buckling on the weakest direction, and design for (a) a length of 15 in, and (b) a length of 8 in with a minimum thickness of $\frac{1}{2}$ in.

Solution

(a) Using Eq. (3-56), we find the limiting slenderness ratio to be

$$\left(\frac{l}{k}\right)_1 = \left(\frac{2\pi^2 CE}{S_y}\right)^{1/2} = \left[\frac{2\pi^2(1)(30)(10^6)}{75(10)^3}\right]^{1/2} = 88.8$$

By using $P_{cr} = n_d P = 4(5000) = 20\,000$ lb, Eqs. (3-64) and (3-66) are solved, using various values of h, to form Table 3-4. The table shows that a cross section of $\frac{5}{8}$ by $\frac{3}{4}$ in, which is marginally suitable, gives the least area.

TABLE 3-4

h	b	A	l/k	TYPE	EQ. NO.
0.375	3.46	1.298	139	Euler	(3-62)
0.500	1.46	0.730	104	Euler	(3-62)
0.625	0.76	0.475	83	Johnson	(3-63)
0.5625	1.03	0.579	92	Euler	(3-62)

(b) An approach similar to that above is used with $l = 8$ in. All trial computations are found to be in the J. B. Johnson region of l/k values. A minimum area occurs when the section is a near square. Thus a cross section of $\frac{1}{2}$ by $\frac{3}{4}$ in is found to be suitable and safe.

3-17   STRUTS, OR SHORT COMPRESSION MEMBERS

A short bar loaded in pure compression by a force P acting along the centroidal axis will shorten in accordance with Hooke's law, until the stress reaches the elastic limit of the material. At this point, permanent set is introduced and usefulness as a machine member may be at an end. If the force P is increased still more, the material either becomes "barrel-like" or fractures. When there is eccentricity in the loading, the elastic limit is encountered at small loads.

A *strut* is a *short compression member* such as the one shown in Fig. 3-24. The compressive stress in the x direction at point D in an intermediate section is the sum of a simple component P/A and a flexural component My/I; that is,

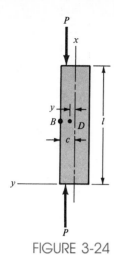

FIGURE 3-24

$$\sigma_c = \frac{P}{A} + \frac{My}{I} = \frac{P}{A} + \frac{PeyA}{IA} = \frac{P}{A}\left(1 + \frac{ey}{k^2}\right) \tag{3-67}$$

where $k = (I/A)^{1/2}$ and is the radius of gyration, y is the coordinate of point D, and e is the eccentricity of loading. The y coordinate of a line parallel to the x axis along which the normal stress is zero is found by setting Eq. (3-67) equal to zero and solving for y. This gives

$$y = -\frac{k^2}{e} \tag{a}$$

As the eccentricity is increased, the line of zero stress moves toward the cross-section centroid. As e is decreased, the line moves far from the section, and the entire section has a compressive normal stress. The largest compressive stress occurs at point B in Fig. 3-24, where $y = c$. Equation (3-67) then becomes

$$\sigma_c = \frac{P}{A}\left(1 + \frac{ec}{k^2}\right) \tag{3-68}$$

Note that the length of the strut does not appear in Eq. (3-68). In order to use the equation for design or analysis, we ought, therefore, to know the range of lengths for which the equation is valid. In other words, how long is a short member?

The difference between the secant formula and Eq. (3-68) is that the secant equation, unlike Eq. (3-68), accounts for an increased bending moment due to bending deflection. Thus the secant equation shows the eccentricity to be magnified by the bending deflection. This difference between the two formulas suggests that one way of differentiating between a "secant column" and a strut, or short compression member, is to say that in a strut, the effect of bending deflection must be limited to a certain small percentage of the eccentricity. If we decide that the limiting percentage is to be 1 percent of e, then the limiting slenderness ratio turns out to be

$$\left(\frac{l}{k}\right)_2 = 0.282\left(\frac{AE}{P_{cr}}\right)^{1/2} \tag{3-69}$$

(see Prob. 3-53). This equation then gives the limiting slenderness ratio for using Eq. (3-68). If the actual slenderness ratio is greater than $(l/k)_2$, then use the secant formula; otherwise, use Eq. (3-68).

EXAMPLE 3-13

Figure 3-25a shows a workpiece clamped to a milling machine table by a bolt tightened to a tension of 2000 lb. The clamp contact is offset from the centroidal axis of the strut by a distance $e = 0.10$ in, as shown in part b of the figure. The strut, or block, is steel, 1 in square and 4 in long, as shown. Determine the maximum compressive stress in the block.

Solution

First we find $A = bh = 1(1) = 1$ in^2, $I = bh^3/12 = 1(1)^3/12 = 0.0833$ in^4, $k^2 = I/A = 0.0833/1 = 0.0833$ in^2, and $l/k = 4/(0.0833)^{1/2} = 13.9$. Equation (3-69) gives the limiting slenderness ratio as

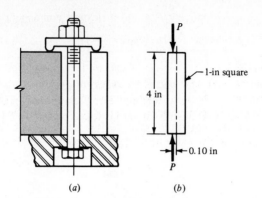

FIGURE 3-25   (a)   (b)

$$\left(\frac{l}{k}\right)_2 = 0.282 \left(\frac{AE}{P_{cr}}\right)^{1/2} = 0.282\left[\frac{1(30)(10^6)}{1000}\right]^{1/2} = 48.8$$

Thus the block could be as long as

$$l = 48.8k = 48.8(0.0833)^{1/2} = 14.1 \text{ in}$$

before it need be treated using the secant formula. So Eq. (3-68) applies and the maximum compressive stress is

Answer $$\sigma_c = \frac{P}{A}\left(1 + \frac{ec}{k^2}\right) = \frac{1000}{1}\left[1 + \frac{0.1(0.5)}{0.0833}\right] = 1600 \text{ psi}$$

PROBLEMS†

3-1   Structures can often be considered to be composed of a combination of tension and torsion members and beams. Each of these members can be analyzed separately to determine its force-deflection relationship and its spring rate. It is possible, then, to obtain the deflection of a structure by considering it as an assembly of springs having various series and parallel relationships.

(a) What is the overall spring rate of three springs in series?
(b) What is the overall spring rate of three springs in parallel?
(c) What is the overall spring rate of a single spring in series with a pair of parallel springs?

3-2   The figure shows a torsion bar *OA* fixed at *O*, supported at *A*, and connected to a cantilever *AB*.

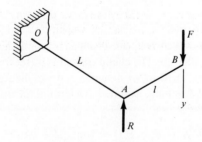

PROBLEM 3-2

---

†An asterisk indicates a design-type problem, a problem with no unique result, or a challenging problem.

The spring rate of the torsion bar is k_T, in newton-meters per radian, and that of the cantilever is k_C, in newtons per meter. What is the overall spring rate based on the deflection y at point B?

3-3 A torsion-bar spring consists of a prismatic bar, usually of round cross section, which is twisted at one end and held fast at the other to form a stiff spring. An engineer needs a stiffer one than usual and so considers building in both ends and applying the torque somewhere in the central portion of the span, as shown in the figure. If the bar is uniform in diameter, that is, if $d = d_1 = d_2$, investigate how the allowable angle of twist, the largest torque, and the spring rate depend on the location x at which the torque is applied.

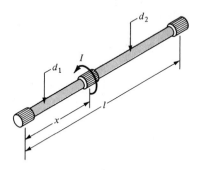

PROBLEM 3-3

3-4 An engineer is forced by geometric considerations to apply the torque on the spring of Prob. 3-3 at the location $x = 0.2l$. For a uniform-diameter spring, this would cause the long leg of the span to be underutilized when both legs have the same diameter. If the diameter of the left leg is reduced sufficiently, the shear stress in the two legs can be made equal. How would this change affect the allowable angle of twist, the largest torque, and the spring rate?

3-5 The figure shows a cantilever consisting of steel angles size $4 \times 4 \times \frac{1}{2}$ in mounted back to back. Find the deflection at B and the maximum stress in the beam.

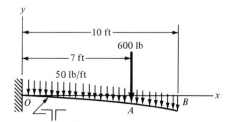

PROBLEM 3-5

3-6 A simply supported beam loaded by two forces is shown in the figure. Select a pair of structural steel channels mounted back to back to support the loads in such a way that the deflection at midspan will not exceed $\frac{1}{16}$ in and the maximum stress will not exceed 6 kpsi.

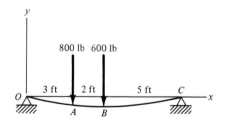

PROBLEM 3-6

3-7   Find the deflection of the steel shaft at *A* in the figure. Find the deflection at midspan. By what percentage do these two values differ?

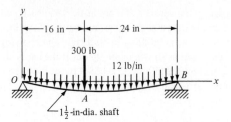

PROBLEM 3-7

3-8   A rectangular steel bar supports the two overhanging loads shown in the figure. Find the deflection at the ends and at the center.

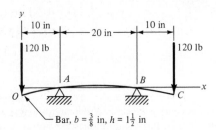

PROBLEM 3-8

3-9   The deflection at *B* of the cantilever shown in the figure can be found by adding the deflection caused by the force at *A* acting alone to that caused by the force at *B* acting alone. This is called the *method of superposition;* it is applicable whenever the deflection is linearly related to the force. Using the formulas in Appendix Table A-9 and superposition, find the deflection of the cantilever at *B* if $I = 13$ in^4 and $E = 30$ Mpsi.

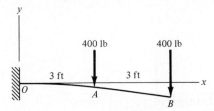

PROBLEM 3-9

3-10   The cantilever shown in the figure is a 4-in, 5.4-lb structural-steel channel. Such sections may be designated by the symbol shown in the figure or by the capital letter *C*. Using the approach suggested in Prob. 3-9, find the deflection at *A*. Where must the load be placed to avoid twisting of the channel section?

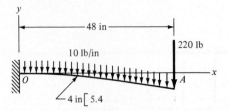

PROBLEM 3-10

3-11   Determine the maximum deflection of the beam shown in the figure. The material is carbon steel.

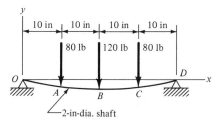

PROBLEM 3-11

3-12*   Illustrated is a rectangular steel bar with simple supports at the ends and loaded by a force F at the middle; the bar is to act as a spring. The ratio of the width to the thickness is to be about $b = 16h$, and the desired spring scale is 2400 lb/in.

(*a*) Find a set of cross-section dimensions, using preferred sizes.

(*b*) What deflection would cause a permanent set in the spring if this is estimated to occur at a normal stress of 90 kpsi?

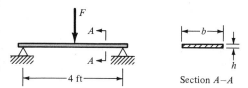

PROBLEM 3-12

3-13*   Illustrated in the figure is a $1\frac{1}{2}$-in-diameter steel countershaft which supports two pulleys. Pulley A delivers power to a machine causing a tension of 600 lb in the tight side of the belt and 80 lb in the loose side, as indicated. Pulley B receives power from a motor. The belt tensions on pulley B have the relation $T_1 = 0.125T_2$. Find the deflection of the shaft in the z direction at pulleys A and B. Assume that the bearings constitute simple supports.

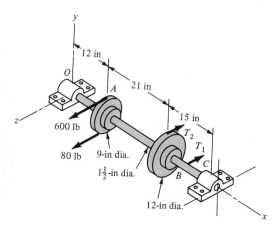

PROBLEM 3-13

3-14*   The figure shows a steel countershaft which supports two pulleys. Pulley C receives power from a motor producing the belt tensions shown. Pulley A transmits this power to another machine through the belt tensions T_1 and T_2 such that $T_1 = T_2/8$.

(a) Find the deflection of the overhanging end of the shaft, assuming simple supports at the bearings.

(b) If roller bearings are used, the slope of the shaft at the bearings should not exceed 0.06° for good bearing life. What shaft diameter is needed to conform to this requirement? Use $\frac{1}{8}$-in increments in any iteration you may make. What is the deflection at pulley C now?

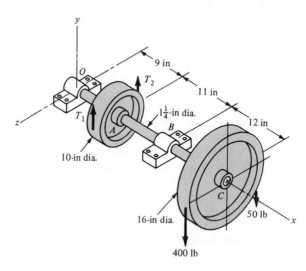

PROBLEM 3-14

400 lb

3-15   A $\frac{5}{8}$-in-diameter plow-steel wire rope has a *rope* modulus of elasticity of $E_r = 12$ Mpsi and is used to support a 500-lb cage and a 2000-lb load in a mine hoist. If the length of the rope between the hoisting drum and the cage is 100 ft, find the stretch in the rope as a mine cart and its load are moved onto the cage platform. The load due to the mine cart is $N = 2000$ lb. The cross-sectional area of the individual rope wire in tension can be estimated from $A_m = 0.4d_r^2$, where d_r is the nominal diameter of the rope.

3-16   When a hoisting cable is long, the weight of the cable itself contributes to the elongation. If a cable has a weight of w newtons per meter, a length of l meters, and a load P attached to the free end, show that the cable elongation is

$$\delta = \frac{Pl}{AE} + \frac{wl^2}{2AE}$$

3-17   For a uniformly loaded beam supported by simple supports, the central deflection is easily found. Suppose an opportunity presented itself to lessen this deflection by moving one support toward the center by a distance a. This would lessen the largest deflection between the supports, but now, the free end would also deflect. What offset a of the right support in the figure would result in a minimum extreme bending moment?

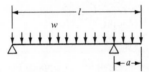

PROBLEM 3-17

3-18   The structure of a diesel-electric locomotive is essentially a composite beam supporting a deck. Above the deck are mounted the diesel prime mover, generator or alternator, radiators, switch gear, and auxiliaries. Beneath the deck are found fuel and lubricant tanks, air reservoirs, and small auxiliaries. This assembly is supported at bolsters by the trucks which house the traction motors and brakes. This equipment is distributed as uniformly as possible in the span between the bolsters. In an approximate way, the loading can be viewed as uniform between the bolsters and simply supported. Because the hoods that shield the equipment from the weather have many rectangular access doors, which are mass-produced, it is important that the hood structure be level and plumb and sit on a flat deck. Aesthetics plays a role too. The center sill beam has a second moment of area of $I = 5450$ in^4, the bolsters are 36 ft apart, and the deck loading is 5000 lb/ft.

(a) What is the camber of the curve to which the deck will be built in order that the service-ready locomotive will have a flat deck?

(b) What equation would you give to locate points on the curve of part (a)?

3-19   The designer of a shaft usually has a slope constraint imposed by the bearings used. This limit will be denoted as ξ. If the shaft shown in the figure is to have a uniform diameter d except in the locality of the bearing mounting, it can be approximated as a uniform beam with simple supports. Show that the diameters to meet the slope constraints at the left and right bearings are, respectively,

$$d_L = \left| \frac{32Fb(b^2 - l^2)}{3\pi El\xi} \right|^{1/4} \qquad d_R = \left| \frac{32Fa(l^2 - a^2)}{3\pi El\xi} \right|^{1/4}$$

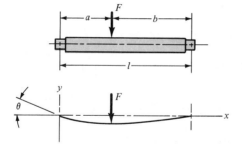

PROBLEM 3-19

3-20   For the shaft shown in the figure, let $a_1 = 4$ in, $b_1 = 12$ in, $a_2 = 10$ in, $F_1 = 100$ lb, $F_2 = 300$ lb, and $E = 30$ Mpsi. The shaft is to be sized so that the maximum slope at either bearing A or bearing B does not exceed 0.001 rad. Determine a suitable diameter d.

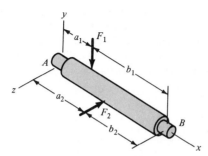

PROBLEM 3-20

3-21   A shaft is to be designed so that it is supported by roller bearings. The basic geometry is shown in the figure. The allowable slope at the bearings is 0.001 mm/mm without bearing life penalty.

For a design factor of 1.28, what uniform-diameter shaft will support the 3.5-kN load 100 mm from the left bearing without penalty? Use $E = 207$ GPa.

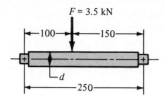

$F = 3.5$ kN

$\leftarrow 100 \rightarrow$ $\leftarrow 150 \rightarrow$

d

$\leftarrow 250 \rightarrow$

PROBLEM 3-21
Dimensions in millimeters.

3-22 Determine the maximum deflection of the shaft of Prob. 3-21.

3-23 See Prob. 3-20 and the accompanying figure. The loads and dimensions are $F_1 = 800$ lb, $F_2 = 600$ lb, $a_1 = 4$ in, $b_1 = 6$ in, and $a_2 = 7$ in. Find the uniform shaft diameter necessary to limit the slope at the bearings to 0.001 in/in. Use a design factor of $n_d = 1.5$ and $E = 29.8$ Mpsi.

3-24 Shown in the figure is a uniform-diameter shaft with bearing shoulders at the ends; the shaft is subjected to a concentrated moment $M = 1200$ lb $\cdot$ in. The shaft is of carbon steel and has $a = 5$ in and $l = 9$ in. The slope at the ends must be limited to 0.002 rad. Find a suitable diameter d.

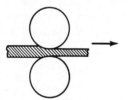

$\leftarrow a \rightarrow$ $\leftarrow b \rightarrow$

M_B

B

$\leftarrow l \rightarrow$

PROBLEM 3-24

3-25 Rolling mills are used to reduce sheet metal cross sections and to process other sheet materials. Rolling devices should provide uniform pressure on the material passing between them. The "window" between them, shown in the figure, should be of constant height. Since the forces can be large, the rollers deflect as the material is rolled. How should a roll be crowned so that after deflection the pressure is uniform and the roller contact surfaces are parallel?

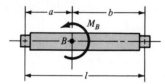

PROBLEM 3-25

3-26 A bar in tension has a circular cross section and includes a conical portion of length l, as shown. The task is to find the spring rate of the entire bar. Equation (3-4) is useful for the outer portions

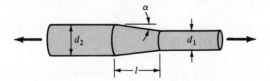

α

d_2 d_1

$\leftarrow l \rightarrow$

PROBLEM 3-26

of diameters d_1 and d_2, but a new relation must be derived for the tapered section. If α is the apex half-angle, as shown, show that the spring rate of the tapered portion of the shaft is

$$k = \frac{EA_1}{l}\left(1 + \frac{2l}{d_1} \tan \alpha\right)$$

3-27   Find expressions for the maximum values of the spring force and deflection y of the impact system shown in the figure. Can you think of a realistic application for this model?

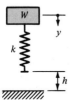

PROBLEM 3-27

3-28   As shown in the figure, the weight W_1 strikes W_2 from a height h. Find the maximum values of the spring force and the deflection of W_2. Name an actual system for which this model might be used.

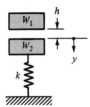

PROBLEM 3-28

3-29   Part a of the figure shows a weight W mounted between two springs. If the free end of spring k_1 is suddenly displaced through the distance $x = a$, as shown in part b, what would be the maximum displacement y of the weight?

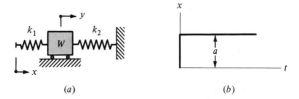

PROBLEM 3-29 (a) (b)

3-30*   In a uniform-cross-section beam with simple supports at the ends loaded by a single concentrated load, the location of the maximum deflection will never exceed $x = l/\sqrt{3} = 0.577l$ regardless of the location of the load along the beam. The importance of this is that you can always get a quick estimate of y_{max} by using $x = l/2$. Prove this statement.

3-31   Use Castigliano's theorem to verify the maximum deflection for the uniformly loaded beam of Appendix Table A-9-7. Neglect shear.

3-32   The rectangular member OAB, shown in the figure, is held horizontal by the round hooked bar

AC. The modulus of elasticity of both parts is 10 Mpsi. Use Castigliano's theorem to find the deflection at *B* due to a force $F = 80$ lb.

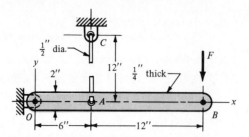

PROBLEM 3-32

3-33 The figure illustrates a torsion-bar spring *OA* having a diameter $d = 12$ mm. The actuating cantilever *AB* also has $d = 12$ mm. Both parts are of carbon steel. Use Castigliano's theorem and find the spring rate k corresponding to a force F acting at B.

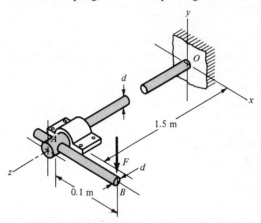

PROBLEM 3-33

3-34 A cable is made using a 16-gauge (0.0625-in) steel wire and three strands of 12-gauge (0.0801-in) copper wire. Find the stress in each wire if the cable is subjected to a tension of 250 lb.

3-35 The figure shows a pressure cylinder of diameter 4 in which uses six SAE grade 5 bolts having a

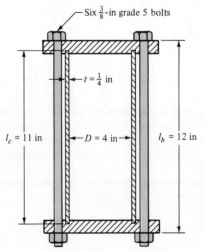

PROBLEM 3-35

grip of 12 in. These bolts have a proof strength (see Chap. 8) of 85 kpsi for this size of bolt. Suppose the bolts are tightened to 90 percent of this strength in accordance with some recommendations.

(a) Find the tensile stress in the bolts and the compressive stress in the cylinder walls.

(b) Repeat part (a), but assume now that a fluid under a pressure of 600 psi is introduced into the cylinder.

3-36   A torsion bar of length L consists of a round core of stiffness $(GJ)_c$ and a shell of stiffness $(GJ)_s$. If a torque T is applied to this composite bar, what percentage of the total torque is carried by the shell?

3-37   The figure shows a $\frac{3}{8}$- by $1\frac{1}{2}$-in rectangular steel bar welded to fixed supports at each end. The bar is axially loaded by the forces $F_A = 10$ kip and $F_B = 5$ kip acting on pins at A and B. Assuming that the bar will not buckle laterally, find the reactions at the fixed supports.

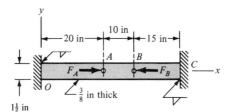

PROBLEM 3-37

3-38   A rectangular aluminum bar $\frac{1}{2}$ in thick and 2 in wide is welded to fixed supports at the ends, and the bar supports a load $W = 800$ lb, acting through a pin as shown. Find the reactions at the supports.

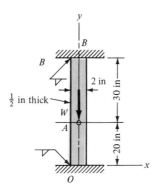

PROBLEM 3-38

3-39   The steel shaft shown in the figure is subjected to a torque T applied at point A. Find the torque reactions at O and B.

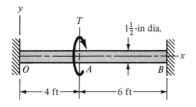

PROBLEM 3-39

3-40   In testing the wear life of gear teeth, the gears are assembled using a pretorsion. In this way, a large torque can exist even though the power input to the tester is small. The arrangement shown in the figure uses this principle. Note the symbol used to indicate the location of the shaft bearings used in the figure. Gears A, B, and C are assembled first, and then gear C is held fixed. Gear D is assembled and meshed with gear C by twisting it through an angle of $4°$ to provide the pretorsion. Find the maximum shear stress in each shaft resulting from this preload.

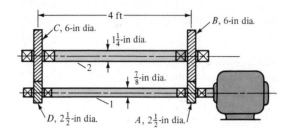

PROBLEM 3-40

3-41   Examine the expression for the deflection of the cantilever beam, end-loaded, shown in Appendix Table A-9-1 for some intermediate point, $x = a$, as

$$y\Big|_{x=a} = \frac{F_1 a^2}{6EI}(a - 3l)$$

In Table A-9-2, for a cantilever with intermediate load, the deflection at the end is

$$y\Big|_{x=l} = \frac{F_2 a^2}{6EI}(a - 3l)$$

These expressions are remarkably similar and become identical when $F_1 = F_2 = 1$. In other words, the deflection at $x = a$ (station 1) due to a unit load at $x = l$ (station 2) is the same as the deflection at station 2 due to a unit load at station 1. Prove that this is true generally for an elastic body even when the lines of action of the loads are not parallel. This is known as a special case of *Maxwell's reciprocal theorem*. (*Hint:* Consider the potential energy of strain when the body is loaded by two forces in either order of application.)

3-42   A steel shaft of uniform 2-in diameter has a bearing span l of 23 in and an overhang of 7 in on which a coupling is to be mounted. A gear is to be attached 9 in to the right of the left bearing and will carry a radial load of 400 lb. We require an estimate of the bending deflection at the coupling. Appendix Table A-9-6 is available, but we can't be sure of how to expand the equation to predict the deflection at the coupling.

(a) Show how Appendix Table A-9-10 and Maxwell's theorem (see Prob. 3-41) can be used to obtain the needed estimate.

(b) Check your work by finding the slope at the right bearing and extending it to the coupling location.

3-43   A thin ring is loaded by two equal and opposite forces F in part a of the figure. A free-body diagram of one quadrant is shown in part b. This is a statically indeterminate problem, because the moment M_A cannot be found by statics. We wish to find the maximum bending moment in the ring due to the forces F. Assume that the radius of the ring is large so that Eq. (3-40) can be used.

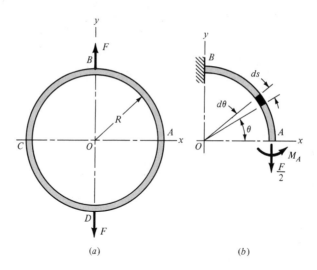

(a)   (b)

3-44 Find the decrease in the diameter of the ring of Prob. 3-43 due to the forces F and along the y axis.

3-45 A cast-iron piston ring has a mean diameter of 81 mm, a radial height $h = 6$ mm, and a thickness $b = 4$ mm. The ring is assembled using an expansion tool which separates the split ends a distance δ by applying a force F as shown. Use Castigliano's theorem and determine the deflection δ as a function of F. Use $E = 131$ GPa and assume Eq. (3-40) applies.

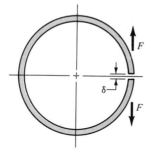

PROBLEM 3-45

3-46 A round tubular column has outside and inside diameters of D and d, respectively, and a diametral ratio of $K = d/D$. Show that buckling will occur when the outside diameter is

$$D = \left[\frac{64 P_{cr} l^2}{\pi^3 C E (1 - K^4)} \right]^{1/4}$$

3-47 For the conditions of Prob. 3-46, show that buckling according to the parabolic formula will occur when the outside diameter is

$$D = 2 \left[\frac{P_{cr}}{\pi S_y (1 - K^2)} + \frac{S_y l^2}{\pi^2 C E (1 + K^2)} \right]^{1/2}$$

3-48   Link 2, shown in the figure, is 1 in wide, has $\frac{1}{2}$-in-diameter bearings at the ends, and is cut from low-carbon steel bar stock having a minimum yield strength of 24 kpsi. The end-condition constants are $C = 1$ and $C = 1.2$ for buckling in and out of the plane of the drawing, respectively.

(a) Using a design factor $n_d = 5$, find a suitable thickness for the link.

(b) Are the bearing stresses at O and B of any significance?

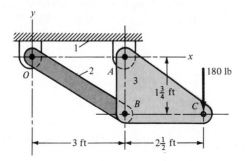

PROBLEM 3-48

3-49*   Link 3, shown schematically in the figure, acts as a brace to support the 1.2-kN load. For buckling in the plane of the figure, the link may be regarded as pinned at both ends. For out-of-plane buckling, the ends are fixed. Select a suitable material and a method of manufacture, such as forging, casting, stamping, or machining, for casual applications of the brace in oil-field machinery. Specify the dimensions of the cross section as well as the ends so as to obtain a strong, safe, well-made, and economical brace.

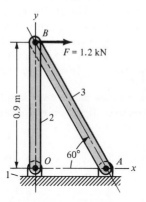

PROBLEM 3-49

3-50   The hydraulic cylinder shown in the figure has a 3-in bore and is to operate at a pressure of 800 psi. With the clevis mount shown, the piston rod should be sized as a column with both ends rounded for any plane of buckling. The rod is to be made of forged AISI 1030 steel without further heat treatment.

(a) Use a design factor $n_d = 3$ and select a preferred size for the rod diameter if the column length is 60 in.

(b) Repeat part (a) but for a column length of 18 in.

(c) What factor of safety actually results for each of the cases above?

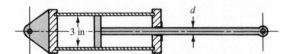

PROBLEM 3-50

3-51   The figure shows a schematic drawing of a vehicular jack which is to be designed to support a maximum weight of 400 kg based on the use of a design factor $n_d = 2.50$. The opposite-handed threads on the two ends of the screw are cut to allow the link angle θ to vary from 15 to 70°. The links are to be machined from AISI 1020 hot-rolled steel bars with a minimum yield strength of 380 MPa. Each of the four links is to consist of two bars, one on each side of the central bearings. The bars are to be 300 mm long and have a bar width of 25 mm. The pinned ends are to be designed to secure an end-condition constant of at least $C = 1.4$ for out-of-plane buckling. Find a suitable preferred thickness and the resulting factor of safety for this thickness.

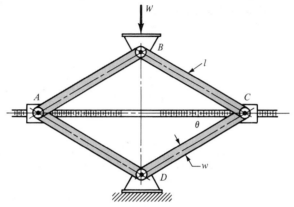

PROBLEM 3-51

3-52   If drawn, a figure for this problem would resemble that for Prob. 3-35. A strut that is a hollow right circular cylinder has an inside diameter of 4 in and a wall thickness of $\frac{3}{8}$ in and is compressed between two circular end plates held by four bolts equally spaced on a bolt circle of 5.68-in diameter. All four bolts are hand-tightened, and then bolt A is tightened to a tension of 2000 lb and bolt C, diagonally opposite, is tightened to a tension of 10 000 lb. The strut axis of symmetry is coincident with the center of the bolt circles. Find the maximum compressive load, the eccentricity of loading, and the largest compressive stress in the strut.

3-53   The secant column equation incorporates the bending deflection, which has the effect of magnifying the eccentricity e at midspan. What is the greatest value of the slenderness ratio for which this influence will not exceed 1 percent of e?

3-54*   Design link CD of the hand-operated toggle press shown in the figure. Specify the cross-section

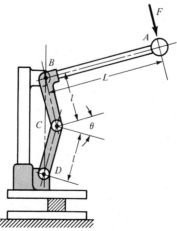

PROBLEM 3-54
$L = 12$ in, $l = 4$ in, $\theta_{min} = 0°$.

dimensions, the bearing size and rod-end dimensions, the material, and the method of processing.

ANSWERS

3-1   (a) $k = 1/[(1/k_1) + (1/k_2) + (1/k_3)]$,   (b) $k = k_1 + k_2 + k_3$

3-5 $y = -0.908$ in, $\sigma = 20.4$ kpsi

3-8   At O, $y = -0.050\ 79$ in; at midspan, $y = 0.019\ 05$ in; at C, $y = -0.050\ 79$ in

3-11 $y_{max} = -0.0130$ in

3-17 $a = 0.293l$

3-21 36.4 mm

3-23 1.57 in

3-27 With $h = 1$ in, $k = 100$ lb/in, and $W = 30$ lb, $y_{max} = 2.130$ in

3-29 $y_{max} = 2k_1a/(k_1 + k_2)$

3-33 $k = 10.60$ kN/m

3-37 $R_O = -3570$ lb, $R_C = -1430$ lb

3-39 $T_{OA} = 0.6T$, $T_{AB} = 0.4T$

3-43 $M_{max} = -Fr/\pi$

3-45 $\delta = 66.4F$ mm

3-48 (a) $t = \frac{1}{2}$ in; (b) no

3-52 $\sigma_c = -9218$ psi, $\sigma_t = 3598$ psi

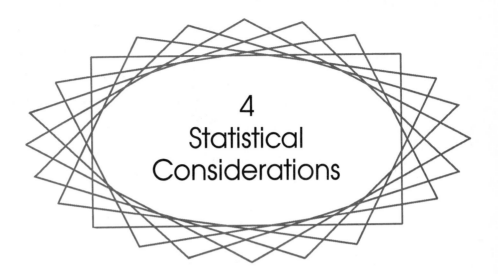

4
Statistical
Considerations

The use of statistics in mechanical design provides a method of dealing with characteristics whose values are variable. Products manufactured in large quantities such as automobiles, watches, lawnmowers, and washing machines, for example, have a life that is variable. One automobile may have so many defects that it must be repaired repeatedly during the first few months of operation; another may operate satisfactorily for years, requiring only minor maintenance.

Methods of quality control are deeply rooted in the use of statistics, and engineering designers need a knowledge of statistics to conform to quality-control standards. The variability inherent in the sizes involved in limits and fits, in stress and strength, in bearing clearances, and in a multitude of other characteristics must be described numerically for proper control. It is no longer satisfactory to say that a product is expected to have a long and trouble-free life. We must now find a way to express such things as product life and product reliability in numerical form in order to achieve a specific quality goal.

A side benefit that you will derive from this chapter is the ability to use the statistical keys on your calculator. These provide quite useful information, even on everyday problems. It is assumed that you are familiar with such topics as permutations, combinations, probability, and probability theorems. These subjects are usually studied in basic college algebra.

4-1 RANDOM VARIABLES*

Consider an experiment to measure strength in a collection of 20 tensile-test specimens that have been machined from a like number of samples selected at random from a carload shipment of, say, UNS G10200 cold-drawn steel. It is reasonable to expect that there will be differences in the ultimate tensile strengths S of each of the individual test specimens. Such differences may occur because of differences in the sizes of the

---

*It may not be necessary to study material introduced by headings in black; see Sec. 1-10.

145

FIGURE 4-1

Sample space showing all possible outcomes of the toss of two dice.

1,1	1,2	1,3	1,4	1,5	1,6
2,1	2,2	2,3	2,4	2,5	2,6
3,1	3,2	3,3	3,4	3,5	3,6
4,1	4,2	4,3	4,4	4,5	4,6
5,1	5,2	5,3	5,4	5,5	5,6
6,1	6,2	6,3	6,4	6,5	6,6

TABLE 4-1

A Probability Distribution

x	2	3	4	5	6	7	8	9	10	11	12
$f(x)$	$\frac{1}{36}$	$\frac{2}{36}$	$\frac{3}{36}$	$\frac{4}{36}$	$\frac{5}{36}$	$\frac{6}{36}$	$\frac{5}{36}$	$\frac{4}{36}$	$\frac{3}{36}$	$\frac{2}{36}$	$\frac{1}{36}$

specimens, in the strength of the material itself, or both. Such an experiment is called a *random experiment,* because the specimens are selected at random. The strength S determined by this experiment is called a *random,* or a *stochastic, variable.* So a random variable is a variable quantity, such as strength, size, or weight, whose value depends upon the outcome of a random experiment.

Let us define a random variable x as the sum of the numbers obtained when two dice are tossed. Either die can display any number from 1 to 6; Fig. 4-1 displays all possible outcomes in what is called the *sample space.* Note that x has a specific value for each possible outcome; for the event 5, 4, $x = 5 + 4 = 9$. It is useful to form a table showing the values of x and the corresponding values of the probability of x, called $p = f(x)$. This is easily done from Fig. 4-1 merely by counting, since there are 36 points, each having a weight $w = \frac{1}{36}$. The results are shown in Table 4-1. Any table, such as this, listing all possible values of a random variable, together with the corresponding probabilities, is called a *probability distribution.*

The values of Table 4-1 are plotted in graphical form in Fig. 4-2. Here it is clear that the probability is a function of x. This *probability function* $p = f(x)$ is often called the

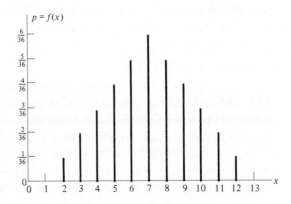

FIGURE 4-2

Frequency distribution.

TABLE 4-2
A Cumulative Probability Distribution

x	2	3	4	5	6	7	8	9	10	11	12
$F(x)$	$\frac{1}{36}$	$\frac{3}{36}$	$\frac{6}{36}$	$\frac{10}{36}$	$\frac{15}{36}$	$\frac{21}{36}$	$\frac{26}{36}$	$\frac{30}{36}$	$\frac{33}{36}$	$\frac{35}{36}$	$\frac{36}{36}$

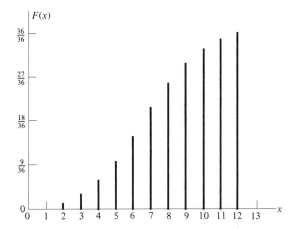

FIGURE 4-3
Cumulative frequency distribution.

frequency function or, sometimes, the *probability density*. The probability that x is less than or equal to a certain value x_i can be obtained from the probability function by summing the probability of all x's up to and including x_i. If we do this with Table 4-1, letting x_i equal 2, then 3, and so on, up to 12, we get Table 4-2, which is called a *cumulative probability distribution*. The function $F(x)$ in Table 4-2 is called a *cumulative probability function*. In terms of $f(x)$ it may be expressed mathematically in the general form

$$F(x_i) = \sum_{x_j \le x_i} f(x_j) \tag{4-1}$$

where $F(x_i)$ is properly called the *distribution function*. The cumulative distribution may also be plotted as a graph (Fig. 4-3).

The variable x of this example is called a *discrete random variable,* because x has only discrete values. A *continuous random variable* is one that can take on any value in a specified interval; for such variables, graphs like Figs. 4-2 and 4-3 would be plotted as continuous curves.

4-2 THE ARITHMETIC MEAN, THE VARIANCE, AND THE STANDARD DEVIATION

In studying the variations in the mechanical properties and characteristics of mechanical elements, we shall generally be dealing with a finite number of elements. The total number of elements, called the *population,* may in some cases be quite large. In such cases it is usually impractical to measure the characteristics of each member of the population, because this involves destructive testing in some cases, and so we select a small part of the group, called a *sample,* for these determinations. Thus the *population* is the entire group, and the *sample* is a part of the population.

The arithmetic mean of a sample, called the *sample mean,* consisting of N elements, is defined by the equation

$$\bar{x} = \frac{x_1 + x_2 + x_3 + \cdots + x_N}{N} = \frac{1}{N}\sum_{1}^{N} x_j \tag{4-2}$$

In a similar manner, a population consisting of N elements has a *population mean* defined by the equation

$$\mu = \frac{x_1 + x_2 + x_3 + \cdots + x_N}{N} = \frac{1}{N}\sum_{1}^{N} x_j \tag{4-3}$$

The *mode* and the *median* are also used as measures of central value. The *mode* is the value that occurs most frequently. The *median* is the middle value if there are an odd number of cases; it is the mean of the two middle values if there are an even number.

Besides the arithmetic mean, it is useful to have another kind of measure which will tell us something about the spread, or dispersion, of the distribution. For any random variable x, the deviation of the ith observation from the mean is $x_i - \bar{x}$. But since the sum of the deviations so defined is always zero, we square them, and define *sample variance* as

$$s_x^2 = \frac{(x_1 - \bar{x})^2 + (x_2 - \bar{x})^2 + \cdots + (x_N - \bar{x})^2}{N - 1} = \frac{1}{N - 1}\sum_{1}^{N} (x_j - \bar{x})^2 \tag{4-4}$$

The *sample standard deviation,* defined as the square root of the variance, is

$$s_x = \left[\frac{1}{N - 1}\sum_{1}^{N} (x_j - \bar{x})^2\right]^{1/2} \tag{4-5}$$

Equations (4-4) and (4-5) are not well suited for use in a calculator. For such purposes, use the alternate form

$$s_x = \left[\frac{\sum x^2 - \dfrac{\left(\sum x\right)^2}{N}}{N - 1}\right]^{1/2} \tag{4-6}$$

for the standard deviation, and

$$s_x^2 = \frac{\sum x^2}{N} - \bar{x}^2 \tag{4-7}$$

for the variance.

It should be observed that some authors define the variance and the standard devia-

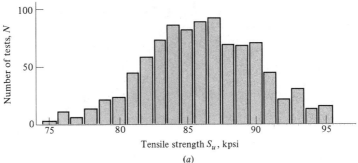

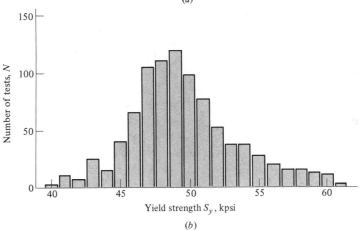

FIGURE 4-4

Distribution of tensile properties of hot-rolled UNS G10350 steel, as rolled. These tests were made from round bars varying in diameter from 1 to 9 in. (*a*) Tensile-strength distributions from 930 heats; $\bar{S}_u = 86.0$ kpsi, $s_{S_u} = 4.94$ kpsi. (*b*) Yield-strength distribution from 899 heats; $\bar{S}_y = 49.5$ kpsi, $s_{S_y} = 5.36$ kpsi. *(By permission, Metals Handbook, vol. 1, 8th ed., American Society for Metals, Metals Park, Ohio, 1961, p. 64.)*

tion using N instead of $N - 1$ in the denominator. For large values of N, there is very little difference. For small values, the denominator $N - 1$ actually gives a better estimate of the variance of the population from which the sample is taken.

Sometimes we are going to be dealing with the standard deviation of the strength of an element. So you must be careful not to be confused by the notation. Note that we are using the *capital letter S* for *strength* and the *lowercase letter s* for *standard deviation*. Figure 4-4 will be useful in visualizing these ideas.

Equations (4-4) to (4-7) apply specifically to the *sample* of a population. When an entire population is considered, the same equations apply, but $\bar{x}$ is replaced with μ, and N weighting is used in the denominators instead of $N - 1$. Also, the resulting standard deviation and variance are designated as $\hat{\sigma}_x$ and $\hat{\sigma}_x^2$; the circumflex accent mark, or "hat," is used to avoid confusion with normal stress.

4-3   DATA PROCESSING

Most small, everyday pocket calculators have statistics keys to be used for the special purpose of computing the mean and the standard deviation. Does your calculator have a $\Sigma+$ key? If so, you have statistical capability. When you enter a number and press $\Sigma+$, the calculator adds this quantity to the total already contained in a memory register. So be sure to clear the memory registers before you start. At the same time

TABLE 4-3

REGISTER		CONTENTS
1	Σy	Dependent variable
2	Σy^2	
3	N	
4	Σx	Independent variable
5	Σx^2	
6	Σxy	

you press $\Sigma+$, the calculator also displays and stores N; the quantity Σx^2 is also computed and stored in a third register. When all the data are entered, press $\bar{x}$, the mean-value key, to display the mean. Depending upon the make of your calculator, other keys can be pressed to recall Σx and Σx^2 from memory. See your user's manual. The manual will also tell you which keys to press to obtain the standard deviation and, with some calculators, the variance.

If you have even a small programmable calculator, the chances are that you have the capability of assimilating data-point pairs. On these, the data are entered in pairs, using an exchange key or an **ENTER** key to separate the terms of each pair. As each data point is keyed in, it is stored in the memory registers as shown in Table 4-3.

The following example can be used to get acquainted with the statistics keys on your calculator.

EXAMPLE 4-1

This example is used to determine the mean and the standard deviation for the grades of a sample of 10 students selected alphabetically. In the table below, x is the final examination grade and y is the course grade. Use your calculator to verify the results shown.

N	1	2	3	4	5	6	7	8	9	10
x	85	74	69	78	93	79	46	75	85	84
y	72.7	52.7	73.2	83.2	78.2	82.3	63.0	82.3	85.7	84.9

$$\sum y = 758.20 \qquad \bar{y} = 75.82 \qquad s_y = 10.75$$

$$\sum x = 768.00 \qquad \bar{x} = 76.80 \qquad s_x = 12.80$$

The standard deviations shown here use $N - 1$ weighting. Some calculators provide the option of using either N or $N - 1$. Also, the results shown above were recorded with the display in the **FIX 2** mode.

4-4 REGRESSION

Statisticians use a process of analysis called *regression* to obtain a curve which best fits a set of data points. The process is called *linear regression* when the best-fitting straight line is to be found. The meaning of the word *best* is open to argument, because there can be many meanings. The usual method, and the one employed here, is that of choosing a best line based on minimizing the squares of the deviations of the data points from the line.

Figure 4-5 shows a set of data points approximated by the line *AB*. The equation of the straight line is

$$y = b + mx \tag{a}$$

where *m* is the slope and *b* is the *y* intercept. The regression equations are

$$m = \frac{\sum xy - \dfrac{\sum x \sum y}{N}}{\sum x^2 - \dfrac{\left(\sum x\right)^2}{N}} \tag{4-8}$$

$$b = \frac{\sum y - m \sum x}{N} \tag{4-9}$$

Examination of Table 4-3 shows that the terms of these equations are available in calculator memory, or can be easily computed.

Once you have established a slope and an intercept, the next point of interest is to discover how well *x* and *y* correlate with each other. If the data points are scattered all over the *xy* plane, there is obviously no correlation. But if all the data points coincide with the regression line, then there is perfect correlation. Most statistical data will be in between these extremes. A *correlation coefficient r*, having the range $-1 \leq r \leq +1$, has been devised to answer this question. The formula is

$$r = \frac{ms_x}{s_y} \tag{4-10}$$

A negative *r* indicates that the regression line has a negative slope. If $r = 0$, there is no correlation; if $r = \pm 1$, there is perfect correlation.

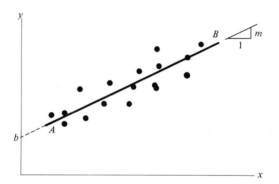

FIGURE 4-5

Set of data points approximated by regression line *AB*.

Equations (4-8) to (4-10) can be solved directly on your calculator if it has the $\Sigma+$ key. There are special calculators for statistics problems in which the constants m, b, and r can be obtained by pressing appropriate keys after the statistical data have been entered. If you have a programmable calculator, the regression equations should be in the calculator library; they can be solved using a software module or a magnetic card or can be programmed directly for solution. See your user's manual. You can use the following example to familiarize yourself with the capability of your own calculator.

EXAMPLE 4-2*   Find the slope, intercept, and correlation coefficient for the following data:

x_i	40.5	38.6	37.9	36.2	35.1	34.6
y_i	104.5	102	100	97.5	95.5	94

What value of y corresponds to $x = 37$?

Solution $y = 33.53 + 1.76x$

Answer $r = 0.995$

At $x = 37$,

Answer $y = 98.65$

Nonlinear regression can be performed merely by operating on one or both of the elements of the data pair with any mathematical function. For example, using natural logarithms, exponentials, and reciprocals, you can obtain logarithmic and exponential curves, as well as others. Apply the transforms to x and y to rectify the data string, perform the linear regression, and then, by inverse process, identify the regression equation.

4-5   NOTATION AND DEFINITIONS

In our use of vectors in this book, we designate them using boldface characters, indicative of the fact that two or three quantities, such as direction and magnitude, are necessary to describe them. This is also true of a random variable which can be characterized by specifying its mean and standard deviation. For this reason we shall adopt the convention of using boldface characters to designate random variables as well as vectors. No confusion between the two is likely to arise.

The terms *stochastic variable* and *variate* are also used to mean a random variable.

A *deterministic quantity* is something that has a single specific value. The mean value of a population is a deterministic quantity, and so is its standard deviation.

A stochastic variable can be described using the mean and the standard deviation, or

---

*From *HP-19C/HP-29C Applications Book*, Hewlett-Packard, Corvallis, Ore., 1977, p. 105.

by using the mean and the *coefficient of variation*. We have not yet defined this coefficient; it is

$$C_x = \frac{\hat{\sigma}_x}{\bar{x}} \tag{4-11}$$

Thus the variate **x** can be expressed in the following two ways:

$$\mathbf{x} = (\bar{x}, \hat{\sigma}_x) = \bar{x}(1, C_x) \tag{4-12}$$

Note that the deterministic quantities $\bar{x}$, $\hat{\sigma}_x$, and C_x are all in lightface.

Symbols containing subscripts to subscripts are difficult to write and to print. For this reason we shall generally employ only a single level of subscripts. Thus the standard deviation of the ultimate-strength variate $\mathbf{S}_u$, for example, will be designated as $\hat{\sigma}_{Su}$; that is, the subscript to S is not set as a subscript.

4-6   THE NORMAL DISTRIBUTION

When Gauss asked the question *What distribution is the most likely parent to a set of data?*, the answer was a distribution which bears his name. The *Gaussian*, or *normal*, *distribution* is an important one whose probability density function is expressed in terms of its mean μ and its standard deviation $\hat{\sigma}$ as

$$f(x) = \frac{1}{\hat{\sigma}\sqrt{2\pi}} \exp\left[-\frac{1}{2}\left(\frac{x - \mu}{\hat{\sigma}}\right)^2\right] \tag{4-13}$$

Since Eq. (4-13) is a probability density function, the area under it, as required, is unity. A plot of Eq. (4-13) is shown in Fig. 4-6 for small and large standard deviations. The bell-shaped curve is taller and narrower for small values of $\hat{\sigma}$ and shorter and broader for large values of $\hat{\sigma}$. The integration of Eq. (4-13) to find the cumulative density function $F(x)$ is not possible in closed form, but must be accomplished numerically. In order to avoid the preparation of many tables for different values of μ and $\hat{\sigma}$, the deviation from the mean is expressed in units of standard deviation by the transform

$$\mathbf{z} = \frac{\mathbf{x} - \mu}{\hat{\sigma}} \tag{4-14}$$

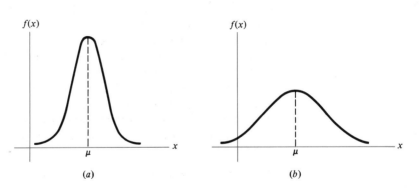

FIGURE 4-6

The shape of the normal distribution curve: (*a*) small $\hat{\sigma}$; (*b*) large $\hat{\sigma}$.

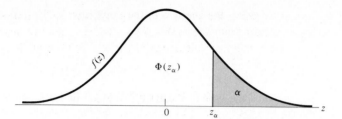

FIGURE 4-7

The standard normal distribution.

and the integral is tabulated in Table A-10 and sketched in Fig. 4-7. The value of the normal cumulative density function is used so often, and manipulated in so many equations, that it has its own particular symbol, $\Phi(z)$. The transformation variate **z** is normally distributed, with a mean of zero and a standard deviation and variance of unity. The probability of an observation less than z is $\Phi(z)$ for negative values of z and $1 - \Phi(z)$ for positive values of z in Table A-10.

EXAMPLE 4-3   In a shipment of 250 connecting rods, the mean tensile strength is found to be 45 kpsi and the standard deviation 5 kpsi.

(*a*) Assuming a normal distribution, how many rods can be expected to have a strength less than 39.5 kpsi?

(*b*) How many are expected to have a strength between 39.5 and 59.5 kpsi?

Solution   (*a*) Substituting in Eq. (4-14) gives the standardized z variable as

$$z_{39.5} = \frac{x - \mu}{\hat{\sigma}} = \frac{S - \bar{S}}{\hat{\sigma}_S} = \frac{39.5 - 45}{5} = -1.10$$

The probability that the strength is less than 39.5 kpsi can be designated as $F(z) = \Phi(-1.10)$. Using Table A-10, we find $\Phi(z_{39.5}) = 0.1357$. So the number having a strength less than 39.5 kpsi is, from Fig. 4-8,

Answer $N\Phi(z_{39.5}) = 250(0.1357) = 33.9 \approx 34$

because $\Phi(z_{39.5})$ represents the proportion of the population N having a strength less than 39.5 kpsi.

(*b*) Corresponding to $S = 59.5$ kpsi, we have

$$z_{59.5} = \frac{59.5 - 45}{5} = 2.90$$

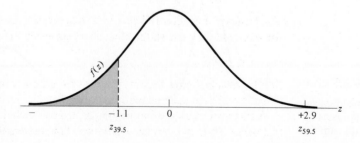

FIGURE 4-8

Referring again to Fig. 4-8, we see that the probability that the strength is less than 59.5 kpsi is $F(z) = \Phi(z_{59.5})$. Since the z variable is positive, we need to find the value complementary to unity. Thus, from Table A-10,

$$\Phi(2.90) = 1 - \Phi(-2.90) = 1 - 0.001\ 87 = 0.998\ 13$$

The probability that the strength lies between 39.5 and 59.5 kpsi is the area between the ordinates at $z_{39.5}$ and $z_{59.5}$ in Fig. 4-8. This probability is found to be

$$p = \Phi(z_{59.5}) - \Phi(z_{39.5}) = \Phi(2.90) - \Phi(-1.10)$$
$$= 0.998\ 13 - 0.1357 = 0.862\ 43$$

Therefore the number of rods expected to have strengths between 39.5 and 59.5 kpsi is

Answer $Np = 250(0.862) = 215.5 \approx 216$

4-7 PROPAGATION OF ERROR

In the equation for axial stress

$$\sigma = \frac{F}{A} \tag{a}$$

suppose both the force F and the area A are random variables. Then Eq. (a) is written as

$$\boldsymbol{\sigma} = \frac{\mathbf{F}}{\mathbf{A}} \tag{b}$$

and we see that the stress $\boldsymbol{\sigma}$ is also a random variable. When Eq. (b) is solved, the errors inherent in $\mathbf{F}$ and in $\mathbf{A}$ are said to be *propagated* to the stress variate $\boldsymbol{\sigma}$. It is not hard to think of many other relations where this will occur.

Suppose we wish to add the two variates $\mathbf{x}$ and $\mathbf{y}$ to form a third variate $\mathbf{z}$. This is written as

$$\mathbf{z} = \mathbf{x} + \mathbf{y} \tag{c}$$

The mean is

$$\mu_z = \mu_x + \mu_y \tag{d}$$

The standard deviation follows the Pythagorean theorem. Thus the standard deviation for *both* addition and subtraction of independent variables is

$$\hat{\sigma}_z = (\hat{\sigma}_x^2 + \hat{\sigma}_y^2)^{1/2} \tag{e}$$

Similar relations have been worked out for a variety of functions and are displayed in Table 4-4. The results shown can easily be combined to form other functions.

An unanswered question in the propagation technique is, What is the distribution that results from the various operations? For answers to this problem, statisticians use

TABLE 4-4

Means, Standard Deviations, and Coefficients of Variation for Simple Operations with Completely Independent (Uncorrelated) Random Variables*

FUNCTION	MEAN VALUE, μ	STANDARD DEVIATION, $\hat{\sigma}$	COEFFICIENT OF VARIATION, C
a	a	0	0
x	$\bar{x}$	$\hat{\sigma}_x$	$\hat{\sigma}/\mu$
$x + a$	$\bar{x} + a$	$\hat{\sigma}_x$	$\hat{\sigma}/\mu$
ax	$a\bar{x}$	$a\hat{\sigma}_x$	$\hat{\sigma}_x/\bar{x}$
$x + y$	$\bar{x} + \bar{y}$	$(\hat{\sigma}_x^2 + \hat{\sigma}_y^2)^{1/2}$	$\hat{\sigma}/\mu$
$x - y$	$\bar{x} - \bar{y}$	$(\hat{\sigma}_x^2 + \hat{\sigma}_y^2)^{1/2}$	$\hat{\sigma}/\mu$
xy	$\overline{xy}$	$C\mu$	$(C_x^2 + C_y^2)^{1/2}$
x/y	$\overline{x/y}$	$C\mu$	$(C_x^2 + C_y^2)^{1/2}$
$1/x$	$1/\bar{x}$	$C_x/\bar{x}$	C_x
x^2	$\bar{x}^2$	$2C_x\bar{x}^2$	$2C_x$
x^3	$\bar{x}^3$	$3C_x\bar{x}^3$	$3C_x$
x^4	$\bar{x}^4$	$4C_x\bar{x}^4$	$4C_x$

*Many of these values are approximate; for a complete listing, see Charles R. Mischke, *Mathematical Model Building*, 2d rev. ed., Iowa State University Press, Ames, 1980, app. C.

closure theorems and the central limit theorem. The results depend upon how the variates are initially distributed. The following guidelines will be helpful:*

1 When normally distributed variables are added or subtracted, the resulting distribution is normal.

2 When a number of random variables of any distribution are added, the resulting distribution approaches normality as the number of distributions involved in the operation increases.

3 When two normally distributed variables are multiplied, the operation is approximately normal, that is to say, *robust*. Suppose **x** and **y** are each normally distributed and that

$$0.1 \leq \frac{\mu_x}{\mu_y} \leq 10$$

$$0.005 \leq C_x \leq 0.20 \tag{4-15}$$

$$0.005 \leq C_y \leq 0.20$$

Then the product is robustly normal if either C_x or C_y is less than 0.075.

4 The quotient of two normally distributed random variables is not strictly normal, but may be robustly normal. The quotient is robustly normal if **x** and **y** are each normally distributed, and if Eq. (4-15) is satisfied.

EXAMPLE 4-4   A round bar subject to a bending load has a diameter **d** = (2.000, 0.002) in. This states that the mean diameter is $\bar{d} = 2.000$ in and the standard deviation is $\hat{\sigma}_d = 0.002$ in. Find the mean and the standard deviation of the second moment of area.

*See E. B. Haugen, *Probabilistic Mechanical Design*, Wiley, New York, 1980, pp. 49–54.

Solution   The second moment of area is given by the equation

$$I = \frac{\pi \mathbf{d}^4}{64}$$

The coefficient of variation of the diameter is

$$C_d = \frac{\hat{\sigma}_d}{\bar{d}} = \frac{0.002}{2.000} = 0.001$$

Then, with Table 4-4, we find

Answer $$\bar{I} = \frac{\pi}{64} \bar{d}^4 = \frac{\pi}{64} (2.000)^4 = 0.785 \text{ in}^4$$

Answer $$\hat{\sigma}_I = 4C_d \bar{x}^4 = 4(0.001)(2.000)^4 = 0.064 \text{ in}^4$$

These results can then be expressed in the form

$$I = (0.785, 0.064) = 0.785(1, 0.0815) \text{ in}^4$$

4-8   LIMITS AND FITS

The subject of limits and fits really deserves a chapter of its own. The subject is included here because the variability inherent in many of the fit classes is so useful in demonstrating the practical application of the statistical ideas presented.

The designer is free to adopt any geometry of fit for shafts and holes that will ensure the intended function. There is sufficient accumulated experience with commonly recurring situations to make standards useful. There are two standards for limits and fits in the United States, one based on inch units and the other based on metric units.* These differ in nomenclature, definitions, and organization. No point would be served by separately studying each of the two systems. The metric version is the newer of the two and is well organized, and so here we present only the metric version but include a set of inch conversions to enable the same system to be used with either system of units.

In using the standard, *capital letters always refer to the hole; lowercase letters are used for the shaft.*

The definitions illustrated in Fig. 4-9 are explained as follows:

• *Basic size* is the size to which limits or deviations are assigned and is the same for both members of the fit.

• *Deviation* is the algebraic difference between a size and the corresponding basic size.

• *Upper deviation* is the algebraic difference between the maximum limit and the corresponding basic size.

Preferred Limits and Fits for Cylindrical Parts, ANSI B4.1-1967; *Preferred Metric Limits and Fits,* ANSI B4.2-1978.

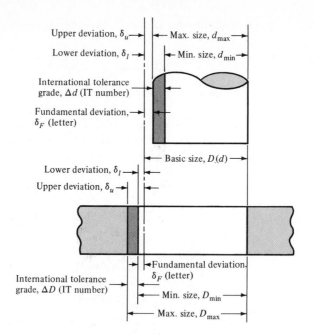

FIGURE 4-9

Definitions applied to a cylindrical fit.

- *Lower deviation* is the algebraic difference between the minimum limit and the corresponding basic size.

- *Fundamental deviation* is either the upper or the lower deviation, depending on which is closer to the basic size.

- *Tolerance* is the difference between the maximum and minimum size limits of a part.

- *International tolerance grade* numbers IT designate groups of tolerances such that the tolerances for a particular IT number have the same relative level of accuracy but vary depending on the basic size.

- *Hole basis* represents a system of fits corresponding to a basic hole size. The fundamental deviation is H.

- *Shaft basis* represents a system of fits corresponding to a basic shaft size. The fundamental deviation is h. The shaft-basis system is not included here.

The magnitude of the tolerance zone is the variation in part size and is the same for both the internal and the external dimensions. The tolerance zones are specified in international tolerance grade numbers, called IT numbers. The smaller grade numbers specify a smaller tolerance zone. These range from IT0 to IT16, but only grades IT6 to IT11 are needed for the preferred fits. These are listed in Tables A-11 to A-13 for basic sizes up to 16 in or 400 mm.

The standard uses *tolerance position letters,* with capital letters for internal dimensions (holes) and lowercase letters for external dimensions (shafts). As shown in Fig. 4-9, the fundamental deviation locates the tolerance zone relative to the basic size.

Table 4-5 shows how the letters are combined with the tolerance grades to establish a preferred fit. The ISO symbol for the hole for a sliding fit with a basic size of 32 mm

TABLE 4-5

Description of Preferred Fits
Using the Basic-Hole System

TYPE OF FIT	DESCRIPTION	SYMBOL
Clearance	*Loose running* fit: for wide commercial tolerances or allowances on external members	H11/c11
	Free running fit: not for use where accuracy is essential, but good for large temperature variations, high running speeds, or heavy journal pressures	H9/d9
	Close running fit: for running on accurate machines and for accurate location at moderate speeds and journal pressures	H8/f7
	Sliding fit: where parts are not intended to run freely, but must move and turn freely and locate accurately	H7/g6
	Locational clearance fit: provides snug fit for location of stationary parts, but can be freely assembled and disassembled	H7/h6
Transition	*Locational transition* fit for accurate location, a compromise between clearance and interference	H7/k6
	Locational transition fit for more accurate location where greater interference is permissible	H7/n6
Interference	*Locational interference* fit: for parts requiring rigidity and alignment with prime accuracy of location but without special bore pressure requirements	H7/p6
	Medium drive fit: for ordinary steel parts or shrink fits on light sections, the tightest fit usable with cast iron	H7/s6
	Force fit: suitable for parts which can be highly stressed or for shrink fits where the heavy pressing forces required are impractical	H7/u6

Source: Preferred Metric Limits and Fits, ANSI B4.2-1978. See also BS 4500.

is 32H7. Inch units are not a part of the standard. However, the designation $(1\frac{3}{8}$ in)H7 includes the same information and is recommended for use here. In both cases, the capital letter H establishes the fundamental deviation and the number 7 defines a tolerance grade of IT7.

For the sliding fit, the corresponding shaft dimensions are defined by the symbol 32g6 [$(1\frac{3}{8}$ in)g6].

The fundamental deviations for shafts are given in Tables A-11 and A-13. For letter codes c, d, f, g, and h,

Upper deviation = fundamental deviation
Lower deviation = upper deviation − tolerance grade

For letter codes k, n, p, s, and u, the deviations for shafts are

Lower deviation = fundamental deviation
Upper deviation = lower deviation + tolerance grade

The lower deviation H (for holes) is zero. For these, the upper deviation equals the tolerance grade.

As shown in Fig. 4-9, we use the following notation:

D = basic size of hole
d = basic size of shaft
δ_u = upper deviation
δ_l = lower deviation
δ_F = fundamental deviation
ΔD = tolerance grade for hole
Δd = tolerance grade for shaft

Note that these quantities are all deterministic. Thus, for the hole,

$$D_{max} = D + \Delta D \qquad D_{min} = D \tag{4-16}$$

For shafts with clearance fits c, d, f, g, and h,

$$d_{max} = d + \delta_F \qquad d_{min} = d + \delta_F - \Delta d \tag{4-17}$$

For shafts with interference fits k, n, p, s, and u,

$$d_{min} = d + \delta_F \qquad d_{max} = d + \delta_F + \Delta d \tag{4-18}$$

EXAMPLE 4-5   Find the shaft and hole dimensions for a loose running fit with a 34-mm basic size.

Solution   From Table 4-5, the ISO symbol is 34H11/c11. From Table A-11, we find that tolerance grade IT11 is 0.160 mm. The symbol 34H11/c11 therefore says that $\Delta D = \Delta d = 0.160$ mm. Using Eq. (4-16) for the hole, we get

Answer $D_{max} = D + \Delta D = 34 + 0.160 = 34.160$ mm

Answer $D_{min} = D = 34.000$ mm

The shaft is designated as a 34c11 shaft. From Table A-12, the fundamental deviation is $\delta_F = -0.120$ mm. Using Eq. (4-17), we get for the shaft dimensions

Answer $d_{max} = d + \delta_F = 34 + (-0.120) = 33.880$ mm

Answer $d_{min} = d + \delta_F - \Delta d = 34 + (-0.120) - 0.160 = 33.720$ mm

---

EXAMPLE 4-6   Find the hole and shaft limits for a medium drive fit using a basic hole size of 2 in.

Solution   The symbol for the fit, from Table 4-5, in inch units is (2 in)H7/s6. For the hole, we use Table A-13 and find the IT7 grade to be $\Delta D = 0.0010$ in. Thus, from Eq. (4-16),

Answer $D_{max} = D + \Delta D = 2 + 0.0010 = 2.0010$ in

Answer $D_{min} = D = 2.0000$ in

The IT6 tolerance for the shaft is $\Delta d = 0.0006$ in. Also, from Table A-14, the fundamental deviation is $\delta_F = 0.0017$ in. Using Eq. (4-18), we get for the shaft that

Answer $d_{min} = d + \delta_F = 2 + 0.0017 = 2.0017$ in

Answer $d_{max} = d + \delta_F + \Delta d = 2 + 0.0017 + 0.0006 = 2.0023$ in

4-9   DIMENSIONING AND TOLERANCING

The following terms are used generally in dimensioning:

- *Nominal size*. The size we use in speaking of an element. For example, we may specify a $\frac{1}{2}$-in bolt or a $1\frac{1}{2}$-in pipe. Either the theoretical size or the actual measured size may be quite different. The bolt, say, may actually measure 0.492 in. And the theoretical size of a $1\frac{1}{2}$-in pipe is 1.900 in for the outside diameter.

- *Basic size*. The exact theoretical size. Limiting dimensions in either the plus or the minus direction begin from the basic dimension.

- *Limits*. The stated maximum and minimum dimensions.

- *Tolerance*. The difference between the two limits.

- *Bilateral tolerance*. The variation in both directions from the basic dimension. That is, the basic size is between the two limits; for example, 1.005 ± 0.002 in. The two parts of the tolerance need not be equal.

- *Unilateral tolerance*. The basic dimension is taken as one of the limits, and variation is permitted in only one direction; for example,

$$1.005 \, {}^{+0.004}_{-0.000} \text{ in}$$

- *Natural tolerance*. A tolerance equal to plus and minus three standard deviations from the mean. For a normal distribution, this ensures that 99.73 percent of production is within the tolerance limits.

- *Clearance*. A general term which refers to the mating of cylindrical parts such as a bolt and a hole. The word *clearance* is used only when the internal member is smaller than the external member. The *diametral clearance* is the measured difference in the two diameters. The *radial clearance* is the difference in the two radii.

- *Interference*. The opposite of clearance, for mating cylindrical parts in which the internal member is larger than the external member.

- *Allowance*. The minimum stated clearance or the maximum stated interference for mating parts.

Variations in dimensions of parts in a production process may occur purely by chance as well as for specific reasons. For example, the operation temperature of a machine tool changes during start-up, and this may have an effect on the dimensions of parts produced during the first 30 min or so of running time. When all such variations in part dimensions have been eliminated, the production process is said to be in *statistical control*. Under these conditions all variations in dimensions occur at random.

EXAMPLE 4-7   Suppose a problem has arisen with the assembly of Example 4-5 and we must reject assemblies when the clearance is less than 0.20 mm and over 0.34 mm. Find the percentage rejected if the tolerances are normal, natural, and the process is centered.

Solution   Using C for clearance, we compute the maximum and minimum clearances as

$$C_{max} = D_{max} - d_{min} = 34.160 - 33.720 = 0.440 \text{ mm}$$
$$C_{min} = D_{min} - d_{max} = 34 - 33.880 = 0.120 \text{ mm}$$

The means of the hole and shaft diameters are easily found. They are

$$\overline{D} = 34.080 \text{ mm} \qquad \overline{d} = 33.800 \text{ mm}$$

Subtracting these from the maxima and minima gives the tolerances as ± 0.080 mm for both the hole and the shaft. Since these are assumed to be natural tolerances, the standard deviations are

$$\hat{\sigma}_D = \hat{\sigma}_d = \frac{0.080}{3} = 0.0267 \text{ mm}$$

Using Table 4-4 for the difference of two variates, we find the standard deviation of the clearance to be

$$\hat{\sigma}_C = (\hat{\sigma}_D^2 + \hat{\sigma}_d^2)^{1/2} = [2(0.0267)^2]^{1/2} = 0.0377 \text{ mm}$$

The standardized variable z is to correspond to a clearance not less than 0.20 mm. Therefore, from Eq. (4-14), we have

$$z_{0.20} = \frac{C - \overline{C}}{\hat{\sigma}_C} = \frac{0.20 - 0.280}{0.0377} = -2.1213$$

From Table A-10, we find the value to be $\Phi(z_{0.20}) = 0.016\ 948$, or 1.69 percent. For clearances greater than 0.34 mm, we have, this time

$$z_{0.34} = \frac{0.34 - 0.280}{0.0377} = 1.5915$$

Since this is positive,

$$\Phi(z_{0.34}) = 1 - \Phi(-1.5915) = 1 - 0.055\ 735 = 0.944\ 265$$

So, by rounding, 94.43 percent of the assemblies have a clearance less than 0.34 mm. This leaves 5.57 percent of them with clearances greater than 0.34 mm. Adding the two reject percentages, we find that a total of 7.26 percent must be rejected.

Tolerance Stacking

An application of propagation-of-error techniques is that of establishing the gap, grip, or interference associated with the assembly of a number of parts each having a size distribution. Consider the parallel vectors representing dimensional random variables in Fig. 4-10. Here the x's are right-tending and the y's left-tending, and w designates the clearance. The vectors are parallel, and so we can write the *scalar equation of random variables* as

$$\mathbf{w} = (\mathbf{x}_1 + \mathbf{x}_3 + \cdots) - (\mathbf{y}_2 + \mathbf{y}_4 + \cdots) = \sum \mathbf{x} - \sum \mathbf{y}$$

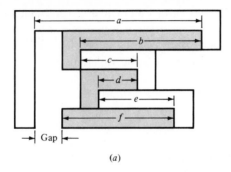

(a)

FIGURE 4-10

An assembly considered as a group of displacement vectors. (a) Six L-shaped blocks A, B, C, D, E, and F are assembled as shown, leaving a gap. The dimensions a, b, c, d, e, and f are toleranced, making them random variables. (b) Corresponding vector diagram showing right-tending displacements as **x** and left-tending as **y** and **w**.

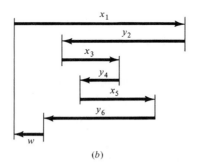

(b)

Then, using Table 4-4, we find the parameters of the assembly to be

$$\overline{w} = \sum \overline{x} - \sum \overline{y} \qquad \hat{\sigma}_w = \left(\sum \hat{\sigma}_x^2 + \sum \hat{\sigma}_y^2 \right)^{1/2}$$

Absolute Tolerance System

The *range numbers* of a gap, such as w in Fig. 4-10, are given by $\overline{w} - t_w$ and $\overline{w} + t_w$. An *absolute tolerance system* is one in which there are *no* instances of the gap w outside the range. It follows that a *statistical tolerance system* is one in which it is possible for some instances of the gap w to lie outside the range.

4-10   THE LOGNORMAL DISTRIBUTION

Sometimes random variables have the following two characteristics:

• The distribution is asymmetrical about the mean.

• The variables have only positive values.

Such characteristics rule out the use of the normal distribution. There are several other distributions that are potentially useful in such situations, one of them being the lognormal (written as a single word) distribution. Especially when life is involved, such as fatigue life under stress, or the wear life of rolling bearings, the lognormal distribution may be a very appropriate one to use.

The *lognormal distribution* is one in which the logarithms of the variate have a normal distribution. Thus the variate itself is said to be lognormally distributed. Let this variate be

$$\mathbf{x} \sim LN(\mu_x, \hat{\sigma}_x) \tag{a}$$

Equation (*a*) states that the random variable **x** is distributed lognormally and that its mean value is μ_x and its standard deviation is $\hat{\sigma}_x$.

Now use the transformation

$$\mathbf{y} = \ln \mathbf{x} \tag{b}$$

Since, by definition, **y** has a normal distribution, we can write

$$\mathbf{y} \sim N(\mu_y, \hat{\sigma}_y) \tag{c}$$

This equation states that the random variable **y** is normally distributed and that its mean value is μ_y and its standard deviation is $\hat{\sigma}_y$.

It is convenient to think of Eq. (*a*) as designating the *parent*, or *principal*, *distribution* while Eq. (*b*) represents the *companion*, or *subsidiary*, *distribution*.

The probability density function (PDF) for **x** can be derived from that for **y**; see Eq. (4-13), and substitute *y* for *x* in that equation. Thus the PDF for the companion distribution is found to be

$$f(x) = \frac{1}{x\hat{\sigma}_y\sqrt{2\pi}} \exp\left[-\frac{1}{2}\left(\frac{y - \mu_y}{\hat{\sigma}_y}\right)^2\right] \tag{4-19}$$

The companion standard deviation $\hat{\sigma}_y$ and mean μ_y in Eq. (4-19) are obtained from

$$\hat{\sigma}_y = [\ln (1 + C_x^2)]^{1/2} \approx C_x \tag{4-20}$$

$$\mu_y = \ln \mu_x - \frac{1}{2} \ln (1 + C_x^2) \approx \ln \mu_x - \frac{C_x^2}{2} \tag{4-21}$$

These equations make it possible to use Table A-10 for statistical computations and eliminate the need for a special table for the lognormal distribution.

EXAMPLE 4-8   A certain brand of light bulb is advertised as having a mean long life of 1.5 kh. The standard deviation is not stated; suppose it is 0.20 kh. If the life is assumed to be lognormally distributed, a likely assumption, then we can designate a parent distribution, the bulb life, as

$$\mathbf{x} \sim LN(1.5, 0.20) \qquad \text{kh}$$

What is the chance of getting a bulb having a life greater than 1000 h?

Solution   Here the parent variate is **x**, with $\hat{\sigma}_x = 0.20$ kh and $\mu_x = 1.5$ kh. The standard deviation and mean of the companion distribution are obtained from Eqs. (4-20) and (4-21) as

$$\hat{\sigma}_y = C_x = \frac{\hat{\sigma}_x}{\mu_x} = \frac{0.20}{1.5} = 0.1333$$

$$\mu_y = \ln \mu_x - \frac{C_x^2}{2} = \ln 1.5 - \frac{(0.1333)^2}{2} = 0.3966$$

The value of the standardized variable z corresponding to a life of 1000 h is

$$z_1 = \frac{y_1 - \mu_y}{\hat{\sigma}_y} = \frac{\ln x_1 - \mu_y}{\hat{\sigma}_y} = \frac{0 - 0.3966}{0.1333} = -2.975$$

Using Table A-10, we find $\Phi(z_1) = 0.0014$. The probability that $x > 1$ kh is then given as

Answer $p(x > 1) = 1 - \Phi(z_1) = 1 - 0.0014 = 0.9986$

Thus there is better than a 99 percent chance of getting a bulb with a life greater than 1000 h.

4-11 THE WEIBULL DISTRIBUTION

The Weibull distribution does not arise from classical statistics and is usually not included in elementary statistics textbooks. It is far more likely to be discussed and used in works dealing with experimental results, particularly reliability. It is a chameleon distribution, asymmetrical, with different values for the mean and the median. It contains within it a good approximation of the normal distribution as well as an exact representation of the exponential distribution. Most reliability information comes from laboratory and field service data, and because of its flexibility, the Weibull distribution is used.

There are two forms of the Weibull distribution, usually simply called the Weibull. These are called the *three-parameter* and the *two-parameter* distributions. The expression for reliability is the value of the cumulative density function complementary to unity. For the Weibull this value is both explicit and simple. It is

$$R(x) = \exp\left[-\left(\frac{x - x_0}{\theta - x_0}\right)^b\right] \qquad x \geq x_0 \tag{4-22}$$

$$R(x) = \exp\left[-\left(\frac{x}{\theta}\right)^b\right] \qquad x \geq 0 \tag{4-23}$$

where Eq. (4-22) is the three-parameter Weibull. The three parameters are

$x_0 = $ the guaranteed value of x $(x_0 \geq 0)$

$\theta = $ a characteristic or scale value $(\theta \geq x_0)$

$b = $ a shape parameter $(b > 0)$

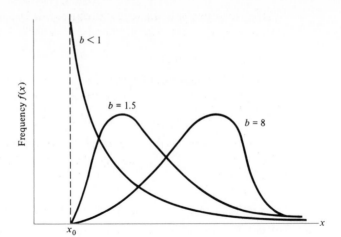

FIGURE 4-11

The density function of the Weibull distribution showing the effect of skewness of the shape parameter b.

Equation (4-23) is the two-parameter Weibull, so named because x_0 is taken as zero.

The Weibull distribution is bounded by x_0 on the lower tail, and hence x_0 is called the *guaranteed* variate.

The characteristic variate θ serves a role similar to the mean and represents a value of x below which lie 63.2 percent of the observations.

The shape parameter b controls the skewness of the distribution. Figure 4-11 shows that large b's skew the distribution to the left and small b's skew it to the right. In the range $3.3 < b < 3.5$, approximate symmetry is obtained along with a good approximation to the normal distribution. When $b = 1$, the distribution is exponential.

To find the probability function, we note that

$$F(x) = 1 - R(x) \tag{a}$$

$$f(x) = \frac{dF(x)}{dx} = -\frac{dR(x)}{dx} \tag{b}$$

Thus, for the Weibull,

$$f(x) = \frac{b}{\theta - x_0}\left(\frac{x - x_0}{\theta - x_0}\right)^{b-1} \exp\left[-\left(\frac{x - x_0}{\theta - x_0}\right)^b\right] \tag{4-24}$$

It is rather difficult to find the mean. The procedure involves integration by parts, then resorting to a table of definite integrals. The result is found to be

$$\mu_x = x_0 + (\theta - x_0)\Gamma\left(1 + \frac{1}{b}\right) \tag{4-25}$$

The standard deviation is then found to be

$$\hat{\sigma}_x = (\theta - x_0)\left[\Gamma\left(1 + \frac{2}{b}\right) - \Gamma^2\left(1 + \frac{1}{b}\right)\right]^{1/2} \tag{4-26}$$

where Γ is the gamma function and may be found in many mathematical tables. With many computers you can store this entire table in memory for recall together with an interpolation routine. The notation for a Weibull distribution is

$$\mathbf{x} \sim W(x_0, \theta, b) \tag{4-27}$$

Two other measures of central tendency can be useful. The *median* is the value of the variate that has a fifty-fifty chance of being exceeded. It is

$$x_{\text{med}} = x_0 + (\theta - x_0)(\ln 2)^{1/b} \tag{4-28}$$

The *mode* is the observation with the highest probability, often the stationary point, of the probability distribution function. For the Weibull it is

$$x_{\text{mode}} = x_0 + (\theta - x_0)\left(\frac{b-1}{b}\right)^{1/b} \tag{4-29}$$

EXAMPLE 4-9 The tensile strength of a cold-drawn SAE 1045 steel bar is given as

$$\mathbf{S}_{ut} \sim W(90.2, 120.5, 4.38) \qquad \text{kpsi}$$

Find the mean and the standard deviation, and compute points so that the probability density function can be plotted if desired.

Solution From Eq. (4-25), the mean of $\mathbf{S}_{ut}$ is

Answer
$$\overline{S}_{ut} = \mu_{Su} = 90.2 + (120.5 - 90.2)\left[\Gamma\left(1 + \frac{1}{4.38}\right)\right] = 117.8 \text{ kpsi}$$

The standard deviation is obtained from Eq. (4-26) as

Answer
$$\hat{\sigma}_{Su} = (120.5 - 90.2)\left[\Gamma\left(1 + \frac{2}{4.38}\right) - \Gamma^2\left(1 + \frac{1}{4.38}\right)\right]^{1/2} = 7.18 \text{ kpsi}$$

Then we use Eq. (4-24) to generate plotting positions. The results are

$x = S_{ut}$	95	100	105	110	115
$f(x)$	0.0003	0.003	0.012	0.029	0.048

$x = S_{ut}$	120	125	130	135	140
$f(x)$	0.054	0.037	0.013	0.002	0.0015

4-12   THE WEIBULL PARAMETERS

To find the three parameters x_0, θ, and b, two graphs are drawn as shown in Fig. 4-12, to rectify the reliability function. The first, Fig. 4-12a, yields x_0; the second, Fig. 4-12b, is the rectification of the first, and is used to verify that the distribution is a Weibull, and to obtain the characteristic parameter θ and the slope b.

In Fig. 4-12a, the data points are plotted under the assumption that $x_0 = 0$. Then,

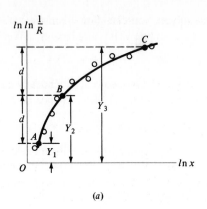

(a)

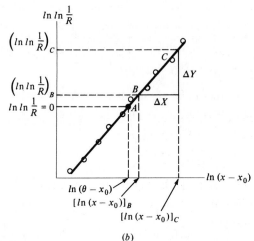

(b)

FIGURE 4-12

Graphs used to estimate the
Weibull parameters from data.
(a) Used with Eq. (4-30) to find x_0.
(b) Used with Eqs. (4-31) and
(4-32) to find θ and b. Note that
$\Delta Y = [\ln \ln (1/R)]_C - [\ln \ln (1/R)]_B$
and that $\Delta X = [\ln (x - x_0)]_B - [\ln (x - x_0)]_C$.

using a french curve, we draw in the best locus. A large graph is preferred to ensure a fairly large radius of curvature at all points. Now we locate any three points, such as A, B, and C, spaced equally along the Y axis. Thus, by construction,

$$Y_2 - Y_1 = Y_3 - Y_2 = d \tag{a}$$

The displacements from the plotting origin at O can be specified in terms of scale factors S_X and S_Y as

$$Y = S_Y \ln \ln \frac{1}{R} \qquad X = S_X \ln (x - x_0) \tag{b}$$

Now we twice take the logarithms of Eq. (4-22). The result can be written as

$$\ln \ln \frac{1}{R} = b \ln (x - x_0) - b \ln (\theta - x_0) \tag{c}$$

Substituting the first part of Eq. (b) in Eq. (a) yields

$$S_Y \ln \ln \frac{1}{R_2} - S_Y \ln \ln \frac{1}{R_1} = S_Y \ln \ln \frac{1}{R_3} - S_Y \ln \ln \frac{1}{R_2}$$

Next, we cancel the scale factor and use Eq. (c) as follows:

$$b \ln (x_2 - x_0) - b \ln (\theta - x_0) - b \ln (x_1 - x_0) + b \ln (\theta - x_0)$$
$$= b \ln (x_3 - x_0) - b \ln (\theta - x_0) - b \ln (x_2 - x_0) + b \ln (\theta - x_0)$$

Then we take the inverse logarithms and solve for x_0. The result is

$$x_0 = x_2 - \frac{(x_3 - x_2)(x_2 - x_1)}{(x_3 - x_2) - (x_2 - x_1)} \tag{4-30}$$

Finally, we use the abscissa coordinates corresponding to points A, B, and C of Fig. 4-12a and compute x_1, x_2, and x_3 and solve Eq. (4-30) for x_0.

Now we compute a new set of data points using $x - x_0$ and replot as in Fig. 4-12b. The plot will straighten if the data are Weibullian and so we will have the x_0 parameter. If the plot is not straight, the data are not Weibullian, we will not have x_0, and we won't care.

If the data are found to be Weibullian, then the plot of Fig. 4-12b can be used to find the remaining parameters. First, we locate point A corresponding to $\ln \ln (1/R) = 0$. Note that

$$\text{inv } \ln (\text{inv } \ln 0) = e = 2.718$$

So point A corresponds to a reliability of

$$R = \frac{1}{e} = \frac{1}{2.718} = 0.368$$

The abscissa point corresponding to A is $\ln (\theta - x_0)$. Therefore the characteristic parameter θ is

$$\theta = \text{inv } \ln (\theta - x_0) + x_0 \tag{4-31}$$

The slope parameter is obtained by scaling the two sides of a right triangle constructed as shown in Fig. 4-12b. Thus the slope is simply

$$b = \frac{\Delta Y}{\Delta X} \tag{4-32}$$

EXAMPLE 4-10   The results of testing many heats of AISI 1020 steel included 1000 separate determinations of the ultimate tensile strength. These ranged from 56 to 72 kpsi. From a histogram, as in Fig. 4-4, of failures, the data in columns 1 and 2 of Table 4-6 were prepared. For each row, the class width is $w = 1$ kpsi. If this is a Weibull distribution, find the three parameters.

Solution   The density f in column 3 of Table 4-6 is the number of strengths of the ith class n_i divided by the product of the sample size N and the class width w.

The cumulative density function in column 4 is the area under the histogram up to the class mark x. It is given by the equation

$$F_i = \frac{f_i w_i}{2} + \sum_{j=1}^{i-1} f_j w_j \tag{1}$$

TABLE 4-6

Computations for Example 4-10 ($x_0 = 55.284$ kpsi)

1 CLASS MARK, x	2 NUMBER OF FAILURES, n	3 $f = \dfrac{n}{Nw}$	4 F	5 $R = 1 - F$	6 $\ln x$	7 $\ln \ln \dfrac{1}{R}$	8 $\ln (x - x_0)$
56.5	2	0.002	0.0010	0.9990	4.034	−6.907	0.196
57.5	18	0.018	0.0110	0.9890	4.052	−4.504	0.796
58.5	23	0.023	0.0315	0.9685	4.069	−3.442	1.168
59.5	31	0.031	0.0585	0.9415	4.086	−2.809	1.439
60.5	83	0.083	0.1155	0.8845	4.103	−2.098	1.652
61.5	109	0.109	0.2115	0.7885	4.119	−1.437	1.827
62.5	138	0.138	0.3350	0.6650	4.135	−0.897	1.976
63.5	151	0.151	0.4795	0.5205	4.151	−0.426	2.106
64.5	139	0.139	0.6245	0.3755	4.167	−0.021	2.221
65.5	130	0.130	0.7590	0.2410	4.182	0.353	2.324
66.5	82	0.082	0.8650	0.1350	4.197	0.694	2.417
67.5	49	0.049	0.9305	0.0695	4.212	0.981	2.503
68.5	28	0.028	0.9690	0.0310	4.227	1.245	2.581
69.5	11	0.011	0.9885	0.0115	4.241	1.496	2.654
70.5	4	0.004	0.9960	0.0040	4.256	1.709	2.722
71.5	2	0.002	0.9990	0.0010	4.270	1.933	2.786

Columns 5, 6, and 7 are obtained next and the results are used to plot Fig. 4-13. In the process of composing this example, this figure was plotted on a large sheet of graph paper so that the curve itself measured 230 mm high and 300 mm along the abscissa. Then the parameter x_0 was determined using three sets of data points along the curve. These three sets are indicated as *ABC*, *DEF*, and *GHI* on the smaller diagram reproduced from the original as Fig. 4-13. Note that *ABC* uses nearly the entire curve, while the range *GHI* includes only the upper two-thirds. The results for these determinations

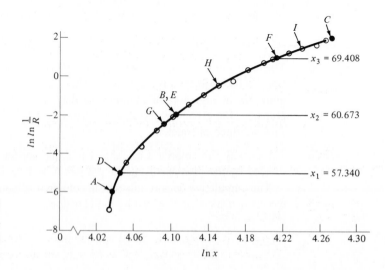

FIGURE 4-13

Graph of ln ln (1/R) versus ln x used to estimate the Weibull parameter x_0.

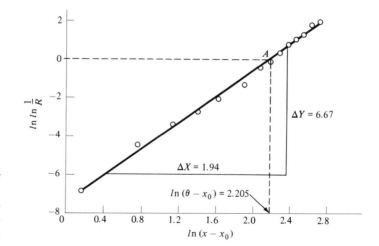

FIGURE 4-14

Graph of ln ln (1/R) versus ln (x − x_0) used to confirm Weibull distribution by rectification and to estimate Weibull parameters θ and b.

are

ABC $x_0 = 54.863$ kpsi

DEF $x_0 = 55.284$ kpsi

GHI $x_0 = 55.328$ kpsi

These are in excellent agreement and testify to the validity of the method.

Using the values on Fig. 4-13 for points DEF and Eq. (4-30), the computation is

Answer $x_0 = 60.673 - \dfrac{(69.408 - 60.673)(60.673 - 57.340)}{(69.408 - 60.673) - (60.673 - 57.340)} = 55.284$ kpsi

With x_0 known, we now compute column 8 and draw Fig. 4-14. Note that the points do indeed fit a straight line quite well. We now draw in this line and locate a point A corresponding to ln ln (1/R) = 0. The corresponding coordinate on the abscissa is used to find the θ parameter. Thus, from Eq. (4-31),

Answer $\theta = \text{inv ln } (\theta - x_0) + x_0 = \text{inv ln } 2.205 + 55.284 = 64.354$ kpsi

Finally, the slope b is found from Eq. (4-32) and the lengths in coordinates of the two right-triangle legs shown in Fig. 4-14:

Answer $b = \dfrac{\Delta Y}{\Delta X} = \dfrac{6.67}{1.94} = 3.44$

Rounding and summarizing these results gives

Answer $S_{ut} = W(x_0, \theta, b) = W(55.3, 64.3, 3.44)$ kpsi

---

4-13   RANKING

Suppose five automobile wheel nuts are to be tightened to a specified wrench torque. For a number of reasons, such as friction, thread form, and lubrication, the resulting

TABLE 4-7

Median Ranks

ORDER NO., j	SAMPLE SIZE, n									
	1	2	3	4	5	6	7	8	9	10
1	0.5000	0.2929	0.2063	0.1591	0.1294	0.1091	0.0943	0.0830	0.0741	0.0670
2		0.7071	0.5000	0.3864	0.3147	0.2655	0.2295	0.2021	0.1806	0.1632
3			0.7937	0.6136	0.5000	0.4218	0.3648	0.3213	0.2871	0.2594
4				0.8409	0.6853	0.5782	0.5000	0.4404	0.3935	0.3557
5					0.8706	0.7345	0.6352	0.5596	0.5000	0.4519
6						0.8909	0.7705	0.6787	0.6065	0.5481
7							0.9057	0.7979	0.7129	0.6443
8								0.9170	0.8194	0.7406
9									0.9259	0.8368
10										0.9330

Source: C. Lipson and N. J. Sheth, *Statistical Design and Analysis of Engineering Experiments*, McGraw-Hill, New York, 1973, p. 482.

clamping force or preload will be different for each of the five fasteners. The statistical concept of *ranking*, based on the cumulative density function, furnishes a method of estimating the median of, say, the highest and lowest clamping force.

Table 4-7 provides values of the median, or 50 percent, CDF corresponding to rank-ordered observations for sample sizes up to $n = 10$. For larger sample sizes, Bernard's approximation can be used. It is

$$\text{CDF}_j = \frac{j - 0.3}{n + 0.4} \tag{4-33}$$

where j is the order number. If mean CDF corresponding to rank-ordered observations is desired, it is simply expressed as

$$\text{CDF}_j = \frac{j}{n + 1} \tag{4-34}$$

If the objective is to estimate parameters of the median CDF locus describing the population, median CDFs are plotted against ordered observations. If population parameters such as mean and standard deviation are desired, the mean CDFs are plotted. An eyeball best-line drawn on coordinates which rectify the CDF is often sufficient. See also the footnote to Prob. 4-53.

PROBLEMS*,†

4-1 At a constant stress level, the cycles-to-failure experience with 70 specimens of 5160H steel from 1.25-in hexagonal bar stock was

---

*As noted in Sec. 1-10, headings for material involving statistical considerations are printed in black type to call your attention to the nature of the content. This practice will also be used here and throughout the book with problem numbers.

†An asterisk indicates a design-type problem, a problem with no unique result, or a challenging problem.

L	60	70	80	90	100	110	120	130	140	150	160	170	180	190	200	210
n	2	1	3	5	8	12	6	10	8	5	2	3	2	1	0	1

where L is the life in 10^3 cycles and n is the number of failures. Estimate the mean and the standard deviation of the population from which the sample was drawn.

4-2   Determinations of the ultimate tensile strength S_{ut} in kpsi of stainless steel sheet (17-7PH, condition TH 1050), in sizes from 0.016 to 0.062 in, in 197 tests divided into 14 classes were

S_{ut}	172	176	180	184	188	192	196	200	204	208	212	216	220	224
n	2	4	2	7	22	22	26	41	31	22	7	5	4	2

where n is the number in the class. Estimate the mean and standard deviation of the population from the sample drawn.

4-3   A total of 58 AISI 1018 cold-drawn steel bars were tested to determine the 0.2 percent offset yield strength S_y in kpsi. The results were:

S_y	64	68	72	76	80	84	88	92
n	2	6	6	9	19	10	4	2

where S_y is the class midpoint and n the class frequency.
(*a*) Estimate the mean and standard deviation and PDF of the population from which the sample came, presuming a normal distribution.
(*b*) Compare the fraction with a yield strength less than 72 kpsi with both the normal fit and the data. Which is a better estimate?

4-4   The base 10 logarithm of 55 cycles-to-failure observations on specimens subjected to a constant stress level in fatigue have been classified as follows:

x	5.625	5.875	6.125	6.375	6.625	6.875	7.125	7.375	7.625	7.875	8.125
n	1	0	0	3	3	6	14	15	10	2	1

Here x is the class midpoint and n the number.
(*a*) Estimate the mean and standard deviation of the population from which the sample was taken and establish the normal PDF.
(*b*) Plot the histogram and superpose the predicted class frequency from the normal fit.

4-5   A $\frac{1}{4}$-in nominal diameter round is formed in an automatic screw machine operation which is initially set to produce a 0.247-in diameter and is reset when tool wear produces diameters in excess of 0.250 in. The stream of parts is mixed and produces a uniform random distribution of diameters.
(*a*) Estimate the mean and standard deviation of a large batch of parts.
(*b*) Find expressions for the PDF and CDF of the population.
(*c*) If, by inspection, the diameters less than 0.248 in are removed, what are the new PDF and CDF as well as the mean and standard deviation of the survivors of the inspection?

4-6   The only detail drawing of a machine part has a dimension smudged beyond legibility. The round in question was created in an automatic screw machine, and 1000 parts are in stock. A random sample of 50 parts gave a mean dimension of $\bar{x} = 0.6241$ in, with a standard deviation estimate of 0.000 581 in. Toleranced dimensions elsewhere on the drawing are given in integral thousandths of an inch. Estimate the missing information on the drawing.

4-7   The CDF of the variate **x** is $F(x) = 0.555x - 33$, where x is in millimeters. Find the PDF, the mean, the standard deviation, and the range numbers a and b of the distribution.

4-8   In the expression $\boldsymbol{\sigma} = \mathbf{F}/\mathbf{A}$, the force is $\mathbf{F} \sim N(3600, 300)$ lb and the area is specified as $\mathbf{A} \sim N(0.112, 0.001)$ in^2. Estimate the mean, the standard deviation, and the coefficient of variation of the corresponding tensile stress.†

4-9   In the expression for uniaxial strain $\boldsymbol{\epsilon} = \boldsymbol{\delta}/\mathbf{l}$, the elongation is specified as $\boldsymbol{\delta} \sim N(0.0015, 0.000\ 092)$ in and the length as $\mathbf{l} \sim U(2.0000, 0.0081)$ in. What are the mean, the standard deviation, and the coefficient of variation of the corresponding strain $\boldsymbol{\epsilon}$ if U is a uniform distribution?

4-10   In Hooke's law for uniaxial stress, $\boldsymbol{\sigma} = \boldsymbol{\epsilon}\mathbf{E}$, the strain is given as $\boldsymbol{\epsilon} \sim N(0.0005, 0.000\ 034)$ and Young's modulus as $\mathbf{E} \sim N(29.5, 0.885)$ Mpsi. Find the mean, the standard deviation, and the coefficient of variation of the corresponding stress $\boldsymbol{\sigma}$ in psi.

4-11   The stretch of a uniform rod in tension is given by the formula $\delta = Fl/AE$. Suppose the terms in this equation are random variables and have parameters as follows:

$\mathbf{F} \sim N(14.7, 1.3)$ kip $\mathbf{A} \sim N(0.226, 0.003)$ in^2

$\mathbf{l} \sim N(1.5, 0.004)$ in $\mathbf{E} \sim N(29.5, 0.885)$ Mpsi

Estimate the mean, the standard deviation, and the coefficient of variation of the corresponding elongation $\boldsymbol{\delta}$ in inches.

4-12   The maximum bending stress in a round bar in flexure occurs in the outer surface and is given by the equation $\sigma = 32M/\pi d^3$. If the moment is specified as $\mathbf{M} \sim N(15\ 000, 1350)$ lb · in and the diameter is $\mathbf{d} \sim N(2.00, 0.005)$ in, find the mean, the standard deviation, and the coefficient of variation of the corresponding stress $\boldsymbol{\sigma}$ in psi.

4-13   Let r be the radial distance to the neutral surface of a curved beam, expressed as the deterministic quantity $r = h/\ln(r_o/r_i)$. When r is given the value $r = 3/\ln(16.5/13.5) = 0.149\ 498\ 659\ 8(10^2)$, expressed as $0.1495(10^2)$, the deterministic quantity r has been rendered stochastic. Suppose that the range numbers are

$0.1495(10^2) - [0.5(10^2)/10^4]$ and $0.1495(10^2) + [0.5(10^2)/10^4]$

All values are equally likely, and so when a deterministic quantity r has been rounded to s, it becomes a uniform random variable with a mean of $\bar{s}$ and a standard deviation of $0.5(10^2)/[\sqrt{3}(10^4)]$. (Confirm this.)

As a generalization, for a number expressed in scientific notation as $s = y(10^k)$, then if m digits are significant, the distribution is uniform and is

$\mathbf{s} \sim U[y(10^k), 0.5(10^k)/\sqrt{3}(10^k)]$

Investigate the nature of the eccentricity e in the relation $e = R - r$ where $r = 15$ in and the

---

†Remember that σ is a deterministic stress, whereas $\boldsymbol{\sigma}$ is a randomly distributed stress, that $\hat{\sigma}$ is used for standard deviation in general, and that $\hat{\sigma}_{\sigma}$ and $\hat{\mu}_{\sigma}$ denote the standard deviation and the mean, respectively, of the stress variate $\boldsymbol{\sigma}$.

computation of r is significant to m digits. Comment on the different values we will get for the inner fiber stress $\sigma_i = -Mc_i/Aer_i$ for $m = 3$, 4, and 5 significant digits.

4-14   A 1-in-diameter steel sphere presses against a $\frac{1}{2}$-in-diameter steel sphere with a load $\mathbf{F} \sim N(50, 4)$ lb. For $\nu = 0.30$ and $E = 30$ Mpsi and the load variation dominating, what are the mean, standard deviation, and coefficient of variation of the maximum Hertzian pressure $\mathbf{p}_{max}$ in psi?

4-15   A 0.5-in steel cylindrical roller 3 in long presses against a 1-in-diameter steel roller with a force $\mathbf{F} \sim N(5000, 400)$ lb. For a Poisson's ratio of 0.30 and a Young's modulus of 30 Mpsi and the load variation dominating, estimate the mean, the standard deviation, and the coefficient of variation of the maximum Hertzian pressure $\mathbf{p}_{max}$ in kpsi.

4-16   The lives of parts are often expressed as the number of cycles of operation that a specified percentage of a population will exceed before experiencing failure. The symbol B is used to designate this definition of life. Thus we can speak of B10 life as the number of cycles to failure exceeded by 90 percent of a population of parts. Using the mean and standard deviation for the data of Prob. 4-1, a normal distribution model, estimate the corresponding B10 life.

4-17   A springmaker is supplying helical-coil springs meeting the requirement for a spring rate k of 10 ± 1 lb/in. The test program of the springmaker shows that the distribution of spring rate is well approximated by a normal distribution. The experience with inspection has shown that 8.1 percent are scrapped with $k < 9$ and 5.5 percent are scrapped with $k > 11$. Estimate the probability density function.

4-18   When a production process is wider than the tolerance interval, inspection rejects a low-end scrap fraction α with $x < x_1$ and an upper-end scrap fraction β with dimensions $x > x_2$. The surviving population has a new density function $g(x)$ related to the original $f(x)$ by a multiplier a. This is because any two observations x_i and x_j will have the same relative probability of occurrence as before. Show that

$$a = \frac{1}{F(x_2) - F(x_1)} = \frac{1}{1 - (\alpha + \beta)}$$

and

$$g(x) = \begin{cases} \dfrac{f(x)}{F(x_2) - F(x_1)} = \dfrac{f(x)}{1 - (\alpha + \beta)} & x_1 \le x \le x_2 \\ 0 & \text{otherwise} \end{cases}$$

4-19   An automatic screw machine produces a run of parts with $\mathbf{d} \sim U[0.748, 0.751]$ in because it was not reset when the diameters reached 0.750 in. The square brackets contain range numbers.
   (a) Estimate the mean, standard deviation, and PDF of the original production run if the parts are thoroughly mixed.
   (b) Using the results of Prob. 4-18, find the new mean, standard deviation, and PDF. Superpose the PDF plots and compare.

4-20   Using the results of Prob. 4-18, show that the mean of a truncated distribution is

$$\hat{\mu}_{x,t} = \int_{x_1}^{x_2} \frac{xf(x)\, dx}{F(x_2) - F(x_1)}$$

and that the variance is

$$\hat{\sigma}_{x,t}^2 = \int_{x_1}^{x_2} \frac{x^2 f(x)\,dx}{F(x_1) - F(x_2)} - \hat{\mu}_{x,t}^2$$

and that the new CDF is

$$G(x_t) = \int_{x_1}^{x_t} \frac{f(x)\,dx}{F(x_2) - F(x_1)}$$

4-21 In Prob. 4-17, what is the new probability density function when 8.1 percent low-end and 5.5 percent high-end scrap are generated in a normally distributed population? The results of Prob. 4-18 will be useful.

4-22 Assemblies of several parts often need a gap (clearance) in order to sustain intended function. The nominal dimension and tolerance of each part can be expressed in the form $\bar{x} \pm t_x$, in terms of the mean size $\bar{x}$, which is midrange, and the equal-amplitude bilateral tolerance t_x. The equation for the mean gap in Fig. 4-10 has been shown to be $\bar{w} = \Sigma \bar{x} - \Sigma \bar{y}$. The *absolute tolerance system* identifies range numbers on the gap, namely $\bar{w} - t_w$ and $\bar{w} + t_w$, such that there are *no* instances of w outside the range. Show that, in this system, t_w is given by

$$t_w = \sum_{\text{all}} t$$

4-23 Three blocks A, B, and C and a grooved block D have dimensions a, b, c, and d as follows:

$a = 1.000 \pm 0.001$ in $b = 2.000 \pm 0.003$ in

$c = 3.000 \pm 0.005$ in $d = 6.020 \pm 0.006$ in

The blocks are assembled as shown in the figure.

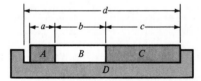

PROBLEM 4-23

(a) Using the absolute tolerance system, determine the nominal gap $\bar{w}$ and its bilateral tolerance amplitude t_w.

(b) Using the statistical tolerance system, determine the nominal gap $\bar{w}$ and its bilateral tolerance amplitude t_w.

4-24 If $x = a \pm \Delta a$, $y = b \pm \Delta b$, and $z = c \pm \Delta c$, show that for the volume V of a rectangular parallelepiped, $V = xyz$,

$$\frac{\Delta V}{\bar{V}} = \frac{\Delta a}{\bar{a}} + \frac{\Delta b}{\bar{b}} + \frac{\Delta c}{\bar{c}}$$

Use this result to place range numbers (absolute tolerance bounds) on the volume of a rectangular parallelepiped with the dimensions

$a = 1.250 \pm 0.001$ in $b = 1.875 \pm 0.002$ in $c = 2.750 \pm 0.003$ in

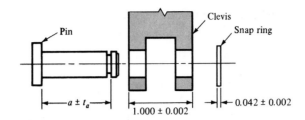

PROBLEM 4-25
Dimensions in inches.

4-25   A pivot in a linkage has a pin as depicted in the figure whose dimension $a \pm t_a$ is to be established. The thickness of the link clevis is 1.000 ± 0.002 in. The designer has concluded that a gap of between 0.004 and 0.14 in will satisfactorily sustain the function of the linkage pivot. (*a*) Determine the dimension a and its tolerance using the absolute tolerance method. (*b*) Establish the dimension a and its tolerance using the statistical tolerance method.

4-26   In Prob. 4-25 the dimensions are

$$\mathbf{a} \sim N(1.051, 0.0015) \text{ in} \qquad \mathbf{b} \sim N(1.000, 0.001) \text{ in}$$

$$\mathbf{c} \sim N(0.042, 0.001) \text{ in}$$

Estimate the probability of observing a gap w of less than 0.004 in. Estimate the probability of observing a gap w less than 0 in.

4-27   The AISI 1040 steel shaft shown in the figure is to mate with an AISI 1040 steel collar, also shown, with a shrink fit. The final dimension of the hole is created by broaching, and the mating surface of the shaft is created in a semiautomatic turret lathe with a grinding attachment. The expected distributions of size are uniformly random when the parts are mixed. The modulus of elasticity is $\mathbf{E} \sim N(30, 0.90)$ Mpsi and the Poisson ratio is $\boldsymbol{\nu} \sim N(0.28, 0.014)$. Estimate the stochastic shrink-fit pressure $\mathbf{p}$.

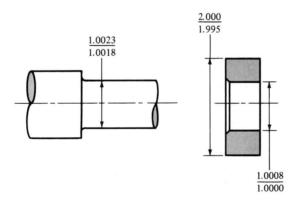

PROBLEM 4-27
Dimensions in inches.

4-28   A guide pin is required to align the assembly of a two-part fixture. The nominal size of the pin is 15 mm. Make the dimensional decisions for a 15-mm basic size locational clearance fit.

4-29   An interference fit of a cast-iron hub of a gear on a steel shaft is required. Make the dimensional decisions for a 45-mm basic size medium drive fit.

4-30   A pin is required for forming a linkage pivot. Find the dimensions required for a 50-mm basic size pin and clevis with a sliding fit.

4-31   A journal bearing and bushing need to be described. The nominal size is 1 in. What dimensions are needed for a 1-in basic size with a close running fit if this is a lightly loaded journal and bushing assembly?

4-32   A circular cross-section O ring has the dimensions shown in the figure. In particular, a No. 240 O ring has an inside diameter D_i and a cross-section diameter W of

$$D_i = 3.734 \pm 0.028 \text{ in} \qquad W = 0.139 \pm 0.004 \text{ in}$$

Using the absolute tolerance system, estimate the mean outside diameter $\overline{D}_o$ and its bilateral tolerance.

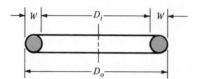

PROBLEM 4-32

4-33   A No. 370 O ring has the dimensions

$$D_i = 208.92 \pm 1.30 \text{ mm} \qquad W = 5.33 \pm 0.13 \text{ mm}$$

Using the absolute tolerance method, estimate the mean outside diameter $\overline{D}_o$ and its bilateral tolerance.

4-34   Estimate the mean outside diameter $\overline{D}_o$ and its bilateral tolerance of the No. 240 O ring of Prob. 4-32 if W is independent of D_i and the statistical tolerancing method is used.

4-35   Find the outside diameter $\overline{D}_o$ and its bilateral tolerance for the No. 370 O ring of Prob. 4-33 if W is independent of D_i and the statistical tolerancing method is used.

4-36   The gland for an O ring under internal pressure is shown in the figure. For a No. 240 O ring, the recommended gland dimensions are

$$G = 0.185 \pm 0.005 \text{ in} \qquad F = 0.106 \pm 0.003 \text{ in}$$

One manufacturer of O rings recommends the gland outer diameter

$$Y_{max} = D_o \qquad Y_{min} = \max [0.99\overline{D}_o, \overline{D}_o - 0.060] \text{ in}$$

The outside diameter of a No. 240 O ring is 4.012 ± 0.036 in and the cross-section diameter is 0.139 ± 0.004 in.

(a) When the end plate is bolted into position, all O rings are compressed. What is the minimum compression (squeeze) of a No. 240 O ring in inches?

(b) When the O ring is placed in the gland prior to the assembly of the end plate, is the ring free or under compression?

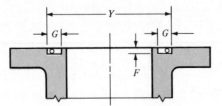

PROBLEMS 4-36 and 4-37

4-37 The gland for an O ring under internal pressure is shown in the figure. For a No. 370 O ring, the recommended gland dimensions are

$$F = 4.32 \pm 0.13 \text{ mm} \qquad G = 7.24 \pm 0.13 \text{ mm}$$

One manufacturer of O rings recommends that the gland's largest diameter be

$$Y_{max} = D_o \qquad Y_{min} = \max (0.99D_o, D_o - 1.52) \text{ mm}$$

The outside diameter of a No. 370 O ring is 219.58 ± 0.34 mm, and the cross-sectional diameter is 5.33 ± 0.13 mm.

(a) When the end plate is bolted in position, the O ring is compressed. What is the minimum compression (squeeze) of a No. 370 O ring in millimeters?

(b) When an O ring is placed in the gland prior to assembly of the end plate, is the ring free or under compression?

4-38 An annealed 70Cu–30Zn brass was reduced in section by cold drawing. This reduction of cross section is called *cold work* and is measured by the formula $W = (A_0 - A_r)/A_0$ (see Chap. 5). The newly worked material was tested for ultimate tensile strength, with the following confirmed results:

W	0	0.1	0.2	0.3	0.4
S'_u	43.0	48.0	53.0	60.0	70.9

where A_0 is the original cross-sectional area, A_r is the reduced value of the area after cold working, W is the fractional reduction in area, and S'_u is the tensile strength of the cold-worked material in kpsi. Note that there is a gain in strength due to cold working. Given the original strength S_u (at zero cold working) of the annealed material as 43.0 kpsi, and calling S_u/S'_u the y variable and the associated fractional cold work W the x variable, perform a regression analysis using the equation $y = a + bx$. How do the results compare with the generally used equation $S_u/S'_u = 1 - W$?

4-39 For a polyethylene material, the true stress–true strain data are

σ	800	1000	1500	2000	2500	3000	3500
ε	0.10	0.15	0.34	0.47	0.60	0.70	0.84

where σ is the true stress in kpsi and ε is the true strain. The experimenter suspects that a relationship of the form $\varepsilon = a + b \ln \sigma$ will rectify the data. Find the regression constants a and b and express σ as a function of ε.

4-40 Two flanges of a bolted joint compress a soft gasket in a controlled manner so that the "squeeze" is between 0.020 in and 0.040 in. Too little squeeze results in leakage and too much in gasket "flow." As shown in the figure, a shouldered cap screw is used to control the gasket compression, and the critical factor in sustaining the function of the joint is the length a of the cap screw. If the dimensions of the members are as shown and gasket production thickness is 0.120 ± 0.005 in, find the length $\bar{a}$ and the tolerance t_a using the absolute tolerance system and the statistical tolerance system.

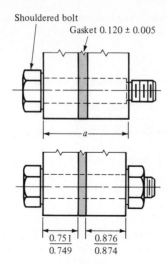

Shouldered bolt

Gasket 0.120 ± 0.005

PROBLEM 4-40
Dimensions in inches.

0.751 / 0.749   0.876 / 0.874

4-41   For Prob. 4-40, conduct a computer simulation with the dimensions

$$a = 1.715 \pm 0.003 \qquad b = 0.750 \pm 0.001$$

$$c = 0.120 \pm 0.005 \qquad d = 0.875 \pm 0.001$$

all in inches, to check the compatibility of the above dimensions with a gasket squeeze in the range 0.020 to 0.040 in. To do this, simulate a gap w from the necessary relation $w = a - b - c - d$ where

$$\mathbf{a} \sim U(1.712, 1.718) \qquad \mathbf{b} \sim U(0.749, 0.751)$$

$$\mathbf{c} \sim U(0.115, 0.125) \qquad \mathbf{d} \sim U(0.874, 0.876)$$

and check the largest and smallest values of w in 1000 evaluations of w. Be sure to recognize the independence of a, b, c, and d by using a different random number for each dimensional simulation that forms a single instance of w. At the same time, histographic information can be assembled so that the general shape of the distribution of w may be seen. Would you expect it to be a uniform random distribution? Why or why not?

4-42*   The tensile 0.2 percent offset yield strength of AISI 1137 cold-drawn steel rounds up to 1 inch in diameter from 2 mills and 25 heats is reported histographically as follows:

S_y	93	95	97	99	101	103	105	107	109	111
n	19	25	38	17	12	10	5	4	4	2

where S_y is the class midpoint in kpsi and n is the number in each class. Presuming the distribution is normal, what is the yield strength exceeded by 99 percent of the population? How can you check on the goodness of fit of a normal distribution to the data?

4-43   Repeat Prob. 4-42, presuming the distribution is lognormal. What is the yield strength exceeded by 99 percent of the population? Check the goodness of fit by plotting the data on appropriate coordinates so as to rectify the ordered data. Compare the normal fit of Prob. 4-42 with the lognormal fit by superposing the PDFs and the histographic PDF.

4-44*   A shipment of 700 helical-coil springs has been received. A random sample of 21 was tested for spring rate k in pounds per inch, and the ordered results were:

7.58	7.60	7.71	7.77	7.77	7.78	7.79
7.85	7.90	7.92	7.96	7.98	7.99	7.99
8.07	8.08	8.14	8.18	8.20	8.30	8.31

Plot the CDF on coordinates that will rectify a normal and a lognormal distribution, that is, k versus z for normal, and $\ln k$ versus z for lognormal. Since the sample is small, use the CDF estimator Eq. (4-33) $F_i = (i - 0.3)/(n + 0.4)$, where i is the order number and n is the sample size. Draw the eyeball best-fit lines in each case, and recover the equation of the CDF. Choose the best fit and estimate the value of k that will be exceeded by 95 percent of the shipment.

4-45   A 5160H steel was tested in fatigue and the distribution of cycles to failure at constant stress level was found to be $\mathbf{n} \sim W(36.9, 133.6, 2.66)$ in 10^3 cycles. Plot the PDF of n and the PDF of the lognormal distribution having the same mean and standard deviation. What is the B10 life (see Prob. 4-16) predicted by both distributions?

4-46*   A material was tested at steady fully reversed loading to determine the number of cycles to failure using 100 specimens. The results were

$(10^{-5})L$	3.05	3.55	4.05	4.55	5.05	5.55	6.05	6.55	7.05	7.55	8.05	8.55	9.05	9.55	10.05
n	3	7	11	16	21	13	13	6	2	0	4	3	0	0	1

where L is the life in cycles and n is the number in each class. Convert to CDF, plot on coordinates that will rectify the ordered data string, draw the eyeball best-fit line, and find the lognormal parameters ($\hat{\mu}$ and $\hat{\sigma}$). Plot the PDF and histographic PDF for comparison.

4-47   The ultimate tensile strength of an AISI 1117 cold-drawn steel is Weibullian, with $\mathbf{S}_u \sim W(70.3, 84.4, 2.01)$. What are the mean, the standard deviation, and the coefficient of variation?

4-48   A 60-45-15 nodular iron has a 0.2 percent yield strength S_y with a mean of 49.0 kpsi, a standard deviation of 4.2 kpsi, and a guaranteed yield strength of 33.8 kpsi. What are the Weibull parameters θ and b?

4-49   A 35018 malleable iron has a 0.2 percent offset yield strength given by the Weibull distribution $\mathbf{S}_y \sim W(34.7, 39.0, 2.93)$ kpsi. What are the mean, the standard deviation, and the coefficient of variation? Is the distribution left- or right-skewed?

4-50   The results of testing many heats of AISI 1020 steel included 1000 separate determinations of the ultimate tensile strength. In Example 4-10 a Weibull fit was made. Using the data in columns 1 and 2 of Table 4-6, fit a normal and a lognormal distribution and plot the PDFs of all three distributions with the histographic PDF superposed for comparison. Which fit seems best?

4-51   The histographic results of a steady load test on a rolling-contact bearing are:

L	1	2	3	4	5	6	7	8	9	10	11	12
n	11	22	38	57	31	19	15	12	11	9	7	5

where L is the life in millions of revolutions and n is the number of failures. Fit a lognormal distribution to these data and plot the PDF with the histographic PDF superposed. From the lognormal distribution, estimate the life at which 10 percent of the bearings under this steady loading will have failed.

4-52 The histographic results of a steady load test on one type of rolling-contact bearing were given in Prob. 4-51. Fit a Weibull distribution to these data and plot the PDF with the histographic PDF superposed. Estimate the life at which 90 percent of the bearings under this steady loading will still be in operation.

4-53*† Six bolts are tightened to exert a mean clamping load of $\bar{F} = 6667$ lb with a standard deviation of 600 lb. Estimate the median values of the least and the greatest clamping loads.

ANSWERS

4-3 (*a*) $\mathbf{S}_y = (78.4, 6.57)$ kpsi
4-6 0.625/0.623 in
4-9 $\hat{\mu}_\epsilon = 0.000\ 750$, $\hat{\sigma}_\epsilon = 0.000\ 046$
4-12 $\boldsymbol{\sigma} = (19\ 099, 1725)$ psi
4-15 $\mathbf{p}_{max} = (229.1, 9.2)$ kpsi
4-19 (*a*) $\mathbf{d} = (0.7495, 0.000\ 86)$ in, (*b*) $\mathbf{d} = (0.749, 0.000\ 577)$ in
4-25 (*a*) $a = 1.051 \pm 0.001$ in, (*b*) $a = 1.051 \pm 0.004$ in
4-28 Hole, 15.018 mm max, 15.000 mm min; shaft, 15.000 mm max, 14.989 mm min
4-31 Hole, $D_{max} = 1.0013$ in, $D_{min} = 1.0000$ in; shaft, $d_{max} = 0.9992$ in, $d_{min} = 0.9984$ in
4-34 $D_o = 4.012 \pm 0.029$ in
4-37 (*a*) 0.75 mm, (*b*) more compressed than free
4-38 $S_u/S_u' = 0.994 - 0.967W$
4-41 In 1000 trials, $w = -0.0210$ max, -0.0391 min.
4-47 $\hat{\mu}_S = 82.8$ kpsi, $\hat{\sigma}_S = 6.62$ kpsi

---

†For help, see Charles R. Mischke, ''A New Approach for the Identification Locus for Estimating CDF-Failure Equations on Rectified Plots,'' *Trans. ASME, J. Vibration, Acoustics, Stress and Reliability in Design*, vol. 109, no. 1, January 1987, pp. 103–112.

PART TWO

FAILURE
PREVENTION

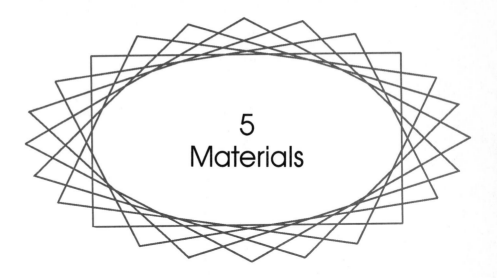

5
Materials

In Chaps. 2 and 3 methods of estimating the stresses in and the deflections of machine members were discussed. It was found that in some methods the elastic properties of the material are used. The stress existing in a machine member has no meaning unless the strength of the material is known; this strength is a property of the particular material in use.

The selection of a material for a machine part or a structural member is one of the decisions the designer is called upon to make. This decision is usually made before the dimensions of the part are determined. After choosing the material and process (the two cannot be divorced), the designer can then proportion the member so that the internal stresses and strains have reasonable and satisfactory values compared with the properties associated with failure of the material.

As important as stress and deflection are in the design of mechanical parts, the selection of a material is not always based upon these factors. Many parts have no loads on them whatever. Parts may be designed merely to fill up space. Members must frequently be designed to resist corrosion. Sometimes temperature effects are more important in design than stress and strain. So many other factors besides stress and strain may govern the design of parts that the designer must have the versatility that comes only with a broad background in materials and processes.

5-1 STATIC STRENGTH

The standard tensile test is used to obtain a variety of characteristics and strengths that are used in design. Figure 5-1 illustrates a typical tension-test specimen and some of the dimensions that are often employed. The original diameter d_0 and length of the gauge l_0, used to measure the strains, are recorded before the test is begun. The specimen is then mounted in the test machine and slowly loaded in tension while the load and strain are observed. At the conclusion of, or during, the test the results are plotted as a *stress-strain diagram* (Fig. 5-2).

FIGURE 5-1

A typical tension-test specimen. Some of the standard values used for d_0 are 2.5, 6.25, and 12.5 mm and 0.5 in, but other sections and sizes are in use. Common gauge lengths l_0 used are 10, 25, and 50 mm and 1 and 2 in.

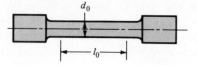

Point P in Fig. 5-2 is called the *proportional limit*. This is the point at which the curve first begins to deviate from a straight line. Point E is called the *elastic limit*. No permanent set will be observable in the specimen if the load is removed at this point. Between P and E the diagram is not a perfectly straight line, even though the specimen is elastic. Thus Hooke's law, which states that stress is proportional to strain, applies only up to the proportional limit.

During the tension test, many materials reach a point at which the strain begins to increase very rapidly without a corresponding increase in stress. This point is called the *yield point*. Not all materials have an obvious yield point. For this reason, *yield strength* S_y is often defined by an *offset method* as shown in Fig. 5-2. Such a yield strength corresponds to a definite or stated amount of permanent set, usually 0.2 or 0.5 percent of the original gauge length, although 0.01, 0.1, and 0.5 percent are sometimes used.

The *ultimate, or tensile, strength* S_u or S_{ut} corresponds to point U in Fig. 5-2 and is the maximum stress reached on the stress-strain diagram. Some materials exhibit a downward trend after the maximum stress is reached. These fracture at point F on the diagram in Fig. 5-2. Others, such as some of the cast irons and high-strength steels, fracture while the stress-strain locus is still rising.

To determine the strain relations for the stress-strain test, let

l_0 = original gauge length

l_i = gauge length corresponding to any load P_i

A_0 = original cross-sectional area

A_i = area of smallest cross section under load P_i

The unit strain from Chap. 2 is

$$\epsilon = \frac{l_i - l_0}{l_0} \tag{5-1}$$

FIGURE 5-2

Stress-strain diagram obtained from the standard tensile test of a ductile material. P marks the proportional limit; E, the elastic limit; Y, the offset yield strength as defined by offset strain OA; U, the maximum or ultimate strength; and F, the fracture strength.

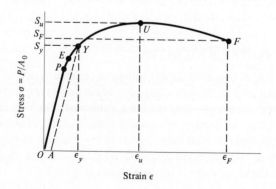

As noted in Sec. 1-8, *strength,* as used in this book, is a built-in property of a material, or of a mechanical element, because of the selection of a particular material or process or both. The strength of a connecting rod, for example, is the same no matter whether it is already an element in an operating machine or whether it is lying on a workbench awaiting assembly with other parts.

On the other hand, stress is something that occurs in a part, usually as a result of being assembled into a machine and loaded. However, stresses may be built into a part by the processing or handling. For example, shot peening produces a compressive *stress* in the outer surface of a part, and also improves the fatigue strength of the part.

In this book we wish to be very careful in distinguishing between *strength,* designated by *S,* and *stress,* designated by σ or τ.

In some books, the terms *proof stress* and *yield stress* are used. The term *proof stress* corresponds to point *Y* in Fig. 5-2 which, in this book, is called the *offset yield strength* S_y or, sometimes, simply the *yield strength.* In most standards the offset is usually specified; usual values are 0.2, 0.5, and 1.0 percent.

The term *yield stress* as used in some books is the stress corresponding to the yield point. Some materials have both an upper and a lower yield point.

The term *true stress* is used to indicate the result obtained when any load used in the tension test is divided by the *true* or *actual* cross-sectional area of the specimen. This means that the load and the cross-sectional area must be measured simultaneously during the test. If the specimen has necked, especial care must be taken to measure the area at the smallest part.

In plotting the true stress-strain diagram it is customary to use a term called *true strain* or, sometimes, *logarithmic strain.* True strain is the sum of the incremental elongations divided by the current length of the filament, or

$$\varepsilon = \int_{l_0}^{l_i} \frac{dl}{l} = \ln \frac{l_i}{l_0} \tag{5-2}$$

where l_0 is the original gauge length and l_i is the gauge length corresponding to load P_i.

The most important characteristic of a true stress-strain diagram (Fig. 5-3) is that the true stress increases all the way to fracture. Thus, as shown in Fig. 5-3, the true fracture stress σ_F is greater than the true ultimate stress σ_u. Contrast this with Fig. 5-2, where the fracture strength S_F is less than the ultimate strength S_u.

Bridgman has pointed out that the true stress-strain diagram of Fig. 5-3 should be corrected because of the triaxial stress state that exists in the neck of the specimen.[*] He observes that the tension is greatest on the axis and smallest on the periphery and that the stress state consists of an axial tension uniform all the way across the neck, plus a hydrostatic tension, which is zero on the periphery and increases to a maximum value on the axis.

Bridgman's correction for the true stress during necking is particularly significant. Designating σ_C as the computed true stress, σ_{ACT} as the corrected or actual stress, *R*

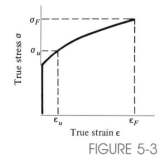

FIGURE 5-3

True stress-strain diagram plotted using Cartesian coordinates.

---

[*]P. W. Bridgman, "The Stress Distribution at the Neck of a Tension Specimen," *ASM,* vol. 32, 1944, p. 553.

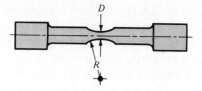

as the radius of the neck (Fig. 5-4), and D as the smallest neck diameter, the equation is

$$\sigma_{\text{ACT}} = \frac{\sigma_C}{\left(1 + \dfrac{4R}{D}\right)\left[\ln\left(1 + \dfrac{D}{4R}\right)\right]}$$ (5-3)

When necking occurs, the engineering strain given by Eq. (5-1) will not be the same at all points within the gauge length. A more satisfactory relation can be obtained by using areas. Since the volume of material remains the same during the test, $A_0 l_0 = A_i l_i$. Consequently, $l_i = l_0(A_0/A_i)$. Substituting this value of l_i in Eq. (5-1) and canceling terms gives

$$\epsilon = \frac{A_0 - A_i}{A_i}$$ (5-4)

But see also Eq. (5-9).

Compression tests are more difficult to make, and the geometry of the test specimens differs from the geometry of those used in tension tests. The reason for this is that the specimen may buckle during testing or it may be difficult to get the stresses distributed evenly. Other difficulties occur because ductile materials will bulge after yielding. However, the results can be plotted on a stress-strain diagram, too, and the same strength definitions can be applied. For many materials the compressive strengths are about the same as the tensile strengths. When substantial differences occur, however, as is the case with the cast irons, the tensile and compressive strengths should be stated separately.

Torsional strengths are found by twisting bars and recording the torque and the twist angle. The results are then plotted as a *torque-twist diagram*. By using the equations in Chap. 2 for torsional stress, both the elastic limit and the *torsional yield strength* S_{sy} may be found. The maximum point on a torque-twist diagram, corresponding to point U on Fig. 5-2, is T_u. The equation

$$S_{su} = \frac{T_u r}{J}$$ (a)

where r is the radius of the bar and J is the polar second moment of area, defines the *modulus of rupture* for the torsion test. Note that the use of Eq. (a) implies that Hooke's law applies to this case. This is not true, because the outermost area of the bar is in a plastic state at the torque T_u. For this reason the quantity S_{su} is called the modulus of rupture. It is incorrect to call S_{su} the ultimate torsional strength.

All of the stresses and strengths defined by the stress-strain diagram of Fig. 5-2 and similar diagrams are specifically known as *engineering stresses* and *strengths* or *nominal stresses* and *strengths*. These are the values normally used in all engineering design. The adjectives *engineering* and *nominal* are used here to emphasize that the stresses are computed using the *original* or *unstressed cross-sectional area* of the specimen. In this book we shall use these modifiers only when we specifically wish to call attention to this distinction.

Sometimes the question arises, Is the compressive strength S_c a negative quantity? The answer is no. All strengths are treated as absolute quantities.

5-2   PLASTIC DEFORMATION

The best current explanation of the relationships between stress and strain is by Datsko.* He describes the plastic region of the true stress–true strain diagram by the equation

$$\sigma = \sigma_0 \varepsilon^m \tag{5-5}$$

where σ = true stress

σ_0 = a strength coefficient, or strain-strengthening coefficient

ε = true plastic strain

m = strain-strengthening exponent

A graph of this equation is a straight line when plotted on log-log paper, as shown in Fig. 5-5. The graph contains three zones of interest: the elastic zone, on line AB, called type I behavior; the plastic zone on line Y_2C, defining type II behavior; and the intermediate zone.

According to Hooke's law, the equation of the elastic portion is

$$\sigma = E\varepsilon \tag{5-6}$$

---

*Joseph Datsko, ''Solid Materials,'' Chap. 7 in Joseph E. Shigley and Charles R. Mischke (eds.), *Standard Handbook of Machine Design*, McGraw-Hill, New York, 1986. See also Joseph Datsko, ''New Look at Material Strength,'' *Machine Design,* vol. 58, no. 3, Feb. 6, 1986, pp. 81–85.

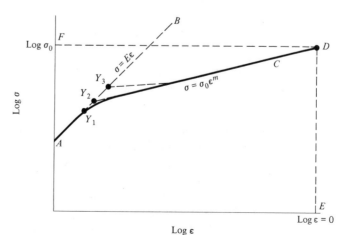

FIGURE 5-5

True stress-strain diagram plotted on log-log paper. Since the values of ε are less than unity, their logarithms are negative. At point E, $\varepsilon = 1$, $\log \varepsilon = 0$, and the ordinate through E locates D and defines the logarithm of the constant σ_0 at F.

where E is the modulus of elasticity. Taking the logarithm of both sides of Eq. (5-6) and recognizing the equation of a straight line, we get

$$\log \sigma = \log E + 1 \log \varepsilon$$

and from Eq. (5-5) we obtain

$$\log \sigma = \log \sigma_0 + m \log \varepsilon$$

From this we conclude that the elastic portion of the line is the same for all materials, having a slope of unity and passing through the point ($\sigma = E$, $\varepsilon = 1$). We also see that the elastic portion has an exponent $m = 1$ and an intercept $\sigma_0 = E$ relative to the true stress-strain equation [Eq. (5-5)].

The constant σ_0 in Eq. (5-5) is the true stress corresponding to a true strain of unity. This constant can be obtained by extending the plastic stress-strain line until it intersects an ordinate through $\varepsilon = 1$ ($\log \varepsilon = 0$). The height of this ordinate is $\log \sigma_0$, as measured parallel to the $\log \sigma$ axis of Fig. 5-5.

The shape of the elastic-plastic zone between the two straight lines varies from one material to another. The three possible yield points Y_1, Y_2, and Y_3 describe the various possibilities that might be observed. An extension of the plastic line would intersect the elastic line at Y_2 and describe an ideal material. Most engineering materials are said to *overyield* to Y_3, because they have a yield strength greater than the ideal value. The alloys of the steels, coppers, brasses, and nickels all have this characteristic. The point Y_1 describes what might be called *underyielding*. Only a few engineering materials have this characteristic, a fully annealed aluminum alloy being one of them.

The relationship between logarithmic strain and unit strain can be obtained by rearranging Eq. (5-1) to read

$$\epsilon = \frac{l_i - l_0}{l_0} = \frac{l_i}{l_0} - 1 \tag{5-7}$$

Then we see that

$$\frac{l_i}{l_0} = \epsilon + 1$$

and so, from Eq. (5-2), we get

$$\varepsilon = \ln (\epsilon + 1) \tag{5-8}$$

A similar relationship can be derived between the true stress and the engineering stress. For convenience in this derivation, let $\sigma =$ true stress and $S =$ engineering stress. Then

$$\sigma = \frac{F_i}{A_i} \quad \text{and} \quad S = \frac{F_i}{A_0}$$

Now, since

$$A_i = A_0 \frac{l_0}{l_i}$$

then

$$\sigma = \frac{F_i}{A_0(l_0/l_i)} = S\frac{l_i}{l_0} = S(\epsilon + 1)$$

But $\epsilon + 1 = \ln^{-1}\varepsilon = e^\varepsilon$. Therefore

$$\sigma = Se^\varepsilon$$

Thus the engineering stress is

$$S = \sigma e^{-\varepsilon} \tag{5-10}$$

or, with Eq. (5-5),

$$S = \sigma_0 \varepsilon^m e^{-\varepsilon} \tag{5-11}$$

Equations (5-7) and (5-8) are used to plot the elastic portion of the diagram. Thus, for this portion of the experiment, the data are acquired in the conventional manner, using an extensometer to obtain the elongation. The extensometer is not used for the plastic portion of the true stress-strain diagram, because the average strain is no longer useful. Its use would also expose an expensive precision instrument to damage when the specimen fractures.

The approach for the plastic region consists in measuring the area of the specimen, being particularly careful to obtain this value at the smallest cross section between the gauge points. Sometimes it is necessary to measure "diameters" of the specimen in two directions, perpendicular to each other, in case the cross section becomes oval-shaped.

From Eq. (5-4), we have

$$\frac{A_0}{A_i} = \epsilon + 1$$

Thus, from Eq. (5-8), we find the logarithmic strain in terms of areas to be

$$\varepsilon = \ln \frac{A_0}{A_i} \tag{5-12}$$

If data are to be corrected in the necking region, then the necking radius R should also be measured and the stress corrected using the Bridgman equation [Eq. (5-3)].

The exponent m represents the slope of the plastic line, as we have seen. This slope is easily obtained after the plastic line has been drawn through the points in the plastic region of the diagram. Another, and easier, method of obtaining the exponent is possible for materials having an ultimate strength greater than the nominal stress at fracture. For these materials the exponent is the same as the logarithmic strain corresponding to the ultimate strength. The proof is as follows:

$$P_i = \sigma A_i = \sigma_0 A_i(\varepsilon)^m \tag{a}$$

where we have used Eq. (5-5). Now, from Eq. (5-12), we have

$$A_i = \frac{A_0}{e^\varepsilon} \tag{b}$$

and so Eq. (*a*) becomes

$$P_i = \frac{\sigma_0 A_0 (\varepsilon)^m}{e^\varepsilon} \qquad\qquad (c)$$

Now the maximum point on the load-deformation diagram, or nominal stress-strain diagram, at least for some materials, is coincident with a zero slope. Thus, for these materials, the derivative of the load with respect to the logarithmic strain must be zero. Thus

$$\frac{dP_i}{d\varepsilon} = \sigma_0 A_0 \frac{d}{d\varepsilon}\,(\varepsilon^m e^{-\varepsilon}) = \sigma_0 A_0 (m\varepsilon^{m-1} e^{-\varepsilon} - \varepsilon^m e^{-\varepsilon}) = 0 \qquad\qquad (d)$$

Solving gives $\varepsilon = m$; but this corresponds to the ultimate load, and so

$$m = \varepsilon_u \qquad\qquad (5\text{-}13)$$

Note again that this relation is valid only if the load-deformation diagram has a point of zero slope.

EXAMPLE 5-1   The first three columns in Table 5-1 list the results obtained from a tensile test of annealed A-40 titanium.

(*a*) Plot the engineering and the true stress-strain diagrams.

(*b*) Find the modulus of elasticity, the yield strength, and the ultimate strength.

(*c*) Find the plastic strain-strengthening coefficient and exponent.

Solution   (*a*) The engineering stress-strain diagram (Fig. 5-6) is plotted from the data in columns 5 and 6 of Table 5-1. The first 11 values in column 6 are obtained from Eq. (5-1). The remaining values in column 6 are found from Eq. (5-4). Note that the strain for the last entry in column 6 does not really mean that the test specimen has elongated

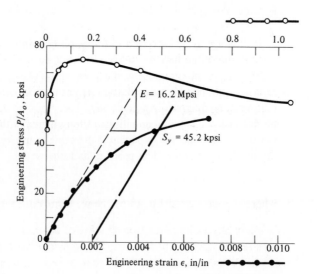

FIGURE 5-6

Engineering stress-strain diagram from test data on annealed A-40 titanium as listed in Table 5-1. Note that two strain scales are used in order to plot all data points.

TABLE 5-1

Results of a Tensile Test of Annealed A-40 Titanium as Reported by Datsko.* (Specimen size is $d_0 = 0.505$ in, $l_0 = 2$ in)

| | | | | ANALYTICAL RESULTS | | | |
| | OBSERVED TEST RESULTS | | | ENGINEERING | | TRUE | |
LOAD P, kip (1)	GAUGE LENGTH l, in (2)	DIAMETER d, in (3)	AREA A, in² (4)	STRESS P/A_0, kpsi (5)	STRAIN ϵ, in/in (6)	STRESS P/A_i, kpsi (7)	STRAIN ε, in/in (8)
0	2.0000	0.505	0.2003	0	0	0	0
1.00	2.0006			5.0	0.000 30	5.0	0.000 30
2.00	2.0012			10.0	0.000 60	10.0	0.000 60
3.00	2.0018			15.0	0.000 90	15.0	0.000 90
4.00	2.0024			20.0	0.001 20	20.0	0.001 20
5.00	2.0035			25.0	0.001 75	25.0	0.001 75
6.00	2.0044			30.0	0.002 20	30.0	0.002 20
7.00	2.0057			34.9	0.002 85	34.9	0.002 85
8.00	2.0070			39.9	0.003 50	39.9	0.003 49
9.00	2.0094	0.504	0.1995	44.9	0.004 70	44.9	0.004 69
10.00	2.0140			49.9	0.007 00	49.9	0.006 98
12.00		0.501	0.1971	59.9	0.016 23	60.9	0.016 10
14.00		0.493	0.1909	69.9	0.049 24	73.3	0.048 07
14.50		0.486	0.1855	72.4	0.079 78	78.2	0.076 76
14.95	2.310	0.470	0.1735	74.6	0.154 47	86.2	0.143 64
14.50		0.442	0.1534	72.4	0.305 74	94.5	0.266 77
14.00		0.425	0.1419	69.9	0.411 56	98.7	0.344 70
11.50	2.480	0.352	0.0973	57.4	1.058 58	118.2	0.722 02

*Data source: Joseph Datsko, *Materials in Design and Manufacturing*, Dept. of Mech. Eng. and Appl. Mech., University of Michigan, Ann Arbor, 1977, Chap. 5.

1.058 58 in/in, because necking has occurred and the result is based on a change of areas. As shown in the table, the final length is actually 2.48 in.

(b) Columns 7 and 8 of Table 5-1 are the computed values of the true stress and strain (Fig. 5-7). The true strain is obtained using Eq. (5-8).

The ultimate strength is $S_{ut} = 74.6$ kpsi; it is the value of the engineering stress in column 5 corresponding to the ultimate load $P = 14.95$ kip.

(c) Figure 5-7 shows that the plastic strain-strengthening coefficient σ_0 is found at the intersection of the true stress-strain line and the ordinate corresponding to $\varepsilon = 1$.

The strain-strengthening exponent is found to be $m = 0.144$, from Eq. (5-13), and corresponds to the ultimate load $P = 14.95$ kip.

5-3 STRENGTH AND COLD WORK

Cold working is the process of stressing or deforming a material in the plastic region of the stress-strain diagram without the deliberate application of heat. Materials can be deformed plastically by the application of heat, as in blacksmithing or hot rolling, but

FIGURE 5-7

True stress-strain diagram plotted from test data on annealed A-40 titanium as listed in Table 5-1. The elastic modulus E is the slope of the dashed line. The strain-strengthening exponent $m = 0.144$ is the slope of the linear portion of the diagram in the plastic region. Note that the logarithmic scale for the ordinate differs from that for the abscissa. These scales were selected in this manner to create a more accurate diagram.

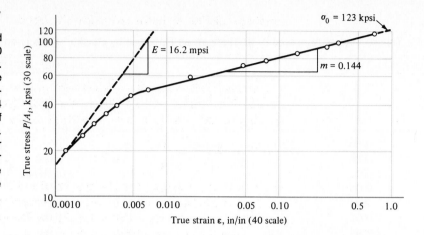

the resulting mechanical properties are quite different from those obtained by cold working. The purpose of this section is to explain what happens to the significant mechanical properties of a material when that material is cold-worked.

Consider the stress-strain diagram of Fig. 5-8a. Here a material has been stressed beyond the yield strength at Y to some point I, in the plastic region, and then the load removed. At this point the material has a permanent plastic deformation ϵ_p. If the load corresponding to point I is now reapplied, the material will be elastically deformed by the amount ϵ_e. Thus at point I the total unit strain consists of the two components ϵ_p and ϵ_e and is given by the equation

$$\epsilon = \epsilon_p + \epsilon_e \qquad (a)$$

This material can be unloaded and reloaded any number of times from and to point I, and it is found that the action always occurs along the straight line which is approxi-

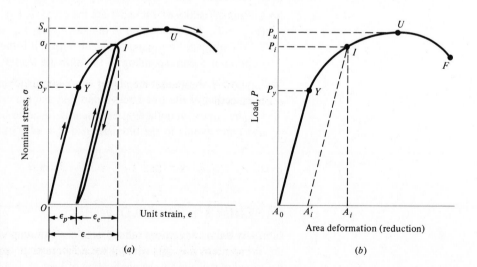

FIGURE 5-8

(a) Stress-strain diagram showing unloading and reloading at point I in the plastic region; (b) analogous load-deformation diagram.

mately parallel to the initial elastic line OY. Thus

$$\epsilon_e = \frac{\sigma_i}{E} \qquad (b)$$

It is possible to construct a similar diagram, as in Fig. 5-8b, where the abscissa is the area deformation and the ordinate is the applied load. The *reduction in area* corresponding to the load P_f, at fracture, is defined as

$$R = \frac{A_0 - A_f}{A_0} = 1 - \frac{A_f}{A_0} \qquad (5\text{-}14)$$

where A_0 is the original area. The quantity R in Eq. (5-14) is usually expressed in percent and tabulated in lists of mechanical properties as a measure of *ductility*. See Appendix Table A-20, for example. Ductility is an important property because it measures the ability of a material to absorb overloads and to be cold-worked. Thus such operations as bending, drawing, heading, and stretch forming are metal-processing operations that require ductile materials.

Figure 5-8b can also be used to define the quantity of cold work. Thus the *cold-work factor* W is

$$W = \frac{A_0 - A_i'}{A_0} \approx \frac{A_0 - A_i}{A_0} \qquad (5\text{-}15)$$

where A_i' corresponds to the area after the load P_i has been released. The approximation in Eq. (5-15) results because of the difficulty of measuring the small diametral changes in the elastic region. If the amount of cold-work is known, then Eq. (5-15) can be solved for the area A_i'. The result is

$$A_i' = A_0(1 - W) \qquad (5\text{-}16)$$

Cold working a material produces a new set of values for the strengths, as can be seen from stress-strain diagrams. If point I is to the left of point U, that is, if $P_i < P_u$, then the new yield strength is

$$S_y' = \frac{P_i}{A_i'} = \sigma_0 \epsilon_i^m \qquad P_i \le P_u \qquad (5\text{-}17)$$

Because of the reduced area, that is, because $A_i' < A_0$, the ultimate strength also changes, and is

$$S_u' = \frac{P_u}{A_i'} \qquad (c)$$

Since $P_u = S_u A_0$, we find, with Eq. (5-13), that

$$S_u' = \frac{S_u A_0}{A_0(1 - W)} = \frac{S_u}{1 - W} \qquad \epsilon_i \le \epsilon_u \qquad (5\text{-}18)$$

which is valid only when point I is to the left of point U.

For points to the right of U, the yield strength is approaching the ultimate strength, and, with small loss in accuracy,

$$S'_u \approx S'_y \approx \sigma_0 \varepsilon_i^m \qquad \varepsilon_i \leq \varepsilon_u \tag{5-19}$$

A little thought will reveal that a bar will have the same ultimate load in tension after being strain-strengthened in tension as it had before. The new strength is of interest to us not because the static ultimate load increases, but—since fatigue strengths are correlated with the local ultimate strengths—because the fatigue strength improves.

EXAMPLE 5-2

An annealed AISI 1018 steel (see Table A-22) has $S_y = 32.0$ kpsi, $S_u = 49.5$ kpsi, $\sigma_f = 91.1$ kpsi, $\sigma_0 = 90$ kpsi, $m = 0.25$, and $\varepsilon_f = 1.05$ in/in. Find the new values of the strengths if the material is given 15 percent cold work.

Solution

From Eq. (5-13), we find the true strain corresponding to the ultimate strength to be

$$\varepsilon_u = m = 0.25$$

The ratio A_0/A_i is, from Eq. (5-15),

$$\frac{A_0}{A_i} = \frac{1}{1 - W} = \frac{1}{1 - 0.15} = 1.176$$

The true strain corresponding to 15 percent cold work is obtained from Eq. (5-12). Thus

$$\varepsilon_i = \ln \frac{A_0}{A_i} = \ln 1.176 = 0.1625$$

Since $\varepsilon_i < \varepsilon_u$, Eqs. (5-17) and (5-18) apply. Therefore,

Answer

$$S'_y = \sigma_0 \varepsilon_i^m = 90(0.1625)^{0.25} = 57.1 \text{ kpsi}$$

Answer

$$S'_u = \frac{S_u}{1 - W} = \frac{49.5}{1 - 0.15} = 58.2 \text{ kpsi}$$

5-4   HARDNESS

The resistance of a material to penetration by a pointed tool is called *hardness*. Though there are many hardness-measuring systems, we shall consider here only the two in greatest use.

    Rockwell hardness measurements are quickly and easily made, they have good reproducibility, and the test machine for them is easy to use. In fact, the hardness number is read directly from a dial. Rockwell hardness scales are designated as A, B, C, . . . , etc. The indenting tools are numbered either 1, 2, or 3. And the load applied is either 60, 100, or 150 kg. Thus the Rockwell B scale, designated R_B, uses a 100-kg load and a No. 2 indenter, which is a $\frac{1}{16}$-in-diameter ball. The Rockwell C scale R_C uses a diamond cone, which is the No. 1 indenter, and a load of 150 kg. Hardness numbers so obtained are relative. Thus a hardness $R_C = 50$ has meaning only in relation to another hardness number using the same scale.

The *Brinell hardness* (note that *Brinell* is spelled with a single letter n and rhymes with the word *bell*) is another test in very general use. In testing, the indenting tool through which force is applied is a ball and the hardness number H_B is found as a number equal to the applied load divided by the spherical surface area of the indentation. Thus the units of H_B are the same as those of stress, though they are seldom used. Brinell hardness testing takes more time, since H_B must be computed from the test data. The primary advantage of the method (both are nondestructive in most cases) is that the Brinell hardness is directly related to the ultimate strength of the material tested. This means that the strength of parts could, if desired, be tested part by part during manufacture.

For steels, the relationship between the minimum ultimate strength and the Brinell hardness number is found to be

$$S_u = \begin{cases} 0.45H_B & \text{kpsi} \\ 3.10H_B & \text{MPa} \end{cases} \tag{5-20}$$

which may also be called the ASTM minimum (see Sec. 1-8).

Similar relationships for cast iron can be derived from data supplied by Krause.[*] Data from 72 tests of gray iron produced by one foundry and poured in two sizes of test bars are reported in graph form. The minimum strength, as defined by the ASTM, is found from these data to be

$$S_u = \begin{cases} 0.23H_B - 12.5 & \text{kpsi} \\ 1.58H_B - 86 & \text{MPa} \end{cases} \tag{5-21}$$

Walton[†] shows a chart from which the SAE minimum strength can be obtained. The result is

$$S_u = 0.2375H_B - 16 \quad \text{kpsi} \tag{5-22}$$

which is even more conservative than the values obtained from Eq. (5-21).

EXAMPLE 5-3   It is necessary to ensure that a certain part supplied by a foundry always meets or exceeds ASTM grade 20 specifications for cast iron (see Table A-24). What hardness should be specified?

Solution   From Eq. (5-21), we have

Answer $$H_B = \frac{S_u + 12.5}{0.23} = \frac{20 + 12.5}{0.23} = 141$$

---

[*]D. E. Krause, "Gray Iron—A Unique Engineering Material," ASTM Special Publication 455, 1969, pp. 3–29, as reported in Charles F. Walton (ed.), *Iron Castings Handbook,* Iron Founders Society, Inc., Cleveland, 1971, pp. 204, 205.

[†]Op. cit.

where S_u is the minimum strength. If the foundry can control the hardness within 20 points, routinely, then specify $145 < H_B < 165$. This imposes no hardship on the foundry and assures the designer that ASTM grade 20 will always be supplied at a predictable cost.

Stochastic Results*

If the hardness is a random variable, then the resulting tensile strength must be treated as a random variable too. The tensile-strength variate corresponding to Eq. (5-20) and based on the same data is

$$\mathbf{S_u} = \begin{cases} (0.5,\ 0.022)\mathbf{H}_B & \text{kpsi} \\ (3.45,\ 0.152)\mathbf{H}_B & \text{MPa} \end{cases} \tag{5-23}$$

for steels.

For cast irons, the relation corresponding to Eq. (5-21) is

$$\mathbf{S_u} = \begin{cases} 0.23\mathbf{H}_B - 9 + (0,\ 1.5) & \text{kpsi} \\ 1.58\mathbf{H}_B - 62 + (0,\ 10) & \text{MPa} \end{cases} \tag{5-24}$$

EXAMPLE 5-4   Brinell hardness tests were made on a random sample of five steel parts during processing. The results were H_B values of 248, 253, 247, 244, and 246.

(a) Estimate the mean and the standard deviation of the ultimate strength in SI units.

(b) The ASTM minimum is established at a level that 99 percent of the population can meet or exceed. On the basis of this definition, what minimum ultimate strength corresponds to this sample testing?

Solution   (a) Using the equations of the normal distribution, we first obtain the mean and standard deviation of the hardness as

$$\overline{H}_B = 247.6 \text{ MPa} \qquad \hat{\sigma}_{HB} = 3.36 \text{ MPa}$$

The ultimate strength is the product of two stochastic quantities. Thus, from Eq. (5-23), we have

$$\mathbf{S}_u = (3.45,\ 0.152)\mathbf{H}_B = (3.45,\ 0.152)(247.6,\ 3.36) \text{ MPa}$$

From Table 4-4 we find the mean of the product of two stochastic quantities to be $\mu_{xy} = \mu_x\mu_y$. Therefore the mean value of $\mathbf{S}_u$ is

Answer $$\bar{S}_u = 3.45(247.6) = 854 \text{ MPa}$$

The coefficient of variation is found from the same table, and is

$$C_{Su} = (C_x^2 + C_y^2)^{1/2} = \left[\left(\frac{0.152}{3.45} \right)^2 + \left(\frac{3.36}{247.6} \right)^2 \right]^{1/2} = 0.0461$$

---

*See Sec. 1-10; headings printed in black are used to indicate that the material deals with statistical methods.

Therefore

Answer $\hat{\sigma}_{Su} = \bar{S}_u C_{Su} = 854(0.0461) = 39.4$ MPa

(b) For $\Phi(z) = 0.01$, we find $z = -2.326$. Thus the minimum ultimate strength (deterministic) is

Answer $S_u = \bar{S}_u + \hat{\sigma}_{Su} z = 854 + 39.4(-2.326) = 762$ MPa

5-5   IMPACT PROPERTIES

An external force applied to a structure or part is called an *impact load* if the time of application is less than one-third the lowest natural period of vibration of the part or structure. Otherwise it is called simply a *static load*.

The *Charpy* and *Izod notched-bar tests* utilize bars of specified geometries to determine brittleness and impact strength. These tests are helpful in comparing several materials and in the determination of low-temperature brittleness. In both tests the specimen is struck by a pendulum released from a fixed height, and the energy absorbed by the specimen, called the *impact value,* is computed from the height of swing after fracture.

The effect of temperature on impact values is shown in Fig. 5-9. Notice the narrow region of critical temperatures where the impact value increases very rapidly. In the low-temperature region the fracture appears as a brittle, shattering type, whereas the appearance is a tough, tearing type above the critical-temperature region. The critical temperature seems to be dependent on both the material and the geometry of the notch. For this reason designers should not rely too heavily on the results of notched-bar tests.

The average strain rate used in obtaining the stress-strain diagram is about 0.001 in/(in · s) or less. When the strain rate is increased, as it is under impact conditions, the strengths increase, as shown in Fig. 5-10. In fact, at very high strain rates the yield strength seems to approach the ultimate strength as a limit. But note that the curves show little change in the elongation. This means that the ductility remains about the same. Also, in view of the sharp increase in yield strength, a mild steel could be expected to behave elastically throughout practically its entire strength range under impact conditions.

The Charpy and Izod tests really provide toughness data under dynamic, rather than

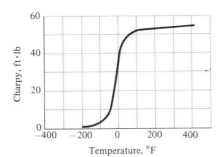

FIGURE 5-9

The effect of temperature on impact values.

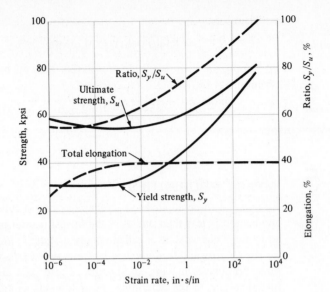

FIGURE 5-10

Influence of strain rate on tensile properties.

static, conditions. It may well be that impact data obtained from these tests are as dependent on the notch geometry as they are on the strain rate. For these reasons it may be better to use the concepts of notch sensitivity, fracture toughness, and fracture mechanics, discussed later in this chapter, to assess the possibility of cracking or fracture.

5-6   TEMPERATURE EFFECTS

Strength and ductility, or brittleness, are properties affected by the temperature of the operating environment.

The effect of temperature on the static properties of steels is typified by the strength versus temperature chart of Fig. 5-11. Note that the tensile strength changes only a

FIGURE 5-11

A plot of the results of 145 tests of 21 carbon and alloy steels showing the effect of operating temperature on the yield strength S_y and the ultimate strength S_{ut}. The ordinate is the ratio of the strength at the operating temperature to the strength at room temperature. The standard deviations were $0.0442 \leq \hat{\sigma} \leq 0.152$ for S_y and $0.099 \leq \hat{\sigma} \leq 0.110$ for S_{ut}. [Data source: E. A. Brandes (ed.), Smithells Metals Reference Book, 6th ed., Butterworth, London, 1983, pp. 22-128 to 22-131.]

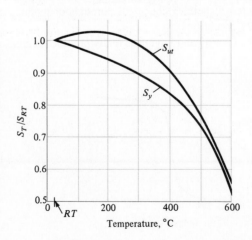

small amount until a certain temperature is reached. At that point it falls off rapidly. The yield strength, however, decreases continuously as the environmental temperature is increased. There is a substantial increase in ductility, as might be expected, at the higher temperatures.

Many tests have been made of ferrous metals subjected to constant loads for long periods of time at elevated temperatures. The specimens were found to be permanently deformed during the tests, even though at times the actual stresses were less than the yield strength of the material obtained from short-time tests made at the same temperature. This continuous deformation under load is called *creep*.

One of the most useful tests to have been devised is the long-time creep test under constant load. Figure 5-12 illustrates a curve which is typical of this kind of test. The curve is obtained at a constant stated temperature. A number of tests are usually run simultaneously at different stress intensities. The curve exhibits three distinct regions. In the first stage are included both the elastic and the plastic deformation. This stage shows a decreasing creep rate which is due to the strain hardening. The second stage shows a constant minimum creep rate caused by the annealing effect. In the third stage the specimen shows a considerable reduction in area, the true stress is increased, and a higher creep eventually leads to fracture.

When the operating temperatures are lower than the transition temperature (Fig. 5-9), the possibility arises that a part could fail by a brittle fracture. This subject will be discussed later in this chapter.

Of course, heat treatment, as will be shown, is used to make substantial changes in the mechanical properties of a material.

Heating due to electric and gas welding also changes the mechanical properties. Such changes may be due to clamping during the welding process, as well as heating; the resulting stresses then remain frozen in when the parts have cooled and the clamps have been removed. Hardness tests can be used to learn whether the strength has been changed by welding, but such tests will not reveal the presence of residual stresses.

5-7 NUMBERING SYSTEMS

The Society of Automotive Engineers (SAE) was the first to recognize the need, and to adopt a system, for the numbering of steels. Later the American Iron and Steel Institute (AISI) adopted a similar system. In 1975 the SAE published the Unified Numbering

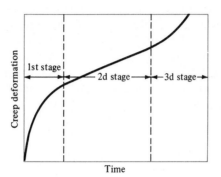

FIGURE 5-12

Creep-time curve.

System for Metals and Alloys (UNS); this system also contains cross-reference numbers for other material specifications.* The UNS uses a letter prefix to designate the material, as, for example, G for the carbon and alloy steels, A for the aluminum alloys, C for the copper-base alloys, and S for the stainless or corrosion-resistant steels. For some materials, not enough agreement has as yet developed in the industry to warrant the establishment of a designation.

For the steels, the first two numbers following the letter prefix indicate the composition, excluding the carbon content. The various compositions used are as follows:

G10	Plain carbon	G46	Nickel-molybdenum
G11	Free-cutting carbon steel with more sulfur or phosphorus	G48	Nickel-molybdenum
		G50	Chromium
G13	Manganese	G51	Chromium
G23	Nickel	G52	Chromium
G25	Nickel	G61	Chromium-vanadium
G31	Nickel-chromium	G86	Chromium-nickel-molybdenum
G33	Nickel-chromium	G87	Chromium-nickel-molybdenum
G40	Molybdenum	G92	Manganese-silicon
G41	Chromium-molybdenum	G94	Nickel-chromium-molybdenum
G43	Nickel-chromium-molybdenum		

The second number pair refers to the approximate carbon content. Thus, G10400 is a plain carbon steel with a carbon content of 0.37 to 0.44 percent. The fifth number following the prefix is used for special situations. For example, the old designation AISI 52100 represents a chromium alloy with about 100 points of carbon. The UNS designation is G52986.

The UNS designations for the stainless steels, prefix S, utilize the older AISI designations for the first three numbers following the prefix. The next two numbers are reserved for special purposes. The first number of the group indicates the approximate composition. Thus 2 is a chromium-nickel-manganese steel, 3 is a chromium-nickel steel, and 4 is a chromium alloy steel. Sometimes stainless steels are referred to by their alloy content. Thus S30200 is often called an 18-8 stainless steel, meaning 18 percent chromium and 8 percent nickel.

The prefix for the aluminum group is the letter A. The first number following the prefix indicates the processing. For example, A9 is a wrought aluminum, while A0 is a casting alloy. The second number designates the main alloy group as shown in Table 5-2. The third number in the group is used to modify the original alloy or to designate the impurity limits. The last two numbers refer to other alloys used with the basic group.

The American Society for Testing and Materials (ASTM) numbering system for cast iron is in widespread use. This system is based on the tensile strength. Thus ASTM No. 30 cast iron has a minimum tensile strength of 30 kpsi. Note from the Appendix, however, that the *typical* tensile strength is 31 kpsi. You should be careful to designate

*Many of the materials discussed in the balance of this chapter are listed in the Appendix tables. Be sure to review these.

TABLE 5-2 Aluminum Alloy Designations		
Aluminum 99.00% pure and greater	Ax1xxx
Copper alloys	Ax2xxx
Manganese alloys	Ax3xxx
Silicon alloys	Ax4xxx
Magnesium alloys	Ax5xxx
Magnesium-silicon alloys	Ax6xxx
Zinc alloys	Ax7xxx

which of the two values is used in design and problem work because of the significance of factor of safety.

5-8   SAND CASTING

Sand casting is a basic low-cost process, and it lends itself to economical production in large quantities with practically no limit to the size, shape, or complexity of the part produced.

In sand casting, the casting is made by pouring molten metal into sand molds. A pattern, constructed of metal or wood, is used to form the cavity into which the molten metal is poured. Recesses or holes in the casting are produced by sand cores introduced into the mold. The designer should make an effort to visualize the pattern and casting in the mold. In this way the problems of core setting, pattern removal, draft, and solidification can be studied.

Steel castings are the most difficult of all to produce, because steel has the highest melting temperature of all materials normally used for casting. This high temperature aggravates all casting problems.

The following rules will be found quite useful in the design of any sand casting:

1 All sections should be designed with a uniform thickness.

2 The casting should be designed so as to produce a gradual change from section to section where this is necessary.

3 Adjoining sections should be designed with generous fillets or radii.

4 A complicated part should be designed as two or more simple castings to be assembled by fasteners or by welding.

Steel, gray iron, brass, bronze, and aluminum are most often used in castings. The minimum wall thickness for any of these materials is about 5 mm; though with particular care, thinner sections can be obtained with some materials.

5-9   SHELL MOLDING

The shell-molding process employs a heated metal pattern, usually made of cast iron, aluminum, or brass, which is placed in a shell-molding machine containing a mixture of dry sand and thermosetting resin. The hot pattern melts the plastic, which, together with the sand, forms a shell about 5 to 10 mm thick around the pattern. The shell is then baked at from 400 to 700°F for a short time while still on the pattern. It is then stripped from the pattern and placed in storage for use in casting.

In the next step the shells are assembled by clamping, bolting, or pasting; they are placed in a backup material, such as steel shot; and the molten metal is poured into the cavity. The thin shell permits the heat to be conducted away so that solidification takes place rapidly. As solidification takes place, the plastic bond is burned and the mold collapses. The permeability of the backup material allows the gases to escape and the casting to air-cool. All this aids in obtaining a fine-grain, stress-free casting.

Shell-mold castings feature a smooth surface, a draft that is quite small, and close tolerances. In general, the rules governing sand casting also apply to shell-mold casting.

5-10   INVESTMENT CASTING

Investment casting uses a pattern which may be made from wax, plastic, frozen mercury, or other material. After the mold is made, the pattern is melted out. Thus a mechanized method of casting a great many patterns is necessary. The mold material is dependent upon the melting point of the cast metal. Thus a plaster mold can be used for some materials while others would require a ceramic mold. After the pattern is melted out, the mold is baked; when baking is finished, the molten metal may be poured in and allowed to cool.

If a number of castings are to be made, then metal or permanent molds may be suitable. Such molds have the advantage that the surfaces are smooth, bright, and accurate, so that little, if any, machining is required. *Metal-mold castings* are also known as *die castings* and *centrifugal castings*.

5-11   POWDER-METALLURGY PROCESS

The powder-metallurgy process is a quantity-production process that uses powders from a single metal, several metals, or a mixture of metals and nonmetals. It consists essentially of mechanically mixing the powders, compacting them in dies at high pressures, and heating the compacted part at a temperature less than the melting point of the major ingredient. The particles are united into a single strong part similar to what would be obtained by melting the same ingredients together. The advantages are (1) the elimination of scrap or waste material, (2) the elimination of machining operations, (3) the low unit cost when mass-produced, and (4) the exact control of composition. Some of the disadvantages are (1) the high cost of dies, (2) the lower physical properties, (3) the higher cost of materials, (4) the limitations on the design, and (5) the limited range of materials which can be used. Parts commonly made by this process are oil-impregnated bearings, incandescent lamp filaments, cemented-carbide tips for tools, and permanent magnets.

5-12   HOT-WORKING PROCESSES

By *hot working* are meant such processes as rolling, forging, hot extrusion, and hot pressing, in which the metal is heated sufficiently to make it plastic and easily worked.

Hot rolling is usually used to create a bar of material of a particular shape and dimensions. Figure 5-13 shows some of the various shapes which are commonly produced by the hot-rolling process. All of them are available in many different sizes as well as in different materials. The materials most available in the hot-rolled bar sizes are steel, aluminum, magnesium, and copper alloys.

Tubing may be manufactured by hot-rolling strip or plate. The edges of the strip are rolled together, creating seams which are either butt-welded or lap-welded. Seamless tubing is manufactured by roll-piercing a solid heated rod with a piercing mandrel.

Extrusion is the process by which great pressure is applied to a heated metal billet or blank, causing it to flow through a restricted orifice. This process is necessarily restricted to materials of low melting point, such as aluminum, copper, magnesium, lead, tin, and zinc.

Forging is the hot working of metal by hammers, presses, or forging machines. In common with other hot-working processes, forging produces a refined grain structure which results in increased strength and ductility. Compared with castings, forgings have greater strength for the same weight. In addition, drop forgings can be made smoother and more accurate than sand castings, so that less machining is necessary. However, the initial cost of the forging dies is usually greater than the cost of patterns for castings, although the greater unit strength rather than the cost is usually the deciding factor between these two processes.

5-13   COLD-WORKING PROCESSES

By *cold working* is meant the forming of the metal while at a low temperature (usually room temperature). In contrast to parts produced by hot working, cold-worked parts have a bright new finish, are more accurate, and require less machining.

Cold-finished bars and shafts are produced by rolling, drawing, turning, grinding,

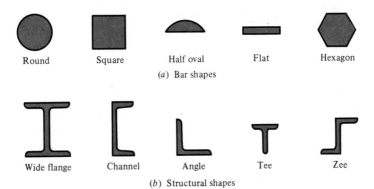

Round   Square   Half oval   Flat   Hexagon

(*a*) Bar shapes

FIGURE 5-13

Common shapes available through hot rolling.

Wide flange   Channel   Angle   Tee   Zee

(*b*) Structural shapes

and polishing. Of these methods, by far the largest percentage of products are made by the cold-rolling and cold-drawing processes. Cold rolling is now used mostly for the production of wide flats and sheets. Practically all cold-finished bars are made by cold drawing but even so are sometimes mistakenly called "cold-rolled bars." In the drawing process, the hot-rolled bars are first cleaned of scale and then drawn by pulling them through a die which reduces the size about $\frac{1}{32}$ to $\frac{1}{16}$ in. This process does not remove material from the bar but reduces, or "draws" down, the size. Many different shapes of hot-rolled bars may be used for cold-drawing.

Cold rolling and cold drawing have the same effect upon the mechanical properties. The cold-working process does not change the grain size but merely distorts it. Cold working results in a large increase in yield strength, an increase in ultimate strength and hardness, and a decrease in ductility. In Fig. 5-14 the properties of a cold-drawn bar are compared with those of a hot-rolled bar of the same material.

Heading is a cold-working process in which the metal is gathered, or upset. This operation is commonly used to make screw and rivet heads and is capable of producing a wide variety of shapes. *Roll threading* is the process of rolling threads by squeezing and rolling a blank between two serrated dies. *Spinning* is the operation of working sheet material around a rotating form into a circular shape. *Stamping* is the term used to describe punch-press operations such as *blanking, coining, forming,* and *shallow drawing.*

5-14   THE HEAT TREATMENT OF STEEL

Heat treatment refers to processes which interrupt or vary the transformation process described by the equilibrium diagram. Other mechanical or chemical operations are sometimes grouped under the heading of heat treatment. The common heat-treating operations are annealing, quenching, tempering, and case hardening.

Annealing   When a material is cold- or hot-worked, residual stresses are built in, and, in addition, the material usually has a higher hardness due to these working operations. These operations change the structure of the material so that it is no longer

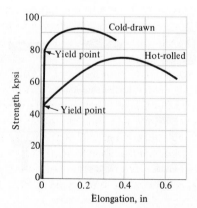

FIGURE 5-14

Stress-strain diagram for hot-rolled and cold-drawn AISI 1035 steel.

represented by the equilibrium diagram. Annealing is a heating operation which permits the material to transform according to the equilibrium diagram. The material to be annealed is heated to a temperature which is approximately 100°F above the critical temperature. It is held at this temperature for a time which is sufficient for the carbon to become dissolved and diffused through the material. The object being treated is then allowed to cool slowly, usually in the furnace in which it was treated. If the transformation is complete, then it is said to have a full anneal. Annealing is used to soften a material and make it more ductile, to relieve residual stresses, and to refine the grain structure.

The term *annealing* includes the process called *normalizing*. Parts to be normalized may be heated to a slightly higher temperature than in full annealing. This produces a coarser grain structure, which is more easily machined if the material is a low-carbon steel. In the normalizing process the part is cooled in still air at room temperature. Since this cooling is more rapid than the slow cooling used in full annealing, less time is available for equilibrium, and the material is harder than fully annealed steel. Normalizing is often used as the final treating operation for steel. The cooling in still air amounts to a slow quench.

Quenching   Eutectoid steel which is fully annealed consists entirely of pearlite, which is obtained from austenite under conditions of equilibrium. A fully annealed hypoeutectoid steel would consist of pearlite plus ferrite, while hypereutectoid steel in the fully annealed condition would consist of pearlite plus cementite. The hardness of steel of a given carbon content depends upon the structure that replaces the pearlite when full annealing is not carried out.

The absence of full annealing indicates a more rapid rate of cooling. The rate of cooling is the factor which determines the hardness. A controlled cooling rate is called *quenching*. A mild quench is obtained by cooling in still air, which, as we have seen, is obtained by the normalizing process. The two most widely used media for quenching are water and oil. The oil quench is quite slow but prevents quenching cracks caused by rapid expansion of the object being treated. Quenching in water is used for carbon steels and for medium-carbon, low-alloy steels.

The effectiveness of quenching depends upon the fact that when austenite is cooled it does not transform into pearlite instantaneously but requires time to initiate and complete the process. Since the transformation ceases at about 800°F, it can be prevented by rapidly cooling the material to a lower temperature. When the material is cooled rapidly to 400°F or less, the austenite is transformed into a structure called *martensite*. Martensite is a supersaturated solid solution of carbon in ferrite and is the hardest and strongest form of steel.

If steel is rapidly cooled to a temperature between 400 and 800°F and held there for a sufficient length of time, the austenite is transformed into a material which is generally called *bainite*. Bainite is a structure which is intermediate between pearlite and martensite. Although there are several structures which can be identified between the temperatures given, depending upon the temperature used, they are collectively known as bainite. By the choice of this transformation temperature, almost any variation of structure may be obtained. These range all the way from coarse pearlite to fine martensite.

Tempering   When a steel specimen has been fully hardened it is very hard and brittle and has high residual stresses. The steel is unstable and tends to contract on aging. This tendency is increased when the specimen is subjected to externally applied loads, because the resultant stresses contribute still more to the instability. These internal stresses can be relieved by a modest heating process called *stress relieving,* or a combination of stress relieving and softening called *tempering* or *drawing.* After the specimen has been fully hardened by being quenched from above the critical temperature, it is reheated to some temperature below the critical temperature for a certain period of time and then allowed to cool in still air. The temperature to which it is reheated depends upon the composition and the degree of hardness or toughness desired.* This reheating operation releases the carbon held in the martensite, forming carbide crystals. The structure obtained is called *tempered martensite.*

The effect of heat-treating operations upon the various mechanical properties of steel is shown graphically in Fig. 5-15.

Case Hardening   The purpose of case hardening is to produce a hard outer surface on a specimen of low-carbon steel while at the same time retaining the ductility and toughness in the core. This is done by increasing the carbon content at the surface.

---

*For the quantitative aspects of tempering in plain carbon and low-alloy steels, see Charles R. Mischke, "The Strength of Cold-Worked and Heat-Treated Steels," chap. 8 in Joseph E. Shigley and Charles R. Mischke (eds.), *Standard Handbook of Machine Design,* McGraw-Hill, New York, 1986.

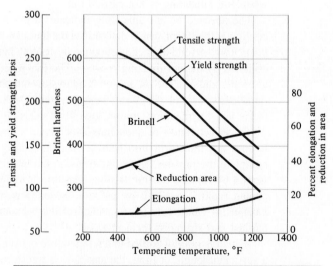

FIGURE 5-15

The effect of thermal-mechanical history on the mechanical properties of AISI 4340 steel. *(Prepared by the International Nickel Company.)*

Condition	Tensile strength, kpsi	Yield strength, kpsi	Reduction in area, %	Elongation in 2 in, %	Brinell hardness, Bhn
Normalized..	200	147	20	10	410
As rolled......	190	144	18	9	380
Annealed.....	120	99	43	18	228

Either solid, liquid, or gaseous carburizing materials may be used. The process consists of introducing the part to be carburized into the carburizing material for a stated time and at a stated temperature, depending upon the depth of case desired and the composition of the part. The part may then be quenched directly from the carburization temperature and tempered, or in some cases it must undergo a double heat treatment in order to ensure that both the core and the case are in proper condition. Some of the more useful case-hardening processes are pack carburizing, gas carburizing, nitriding, cyaniding, induction hardening, and flame hardening.

5-15 ALLOY STEELS

Although a plain carbon steel is an alloy of iron and carbon with small amounts of manganese, silicon, sulfur, and phosphorus, the term *alloy steel* is applied when one or more elements other than carbon are introduced in sufficient quantities to modify its properties substantially. The alloy steels not only possess more desirable physical properties but also permit a greater latitude in the heat-treating process.

Chromium   The addition of chromium results in the formation of various carbides of chromium which are very hard, yet the resulting steel is more ductile than a steel of the same hardness produced by a simple increase in carbon content. Chromium also refines the grain structure so that these two combined effects result in both increased toughness and increased hardness. The addition of chromium increases the critical range of temperatures and moves the eutectoid point to the left. Chromium is thus a very useful alloying element.

Nickel   The addition of nickel to steel also causes the eutectoid point to move to the left and increases the critical range of temperatures. Nickel is soluble in ferrite and does not form carbides or oxides. This increases the strength without decreasing the ductility. Case hardening of nickel steels results in a better core than can be obtained with plain carbon steels. Chromium is frequently used in combination with nickel to obtain the toughness and ductility provided by the nickel and the wear resistance and hardness contributed by the chromium.

Manganese   Manganese is added to all steels as a deoxidizing and desulfurizing agent, but if the sulfur content is low and the manganese content is over 1 percent, the steel is classified as a manganese alloy. Manganese dissolves in the ferrite and also forms carbides. It causes the eutectoid point to move to the left and lowers the critical range of temperatures. It increases the time required for transformation so that oil quenching becomes practicable.

Silicon   Silicon is added to all steels as a deoxidizing agent. When added to very-low-carbon steels, it produces a brittle material with a low hysteresis loss and a high magnetic permeability. The principal use of silicon is with other alloying elements, such as manganese, chromium, and vanadium, to stabilize the carbides.

Molybdenum   While molybdenum is used alone in a few steels, it finds its greatest use when combined with other alloying elements, such as nickel, chromium, or both. Molybdenum forms carbides and also dissolves in ferrite to some extent, so that it adds both hardness and toughness. Molybdenum increases the critical range of temperatures and substantially lowers the transformation point. Because of this lowering of the transformation point, molybdenum is most effective in producing desirable oil-hardening and air-hardening properties. Except for carbon, it has the greatest hardening effect, and because it also contributes to a fine grain size, this results in the retention of a great deal of toughness.

Vanadium   Vanadium has a very strong tendency to form carbides; hence it is used only in small amounts. It is a strong deoxidizing agent and promotes a fine grain size. Since some vanadium is dissolved in the ferrite, it also toughens the steel. Vanadium gives a wide hardening range to steel, and the alloy can be hardened from a higher temperature. It is very difficult to soften vanadium steel by tempering; hence it is widely used in tool steels.

Tungsten   Tungsten is widely used in tool steels because the tool will maintain its hardness even at red heat. Tungsten produces a fine, dense structure and adds both toughness and hardness. Its effect is similar to that of molybdenum, except that it must be added in greater quantities.

5-16   CORROSION-RESISTANT STEELS

Iron-base alloys containing at least 12 percent chromium are called *stainless steels*. The most important characteristic of these steels is their resistance to many, but not all, corrosive conditions. The four types available are the ferritic chromium steels, the austenitic chromium-nickel steels, and the martensitic and precipitation-hardenable stainless steels.

The ferritic chromium steels have a chromium content ranging from 12 to 27 percent. Their corrosion resistance is a function of the chromium content, so that alloys containing less than 12 percent still exhibit some corrosion resistance, although they may rust. The quench-hardenability of these steels is a function of both the chromium and the carbon content. The very-high-carbon steels have good quench-hardenability up to about 18 percent chromium, while in the lower carbon ranges it ceases at about 13 percent. If a little nickel is added, these steels retain some degree of hardenability up to 20 percent chromium. If the chromium content exceeds 18 percent, they become difficult to weld, and at the very high chromium levels the hardness becomes so great that very careful attention must be paid to the service conditions. Since chromium is expensive, the designer will choose the lowest chromium content consistent with the corrosive conditions.

The chromium-nickel stainless steels retain the austenitic structure at room temperature; hence they are not amenable to heat treatment. The strength of these steels can be greatly improved by cold working. They are not magnetic unless cold-worked. Their work-hardenability properties also cause them to be difficult to machine. All the chromium-nickel steels may be welded. They have greater corrosion-resistant properties than the plain chromium steels. When more chromium is added for greater corrosion resistance, more nickel must also be added if the austenitic properties are to be retained.

5-17   CASTING MATERIALS

Gray Cast Iron   Of all the cast materials, gray cast iron is the most widely used. This is because it has a very low cost, is easily cast in large quantities, and is easy to machine. The principal objections to the use of gray cast iron are that it is brittle and that it is weak in tension. In addition to a high carbon content (over 1.7 percent and usually greater than 2 percent), cast iron also has a high silicon content, with low percentages of sulfur, manganese, and phosphorus. The resultant alloy is composed of pearlite, ferrite, and graphite, and under certain conditions the pearlite may decompose into graphite and ferrite. The resulting product then contains all ferrite and graphite. The graphite, in the form of thin flakes distributed evenly throughout the structure, darkens it; hence, the name *gray cast iron*.

Gray cast iron is not readily welded, because it may crack, but this tendency may be reduced if the part is carefully preheated. Although the castings are generally used in the as-cast condition, a mild anneal reduces cooling stresses and improves the machinability. The tensile strength of gray cast iron varies from 100 to 400 MPa (15 to 60 kpsi), and the compressive strengths are 3 to 4 times the tensile strengths. The modulus of elasticity varies widely, with values extending all the way from 75 to 150 GPa (11 to 22 Mpsi).

White Cast Iron   If all the carbon in cast iron is in the form of cementite and pearlite, with no graphite present, the resulting structure is white and is known as *white cast iron*. This may be produced in two ways. The composition may be adjusted by keeping the carbon and silicon content low, or the gray-cast-iron composition may be cast against chills in order to promote rapid cooling. By either method a casting with large amounts of cementite is produced, and as a result the product is very brittle and hard to machine but also very resistant to wear. A chill is usually used in the production of gray-iron castings in order to provide a very hard surface within a particular area of the casting, while at the same time retaining the more desirable gray structure within the remaining portion. This produces a relatively tough casting with a wear-resistant area.

Malleable Cast Iron   If white cast iron within a certain composition range is annealed, a product called *malleable cast iron* is formed. The annealing process frees the carbon so that it is present as graphite, just as in gray cast iron but in a different form. In gray cast iron the graphite is present in a thin flake form, while in malleable cast iron it has a nodular form and is known as *temper carbon*. A good grade of malleable cast iron may have a tensile strength of over 350 MPa (50 kpsi), with an elongation of as much as 18 percent. The percentage elongation of a gray cast iron, on the other hand, is seldom over 1 percent. Because of the time required for annealing (up to 6 days for large and heavy castings), malleable iron is necessarily somewhat more expensive than gray cast iron.

Ductile and Nodular Cast Iron Because of the lengthy heat treatment required to produce malleable cast iron, a cast iron has long been desired which would combine the ductile properties of malleable iron with the ease of casting and machining of gray iron and at the same time would possess these properties in the as-cast conditions. A process for producing such a material using cesium with magnesium seems to fulfill these requirements.

Ductile cast iron, or *nodular cast iron,* as it is sometimes called, is essentially the same as malleable cast iron, because both contain graphite in the form of spheroids. However, ductile cast iron in the as-cast condition exhibits properties very close to those of malleable iron, and if a simple 1-h anneal is given and is followed by a slow cool, it exhibits even more ductility than the malleable product. Ductile iron is made by adding magnesium to the melt; since magnesium boils at this temperature, it is necessary to alloy it with other elements before it is introduced.

Ductile iron has a high modulus of elasticity (172 GPa or 25 Mpsi) as compared with gray cast iron, and it is elastic in the sense that a portion of the stress-strain curve is a straight line. Gray cast iron, on the other hand, does not obey Hooke's law, because the modulus of elasticity steadily decreases with increase in stress. Like gray cast iron, however, nodular iron has a compressive strength which is higher than the tensile strength, although the difference is not as great. Since this is a new product, its full range of application has not as yet developed, but an obvious area of application is for those castings requiring shock and impact resistance.

Alloy Cast Irons Nickel, chromium, and molybdenum are the most common alloying elements used in cast iron. Nickel is a general-purpose alloying element, usually added in amounts up to 5 percent. Nickel increases the strength and density, improves the wearing qualities, and raises the machinability. If the nickel content is raised to 10 to 18 percent, an austenitic structure with valuable heat- and corrosion-resistant properties results. Chromium increases the hardness and wear resistance and, when used with a chill, increases the tendency to form white iron. When chromium and nickel are both added, the hardness and strength are improved without a reduction in the machinability rating. Molybdenum added in quantities up to 1.25 percent increases the stiffness, hardness, tensile strength, and impact resistance. It is a widely used alloying element.

Cast Steels The advantage of the casting process is that parts having complex shapes can be manufactured at costs less than fabrication by other means, such as welding. Thus the choice of steel castings is logical when the part is complex and when it must also have a high strength. The higher melting temperatures for steels do aggravate the casting problems and require closer attention to such details as core design, section thicknesses, fillets, and the progress of cooling. The same alloying elements used for the wrought steels can be used for cast steels to improve the strength and other mechanical properties. Cast-steel parts can also be heat-treated to alter the mechanical properties, and, unlike the cast irons, they can be welded.

5-18 NONFERROUS METALS

Aluminum The outstanding characteristics of aluminum and its alloys are their strength-weight ratio, their resistance to corrosion, and their high thermal and electrical conductivity. The density of aluminum is about 2770 kg/m^3 (0.10 lb/in^3), compared with 7750 kg/m^3 (0.28 lb/in^3) for steel. Pure aluminum has a tensile strength of about 90 MPa (13 kpsi), but this can be improved considerably by cold working and also by alloying with other materials. The modulus of elasticity of aluminum, as well as of its alloys, is 71 GPa (10.3 Mpsi), which means that it has about one-third the stiffness of steel.

Considering the cost and strength of aluminum and its alloys, they are among the most versatile materials from the standpoint of fabrication. Aluminum can be processed by sand casting, die casting, hot or cold working, or extruding. Its alloys can be machined, press-worked, soldered, brazed, or welded. Aluminum melts at 660°C (1215°F), which makes it very desirable for the production of either permanent or sand-mold castings. It is commercially available in the form of plate, bar, sheet, foil, rod, and tube and in structural and extruded shapes. Certain precautions must be taken in joining aluminum by soldering, brazing, or welding; these joining methods are not recommended for all alloys.

The corrosion resistance of the aluminum alloys depends upon the formation of a thin oxide coating. This film forms spontaneously because aluminum is inherently very reactive. Constant erosion or abrasion removes this film and allows corrosion to take place. An extra-heavy oxide film may be produced by the process called *anodizing*. In this process the specimen is made to become the anode in an electrolyte, which may be chromic acid, oxalic acid, or sulfuric acid. It is possible in this process to control the color of the resulting film very accurately.

The most useful alloying elements for aluminum are copper, silicon, manganese, magnesium, and iron. Aluminum alloys are classified as *casting alloys* or *wrought alloys*. The casting alloys have greater percentages of alloying elements to facilitate casting, but this makes cold working difficult. Many of the casting alloys, and some of the wrought alloys, cannot be hardened by heat treatment. The alloys that are heat-treatable use an alloying element which dissolves in the aluminum. The heat treatment consists of heating the specimen to a temperature which permits the alloying element to pass into solution, then quenching so rapidly that the alloying element is not precipitated. The aging process may be accelerated by heating slightly, which results in even greater hardness and strength. One of the better-known heat-treatable alloys is duraluminum, or 2017 (4 percent Cu, 0.5 percent Mg, 0.5 percent Mn). This alloy hardens in 4 days at room temperature. Because of this rapid aging, the alloy must be stored under refrigeration after quenching and before forming, or it must be formed immediately after quenching. Other alloys (such as 5053) have been developed which age-harden much more slowly, so that only mild refrigeration is required before forming. After forming, they are artificially aged in a furnace and possess approximately the same strength and hardness as the 2024 alloys. Those alloys of aluminum which cannot be heat-treated can be hardened only by cold working. Both work hardening and the hardening produced by heat treatment may be removed by an annealing process.

Magnesium   The density of magnesium is about 1800 kg/m^3 (0.065 lb/in^3), which is two-thirds that of aluminum and one-fourth that of steel. Since it is the lightest of all commercial metals, its greatest use is in the aircraft industry, but uses are now being found for it in other applications. Although the magnesium alloys do not have great strength, because of their light weight the strength-weight ratio compares favorably with the stronger aluminum and steel alloys. Even so, magnesium alloys find their greatest use in applications where strength is not an important consideration. Magnesium will not withstand elevated temperatures; the yield point is definitely reduced when the temperature is raised to that of boiling water.

Magnesium and its alloys have a modulus of elasticity of 45 GPa (6.5 Mpsi) in tension and in compression, although some alloys are not as strong in compression as in tension. Curiously enough, cold working reduces the modulus of elasticity.

Copper-base Alloys When copper is alloyed with zinc, it is usually called *brass*. If it is alloyed with another element, it is often called *bronze*. Sometimes the other element is specified too, as, for example, *tin bronze* or *phosphor bronze*. There are hundreds of variations in each category.

Brass with 5 to 15 Percent Zinc The low-zinc brasses are easy to cold work, especially those with the higher zinc content. They are ductile but often hard to machine. The corrosion resistance is good. Alloys included in this group are *gilding brass* (5 percent Zn), *commercial bronze* (10 percent Zn), and *red brass* (15 percent Zn). Gilding brass is used mostly for jewelry and articles to be gold-plated; it has the same ductility as copper but greater strength, accompanied by poor machining characteristics. Commercial bronze is used for jewelry and for forgings and stampings, because of its ductility. Its machining properties are poor, but it has excellent cold-working properties. Red brass has good corrosion resistance as well as high-temperature strength. Because of this it is used a great deal in the form of tubing or piping to carry hot water in such applications as radiators or condensers.

Brass with 20 to 36 Percent Zinc Included in the intermediate-zinc group are *low brass* (20 percent Zn), *cartridge brass* (30 percent Zn), and *yellow brass* (35 percent Zn). Since zinc is cheaper than copper, these alloys cost less than those with more copper and less zinc. They also have better machinability and slightly greater strength; this is offset, however, by poor corrosion resistance and the possibility of season cracking at points of residual stresses. Low brass is very similar to red brass and is used for articles requiring deep-drawing operations. Of the copper-zinc alloys, cartridge brass has the best combination of ductility and strength. Cartridge cases were originally manufactured entirely by cold working; the process consisted in a series of deep draws, each draw being followed by an anneal to place the material in condition for the next draw; hence, the name cartridge brass. Although the hot-working ability of yellow brass is poor, it can be used in practically any other fabricating process and is therefore employed in a large variety of products.

When small amounts of lead are added to the brasses, their machinability is greatly improved and there is some improvement in their abilities to be hot-worked. The addition of lead impairs both the cold-working and welding properties. In this group are *low-leaded brass* ($32\frac{1}{2}$ percent Zn, $\frac{1}{2}$ percent Pb), *high-leaded brass* (34 percent Zn, 2 percent Pb), and *free-cutting brass* ($35\frac{1}{2}$ percent Zn, 3 percent Pb). The low-leaded brass is not only easy to machine but has good cold-working properties. It is used for various screw-machine parts. High-leaded brass, sometimes called *engraver's brass,* is used for instrument, lock, and watch parts. Free-cutting brass is also used for screw-machine parts and has good corrosion resistance with excellent mechanical properties.

Admiralty metal (28 percent Zn) contains 1 percent tin, which imparts excellent corrosion resistance, especially to saltwater. It has good strength and ductility but only fair machining and working characteristics. Because of its corrosion resistance it is used in power-plant and chemical equipment. *Aluminum brass* (22 percent Zn) contains 2 percent aluminum and is used for the same purposes as admiralty metal, because it has nearly the same properties and characteristics. In the form of tubing or piping, it is favored over admiralty metal, because it has better resistance to erosion caused by high-velocity water.

Brass with 36 to 40 Percent Zinc   Brasses with more than 38 percent zinc are less ductile than cartridge brass and cannot be cold-worked as severely. They are frequently hot-worked and extruded. *Muntz metal* (40 percent Zn) is low in cost and mildly corrosion-resistant. *Naval brass* has the same composition as Muntz metal except for the addition of 0.75 percent tin, which contributes to the corrosion resistance.

Bronze   *Silicon bronze,* containing 3 percent silicon and 1 percent manganese in addition to the copper, has mechanical properties equal to those of mild steel, as well as good corrosion resistance. It can be hot- or cold-worked, machined, or welded. It is useful wherever corrosion resistance combined with strength is required.

Phosphor bronze, made with up to 11 percent tin and containing small amounts of phosphorus, is especially resistant to fatigue and corrosion. It has a high tensile strength and a high capacity to absorb energy, and it is also resistant to wear. These properties make it very useful as a spring material.

Aluminum bronze is a heat-treatable alloy containing up to 12 percent aluminum. This alloy has strength and corrosion-resistance properties which are better than brass, and in addition, its properties may be varied over a wide range by cold working, heat treating, or changing the composition. When iron is added in amounts up to 4 percent, the alloy has a high endurance limit, a high shock resistance, and excellent wear resistance.

Beryllium bronze is another heat-treatable alloy, containing about 2 percent beryllium. This alloy is very corrosion-resistant and has high strength, hardness, and resistance to wear. Although it is expensive, it is used for springs and other parts subjected to fatigue loading where corrosion resistance is required.

5-19   PLASTICS

The term *thermoplastics* is used to mean any plastic that flows or is moldable when heat is applied to it; the term is sometimes applied to plastics moldable under pressure. Such plastics can be remolded when heated.

A *thermoset* is a plastic for which the polymerization process is finished in a hot molding press where the plastic is liquefied under pressure. Thermoset plastics cannot be remolded.

Table 5-3 lists some of the most widely used thermoplastics, together with some of their characteristics and the range of their properties. Table 5-4, listing some of the thermosets, is similar. These tables are presented for information only and should not be used to make a final design decision. The range of properties and characteristics that

TABLE 5-3

The Thermoplastics

NAME	S_u, kpsi	E, Mpsi	HARDNESS ROCKWELL	ELONGA-TION, %	DIMEN-SIONAL STABILITY	HEAT RESISTANCE	CHEMICAL RESISTANCE	PROCESSING
ABS group	2–8	0.10–0.37	60–110R	3–50	Good	*	Fair	EMST
Acetal group	8–10	0.41–0.52	80–94M	40–60	Excellent	Good	High	M
Acrylic	5–10	0.20–0.47	92–100M	3–75	High	*	Fair	EMS
Fluoroplastic group	0.50–7	···	50–80D	100–300	High	Excellent	Excellent	MPR†
Nylon	8–14	0.18–0.45	112–120R	10–200	Poor	Poor	Good	CEM
Phenylene oxide	7–18	0.35–0.92	115R, 106L	5–60	Excellent	Good	Fair	EFM
Polycarbonate	8–16	0.34–0.86	62–91M	10–125	Excellent	Excellent	Fair	EMS
Polyester	8–18	0.28–1.6	65–90M	1–300	Excellent	Poor	Excellent	CLMR
Polyimide	6–50	···	88–120M	Very low	Excellent	Excellent	Excellent†	CLMP
Polyphenylene sulfide	14–19	0.11	122R	1.0	Good	Excellent	Excellent	M
Polystyrene group	1.5–12	0.14–0.60	10–90M	0.5–60	···	Poor	Poor	EM
Polysulfone	10	0.36	120R	50–100	Excellent	Excellent	Excellent†	EFM
Polyvinyl chloride	1.5–7.5	0.35–0.60	65–85D	40–450	···	Poor	Poor	EFM

*Heat-resistant grades available.

†With exceptions.

C	Coatings	L	Laminates
E	Extrusions	M	Moldings
F	Foams	P	Press and sinter methods

R Resins
S Sheet
T Tubing

Source: These data have been obtained from the *Machine Design Materials Reference Issue,* published by Penton/IPC, Cleveland. These reference issues are published about every 2 years and constitute an excellent source of data on a great variety of materials.

TABLE 5-4

The Thermosets

NAME	S_u, MPa	E, MPa	HARDNESS ROCKWELL	ELONGA-TION, %	DIMEN-SIONAL STABILITY	HEAT RESISTANCE	CHEMICAL RESISTANCE	PROCESSING
Alkyd	3–9	0.05–0.30	99M*	···	Excellent	Good	Fair	M
Allylic	4–10	···	105–120M	···	Excellent	Excellent	Excellent	CM
Amino group	5–8	0.13–0.24	110–120M	0.30–0.90	Good	Excellent*	Excellent*	LR
Epoxy	5–20	0.03–0.30*	80–120M	1–10	Excellent	Excellent	Excellent	CMR
Phenolics	5–9	0.10–0.25	70–95E	···	Excellent	Excellent	Good	EMR
Silicones	5–6	···	80–90M	···	···	Excellent	Excellent	CLMR

*With exceptions.

C	Coatings	L	Laminates
E	Extrusions	M	Moldings
F	Foams	P	Press and sinter methods

R Resins
S Sheet
T Tubing

Source: These data have been obtained from the *Machine Design Materials Reference Issue,* published by Penton/IPC, Cleveland. These reference issues are published about every 2 years and constitute an excellent source of data on a great variety of materials.

can be obtained with plastics is very great. The influence of many factors, such as cost, moldability, coefficient of friction, weathering, impact strength, and the effect of fillers and reinforcements, must be considered. Manufacturers' catalogs will be found quite helpful in making possible selections.

5-20   NOTCH SENSITIVITY

In Sec. 2-14 it was pointed out that the existence of irregularities or discontinuities, such as holes, grooves, or notches, in a part increases the theoretical stresses significantly in the immediate vicinity of the discontinuity. And Eq. (2-48) defined a stress concentration factor K_t which is used with the nominal stress to obtain the maximum resulting stress due to the irregularity or defect. It turns out that some materials are not fully sensitive to the presence of notches and hence, for these, a reduced value of K_t can be used. For these materials, the maximum stress is, in fact,

$$\sigma_{\max} = K_f \sigma_0 \tag{5-25}$$

where K_f is a reduced value of K_t and σ_0 is the nominal stress. The factor K_f is commonly called a *fatigue stress-concentration factor*, and hence the subscript f; but we shall find many instances when its use is indicated where only static stresses are present. So it is convenient to think of K_f as a stress-concentration factor reduced from K_t because of lessened sensitivity to notches. The resulting factor is defined by the equation

$$K_f = \frac{\text{maximum stress in notched specimen}}{\text{stress in notch-free specimen}} \tag{a}$$

Notch sensitivity q is defined by the equation

$$q = \frac{K_f - 1}{K_t - 1} \tag{b}$$

where q is usually between zero and unity. Equation (b) shows that if $q = 0$, then $K_f = 1$, and the material has no sensitivity to notches at all. On the other hand, if $q = 1$, then $K_f = K_t$, and the material has full notch sensitivity. In analysis or design work, find K_t first, from the geometry of the part. Then specify the material, find q, and solve for K_f from the equation

$$K_f = 1 + q(K_t - 1) \tag{5-26}$$

For steels and 2024 aluminum alloys, use Fig. 5-16 to find q for bending and axial loading. For shear loading, use Fig. 5-17. In using these charts it is well to know that the actual test results from which the curves were derived exhibit a large amount of scatter. Because of this scatter it is always safe to use $K_f = K_t$ if there is any doubt about the true value of q. Also, note that q is not far from unity for large notch radii.

The notch sensitivity of the cast irons is very low, varying from 0 to about 0.20, depending upon the tensile strength. To be on the conservative side, it is recommended that the value $q = 0.20$ be used for all grades of cast iron.

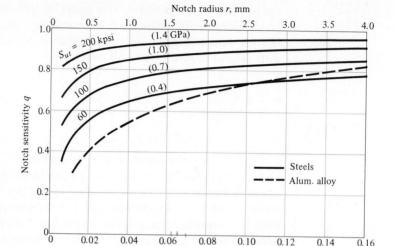

FIGURE 5-16

Notch-sensitivity charts for steels and UNS A92024-T wrought aluminum alloys subjected to reversed bending or reversed axial loads. For larger notch radii, use the values of q corresponding to $r = 0.16$ in (4 mm). *[Reproduced by permission from George Sines and J. L. Waisman (eds.), Metal Fatigue, McGraw-Hill, New York, 1959, pp. 296, 298.]*

Statistical Results

When the notch sensitivity q is obtained from Figs. 5-16 and 5-17, the resulting value of K_f from Eq. (5-26) may be treated as the mean value. The coefficient of variation then depends upon the type of discontinuity. Table 5-5 can be used to find values for the steels.

5-21   INTRODUCTION TO FRACTURE MECHANICS

The use of elastic stress-concentration factors provides an indication of the average load required on a part for the onset of plastic deformation, or yielding; these factors

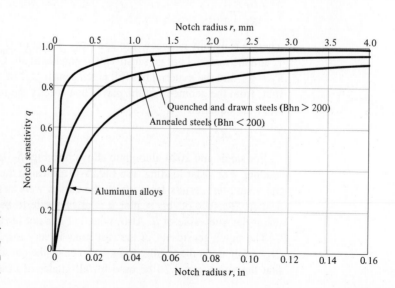

FIGURE 5-17

Notch-sensitivity curves for materials in reversed torsion. For larger notch radii, use the values of q corresponding to $r = 0.16$ in (4 mm).

TABLE 5-5

Coefficients of Variation C_K for Steels

NOTCH TYPE	COEFFICIENT OF VARIATION C_K
Transverse hole	0.11
Shoulder	0.08
Groove	0.13
Others	0.11

are also useful for analysis of the loads on a part that will cause fatigue fracture. However, stress-concentration factors are limited to structures for which all dimensions are precisely known, particularly the radius of curvature in regions of high stress concentration. When there exists a crack, flaw, inclusion, or defect of unknown small radius in a part, the elastic stress-concentration factor approaches infinity as the root radius approaches zero, thus rendering the stress-concentration factor useless. Furthermore, even if the radius of curvature of the flaw tip is known, the high local stresses there will lead to local plastic deformation surrounded by a region of elastic deformation. Elastic stress-concentration factors are no longer valid for this situation, so analysis from the point of view of stress-concentration factors does not lead to criteria useful for design when very sharp cracks are present.

By combining analysis of the gross elastic changes in a structure or part that occur as a sharp brittle crack grows with measurements of the energy required to produce new fracture surfaces, it is possible to calculate the average stress (if no crack were present) which will cause crack growth in a part. Such calculation is possible only for parts with cracks for which the elastic analysis has been completed, and for materials that crack in a relatively brittle manner and for which the fracture energy has been carefully measured. The term *relatively brittle* is rigorously defined in the test procedures,* but it means, roughly, *fracture without yielding occurring throughout the fractured cross section.*

Thus glass, hard steels, strong aluminum alloys, and even low-carbon steel below the ductile-to-brittle transition temperature can be analyzed in this way. Fortunately, ductile materials blunt sharp cracks, as we have previously discovered, so that fracture occurs at average stresses of the order of the yield strength, and the designer is prepared for this condition. The middle ground of materials that lie between "relatively brittle" and "ductile" is now being actively analyzed, but exact design criteria for these materials are not yet available and will not be covered here.

5-22   STRESS STATE IN A CRACK

Suppose a sharp transverse full-thickness crack of length $2a$ is located in the center of a rectangular plate of material, as in Fig. 5-18. An average axial tensile stress σ is applied to both ends of the plate. If the plate length $2h$ is large compared with the width $2b$, and width $2b$ is large compared with the crack length $2a$, elastic analysis shows

---

*BS 5447:1977 and ASTM E399-78.

FIGURE 5-18

Plate of length 2h, width 2b, containing a central crack of length 2a; tensile stress σ acts in longitudinal direction.

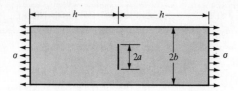

that the conditions for crack growth are controlled by the magnitude of the elastic stress intensity factor K, and that for this case

$$K_0 = \sigma\sqrt{\pi a} \tag{5-27}$$

We shall employ the units of MPa $\sqrt{m}$ for the factor K_0. If, for example, $h/b = 1$ and $a/b = 0.5$, then the magnitude of K_0 must be modified, in this case by a factor of 1.32, and so then

$$K_I = 1.32\sigma\sqrt{\pi a} \tag{5-28}$$

Thus it can be seen that K_I is a function of the average axial stress and the geometry of the part. Solutions for this particular problem over a wide range of ratios h/b and a/b have been calculated and are given graphically in Fig. 5-19, where K_I is the desired value and K_0 is the base value calculated from Eq. (5-27).

5-23 FRACTURE TOUGHNESS

The previous section describes the conditions in a part under a given applied stress by calculation of the stress-intensity factor. This value is, from the designer's point of view, a condition analogous to *stress*. This section discusses the other half of the design equation, the value analogous to *strength* of the material, that is, the *critical stress-intensity factor*, also called *fracture toughness*, and designated by the symbol K_c.

Through carefully controlled testing of a given material, the stress-intensity factor at which a crack will propagate is measured. This is the critical stress-intensity factor K_c.

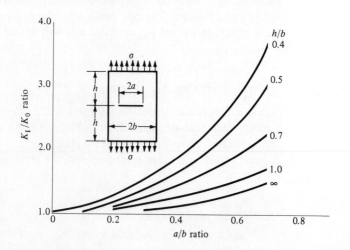

FIGURE 5-19

Plate containing a central crack loaded in longitudinal tension.

Thus for a known applied stress σ acting on a part of known or assumed crack length $2a$, when the magnitude of K reaches K_c, the crack will propagate. For the designer, the factor of safety n is thus

$$n = \frac{K_c}{K} \tag{5-29}$$

The enormous power of this method of analysis is that it enables the designer to use the value of K_c (usually measured in a single edge-notch specimen) for a given material in the design of a part that may be much more complex than the original test specimen.

EXAMPLE 5-5   A steel ship deck that is 30 mm thick, 12 m wide, and 20 m long (in the tensile-stress direction) is operated below its ductile-to-brittle transition temperature (with $K_c = 28.3$ MPa$\sqrt{m}$). If a 65-mm-long central transverse crack is present, calculate the tensile stress for catastrophic failure. Compare this stress with the yield strength of 240 MPa for this steel.

Solution   From Fig. 5-18, $2a = 65$ mm, $2b = 12$ m, and $2h = 20$ m. Thus $a/b = 32.5/6(10)^3 = 0.005$ and $h/b = 10/6 = 1.67$. Since a/b is so small, this may be considered as an infinite plate, and so Eq. (5-27) need not be modified. Solving for the stress then gives $\sigma = K_I/\sqrt{\pi a}$. Since fracture will occur when $K_I = K_{Ic}$, we have

$$\sigma = \frac{K_{Ic}}{\sqrt{\pi a}} = \frac{28.3}{\sqrt{\pi(32.5)(10)^{-3}}} = 88.6 \text{ MPa}$$

Thus catastrophic fracture will occur at a strength-stress ratio of

$$\frac{S_y}{\sigma} = \frac{240}{88.6} = 2.71$$

5-24   FRACTURE CONDITIONS

One of the first problems facing the designer is that of deciding whether the conditions exist, or not, for a brittle fracture. Low-temperature operation, that is, operation below room temperature, is a key indicator that brittle fracture is a possible failure mode. Tables of transition temperatures for various materials have not been published, possibly because of the wide variation in values, even with a single material. Thus, in many situations, laboratory testing may give the only clue to the possibility of brittle fracture.

Another key indicator of the possibility of fracture is the ratio of the yield strength to the ultimate strength. A high ratio S_y/S_u indicates there is only a small ability to absorb energy in the plastic region and hence there is a likelihood of brittle fracture.

High Charpy impact values can be used as a rough indicator of the possibility of a brittle fracture, though it is impossible to specify a transition value.*

---

*Charpy values are available for a wide range of materials. See, for example, Eric A. Brandes (ed.), *Smithell's Metals Reference Book,* 6th ed., Butterworth, London, 1983, Chap. 22.

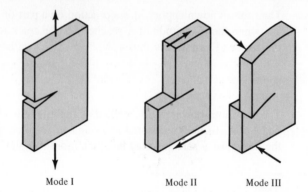

FIGURE 5-20

Deformation modes: mode I is tension; modes II and III are both shear modes.

Mode I   Mode II   Mode III

The three possible ways of separating a plate are shown in Fig. 5-20. Note that modes II and III are fundamentally shear modes of fracture, but mode II stresses and deformation stay within the plane of the plate. Mode III is out-of-plane shear.

Some stress analyses and fewer critical stress-intensity factor measurements have been made for modes I and III, but they are still limited in scope. The nomenclature K_I for stress-intensity factor and K_{Ic} for critical stress-intensity factor under mode I conditions are in general use, so for clarity the subscript I will be added here. When analysis of K and measurements of K_c are made generally available for modes II and III, then more design can be extended to these modes. The procedure is exactly the same as for mode I analysis.

In general, increasing the thickness of a part leads to a decrease in K_{Ic}. The value of K_{Ic} becomes asymptotic to a minimum value with increasing thickness; this minimum value is called the *plane strain critical stress-intensity factor,* because deformation in the thickness direction at the tip of a crack is constrained by the surrounding elastic material so that most of the strain occurs in the two directions that lie in the plane of the plate. The test requirements* for measuring K_{Ic} provide for essentially plane strain conditions, so the published values of K_{Ic} are usually plane strain values. Since use of the minimum (plane strain) value of K_{Ic} will lead to a conservative design for thinner parts, the plane strain K_{Ic} is usually used. However, if the designer has available a reliable value of K_{Ic} for the thickness of the part to be designed, this value should be used.

In practice, the crack length and location assumed in design are the worst combination of crack size and location, leading to the weakest structure. Thus $2a$ is the longest crack for the part that will not be discovered by the crack detection methods used in manufacture and in service. The location and orientation of the assumed crack or cracks must be selected as the worst conceivable. Sometimes more than one location might be critical, so analysis of the part with cracks in any or all locations must be made.

The failure analyst often has a simpler task than the designer, if fractography can establish accurately the location and size of the crack that led to the final fracture. It is then a matter of determining K_I as a function of stress (which may not be known) and comparing K_I with K_{Ic} measured for the material to obtain an estimate of the stress at the time of final fracture.

---

*ASTM Standard E399-72.

5-25   STRESS-INTENSITY FACTORS

A substantial number of geometries for stress-intensity factors have been compiled in recent years.* Some of these are included here as Figs. 5-21 to 5-26. If K_I is needed for a configuration not included in the literature, the designer's only recourse is to carry out the complete analysis alone. A large body of literature on this subject is summarized in a form useful to the designer,† and typical values of K_{Ic} are listed in Table 5-6.‡

Note carefully in Table 5-6 the general inverse relationship between yield strength

---

*H. Tada, P. C. Paris, and G. R. Irwin, *The Stress Analysis of Cracks Handbook,* Del Research, Hellertown, Pa., 1973; G. C. M. Sih, *Handbook of Stress Intensity Factors,* Lehigh University, Bethlehem, Pa., 1973; D. P. Rooke and D. J. Cartwright, *Compendium of Stress Intensity Factors,* H.M.S.O., Hillingdon Press, Uxbridge, England, 1976.

†David K. Felbeck and Anthony G. Atkins, *Strength and Fracture of Engineering Solids,* Prentice-Hall, Englewood Cliffs, N.J., 1984; Kåre Hellan, *Introduction to Fracture Mechanics,* McGraw-Hill, New York, 1984.

‡For an extensive compilation of K_c values, see *Damage Tolerant Design Handbook,* MCIC-HB-01, Air Force Materials Laboratory, Wright-Patterson Air Force Base, Ohio, December 1972 and supplements.

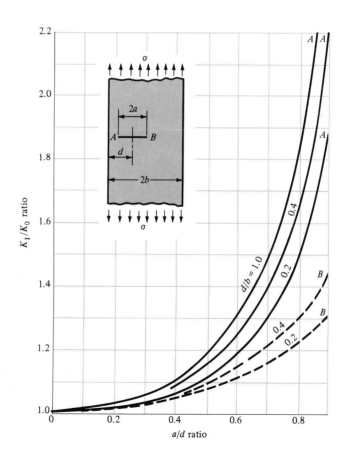

FIGURE 5-21

Off-center crack in a plate in longitudinal tension; solid curves are for the crack tip at A; dashed curves for tip at B.

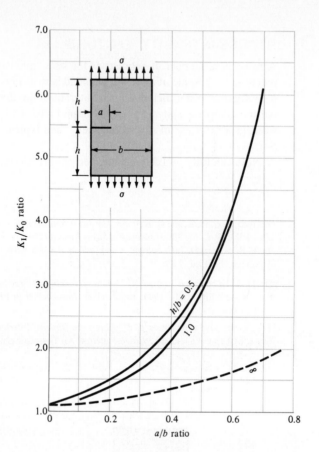

FIGURE 5-22

Plate loaded in longitudinal tension with a crack at the edge; for the solid curve there are no constraints to bending; the dashed curve was obtained with bending constraints added.

TABLE 5-6

Values of K_{Ic} for Some Engineering Materials

MATERIAL	K_{Ic}, MPa $\sqrt{m}$	S_y, MPa
Aluminum		
2024	26	455
7075	24	495
7178	33	490
Titanium		
Ti-6AL-4V	115	910
Ti-6AL-4V	55	1035
Steel		
4340	99	860
4340	60	1515
52100	14	2070

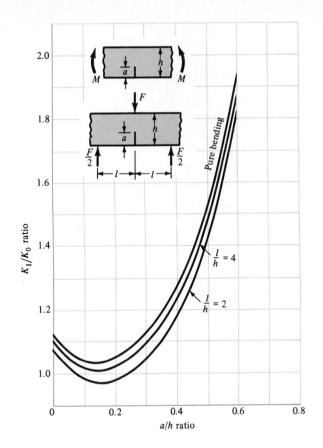

FIGURE 5-23

Beams of rectangular cross section having an edge crack.

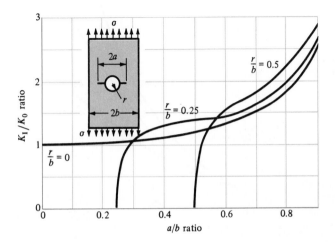

FIGURE 5-24

Plate in tension containing a circular hole with two cracks.

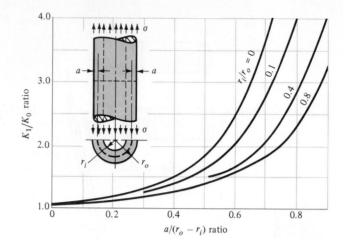

FIGURE 5-25

A cylinder loaded in axial tension having a radial crack of depth a extending completely around the circumference of the cylinder.

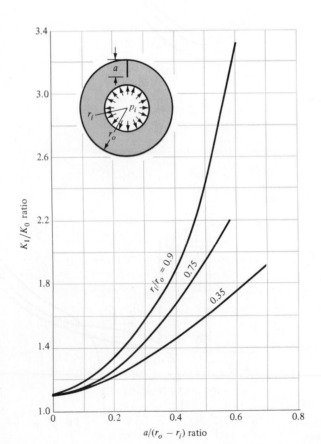

FIGURE 5-26

Cylinder subjected to internal pressure p_i having a radial crack in the longitudinal direction of depth a. Use Eq. (2-51) for the tangential stress at $r = r_o$.

and K_{Ic}. This often leads to the choice of a material of lower yield strength and higher K_{Ic}, as is shown in the example that follows.

EXAMPLE 5-6

A plate of width 1.4 m and length 2.8 m is required to support a tensile force in the 2.8-m direction of 4.0 MN. Inspection procedures will only detect through-thickness edge cracks larger than 2.7 mm. The two Ti-6AL-4V alloys in Table 5-6 are being considered for this application, for which the safety factor must be 1.3 and minimum weight is important. Which alloy should be used?

Solution

(*a*) We elect first to determine the thickness required to resist yielding. Since $\sigma = P/wt$, we have $t = P/w\sigma$. But

$$\sigma_{all} = \frac{S_y}{n} = \frac{910}{1.3} = 700 \text{ MPa}$$

Thus

$$t = \frac{P}{w\sigma_{all}} = \frac{4.0(10)^3}{1.4(700)} = 4.08 \text{ mm}$$

where we have $S_y = 910$ MPa for the weaker titanium alloy. For the stronger alloy, we have, from Table 5-6,

$$\sigma_{all} = \frac{1035}{1.3} = 796 \text{ MPa}$$

and so the thickness is

Answer

$$t = \frac{P}{w\sigma_{all}} = \frac{4.0(10)^3}{1.4(796)} = 3.59 \text{ mm}$$

(*b*) Now let us find the thickness required to prevent crack growth. Using Fig. 5-22, we have

$$\frac{h}{b} = \frac{2.8/2}{1.4} = 1 \qquad \frac{a}{b} = \frac{2.7}{1.4(10)^3} = 0.001\ 93$$

Corresponding to these ratios we find from Fig. 5-22 that $K_I/K_0 = 1.1$. Thus $K_I = 1.1\sigma\sqrt{\pi a}$. From Table 5-6 we next find $K_{Ic} = 115$ MPa$\sqrt{m}$ for the weaker of the two alloys. The stress at fracture will be

$$\sigma = \frac{K_{Ic}}{1.1\sqrt{\pi a}} = \frac{115\sqrt{(10)^3}}{1.1\sqrt{\pi(2.7)}} = 1135 \text{ MPa}$$

This stress is larger than the yield strength, and so yielding governs the design when the weaker of the two alloys is used.

For the stronger alloy, we see from Table 5-6 that $K_{Ic} = 55$. Thus

$$\sigma = \frac{K_{Ic}}{1.1\sqrt{\pi a}} = \frac{5.5\sqrt{(10)^3}}{1.1\sqrt{\pi(2.7)}} = 543 \text{ MPa}$$

Then the allowable stress is $\sigma_{all} = \sigma/n = 543/1.3 = 418$ MPa. Thus, the required thickness is

Answer

$$t = \frac{P}{w\sigma_{all}} = \frac{4.0(10)^3}{1.4(418)} = 6.84 \text{ mm}$$

This example shows that the fracture toughness K_{Ic} limits the design when the stronger alloy is used and so a thickness of 6.84 mm is required. When a weaker alloy is used, the design is limited by its yield strength, giving a thickness of only 4.08 mm. Thus the weaker alloy leads to a thinner and lighter-weight choice.

5-26   STRESS-CORROSION CRACKING

Parts subjected to continuous static loads in certain corrosive environments may, over a period of time, develop serious cracks. This phenomenon is known as *stress-corrosion cracking*. Examples of such parts are door-lock springs, watch springs, lock washers, marine and bridge cables, and other highly stressed parts subjected to atmospheric or other corrosive surroundings. The stress, environment, time, and alloy structure of the part all seem to have an influence on the cracking, with each factor speeding up the influence of the others.

Stress-time tests* can be made on specimens in a corrosive environment in order to determine the limiting value of the fracture toughness. The curve shown in Fig. 5-27 typifies the results of many of these experiments. The tests must be run on a number of specimens, each subject to a constant but different load and each having the same size initial crack. It will then be found that the rate of crack growth depends both upon stress and upon time. When the times to fracture corresponding to the values of K_I are noted and plotted, a curve like that of Fig. 5-27 will be obtained. The limiting value of the stress-intensity factor is here designated as K'_{Ic}, corresponding to point C on the curve. Crack growth will not be obtained for stress-intensity factors less than this value, no matter how long the loaded specimen remains in the environment. Unfortu-

---

*See H. O. Fuchs and R. I. Stephens, *Metal Fatigue in Engineering*, Wiley, New York, 1980, p. 218.

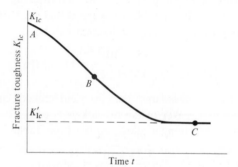

FIGURE 5-27

Change in fracture toughness with time.

nately, these tests require a great deal of time for completion, usually not less than 1000 h.*

PROBLEMS

5-1   A specimen of medium-carbon steel having an initial diameter of 0.503 in was tested in tension using a gauge length of 2 in. The following data were obtained for the elastic and plastic states:

ELASTIC STATE		PLASTIC STATE	
LOAD P, lb	ELONGATION, in	LOAD P, lb	AREA A_i, in²
1 000	0.0004	8 800	0.1984
2 000	0.0006	9 200	0.1978
3 000	0.0010	9 100	0.1963
4 000	0.0013	13 200	0.1924
7 000	0.0023	15 200	0.1875
8 400	0.0028	17 000	0.1563
8 800	0.0036	16 400	0.1307
9 200	0.0089	14 800	0.1077

Note that there is some overlap in the data. Plot the engineering or nominal stress-strain diagram using two scales for the unit strain ϵ, one from zero to about 0.02 in/in and the other from zero to maximum strain. From this diagram find the modulus of elasticity, the 0.2 percent offset yield strength, the ultimate strength, and the percent reduction in area.

5-2   Compute the true stress and the logarithmic strain using the data of Prob. 5-1 and plot the results on log-log paper. Then find the plastic strength coefficient σ_0 and the strain-strengthening exponent m. Find also the yield strength and the ultimate strength after the specimen has had 20 percent cold work.

5-3   The stress-strain data from a tensile test on a cast-iron specimen are

Engineering stress, kpsi	5	10	16	19	26	32	40	46	49	54
Engineering strain, $\epsilon \cdot 10^{-3}$ in/in	0.20	0.44	0.80	1.0	1.5	2.0	2.8	3.4	4.0	5.0

Plot the stress-strain locus and find the 0.1 percent offset yield strength, and the tangent modulus of elasticity at zero stress and at 20 kpsi.

5-4   Having completed the task of Prob. 5-3, an engineer is unsatisfied with the accuracy of the result, depending as it does on drawing tangents to unknown curves. Noting that the stress-strain

---

*For some values of the stress-intensity factors K'_{1c}, see *Damage Tolerant Handbook*, Metals and Ceramics Information Center, Battelle, Columbus, Ohio, 1975.

locus on Cartesian coordinates looks smooth and parabolic, the engineer writes

$$\epsilon = c_0 + c_1\sigma + c_2\sigma^2$$

and then notes the necessity for $c_0 = 0$ and replots the data with ϵ/σ as the ordinate and σ as the abscissa. If the plotted locus is satisfactorily straight, find the constants c_1 and c_2 by regression. From this regression equation, determine the tangent modulus at zero load and at 20 kpsi, as well as the 0.1 percent offset yield strength. How do these determinations compare with the technique of Prob. 5-3? What can you say about the error?

5-5   The true stress–true strain data from a tensile test of a polyethylene plastic are

True stress, kpsi	1.0	1.5	2.0	2.5	3.0	3.5
True strain, in/in	0.15	0.32	0.46	0.60	0.70	0.82

Use these data from the plastic range and by regression determine the strain-strengthening exponent σ_0 and the strain-strengthening exponent m. Can you say anything about the error in either of these determinations?

5-6   A straight bar of arbitrary cross section and thickness h is cold-formed to an inner radius R about an anvil as shown in the figure. Some surface at distance N having an original length L_{AB} will remain unchanged in length after bending. This length is

$$L_{AB} = L_{AB'} = \frac{\pi(R + N)}{2}$$

The lengths of the outer and inner surfaces, after bending, are

$$L_o = \frac{\pi}{2}(R + h) \qquad L_i = \frac{\pi}{2}R$$

Using Eq. (5-8), we then find the true strains to be

$$\varepsilon_o = \ln\frac{R + h}{R + N} \qquad \varepsilon_i = \ln\frac{R}{R + N}$$

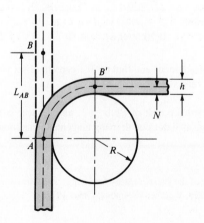

PROBLEM 5-6

Tests show that $|\varepsilon_o| = |\varepsilon_i|$. Show that

$$N = R\left[\left(1 + \frac{h}{R}\right)^{1/2} - 1\right]$$

and

$$\varepsilon_o = \ln\left(1 + \frac{h}{R}\right)^{1/2}$$

5-7   A 12-gauge (0.1094-in-thick) strap is to be cold-bent around a $\frac{1}{8}$-in-radius anvil. If the material is annealed AISI 1018 steel, use the results of Prob. 5-6 to
(*a*) Estimate the plastic strain at the surfaces.
(*b*) Estimate the new ultimate strength at the inner and outer surfaces.
(*c*) Estimate the new yield strength at the surfaces of the curve.

5-8   The strain-strengthening exponent for some 1000-series steels has been measured as follows:

Steel	1002	1008	1010	1018	1020	1045
m	0.29	0.24	0.26	0.25	0.22	0.14
	0.27	0.24	0.23			
			0.23			
			0.23			

Is there a correlation between carbon content in percent (or in points) and the strain-strengthening exponent m?

5-9   For heat-treated AISI 1045 steel, the following Brinell hardnesses and strain-strengthening exponents have been observed:

H_B	225	390	410	450	500	595
m	0.61	0.45	0.60	0.35	0.25	0.07
	1.00					

Is m a decreasing monotone with hardness? What is the regression relation?

5-10   In the torsion of a circular shaft of radius r, length l, plastically twisted through the angle θ, Datsko* has shown that at some angle α to the axial direction there exists a surface filament having the largest tensile strain. This angle is given by the equation

$$\alpha = \frac{1}{2}\tan^{-1}\left(\frac{2l}{r\theta}\right)$$

and the true strain by

$$\varepsilon = \ln\left\{\cos\alpha\left[1 + \left(\tan\alpha + \frac{r\theta}{l}\right)^2\right]^{1/2}\right\}$$

---

*Joseph Datsko, *Materials in Design and Manufacture,* Joseph Datsko Consultants, Ann Arbor, Mich., 1978, pp. 7-19 to 7-23.

A shaft has a diameter of 1.0 in, is 10 in long, and is twisted through an angle of 10 rad.
(a) Find the true strain of an axial filament.
(b) Find the maximum true strain at the surface.
(c) Determine whether a shaft of 2011-T6 aluminum alloy can be cold-worked to this extent.

ANSWERS

5-1 $E = 30$ Mpsi, $S_y = 45.5$ kpsi, $S_u = 85.5$ kpsi, $R = 45.8$ percent

5-4 $E_0 = 27.6$ Mpsi, $E_{20} = 18.9$ Mpsi; $\varepsilon/\sigma = 0.362(10^{-7}) + 0.835(10^{-12})\sigma$; E_0 is between 26.8 and 28.4 Mpsi with 95 percent confidence

5-7 (a) $|\varepsilon_i| = |\varepsilon_o| = 0.314$ in/in; (b) 67.8 kpsi; (c) 67.8 kpsi

5-10 (b) 0.247 in/in

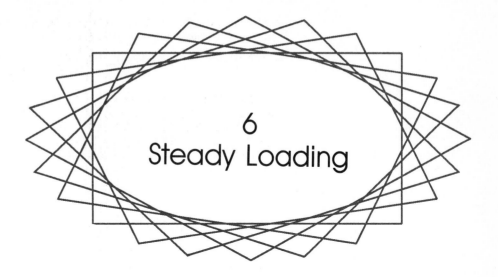

6
Steady Loading

In Chap. 1 we learned that *strength is a property or characteristic of a material or of a mechanical element*. This property may be inherent in a material or may result from the treatment and processing of the material. The strength of a mechanical part is a property completely independent of whether or not that part is subjected to load or force. In fact, this strength property is a characteristic of the element even before it is assembled with other elements into a machine or system.

In addition to considering the strength of a single part, in this chapter (in the optional sections; see Sec. 1-10) we shall also consider the strength of a collection of mechanical parts all alike. Each part is quite likely to deviate slightly from others in the collection with respect to dimensions, processing, and machining or forming. Thus strength as a property of an entire collection of parts is a stochastic quantity, characterized by a mean and a standard deviation.

A *static load* is a stationary force or moment acting on a member. To be stationary, the force or moment must be unchanging in magnitude, point or points of application, and direction. A static load can be axial compression or tension, a shear load, a bending load, a torsional load, or any combination of these. But the load cannot change in any manner if it is still to be considered static. (Sometimes a load is assumed to be static when it is known that some variation is to be expected. This assumption is usually made to get a rough idea of the component dimensions and to simplify design computations when variations in load are few or minor in nature.)

In this chapter we shall examine the relations between the strength of a part and its anticipated static loading in order to select the optimum material and dimensions for satisfying the requirement that the part must not fail in service. In Chap. 1 it was observed that there are two distinct and separate approaches for accomplishing these objectives:

1. The *deterministic,* or *factor-of-safety, approach*. In this method the maximum stress or stresses in a part are kept below the minimum strength by a suitable design factor or margin of safety, to ensure that the part will not fail. Of course, this involves a consideration of the material, the processing, and the dimensions.

2. The *stochastic,* or *reliability, approach.** This method involves the selection of materials, processing, and dimensions such that the probability of failure is always less than a preselected value.

Figures 6-1 to 6-14 are photographs of several failed parts. They exemplify the need to be well versed in the principles of design for satisfactory performance. So we shall develop failure and safety criteria for one-, two-, and three-dimensional stress states, with and without stress concentration, for both ductile and brittle materials.

6-1   STATIC STRENGTH

Ideally, in designing any machine element, the engineer should have at his or her disposal the results of a great many strength tests of the particular material chosen. These tests should be made on specimens having the same heat treatment, surface finish, and size as the element the engineer proposes to design; and the tests should be made under exactly the same loading conditions as the part will experience in service. This means that if the part is to experience a bending load, it should be tested with a bending load. If it is to be subjected to combined bending and torsion, it should be tested under combined bending and torsion. If it is made of heat-treated AISI 1040 steel drawn at 500°C with a ground finish, the specimens tested should be of the same material prepared in the same manner. Such tests will provide very useful and precise information. Whenever such data are available for design purposes, the engineer can be assured that he or she is doing the best possible job of engineering.

The cost of gathering such extensive data prior to design is justified if failure of the part may endanger human life or if the part is manufactured in sufficiently large quantities. Refrigerators and other appliances, for example, have very good reliabilities because the parts are made in such large quantities that they can be thoroughly tested in

---

*Sections dealing with this approach may be optional; see Sec. 1-10.

FIGURE 6-1

(*a*) Failure of a truck drive-shaft spline due to corrosion fatigue. Note that it was necessary to use clear tape to hold the pieces in place. (*b*) Direct end view of failure.

(*a*)   (*b*)

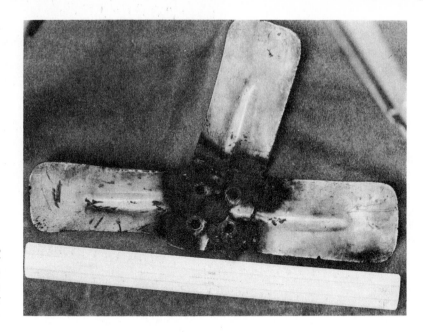

FIGURE 6-2

Fatigue failure of an automotive cooling fan due to vibrations caused by a defective water pump.

advance of manufacture. The cost of making these tests is very low when it is divided by the total number of parts manufactured.

You can now appreciate the following four design categories:

1 Failure of the part would endanger human life, or the part is made in extremely large quantities; consequently, an elaborate testing program is justified during design.

2 The part is made in large enough quantities that a moderate series of tests is feasible.

FIGURE 6-3

Typical failure of a stamped steel alternator bracket after about 40 000 km. The failure was probably due to residual stresses caused by the cold-forming operation. The high failure rate prompted the manufacturer to redesign the bracket as a die casting.

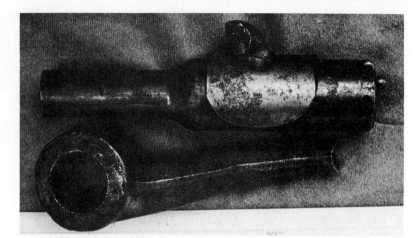

FIGURE 6-4

Failure of an automotive drag link. This failure occurred after about 225 000 km. Fortunately the car was in park and against a curb. Such a failure results in total disconnect of the steering wheel from the steering mechanism.

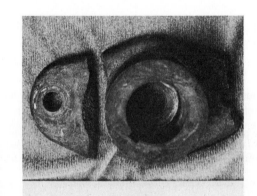

FIGURE 6-5

Impact failure of a lawn-mower blade driver hub. The blade impacted a surveying pipe marker.

FIGURE 6-6

Failure of an overhead-pulley retaining bolt on a weight-lifting machine. A manufacturing error caused a gap that forced the bolt to take the entire moment load.

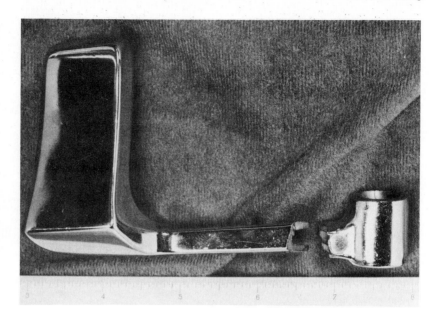

FIGURE 6-7

Failure of an interior die-cast car-door handle. Failure occurred after about every 72 000 km. Probable causes were the electroplating material, stress concentration, the long lever arm required to operate a "sticky" door-release mechanism, and the high actuation forces.

3 The part is made in such small quantities that testing is not justified at all; or the design must be completed so rapidly that there is not enough time for testing.

4 The part has already been designed, manufactured, and tested and found to be unsatisfactory. Analysis is required to understand why the part is unsatisfactory and what to do to improve it.

More often than not it is necessary to design using only published values of yield strength, ultimate strength, percentage reduction in area, and percentage elongation, such as those listed in the Appendix. How can one use such meager data to design

FIGURE 6-8

Chain test fixture that failed in one cycle. To alleviate complaints of excessive wear, the manufacturer decided to case-harden the material. (*a*) Two halves showing fracture; this is an excellent example of brittle fracture initiated by stress concentration. (*b*) Enlarged view of one portion to show cracks induced by stress concentration at the support-pin holes.

(*a*)

(*b*)

FIGURE 6-9

Automotive rocker-arm
articulation-joint fatigue failure.

against both static and dynamic loads, biaxial and triaxial stress states, high and low temperatures, and very large and very small parts? These and similar questions will be addressed in this chapter and those to follow; but think how much better it would be to have data available that duplicate the actual design situation.

FIGURE 6-10

Valve-spring failure caused by
spring surge in an overrevved
engine. The fractures exhibit the
classic 45° shear failure.

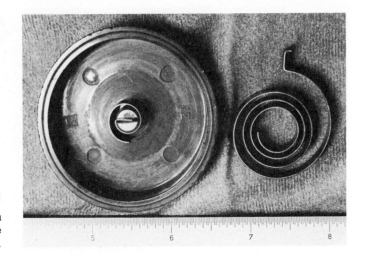

FIGURE 6-11

Torsional spring failure. This is a
Corvair choke spring; a very rare
failure.

6-2   STRESS CONCENTRATION

The reduced value K_f of the stress-concentration factor (see Sec. 5-20) must sometimes
be used when parts are loaded statically, but not always.

Stress concentration is a highly localized effect; in some instances it may be due
only to a surface scratch. If the material is ductile, even a normal load will cause
yielding in the immediate vicinity of the notch. This yielding is the same as cold

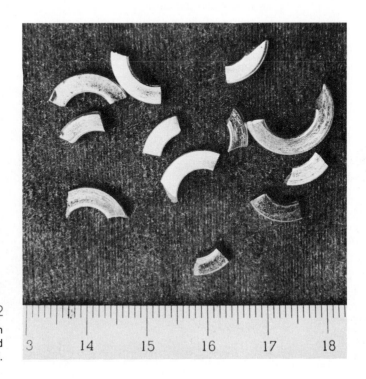

FIGURE 6-12

Brittle fracture of a lock washer in
one-half cycle. The washer failed
when it was installed.

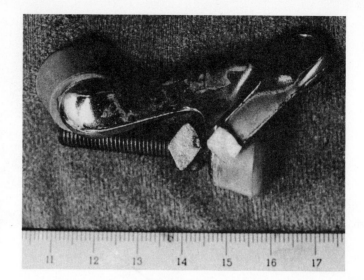

FIGURE 6-13

Fatigue failure of a die-cast residence door bumper. This bumper is installed on the door hinge to prevent the doorknob from impacting the wall.

working the material and, according to Sec. 5-3, increases the strength significantly at the notch. Since the loads are static, the part will carry them quite satisfactorily, and stress concentration need not be considered in the design or analysis.

When using this rule for ductile materials with static loads, be very careful to assure yourself that the material really is ductile and that the possibility of a brittle fracture (see Secs. 5-24 and 5-25) can be ruled out. Generally, a material may be considered ductile if the tension test reveals a true strain at fracture greater than 5 percent. But this is not a hard-and-fast rule.

Of course, if the material is brittle or acts as a brittle material, then K_t (the "full value" of K_f) should be used in computing the stress.

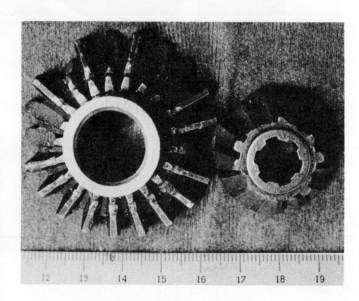

FIGURE 6-14

A gear failure from a $7\frac{1}{2}$-hp (5.6-kW) American-made outboard motor. The large gear has a $1\frac{7}{8}$-in (47.6-mm) outside diameter and had 21 teeth; 6 are broken. The pinion had 14 teeth; all are broken. Failure occurred when the propeller struck a steel auger placed in the lake bottom as an anchorage. Owner had replaced the shear pin with a substitute pin.

6-3   FAILURE THEORIES

When a part is loaded so that the stress state is uniaxial, then the stress and the strength can be compared directly to determine the degree of safety, or to learn whether the part will fail. The method is simple, because there is only one value of stress and there is only one value of strength, be it yield strength, ultimate strength, shear strength, or whatever, as appropriate.

The problem becomes more complicated when the stress state is biaxial or triaxial. In such cases there are a multitude of stresses, but still only one significant strength. So how do we learn whether the part is safe or not, and, if so, how safe? A number of failure theories have been proposed to help answer this question. In the next several sections we shall present some of these theories. Then, in the sections that follow, we shall indicate which of the theories are most useful and how they are used in analysis and design.

6-4   THE MAXIMUM-NORMAL-STRESS THEORY

The maximum-normal-stress theory states that *failure occurs whenever one of the three principal stresses equals the strength.* Suppose we arrange the three principal stresses for any stress state in the ordered form

$$\sigma_1 > \sigma_2 > \sigma_3 \tag{6-1}$$

Then this theory predicts that failure occurs whenever

$$\sigma_1 = S_t \qquad \text{or} \qquad \sigma_3 = -S_c \tag{6-2}$$

where S_t and S_c are tensile and compressive strengths, usually either the yield or ultimate, respectively. Figures 6-15 and 6-16 illustrate stress states associated with safety and with failure.

FIGURE 6-15

The maximum-normal-stress theory in three dimensions. The right rectangular prism encloses all safe values of any combination of stress components. The compressive strength S_c need not equal the tensile strength S_t. For this theory, these may be either yield or ultimate strengths. Note, too, that strengths are always positive quantities, but stresses may be either positive or negative.

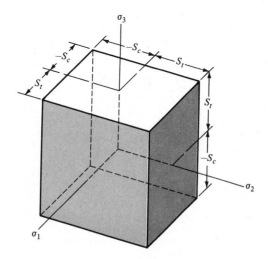

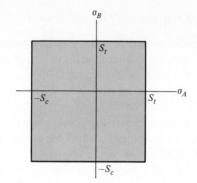

FIGURE 6-16

Graph of the maximum-normal-stress theory of failure for biaxial stress states using $S_c > S_t$. Stress states inside the graph are safe.

6-5   THE MAXIMUM-NORMAL-STRAIN THEORY

The *maximum-strain theory*, also called *Saint-Venant's theory*, applies only in the elastic range of stresses. This theory states that *yielding occurs when the largest of the three principal strains becomes equal to the strain corresponding to the yield strength.* If we assume that the yield strengths in tension and compression are equal, then the strains due to the stresses can be equated to the strain corresponding to the yield strength. Using Table 2-1 in this manner gives the condition for yielding as

$$\sigma_1 - \nu(\sigma_2 + \sigma_3) = \pm S_y$$
$$\sigma_2 - \nu(\sigma_3 + \sigma_1) = \pm S_y \tag{6-3}$$
$$\sigma_3 - \nu(\sigma_1 + \sigma_2) = \pm S_y$$

If one of the three principal stresses is zero and the remaining two are designated as σ_A and σ_B, then, for biaxial stress states, the criterion for yielding is written

$$\sigma_A - \nu\sigma_B = \pm S_y$$
$$\sigma_B - \nu\sigma_A = \pm S_y \tag{6-4}$$

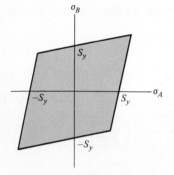

FIGURE 6-17

Graph of the maximum-normal-strain theory for biaxial stress states; based on $\nu = 0.30$. Note that compressive strengths are treated as positive quantities.

It is possible to construct a prism,* as in Fig. 6-15, to illustrate the maximum-strain-energy theory in three dimensions, but the result isn't as informative as a two-dimensional plot. Figure 6-17 shows such a plot for a Poisson ratio of 0.30 and for equal tensile and compressive yield strengths.

6-6   THE MAXIMUM-SHEAR-STRESS THEORY

The maximum-shear-stress theory states that *yielding begins whenever the maximum shear stress in any element becomes equal to the maximum shear stress in a tension-test specimen of the same material when that specimen begins to yield.*

---

*The resulting three-dimensional prism can be found in J. A. Collins, *Failure of Materials in Mechanical Design*, Wiley, New York, 1981, p. 134.

If we order the principal normal stresses as $\sigma_1 > \sigma_2 > \sigma_3$, then the maximum-shear-stress theory predicts that yielding will occur whenever

$$\tau_{\max} \geq \frac{S_y}{2} \qquad \text{or} \qquad \sigma_1 - \sigma_3 \geq S_y \tag{6-5}$$

Note that this theory also states that the yield strength in shear is given by the equation

$$S_{sy} = 0.50 S_y \tag{6-6}$$

To develop an even better understanding of this theory, we repeat Eq. (2-11) for the three principal shear stresses here. These are

$$\tau_{1/2} = \frac{\sigma_1 - \sigma_2}{2} \qquad \tau_{2/3} = \frac{\sigma_2 - \sigma_3}{2} \qquad \tau_{1/3} = \frac{\sigma_1 - \sigma_3}{2} \tag{6-7}$$

Equation (6-7) shows that failure is predicted when any one of these three shear stresses is maximum. Suppose we decompose the normal principal stresses into the components

$$\sigma_1 = \sigma_1' + \sigma_1''$$
$$\sigma_2 = \sigma_2' + \sigma_2'' \tag{a}$$
$$\sigma_3 = \sigma_3' + \sigma_3''$$

such that

$$\sigma_1'' = \sigma_2'' = \sigma_3'' \tag{b}$$

The stresses in Eq. (b) are called the *hydrostatic components* since they are equal. If it should happen that $\sigma_1' = \sigma_2' = \sigma_3' = 0$, then the three shear stresses, given by Eq. (6-7), would all be zero and there could be no yielding regardless of the magnitudes of the hydrostatic stresses. Thus the hydrostatic components have no effect on the size of the Mohr's circle but merely serve to shift it along the normal-stress axis. It is for this reason that the yielding criterion for the general stress state can be represented by the oblique regular hexagonal cylinder of Fig. 6-18. Figure 6-19 illustrates the theory for biaxial stresses.

6-7   THE STRAIN-ENERGY THEORIES

The *maximum-strain-energy theory* predicts that failure by *yielding occurs when the total strain energy in a unit volume reaches or exceeds the strain energy in the same volume corresponding to the yield strength in tension or in compression.*

The strain energy stored in a unit volume when stressed uniaxially to the yield strength can be found from Eq. (3-33). Thus

$$u_S = \frac{S_y^2}{2E} \tag{a}$$

With the help of the triaxial stress-strain relations in Table 2-1, we find the total strain

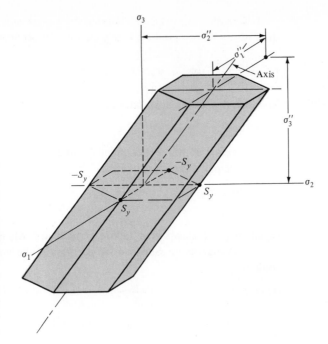

FIGURE 6-18

The maximum-shear-stress theory graphically represented in three dimensions. The hexagonal cylinder encloses all safe (free of yielding) values of the general stress state given by σ_1, σ_2, and σ_3. The axis of the cylinder is inclined equally to each of the three principal directions and is the locus of the points described by the triad of hydrostatic components σ''_1, σ''_2, and σ''_3.

energy in a unit volume subjected to combined stresses to be

$$u_\sigma = \frac{\epsilon_1 \sigma_1}{2} + \frac{\epsilon_2 \sigma_2}{2} + \frac{\epsilon_3 \sigma_3}{2}$$

$$= \frac{1}{2E}[\sigma_1^2 + \sigma_2^2 + \sigma_3^2 - 2\nu(\sigma_1\sigma_2 + \sigma_2\sigma_3 + \sigma_3\sigma_1)] \qquad (b)$$

Since this theory is no longer favored, no graph of the theory is shown here and the biaxial stress equations are not given. You may wish, however, to determine these to satisfy your own curiosity.

The *distortion-energy theory* originated because of the observation that ductile materials stressed hydrostatically exhibited yield strengths greatly in excess of the values given by the simple tension test. Therefore it was postulated that yielding was not a simple tensile or compressive phenomenon at all, but, rather, that it was related somehow to the angular distortion of the stressed element. To develop the theory, note, in Fig. 6-20a, the unit volume subjected to any three-dimensional stress state designated by the stresses σ_1, σ_2, and σ_3. The stress state shown in Fig. 6-20b is one of hydrostatic tension due to the stresses σ_{av} acting in each of the same principal directions as in Fig. 6-20a. The formula for σ_{av} is

$$\sigma_{av} = \frac{\sigma_1 + \sigma_2 + \sigma_3}{3} \qquad (c)$$

Thus the element in Fig. 6-20b undergoes pure volume change, that is, no angular distortion. If we regard σ_{av} as a component of σ_1, σ_2, and σ_3, then this component can

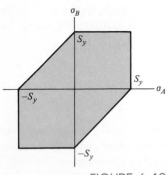

FIGURE 6-19

The maximum-shear-stress theory for biaxial stresses. σ_A and σ_B are the two nonzero principal stresses. Note that in the first and third quadrants, this theory is the same as the maximum-normal-stress theory.

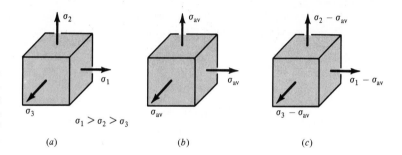

FIGURE 6-20

(a) Element with triaxial stresses; this element undergoes both volume change and angular distortion. (b) Element under hydrostatic tension undergoes only volume change. (c) Element has angular distortion without volume change.

(a) (b) (c)

be subtracted from them, resulting in the stress state shown in Fig. 6-20c. This element is subjected to pure angular distortion, that is, no volume change.

Equation (b) gives the total strain energy for the element of Fig. 6-20a. The strain energy for producing only volume change can be obtained by substituting σ_{av} for σ_1, σ_2, and σ_3 in Eq. (b). The result is

$$u_v = \frac{3\sigma_{av}^2}{2E}(1 - 2\nu) \tag{d}$$

If we now substitute the square of Eq. (c) in Eq. (d) and simplify the expression, we get

$$u_v = \frac{1 - 2\nu}{6E}(\sigma_1^2 + \sigma_2^2 + \sigma_3^2 + 2\sigma_1\sigma_2 + 2\sigma_2\sigma_3 + 2\sigma_3\sigma_1) \tag{6-8}$$

Then the distortion energy is obtained by subtracting Eq. (6-8) from Eq. (b). This gives

$$u_d = u_\sigma - u_v = \frac{1 + \nu}{3E}\left[\frac{(\sigma_1 - \sigma_2)^2 + (\sigma_2 - \sigma_3)^2 + (\sigma_3 - \sigma_1)^2}{2}\right] \tag{6-9}$$

Note that the distortion energy is zero if $\sigma_1 = \sigma_2 = \sigma_3$.

In words, *the distortion-energy theory predicts that yielding will occur whenever the distortion energy in a unit volume equals the distortion energy in the same volume when uniaxially stressed to the yield strength.* For the simple tension test, let $\sigma_1 = \sigma'$, $\sigma_2 = \sigma_3 = 0$. The distortion energy is

$$u_d = \frac{1 + \nu}{3E}\sigma'^2 \tag{6-10}$$

Setting Eqs. (6-9) and (6-10) equal to each other gives

$$\sigma' = \left[\frac{(\sigma_1 - \sigma_2)^2 + (\sigma_2 - \sigma_3)^2 + (\sigma_1 - \sigma_3)^2}{2}\right]^{1/2} \tag{6-11}$$

Therefore yielding is predicted to occur when

$$\sigma' \geq S_y \tag{6-12}$$

The stress σ' should be called by a special name, because it represents the entire stress state σ_1, σ_2, and σ_3. The preferred names are the *effective stress* and the *von Mises stress,* after Dr. R. von Mises, who contributed to the theory.

For the biaxial stress state, let σ_A and σ_B be the two nonzero principal stresses. Then, from Eq. (6-11), we get

$$\sigma' = (\sigma_A^2 - \sigma_A\sigma_B + \sigma_B^2)^{1/2} \tag{6-13}$$

The distortion-energy theory is also called:

- The *shear-energy theory*
- The *von Mises–Hencky theory*
- The *octahedral-shear-stress theory*

Under the name of the octahedral-shear-stress theory, *failure is assumed to occur whenever the octahedral shear stress for any stress state equals or exceeds the octahedral shear stress for the simple tension-test specimen at failure.* Equation (2-12) is

$$\tau_{\text{oct}} = \tfrac{1}{3}[(\sigma_1 - \sigma_2)^2 + (\sigma_2 - \sigma_3)^2 + (\sigma_1 - \sigma_3)^2]^{1/2} \tag{e}$$

Using the tension test results $\sigma_1 = \sigma'$, $\sigma_2 = \sigma_3 = 0$, as before, we find from Eq. (*e*) that

$$\tau_{\text{oct}} = \frac{\sigma'}{3}(2)^{1/2} \tag{f}$$

Solving Eqs. (*e*) and (*f*) for σ' yields

$$\sigma' = \left[\frac{(\sigma_1 - \sigma_2)^2 + (\sigma_2 - \sigma_3)^2 + (\sigma_1 - \sigma_3)^2}{2}\right]^{1/2} \tag{g}$$

which is identical with Eq. (6-11).

Figure 6-21 shows the distortion-energy theory for triaxial stress states. Notice that the hydrostatic components σ_1'', σ_2'', and σ_3'', as defined by Eq. (*a*) in Sec. 6-6, always lie on the axis of the cylinder no matter how far it is extended from the origin. The representation for biaxial stress states is shown in Fig. 6-22. This is a truer representation of the ellipse because of the distortion inherent in pictorial representation.*

The mathematical manipulation involved in the development of a theory often tends to obscure the real value and usefulness of the result. Equations (6-11) and (6-13) mean that a complex stress situation can be represented by a single value. Just think about it. The von Mises stress can be used to represent the most complicated stress situation you can think of! For example, the stress state σ_x, σ_y, σ_z, τ_{xy}, τ_{xz}, τ_{yz} can be represented by the single value σ'.

6-8   THE INTERNAL-FRICTION THEORY

The compressive strength of a material that fails in compression by a sudden shattering fracture has a specific value. However, if the material is not one that fails in this

---

*Figure 6-21 was drafted using oblique pictorial drawing. The receding axis is scaled at 50 percent. In an isometric drawing, the scales of the three axes are equal. But isometric drawing could not be used in this case because the axis of the ellipse would overlap a coordinate axis.

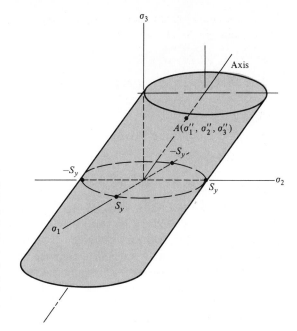

FIGURE 6-21

The distortion-energy theory graphically represented in three dimensions. The oblique elliptical cylinder encloses all safe values (free of yielding) of the general stress state given by σ_1, σ_2, and σ_3. The axis of the cylinder is inclined equally to each of the three principal directions and is the locus of the points described by the triad of hydrostatic components σ_1'', σ_2'', and σ_3''.

manner, then the compressive strength must be defined in some arbitrary manner corresponding to a specified allowed distortion.

Not all materials have compressive strengths equal to their corresponding tensile values. For example, the yield strength of magnesium alloys in compression may be as little as 50 percent of their yield strength in tension. And the ultimate strength of the gray cast irons in compression varies from about 3 to 4 times greater than the ultimate tensile strength. So, in this section, we are primarily interested in those theories that can be used to predict failure for materials whose strengths in tension and compression are not equal.

One such theory is the Mohr theory, which is similar to the maximum-shear-stress theory because it predicts failure only on the basis of the largest of the three principal

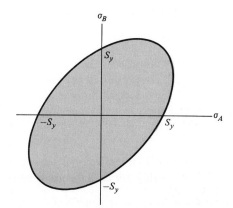

FIGURE 6-22

The distortion-energy theory for biaxial stress states. This is a true plot of points obtained from Eq. (6-12) with $\sigma = S_y$.

FIGURE 6-23

Three Mohr's circles, one for the uniaxial compression test, one for a test in pure shear, and one for the uniaxial tension test, are used to define failure by the Mohr theory. The strengths S_c and S_t are the compressive and tensile strengths, respectively; they can be used either for yield or for the ultimate strength.

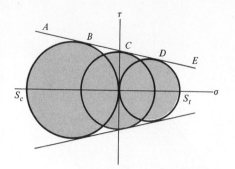

shear stresses (see Sec. 2-3). Thus, in Fig. 2-5, the shear stress $\tau_{1/3}$ is the critical one and σ_1, σ_2, and σ_3 need not be considered.

The basis of Mohr's theory of failure is shown in Fig. 6-23. Three circles are shown, one representing the strength S_c from the uniaxial compression test, one representing the strength S_t from the uniaxial tensile test, and the third, the central circle, taken from a test of pure shear at failure. The Mohr theory predicts failure for any other stress state in which the largest of the three Mohr's circles corresponding to σ_1, σ_2, and σ_3 is tangent to the line AE in Fig. 6-23. The theory can be used to predict either the beginning of yield or the beginning of fracture.

A graphical representation of Mohr's theory is similar to Figs. 6-18 and 6-19, but the hexagonal sides are of different lengths. The theory is also valid for hydrostatic stress states.

A variation of Mohr's theory is called the *Coulomb-Mohr theory* or the *internal-friction theory*. This theory is based on the assumption that line BCD in Fig. 6-23 is straight. Order the three principal stresses so that $\sigma_1 > \sigma_2 > \sigma_3$. Then for any stress state producing a circle tangent to line BCD, between points B and D, it is true that σ_1 and σ_3 have opposite signs. For this state of stress the Mohr theory applies and the two stresses and the strengths are related by the equation

$$\frac{\sigma_1}{S_t} - \frac{\sigma_3}{S_c} = 1 \qquad \sigma_1 \geq 0, \ \sigma_3 \leq 0 \tag{6-14}$$

For biaxial stress states in which σ_1 and σ_3 have like signs, the internal-friction theory is the same as the maximum-normal-stress theory and failure is predicted by

$$\sigma_1 = S_t \qquad \sigma_1 > 0$$
$$\sigma_3 = -S_c \qquad \sigma_3 < 0 \tag{6-15}$$

Either yield strength or ultimate strength can be used with Eqs. (6-14) and (6-15). Note again that strengths are always treated as positive values. The internal-friction theory is shown in Fig. 6-24 for a biaxial stress state. The nonzero stresses are σ_A and σ_B. This figure is plotted for a material, such as gray cast iron, in which $S_{uc} > S_{ut}$.

EXAMPLE 6-1   A material has a minimum yield strength in tension and compression of $S_y = 100$ kpsi. Compute the factor of safety for each theory discussed in the previous sections for the following stress states:

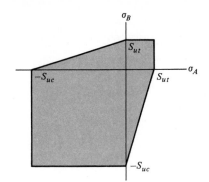

FIGURE 6-24

Plot of the internal-friction, or Coulomb-Mohr, theory of failure for biaxial stress states with $S_{uc} = 3S_{ut}$.

(a) $\sigma_1 = 70$ kpsi, $\sigma_2 = 70$ kpsi, $\sigma_3 = 0$

(b) $\sigma_1 = 70$ kpsi, $\sigma_2 = 30$ kpsi, $\sigma_3 = 0$

(c) $\sigma_1 = 70$ kpsi, $\sigma_2 = 0$, $\sigma_3 = -30$ kpsi

(d) $\sigma_1 = 0$, $\sigma_2 = -30$ kpsi, $\sigma_3 = -70$ kpsi

(e) $\sigma_1 = 30$ kpsi, $\sigma_2 = 30$ kpsi, $\sigma_3 = 30$ kpsi

Solution   The results, rounded off, are displayed in Table 6-1.

TABLE 6-1

Factors of Safety n

THEORY OF FAILURE	PART				
	(a)	(b)	(c)	(d)	(e)
Maximum-normal-stress	1.43	1.43	1.43	1.43	3.33
Maximum-shear-stress	1.43	1.43	1.00	1.43	∞
Distortion-energy	1.43	1.64	1.13	1.64	∞
Maximum-strain-energy	1.21	1.48	1.19	1.48	3.04
Coulomb-Mohr	1.43	1.43	1.00	1.43	ND*

*Not defined

6-9   FAILURE OF DUCTILE MATERIALS

Having studied some of the various theories of failure, we shall now evaluate them and show how they are applied in design and analysis. In this section we limit our studies to materials and parts that are known to fail in a ductile manner. Materials that fail in a brittle manner will be considered separately because these require different failure theories.

To decide on appropriate and workable theories of failure, Marin* collected data from many sources. Some of the data points used to select failure theories for ductile materials are shown in Fig. 6-25†. Marin also collected many data for copper and

---

*Joseph Marin was one of the pioneers in the collection, development, and dissemination of material on the failure of engineering elements. He has published many books and papers on the subject. Here the reference used is Joseph Marin, *Engineering Materials*, Prentice-Hall, Englewood Cliffs, N.J., 1952.

†Marin, op. cit., pp. 156, 157.

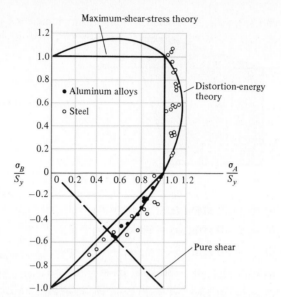

FIGURE 6-25

Graph of two failure theories for biaxial stress showing how test results confirm the predictions of the two theories. Any stress state is considered safe if its coordinates define a point within the safety region defined by each respective graph.

nickel alloys; if shown, the data points for these would be mingled with those already diagrammed.

Figure 6-25 shows that either the maximum-shear-stress theory or the distortion-energy theory is acceptable for design and analysis of materials that would fail in a ductile manner. You may wish to plot other theories using a red or blue pencil on Fig. 6-25 to show why they are not acceptable or are not used.

The selection of one or the other of these two theories is something that you, the engineer, must decide. For design purposes the maximum-shear-stress theory is easy and quick to use. If the problem is to learn why a part failed, then the distortion-energy theory may be the best to use; Fig. 6-25 shows that the locus of the distortion-energy theory passes closer to the central area of the data points, and thus may be a better predictor of failure.

If the principal stresses are ordered $\sigma_1 > \sigma_2 > \sigma_3$, then, for the simple tension test, $\sigma_1 = S_y$ and $\sigma_2 = \sigma_3 = 0$. The maximum shear stress at yielding is $\sigma_1/2$. Thus, the maximum-shear-stress theory predicts the yield strength in shear to be $S_{sy} = S_y/2$. This can be seen by the pure-shear line in Fig. 6-25, which defines all stress states having $\sigma_B = -\sigma_A$.

To determine the yield strength in shear as predicted by the distortion-energy theory, replace σ' with S_y in Eq. (6-13). For pure shear, $\sigma_B = -\sigma_A$ and $\tau = \sigma_A$; consequently, $S_{sy} = S_y/\sqrt{3} = 0.577S_y$. This equation helps to explain the use of $S_{sy} = 0.60S_y$ in some design codes.

These two relations can be expressed as the equation pair

$$S_{sy} = \begin{cases} 0.5S_y & \text{maximum-shear-stress theory} \\ 0.577S_y & \text{distortion-energy theory} \end{cases} \tag{6-16}$$

It is possible to bypass a Mohr's circle analysis for the special case of combined bending and torsion when finding σ_A and σ_B for use with failure theories. Let the two stresses obtained from combined bending and torsion be σ_x and τ_{xy}. Then a Mohr's circle for this stress state will reveal the two nonzero principal stresses to be

$$\sigma_A, \ \sigma_B = \frac{\sigma_x}{2} \pm \left[\left(\frac{\sigma_x}{2}\right)^2 + \tau_{xy}^2\right]^{1/2} \tag{6-17}$$

Equation (6-17) can be used with Eq. (6-13) to obtain the von Mises stress for combined bending and torsion. The result is

$$\sigma' = (\sigma_x^2 + 3\tau_{xy}^2)^{1/2} \tag{6-18}$$

EXAMPLE 6-2

This example illustrates the use of a failure theory to determine the strength of a mechanical element or component. The example may also clear up any confusion existing between the phrases *strength of a machine part* and *strength of a material*, or *strength of a part at a point*.

A certain force F applied at D near the end of the 15-in lever shown in Fig. 6-26, which is quite similar to a socket wrench, results in certain stresses in the cantilevered bar *OABC*. This bar *(OABC)* is of AISI 1035 steel, forged and heat-treated so that it has a minimum (ASTM) yield strength of 81 kpsi. We presume that this component would be of no value after yielding. Thus the force F required to initiate yielding can be regarded as the strength of the component part. Find this force.

Solution

We assume that the lever *DC* is strong and hence not a part of the problem. A 1035

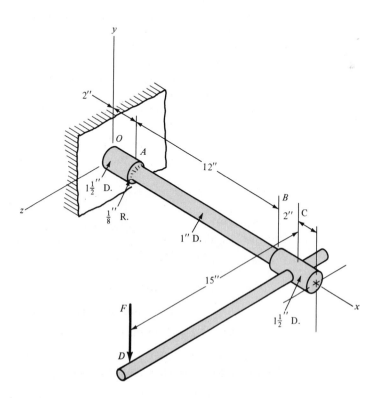

FIGURE 6-26

steel, heat-treated, will have a reduction in area of 50 percent or more and hence is a ductile material at normal temperatures. This also means that stress concentration at shoulder A need not be considered. A stress element at A on the top surface will be subjected to a tensile bending stress and a torsional stress. This point, on the 1-in-diameter section, is the weakest section, and governs the strength of the assembly. The two stresses are

$$\sigma_x = \frac{M}{I/c} = \frac{32M}{\pi d^3} = \frac{32(14F)}{\pi(1^3)} = 142.6F \qquad \text{psi}$$

$$\tau_{xz} = \frac{Tr}{J} = \frac{16T}{\pi d^3} = \frac{16(15F)}{\pi(1^3)} = 76.4F \qquad \text{psi}$$

Employing the distortion-energy theory, we find, from Eq. (6-18), that

$$\sigma' = (\sigma_x^2 + 3\tau_{xz}^2)^{1/2} = [(142.6F)^2 + 3(76.4F)^2]^{1/2} = 195F \qquad \text{psi}$$

Equating the von Mises stress to S_y, we solve for F and get

Answer $$F = \frac{S_y}{195} = \frac{81\ 000}{195} = 415 \text{ lb}$$

In this example the strength of the *material* at point A is $S_y = 81$ kpsi. The strength of the *assembly* or *component* is $F = 415$ lb.

EXAMPLE 6-3   The cantilevered tube shown in Fig. 6-27 is to be made of 2014 aluminum alloy treated to obtain a specified minimum yield strength of 276 MPa. We wish to select a stock-size tube from Table A-8 using a design factor $n_d = 4$. The bending load is $F = 1.75$ kN, the axial tension is $P = 9.0$ kN, and the torsion is $T = 72$ N · m. What is the realized factor of safety?

Solution   The area properties are

$$A = \frac{\pi}{4}(d^2 - d_i^2) \qquad \frac{I}{c} = \frac{\pi}{32}\frac{d^4 - d_i^4}{d}$$

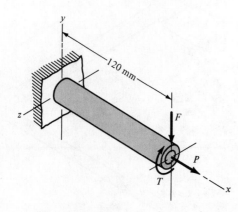

FIGURE 6-27

and

$$\frac{J}{r} = \frac{\pi}{16} \frac{d^4 - d_i^4}{d}$$

Since the maximum bending moment is $M = 120F$, the normal stress, for an element on the top surface of the tube at the origin, is

$$\sigma_x = \frac{P}{A} + \frac{M}{I/c} = \frac{4P}{\pi(d^2 - d_i^2)} + \frac{32(120)Fd}{\pi(d^4 - d_i^4)}$$

$$= \frac{11.46}{d^2 - d_i^2} + \frac{1222d}{d^4 - d_i^4} \tag{1}$$

The torsional stress at the same point is

$$\tau_{xz} = \frac{T}{J/r} = \frac{72(16)d}{\pi(d^4 - d_i^4)} = \frac{366.7d}{d^4 - d_i^4} \tag{2}$$

We next choose the maximum-shear-stress theory as the design basis. The maximum shear stress can be found from a Mohr's circle diagram. It is

$$\tau_{\max} = \left[\left(\frac{\sigma_x}{2}\right)^2 + \tau_{xz}^2 \right]^{1/2} \tag{3}$$

On the basis of the given design factor, this shear stress must conform to

$$\tau_{\max} \leq \frac{S_{sy}}{n_d} \tag{4}$$

Here $S_{sy} = S_y/2 = 0.276/2 = 0.138$ GPa. We have used gigapascals in this relation so that diameters in millimeters can be used in Eqs. (1) and (2). Substituting the value of S_{sy} and $n_d = 4$ in Eq. (4), we find the goal for $\tau_{\max}$ to be

$$\tau_{\max} \leq \frac{0.138}{4} = 0.0345 \text{ GPa} \tag{5}$$

Substituting metric sizes from Table A-8 reveals that a 50- × 5-mm tube is satisfactory. The maximum shear stress is found to be $\tau_{\max} = 0.0297$ GPa for this size. Thus the realized factor of safety is

Answer $$n = \frac{S_{sy}}{\tau_{\max}} = \frac{0.138}{0.0297} = 4.65$$

The solution also reveals for a 50- × 4-mm tube, the next size smaller, a factor of safety of

$$n = \frac{0.138}{0.0355} = 3.89$$

6-10   FAILURE OF BRITTLE MATERIALS

This section is concerned with the failure, or strength, of brittle materials in the usual meaning of the word *brittle*. But it is also concerned with the failure of materials that are usually considered ductile but for some reason tend to fail in a brittle manner. We have already learned, for example, that a normally ductile material may develop a brittle fracture, or crack, if used below the transition temperature.

In order to select suitable theories for use in the analysis of brittle failure, we collect a quantity of data and compare it with the various theories. This has been done in Fig. 6-28. The data are from a large number of biaxial-stress tests of gray cast iron. Some of the tests had their values normalized so that they could all be plotted at the same scale. Not plotted were a large number of results for ASTM grade 40 cast iron from the same source; these data exhibited the same trend as in Fig. 6-28.

Among the various theories already described in this chapter, three are not limited to the prediction of yielding. The graphs for two of these, the maximum-normal-stress theory and the Coulomb-Mohr theory, have been added to Fig. 6-28. In addition, a third theory, here called the *modified Mohr theory,* has been added to the figure.

In the first quadrant, where the two biaxial stresses are positive, the theories are alike and give the same results as the maximum-normal-stress theory. This is expressed in Table 6-2 in the column for the first quadrant. It is in the fourth quadrant, where the stresses have opposite senses, that the theories differ. For example, a line drawn at a slope $\sigma_B/\sigma_A = -1$ defines the shear strength at its intersection with each failure theory.

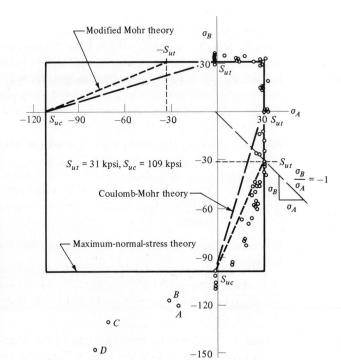

FIGURE 6-28

A plot of experimental data points obtained from tests of cast iron. Shown also are the graphs of three failure theories of possible usefulness for brittle materials. Note points *A, B, C,* and *D.* To avoid congestion in the first quadrant, points have been plotted for $\sigma_A > \sigma_B$ as well as for the opposite sense. [Source of data: Charles F. Walton (ed.), Iron Castings Handbook, Iron Founders' Society, 1971, pp. 215, 216. Cleveland, Ohio]

TABLE 6-2

Recommended Formulas for Prediction of Safety When a Brittle Failure is Considered. (Factor of Safety $= n$; Use $n = 1$ to Predict Failure. For Biaxial Stress States with σ_A and σ_B the Two Nonzero Stresses, and for $\sigma_A \geq \sigma_B$. Note that S_{uc} is Always Treated as Positive)

FAILURE THEORY	FIRST QUADRANT $\sigma_A \geq 0,\ \sigma_B \geq 0$	FOURTH QUADRANT $\sigma_A \geq 0,\ \sigma_B < 0$		EQUATION NO.
Coulomb-Mohr	$\sigma_A = \dfrac{S_{ut}}{n}$	$\dfrac{\sigma_A}{S_{ut}} - \dfrac{\sigma_B}{S_{uc}} = \dfrac{1}{n}$		(6-19)
Modified-Mohr	$\sigma_A = \dfrac{S_{ut}}{n}$	$\sigma_A = \dfrac{S_{ut}}{n}$	$\sigma_B \geq -S_{ut}$	(6-20)
		$\sigma_A - \dfrac{S_{ut}\sigma_B}{S_{uc} - S_{ut}} = \dfrac{S_{uc}S_{ut}}{n(S_{uc} - S_{ut})}$	$\sigma_B < -S_{ut}$	

This intersection yields

$$S_{su} = S_{ut} \tag{a}$$

for the maximum-normal-stress theory and for the modified Mohr theory. But the Coulomb-Mohr theory gives

$$S_{su} = \frac{S_{ut}}{1 + (S_{ut}/S_{uc})} \tag{b}$$

For the cast iron of Fig. 6-28, this works out to

$$S_{su} = 0.78 S_{ut} \tag{c}$$

Equations (6-19) and (6-20) in Table 6-2 can be deduced by writing equations for a straight line and solving for the constants using the coordinates of the line ends.

The data plotted in Fig. 6-28 show that either the Coulomb-Mohr theory or the modified Mohr theory is acceptable for design purposes. In analysis, however, the modified Mohr theory appears to be a better predictor of actual failure. Both theories are easily programmed using Table 6-2.

The small amount of data shown in the third quadrant of Fig. 6-28 is insufficient to develop a recommendation. The use of the maximum-normal-stress theory, when both of the biaxial stresses are negative, appears to be satisfactory.

EXAMPLE 6-4   Figure 6-29 shows a torsion-bar spring having maximum loads of $F = 35$ N and $T = 8$ N · m. The material is ASTM grade 30 cast iron. The theoretical stress-concentration factors are 1.68 for bending and 1.42 for torsion. Determine a suitable diameter d using a design factor of 4.

Solution   Equation (5-26) for the reduced value of the stress-concentration factor cannot be solved until the fillet radius is known. Therefore we shall use the full values of the

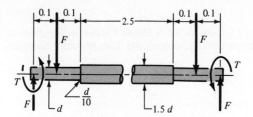

FIGURE 6-29

Dimensions in meters.

stress-concentration factors for preliminary calculation. The bending moment at the shoulders is

$$M = 100(35) = 3500 \text{ N} \cdot \text{m}$$

So the bending stress is

$$\sigma_x = K_t \frac{32M}{\pi d^3} = \frac{1.68(32)(3500)}{\pi d^3} = \frac{59.9(10^3)}{d^3} \text{ MPa}$$

The torsional stress is

$$\tau_{xz} = K_{ts} \frac{16T}{\pi d^3} = \frac{1.42(16)(8)(10^3)}{\pi d^3} = \frac{57.9(10^3)}{d^3} \text{ MPa}$$

The corresponding nonzero principal stresses are found to be

$$\sigma_A = \frac{95.1(10^3)}{d^3} \text{ MPa} \qquad \sigma_B = -\frac{35.2(10^3)}{d^3} \text{ MPa}$$

From Table A-24 we find the minimum tensile strength to be $S_{ut} = 30(6.89) = 207$ MPa. The compressive strength is listed as 109 kpsi; but this is the typical value, not the minimum. So we select $S_{uc} = 105$ kpsi as the minimum and plan to insist that it be used in the purchase specifications. Thus, in SI units, $S_{uc} = 6.89(105) = 723$ MPa.

We now opt for the Coulomb-Mohr theory. Equation (6-19) in Table 6-2 applies. Substituting gives

$$\frac{95.1(10^3)}{207d^3} - \frac{-35.2(10^3)}{723d^3} = \frac{1}{4}$$

Solving gives $d = 12.7$ mm. The next-largest preferred size in Table A-17 is 14 mm. But let us try $d = 12$ mm first; it just might be safe.

For cast iron, $q = 0.20$. Therefore, from Eq. (5-26), the reduced value of the stress-concentration factor for bending is

$$K_f = 1 + q(K_t - 1) = 1 + 0.20(1.68 - 1) = 1.14$$

and, for torsion,

$$K_{sf} = 1 + 0.20(1.42 - 1) = 1.08$$

Using these values and $d = 12$ mm, the stresses are recalculated and found to be

$$\sigma_x = 23.5 \text{ MPa} \qquad \tau_{xz} = 25.5 \text{ MPa}$$

$$\sigma_A = 39.8 \text{ MPa} \qquad \sigma_B = -16.3 \text{ MPa}$$

Solving Eq. (6-19) for the factor of safety this time gives $n = 4.66$. Thus a diameter of 12 mm is quite satisfactory.

6-11   STOCHASTIC ANALYSIS—INTRODUCTION

The purpose of stochastic analysis, when stress and strength are involved, is to determine the reliability with both stress and strength distributions known at the critical location in a part. We shall simplify the notation for a while by not qualifying the strength S or the stress σ with subscripts. Remembering from Chap. 4 that boldface characters are used for stochastic variables, we assume that both the stress and the strength are Gaussian, and hence that

$$\boldsymbol{\sigma} \sim N(\hat{\mu}_\sigma, \hat{\sigma}_\sigma) \qquad \mathbf{S} \sim N(\hat{\mu}_S, \hat{\sigma}_S)$$

where $\hat{\mu}$ and $\hat{\sigma}$ are the mean and the standard deviation, respectively.

It is convenient to define the term *stress margin* **m,** which is the difference between the strength and the stress. Thus, for specific values of S and σ,

$$m = S - \sigma \tag{6-21}$$

Reliability is the probability that the strength exceeds the stress, or, alternatively, the probability that the stress margin is greater than zero. This can be expressed in the form

$$R = p(S > \sigma) = p[(S - \sigma) > 0] = p(m > 0)$$

The central limit theorem of statistics says, in part, that the difference between variables that each have a normal distribution is likewise normally distributed. Therefore

$$\mathbf{m} \sim N(\hat{\mu}_m, \hat{\sigma}_m)$$

The distribution of m is shown in Fig. 6-30.

Referring next to Table 4-4, for the difference between two independent random variables, we find

$$\hat{\mu}_m = \hat{\mu}_S - \hat{\mu}_\sigma \qquad \hat{\sigma}_m = (\hat{\sigma}_S^2 + \hat{\sigma}_\sigma^2)^{1/2} \tag{a}$$

To find the probability that $m > 0$, we find the z variable corresponding to $m = 0$ in Fig. 6-30. Using Eq. (4-14), this gives

$$z = \frac{m - \hat{\mu}_m}{\hat{\sigma}_m} = \frac{0 - \hat{\mu}_m}{\hat{\sigma}_m} = -\frac{\hat{\mu}_m}{\hat{\sigma}_m} \tag{b}$$

Substituting equation pair (a) in Eq. (b) yields

$$z = -\frac{\hat{\mu}_S - \hat{\mu}_\sigma}{(\hat{\sigma}_S^2 + \hat{\sigma}_\sigma^2)^{1/2}} \tag{6-22}$$

This is called the *coupling equation,* because it relates the reliability, through z, to the statistical parameters of the normally distributed strength and stress. The reliability is

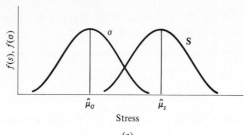

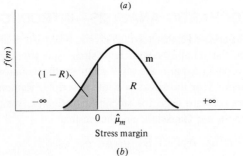

FIGURE 6-30

Plot of density functions showing how the interference of **S** and σ is used to obtain the stress margin **m**. (*a*) Stress and strength distributions. (*b*) Distribution of interference; the reliability R is the area of the density function for all m's greater than zero; the interference is the area $(1 - R)$.

given by

$$R = \int_{z}^{\infty} \frac{1}{\sqrt{2\pi}} \exp\left(-\frac{u^2}{2}\right) du \qquad (6\text{-}23)$$

or by $(1 - \text{tabulation})$ in Table A-10 when z is negative (that is, when high reliability exists).

EXAMPLE 6-5

The load-induced stress at the critical location of a part in simple tension is $\sigma \sim N(40, 5)$ kpsi, and the yield strength is $\mathbf{S} \sim N(60, 4)$ kpsi. Estimate the reliability against yielding.

Solution

$$z = -\frac{\hat{\mu}_S - \hat{\mu}_\sigma}{(\hat{\sigma}_S^2 + \hat{\sigma}_\sigma^2)^{1/2}} = -\frac{60 - 40}{(4^2 + 5^2)^{1/2}} = -3.12$$

From Table A-10 we obtain

Y	3.10	3.12	3.20
X	0.000 968	x	0.000 687

Using the method of Prob. 1-2, we obtain $x = 0.000\ 912$. Therefore the reliability is

Answer

$$R = R(z) = R(-3.12) = 1 - \text{tabulation} = 1 - 0.000\ 912 = 0.9991$$

EXAMPLE 6-6

Find the diameter of a tension member if the yield strength is $\mathbf{S} \sim N(60, 4)$ kpsi, the load is $\mathbf{F} \sim N(50, 5)$ kip, and the desired reliability is 0.999.

Solution

Since $\sigma = 4F/\pi d^2$, we have, from Table 4-4, that the mean and standard deviation of the stress are

$$\hat{\mu}_\sigma = \frac{4}{\pi d^2}\hat{\mu}_F = \frac{4(50)}{\pi d^2} = \frac{63.7}{d^2} \quad \text{kpsi}$$

$$\hat{\sigma}_\sigma = \frac{4}{\pi d^2}\hat{\sigma}_F = \frac{4(5)}{\pi d^2} = \frac{6.37}{d^2} \quad \text{kpsi}$$

Corresponding to $R = 0.999$, $z = -3.12$. Rearranging Eq. (6-22), we have

$$z(\hat{\sigma}_S^2 + \hat{\sigma}_\sigma^2)^{1/2} = -\hat{\mu}_S + \hat{\mu}_\sigma \qquad (1)$$

Substituting appropriate values next gives the equation

$$-3.12\left[(4)^2 + \left(\frac{6.37}{d^2}\right)^2\right]^{1/2} = -60 + \frac{63.7}{d^2} \qquad (2)$$

or

$$(4)^2 + \left(\frac{6.37}{d^2}\right)^2 = \left(\frac{-60}{-3.12} + \frac{63.7}{-3.12d^2}\right)^2$$

This reduces to

$$d^4 - 2.219d^2 + 1.063 = 0$$

Solving Eq. (3) as a quadratic equation gives $d^2 = 1.519$, and so

Answer $d = (1.519)^{1/2} = 1.232$ in

6-12 FACTOR OF SAFETY—A NOTE

One of the questions that sometimes arise among engineers when they study reliability design for the first time is, But what is the factor of safety? The answer is that the factor of safety is simply not pertinent to the reliability approach. An element or part may be analyzed to determine its reliability, or it may be designed to a reliability specification. The factor of safety does not enter into either of these approaches.

Recalling that the factor of safety is the ratio of the minimum strength to the maximum stress, there is a way to acquire some additional insight into the problem. For example, we can be pretty certain that the minimum strength is around 3 or 4 standard deviations less than the mean strength. In the same way, the maximum stress will probably not exceed the sum of the mean stress and 3 or 4 standard deviations. We assume that the possibility of catastrophic situations has been accounted for in the statistical parameters. If, now, we are not too fussy, we might select 3 standard deviations for both and express the factor of safety as

$$n = \frac{\hat{\mu}_S - 3\hat{\sigma}_S}{\hat{\mu}_\sigma + 3\hat{\sigma}_\sigma} \qquad (6\text{-}24)$$

As an example of the use of this equation, the mean stress in Example 6-5 is found to be 40 kpsi and the standard deviation 5 kpsi. Thus, for this example, the factor of safety turns out to be

$$n = \frac{60 - (3)(4)}{40 + (3)(5)} = 0.873$$

Equation (6-24) will not be used in this book. It is presented so as to give you some insight into the two methods of design. Neither method is perfect. Both have their advantages and disadvantages. Safety and economy in design are strange and often antagonistic bedfellows. A thorough knowledge of both approaches is the surest guarantee of design success.

6-13   LOGNORMAL INTERFERENCE

Suppose we have an interference between a strength and a stress which are both lognormal and are given by $\mathbf{S} \sim LN(\hat{\mu}_S, \hat{\sigma}_S)$ and $\boldsymbol{\sigma} \sim LN(\hat{\mu}_\sigma, \hat{\sigma}_\sigma)$. Since both are lognormal, we can write

$$\mathbf{S}_{\ln} = \ln \mathbf{S} \qquad \boldsymbol{\sigma}_{\ln} = \ln \boldsymbol{\sigma} \tag{6-25}$$

Also, from Eqs. (4-20) and (4-21),

$$\hat{\sigma}_{\ln S} = C_S = \frac{\hat{\sigma}_S}{\hat{\mu}_S} \qquad \hat{\sigma}_{\ln \sigma} = C_\sigma = \frac{\hat{\sigma}_\sigma}{\hat{\mu}_\sigma} \tag{6-26}$$

and

$$\hat{\mu}_{\ln S} = \ln \hat{\mu}_S - \tfrac{1}{2} C_S^2$$
$$\hat{\mu}_{\ln \sigma} = \ln \hat{\mu}_\sigma - \tfrac{1}{2} C_\sigma^2 \tag{6-27}$$

With these transformations, we can write the coupling equation [Eq. (6-22)] as

$$z = -\frac{\hat{\mu}_{\ln S} - \hat{\mu}_{\ln \sigma}}{(\hat{\sigma}_{\ln S}^2 + \hat{\sigma}_{\ln \sigma}^2)^{1/2}} \tag{6-28}$$

6-14   INTERFERENCE—GENERAL

In the previous sections, we have employed interference theory to estimate reliability when the distributions are both normal and when they are both lognormal. Sometimes, however, it turns out that the strength has, say, a Weibull distribution while the stress is distributed lognormally. In fact, stresses are quite likely to have a lognormal distribution, because the multiplication of variates that are normally distributed produces a result that approaches lognormal. What all this means is that we must expect to encounter interference problems involving mixed distributions and we need a general method to handle the problem.

It is quite likely that we will use interference theory for problems involving distributions other than strength and stress. For this reason we employ the subscript 1 to

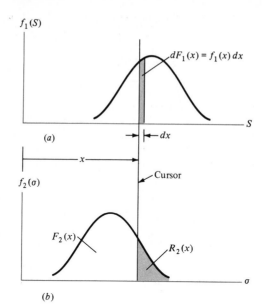

FIGURE 6-31

(a) PDF of the strength distribution; (b) PDF of the load-induced stress distribution.

designate the strength distribution and the subscript 2 to designate the stress distribution. Figure 6-31 shows these two distributions aligned so that a single cursor x can be used to identify points on both distributions. We can now write

$$\begin{pmatrix} \text{Probability that} \\ \text{stress is less} \\ \text{than strength} \end{pmatrix} = dp(\sigma < x) = dR = F_2(x)\, dF_1(x)$$

By substituting $1 - R_2$ for F_2 and $-dR_1$ for dF_1, we have

$$dR = -[1 - R_2(x)]\, dR_1(x)$$

The reliability for all possible locations of the cursor is obtained by integrating x from $-\infty$ to ∞; but this corresponds to an integration from 1 to 0 on the reliability R_1. Therefore

$$R = -\int_1^0 [1 - R_2(x)]\, dR_1(x)$$

which can be written

$$R = 1 - \int_0^1 R_2\, dR_1 \tag{6-29}$$

where

$$R_1(x) = \int_x^\infty f_1(S)\, dS \tag{6-30}$$

$$R_2(x) = \int_x^\infty f_2(\sigma)\, d\sigma \tag{6-31}$$

FIGURE 6-32

Curve shapes of the R_1-R_2 relation. In each case the shaded area is equal to $1 - R$ and is obtained by numerical integration. (a) Typical curve for asymptotic distribution; (b) curve shape obtained for lower truncated distributions such as Weibull.

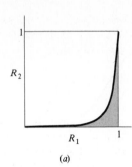

(a)

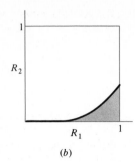

(b)

For the usual distributions encountered, plots of R_1 versus R_2 appear as shown in Fig. 6-32. Both of the cases shown are amenable to numerical integration and computer solution. When the reliability is high, the bulk of the integration area is under the right-hand spike of Fig. 6-32a. A system of increasingly finer Simpson's rule ordinate spacing is in order.

PROBLEMS*

6-1   A hot-rolled bar has a minimum yield strength in tension and compression of 44 kpsi. Find the factors of safety for each applicable theory of failure for the following stress states:
(a) $\sigma_x = 9$ kpsi, $\sigma_y = -5$ kpsi
(b) $\sigma_x = 12$ kpsi, $\tau_{xy} = 3$ kpsi ccw
(c) $\sigma_x = -4$ kpsi, $\sigma_y = -9$ kpsi, $\tau_{xy} = 5$ kpsi cw
(d) $\sigma_x = 11$ kpsi, $\sigma_y = 4$ kpsi, $\tau_{xy} = 1$ kpsi cw

6-2   Repeat Prob. 6-1 but with $S_y = 390$ MPa for:
(a) $\sigma_A = 65$ MPa, $\sigma_B = 72$ MPa
(b) $\sigma_A = 72$ MPa, $\sigma_B = 40$ MPa
(c) $\sigma_A = 64$ MPa, $\sigma_B = -64$ MPa
(d) $\sigma_A = 0$ MPa, $\sigma_B = -72$ MPa

6-3   Repeat Prob. 6-1 but with $S_y = 420$ MPa for:
(a) $\sigma_x = 180$ MPa, $\sigma_y = 180$ MPa
(b) $\sigma_x = 140$ MPa, $\tau_{xy} = 80$ MPa cw
(c) $\sigma_y = -80$ MPa, $\tau_{xy} = 120$ MPa ccw
(d) $\tau_{xy} = 200$ MPa

6-4   Repeat Prob. 6-1 for:
(a) $\sigma_A = 10$ kpsi, $\sigma_B = 10$ kpsi
(b) $\sigma_A = 10$ kpsi, $\sigma_B = 5$ kpsi
(c) $\sigma_A = 10$ kpsi, $\sigma_B = -10$ kpsi
(d) $\sigma_A = 0$ kpsi, $\sigma_B = -10$ kpsi

6-5   An ASTM 30 cast iron has minimum ultimate strengths of 30 kpsi in tension and 100 kpsi in compression. Find the factors of safety using the maximum-normal-stress theory and the

---

*The asterisk indicates a problem that may not have a unique result, a particularly challenging problem, or a design-type problem.

Coulomb-Mohr theory for each of the stress states that follow. Plot a $\sigma_A - \sigma_B$ diagram to scale as in Figs. 6-16 and 6-24 and locate the coordinates of each stress state.

(a) $\sigma_A = 20$ kpsi, $\sigma_B = 20$ kpsi
(b) $\tau_{xy} = 15$ kpsi
(c) $\sigma_A = \sigma_B = -80$ kpsi
(d) $\sigma_A = 20$ kpsi, $\sigma_B = -10$ kpsi

6-6   This problem illustrates that the factor of safety for a machine element depends on the particular point selected for analysis. Here you are to compute factors of safety, based upon the distortion-energy theory, for stress elements at A and B of the member shown in the figure. This bar is made of AISI 1006 cold-drawn steel and is loaded by the forces $F = 0.55$ kN, $P = 8.0$ kN, and $T = 30$ N · m.

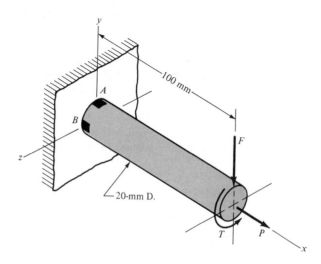

PROBLEM 6-6

6-7*   Design the lever arm CD of Fig. 6-26 by specifying a suitable size and material.

6-8   The figure shows a crank loaded by a force $F = 300$ lb which causes twisting and bending of the $\frac{3}{4}$-in-diameter shaft fixed to a support at the origin of the reference system. In actuality, the support may be an inertia which we wish to rotate, but for the purposes of a strength analysis we

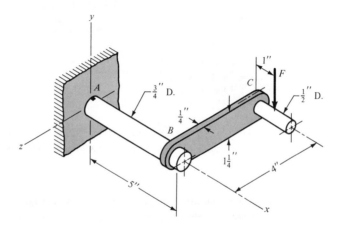

PROBLEM 6-8

can consider this to be a statics problem. The material of the shaft AB is hot-rolled AISI 1018 steel (Table A-20). Using the maximum-shear-stress theory, find the factor of safety based on the stress state at point A.

6-9   A spherical pressure vessel is formed of 18-gauge (0.05-in) cold-drawn AISI 1018 sheet steel. If the vessel has a diameter of 8 in, estimate the pressure necessary to initiate yielding. What is the estimated bursting pressure?

6-10   This problem illustrates that the strength of a machine part can sometimes be measured in units other than those of force or moment. For example, the maximum speed that a flywheel can reach without yielding or fracturing is a measure of its strength. In this problem you have a rotating ring made of hot-forged AISI 1020 steel; the ring has a 6-in inside diameter and a 10-in outside diameter and is 1.5 in thick. What speed in revolutions per minute would cause the ring to yield? At what radius would yielding begin? [*Note:* The maximum radial stress occurs at $r = (r_o r_i)^{1/2}$; see Eq. (2-56).]

6-11   A light pressure vessel is made of 2024-T3 aluminum alloy tubing with suitable end closures. This cylinder has a $3\frac{1}{2}$-in OD, a 0.065-in wall thickness, and $\nu = 0.334$. The purchase order specifies a minimum yield strength of 46 kpsi. What is the factor of safety if the pressure-release valve is set at 500 psi?

6-12   A cold-drawn AISI 1015 steel tube is 300 mm OD by 200 mm ID and is to be subjected to an external pressure caused by a shrink fit. What maximum pressure would cause the material of the tube to yield?

6-13   What speed would cause fracture of the ring of Prob. 6-10 if it were made of grade 30 cast iron?

6-14   The figure shows a shaft mounted in bearings at A and D and having pulleys at B and C. The forces shown acting on the pulley surfaces represent the belt pulls. The shaft is to be made of ASTM grade 25 cast iron using a design factor $n_d = 2.8$. What diameter should be used for the shaft?

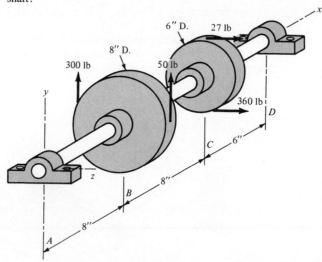

PROBLEM 6-14

6-15   By modern standards, the shaft design of Prob. 6-14 is poor because it is so long. Suppose it is redesigned by halving the length dimensions. Using the same material and design factor as in Prob. 6-14, find the new shaft diameter.

6-16   The clevis pin shown in the figure is 12 mm in diameter and has the dimensions $a = 12$ mm and $b = 18$ mm. The pin is machined from AISI 1018 hot-rolled steel (Table A-20) and is to be loaded to no more than 4.4 kN. Determine whether or not the assumed loading of c yields a factor of safety any different from that of d. Use the maximum-shear-stress theory.

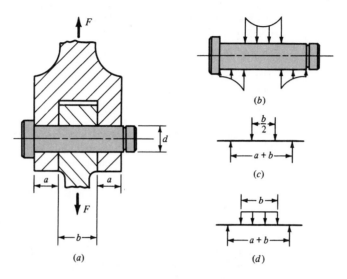

PROBLEM 6-16   (a)   (b)   (c)   (d)

6-17   Repeat Prob. 6-16, but this time use the distortion-energy theory.

6-18   A stochastic force $F \sim N(410, 45)$ lb applied at D near the end of the 15-in lever shown in the figure results in certain stresses in the cantilevered bar $OABC$.

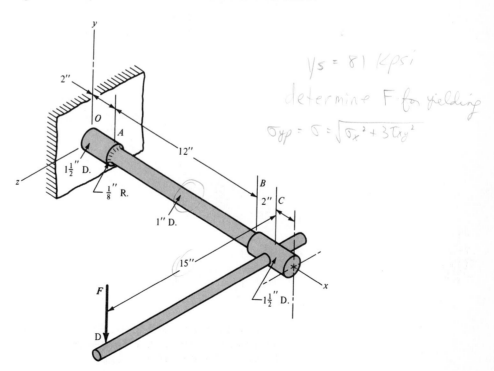

PROBLEM 6-18

(a) Determine the mean and standard deviation of the principal stresses for a stress element on the upper surface of the 1-in section at A.

(b) What are the mean and standard deviation of the von Mises stresses corresponding to the stress element at A?

6-19   A carbon steel collar of length 1 in is to be machined to inside and outside diameters, respectively, of

$$D_i = 0.750 \pm 0.0004 \text{ in} \qquad D_o = 1.125 \pm 0.002 \text{ in}$$

This collar is to be shrink-fitted to a hollow steel shaft having inside and outside diameters, respectively, of

$$d_i = 0.375 \pm 0.002 \text{ in} \qquad d_o = 0.752 \pm 0.0004 \text{ in}$$

These tolerances are assumed to have a normal distribution, to be centered in the spread interval, and to have a total spread of ± 4 standard deviations. Determine the means and the standard deviations of the tangential stress components for both cylinders at the interface.

6-20   Suppose the collar of Prob. 6-19 has a yield strength of $\mathbf{S}_y \sim N(95.5, 6.59)$ kpsi. What is the probability that the material will not yield?

6-21   The clevis pin of Prob. 6-16 is to be made of a malleable cast iron with a tensile strength of $\mathbf{S}_{ut} \sim N(44.5, 4.34)$ kpsi. The load is given as $\mathbf{F} \sim N(1500, 150)$ lb.

(a) Determine the reliability for a point at midspan and on the bottom of the pin. Note at this point that the direct shear is zero.

(b) Repeat part (a), but with a strength distribution that is Weibullian and is given as $\mathbf{S}_{ut} \sim W(27.7, 46.2, 4.32)$ kpsi.

6-22   A carbon steel tube has an outside diameter of 1 in and a wall thickness of $\frac{1}{8}$ in. The tube is to carry an internal hydraulic pressure given as $\mathbf{p} \sim N(6000, 500)$ psi. The material of the tube has a yield strength of $\mathbf{S}_y \sim N(50, 4.1)$ kpsi. Find the reliability using thin-wall theory.

6-23*   A split-ring clamp-type shaft collar is shown in the figure. The collar is 2 in OD by 1 in ID by $\frac{1}{2}$ in wide. The screw is designated as $\frac{1}{4}$-28 UNF. The relation between the screw tightening torque T, the nominal screw diameter d, and the tension in the screw F_i is approximately $T = 0.2F_i d$. The shaft is sized to obtain a close running fit. Find the axial holding force F_x of the collar as a function of the coefficient of friction and the screw torque.

PROBLEM 6-23

6-24*   Suppose the collar of Prob. 6-23 is tightened using a screw torque of 190 lb · in. The collar material is AISI 1040 steel heat-treated to a minimum tensile yield strength of 63 kpsi.

(a) Estimate the tension in the screw.

(b) By relating the tangential stress to the hoop tension, find the internal pressure of the shaft on the ring.

(c) Find the tangential and radial stresses in the ring at the inner surface.

(d) Determine the maximum shear stress and the von Mises stress.

(e) What are the factors of safety based on the maximum-shear-stress theory and the distortion-energy theory?

6-25   In Prob. 6-23, the role of the screw was to induce the hoop tension which produces the clamping. The screw should be placed so that no moment is induced in the ring. Just where should the screw be located?

6-26   A tube has another tube shrunk over it. The specifications are:

	INNER MEMBER	OUTER MEMBER
ID	1.000 ± 0.002 in	1.999 ± 0.0004 in
OD	2.000 ± 0.0004 in	3.000 ± 0.004 in

Both tubes are made of a plain carbon steel.
(a) Find the shrink-fit pressure and the stresses at the fit surface.
(b) If the inner tube is changed to solid shafting with the same outside dimensions, find the shrink-fit pressure and the stresses at the fit surface.

6-27   Steel tubes with a Young's modulus of 207 GPa have the specifications:

	INNER TUBE	OUTER TUBE
ID	25 ± 0.050 mm	49.98 ± 0.010 mm
OD	50 ± 0.010 mm	75 ± 0.10 mm

These are shrink-fitted together. Find the shrink-fit pressure and the von Mises stress in each body at the fit surface.

6-28   A 2-in-diameter solid steel shaft has a gear with ASTM grade 20 cast-iron hub ($E = 14.5$ Mpsi) shrink-fitted to it. The specifications for the shaft are

$$2.000 \begin{array}{c} + \ 0.0000 \\ - \ 0.0004 \end{array} \text{in}$$

The hole in the hub is sized at 1.999 ± 0.0004 in and an OD of $4.00 \pm \frac{1}{32}$ in. Use the modified Mohr theory and estimate the factor of safety guarding against fracture in the gear hub due to the shrink fit.

6-29   A steel shaft with a diameter of

$$1.875 \begin{array}{c} + \ 0.0000 \\ - \ 0.0004 \end{array} \text{in}$$

is shrink-fitted to the hub of a pulley with a bore of 1.870 ± 0.002 in and a hub diameter of $3.25 \pm \frac{1}{32}$ in. The pulley hub, which is grade 40 cast iron, has an ultimate tensile strength of $\mathbf{S}_{ut} \sim N(44.5, 4.34)$ kpsi, a compressive ultimate strength of $\mathbf{S}_{uc} \sim N(140, 13.1)$ kpsi, and a modulus of elasticity $E = 17$ Mpsi. Estimate the reliability of the hub against fracture due to the shrink-fit pressure.

6-30   This problem is an exercise in using Eq. (6-29) to estimate reliability. A tensile strength is uniformly distributed in the interval $46 \le S_{ut} \le 54$ kpsi; that is, $\mathbf{S} \sim U[48, 54]$ kpsi. The load-induced stress is $\boldsymbol{\sigma} \sim U[40, 48]$ kpsi. Estimate the reliability and check by drawing a figure similar to Fig. 6-32.

6-31   Solve Prob. 6-30 using the idea of stress margin of Eq. (6-21), taking careful note that the distributions of **S** and **σ** are uniform random. The stress margin is the difference between two uniform random variables which overlap. The distribution of **m** is isosceles triangular between the range numbers of **m**. Determine the reliability and note the factor of safety.

6-32   Write a computer program to simulate conditions of Prob. 6-30 using a uniform random number generator supplied to your machine by the computer manufacturer. Generate independent random instances of the stress and strength in order to create random instances of the stress margin. Keep track of the number of instances in which $m > 0$. The reliability is this number divided by the number of trials. (Four instances of 10 000 trials estimate the reliability as 0.9718, 0.9671, 0.9682, and 0.9690. Do you agree?)

ANSWERS   **6-3** For parts (*a*), (*b*), (*c*), and (*d*), respectively, the results are:
Maximum-normal-stress theory: 2.33, 2.39, 2.52, 2.10
Maximum-shear-stress theory: 2.33, 1.98, 1.66, 1.05
Distortion-energy theory: 2.33, 2.14, 1.89, 1.21
Coulomb-Mohr theory: 2.33, 1.98, 1.66, 1.05
6-6 At *A*, $n = 3.27$; at *B*, $n = 7.33$
6-9 1309 psi to yield using the distortion-energy theory; 1590 psi to burst using the maximum-normal-stress theory
6-12 88.9 MPa
6-15 $1\frac{1}{8}$ in
6-18 (*a*) $\sigma_1 \sim N(72.1, 7.93)$ kpsi, $\sigma_3 \sim N(-13.6, 1.50)$ kpsi; (*b*) $\sigma' \sim N(79.8, 8.77)$ kpsi
6-21 (*a*) $R = 0.9998$; (*b*) $R = 0.9997$
6-23 $F_x = 2\pi\mu T/(0.2d)$
6-27 $p = 19.4$ MPa; inner tube, $\sigma' = 28.4$ MPa; outer tube, $\sigma' = 62.5$ MPa
6-29 $R = 0.46$

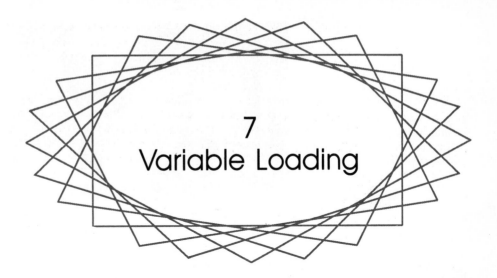

7
Variable Loading

In Chap. 6 we considered the analysis and design of parts subjected to static loads. It is an entirely different matter when the parts are subjected to time-varying, or nonstatic, loads. In this chapter we shall examine how parts fail under nonstatic conditions and how to design them to resist such conditions.

7-1   INTRODUCTION

In most testing of those properties of materials that relate to the stress-strain diagram, the load is applied gradually, to give sufficient time for the strain to fully develop. Furthermore, the specimen is tested to destruction, and so the stresses are applied only once. Testing of this kind is applicable, then, to what are known as *static conditions;* such conditions closely approximate the actual conditions to which many structural and machine members are subjected.

The condition frequently arises, however, in which the stresses vary or they fluctuate between levels. For example, a particular fiber on the surface of a rotating shaft subjected to the action of bending loads undergoes both tension and compression for each revolution of the shaft. If the shaft is part of an electric motor rotating at 1725 rev/min, the fiber is stressed in tension and compression 1725 times each minute. If, in addition, the shaft is also axially loaded (as it would be, for example, by a helical or worm gear), an axial component of stress is superposed upon the bending component. In this case, some stress is always present in any one fiber, but now the *level* of stress is fluctuating. These and other kinds of loading occurring in machine members produce stresses which are called *repeated, alternating,* or *fluctuating* stresses.

Often, machine members are found to have failed under the action of repeated or fluctuating stresses; yet the most careful analysis reveals that the actual maximum stresses were below the ultimate strength of the material, and quite frequently even below the yield strength. The most distinguishing characteristic of these failures is that the stresses have been repeated a very large number of times. Hence the failure is called a *fatigue failure*.

FIGURE 7-1

A fatigue failure of a $7\frac{1}{2}$-in-diameter forging at a press fit. The specimen is UNS G10450 steel, normalized and tempered and has been subjected to rotating bending. *(Courtesy of The Timken Company.)*

A fatigue failure begins with a small crack. The initial crack is so minute that it cannot be detected by the naked eye and is even quite difficult to locate in a Magnaflux or x-ray inspection. The crack will develop at a point of discontinuity in the material, such as a change in cross section, a keyway, or a hole. Less obvious points at which fatigue failures are likely to begin are inspection or stamp marks, internal cracks, or even irregularities caused by machining. Once a crack is initiated, the stress-concentration effect becomes greater and the crack progresses more rapidly. As the stressed area decreases in size, the stress increases in magnitude until, finally, the remaining area fails suddenly. A fatigue failure, therefore, is characterized by two distinct regions (Fig. 7-1). The first of these is due to the progressive development of the crack, while the second is due to the sudden fracture. The zone of sudden fracture is very similar in appearance to the fracture of a brittle material, such as cast iron, that has failed in tension.

When machine parts fail statically, they usually develop a very large deflection, because the stress has exceeded the yield strength; and the part is replaced before fracture actually occurs. Thus many static failures give visible warning in advance. But a fatigue failure gives no warning! It is sudden and total, and hence dangerous. It is relatively simple to design against a static failure, because our knowledge is comprehensive. Fatigue is a much more complicated phenomenon, only partially understood, and the engineer seeking competence must acquire as much knowledge of the subject as possible. Anyone who lacks knowledge of fatigue can double or triple design factors and formulate a design that will not fail. However, such designs cannot compete in the marketplace. Neither can the engineers who produce them.

7-2   THE STRAIN-LIFE THEORY OF FATIGUE FAILURE

The best theory yet advanced to explain the nature of fatigue failure is called by some the *strain-life theory*. The theory can be used to estimate fatigue strengths, but when it is so used it is necessary to compound several idealizations, and so some uncertainties will exist in the results. For this reason, the theory is presented here only because of its value in explaining the nature of fatigue.

A fatigue failure almost always begins at a local discontinuity such as a notch, crack, or other area of stress concentration. When the stress at the discontinuity ex-

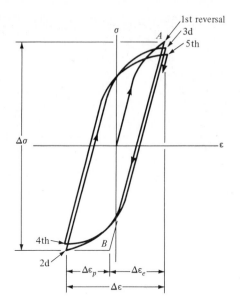

FIGURE 7-2

True stress–true strain hysteresis loops showing the first five stress reversals of a cyclic-softening material. The graph is slightly exaggerated for clarity. Note that the slope of line AB is the modulus of elasticity E. The stress range is $\Delta\sigma$; $\Delta\varepsilon_p$ is the plastic-strain range, and $\Delta\varepsilon_e$ is the elastic-strain range. The total-strain range is $\Delta\varepsilon = \Delta\varepsilon_p + \Delta\varepsilon_e$.

ceeds the elastic limit, plastic strain occurs. If a fatigue fracture is to occur, there must exist cyclic plastic strains. Thus we shall need to investigate the behavior of materials subject to cyclic deformation.

In 1910, Bairstow verified by experiment Bauschinger's theory that the elastic limits of iron and steel can be changed, either up or down, by the cyclic variations of stress.[*] In general, the elastic limits of annealed steels are likely to increase when subjected to cycles of stress reversals, while cold-drawn steels exhibit a decreasing elastic limit.

Test specimens subjected to reversed bending are not suitable for strain cycling, because of the difficulty of measuring plastic strains. Consequently, most of the research has been done using axial specimens. By using electrical transducers, it is possible to generate signals that are proportional to the stress and strain, respectively. These signals can then be displayed on an oscilloscope or plotted on an *XY* plotter. R. W. Landgraf has investigated the low-cycle fatigue behavior of a large number of very high-strength steels, and during his research he made many cyclic stress-strain plots.[†] Figure 7-2 has been constructed to show the general appearance of these plots for the first few cycles of controlled cyclic strain. In this case the strength decreases with stress repetitions, as evidenced by the fact that the reversals occur at ever-smaller stress levels. As previously noted, other materials may be strengthened, instead, by cyclic stress reversals.

Slightly different results may be obtained if the first reversal occurs in the compressive region; this is probably due to the fatigue-strengthening effect of compression.

Landgraf's paper contains a number of plots that compare the monotonic stress-

[*]L. Bairstow, "The Elastic Limits of Iron and Steel under Cyclic Variations of Stress," *Philosophical Transactions* Series A, vol. 210, Royal Society of London, 1910, pp. 35–55.

[†]R. W. Landgraf, *Cyclic Deformation and Fatigue Behavior of Hardened Steels,* Report no. 320, Department of Theoretical and Applied Mechanics, University of Illinois, Urbana, 1968, pp. 84–90.

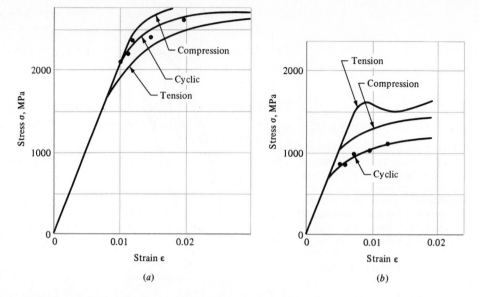

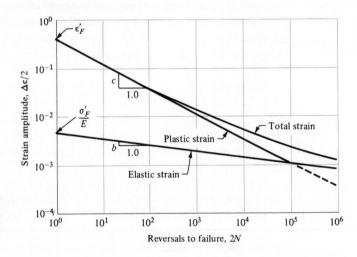

FIGURE 7-3

Monotonic and cyclic stress-strain
results. (a) Ausformed H-11 steel;
660 Bhn; (b) SAE 4142 steel,
400 Bhn.

strain relations in both tension and compression with the cyclic stress-strain curve.*
Two of these have been redrawn and are shown in Fig. 7-3. The importance of these is
that they emphasize the difficulty of attempting to predict the fatigue strength of a
material from known values of monotonic yield or ultimate strengths in the low-cycle
region.

The SAE Fatigue Design and Evaluation Steering Committee released a report in
1975 in which the life in reversals to failure is related to the strain amplitude.† The
report contains a plot of this relationship for SAE 1020 hot-rolled steel; the graph has
been reproduced as Fig. 7-4. To explain the graph, we first define the following terms:

FIGURE 7-4

A log-log plot showing how the
fatigue life is related to the true-
strain amplitude for hot-rolled
SAE 1020 steel. (Reproduced
from Tech. Rep. SAE J1099, by
permission.)

*Ibid., pp. 58–62.

†*Technical Report on Fatigue Properties*, SAE J1099, 1975.

Fatigue ductility coefficient ε'_F is the true strain corresponding to fracture in one reversal (point *A* in Fig. 7-2). The plastic-strain line begins at this point in Fig. 7-4.

Fatigue strength coefficient σ'_F is the true stress corresponding to fracture in one reversal (point *A* in Fig. 7-2). Note in Fig. 7-4 that the elastic-strain line begins at σ'_F/E.

Fatigue ductility exponent c is the slope of the plastic-strain line in Fig. 7-4 and is the power to which the life 2*N* must be raised to be proportional to the true plastic-strain amplitude.

Fatigue strength exponent b is the slope of the elastic-strain line, and is the power to which the life 2*N* must be raised to be proportional to the true-stress amplitude.

Now, from Fig. 7-2, we see that the total strain is the sum of the elastic and plastic components. Therefore the total strain amplitude is

$$\frac{\Delta\varepsilon}{2} = \frac{\Delta\varepsilon_e}{2} + \frac{\Delta\varepsilon_p}{2} \tag{a}$$

The equation of the plastic-strain line in Fig. 7-4 is

$$\frac{\Delta\varepsilon_p}{2} = \varepsilon'_F(2N)^c \tag{7-1}$$

The equation of the elastic strain line is

$$\frac{\Delta\varepsilon_e}{2} = \frac{\sigma'_F}{E}(2N)^b \tag{7-2}$$

Therefore, from Eq. (*a*), we have for the total-strain amplitude

$$\frac{\Delta\varepsilon}{2} = \frac{\sigma'_F}{E}(2N)^b + \varepsilon'_F(2N)^c \tag{7-3}$$

which is the Manson-Coffin relationship between fatigue life and total strain.* Some values of the coefficients and exponents are listed in Table 7-1. Many more are included in the SAE J1099 report.

Though Eq. (7-3) is a perfectly legitimate equation for obtaining the fatigue life of a part when the strain and other cyclic characteristics are given, it appears to be of little use to the designer. The question of how to determine the total strain at the bottom of a notch or discontinuity has not been answered. There are no tables or charts of strain concentration factors in the literature. It is possible that strain concentration factors will become available in research literature very soon because of the increase in the use of finite-element analysis. Moreover, finite-element analysis can of itself approximate the strains that will occur at all points in the subject structure. Lacking this tool, many

---

*J. F. Tavernelli and L. F. Coffin, Jr., "Experimental Support for Generalized Equation Predicting Low Cycle Fatigue," and S. S. Manson, discussion, *Trans. ASME, J. Basic Eng.*, vol. 84, no. 4, pp. 533–537.

TABLE 7-1

Cyclic Properties of Some High-Strength Steels

AISI NUMBER	PROCESSING	BRINELL HARDNESS H_B	CYCLIC YIELD STRENGTH S'_y, kpsi	FATIGUE STRENGTH COEFFICIENT σ'_F, kpsi	FATIGUE DUCTILITY COEFFICIENT ε'_F	FATIGUE STRENGTH EXPONENT b	FATIGUE DUCTILITY EXPONENT c	FATIGUE STRAIN-HARDENING EXPONENT m
1045	Q & T 80°F	705	$\cdots$	310	$\cdots$	−0.065	−1.0	0.10
1045	Q & T 360°F	595	250	395	0.07	−0.055	−0.60	0.13
1045	Q & T 500°F	500	185	330	0.25	−0.08	−0.68	0.12
1045	Q & T 600°F	450	140	260	0.35	−0.07	−0.69	0.12
1045	Q & T 720°F	390	110	230	0.45	−0.074	−0.68	0.14
4142	Q & T 80°F	670	300	375	$\cdots$	−0.075	−1.0	0.05
4142	Q & T 400°F	560	250	385	0.07	−0.076	−0.76	0.11
4142	Q & T 600°F	475	195	315	0.09	−0.081	−0.66	0.14
4142	Q & T 700°F	450	155	290	0.40	−0.080	−0.73	0.12
4142	Q & T 840°F	380	120	265	0.45	−0.080	−0.75	0.14
4142*	Q & D 550°F	475	160	300	0.20	−0.082	−0.77	0.12
4142	Q & D 650°F	450	155	305	0.60	−0.090	−0.76	0.13
4142	Q & D 800°F	400	130	275	0.50	−0.090	−0.75	0.14

*Deformed 14 percent.

Source: Data from R. W. Landgraf, *Cyclic Deformation and Fatigue Behavior of Hardened Steels,* Report no. 320, Department of Theoretical and Applied Mechanics, University of Illinois, Urbana, 1968.

engineers will not find the strain-life analysis a very useful tool for estimating fatigue strengths.

7-3 STRESS-LIFE DEFINITIONS

To determine the strength of materials under the action of fatigue loads, specimens are subjected to repeated or varying forces of specified magnitudes while the cycles or stress reversals are counted to destruction. The most widely used fatigue-testing device is the R. R. Moore high-speed rotating-beam machine. This machine subjects the specimen to pure bending (no transverse shear) by means of weights. The specimen, shown in Fig. 7-5, is very carefully machined and polished, with a final polishing in an axial direction to avoid circumferential scratches. Other fatigue-testing machines are available for applying fluctuating or reversed axial stresses, torsional stresses, or combined stresses to the test specimens.

To establish the fatigue strength of a material, quite a number of tests are necessary because of the statistical nature of fatigue. For the rotating-beam test, a constant bend-

FIGURE 7-5

Test-specimen geometry for the R. R. Moore rotating-beam machine. The bending moment is uniform over the curved portion so that fracture into two equal halves indicates failure at the highest-stressed portion, a valid test of material; whereas a fracture elsewhere (not at the highest-stress level) is grounds for suspicion of material flaw.

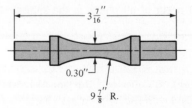

$3\frac{7}{16}''$

$0.30''$

$9\frac{7}{8}''$ R.

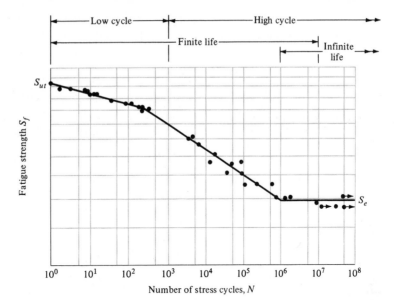

FIGURE 7-6

An *S-N* diagram plotted from the results of completely reversed axial fatigue tests. Material: UNS G41300 steel, normalized; S_{ut} = 116 kpsi; maximum S_{ut} = 125 kpsi. *(Data from NACA Tech. Note 3866, December 1966.)*

ing load is applied, and the number of revolutions (stress reversals) of the beam required for failure is recorded. The first test is made at a stress which is somewhat under the ultimate strength of the material. The second test is made at a stress which is less than that used in the first. This process is continued, and the results are plotted as an *S-N* diagram (Fig. 7-6). This chart may be plotted on semilog paper or on log-log paper. In the case of ferrous metals and alloys, the graph becomes horizontal after the material has been stressed for a certain number of cycles. Plotting on log paper emphasizes the bend in the curve, which might not be apparent if the results were plotted by using Cartesian coordinates.

The ordinate of the *S-N* diagram is called the *fatigue strength* S_f; a statement of this strength must always be accompanied by a statement of the number of cycles *N* to which it corresponds.

Soon we shall learn that *S-N* diagrams can be determined either for a test specimen or for an actual mechanical element. Even when the material of the test specimen and that of the mechanical element are identical, there will be significant differences between the diagrams for the two.

In the case of the steels, a knee occurs in the graph, and beyond this knee failure will not occur, no matter how great the number of cycles. The strength corresponding to the knee is called the *endurance limit* S_e, or the *fatigue limit*. The graph of Fig. 7-6 never does become horizontal for nonferrous metals and alloys, and hence these materials do not have an endurance limit.

We note that a stress cycle (*N* = 1) constitutes a single application and removal of a load and then another application and removal of the load in the opposite direction. Thus $N = \frac{1}{2}$ means the load is applied once and then removed, which is the case with the simple tension test.

The body of knowledge available on fatigue failure from *N* = 1 to *N* = 1000 cycles is generally classified as *low-cycle fatigue*, as indicated in Fig. 7-6. *High-cycle fatigue*, then, is concerned with failure corresponding to stress cycles greater than 10^3 cycles.

We also distinguish a *finite-life region* and an *infinite-life region* in Fig. 7-6. The boundary between these regions cannot be clearly defined except for a specific material; but it lies somewhere between 10^6 and 10^7 cycles for steels, as shown in Fig. 7-6.

As noted previously, it is always good engineering practice to conduct a testing program on the materials to be employed in design and manufacture. This, in fact, is a requirement, not an option, in guarding against the possibility of a fatigue failure. *Because of this necessity for testing, it would really be unnecessary for us to proceed any further in the study of fatigue failure except for one important reason: the desire to know why fatigue failures occur so that the most effective method or methods can be used to improve fatigue strength.* Thus our primary purpose in studying fatigue is to understand why failures occur so that we can guard against them in an optimum manner. For this reason, the analytical design approaches presented in this book, or in any other book, for that matter, do not yield absolutely precise results. The results should be taken as a guide, as something which indicates what is important and what is not important in designing against fatigue failure.

The methods of fatigue-failure analysis represent a combination of engineering and science. Often science fails to provide the answers which are needed. But the airplane must still be made to fly—safely. And the automobile must be manufactured with a reliability that will ensure a long and trouble-free life and at the same time produce profits for the stockholders of the industry. Thus, while science has not yet completely explained the actual mechanism of fatigue, the engineer must still design things that will not fail. In a sense this is a classic example of the true meaning of engineering as contrasted with science. Engineers use science to solve their problems *if* the science is available. But available or not, the problem must be solved, and whatever form the solution takes under these conditions is called engineering.

7-4   PRELIMINARY OBSERVATIONS

In the material to follow we shall learn that fatigue strength has an observable relation with tensile strength. *Minimum* values of the tensile strength, designated S_{ut}, are always used with the factor-of-safety method. Table A-20 lists minimum values for steels. Other tables in the Appendix list *typical values;* these are neither the means nor the minima, but are attainable.

When stochastic methods of analysis are used, the *mean* tensile strength, usually designated $\bar{S}_{ut}$, is needed for analysis.

7-5   THE ENDURANCE LIMIT

The determination of endurance limits by fatigue testing is now routine, though a lengthy procedure. Generally, stress testing is preferred to strain testing for endurance limits.

For preliminary and prototype design and for some failure analysis as well, a quick method of estimating endurance limits is needed. There are great quantities of data in the literature on the results of rotating-beam tests and simple tension tests of specimens taken from the same bar or ingot. By plotting these as in Fig. 7-7, it is possible to see

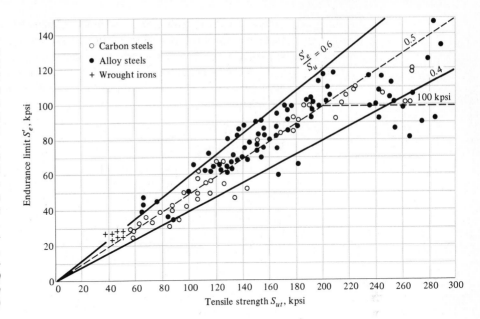

FIGURE 7-7

Graph of endurance limits versus tensile strengths from actual test results for a large number of wrought irons and steels. Ratios of S'_e/S_{ut} = 0.60, 0.50, and 0.4 are shown by the solid and dashed lines. Note also the horizontal dashed line for S'_e = 100 kpsi. Points shown having a tensile strength greater than 200 kpsi have a mean endurance limit $\overline{S}'_e$ = 107 kpsi and a standard deviation of 13.5 kpsi. *(Collated from data compiled by H. J. Grover, S. A. Gordon, and L. R. Jackson in Fatigue of Metals and Structures, Bureau of Naval Weapons Document NAVWEPS 00-25-534, 1960 rev.; and from Fatigue Design Handbook, SAE, 1968, p. 42.)*

whether there is any correlation between the two sets of results. The graph appears to suggest that the endurance limit ranges from about 40 to 60 percent of the tensile strength for steels up to about 200 kpsi (1400 MPa). Beginning at about S_{ut} = 200 kpsi (1400 MPa), the scatter appears to increase, but the trend seems to level off, as suggested by the dashed horizontal line at S'_e = 100 kpsi (700 MPa).

Another series of tests, this time for various microstructures, is shown in Table 7-2. In this table the endurance limits vary from about 23 to 63 percent of the tensile strength.*

Now, it is important to observe that the dispersion of the endurance limit is *not* due to a dispersion in the tensile strengths of the specimen, but rather that the spread occurs even when the tensile strengths of a large number of specimens remain exactly the same. Keep this in mind when choosing factors of safety.

We wish now to present a method for estimating endurance limits. Note that estimates obtained from quantities of data obtained from many sources probably have a large spread and might deviate significantly from the results of actual laboratory tests

---

*But see H. O. Fuchs and R. I. Stephens, *Metal Fatigue in Engineering,* Wiley, New York, 1980, pp. 69–71, which reports a range of 35 to 60 percent for steels having $S_{ut} <$ 1400 MPa and as low as 20 percent for high-strength steels.

TABLE 7-2

Endurance-Limit Ratio S'_e/S_{ut} for Various Steel Microstructures

	FERRITE		PEARLITE		MARTENSITE	
	RANGE	AVERAGE	RANGE	AVERAGE	RANGE	AVERAGE
Carbon steel	0.57–0.63	0.60	0.38–0.41	0.40	⋯	0.25
Alloy steel	⋯	⋯	⋯	⋯	0.23–0.47	0.35

Source: Adapted from L. Sors, *Fatigue Design of Machine Components,* Pergamon Press, Oxford, England, 1971.

of the mechanical properties of specimens obtained through strict purchase-order specifications. Since the area of uncertainty is greater, compensation must be made by the employment of larger design factors than would be used for static design.

Mischke* has analyzed a great deal of actual test data from several sources and concluded that endurance limit can, indeed, be related to tensile strength. For steels, the relationship is

$$S_e' = \begin{cases} 0.504S_{ut} & S_{ut} \le 200 \text{ kpsi (1400 MPa)} \\ 100 \text{ kpsi} & S_{ut} > 200 \text{ kpsi} \\ 700 \text{ MPa} & S_{ut} > 1400 \text{ MPa} \end{cases} \tag{7-4}$$

where S_{ut} is the minimum tensile strength. The prime mark on S_e' in this equation refers to the rotating-beam specimen itself. We wish to reserve the unprimed symbol S_e for the endurance limit of any particular machine element subjected to any kind of loading. Soon we shall learn that the two strengths may be quite different.

The data of Table 7-2 emphasize the difficulty of attempting to provide a single rule for deriving the endurance limit from the tensile strength. The table also shows a part of the cause of this difficulty. Steels treated to give different microstructures have different S_e'/S_{ut} ratios. It appears that the more ductile microstructures have a higher ratio. Martensite has a very brittle nature and is highly susceptible to fatigue-induced cracking; thus the ratio is low. When designs include detailed heat-treating specifications to obtain specific microstructures, it is possible to use an estimate of the endurance limit based on test data for the particular microstructure; such estimates are much more reliable and indeed should be used.

The endurance limits for various classes of cast irons, polished or machined, are given in Table A-24. Aluminum alloys do not have an endurance limit. The fatigue strengths of some aluminum alloys at $50(10^7)$ cycles of reversed stress are given in Table A-23.

7-6   THE FATIGUE STRENGTH

As shown in Fig. 7-6, the region of low-cycle fatigue extends from $N = 1$ to about $N = 1000$ cycles. In this region the fatigue strength S_f is only slightly less than the tensile strength S_{ut}. If a more precise estimate of S_f is needed in this region, then laboratory tests should not be too lengthy or expensive, provided the parts are not too large in size; usually no more than 1000 cycles of stress reversal is needed.

Figure 7-6 shows that the high-cycle fatigue domain extends from 10^3 cycles, for steels, to the endurance-limit life N_e, which is about 10^6 cycles or only slightly more. The purpose of this section is to develop methods of approximating the S-N diagram when information may be as sparse as to include only the results of the simple tension test.

To develop an analytical approach, let the equation of the S-N line be

$$S_f = aN^b \tag{7-5}$$

---

*Charles R. Mischke, "Prediction of Stochastic Endurance Strength," *Trans. of ASME, J. Vibration, Acoustics, Stress, and Reliability in Design*, vol. 109, no. 1, pp. 113–122, January 1987.

Then, at 10^3 cycles,

$$(S_f)_{10^3} = a(10^3)^b = a(10)^{3b} = fS_{ut}$$

Solving for the factor f gives

$$f = \frac{a}{S_{ut}} (10)^{3b} \tag{a}$$

which is *not* constant. Now, in high-cycle fatigue, with stress levels below the proportional limit, the strain is predominantly elastic. Therefore, with $\sigma_a = \Delta\varepsilon_e E/2$, Eq. (7-2) becomes

$$\sigma_a = \sigma_f'(2N)^b \tag{b}$$

Solving for the exponent b gives

$$b = -\frac{1}{\log 2N} \log \frac{\sigma_f'}{\sigma_a} = -\frac{\log (\sigma_f'/S_e)}{\log 2N_e} \tag{c}$$

Multiplying both sides by 3 and substituting $N_e = 10^6$ cycles gives

$$3b = \log \left(\frac{\sigma_f'}{S_e}\right)^{-1/2.1} \tag{d}$$

which makes f in Eq. (a) become

$$f = \frac{a}{S_{ut}} (10)^{3b} = \frac{2^b \sigma_f'}{S_{ut}} \left(\frac{\sigma_f'}{S_e}\right) \tag{e}$$

The fatigue strength coefficient σ_f' is approximately given by the equation

$$\sigma_f' = S_{ut} + 58.8 \qquad \text{kpsi} \tag{f}$$

Thus Eq. (e) can be solved when S_{ut} and S_e are given, because b can be found from Eq. (c). As an example solution, let $S_{ut} = 105$ kpsi and $S_e = 62$ kpsi; these are data for an AISI 1045 steel. Then Eq. (f) gives $\sigma_f' = 163.8$ kpsi, and Eq. (c) gives $b = -0.0670$. Solving Eq. (e) yields the result $f = 0.94$. Note that the use of $S_e' = 0.5S_{ut}$, for the rotating-beam endurance limit, produces some simplification in Eq. (e) but also roughens the approximation.

In the factor-of-safety, or deterministic, approach to design, the strengths used, as indicated in Chap. 1, are always the *minimum* expected values. No specific designation is applied to any strength S to indicate this fact, because it is always so used.

On the other hand, when statistical methods of design are used, the *mean strengths* $\overline{S}$ or $\hat{\mu}_S$ are used.

Keeping the above facts in mind, we shall approximate the S-N diagram with a line on the log S–log N chart joining $0.9S_{ut}$ at 10^3 cycles and S_e at 10^6 cycles to define the fatigue strength S_f corresponding to any life N between 10^3 and 10^6 cycles. We have noted that S_{ut} is a minimum expected value of the tensile strength. This is also true of S_e at the lower end of the S-N line, because most tests indicate that N_e is somewhat more than 10^6 cycles.

Another way to obtain the finite-life fatigue strength S_f is to plot the line on log-log paper. Then the result can be read off. A disadvantage of this approach is that the slope of the *S-N* line on standard log-log graph paper is so small that accuracy is difficult to obtain.

To avoid the use of log-log paper, write Eq. (7-5) as

$$\log S_f = \log a + b \log N \tag{g}$$

This line is to intersect 10^6 cycles at S_e and 10^3 cycles at $0.9S_{ut}$. When these are substituted into Eq. (g), the resulting equations can be solved for a and b. The results are

$$a = \frac{(0.9S_{ut})^2}{S_e}$$

$$b = -\frac{1}{3} \log \frac{0.9S_{ut}}{S_e} \tag{7-6}$$

Note that the constant a depends upon the units used. Units of MPa (N/mm²) or kpsi are most convenient for these equations, but any consistent units can be used.

Suppose a completely reversed stress σ_a is given. The number of cycles of life corresponding to this stress can be found from Eq. (7-5) by substituting σ_a for S_f. The result is

$$N = \left(\frac{\sigma_a}{a}\right)^{1/b} \tag{7-7}$$

It is probably worth noting that either S_e or S_e' can be used in Eq. (7-6).

EXAMPLE 7-1   An AISI 1045 steel has a minimum tensile strength of 95 kpsi and a minimum expected yield strength of 74 kpsi.

(a) Estimate the rotating-beam endurance limit.
(b) Estimate the fatigue strength corresponding to 10^4 cycles of life.
(c) Estimate the expected life corresponding to a completely reversed stress of 55 kpsi.

Solution   (a) From Eq. (7-4), we find

Answer $S_e' = 0.5S_{ut} = 0.5(95) = 47.5$ kpsi

(b) Using equation pair (7-6), we obtain

$$a = \frac{(0.9S_{ut})^2}{S_e} = \frac{[0.9(95)]^2}{47.5} = 153.9 \text{ kpsi}$$

$$b = -\frac{1}{3} \log \frac{0.9S_{ut}}{S_e} = -\frac{1}{3} \log \frac{0.9(95)}{47.5} = -0.0851$$

Then, from Eq. (7-5),

Answer $S_f = aN^b = 153.9(10^4)^{-0.0851} = 70.3$ kpsi

(c) Equation (7-7), for $\sigma_a = 55$ kpsi, gives

Answer $N = \left(\frac{\sigma_a}{a}\right)^{1/b} = \left(\frac{55}{153.9}\right)^{-1/0.0851} = 1.78(10^5)$ cycles

7-7   THE ENDURANCE-LIMIT AND FATIGUE-STRENGTH VARIATES*

When reliability is important, then fatigue testing must certainly be undertaken. There is no other way. Consequently, the methods of stochastic analysis presented here and in other sections of this book constitute guidelines which enable the designer to obtain a good understanding of the various issues involved and help in the development of a safe and reliable design.

In statistical analysis, a variate $\mathbf{x}$ can be expressed in the following two forms:

$$\mathbf{x} = \begin{cases} (\hat{\mu}_x, \hat{\sigma}_x) \\ \hat{\mu}_x(1, C_x) \end{cases} \tag{7-8}$$

where the coefficient of variation is $C_x = \hat{\sigma}_x/\hat{\mu}_x$. The second form of Eq. (7-8) is often more convenient and will be employed frequently in the material to follow.

The stochastic version of Eq. (7-4) is

$$\mathbf{S}'_e = \begin{cases} 0.504(1, 0.146)\overline{S}_{ut} & \overline{S}_{ut} \leq 200 \text{ kpsi (1400 MPa)} \\ 100(1, 0.146) \quad \text{kpsi} & \overline{S}_{ut} > 200 \text{ kpsi} \\ 700(1, 0.146) \quad \text{MPa} & \overline{S}_{ut} > 1400 \text{ MPa} \end{cases} \tag{7-9}$$

where $\mathbf{S}'_e$ is the standard rotating-beam endurance limit expressed as a random variable and $\overline{S}_{ut}$ is the mean tensile strength. For more on Eq. (7-9), see Sec. 7-10.

Let us consider next the problem of finding the fatigue-strength variate $\mathbf{S}_f$ corresponding to a deterministic value of N when the variates $\mathbf{S}_{ut}$ and $\mathbf{S}_e$ are given. First, let $\mathbf{S}_e = (\hat{\mu}_{Se}, \hat{\sigma}_{Se})$ and $\mathbf{S}_{ut} = (\hat{\mu}_{Sut}, \hat{\sigma}_{Sut})$. Then the corresponding coefficients of variation are

$$C_{Se} = \frac{\hat{\sigma}_{Se}}{\hat{\mu}_{Se}} \qquad C_{Su} = \frac{\hat{\sigma}_{Sut}}{\hat{\mu}_{Sut}} \tag{a}$$

In order to obtain C_{Sf}, we must interpolate logarithmically. If $C = c + d \log N$, then

$$C_{Su} = c + d \log 10^3 = c + 3d$$
$$C_{Se} = c + d \log 10^6 = c + 6d \tag{b}$$

and so

$$C_{Sf} = c + d \log N \tag{c}$$

We can solve the two equations of set (b) simultaneously and substitute the results in Eq. (c). Thus

$$C_{Sf} = 2C_{Su} - C_{Se} + \tfrac{1}{3}(C_{Se} - C_{Su}) \log N \tag{7-10}$$

Using the definition of the coefficient of variation, we recognize that

$$\hat{\sigma}_{Sf} = C_{Sf}\hat{\mu}_{Sf} = C_{Sf}\overline{S}_f \tag{7-11}$$

Since N is deterministic, the mean fatigue strength $\overline{S}_f$ can be obtained from Eq. (7-5). Thus

$$\overline{S}_f = aN^b \tag{7-12}$$

---

*This section may be omitted; see Sec. 1-10.

where the constants a and b are found from Eq. (7-6) using $\overline{S}_{ut}$ and $\overline{S}_e$. Since $\mathbf{S}_f = (\overline{S}_f, \hat{\sigma}_{Sf})$, the procedure is to solve Eqs. (7-12), (7-10), and (7-11), in that order.

The reverse problem is that in which $\mathbf{S}_e$ and $\mathbf{S}_{ut}$ are given as before. With an alternating stress variate $\boldsymbol{\sigma}_a$ given, the problem is to find the corresponding value of the life $\mathbf{N}$. It turns out that $\mathbf{N}$ has a lognormal distribution and, in many cases, occupies a large portion of the log N axis. This problem can be solved graphically or analytically.

7-8 ENDURANCE-LIMIT MODIFYING FACTORS

We have seen that the rotating-beam specimen used in the laboratory to determine endurance limits is prepared very carefully and tested under closely controlled conditions. It is unrealistic to expect the endurance limit of a mechanical or structural member to match the values obtained in the laboratory.

Marin* classifies some of the factors that modify the endurance limit, and these are shown in Table 7-3. To account for the most important of these conditions, we employ a variety of modifying factors, each of which is intended to account for a single effect. Using this idea, we may write

$$S_e = k_a k_b k_c k_d k_e S'_e \qquad (7\text{-}13)$$

where S_e = endurance limit of mechanical element
S'_e = endurance limit of test specimen
k_a = surface factor
k_b = size factor
k_c = load factor
k_d = temperature factor
k_e = miscellaneous-effects factor

Surface Factor k_a

The surface of the rotating-beam specimen is highly polished, with a final polishing in the axial direction to smooth out any circumferential scratches. The modification factors depend upon the quality of the finish and upon the tensile strength. To derive a formula for k_a, a total of 39 data points† for various surface finishes were examined.

*Joseph Marin, *Mechanical Behavior of Engineering Materials,* Prentice-Hall, Englewood Cliffs, N.J., 1962, p. 224.

†*Data source:* C. G. Noll and C. Lipson, "Allowable Working Stresses," *Society for Experimental Stress Analysis,* vol. III, no. 2, 1946, p. 49.

TABLE 7-3

Conditions Affecting the Endurance Limit

Material: Chemical composition, basis of failure, variability
Manufacturing: Method of manufacture, heat treatment, fretting corrosion, surface condition, stress concentration
Environment: Corrosion, temperature, stress state, relaxation times
Design: Size, shape, life, stress state, stress concentration, speed, fretting, galling

TABLE 7-4	SURFACE	FACTOR a		EXPONENT
Surface Finish Factors	FINISH	kpsi	MPa	b
	Ground	1.34	1.58	−0.085
	Machined or cold-drawn	2.70	4.51	−0.265
	Hot-rolled	14.4	57.7	−0.718
	As forged	39.9	272.	−0.995

This formula is

$$k_a = a S_{ut}^b \tag{7-14}$$

where S_{ut} is the minimum tensile strength and a and b are to be found in Table 7-4.

Size Factor k_b

The size factor has been evaluated using 133 sets of data points.[*]
The results for bending and torsion may be expressed as

$$k_b = \begin{cases} \left(\dfrac{d}{0.3}\right)^{-0.1133} & \text{in} & 0.11 \le d \le 2 \text{ in} \\ \left(\dfrac{d}{7.62}\right)^{-0.1133} & \text{mm} & 2.79 \le d \le 51 \text{ mm} \end{cases} \tag{7-15}$$

For larger sizes, k_b varies from 0.60 to 0.75 for bending and torsion.
  For axial loading there is no size effect. Therefore, use

$$k_b = 1 \tag{7-16}$$

  One of the problems that arise in using Eq. (7-15) is what to do when a round bar in bending is not rotating, or when a noncircular cross section is used. For example, what is the size factor for a bar 6 mm thick and 40 mm wide? The approach to be used here employs an *effective dimension* d_e obtained by equating the volume of material stressed at and above 95 percent of the maximum stress to the same volume in the rotating-beam specimen.[†] It turns out that when these two volumes are equated, the lengths cancel, and so we need only consider the areas. For a rotating round section, the 95 percent stress area is the area in a ring having an outside diameter d and an inside diameter of $0.95d$. So, designating the 95 percent stress area $A_{0.95\sigma}$, we have

$$A_{0.95\sigma} = \frac{\pi}{4}[d^2 - (0.95d)^2] = 0.0766 d^2 \tag{7-17}$$

This equation is also valid for a rotating hollow round. For nonrotating solid or hollow rounds, the 95 percent stress area is twice the area outside of two parallel chords having

[*]Mischke, op. cit., Table 3.

[†]See R. Kuguel, ''A Relation between Theoretical Stress Concentration Factor and Fatigue Notch Factor Deduced from the Concept of Highly Stressed Volume,'' *Proc. ASTM*, vol. 61, 1961, pp. 732–748.

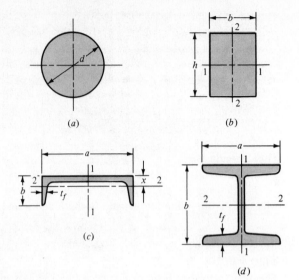

FIGURE 7-8

(a) Solid round; (b) rectangular
section; (c) channel section;
(d) wide-flange section.

a spacing of $0.95D$, where D is the diameter. Using an exact computation, this is

$$A_{0.95\sigma} = 0.0105D^2 \tag{a}$$

when rounded. Setting Eqs. (7-17) and (a) equal to each other enables us to solve for the effective diameter. This gives

$$d_e = 0.370D \tag{7-18}$$

as the effective size of round corresponding to a nonrotating solid or hollow round.

A rectangular section of dimensions $h \times b$ has $A_{0.95\sigma} = 0.05hb$. Using the same approach as before, we have

$$d_e = 0.808(hb)^{1/2} \tag{7-19}$$

These sections are shown in Fig. 7-8 together with a channel and a wide-flange or I-beam section. For the channel,

$$A_{0.95\sigma} = \begin{cases} 0.05ab & \text{axis 1-1} \\ 0.052xa + 0.1t_f(b - x) & \text{axis 2-2} \end{cases} \tag{7-20}$$

The 95 percent stress area for the wide flange is

$$A_{0.95\sigma} = \begin{cases} 0.10at_f & \text{axis 1-1} \\ 0.05ba \quad t_f > 0.025a & \text{axis 2-2} \end{cases} \tag{7-21}$$

Load Factor k_c

The load factor is given by the equation

$$k_c = \begin{cases} 0.923 & \text{axial loading} & S_{ut} \le 220 \text{ kpsi (1520 MPa)} \\ 1 & \text{axial loading} & S_{ut} > 220 \text{ kpsi (1520 MPa)} \\ 1 & \text{bending} \\ 0.577 & \text{torsion and shear} \end{cases} \tag{7-22}$$

Though there is no apparent size effect for specimens tested in axial or push-pull fatigue, there is a definite difference between the axial fatigue limit and that in reversed bending. A very extensive collection of data has been made by R. W. Landgraf, now of Ford Motor Company, on axial fatigue.* These results were analyzed, resulting in the values shown for axial loading in Eq. (7-22), above.

A collection of 52 data points† comparing the torsional endurance limit with the bending endurance limit yielded a load factor for torsion of 0.565. Using a different set of data points, Mischke‡ obtained the result $k_c = 0.585$. Both of these are very close to the value of 0.577 shown in Eq. (7-22), which can be obtained from the distortion-energy theory.

Temperature Factor k_d

When operating temperatures are below room temperature, brittle fracture is a strong possibility and should be investigated first. When the operating temperatures are higher than room temperature, yielding should be investigated first because the yield strength drops off so rapidly with temperature; see Fig. 5-11. Any stress will induce creep in a material operating at high temperatures; so this factor must be considered too. Finally, it may be true that there is no fatigue limit for materials operating at high temperatures. Because of the reduced fatigue resistance, the failure process is, to some extent, dependent on time.

The limited amount of data available show that the endurance limit for steels increases slightly as the temperature rises and then begins to fall off in the 400 to 700°F range, not unlike the behavior of the tensile strength shown in Fig. 5-11. For this reason it is probably true that the endurance limit is related to tensile strength at elevated temperatures in the same manner as at room temperature.§ It seems quite logical, therefore, to employ the same relations to predict endurance limit at elevated temperatures as are used at room temperature, at least until more comprehensive data become available. At the very least, this practice will provide a useful standard against which the performance of various materials can be compared.

Table 7-5 has been obtained from Fig. 5-11 using only the tensile-strength data. Note that the table represents 145 tests of 21 different carbon and alloy steels and that the maximum standard deviation is just 0.110.

Two types of problems arise when temperature is a consideration. If the rotating-beam endurance limit is known at room temperature, then use

$$k_d = \frac{S_T}{S_{RT}} \tag{7-23}$$

from Table 7-5 and proceed as usual. If the rotating-beam endurance limit is not given, then compute it using Eq. (7-4) and the temperature-corrected tensile strength obtained using the factor from Table 7-5. Then use $k_d = 1$.

---

*Landgraf, *op. cit.* and by personal communication.

†*Source*: Thomas J. Dolan, "Physical Properties," in Oscar J. Horger (ed.), *ASME Handbook—Metals Engineering Design*, McGraw-Hill, New York, 1953, p. 97.

‡Op. cit., table 6.

§For more, see Table 2 of ANSI/ASME B106.1M-1985 shaft standard; and E. A. Brandes (ed.), *Smithell's Metals Reference Book*, 6th ed., Butterworth, London, 1983, pp. 22-134 to 22-136, where endurance limits from 100 to 650°C are tabulated.

TABLE 7-5

Effect of Operating Temperature on the Tensile Strength of Steel.* (S_T = Tensile Strength at Operating Temperature; S_{RT} = Tensile Strength at Room Temperature; $0.099 \leq \hat{\sigma} \leq 0.110$)

TEMPERATURE, °C	S_T/S_{RT}	TEMPERATURE, °F	S_T/S_{RT}
20	1.000	70	1.000
50	1.010	100	1.008
100	1.020	200	1.020
150	1.025	300	1.024
200	1.020	400	1.018
250	1.000	500	0.995
300	0.975	600	0.963
350	0.943	700	0.927
400	0.900	800	0.872
450	0.840	900	0.797
500	0.766	1000	0.698
550	0.670	1100	0.567
600	0.546		

Data source: Fig. 5-11.

7-9   MISCELLANEOUS-EFFECTS FACTOR k_e

Though the factor k_e is intended to account for the reduction in endurance limit due to all other effects, it is really intended as a reminder that these must be accounted for, because actual values of k_e are not always available.

Residual stresses may either improve the endurance limit or affect it adversely. Generally, if the residual stress in the surface of the part is compression, the endurance limit is improved. Fatigue failures appear to be tensile failures, or at least to be caused by tensile stress, and so anything which reduces tensile stress will also reduce the possibility of a fatigue failure. Operations such as shot peening, hammering, and cold rolling build compressive stresses into the surface of the part and improve the endurance limit significantly. Of course, the material must not be worked to exhaustion.

The endurance limits of parts which are made from rolled or drawn sheets or bars, as well as parts which are forged, may be affected by the so-called *directional characteristics* of the operation. Rolled or drawn parts, for example, have an endurance limit in the transverse direction which may be 10 to 20 percent less than the endurance limit in the longitudinal direction.

Parts which are case-hardened may fail at the surface or at the maximum core radius, depending upon the stress gradient. Figure 7-9 shows the typical triangular stress distribution of a bar under bending or torsion. Also plotted as a heavy line in this figure are the endurance limits S_e for the case and core. For this example the endurance limit of the core rules the design because the figure shows that the stress σ or τ, whichever applies, at the outer core radius, is appreciably larger than the core endurance limit.

Of course, if stress concentration is also present, the stress gradient is much steeper, and hence failure in the core is unlikely.

Corrosion   It is to be expected that parts which operate in a corrosive atmosphere will have a lowered fatigue resistance. This is, of course, true, and it is due to the roughening or pitting of the surface by the corrosive material. But the problem is not so

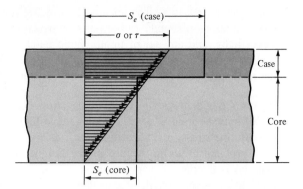

FIGURE 7-9

The failure of a case-hardened part in bending or torsion. In this example, failure occurs in the core.

simple as the one of finding the endurance limit of a specimen which has been corroded. The reason for this is that the corrosion and the stressing occur at the same time. Basically, this means that in time any part will fail when subjected to repeated stressing in a corrosive atmosphere. There is no fatigue limit. Thus the designer's problem is to attempt to minimize the factors that affect the fatigue life; these are:

- Mean or static stress
- Alternating stress
- Electrolyte concentration
- Dissolved oxygen in electrolyte
- Material properties and composition
- Temperature
- Cyclic frequency
- Fluid flow rate around specimen
- Local crevices

Electrolytic Plating   Metallic coatings, such as chromium plating, nickel plating, or cadmium plating, reduce the endurance limit by as much as 50 percent. In some cases the reduction by coatings has been so severe that it has been necessary to eliminate the plating process. Zinc plating does not affect the fatigue strength. Anodic oxidation of light alloys reduces bending endurance limits by as much as 39 percent but has no effect on the torsional endurance limit.

Metal Spraying   Metal spraying results in surface imperfections that can initiate cracks. Limited tests show reductions of 14 percent in the fatigue strength.

Cyclic Frequency   If, for any reason, the fatigue process becomes time-dependent, then it also becomes frequency-dependent. Under normal conditions, fatigue failure is independent of frequency. But when corrosion or high temperatures, or both, are encountered, the cyclic rate becomes important. The slower the frequency and the higher the temperature, the higher the crack propagation rate and the shorter the life at a given stress level.

Frettage Corrosion   The phenomenon of frettage corrosion is the result of microscopic motions of tightly fitting parts or structures. Bolted joints, bearing-race fits, wheel hubs, and any set of tightly fitted parts are examples. The process involves surface discoloration, pitting, and eventual fatigue. The frettage factor k_e depends upon the material of the mating pairs and ranges from 0.24 to 0.90.

Stress Concentration   The fatigue stress-concentration factor K_f given by Eq. (5-26) must be used when designing against fatigue. Normally, this factor is used to increase the stress, as indicated by Eq. (5-25).

A problem arises in the use of the fatigue stress-concentration factor when the material is ductile, or performs as a ductile material, and we are interested in the finite-life fatigue strength. Recall that a stress-concentration factor need not be used with ductile materials when they are subject only to static loads, because yielding will relieve the stress concentration. This means that at $N = 10^3$ cycles, the load is practically static and so a stress-concentration factor need not be employed. Since we must use K_f at 10^6 cycles, what should we do about lives between 10^3 and 10^6 cycles?

A widely used approach is to use the miscellaneous-effects factor k_e as a strength-reduction factor for cases of this nature, thus reducing only the fatigue limit. With this approach we define

$$k_e = \frac{1}{K_f} \tag{7-24}$$

as the *fatigue-strength-reduction factor*. Thus, on the *S-N* diagram, k_e reduces the endurance limit at 10^6 cycles but has no effect at 10^3 cycles.

An alternative approach is to use a reduced value of K_f, designated as K_f', for lives less than $N = 10^6$. To do this, let $K_f' = K_f$ at 10^6 cycles and $K_f' = 1$ at 10^3 cycles, and write

$$K_f' = aN^b \tag{7-25}$$

this equation can be solved for a and b in the same manner as Eq. (7-5), giving

$$a = \frac{1}{K_f} \qquad b = -\frac{1}{3} \log \frac{1}{K_f} \tag{7-26}$$

EXAMPLE 7-2   An AISI 1015 cold-drawn steel bar has a diameter of 1 in.

(*a*) Estimate the endurance limit.

(*b*) Estimate the endurance limit for reversed bending without rotation.

(*c*) Estimate the fatigue strength at $N = 70(10^3)$ cycles at an operating temperature of 550°F.

Solution   (*a*) From Table A-20, find the minimum tensile strength as $S_{ut} = 56$ kpsi. Then from Eq. (7-4) the endurance limit of a test specimen is

$$S_e' = 0.504 S_{ut} = 0.504(56) = 28.2 \text{ kpsi}$$

We use Eq. (7-14) to get the surface factor k_a. Using $a = 2.70$ and $b = -0.265$ for cold-drawn steel gives

$$k_a = aS_{ut}^b = 2.70(56)^{-0.265} = 0.929$$

From Eq. (7-15) we find the size factor as

$$k_b = \left(\frac{d}{0.3}\right)^{-0.1133} = \left(\frac{1}{0..3}\right)^{-0.1133} = 0.872$$

Also, $k_c = k_d = k_e = 1$. Therefore, from Eq. (7-13), we have

Answer $\qquad$ $S_e = 0.929(0.872)(28.2) = 22.8$ kpsi

(b) From Eq. (7-18), find

$$d_e = 0.370D = 0.370(1) = 0.370 \text{ in}$$

Then Eq. (7-15) gives

$$k_b = \left(\frac{d}{0.3}\right)^{-0.1133} = \left(\frac{0.370}{0.3}\right)^{-0.1133} = 0.977$$

Thus the endurance limit for nonrotating reversed bending is

Answer $\qquad$ $S_e = 0.929(0.977)(28.2) = 25.6$ kpsi

Note that nonrotating reversed bending means that the bending force changes direction $(+ \text{ or } -)$. It is not the same as one-way or push-push bending.

(c) From Table 7-5, by linear interpolation, find $S_T/S_{RT} = 0.979$ at 550°F. Then

$$S_{utT} = \frac{S_T}{S_{RT}} S_{ut} = 0.979(56) = 54.8 \text{ kpsi}$$

Thus $S_e' = 0.504S_{utT} = 0.504(54.8) = 27.6$ kpsi. The surface and size factors are the same as in part (a), and so

$$S_e = 0.929(0.872)(27.6) = 22.4 \text{ kpsi}$$

Next, using Eq. (7-6), find

$$a = \frac{(0.9S_{ut})^2}{S_e} = \frac{[0.9(54.8)]^2}{22.4} = 108.6$$

$$b = -\frac{1}{3}\log\frac{0.9S_{ut}}{S_e} = -\frac{1}{3}\log\frac{0.9(54.8)}{22.4} = -0.1143$$

Then Eq. (7-5) gives the corresponding fatigue strength as

Answer $\qquad$ $S_f = aN^b = 108.6[70(10^3)]^{-0.1143} = 30.3$ kpsi

Notice how conveniently this approach handles the temperature effect at both ends of the *S-N* diagram.

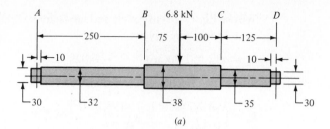

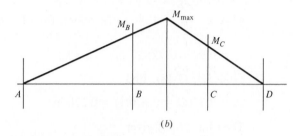

FIGURE 7-10

(a) Shaft drawing; all dimensions in millimeters; all fillets 3-mm radius. The shaft rotates and the load is stationary; material is machined from AISI 1050 cold-drawn steel. (b) Bending-moment diagram.

EXAMPLE 7-3

Figure 7-10a shows a rotating shaft supported in ball bearings at A and D and loaded by the nonrotating force F. Estimate the life of the part.

Solution

From Fig. 7-10b we learn that failure will probably occur at B rather than at C or at the point of maximum moment. Point B has a smaller cross section, a higher bending moment, and a higher stress-concentration factor than C, and the location of maximum moment has a larger size and no stress-concentration factor.

We shall solve the problem by first estimating the strength at point B, since the strength will be different elsewhere, and comparing this strength with the stress at the same point.

From Table A-20 we find $S_{ut} = 690$ MPa and $S_y = 580$ MPa. Therefore

$$S'_e = 0.504(690) = 347.8 \text{ MPa}$$

Using Eq. (7-14) with $a = 4.51$ and $b = -0.265$, we find the surface factor to be

$$k_a = aS^b_{ut} = 4.51(690)^{-0.265} = 0.798$$

The size factor is found from Eq. (7-15) to be

$$k_b = \left(\frac{d}{7.62}\right)^{-0.1133} = \left(\frac{32}{7.62}\right)^{-0.1133} = 0.850$$

To find the stress-concentration factor, we enter Table A-15-9 with the values

$$\frac{D}{d} = \frac{38}{32} = 1.1875 \qquad \text{and} \qquad \frac{r}{d} = \frac{3}{32} = 0.093\ 75$$

and find $K_t = 1.65$. From Fig. 5-16 we obtain $q = 0.82$, corresponding to $r = 3$ mm and $S_{ut} = 690$ MPa. Using Eq. (5-26) gives

$$K_f = 1 + q(K_t - 1) = 1 + 0.82(1.65 - 1) = 1.53$$

We are asked to find the life of a ductile material, and so we elect to use Eq. (7-24), treating stress concentration as a fatigue-strength-reduction effect. Thus

$$k_e = \frac{1}{K_f} = \frac{1}{1.53} = 0.654$$

The remaining modifying factors are unity, and so Eq. (7-13) gives

$$S_e = 0.798(0.850)(0.654)(347.8) = 154.3 \text{ MPa}$$

The next step is to determine the stress at point B; the bending moment is

$$M_B = 250\frac{225F}{550} = 250\frac{(225)(6.8)}{550} = 695 \text{ N} \cdot \text{m}$$

The section modulus is

$$\frac{I}{c} = \frac{\pi d^3}{32} = \frac{\pi(3.2)^3}{32} = 3.22 \text{ cm}^3$$

when rounded. Therefore the bending stress is

$$\sigma = \frac{M}{I/c} = \frac{695}{3.22} = 216 \text{ MPa}$$

This stress is greater than the endurance limit, and so the part will have only a finite life. Going to Eq. (7-6), we find

$$a = \frac{[0.9(690)]^2}{154.3} = 2499 \qquad b = -\frac{1}{3} \log \frac{0.9(690)}{154.3} = -0.2016$$

Then we use Eq. (7-7) to estimate the life. This gives

Answer $$N = \left(\frac{\sigma_a}{a}\right)^{1/b} = \left(\frac{216}{2499}\right)^{1/(-0.2016)} = 188(10^3) \text{ cycles}$$

7-10 THE ENDURANCE LIMIT AS A RANDOM VARIABLE

If we wish to describe the endurance limit stochastically, then Eq. (7-13) must be written in the form

$$\mathbf{S}_e = \mathbf{k}_a k_b \mathbf{k}_c \mathbf{k}_d \mathbf{k}_e \mathbf{S}'_e \tag{7-27}$$

where all the terms except k_b are random variables and hence are in boldface type.

Surface Factor $\mathbf{k}_a$

The relation for $\mathbf{k}_a$ is similar to Eq. (7-14) and is written

$$\mathbf{k}_a = a\bar{S}_{ut}^b(1, C) \tag{7-28}$$

where the *mean* strength $\bar{S}_{ut}$ is used and the coefficient of variation C is obtained from Table 7-6. Use Table 7-4 for factors a and b.

SURFACE FINISH	FACTOR C
Ground	0.13
Machine or cold-drawn	0.06
Hot-rolled	0.11
As forged	0.08

TABLE 7-6

Coefficient of Variation C for Surface-Finish Factors

Size Factor k_b

The size factor is always deterministic; therefore, use Eq. (7-15) or (7-16).

Load Factor k_c

In analyzing data to determine values for the load factor, it was possible to obtain the coefficients of variation too. The results are summarized in the equation

$$k_c = \begin{cases} 0.923(1,\ 0.044) & \text{axial} & \bar{S}_{ut} \le 220 \text{ kpsi (1520 MPa)} \\ 1.0(1,\ 0) & \text{axial} & \bar{S}_{ut} > 220 \text{ kpsi (1520 MPa)} \\ 1.0(1,\ 0) & \text{bending} \\ 0.577(1,\ 0.11) & \text{torsion and shear} \end{cases} \quad (7\text{-}29)$$

Temperature Factor k_d

Use Table 7-5 and Eq. (7-23) for the temperature factor. The standard deviation is estimated to be $\hat{\sigma}_k = 0.110$.

EXAMPLE 7-4

The bar shown in Fig. 7-11 is machined from a cold-drawn flat having a mean tensile strength of 80 kpsi. The axial load shown is completely reversed. Find the reliability, assuming both the stress and the strength have normal distributions.

Solution

The rotating-beam endurance limit is first found using Eq. (7-9):

$$\mathbf{S}'_e = 0.504(1,\ 0.146)\bar{S}_{ut} = 0.504(1,\ 0.146)(80)$$

$$= 40.3(1,\ 0.146) \quad \text{kpsi}$$

Then, from Tables 7-4 and 7-6, find $a = 2.70$, $b = -0.265$, and $C = 0.06$ for a machined surface. Using Eq. (7-28), we compute the surface-finish factor as

$$\mathbf{k}_a = a\bar{S}_{ut}^b(1,\ C) = 2.70(1,\ 0.06)(80)^{-0.265} = 0.845(1,\ 0.06)$$

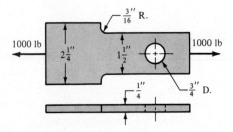

FIGURE 7-11

The size factor k_b is unity for axial loading. The load factor for axial loading is found from Eq. (7-29) to be

$$\mathbf{k}_c = 0.923(1, 0.044)$$

The remaining factors are all unity. The endurance limit of the bar is therefore given by the stochastic equation

$$\mathbf{S}_e = \mathbf{k}_a\mathbf{k}_b\mathbf{k}_c\mathbf{k}_d\mathbf{k}_e\mathbf{S}'_e$$
$$= 0.845(1, 0.06)(1)[0.923(1, 0.044)](1)(1)[40.3(1, 0.146)]$$

Therefore the mean value of the endurance limit is

$$\bar{S}_e = 0.845(0.923)(40.3) = 31.4 \text{ kpsi}$$

The coefficient of variation is

$$C_{Se} = (C_{ka}^2 + C_{kc}^2 + C_{S'e}^2)^{1/2}$$
$$= [(0.06)^2 + (0.044)^2 + (0.146)^2]^{1/2} = 0.164$$

Now we can express the endurance limit as

$$\mathbf{S}_e = 31.4(1, 0.164) \qquad \text{kpsi}$$

In computing the stress, the section at the hole governs. Using the terminology of Table A-15-1, we find $d/w = 0.50$. Therefore $K_t = 2.18$. The notch sensitivity q is found at the end of the chart (Fig. 5-16) as 0.81. Then the fatigue stress-concentration factor is found to be

$$K_f = 1 + q(K_t - 1) = 1 + 0.81(2.18 - 1) = 1.96$$

From Table 5-5, we note that $C_K = 0.11$. Therefore

$$\mathbf{K}_f = 1.96(1, 0.11)$$

The stress is

$$\boldsymbol{\sigma} = \mathbf{K}_f\frac{F}{A} = 1.96(1, 0.11)\frac{1000}{0.25(0.75)}(10^{-3}) = 10.5(1, 0.11) \qquad \text{kpsi}$$

The standard deviations of the strength and stress are now found to be

$$\hat{\sigma}_{Se} = 31.4(0.164) = 5.15 \text{ kpsi}$$
$$\hat{\sigma}_\sigma = 10.5(0.11) = 1.155 \text{ kpsi}$$

Using the coupling equation for the interference of two normal distributions gives the standardized variable as

$$z = -\frac{\mu_{Se} - \mu_\sigma}{(\hat{\sigma}_{Se}^2 + \hat{\sigma}_\sigma^2)^{1/2}} = -\frac{31.4 - 10.5}{[(5.15)^2 + (1.155)^2]^{1/2}} = -3.960$$

Answer Using Table A-10 and linear interpolation, we find this corresponds to a reliability of 0.999 962.

7-11

THE DISTRIBUTIONS

We have given both deterministic and stochastic substance to the fatigue modification factors, given by Eq. (7-27) and also called the *Marin modification factors*. Their multiplicative nature is supported by an extensive statistical analysis of a UNS G43400 (electric furnace or aircraft quality) steel which resulted in a correlation coefficient of 0.85 for a multiplicative form and 0.40 for an additive form. A theorem exists to show that the most likely distribution of data about the mean (or regression line) for S_e is Gaussian (normal). Note that this theorem does not prove that the distribution *is* Gaussian, but merely that it is most likely so. It turns out that the surface factor k_a is approximately lognormal, while the load and temperature factors k_c and k_d can be viewed as normal.

The question of the distribution of S_e is answered by the central limit theorem of statistics. The individual distributional nature of the elements of the product is not vital, for the distribution of the multiple-element product approaches lognormal regardless. Since each element is (1) continuous, (2) not seriously skewed, and (3) several in number, it would be difficult to detect deviation from lognormality. The presumption of lognormality in this circumstance is said to be *robust*. On account of the asymptotic approach of the distribution to its final lognormal form, there is some error introduced by the presumption. For example, in addition, even if the underlying distribution is rectangular or triangular, the means of a sample of 4 are approximately normally distributed, indicating rapid convergence to asymptotic form. The presumption of normality here is said to be robust. If the elements of a product are lognormal themselves, then the product is lognormal exactly; that is, for the operation, the result is *closed*. This indicates that if some of the product elements are individually lognormal, the convergence is accelerated. For finite-life fatigue, $S_f \sim LN$ in the interval 10^3 to 10^6 cycles.

In a strength-limited design, a load-induced stress is related to a fatigue strength. When the problem involves an axial tension of a notched round rod, the load-induced stress σ can be expressed as

$$\sigma = K_f \sigma_0 = K_f \frac{F}{A} = \frac{4}{\pi} K_f F \left(\frac{1}{d}\right)^2 \tag{a}$$

or, for bending of a notched flat bar,

$$\sigma = K_f \sigma_0 = K_f \frac{6M}{td^2} = 6K_f \frac{Fx}{td^2} \tag{b}$$

Note that there are three stochastic factors in the first case and five in the second. In these cases, lognormality of σ is a viable estimate, that is to say, robust. In the load-induced stress σ, when machining controls cause the geometric factors to have small coefficients of variation, the load dominates and its distribution persists, because of the controlling coefficient of variation.

7-12

FLUCTUATING STRESSES

Quite frequently it is necessary to determine the strength of parts corresponding to stress situations other than complete reversals. Many times in design the stresses fluc-

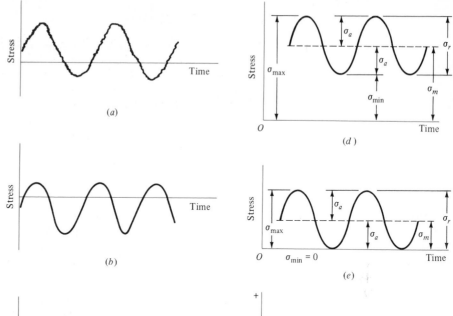

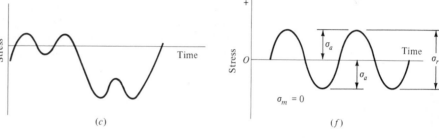

FIGURE 7-12

Some stress-time relations: (a) fluctuating stress with high-frequency ripple; (b and c) nonsinusoidal fluctuating stress; (d) sinusoidal fluctuating stress; (e) repeated stress; (f) completely reversed sinusoidal stress.

tuate without passing through zero. Figure 7-12 illustrates some of the various stress-time relationships which may occur. The components of stress with which we must deal, some of which are shown in Fig. 7-12d, are

σ_{min} = minimum stress σm = midrange or mean stress

σ_{max} = maximum stress σ_r = stress range

σ_a = stress amplitude σ_s = steady, or static, stress

The steady, or static, stress is *not* the same as the mean stress; in fact, it may have any value between σ_{min} and σ_{max}. The steady stress exists because of a fixed load or preload applied to the part, and it is usually independent of the varying portion of the load. A helical compression spring, for example, is always loaded into a space shorter than the free length of the spring. The stress created by this initial compression is called the steady, or static, component of the stress. It is not the same as the mean stress.

We shall have occasion to apply the subscripts of these components to shear stresses as well as normal stresses.

The following relations are evident from Fig. 7-12:

$$\sigma_m = \frac{\sigma_{max} + \sigma_{min}}{2} \qquad (7\text{-}30)$$

$$\sigma_a = \frac{\sigma_{max} - \sigma_{min}}{2} \qquad (7\text{-}31)$$

Although some of the stress components have been defined by using a sine stress-time relation, the exact shape of the curve does not appear to be of particular significance.

In addition, the stress ratios

$$R = \frac{\sigma_{min}}{\sigma_{max}} \qquad (7\text{-}32)$$

and

$$A = \frac{\sigma_a}{\sigma_m} \qquad (7\text{-}33)$$

are often used in describing fluctuating stresses.

7-13 FATIGUE STRENGTH UNDER FLUCTUATING STRESSES

Now that we have defined the various components of stress associated with a part subjected to fluctuating stress, we want to vary both the mean stress and the stress amplitude, or alternating component, to learn something about the fatigue resistance of parts when subjected to such situations. Three methods of plotting the results of such tests are in general use and are shown in Figs. 7-13, 7-14, and 7-15.

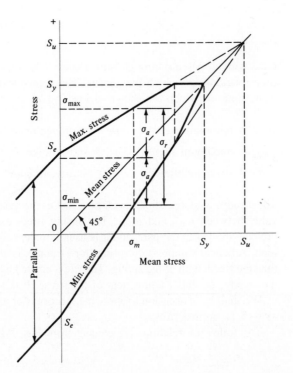

FIGURE 7-13

Modified Goodman diagram showing all the strengths and the limiting values of all stress components for a particular mean stress.

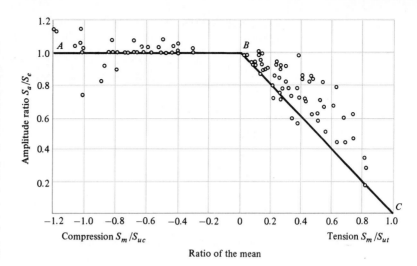

FIGURE 7-14

Plot of fatigue failures for mean stresses in both the tensile and compressive regions. Normalizing the data by using the ratios of mean strength to tensile strength S_m/S_{ut}, mean strength to compressive strength S_m/S_{uc}, and strength amplitude to endurance limit S_a/S_e enables a plot of experimental results for a variety of steels. *[Data source: Thomas J. Dolan, "Stress Range," sec. 6.2 in O. J. Horger (ed.), ASME Handbook— Metals Engineering Design, McGraw-Hill, New York, 1953.]*

The *modified Goodman diagram* of Fig. 7-13 has the mean stress plotted along the abscissa and all other components of stress plotted on the ordinate, with tension in the positive direction. The endurance limit, fatigue strength, or finite-life strength, which-ever applies, is plotted on the ordinate above and below the origin. The mean-stress line is a 45° line from the origin to the tensile strength of the part. The modified Goodman diagram consists of the lines constructed to S_e (or S_f) above and below the origin. Note that the yield strength is also plotted on both axes, because yielding would be the criterion of failure if σ_{max} exceeded S_y.

Another way to display test results is shown in Fig. 7-14. Here the abscissa repre-

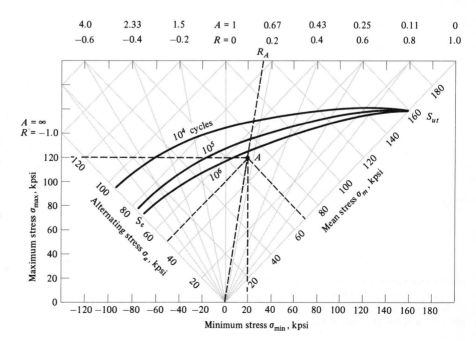

FIGURE 7-15

Master fatigue diagram created for AISI 4340 steel having S_{ut} = 158 kpsi and S_y = 147 kpsi. The stress components at A are σ_{min} = 20, σ_{max} = 120, σ_m = 70, and σ_a = 50, all in kpsi. *(Source: H. J. Grover, Fatigue of Aircraft Structures, U.S. Government Printing Office, Washington, D.C., 1966, pp. 317, 322. See also J. A. Collins, Failure of Materials in Mechanical Design, Wiley, New York, 1981, p. 216.)*

sents the ratio of the mean strength to the ultimate strength, with tension plotted to the right and compression to the left. The ordinate is the ratio of the alternating strength to the endurance limit. The line BC then represents the modified Goodman criterion of failure. Note that the existence of mean stress in the compressive region has little effect on the endurance limit.

The very clever diagram of Fig. 7-15 is unique in that it displays four of the stress components as well as the two stress ratios. A curve representing the endurance limit for values of R beginning at $R = -1$ and ending with $R = 1$ begins at S_e on the σ_a axis and ends at S_{ut} on the σ_m axis. Constant-life curves for $N = 10^5$ and $N = 10^4$ cycles have been drawn too. Any stress state, such as the one at A, can be described by the minimum and maximum components, or by the mean and alternating components. And safety is indicated whenever the point described by the stress components lies below the constant-life line.

When the mean stress is compression, failure occurs whenever $\sigma_a = S_e$ or whenever $\sigma_{\max} = S_{yc}$, as indicated by the left-hand side of Fig. 7-14. Neither a fatigue diagram nor any other failure criteria need be developed.

In Fig. 7-16, the tensile side of Fig. 7-14 has been redrawn using strengths, instead of strength ratios, with the same modified Goodman criterion together with three additional criteria of failure. Such diagrams are often constructed for analysis and design purposes; they are easy to use and the results can be scaled off directly.

Either the fatigue limit S_e or the finite-life strength S_f is plotted on the ordinate of Fig. 7-16. These values will have already been corrected using the Marin factors of Eq. (7-13). Note that the yield strength S_{yt} is plotted on the ordinate too. This serves as a reminder that yielding rather than fatigue might be the criterion of failure.

The mean-stress axis of Fig. 7-16 has the yield strength S_{yt} and the tensile strength S_{ut} plotted along it.

Four criteria of failure are diagrammed in Fig. 7-16: the Soderberg, the modified Goodman, the Gerber, and yielding. The diagram shows that only the Soderberg criterion guards against yielding.

The linear theories of Fig. 7-16 can be placed in equation form for machine compu-

FIGURE 7-16

Fatigue diagram showing various criteria of failure. For each criterion, points on or outside the respective line indicate failure. Some point A on the Goodman line, for example, gives the strength S_m as the limiting value of σ_m corresponding to the strength S_a, which, paired with σ_m, is the limiting value of σ_a.

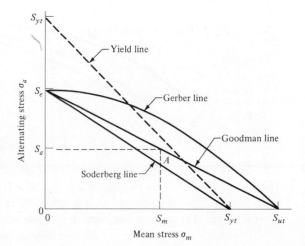

tation by writing the equation of a straight line in intercept form. This form is

$$\frac{x}{a} + \frac{y}{b} = 1 \qquad (a)$$

where a and b are the x and y intercepts, respectively. This equation for the Soderberg line is

$$\frac{S_a}{S_e} + \frac{S_m}{S_{yt}} = 1 \qquad (7\text{-}34)$$

Similarly, we find the modified Goodman relation to be

$$\frac{S_a}{S_e} + \frac{S_m}{S_{ut}} = 1 \qquad (7\text{-}35)$$

Examination of Fig. 7-14 shows that the line representing the Gerber theory has a better chance of passing through the central portion of the failure points and hence should be a better predictor. This theory is also called the *Gerber parabolic relation,* because the equation is

$$\frac{S_a}{S_e} + \left(\frac{S_m}{S_{ut}}\right)^2 = 1 \qquad (7\text{-}36)$$

Though unnecessary, we can complete the picture by defining yielding on the first cycle by the equation

$$\frac{S_a}{S_{yt}} + \frac{S_m}{S_{yt}} = 1 \qquad (7\text{-}37)$$

The methods of Chap. 6, however, are usually preferred over this equation.

The stresses σ_a and σ_m can replace S_a and S_m in Eqs. (7-34) to (7-36) if each strength is divided by a factor of safety n. When this is done, the Soderberg equation becomes

$$\frac{\sigma_a}{S_e} + \frac{\sigma_m}{S_{yt}} = \frac{1}{n} \qquad (7\text{-}38)$$

The modified Goodman relation is

$$\frac{\sigma_a}{S_e} + \frac{\sigma_m}{S_{ut}} = \frac{1}{n} \qquad (7\text{-}39)$$

and the Gerber equation is

$$\frac{n\sigma_a}{S_e} + \left(\frac{n\sigma_m}{S_{ut}}\right)^2 = 1 \qquad (7\text{-}40)$$

The meaning of these equations is illustrated in Fig. 7-17, using the modified Goodman theory as an example. While Eqs. (7-38) to (7-40) represent the usual approach to the determination of factor of safety, other methods can be developed, some based on the concept of a *load line.* Examples which follow will illustrate some of these alternate approaches.

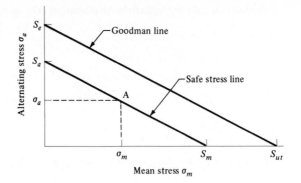

FIGURE 7-17

The safe-stress line through *A* is constructed parallel to the modified Goodman line. Note that the safe-stress line is the locus of all sets of σ_a-σ_m stresses having a factor of safety *n* and that $S_m = n\sigma_m$ and $S_a = n\sigma_a$.

Many authors utilize Eqs. (7-34) to (7-40) to derive an entire series of formulas for the design and analysis of common problems which arise frequently. But this proliferation of formulas based on, say, the Goodman line detracts from the basic simplicity of the graphic and algebraic approaches presented here. The graphic approach is easy and quick and requires only a shirt-pocket or purse-size 15-cm scale. Furthermore, this approach to the development of a sound understanding of a few basic principles will enable you to develop your own specialized formulas to solve almost any problem situation you can conceive.

EXAMPLE 7-5

A 1.5-in tension bar has been machined from AISI 1050 cold-drawn round bar. This part is to withstand a fluctuating tensile load varying from 0 to 16 kip. Because of the design of the ends and the fillet radius, a fatigue stress-concentration factor $K_f = 1.85$ exists. The remaining Marin factors have been worked out, and are $k_a = 0.797$, $k_b = k_d = 1$, and $k_c = 0.923$.

(a) Use K_f as a fatigue-strength-reduction factor with the modified Goodman line, and determine the factor of safety using a complete graphical solution under the assumption that σ_m remains fixed.

(b) The same as part (a), except that σ_a remains fixed.

(c) The same as part (a), except that the ratio σ_a/σ_m is a constant.

(d) The same as part (a), but use K_f as a stress-concentration factor.

(e) The same as part (d), except that σ_a remains fixed.

(f) The same as part (d), except that the ratio σ_a/σ_m is a constant.

(g) Use the methods of Chap. 6 and find the factor of safety guarding against yielding.

Solution

(a) From Table A-20, find $S_y = 84$ kpsi, $S_{ut} = 100$ kpsi; therefore $S'_e = 0.504(100) = 50.4$ kpsi. Treating K_f as a fatigue-strength-reduction factor gives k_e as

$$k_e = \frac{1}{K_f} = \frac{1}{1.85} = 0.541$$

Therefore the corrected endurance limit is

$$S_e = k_a k_b k_c k_d k_e S'_e = 0.797(1)(0.923)(1)(0.541)(50.4) = 20.0 \text{ kpsi}$$

The stresses are found to be

$$\sigma_a = \frac{4F_a}{\pi d^2} = \frac{4(8)}{\pi(1.5)^2} = 4.53 \text{ kpsi}$$

$$\sigma_m = \frac{4F_m}{\pi d^2} = \frac{4(8)}{\pi(1.5)^2} = 4.53 \text{ kpsi}$$

These data are now plotted to scale in Fig. 7-18a. With σ_m fixed, the danger of an overload is with σ_a. Therefore the line AB in Fig. 7-18a is the *load line*, giving $S_a = 19.1$ kpsi. The factor of safety is

Answer $$n = \frac{S_a}{\sigma_a} = \frac{19.1}{4.53} = 4.22$$

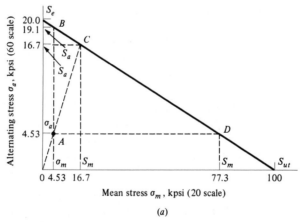

(a)

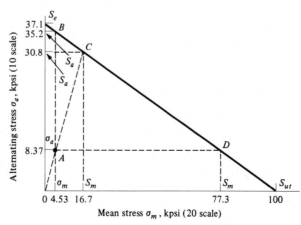

FIGURE 7-18

Modified Goodman diagrams used in graphical analysis. (a) K_f used as fatigue-strength-reduction factor; (b) K_f used as stress-increasing factor.

(b)

(b) With σ_a fixed, the danger of an overload is with σ_m. Therefore the line AD in Fig. 7-18a is the load line, giving $S_m = 77.3$ kpsi. The factor of safety is

Answer
$$n = \frac{S_m}{\sigma_m} = \frac{77.3}{4.53} = 17.07$$

(c) If σ_a and σ_m always have the same constant relation to each other, then both may overload proportionally. Therefore the load line is AC in Fig. 7-18a and $S_a = S_m = 16.7$ kpsi. The factor of safety is

Answer
$$n = \frac{S_a}{\sigma_a} = \frac{S_m}{\sigma_m} = \frac{16.7}{4.53} = 3.68$$

(d, e, f) When K_f is used as a stress-concentration factor, the stresses to be used in the fatigue diagram are

$$\sigma_a = K_f \frac{4F_a}{\pi d^2} = 1.85 \frac{4(8)}{\pi(1.5)^2} = 8.37 \text{ kpsi}$$

$$\sigma_m = \frac{4F_m}{\pi d^2} = 4.53 \text{ kpsi}$$

Also, $k_e = 1$, and so the endurance limit is

$$S_e = 0.797(1)(0.923)(1)(1)(50.4) = 37.1 \text{ kpsi}$$

These data are plotted in Fig. 7-18b. Note that the ordinate scale has been chosen differently from that in Fig. 7-18a. The factors of safety are obtained from this graph exactly the same as before. The results are the same. Try it.

(g) The factor of safety guarding against yielding can be obtained graphically by plotting the yield line of Fig. 7-16 in Fig. 7-18a and following the procedure of part (a). This cannot be done with the graph of Fig. 7-18b, however, because the stress σ_a on that graph has been augmented by K_f. This is because a stress-concentration factor is not used with ductile materials subjected to static loads. This problem with K_f can be sidetracked by always using the analytic methods of Chap. 6 to predict static safety (or failure). Thus the factor of safety guarding against yielding is

Answer
$$n = \frac{S_y}{\sigma_a + \sigma_m} = \frac{84}{4.53 + 4.53} = 9.28$$

EXAMPLE 7-6   Solve parts (a), (b), and (c) of Example 7-5 analytically, using the modified Goodman equations.

Solution   (a) Use $S_m = \sigma_m = 4.53$ kpsi, $S_e = 20.0$ kpsi, and $S_{ut} = 100$ kpsi. Then Eq. (7-35) yields

$$S_a = S_e\left(1 - \frac{S_m}{S_{ut}}\right) = 20.0\left(1 - \frac{4.53}{100}\right) = 19.1 \text{ kpsi}$$

The factor of safety is

Answer $$n = \frac{S_a}{\sigma_a} = \frac{19.1}{4.53} = 4.22$$

(b) With the overload on σ_m, solve Eq. (7-35) for S_m and then substitute $S_a = \sigma_a = 4.53$ kpsi. Thus

$$S_m = S_{ut}\left(1 - \frac{S_a}{S_e}\right) = 100\left(1 - \frac{4.53}{20.0}\right) = 77.3 \text{ kpsi}$$

The factor of safety is

Answer $$n = \frac{S_m}{\sigma_m} = \frac{77.3}{4.53} = 17.07$$

(c) Equation (7-39) applies, but here we choose an alternate approach for illustrative purposes. First, from Eq. (7-33),

$$\frac{\sigma_a}{\sigma_m} = \frac{S_a}{S_m} = A \tag{7-41}$$

Then Eq. (7-35), for the modified Goodman relation, becomes

$$S_m = \frac{S_e S_{ut}}{S_e + A S_{ut}} \tag{7-42}$$

For this problem, $\sigma_a/\sigma_m = 4.53/4.53 = 1$. Then, from Eq. (7-42),

$$S_m = \frac{20.0(100)}{20.0 + 1(100)} = 16.7 \text{ kpsi}$$

and so

Answer $$n = \frac{S_m}{\sigma_m} = \frac{16.7}{4.53} = 3.68$$

EXAMPLE 7-7   A flat leaf spring is used to retain a roller-follower mechanism in contact with a plate cam. The range of motion of the follower is fixed, and so the alternating component of the bending moment in the spring is fixed, too. However, the spring preload can be adjusted to accommodate various cam speeds. When the cam is run at a high speed, the preload must be increased so as to prevent follower float or jump. For lower speeds, the preload should be decreased to obtain a longer life for the cam and roller contact surfaces.

The spring is a steel cantilever 32 in long, 2 in wide, and $\frac{1}{4}$ in thick, as shown in Fig. 7-19. The strengths are $S_u = 150$ kpsi, $S_y = 127$ kpsi, and $S_e = 28$ kpsi fully corrected. The total cam motion is 2 in and we wish to preload the spring by deflecting it 2 in for slow speeds and up to 5 in for high speeds. Determine factors of safety for each of these conditions and also check for the possibility of a static failure by yielding. Use the modified Goodman criterion.

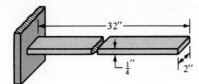

FIGURE 7-19

A cantilever cam-follower spring.

Solution The second moment of area is

$$I = \frac{bh^3}{12} = \frac{2(0.25)^3}{12} = 0.0026 \text{ in}^4$$

The deflection is related to the force by the equation $y = Fl^3/3EI$. Therefore the spring rate is

$$k = \frac{F}{y} = \frac{3EI}{l^3} = \frac{3(30)(10^6)(0.0026)}{(32)^3} = 7.14 \text{ lb/in}$$

The bending stress caused by a 1-in deflection is

$$\sigma = \frac{Mc}{I} = \frac{7.14(32)(0.125)}{0.0026}(10^{-3}) = 11 \text{ kpsi}$$

Thus $\sigma_a = 11$ kpsi. Now express σ_m in parameter form as

$$\sigma_m = 11\delta$$

where δ = deflection. Suppose we designate δ^* as the deflection necessary to produce failure. Substituting in Eq. (7-35) gives

$$\frac{11}{28} + \frac{11\delta^*}{150} = 1$$

or

$$\delta^* = \tfrac{150}{11} (1 - \tfrac{11}{28}) = 8.28 \text{ in}$$

For slow speeds, $\delta = 2$ in and the factor of safety is

Answer $$n = \frac{\delta^*}{\delta} = \frac{8.28}{2} = 4.14$$

Thus, at slow speeds, a 4.14-fold increase in preset produces a failure. This is what $n = 4.14$ means.

At high speeds, $\delta = 5$ in and the factor of safety is

Answer $$n = \frac{\delta^*}{\delta} = \frac{8.28}{5} = 1.66$$

Therefore, in this case, $n = 1.66$ means that a failure is produced if the preset is increased 1.66 times.

The preset required to produce yielding on the first cycle can be found from the equation

$$\sigma_a + \sigma_m = S_{yt}$$

or, substituting,

$$11 + 11\delta^* = 127 \qquad \delta^* = \frac{127 - 11}{11} = 10.5 \text{ in}$$

Then, for $\delta = 5$ in,

Answer $\qquad n = \dfrac{\delta^*}{\delta} = \dfrac{10.5}{5} = 2.1$

Thus a 2.1-fold increase in preset produces first-cycle yielding. This is what n means in this case.

All of the above values result when the factor of safety is obtained on the basis of *variations in stress*. Let us now see how the factor of safety is found on the basis of *variations in strength*. Using Eq. (7-39), we have

$$\frac{\sigma_a}{S_e} + \frac{\sigma_m}{S_{ut}} = \frac{1}{n} = \frac{11}{28} + \frac{11\delta}{150}$$

For $\delta = 5$ in, we obtain

$$n = \frac{1}{11/28 + 11\delta/150} = \frac{1}{11/28 + 11(5)/150} = 1.32$$

The meaning of n here is that failure would occur if both S_e and S_{ut} were reduced by the factor 1.32.

For yielding, $\sigma_a + \sigma_m = S_{yt}/n$, or

$$n = \frac{S_{yt}}{\sigma_a + \sigma_m} = \frac{127}{11 + 11\delta} = \frac{127}{11 + 11(5)} = 1.92$$

This means that failure by yielding would occur if the yield strength were to be reduced by the factor 1.92.

Additional insight can be obtained for this algebraic solution by sketching a fatigue diagram and plotting the stress points relative to the Goodman line.

7-14   TORSIONAL FATIGUE STRENGTH UNDER PULSATING STRESSES

Extensive tests by Smith* provide some very interesting results on pulsating torsional fatigue. Smith's first result, based on 72 tests, shows that the existence of a torsional mean stress not more than the torsional yield strength has no effect on the torsional endurance limit, provided the material is *ductile, polished, notch-free, and cylindrical*.

Smith's second result applies to materials with stress concentration, notches, or surface imperfections. In this case, he finds that the torsional fatigue limit decreases steadily with torsional mean stress, not unlike the Goodman line. Since the great

---

*James O. Smith, "The Effect of Range of Stress on the Fatigue Strength of Metals," *Univ. of Ill., Eng. Exp. Sta. Bull. 334,* 1942.

majority of parts will have surfaces that are less than perfect, this result indicates that the modified Goodman theory applies to torsion too.

Joerres,* of Associated Spring–Barnes Group, confirms Smith's results and also recommends the use of the modified Goodman relation for pulsating torsion. In constructing the Goodman diagram, Joerres uses

$$S_{su} = 0.67S_{ut} \tag{7-43}$$

Also, from Chap. 6, $S_{sy} = 0.577S_{yt}$, from the distortion-energy theory; and the load factor k_c is given in Eq. (7-22) as 0.577 for torsion.

7-15   COMBINATIONS OF LOADING MODES

In Sec. 7-8 we learned that a load factor k_c is used to obtain the endurance limit and hence that the result is dependent on whether the loading is axial, bending, or torsion. In this section we want to answer the question, How do we proceed when the loading is a mixture of, say, axial, bending, and torsional loads? In addition to the complication introduced by the fact that a separate endurance limit is associated with each mode of loading, there may also be multiple stress-concentration factors, one also for each mode of loading. Fortunately, the answer turns out to be rather simple:

1 For the strength, use the fully corrected endurance limit for bending, S_e or $\mathbf{S}_e$.

2 Apply the appropriate fatigue stress-concentration factors to the alternating components of the torsional stress, the bending stress, and the axial stress components.

3 Multiply any alternating axial stress components by the factor $1/k_{c,\text{ax}} = 1/0.923 = 1.083$.

4 Enter the resultant stresses into a Mohr's circle analysis and find the principal stresses.

5 Using the results of step 4, find the von Mises alternating stress σ'_a or the stress variate $\boldsymbol{\sigma}'_a$.

6 Compare σ'_a with S_e to find the factor of safety, or interfere $\boldsymbol{\sigma}'_a$ with $\mathbf{S}_e$ to find the reliability.

This approach is based on the assumption that all stress components are completely reversed and are always in time phase with each other. If they are not in phase but have the same frequency, the maxima can be found by expressing each component in trigonometric terms, using phase angles, and then finding the sum. If two or more stress components have differing frequencies, the problem is difficult; one solution is to assume that the two (or more) components often reach an in-phase condition, so that their magnitudes are additive.

If mean stresses are also present, then steps 4 and 5 can be repeated for them and the

---

*Robert E. Joerres, "Springs," chap. 24 in Joseph E. Shigley and Charles R. Mischke (eds.), *Standard Handbook of Machine Design*, McGraw-Hill, New York, 1986. See also *Design Handbook*, Associated Spring–Barnes Group, Bristol, Conn., 1981, pp. 35–38.

resulting von Mises mean stress σ'_m used with σ'_a in forming a modified Goodman solution. Note that mean stresses are not augmented by the fatigue stress-concentration factor K_f or K_{sf} unless they behave as brittle materials. Also, the factor $1/k_{c,ax}$ should not be applied to axial mean stresses, since they are regarded as static.

It is worth noting that the analysis outlined above assumes a size factor for axial loading the same as for bending and torsion. When bending is present, the existence of an axial component is usually relatively small, and so in most cases this loss of accuracy is small and always conservative.

EXAMPLE 7-8   A rotating shaft is made of 42- × 4-mm AISI 1018 cold-drawn steel tubing and has a 6-mm diameter hole drilled transversely through it. Estimate the factor of safety guarding against fatigue and static failures for the following loading conditions:

(a) The shaft is subjected to a completely reversed torque of 120 N · m in phase with a completely reversed bending moment of 150 N · m.

(b) The shaft is subjected to a pulsating torque fluctuating from 20 to 160 N · m and a steady bending moment of 150 N · m.

Solution   Here we follow the procedure of estimating the strengths and then the stresses, followed by relating the two.

From Table A-20 we find the minimum strengths to be $S_{ut} = 440$ MPa and $S_{yt} = 370$ MPa. Thus the endurance limit of the rotating-beam specimen is $S'_e = 0.504(440) = 222$ MPa, from Eq. (7-4). The surface factor is obtained from Eq. (7-14) with $a = 1.58$ and $b = -0.085$. The result is

$$k_a = aS_{ut}^b = 1.58(440)^{-0.085} = 0.942$$

The size factor is obtained from Eq. (7-15) as

$$k_b = \left(\frac{d}{7.62}\right)^{-0.1133} = \left(\frac{42}{7.62}\right)^{-0.1133} = 0.824$$

The remaining Marin factors are all unity, and so

$$S_e = 0.942(0.824)(222) = 172 \text{ MPa}$$

(a) Theoretical stress-concentration factors are found from Table A-16. Using $a/D = 6/42 = 0.143$ and $d/D = 34/42 = 0.810$, and using linear interpolation, we obtain $A = 0.798$ and $K_t = 2.366$ for bending; and $A = 0.89$ and $K_{ts} = 1.75$ for torsion. Thus, for bending,

$$Z_{net} = \frac{\pi A}{32D}(D^4 - d^4) = \frac{\pi(0.798)}{32(4.2)}[(4.2)^4 - (3.4)^4] = 3.31 \text{ cm}^3$$

and for torsion,

$$J_{net} = \frac{\pi A}{32}(D^4 - d^4) = \frac{\pi(0.89)}{32}[(4.2)^4 - (3.4)^4] = 15.5 \text{ cm}^4$$

Next, using Figs. 5-16 and 5-17 we find the notch sensitivities to be 0.78 for bending and 0.96 for torsion. The two corresponding fatigue stress-concentration factors are obtained from Eq. (5-26) as

$$K_f = 1 + q(K_t - 1) = 1 + 0.78(2.366 - 1) = 2.07$$
$$K_{fs} = 1 + 0.96(1.75 - 1) = 1.72$$

The bending stress is now found to be

$$\sigma_x = K_f \frac{M}{Z_{net}} = 2.07 \frac{150}{3.31} = 93.8 \text{ MPa}$$

and the torsional stress is

$$\tau_{xy} = K_{fs} \frac{TD}{2J_{net}} = 1.72 \frac{120(4.2)}{2(15.5)} = 28.0 \text{ MPa}$$

The two nonzero principal stresses are found from a Mohr's circle analysis to be

$$\sigma_A, \sigma_B = \frac{\sigma_x}{2} \pm \left[\left(\frac{\sigma_x}{2}\right)^2 + \tau_{xy}^2\right]^{1/2}$$
$$= \frac{93.8}{2} \pm \left[\left(\frac{93.8}{2}\right)^2 + (28.0)^2\right]^{1/2} = 101.5, -7.7 \text{ MPa}$$

From Eq. (6-13), the von Mises stress is

$$\sigma' = (\sigma_A^2 - \sigma_A\sigma_B + \sigma_B^2)^{1/2}$$
$$= [(101.5)^2 - (101.5)(-7.7) + (-7.7)^2]^{1/2} = 105.6 \text{ MPa}$$

Finally, we estimate the factor of safety guarding against a fatigue failure to be

Answer

$$n = \frac{S_e}{\sigma'} = \frac{172}{105.6} = 1.63$$

To find the factor of safety guarding against yielding, we omit the use of K_f and K_{fs}, because the material is ductile, and find $\sigma_x = 45.3$ MPa and $\tau_{xy} = 16.3$ MPa. With these components and a shortcut formula, we find the von Mises stress to be

$$\sigma' = (\sigma_x^2 + 3\tau_{xy}^2)^{1/2}$$
$$= [(45.3)^2 + 3(16.3)^2]^{1/2} = 53.4 \text{ MPa} \tag{7-44}$$

and so

Answer

$$n = \frac{S_{yt}}{\sigma'} = \frac{370}{53.4} = 6.93$$

(b) This part asks us to find factors of safety when the alternating component is due to a pulsating torsion and the mean component is due to both torsion and bending. We have $T_a = (160 - 20)/2 = 70$ N · m and $T_m = 20 + 70 = 90$ N · m. The corresponding alternating and mean stress components are

$$\tau_{xya} = K_{fs} \frac{T_a D}{2J_{net}} = 1.72 \frac{70(4.2)}{2(15.5)} = 16.3 \text{ MPa}$$
$$\tau_{xym} = \frac{T_m D}{2J_{net}} = \frac{90(4.2)}{2(15.5)} = 12.2 \text{ MPa}$$

The bending stress is

$$\sigma_{xm} = \frac{M_m}{Z_{net}} = \frac{150}{3.31} = 45.3 \text{ MPa}$$

We next employ these components to find σ'_a and σ'_m. Using Eq. (7-44) for both, we get

$$\sigma'_a = [3(16.3)^2]^{1/2} = 28.2 \text{ MPa}$$

$$\sigma'_m = [(45.3)^2 + 3(12.2)^2]^{1/2} = 50.0 \text{ MPa}$$

To use the modified Goodman relation, rearrange Eq. (7-39) and solve. This gives

Answer
$$n = \frac{S_e S_{ut}}{\sigma_a S_{ut} + \sigma_m S_e} = \frac{172(440)}{28.2(440) + 50.0(172)} = 3.60 \qquad (7\text{-}45)$$

To check failure by yielding, we observe that $T_{max} = 160 \text{ N} \cdot \text{m}$ and so

$$\tau_{xy} = \frac{T_{max}D}{2J_{net}} = \frac{160(4.2)}{2(15.5)} = 21.7 \text{ MPa}$$

Since $\sigma_x = \sigma_{xm} = 45.3$ MPa, we have

$$\sigma' = [(45.3)^2 + 3(21.7)^2]^{1/2} = 58.9 \text{ MPa}$$

and so the factor of safety guarding against yielding is

Answer
$$n = \frac{S_{yt}}{\sigma'} = \frac{370}{58.9} = 6.28$$

7-16   CUMULATIVE FATIGUE DAMAGE

Instead of a single reversed stress σ for n cycles, suppose a part is subjected to σ_1 for n_1 cycles, σ_2 for n_2 cycles, etc. Under these conditions our problem is to estimate the fatigue life of a part subjected to these reversed stresses, or to estimate the factor of safety if the part has an infinite life. A search of the literature reveals that this problem has not been solved completely. Therefore the results obtained using either of the approaches presented here should be employed as guides to indicate how one might seek improvement. An approach consistently in agreement with experiment has not yet been reported in the literature of the subject.

The theory which is in greatest use at the present time to explain cumulative fatigue damage is the *Palmgren-Miner cycle-ratio summation theory*, also called *Miner's rule*.* Mathematically, this theory is stated as

---

*A. Palmgren, "Die Lebensdauer von Kugellagern," *ZVDI*, vol. 68, pp. 339–341, 1924; M. A. Miner, "Cumulative Damage in Fatigue," *J. Appl. Mech.*, vol. 12, *Trans. ASME*, vol. 67, pp. A159–A164, 1945.

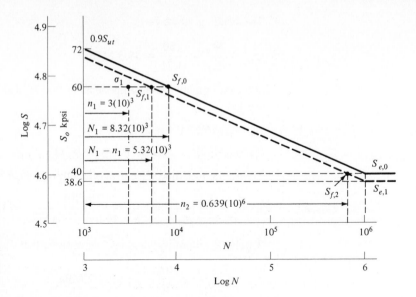

FIGURE 7-20

Use of Miner's rule to predict the endurance limit of a material which has been overstressed for a finite number of cycles.

$$\frac{n_1}{N_1} + \frac{n_2}{N_2} + \cdots + \frac{n_i}{N_i} = C \tag{7-46}$$

where n is the number of cycles of stress σ applied to the specimen and N is the life corresponding to σ. The constant C is determined by experiment and is usually found in the range

$$0.7 \le C \le 2.2$$

Many authorities recommend using $C = 1$, and then Eq. (7-46) may be written

$$\sum \frac{n}{N} = 1 \tag{7-47}$$

To illustrate the use of Miner's rule, let us choose a steel having the properties $S_{ut} = 80$ kpsi and $S'_{e,0} = 40$ kpsi, where we have used the designation $S'_{e,0}$ instead of the more usual S'_e to indicate the endurance limit of the *virgin*, or *undamaged, material*. The log S–log N diagram for this material is shown in Fig. 7-20 by the heavy solid line. Now apply, say, a reversed stress $\sigma_1 = 60$ kpsi for $n_1 = 3000$ cycles. Since $\sigma_1 > S'_{e,0}$, the endurance limit will be damaged, and we wish to find the new endurance limit $S'_{e,1}$ of the damaged material using Miner's rule. The figure shows that the material has a life $N_1 = 8320$ cycles, and consequently, after the application of σ_1 for 3000 cycles, there are $N_1 - n_1 = 5320$ cycles of life remaining. This locates the finite-life strength $S_{f,1}$ of the damaged material, as shown in Fig. 7-20. To get a second point, we ask the question, With n_1 and N_1 given, how many cycles of stress $\sigma_2 = S'_{e,0}$ can be applied before the damaged material fails? This corresponds to n_2 cycles of stress reversal, and hence, from Eq. (7-47), we have

$$\frac{n_1}{N_1} + \frac{n_2}{N_2} = 1 \tag{a}$$

or

$$n_2 = \left(1 - \frac{n_1}{N_1}\right)N_2 \qquad\qquad (b)$$

Then

$$n_2 = \left[1 - \frac{3(10)^3}{8.32(10)^3}\right](10)^6 = 0.639(10)^6 \ cycles$$

This corresponds to the finite-life strength $S_{f,2}$ in Fig. 7-20. A line through $S_{f,1}$ and $S_{f,2}$ is the log S–log N diagram of the damaged material according to Miner's rule. The new endurance limit is $S_{e,1} = 38.6$ kpsi.

Though Miner's rule is quite generally used, it fails in two ways to agree with experiment. First, note that this theory states that the static strength S_{ut} is damaged, that is, decreased, because of the application of σ_1; see Fig. 7-20 at $N = 10^3$ cycles. Experiments fail to verify this prediction.

Miner's rule, as given by Eq. (7-47), does not account for the order in which the stresses are applied, and hence ignores any stresses less than $S'_{e,0}$. But it can be seen in Fig. 7-20 that a stress σ_3 in the range $S'_{e,1} < \sigma_3 < S'_{e,0}$ would cause damage if applied after the endurance limit had been damaged by the application of σ_1.

*Manson's** approach overcomes both of the deficiencies noted for the Palmgren-Miner method; historically it is a much more recent approach, and it is just as easy to use. Except for a slight change, we shall use and recommend the Manson method in this book. Manson plotted the S-log N diagram instead of a log S–log N plot as is recommended here. Manson also resorted to experiment to find the point of convergence of the S-log N lines corresponding to the static strength, instead of arbitrarily selecting the intersection of $N = 10^3$ cycles with $S = 0.9S_{ut}$ as is done here. Of course, it is always better to use experiment, but our purpose in this book has been to use the simple tension-test data to learn as much as possible about fatigue failure.

The method of Manson, as presented here, consists in having all log S–log N lines, that is, lines for both the damaged and the virgin material, converge to the same point, $0.9S_{ut}$ at 10^3 cycles. In addition, the log S–log N lines must be constructed in the same historical order in which the stresses occur.

The data from the preceding example are used for illustrative purposes. The results are shown in Fig. 7-21. Note that the strength $S_{f,1}$ corresponding to $N_1 - n_1 = 5.32(10)^3$ cycles is found in the same manner as before. Through this point and through $0.9S_{ut}$ at 10^3 cycles, draw the heavy dashed line to meet $N = 10^6$ cycles and define the endurance limit $S'_{e,1}$ of the damaged material. In this case the new endurance limit is 34.1 kpsi, somewhat less than that found by Miner's method.

It is now easy to see from Fig. 7-21 that a reversed stress $\sigma = 36$ kpsi, say, would not harm the endurance limit of the virgin material, no matter how many cycles it might be applied. However, if $\sigma = 36$ kpsi should be applied *after* the material was damaged by $\sigma_1 = 60$ kpsi, then additional damage would be done.

---

*S. S. Manson, A. J. Nachtigall, C. R. Ensign, and J. C. Freche, "Further Investigation of a Relation for Cumulative Fatigue Damage in Bending," *Trans. ASME, J. Eng. Ind.,* ser. B, vol. 87, no. 1, pp. 25–35, February 1965.

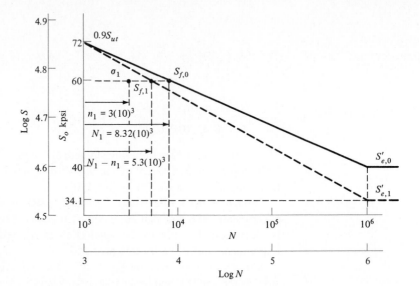

FIGURE 7-21

Use of Manson's method to predict the endurance limit of a material which has been overstressed for a finite number of cycles.

7-17 THE FRACTURE-MECHANICS APPROACH

Here we present a method of estimating the remaining life in a specimen after discovery of a crack. The method is known as *linear elastic fracture mechanics,* and it requires the assumption that plane strain conditions prevail.*

Assuming an elastic isotropic material, the growth of a fatigue crack is approximated by the equation

$$\frac{da}{dN} = \frac{C(\Delta K - \Delta K_{\text{th}})^m}{(1 - R)K_c - \Delta K} \tag{7-48}$$

where C and m are empirical constants. The stress ratio R is given by Eq. (7-32), and K_c is the critical stress-intensity factor as defined in Sec. 5-23. The quantity ΔK depends upon the stress range, and is

$$\Delta K = \sigma_r(\pi a)^{1/2}\left(\frac{K_{\text{I}}}{K_0}\right) \tag{7-49}$$

where σ_r is the stress range as defined in Sec. 7-12. The ratio $K_{\text{I}}K_0$ corrects for the geometry of the problem and is the ordinate in Figs. 5-19 and 5-21 to 5-27.

The quantity ΔK_{th} in Eq. (7-48) is called the threshold value; crack growth is not expected for lesser values of ΔK.

We can bypass much of the complexity of Eq. (7-48) by using the power equation

$$\frac{da}{dN} = C(\Delta K_{\text{I}})^m \tag{a}$$

*Recommended references are: J. A. Collins, *Failure of Materials in Mechanical Design,* Wiley, New York 1981; H. O. Fuchs and R. I. Stephens, *Metal Fatigue in Engineering,* Wiley, New York, 1980; and Harold S. Reemsnyder, "Constant Amplitude Fatigue Life Assessment Models," *SAE Trans. 820688,* vol. 91, November 1983.

Values of the Factor C and Exponent m in Eq. (a)

MATERIAL	EQUATION AND UNITS
Ferritic-pearlitic steels	$\frac{da}{dN}$(m/cycle) $= 6.9(10^{-12})(\Delta K\ \text{MPa}\sqrt{m})^{3.0}$
	$\frac{da}{dN}$(in/cycle) $= 3.6(10^{-10})(\Delta K\ \text{kpsi}\sqrt{in})^{3.0}$
Martensitic steels	$\frac{da}{dN}$(m/cycle) $= 1.35(10^{-10})(\Delta K\ \text{MPa}\sqrt{m})^{2.25}$
	$\frac{da}{dN}$(in/cycle) $= 6.6(10^{-9})(\Delta K\ \text{kpsi}\sqrt{in})^{2.25}$
Austenitic stainless steels	$\frac{da}{dN}$(m/cycle) $= 5.6(10^{-12})(\Delta K\ \text{MPa}\sqrt{m})^{3.25}$
	$\frac{da}{dN}$(in/cycle) $= 3.0(10^{-10})(\Delta K\ \text{kpsi}\sqrt{in})^{3.25}$

Source: Fuchs and Stephens, op. cit., pp. 85, 86.

Values of C and m for several classes of materials are listed in Table 7-7. By rearranging and integrating both sides of Eq. (7-50), we have

$$\int_{a_0}^{a_f} da = \int_0^{N_f} C(\Delta K_I)^m\ dN \tag{7-50}$$

Here a_0 is the initial crack length, a_f is the final crack length corresponding to failure, and N_f is the estimated number of cycles required to produce a failure. Note that the ratio K_I/K_0 may vary in the integration interval. If this should happen, then Reemsnyder* suggests the use of numerical integration employing the algorithm

$$\delta a_j = C(\Delta K)_j^m(\delta N)_j$$
$$a_{j+1} = a_j + \delta a_j$$
$$N_{j+1} = N_j + \delta N_j \tag{7-51}$$
$$N_j = \sum \delta N_j$$

Here δa_j and δN_j are increments of the crack length and the number of cycles. The procedure is to select a value of δN_j, compute ΔK using a_0, and then find the next value of a. Repeat until $a = a_f$. The procedure can be illustrated using a simple example.

EXAMPLE 7-9

The bar shown in Fig. 7-22 is subjected to a repeated moment $M_{\max} = 208$ lb · in. The bar is hot-rolled steel with $S_{ut} = 60$ kpsi and $S_y = 33$ kpsi. As shown, a nick of size 0.004 in has been discovered on the bottom of the bar. Estimate the number of cycles of life remaining.

*Op. cit.

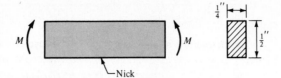

FIGURE 7-22

— Nick

Solution

The stress range σ_r for use in Eq. (7-49) is always computed using the nominal (uncracked) area. Thus

$$\frac{I}{c} = \frac{bh^2}{6} = \frac{0.25(0.5)^2}{6} = 0.0104 \text{ in}^3$$

Therefore the stress range is

$$\sigma_r = \frac{M}{I/c} = \frac{208}{0.0104}(10^{-3}) = 20 \text{ kpsi}$$

If the crack grows, it will eventually become long enough that the bar will fail by yielding. Designate this crack length a_f. The section modulus corresponding to this length is

$$\left(\frac{I}{c}\right)_f = \frac{b(0.5 - a_f)^2}{6}$$

Since $\sigma_{max} = S_y = M/(I/c)_f$, we have $M/S_y = (I/c)_f$, or

$$\frac{208}{33(10^3)} = \frac{0.25}{6}(0.5 - a_f)^2 \tag{1}$$

The solution to this equation is $a_f = 0.111$ in.

Table 7-7 gives $C = 3.6(10^{-10})$ and $m = 3.0$ for use in Eq. (a). Referring next to Fig. 5-23, we compute the largest a/h ratio as

$$\frac{a}{h} = \frac{0.111}{0.5} = 0.222$$

Thus a/h varies from near zero to 0.222, and for this range $K_I/K_0 = 1.06$ and is nearly a constant. We shall assume it to be so, and integrate Eq. (7-50) analytically. Thus

$$K_I = \sigma_r(\pi a)^{1/2}\left(\frac{K_I}{K_0}\right) = [20(\pi)^{1/2}(1.06)]^{3.0}a^{1.5}$$

Thus Eq. (7-50) becomes

$$\int_{a_0}^{a_f} \frac{da}{a^{1.5}} = \int_0^{N_f} 3.6(10^{-10})[20(\pi)^{1/2}(1.06)]^{3.0} \, dN$$

Performing the indicated operations, we have

$$\frac{-2}{\sqrt{a}}\bigg|_{0.004}^{0.111} = 1.91(10^{-5})N\bigg|_0^{N_f}$$

Answer from which we find $N_f = 1.34(10^6)$ cycles, which is the estimated remaining life.

7-18   SURFACE STRENGTH

Our studies thus far have dealt with the failure of a machine element by yielding, by fracture, and by fatigue. The endurance limit obtained by the rotating-beam test is frequently called the *flexural endurance limit,* because it is a test of a rotating beam. In this section we shall study a property of *mating materials* called the *surface endurance shear.* The design engineer must frequently solve problems in which two machine elements mate with one another by rolling, sliding, or a combination of rolling and sliding contact. Obvious examples of such combinations are the mating teeth of a pair of gears, a cam and follower, a wheel and rail, and a chain and sprocket. A knowledge of the surface strength of materials is necessary if the designer is to create machines having a long and satisfactory life.

When two surfaces roll or roll and slide against one another with sufficient force, a pitting failure will occur after a certain number of cycles of operation. Authorities are not in complete agreement on the exact mechanism of the pitting; although the subject is quite complicated, they do agree that the Hertz stresses, the number of cycles, the surface finish, the hardness, the degree of lubrication, and the temperature all influence the strength. In Sec. 2-19 it was learned that, when two surfaces are pressed together, a maximum shear stress is developed slightly below the contacting surface. It is postulated by some authorities that a surface fatigue failure is initiated by this maximum shear stress and then is propagated rapidly to the surface. The lubricant then enters the crack which is formed and, under pressure, eventually wedges the chip loose.

To determine the surface fatigue strength of mating materials, Buckingham designed a simple machine for testing a pair of contacting rolling surfaces in connection with his investigation of the wear of gear teeth. Buckingham and, later, Talbourdet gathered large numbers of data from many tests so that considerable design information is now available. To make the results useful for designers, Buckingham defined a *load-stress factor,* also called a *wear factor,* which is derived from the Hertz equations. Equations (2-93) and (2-94) for contacting cylinders are found to be

$$b = \sqrt{\frac{2F}{\pi l} \frac{(1 - \nu_1^2)/E_1 + (1 - \nu_2^2)/E_2}{(1/d_1) + (1/d_2)}} \tag{7-52}$$

$$p_{max} = \frac{2F}{\pi bl} \tag{7-53}$$

where b = half width of rectangular contact area
   F = contact force
   l = width of cylinders
   ν = Poisson's ratio
   E = modulus of elasticity
   d = cylinder diameter

On the average, $\nu = 0.30$ for engineering materials. Thus, let $\nu = \nu_1 = \nu_2 = 0.30$. Also, it is more convenient to use the cylinder radius; so let $2r = d$. If we then designate the width of the cylinders as w instead of l and remove the square root sign, Eq. (7-52) becomes

$$b^2 = 1.16 \frac{F}{w} \frac{(1/E_1) + (1/E_2)}{(1/r_1) + (1/r_2)} \tag{7-54}$$

Next, define a new kind of endurance property called *surface strength,* which is qualified by the number of cycles at which the first tangible evidence of fatigue is observed. Using Eq. (7-53), this strength is

$$S_C = \frac{2F}{\pi bw} \tag{7-55}$$

which may also be called the *contact strength,* the *contact fatigue strength,* or the *Hertzian endurance strength.* This strength is the contacting pressure which, after a specified number of cycles, will cause failure of the surface. Such failures are often called *wear* because they occur after a very long time. They should not be confused with abrasive wear, however. By substituting the value of b from Eq. (7-54) in (7-55) and rearranging, we obtain

$$2.857S_C^2\left(\frac{1}{E_1} + \frac{1}{E_2}\right) = \frac{F}{w}\left(\frac{1}{r_1} + \frac{1}{r_2}\right) \tag{7-56}$$

The left side of this equation contains E_1, E_2, and S_C, constants which come about because of the selection of a certain material for each element of the pair. We call this K_1, *Buckingham's load-stress factor.* Once the two materials have been selected, K_1 is computed from the equation

$$K_1 = 2.857S_C^2\left(\frac{1}{E_1} + \frac{1}{E_2}\right) \tag{7-57}$$

With K_1 known, we now write the design equation as

$$K_1 = \frac{F}{w}\left(\frac{1}{r_1} + \frac{1}{r_2}\right) \tag{7-58}$$

which, if satisfied, defines a surface fatigue failure in 10^8 cycles of repeated stress, according to Talbourdet's experiments. Since we usually want to define factor of safety n instead of failure, we would write Eq. (7-58) in the form

$$\frac{K_1}{n} = \frac{F}{w}\left(\frac{1}{r_1} + \frac{1}{r_2}\right) \tag{7-59}$$

Values of the surface fatigue strength for steels can be obtained from the equation

$$S_C = \begin{cases} 0.4H_B - 10 \text{ kpsi} \\ 2.76H_B - 70 \text{ MPa} \end{cases} \tag{7-60}$$

where H_B is the Brinell hardness number and where it is understood that these strengths are valid only for 10^8 cycles of repeated contact stress. If the two materials have different hardnesses, the lesser value is generally, though not always, used. The results of this procedure agree with the values of the load-stress factors recommended by Buckingham.

PROBLEMS*

7-1 Estimate the fatigue strength of a rotating-beam specimen made of AISI 1020 hot-rolled steel corresponding to a life of 12.5 kcycles of stress reversal. Estimate the life of the specimen corresponding to a stress amplitude of 36 kpsi.

*The asterisk indicates a problem that may not have a unique solution.

7-2   Derive Eqs. (7-6) and (7-7).

7-3   Write a program for solving Eqs. (7-6), (7-5), and (7-7), in that order.

7-4   A rotating-beam specimen has been made from an AISI 1137 cold-drawn steel bar and has $\bar{S}_{ut} = 734$ MPa and $\hat{\sigma}_{Sut} = 4.24$ MPa.

(a) Estimate the endurance limit and the standard deviation; use SI units.

(b) Estimate the fatigue strength corresponding to a desired life of 130 kcycles.

7-5   A $\frac{3}{16}$-in drill rod was heat-treated and ground; the measured hardness was then found to be 490 Bhn. Estimate the endurance limit if the rod is used in rotating bending.

7-6   Solve Prob. 7-5 if the hardness of production pieces was found to be $\mathbf{H}_B = 490(1, 0.03)$.

7-7   Estimate the endurance limit of a 32-mm bar of AISI 1035 steel having a machined finish and heat-treated to a minimum tensile strength of 710 MPa.

7-8   Two steels are being considered for manufacture as forged connecting rods. One is AISI 4340 Cr-Mo-Ni steel capable of being heat-treated to a minimum tensile strength of 260 kpsi. The other is a plain carbon AISI 1040 steel with an attainable minimum S_{ut} of 113 kpsi. If each rod is to have a size of $\frac{3}{4}$ in, is there any advantage in using the alloy steel? Why?

7-9   A rectangular bar is cut from an AISI 1018 cold-drawn steel flat. The bar is 60 mm wide by 10 mm thick and has a 12-mm-diameter hole drilled through the center. The bar is loaded in push-pull fatigue by axial forces F uniformly distributed across the width. On the basis of a design factor $n_d = 1.8$, estimate the maximum allowable force F that can be applied.

7-10   The figure is an idealized representation of a machine member subjected to the action of an alternating force F which places the member in completely reversed bending. The material is AISI 1050 steel, heat-treated and tempered to 800°F with a ground finish. Assume the strengths listed in Table A-21 are attainable as mean values. On the basis of a realiability of 50 percent, estimate the force that can be applied.

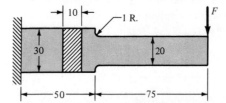

PROBLEM 7-10
Dimensions in millimeters.

7-11   Bearing reactions are located at A and B of the countershaft shown in the figure. The bending forces are $F_1 = 2.1$ kN and $F_2 = 4.5$ kN. The steel used has minimum strengths of $S_{ut} =$

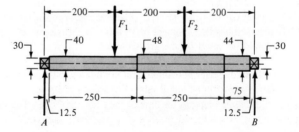

PROBLEM 7-11
Dimensions in millimeters; all fillets 1.6-mm radius.

610 MPa and S_{yt} = 490 MPa. Estimate the factor of safety based on a machined finish for the shaft.

7-12   The steel shaft shown in the figure has a ground finish and a minimum tensile strength of 89 kpsi. The shaft rotates at 1720 rev/min and is supported in rolling bearings at A and B. Estimate the life of the part if the forces are F_1 = 2000 lb and F_2 = 3000 lb.

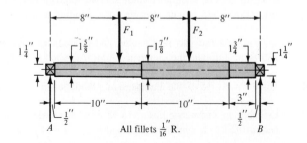

PROBLEM 7-12

7-13   Illustrated is a shaft machined from a bar of cold-drawn steel having a mean tensile strength of 550 MPa. Determine the reliability if the force is $F \sim N(6.0, 0.1)$ kN. Use *LNLN* for the interference.

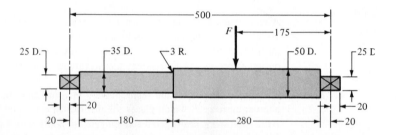

PROBLEM 7-13
Dimensions in millimeters.

7-14   Bearing reactions R_1 and R_2 are exerted on the shaft shown in the figure, which rotates at 1150 rev/min and supports the 10-kip bending force. The specifications call for a ductile steel having minimum strengths of S_{ut} = 120 kpsi and S_{yt} = 90 kpsi. The shaft is to have a machined finish and to have a life of 75 min. Use a design factor of 1.60 and find an appropriate diameter.

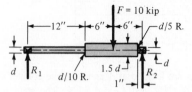

PROBLEM 7-14

7-15   A solid round steel bar is machined to a diameter of 1.25 in. A groove $\frac{1}{8}$ in deep and with radius $\frac{1}{8}$ in is cut into the bar. The material has a mean tensile strength of 110 kpsi. A completely reversed bending moment M = 1400 lb · in is applied. Estimate the reliability. The size factor should be based on the gross diameter.

7-16   A $1\frac{1}{4}$-in-diameter hot-rolled steel bar has a $\frac{1}{8}$-in diameter hole drilled transversely through it. The bar is nonrotating and is subject to a completely reversed bending moment of M = 1600 lb · in in the same plane as the axis of the transverse hole. The material has a mean tensile strength of

58 kpsi. Estimate the reliability. The size factor should be based on the gross size. Use Table A-16 for K_t.

7-17   The same as Prob. 7-15, except that a completely reversed torsional moment of $T = 1400$ lb · in is applied.

7-18   The same as Prob. 7-16, except that the bar is subjected to a completely reversed torsional moment of 2400 lb · in.

7-19*   The figure shows modified Goodman fatigue diagrams containing six different load lines. For each case, make a sketch of the geometry of an actual physical case and explain how the load line is produced.

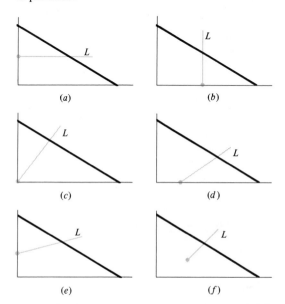

(a)   (b)

(c)   (d)

PROBLEM 7-19   (e)   (f)

7-20   The cold-drawn AISI 1018 steel bar shown in the figure is subjected to a tension load fluctuating between 800 and 3000 lb. Estimate the factors of safety guarding against failure by yielding and by fatigue action.

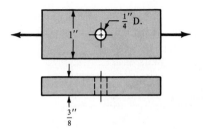

PROBLEM 7-20

7-21   The same as Prob. 7-20, except that the load fluctuates between −800 and 3000 lb. Assume no buckling.

7-22   The same as Prob. 7-21, except that the load fluctuates between 800 and −3000 lb. Assume no buckling.

7-23   The flat steel spring illustrated is loaded in bending by the force F. The spring supports a static weight of exactly 9.36 kN. During operation, the total load on the spring is estimated to fluctuate up to 10.67 kN maximum. The spring is forged of a 95-point carbon steel and after heat treatment has the following minimum properties: $S_{ut} = 1400$ MPa, $S_{yt} = 950$ MPa, $H_B = 399$, and 32 percent reduction in area. Estimate the factor of safety if the spring is 18 mm thick.

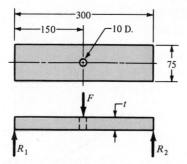

PROBLEM 7-23
Dimensions in millimeters.

7-24   The figure shows a formed round-wire cantilever spring subjected to a varying force. A hardness test made on 25 springs gave a minimum hardness of 380 Bhn. It is apparent from the mounting details that there is no stress concentration. A visual inspection of the springs indicates that the surface finish corresponds closely to a hot-rolled finish. What number of load applications is likely to cause failure?

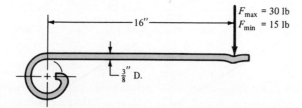

PROBLEM 7-24

7-25*   The figure is a drawing of a 3- × 18-mm latching spring. A preload is obtained during assembly by shimming under the bolts so as to obtain an estimated initial deflection of 2 mm. The latching operation itself requires an additional deflection of exactly 4 mm. The material is ground high-carbon steel, bent and then hardened and tempered to a minimum hardness of 490 Bhn. The radius of the bend is 3 mm.
(a) Find the maximum and minimum latching forces.
(b) Is it likely that the spring will fail in fatigue?

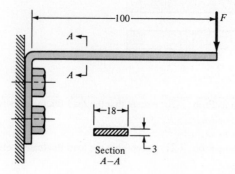

PROBLEM 7-25
Dimensions in millimeters.

7-26 The figure shows the free-body diagram of a connecting-link portion having stress concentration at three places. The dimensions are: $r = 0.25$ in, $d = 0.75$ in, $h = 0.50$ in, $w_1 = 3.75$ in, and $w_2 = 2.5$ in. The forces F fluctuate between a tension of 4 kip and a compression of 16 kip. Neglect column action and find the least factor of safety. Material is 1018 CD steel.

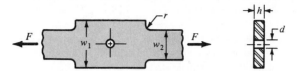

PROBLEM 7-26

7-27 A bar of steel has the minimum properties $S_e = 276$ MPa, $S_y = 413$ MPa, and $S_u = 551$ MPa. For each of the cases below, find the factor of safety guarding against a static failure and either the factor of safety guarding against a fatigue failure or the expected life of the part.
(*a*) A steady torsional stress of 103 MPa and an alternating bending stress of 172 MPa
(*b*) A steady torsional stress of 138 MPa and an alternating torsional stress of 69 MPa
(*c*) A steady torsional stress of 103 MPa, an alternating torsional stress of 69 MPa, and an alternating bending stress of 83 MPa
(*d*) An alternating torsional stress of 207 MPa
(*e*) An alternating torsional stress of 103 MPa and a steady tensile stress of 103 MPa

7-28 A spherical pressure vessel 600 mm in diameter is made of cold-drawn sheet steel having $S_u = 440$ MPa, $S_y = 370$ MPa, and a thickness of 3 mm. The vessel is to withstand an infinite number of pressure fluctuations from 0 to p_{max}.
(*a*) What maximum pressure will cause static yielding?
(*b*) What maximum pressure will eventually cause a fatigue failure? In any case, assume that the joints and connections are adequately reinforced and do not weaken the vessel.

7-29 The figure shows the head end of a long power screw. Such screws are usually designed for compression loading; but when they are long, tension loading avoids the buckling problem. To obtain a good resistance against wear, the screw is made of a medium-carbon steel and heat-treated to obtain the minimum properties of $S_{ut} = 148$ kpsi and $S_{yt} = 112$ kpsi. All important

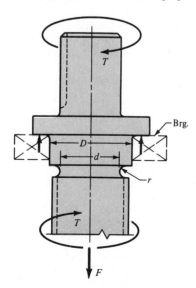

PROBLEM 7-29

surfaces have been finished by grinding. On the drive stroke, $F = 15$ kip and $T = 3000$ lb · in. On the return stroke, $F = 2$ kip and $T = -300$ lb · in. In this problem, we wish to investigate the safety of the design based on the stresses in the thread run-out groove. The important dimensions are $d = 1.25$ in, $D = 1.75$ in, and $r = \frac{1}{8}$ in. For stress-concentration purposes, treat the run-out groove as a shoulder fillet, rather than as a simple groove, even though the threads do add some strengthening.

7-30 A $\frac{1}{4}$- × $1\frac{1}{2}$-in steel bar has a $\frac{3}{4}$-in-diameter drilled hole located in the center. The bar is subjected to a completely reversed axial load with a deterministic amplitude of 2100 lb. The material has a mean ultimate tensile strength of $S_{ut} = 80$ kpsi.

(a) Estimate the reliability using Eqs. (6-22) and (6-28).

(b) Decide which estimate is better.

(c) Confirm your decision in part (b) with a computer simulation treating the random variables as normal variates.

ANSWERS

7-4 (a) $\mathbf{S}'_e = 370(1, 0.146)$ MPa; (b) $\mathbf{S}_f = 439(1, 0.1199)$ MPa

7-9 $S_e = 184$ MPa, $K_f = 2.17$, $F = 22.6$ kN

7-12 $S_e = 18.2$ kpsi, $\sigma = 46.27$ kpsi, $t = 7.5$ min

7-13 $R = 0.8852$

7-16 $R = 0.7379$

7-20 5.06, 2.39

7-23 3.54

7-26 1.80 at the fillet

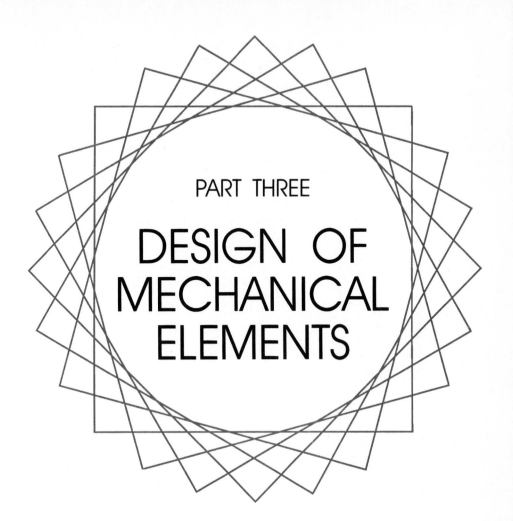

PART THREE

DESIGN OF MECHANICAL ELEMENTS

8
The Design of Screws, Fasteners, and Connections

8-1 THREAD STANDARDS AND DEFINITIONS

The terminology of screw threads, illustrated in Fig. 8-1, is explained as follows:

The *pitch* is the distance between adjacent thread forms measured parallel to the thread axis. The pitch in U.S. units is the reciprocal of the number of thread forms per inch N.

The *major diameter d* is the largest diameter of a screw thread.

The *minor diameter d_r or d_1* is the smallest diameter of a screw thread.

The lead l, not shown, is the distance the nut moves parallel to the screw axis when the nut is given one turn. For a single thread, as in Fig. 8-1, the lead is the same as the pitch.

A *multiple-threaded* product is one having two or more threads cut beside each other (imagine two or more strings wound side by side around a pencil). Standardized products such as screws, bolts, and nuts all have single threads; a *double-threaded* screw has a lead equal to twice the pitch, a *triple-threaded* screw has a lead equal to 3 times the pitch, and so on.

All threads are made according to the *right-hand rule* unless otherwise noted.

The *American National (Unified)* thread standard has been approved in this country and in Great Britain for use on all standard threaded products. The thread angle is 60° and the crests of the thread may be either flat or rounded.

Figure 8-2 shows the thread geometry of the metric M and MJ profiles. The M profile replaces the inch class and is the basic ISO 68 profile with 60° symmetric threads. The MJ profile has a rounded fillet at the root of the external thread and a larger minor diameter of both the internal and external threads. This profile is especially useful where high fatigue strength is required.

Tables 8-1 and 8-2 will be useful in specifying and designing threaded parts. Note that the thread size is specified by giving the pitch p for metric sizes and by giving the number of threads per inch N for the Unified sizes. The screw sizes in Table 8-2 with diameter under $\frac{1}{4}$ in are numbered or gauge sizes. The second column in Table 8-2 shows that a No. 8 screw has a nominal major diameter of 0.1640 in.

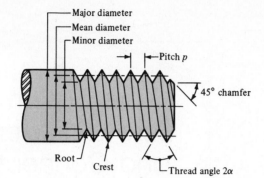

FIGURE 8-1

Terminology of screw threads. Sharp vee threads shown for clarity; the crests and roots are actually flattened or rounded during the forming operation.

A great many tensile tests of threaded rods have shown that an unthreaded rod having a diameter equal to the mean of the pitch diameter and minor diameter will have the same tensile strength as the threaded rod. The area of this unthreaded rod is called the tensile-stress area A_t of the threaded rod; values of A_t are listed in both tables.

Two major Unified thread series are in common use: UN and UNR. The difference between these is simply that a root radius must be used in the UNR series. Because of reduced thread stress-concentration factors, UNR series threads have improved fatigue strengths. Unified threads are specified by stating the nominal major diameter, the number of threads per inch, and the thread series, for example, $\frac{5}{8}''$-18 UNRF or 0.625″-18 UNRF.

Metric threads are specified by writing the diameter and pitch in millimeters, in that order. Thus, M12 × 1.75 is a thread having a nominal major diameter of 12 mm and a pitch of 1.75 mm. Note that the letter M, which precedes the diameter, is the clue to the metric designation.

Square and Acme threads, shown in Fig 8-3a and b, respectively, are used on screws when power is to be transmitted. Table 8-3 lists the preferred pitches for inch-series Acme threads. However, other pitches can be and often are used, since the need for a standard for such threads is not great.

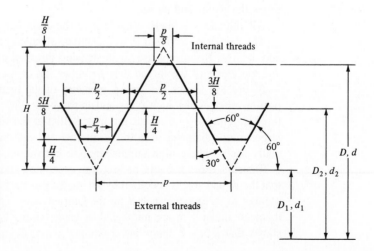

FIGURE 8-2

Basic thread profile for metric M and MJ threads. D (d) = basic major diameter of internal (external) thread; D_1 (d_1) = basic minor diameter of internal (external) thread; D_2 (d_2) = basic pitch diameter of internal (external) thread; p = pitch; $H = 0.5(3)^{1/2}\,p$.

TABLE 8-1

Diameters and Areas of
Course-Pitch and Fine-Pitch
Metric Threads. (All
Dimensions in Millimeters)*

NOMINAL MAJOR DIAMETER d	COARSE-PITCH SERIES			FINE-PITCH SERIES		
	PITCH p	TENSILE-STRESS AREA A_t	MINOR-DIAMETER AREA A_r	PITCH p	TENSILE-STRESS AREA A_t	MINOR-DIAMETER AREA A_r
1.6	0.35	1.27	1.07			
2	0.40	2.07	1.79			
2.5	0.45	3.39	2.98			
3	0.5	5.03	4.47			
3.5	0.6	6.78	6.00			
4	0.7	8.78	7.75			
5	0.8	14.2	12.7			
6	1	20.1	17.9			
8	1.25	36.6	32.8	1	39.2	36.0
10	1.5	58.0	52.3	1.25	61.2	56.3
12	1.75	84.3	76.3	1.25	92.1	86.0
14	2	115	104	1.5	125	116
16	2	157	144	1.5	167	157
20	2.5	245	225	1.5	272	259
24	3	353	324	2	384	365
30	3.5	561	519	2	621	596
36	4	817	759	2	915	884
42	4.5	1120	1050	2	1260	1230
48	5	1470	1380	2	1670	1630
56	5.5	2030	1910	2	2300	2250
64	6	2680	2520	2	3030	2980
72	6	3460	3280	2	3860	3800
80	6	4340	4140	1.5	4850	4800
90	6	5590	5360	2	6100	6020
100	6	6990	6740	2	7560	7470
110				2	9180	9080

*The equations and data used to develop this table have been obtained from ANSI B1.1-1974 and B18.3.1-1978. The minor diameter was found from the equation $d_r = d - 1.226\ 869p$, and the pitch diameter from $d_m = d - 0.649\ 519p$. The mean of the pitch diameter and the minor diameter was used to compute the tensile-stress area.

Modifications are frequently made to both Acme and square threads. For instance, the square thread is sometimes modified by cutting the space between the teeth so as to have an included thread angle of 10 to 15°. This is not difficult, since these threads are usually cut with a single-point tool anyhow; the modification retains most of the high efficiency inherent in square threads and makes the cutting simpler. Acme threads are sometimes modified to a stub form by making the teeth shorter. This results in a larger minor diameter and a somewhat stronger screw.

8-2   THE MECHANICS OF POWER SCREWS

A power screw is a device used in machinery to change angular motion into linear motion, and, usually, to transmit power. Familiar applications include the lead screws of lathes, and the screws for vises, presses, and jacks.

TABLE 8-2

Diameters and Area of Unified Screw Threads UNC and UNF*

SIZE DESIGNATION	NOMINAL MAJOR DIAMETER, in	COARSE SERIES—UNC			FINE SERIES—UNF		
		THREADS PER INCH N	TENSILE-STRESS AREA A_t, in^2	MINOR-DIAMETER AREA A_r, in^2	THREADS PER INCH N	TENSILE-STRESS AREA A_t, in^2	MINOR-DIAMETER AREA A_r, in^2
0	0.0600				80	0.001 80	0.001 51
1	0.0730	64	0.002 63	0.002 18	72	0.002 78	0.002 37
2	0.0860	56	0.003 70	0.003 10	64	0.003 94	0.003 39
3	0.0990	48	0.004 87	0.004 06	56	0.005 23	0.004 51
4	0.1120	40	0.006 04	0.004 96	48	0.006 61	0.005 66
5	0.1250	40	0.007 96	0.006 72	44	0.008 80	0.007 16
6	0.1380	32	0.009 09	0.007 45	40	0.010 15	0.008 74
8	0.1640	32	0.014 0	0.011 96	36	0.014 74	0.012 85
10	0.1900	24	0.017 5	0.014 50	32	0.020 0	0.017 5
12	0.2160	24	0.024 2	0.020 6	28	0.025 8	0.022 6
$\frac{1}{4}$	0.2500	20	0.031 8	0.026 9	28	0.036 4	0.032 6
$\frac{5}{16}$	0.3125	18	0.052 4	0.045 4	24	0.058 0	0.052 4
$\frac{3}{8}$	0.3750	16	0.077 5	0.067 8	24	0.087 8	0.080 9
$\frac{7}{16}$	0.4375	14	0.106 3	0.093 3	20	0.118 7	0.109 0
$\frac{1}{2}$	0.5000	13	0.141 9	0.125 7	20	0.159 9	0.148 6
$\frac{9}{16}$	0.5625	12	0.182	0.162	18	0.203	0.189
$\frac{5}{8}$	0.6250	11	0.226	0.202	18	0.256	0.240
$\frac{3}{4}$	0.7500	10	0.334	0.302	16	0.373	0.351
$\frac{7}{8}$	0.8750	9	0.462	0.419	14	0.509	0.480
1	1.0000	8	0.606	0.551	12	0.663	0.625
$1\frac{1}{4}$	1.2500	7	0.969	0.890	12	1.073	1.024
$1\frac{1}{2}$	1.5000	6	1.405	1.294	12	1.581	1.521

*This table was compiled from ANSI B1.1-1974. The minor diameter was found from the equation $d_r = d - 1.299\ 038p$, and the pitch diameter from $d_m = d - 0.649\ 519p$. The mean of the pitch diameter and the minor diameter was used to compute the tensile-stress area.

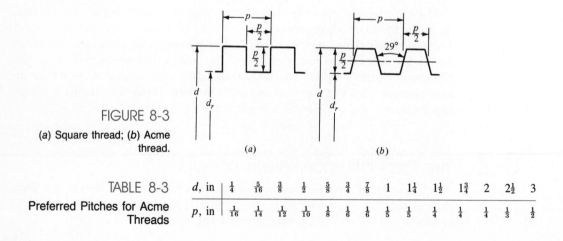

FIGURE 8-3

(a) Square thread; (b) Acme thread.

(a)   (b)

TABLE 8-3

Preferred Pitches for Acme Threads

d, in	$\frac{1}{4}$	$\frac{5}{16}$	$\frac{3}{8}$	$\frac{1}{2}$	$\frac{5}{8}$	$\frac{3}{4}$	$\frac{7}{8}$	1	$1\frac{1}{4}$	$1\frac{1}{2}$	$1\frac{3}{4}$	2	$2\frac{1}{2}$	3
p, in	$\frac{1}{16}$	$\frac{1}{14}$	$\frac{1}{12}$	$\frac{1}{10}$	$\frac{1}{8}$	$\frac{1}{6}$	$\frac{1}{6}$	$\frac{1}{5}$	$\frac{1}{5}$	$\frac{1}{4}$	$\frac{1}{4}$	$\frac{1}{4}$	$\frac{1}{3}$	$\frac{1}{2}$

An application of power screws to a power-driven jack is shown in Fig. 8-4. You should be able to identify the worm, the worm gear, the screw, and the nut. Is the worm gear supported by one bearing or two?

In Fig. 8-5 a square-threaded power screw with single thread having a mean diameter d_m, a pitch p, a lead angle λ, and a helix angle ψ is loaded by the axial compressive force F. We wish to find an expression for the torque required to raise this load, and another expression for the torque required to lower the load.

First, imagine that a single thread of the screw is unrolled or developed (Fig. 8-6) for exactly the single turn. Then one edge of the thread will form the hypotenuse of a right triangle whose base is the circumference of the mean-thread-diameter circle and whose height is the lead. The angle λ, in Figs. 8-5 and 8-6, is the lead angle of the thread. We represent the summation of all the unit axial forces acting upon the normal thread area by F. To raise the load, a force P acts to the right (Fig. 8-6a), and to lower

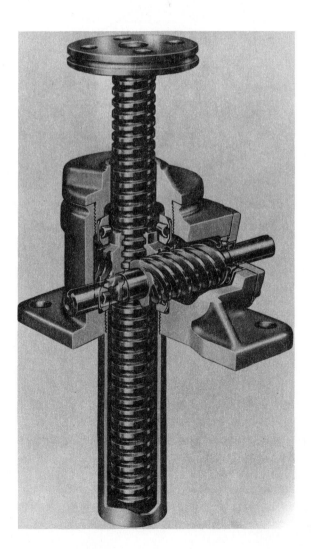

FIGURE 8-4

The Joyce worm-gear screw jack.
(Courtesy Joyce/Dayton Corp.,
Dayton, Ohio.)

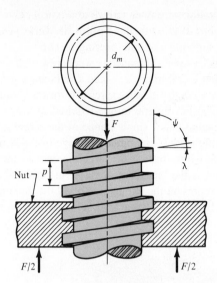

FIGURE 8-5

Portion of a power screw.

the load, P acts to the left (Fig. 8-6b). The friction force is the product of the coefficient of friction μ with the normal force N, and acts to oppose the motion. The system is in equilibrium under the action of these forces, and hence, for raising the load, we have

$$\sum F_H = P - N \sin \lambda - \mu N \cos \lambda = 0$$
$$\sum F_V = F + \mu N \sin \lambda - N \cos \lambda = 0 \qquad (a)$$

In a similar manner, for lowering the load, we have

$$\sum F_H = -P - N \sin \lambda + \mu N \cos \lambda = 0$$
$$\sum F_V = F - \mu N \sin \lambda - N \cos \lambda = 0 \qquad (b)$$

Since we are not interested in the normal force N, we eliminate it from each of these sets of equations and solve the result for P. For raising the load, this gives

$$P = \frac{F(\sin \lambda + \mu \cos \lambda)}{\cos \lambda - \mu \sin \lambda} \qquad (c)$$

and for lowering the load,

$$P = \frac{F(\mu \cos \lambda - \sin \lambda)}{\cos \lambda + \mu \cos \lambda} \qquad (d)$$

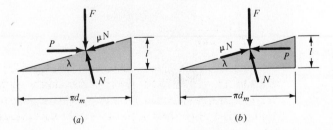

FIGURE 8-6

Force diagrams: (a) lifting the load; (b) lowering the load.

Next, divide the numerator and the denominator of these equations by $\cos \lambda$ and use the relation $\tan \lambda = l/\pi d_m$ (Fig. 8-6). We then have, respectively,

$$P = \frac{F[(l/\pi d_m) + \mu]}{1 - (\mu l/\pi d_m)} \tag{e}$$

$$P = \frac{F[\mu - (l/\pi d_m)]}{1 + (\mu l/\pi d_m)} \tag{f}$$

Finally, noting that the torque is the product of the force P and the mean radius $d_m/2$, for raising the load we can write

$$T = \frac{Fd_m}{2}\left(\frac{l + \pi\mu d_m}{\pi d_m - \mu l}\right) \tag{8-1}$$

where T is the torque required for two purposes: to overcome thread friction and to raise the load.

The torque required to lower the load, from Eq. (f), is found to be

$$T = \frac{Fd_m}{2}\left(\frac{\pi\mu d_m - l}{\pi d_m + \mu l}\right) \tag{8-2}$$

This is the torque required to overcome a part of the friction in lowering the load. It may turn out, in specific instances where the lead is large or the friction is low, that the load will lower itself by causing the screw to spin without any external effort. In such cases, the torque T from Eq. (8-2) will be negative or zero. When a positive torque is obtained from this equation, the screw is said to be *self-locking*. Thus the condition for self-locking is

$$\pi\mu d_m > l$$

Now divide both sides of this inequality by πd_m. Recognizing that $l/\pi d_m = \tan \lambda$, we get

$$\mu > \tan \lambda \tag{8-3}$$

This relation states that self-locking is obtained whenever the coefficient of thread friction is equal to or greater than the tangent of the thread lead angle.

An expression for efficiency is also useful in the evaluation of power screws. If we let $\mu = 0$ in Eq. (8-1), we obtain

$$T_0 = \frac{Fl}{2\pi} \tag{g}$$

which, since thread friction has been eliminated, is the torque required only to raise the load. The efficiency is therefore

$$e = \frac{T_0}{T} = \frac{Fl}{2\pi T} \tag{8-4}$$

The preceding equations have been developed for square threads where the normal thread loads are parallel to the axis of the screw. In the case of Acme or other threads, the normal thread load is inclined to the axis because of the thread angle 2α and the lead angle λ. Since lead angles are small, this inclination can be neglected and only the

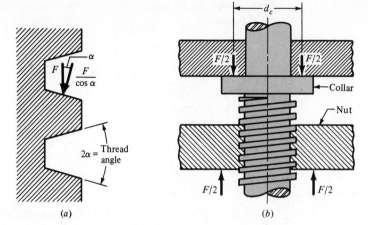

FIGURE 8-7

(a) Normal thread force is increased because of angle α; (b) thrust collar has frictional diameter d_c.

effect of the thread angle (Fig. 8-7a) considered. The effect of the angle α is to increase the frictional force by the wedging action of the threads. Therefore the frictional terms in Eq. (8-1) must be divided by cos α. For raising the load, or for tightening a screw or bolt, this yields

$$T = \frac{Fd_m}{2}\left(\frac{l + \pi\mu d_m \sec \alpha}{\pi d_m - \mu l \sec \alpha}\right) \qquad (8\text{-}5)$$

In using Eq. (8-5), remember that it is an approximation because the effect of the lead angle has been neglected.

For power screws, the Acme thread is not as efficient as the square thread, because of the additional friction due to the wedging action, but it is often preferred because it is easier to machine and permits the use of a split nut, which can be adjusted to take up for wear.

Usually a third component of torque must be applied in power-screw applications. When the screw is loaded axially, a thrust or collar bearing must be employed between the rotating and stationary members in order to carry the axial component. Figure 8-7b shows a typical thrust collar in which the load is assumed to be concentrated at the mean collar diameter d_c. If μ_c is the coefficient of collar friction, the torque required is

$$T_c = \frac{F\mu_c d_c}{2} \qquad (8\text{-}6)$$

For large collars, the torque should probably be computed in a manner similar to that employed for disk clutches.

EXAMPLE 8-1   A square-thread power screw has a major diameter of 32 mm and a pitch of 4 mm with double threads, and it is to be used in an application similar to that of Fig. 8-4. The given data include $\mu = \mu_c = 0.08$, $d_c = 40$ mm, and $F = 6.4$ kN per screw.

(a) Find the thread depth, thread width, mean or pitch diameter, minor diameter, and lead.

(b) Find the torque required to rotate the screw "against" the load.

(c) Find the torque required to rotate the screw "with" the load.

(d) Find the overall efficiency.

Solution   (a) From Fig. 8-3a the thread depth and width are the same and equal to half the pitch, or 2 mm. Also,

$$d_m = d - \frac{p}{2} = 32 - 2 = 30 \text{ mm}$$

$$d_r = d - p = 32 - 4 = 28 \text{ mm}$$

$$l = np = 2(4) = 8 \text{ mm}$$

(b) Using Eqs. (8-1) and (8-6), the torque required to turn the screw against the load is

$$T = \frac{Fd_m}{2}\left(\frac{l + \pi\mu d_m}{\pi d_m - \mu l}\right) + \frac{F\mu_c d}{2}$$

$$= \frac{6.4(30)}{2}\left[\frac{8 + \pi(0.08)\,(30)}{\pi(30) - 0.08(8)}\right] + \frac{6.4(0.08)\,(40)}{2}$$

Answer $= 15.94 + 10.24 = 26.18 \text{ N} \cdot \text{m}$

(c) The torque required to lower the load, that is, to rotate the screw with the aid of the load, is obtained using Eqs. (8-2) and (8-6). Thus

$$T = \frac{Fd_m}{2}\left(\frac{\pi\mu d_m - l}{\pi d_m + \mu l}\right) + \frac{F\mu_c d_c}{2}$$

$$= \frac{6.4(30)}{2}\left[\frac{\pi(0.08)\,(30) - 8}{\pi(30) + (0.08)\,(8)}\right] + \frac{6.4(0.08)\,(40)}{2}$$

Answer $= -0.466 + 10.24 = 9.77 \text{ N} \cdot \text{m}$

The minus sign in the first term indicates that the screw alone is not self-locking and would rotate under the action of the load except for the fact that collar friction is present and must be overcome too. Thus the torque required to rotate the screw "with" the load is less than is necessary to overcome collar friction alone.

(d) The overall efficiency is

Answer $$e = \frac{Fl}{2\pi T} = \frac{6.4(8)}{2\pi(26.18)} = 0.311$$

8-3   THREADED FASTENERS

Figure 8-8 is a drawing of a standard hexagon-head bolt. Points of stress concentration are at the fillet and at the start of the threads. See Table A-26 for dimensions. The diameter of the washer face is the same as the width across flats of the hexagon. The

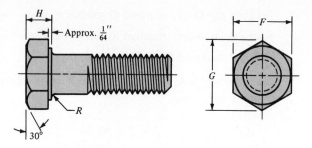

FIGURE 8-8

Hexagon-head bolt; note the washer face, the fillet under the head, the start of threads, and the chamfer on both ends. Bolt lengths are always measured underneath the head.

thread length of inch-series bolts is

$$L_T = \begin{cases} 2D + \frac{1}{4} \text{ in} & L \le 6 \text{ in} \\ 2D + \frac{1}{2} \text{ in} & L > 6 \text{ in} \end{cases} \tag{8-7}$$

and for metric bolts is

$$L_T = \begin{cases} 2D + 6 & L \le 125 \qquad D \le 48 \\ 2D + 12 & 125 < L \le 200 \\ 2D + 25 & L > 200 \end{cases} \tag{8-8}$$

where the dimensions are in millimeters. The ideal bolt length is one in which only one or two threads project from the nut after it is tightened. Bolt holes may have burrs or sharp edges after drilling. These could bite into the fillet and increase stress concentration. Therefore, washers must always be used under the bolt head to prevent this. They should be of hardened steel and loaded onto the bolt so that the rounded edge of the stamped hole faces the washer face of the bolt. Sometimes it is necessary to use washers under the nut too.

The purpose of a bolt is to clamp two or more parts together. The clamping load stretches or elongates the bolt; the load is obtained by twisting the nut until the bolt has elongated almost to the elastic limit. If the nut does not loosen, this bolt tension remains as the preload or clamping force. When tightening, the mechanic should, if possible, hold the bolt head stationary and twist the nut; in this way the bolt shank will not feel the thread-friction torque.

The head of a hexagon-head cap screw is slightly thinner than that of a hexagon-head bolt. Dimensions of hexagon-head cap screws are listed in Table A-27. Hexagon-head cap screws are used in the same applications as bolts and also in applications in which one of the clamped members is threaded. Three other common cap-screw head styles are shown in Fig. 8-9.

A variety of machine-screw head styles are shown in Fig. 8-10. Inch-series machine screws are generally available in sizes from No. 0 to about $\frac{3}{8}$ in.

Several styles of hexagonal nuts are illustrated in Fig. 8-11; their dimensions are given in Table A-28. The material of the nut must be selected carefully to match that of the bolt. During tightening, the first thread of the nut tends to take the entire load; but yielding occurs, with some strengthening due to the cold work that takes place, and the load is eventually divided over about three nut threads. For this reason you should never reuse nuts; in fact, it can be dangerous to do so.

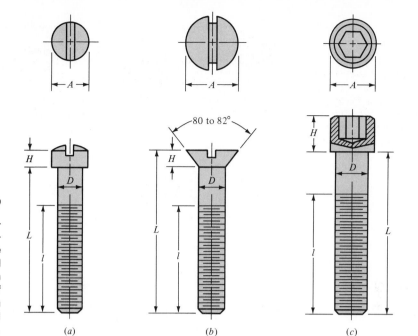

FIGURE 8-9

Typical cap-screw heads: (a) fillister head; (b) flat head; (c) hexagonal socket head. Cap screws are also manufactured with hexagonal heads similar to the one shown in Fig. 8-8, as well as a variety of other head styles. This illustration uses one of the conventional methods of representing threads.

8-4 TENSION CONNECTIONS—THE FASTENER

When a connection is desired which can be disassembled without destructive methods and which is strong enough to resist external tensile loads, moment loads, and shear loads, or a combination of these, then the simple bolted joint using hardened-steel washers is a good solution. Such a joint can also be dangerous unless it is properly designed and assembled by a *trained* mechanic.

A section through a tension-loaded bolted joint is illustrated in Fig. 8-12. Notice the clearance space provided by the bolt holes. Notice, too, how the bolt threads extend into the body of the connection.

As noted previously, the purpose of the bolt is to clamp the two, or more, parts together. Twisting the nut stretches the bolt to produce the clamping force. This clamping force is called the *pre-tension* or *bolt preload*. It exists in the connection after the nut has been properly tightened no matter whether the external tensile load P is exerted or not.

Of course, since the members are being clamped together, the clamping force which produces tension in the bolt induces compression in the members.

Figure 8-13 shows another tension-loaded connection. This joint uses cap screws threaded into one of the members. An alternative approach to this problem (of not using a nut) would be to use studs. A stud is a rod threaded on both ends. The stud is screwed into the lower member first; then the top member is positioned and fastened down with hardened washers and nuts. The studs are regarded as permanent, and so the joint can be disassembled merely by removing the nut and washer. Thus the threaded part of the lower member is not damaged by reusing the threads.

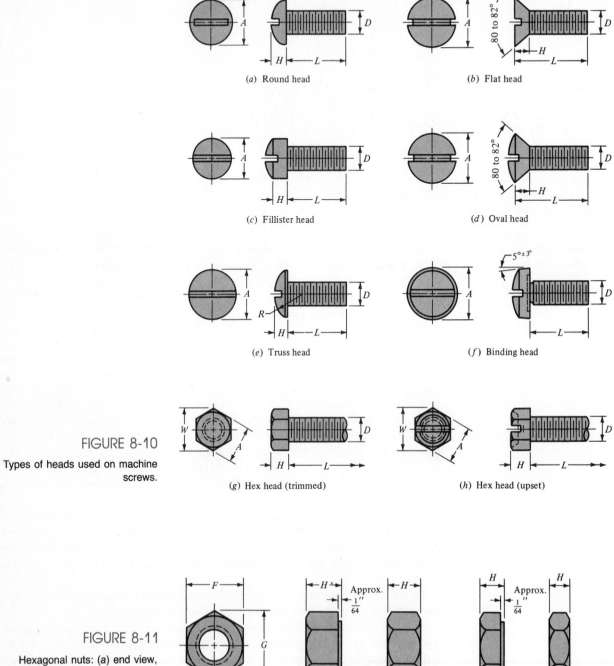

FIGURE 8-10

Types of heads used on machine screws.

(a) Round head

(b) Flat head

(c) Fillister head

(d) Oval head

(e) Truss head

(f) Binding head

(g) Hex head (trimmed)

(h) Hex head (upset)

FIGURE 8-11

Hexagonal nuts: (a) end view, general; (b) washer-faced regular nut; (c) regular nut chamfered on both sides; (d) jam nut with washer face; (e) jam nut chamfered on both sides.

(a)   (b)   (c)   (d)   (e)

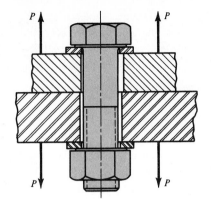

FIGURE 8-12

A bolted connection loaded in tension by the forces P. Note the use of two washers. A simplified conventional method is used here to represent the screw threads. Note how the threads extend into the body of the connection. This is usual and is desired.

The *spring constant,* or *stiffness constant,* of an elastic member such as a bolt, as we learned in Chap. 3, is the ratio between the force applied to the member and the deflection produced by that force. We can use Eq. (3-4) and the results of Prob. 3-1 to find the stiffness constant of a fastener in any bolted connection.

The *grip* of a connection is the total thickness of the clamped material. In Fig. 8-12 the grip is the sum of the thicknesses of both members and both washers. In Fig. 8-13 the grip is the thickness of the top member plus that of the washer.

The stiffness of the portion of a bolt or screw within the clamped zone will generally consist of two parts, that of the unthreaded shank portion and that of the threaded portion. Thus the stiffness constant of the bolt is equivalent to the stiffnesses of two springs in series. Using the results of Prob. 3-1, we find

$$\frac{1}{k} = \frac{1}{k_1} + \frac{1}{k_2} \qquad \text{or} \qquad k = \frac{k_1 k_2}{k_1 + k_2} \tag{8-9}$$

for two springs in series. From Eq. (3-4), the spring rates of the threaded and unthreaded portions of the bolt in the clamped zone are, respectively,

$$k_T = \frac{A_t E}{l_T} \qquad k_d = \frac{A_d E}{l_d} \tag{8-10}$$

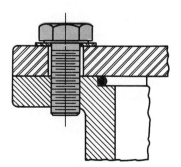

FIGURE 8-13

Section of a cylindrical pressure vessel. Hexagon-head cap screws are used to fasten the cylinder head to the body. Note the use of the O-ring seal.

where A_t = tensile-stress area (Tables 8-1, 8-2)

$\quad\quad l_T$ = length of threaded portion of grip

$\quad\quad A_d$ = major-diameter area of fastener

$\quad\quad l_d$ = length of unthreaded portion in grip

Substituting these stiffnesses in Eq. (8-9) gives

$$k_b = \frac{A_d A_T E}{A_d l_T + A_T l_d} \tag{8-11}$$

where k_b is the estimated effective stiffness of the bolt or cap screw in the clamped zone. For short fasteners, the one in Fig. 8-13, for example, the unthreaded area is small and so the first of the expressions in Eq. (8-10) can be used to find k_b; and in the case of long fasteners, the threaded area is relatively small, and so the second expression in Eq. (8-10) can be used.

8-5 TENSION CONNECTIONS—THE MEMBERS

In the previous section, we determined the stiffness of the fastener in the clamped zone. In this section, we wish to study the stiffness of the members in the clamped zone. Both of these stiffnesses must be known in order to learn what happens when the assembled connection is subjected to an external tensile loading.

There may be more than two members included in the grip of the fastener. All together these act like compressive springs in series, and hence the total spring rate of the members is

$$\frac{1}{k_m} = \frac{1}{k_1} + \frac{1}{k_2} + \frac{1}{k_3} + \cdots + \frac{1}{k_i} \tag{8-12}$$

If one of the members is a soft gasket, its stiffness relative to the other members is usually so small that for all practical purposes the others can be neglected and only the gasket stiffness used.

If there is no gasket, the stiffness of the members is rather difficult to obtain, except by experimentation, because the compression spreads out between the bolt head and the nut and hence the area is not uniform. There are, however, some cases in which this area can be determined.

Ito[*] has used ultrasonic techniques to determine the pressure distribution at the member interface. The results show that the pressure stays high out to about 1.5 bolt radii. The pressure, however, falls off farther away from the bolt. Thus Ito suggests the use of Rotscher's pressure-cone method for stiffness calculations with a variable cone angle. This method is quite complicated, and so here we choose to use a simpler approach using a fixed cone angle.

Figure 8-14b illustrates the general cone geometry using a half-apex angle α. An angle $\alpha = 45°$ has been used, but Little[*] reports that this overestimates the clamping

---

[*]Y. Ito, J. Toyoda, and S. Nagata, "Interface Pressure Distribution in a Bolt-Flange Assembly," ASME paper no. 77-WA/DE-11, 1977.

[*]R. E. Little, "Bolted Joints: How Much Give?," *Machine Design*, Nov. 9, 1967.

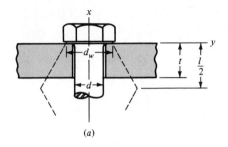

(a)

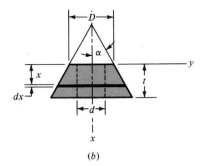

FIGURE 8-14

Compression of a member assumed to be confined to the frustum of a hollow cone.

(b)

stiffness. When loading is restricted to a washer-face annulus (hardened steel, cast iron, or aluminum), the proper apex angle is smaller. Osgood[†] reports a range of $25° \le \alpha \le 33°$ for most combinations. In this book we shall use $\alpha = 30°$ except in cases in which the material is insufficient.

Referring now to Fig. 8-14, the elongation of an element of the cone of thickness dx subjected to a tensile force P is, from Eq. (3-3),

$$d\delta = \frac{P\,dx}{EA} \tag{a}$$

The area of the element is

$$A = \pi(r_o^2 - r_i^2) = \pi\left[\left(x \tan \alpha + \frac{D}{2}\right)^2 - \left(\frac{d}{2}\right)^2\right]$$

$$= \pi\left(x \tan \alpha + \frac{D+d}{2}\right)\left(x \tan \alpha + \frac{D-d}{2}\right) \tag{b}$$

Substituting this in Eq. (a) and integrating the left side gives the elongation as

$$\delta = \frac{P}{\pi E} \int_0^t \frac{dx}{[x \tan \alpha + (D+d)/2][x \tan \alpha + (D-d)/2]} \tag{c}$$

Using a table of integrals, we find the result to be

---

† C. C. Osgood, "Saving Weight on Bolted Joints," *Machine Design*, Oct. 25, 1979.

$$\delta = \frac{P}{\pi E d \tan \alpha} \ln \frac{(2t \tan \alpha + D - d)(D + d)}{(2t \tan \alpha + D + d)(D - d)} \tag{d}$$

Thus the spring rate or stiffness of this frustum is

$$k = \frac{P}{\delta} = \frac{\pi E d \tan \alpha}{\ln \dfrac{(2t \tan \alpha + D - d)(D + d)}{(2t \tan \alpha + D + d)(D - d)}} \tag{8-13}$$

With $\alpha = 30°$, this becomes

$$k = \frac{0.577 \pi E d}{\ln \dfrac{(1.15t + D - d)(D + d)}{(1.15t + D + d)(D - d)}} \tag{8-14}$$

Equation (8-14), or (8-13), must be solved separately for each frustum in the joint. Then individual stiffnesses are assembled to obtain k_m using Eq. (8-12).

If the members of the joint have the same Young's modulus E with symmetrical frusta back to back, then they act as two identical springs in series. From Eq. (8-12) we learn that $k_m = k/2$. Using the grip as $l = 2t$ and d_w as the diameter of the washer face, we find the spring rate of the members to be

$$k_m = \frac{\pi E d \tan \alpha}{2 \ln \dfrac{(l \tan \alpha + d_w - d)(d_w + d)}{(l \tan \alpha + d_w + d)(d_w - d)}} \tag{8-15}$$

The diameter of the washer face is about 50 percent greater than the fastener diameter for standard hexagon-head bolts and cap screws. Thus we can simplify Eq. (8-15) by letting $d_w = 1.5d$. If we also use $\alpha = 30°$, then Eq. (8-15) can be written as

$$k_m = \frac{0.577 \pi E d}{2 \ln \left(5 \dfrac{0.577l + 0.5d}{0.577l + 2.5d} \right)} \tag{8-16}$$

It is easy to program the numbered equations in this section, and you should do so. The time spent in programming will save many hours of formula plugging.

8-6   BOLT STRENGTH

The bolt strength is the key factor in the design or analysis of bolted connections. In the specification standards for bolts, the strength is specified by stating the *minimum proof strength*, or *minimum proof load*, and the *minimum tensile strength*.

The *proof load* is the maximum load (force) that a bolt can withstand without acquiring a permanent set. The *proof strength* is the quotient of the proof load and the tensile-stress area. The proof strength thus corresponds roughly to the yield-point strength and is about 90 percent of the 0.2 percent offset yield strength.

The values of the mean proof strength, the mean tensile strength, and the corresponding standard deviations are not a part of the specification codes, and so it is the designer's responsibility to obtain these values, perhaps by laboratory testing, before designing to a reliability specification.

The SAE specifications are found in Table 8-4. The bolt grades are numbered according to the tensile strengths, with decimals used for variations at the same strength level. Bolts and screws are available in all grades listed. Studs are available in grades 1, 2, 4, 5, 8, and 8.1. Grade 8.1 is not listed.

TABLE 8-4

SAE Specifications for Steel Bolts

SAE GRADE NO.	SIZE RANGE, INCLUSIVE, in	MINIMUM PROOF STRENGTH, kpsi	MINIMUM TENSILE STRENGTH, kpsi	MINIMUM YIELD STRENGTH, kpsi	MATERIAL	HEAD MARKING
1	$\frac{1}{4}-1\frac{1}{2}$	33	60	36	Low or medium carbon	
2	$\frac{1}{4}-\frac{3}{4}$	55	74	57	Low or medium carbon	
	$\frac{7}{8}-1\frac{1}{2}$	33	60	36		
4	$\frac{1}{4}-1\frac{1}{2}$	65	115	100	Medium carbon, cold-drawn	
5	$\frac{1}{4}-1$	85	120	92	Medium carbon, Q&T	
	$1\frac{1}{8}-1\frac{1}{2}$	74	105	81		
5.2	$\frac{1}{4}-1$	85	120	92	Low-carbon martensite, Q&T	
7	$\frac{1}{4}-1\frac{1}{2}$	105	133	115	Medium-carbon alloy, Q&T	
8	$\frac{1}{4}-1\frac{1}{2}$	120	150	130	Medium-carbon alloy, Q&T	
8.2	$\frac{1}{4}-1$	120	150	130	Low-carbon martensite, Q&T	

TABLE 8-5

ASTM Specifications for Steel Bolts

ASTM DESIGNATION NO.	SIZE RANGE, INCLUSIVE, in	MINIMUM PROOF STRENGTH, kpsi	MINIMUM TENSILE STRENGTH, kpsi	MINIMUM YIELD STRENGTH, kpsi	MATERIAL	HEAD MARKING
A307	$\frac{1}{4}$–$1\frac{1}{2}$	33	60	36	Low carbon	
A325, type 1	$\frac{1}{2}$–1 $1\frac{1}{8}$–$1\frac{1}{2}$	85 74	120 105	92 81	Medium carbon, Q&T	A325
A325, type 2	$\frac{1}{2}$–1 $1\frac{1}{8}$–$1\frac{1}{2}$	85 74	120 105	92 81	Low-carbon martensite, Q&T	A325
A325, type 3	$\frac{1}{2}$–1 $1\frac{1}{8}$–$1\frac{1}{2}$	85 74	120 105	92 81	Weathering steel, Q&T	A325
A354, grade BC					Alloy-steel, Q&T	BC
A354, grade BD	$\frac{1}{4}$–4	120	150	130	Alloy steel, Q&T	
A449	$\frac{1}{4}$–1 $1\frac{1}{8}$–$1\frac{1}{2}$ $1\frac{3}{4}$–3	85 74 55	120 105 90	92 81 58	Medium-carbon, Q&T	
A490, type 1	$\frac{1}{2}$–$1\frac{1}{2}$	120	150	130	Alloy steel, Q&T	A490
A490, type 3					Weathering steel, Q&T	A490

ASTM specifications are listed in Table 8-5. ASTM threads are shorter because ASTM deals mostly with structures; structural connections are generally loaded in shear, and the decreased thread length provides more shank area.

Specifications for metric fasteners are given in Table 8-6.

It is worth noting that all specification-grade bolts made in this country bear a manufacturer's mark or logo, in addition to the grade marking, on the bolt head. Such marks confirm that the bolt meets or exceeds specifications. If such marks are missing, the bolt may be imported; for imported bolts there is no obligation to meet specifications.

TABLE 8-6

Metric Mechanical-Property Classes for Steel Bolts, Screws, and Studs*

PROPERTY CLASS	SIZE RANGE, INCLUSIVE	MINIMUM PROOF STRENGTH, MPa	MINIMUM TENSILE STRENGTH, MPa	MINIMUM YIELD STRENGTH, MPa	MATERIAL	HEAD MARKING
4.6	M5–M36	225	400	240	Low or medium carbon	4.6
4.8	M1.6–M16	310	420	340	Low or medium carbon	4.8
5.8	M5–M24	380	520	420	Low or medium carbon	5.8
8.8	M16–M36	600	830	660	Medium carbon, Q&T	8.8
9.8	M1.6–M16	650	900	720	Medium carbon, Q&T	9.8
10.9	M5–M36	830	1040	940	Low-carbon martensite, Q&T	10.9
12.9	M1.6–M36	970	1220	1100	Alloy, Q&T	12.9

*The thread length for bolts and cap screws is

$$L_T = \begin{cases} 2d + 6 & L \le 125 \\ 2d + 12 & 125 < L \le 200 \\ 2d + 25 & L > 200 \end{cases}$$

where L is the bolt length. The thread length for structural bolts is slightly shorter than given above.

8-7   TENSION CONNECTIONS—THE EXTERNAL LOAD

Let us now consider what happens when an external tensile load P, as in Fig. 8-12, is applied to a bolted connection. It is to be assumed, of course, that the clamping force, which we will call the *preload* F_i, has been correctly applied by tightening the nut *before* P is applied. The nomenclature used is:

F_i = preload

P = external tensile load

P_b = portion of P taken by bolt

P_m = portion of P taken by members

$F_b = P_b + F_i$ = resultant bolt load

$F_m = P_m - F_i$ = resultant load on members

The load P is tension, and it causes the connection to stretch, or elongate, through some distance δ. We can relate this elongation to the stiffnesses by recalling that k is the force divided by the deflection. Thus

$$\delta = \frac{P_b}{k_b} \quad \text{and} \quad \delta = \frac{P_m}{k_m} \tag{a}$$

or

$$P_b = P_m \frac{k_b}{k_m} \tag{b}$$

Since $P = P_b + P_m$, we have

$$P_b = \frac{k_b P}{k_b + k_m} \tag{c}$$

Therefore the resultant bolt load is

$$F_b = P_b + F_i = \frac{k_b P}{k_b + k_m} + F_i \qquad F_m < 0 \tag{8-17}$$

and the resultant load on the connected members is

$$F_m = P_b - F_i = \frac{k_m P}{k_b + k_m} - F_i \qquad F_m < 0 \tag{8-18}$$

Of course, these results are valid only as long as some clamping load remains in the members; this is indicated by the qualifier in the equations.

Table 8-7 is included to provide some information on the relative values of the stiffnesses encountered. The grip contains only two members, both of steel, and no washers. The ratios C and $1 - C$ are the coefficients of P in Eqs. (8-17) and (8-18), respectively. They describe the proportion of the external load taken by the bolt and by the members, respectively. In all cases, the members take over 80 percent of the external load. Think how important this is when fatigue loading is present. Note also that making the grip longer causes the members to take an even greater percentage of the external load.

TABLE 8-7

Computation of Bolt and Member Stiffnesses. Steel members clamped using a $\frac{1''}{2}$-13 NC steel bolt.

$$C = \frac{k_b}{k_b + k_m}$$

BOLT GRIP, in	STIFFNESSES, Mlb/in		C	1 − C
	k_b	k_m		
2	2.57	12.69	0.168	0.832
3	1.79	11.33	0.136	0.864
4	1.37	10.63	0.114	0.886

8-8   TORQUE REQUIREMENTS

Having learned that a high preload is very desirable in important bolted connections, we must next consider means of ensuring that the preload is actually developed when the parts are assembled.

If the overall length of the bolt can actually be measured with a micrometer when it is assembled, the bolt elongation due to the preload F_i can be computed using the formula $\delta = F_i l/(AE)$. Then the nut is simply tightened until the bolt elongates through the distance δ. This ensures that the desired preload has been attained.

The elongation of a screw cannot usually be measured, because the threaded end is often in a blind hole. It is also impractical in many cases to measure bolt elongation. In such cases the wrench torque required to develop the specified preload must be estimated. Then torque wrenching, pneumatic-impact wrenching, or the turn-of-the-nut method may be used.

The torque wrench has a built-in dial which indicates the proper torque.

With impact wrenching, the air pressure is adjusted so that the wrench stalls when the proper torque is obtained, or in some wrenches, the air automatically shuts off at the desired torque.

The turn-of-the-nut method requires that we first define the meaning of snug-tight. The *snug-tight* condition is the tightness attained by a few impacts of an impact wrench, or the full effort of a person using an ordinary wrench. When the snug-tight condition is attained, all additional turning develops useful tension in the bolt. The turn-of-the-nut method requires that you compute the fractional number of turns necessary to develop the required preload from the snug-tight condition. For example, for heavy hexagon structural bolts, the turn-of-the-nut specification states that the nut should be turned a minimum of 180° from the snug-tight condition under optimum conditions. Note that this is also about the correct rotation for the wheel nuts of a passenger car.

Although the coefficients of friction may vary widely, we can obtain a good estimate of the torque required to produce a given preload by combining Eqs. (8-5) and (8-6):

$$T = \frac{F_i d_m}{2}\left(\frac{l + \pi \mu d_m \sec \alpha}{\pi d_m - \mu l \sec \alpha}\right) + \frac{F_i \mu_c d_c}{2} \qquad (a)$$

Since $\tan \lambda = l/\pi d_m$, we divide the numerator and denominator of the first term by πd_m and get

$$T = \frac{F_i d_m}{2}\left(\frac{\tan \lambda + \mu \sec \alpha}{l - \mu \tan \lambda \sec \alpha}\right) + \frac{F_i \mu_c d_c}{2} \qquad (b)$$

The diameter of the washer face of a hexagonal nut is the same as the width across flats and equal to $1\frac{1}{2}$ times the nominal size. Therefore the mean collar diameter is $d_c = (d + 1.5d)/2 = 1.25d$. Equation ($b$) can now be arranged to give

$$T = \left[\left(\frac{d_m}{2d} \right) \left(\frac{\tan \lambda + \mu \sec \alpha}{1 - \mu \tan \lambda \sec \alpha} \right) + 0.625 \mu_c \right] F_i d \qquad (c)$$

We now define a *torque coefficient K* as the term in brackets, and so

$$K = \left(\frac{d_m}{2d} \right) \left(\frac{\tan \lambda + \mu \sec \alpha}{1 - \mu \tan \lambda \sec \alpha} \right) + 0.625 \mu_c \qquad (8\text{-}19)$$

Equation (c) can now be written

$$T = KF_i d \qquad (8\text{-}20)$$

The coefficient of friction depends upon the surface smoothness, accuracy, and degree of lubrication. On the average, both μ and μ_c are about 0.15. The interesting fact about Eq. (8-19) is that $K \approx 0.20$ for $\mu = \mu_c = 0.15$ no matter what size bolts are employed and no matter whether the threads are coarse or fine.

Blake and Kurtz have published results of numerous tests of the torquing of bolts.* By subjecting their data to a statistical analysis, we can learn something about the distribution of the torque coefficients and the resulting preload. Blake and Kurtz determined the preload in quantities of unlubricated and lubricated bolts of size $\frac{1}{2}''$-20 UNF when torqued to 800 lb · in. This corresponds roughly to an M12 × 1.25 bolt torqued to 90 N · m. The statistical analyses of these two groups of bolts, converted to SI units, are displayed in Tables 8-8 and 8-9.

We first note that both groups have about the same mean preload, 34 kN. The unlubricated bolts have a standard deviation of 4.9 kN, which is about 15 percent of the mean. The lubricated bolts have a standard deviation of 3 kN, or about 9 percent of the mean, a substantial reduction. These deviations are quite large, though, and emphasize the necessity for quality-control procedures throughout the entire manufacturing and assembly process to ensure uniformity.

---

*J. C. Blake and H. J. Kurtz, "The Uncertainties of Measuring Fastener Preload," *Machine Design*, vol. 37, Sept. 30, 1965, pp. 128–131.

TABLE 8-8

Distribution of Preload F_i for 20 Tests of Unlubricated Bolts Torqued to 90 N · m

23.6,	27.6,	28.0,	29.4,	30.3,	30.7,	32.9,	33.8,	33.8,	33.8,
34.7,	35.6,	35.6,	37.4,	37.8,	37.8,	39.2,	40.0,	40.5,	42.7

Mean value, $\overline{F}_i = 34.3$ kN. Standard deviation, $\hat{\sigma} = 4.91$ kN.

TABLE 8-9

Distribution of Preload F_i for 10 Tests of Lubricated Bolts Torqued to 90 N · m

30.3,	32.5,	32.5,	32.9,	32.9,	33.8,	34.3,	34.7,	37.4,	40.5

Mean value, $\overline{F}_i = 34.18$ kN. Standard deviation, $\hat{\sigma} = 2.88$ kN.

TABLE 8-10
Torque Factors K for Use with
Eq. (8-20)

BOLT CONDITION	K
Nonplated, black finish	0.30
Zinc-plated	0.20
Lubricated	0.18
Cadmium-plated	0.16
With Bowman Anti-Seize	0.12
With Bowma-Grip nuts	0.09

The means obtained from the two samples are nearly identical, approximately 34 kN; using Eq. (8-20), we find, for both samples, $K = 0.208$.

Bowman Distribution,* a large manufacturer of fasteners, recommends the values shown in Table 8-10. In this book we shall use these values and use $K = 0.2$ when the bolt condition is not stated.

8-9   BOLT PRELOAD—STATIC LOADING

Let us write Eq. (8-17) as

$$F_b = \frac{k_b P}{k_b + k_m} + F_i = CP + F_i \qquad (a)$$

where C is called the joint constant and is defined in Eq. (a) as

$$C = \frac{k_b}{k_b + k_m} \qquad (8\text{-}21)$$

Then Eq. (8-18) will become

$$F_m = (1 - C)P - F_i \qquad (8\text{-}22)$$

The tensile stress in the bolt can be found by dividing both terms of Eq. (a) by the tensile-stress area A_t. This yields

$$\sigma_b = \frac{CP}{A_t} + \frac{F_i}{A_t} \qquad (b)$$

But the limiting value of σ_b is the proof strength S_b. Thus, with the introduction of a load factor n, Eq. (b) becomes

$$\frac{CnP}{A_t} + \frac{F_i}{A_t} = S_P \qquad (c)$$

or

$$n = \frac{S_p A_t - F_i}{CP} \qquad (8\text{-}23)$$

---

*Bowman Distribution–Barnes Group, *Fastener Facts,* Cleveland, 1985, p. 90.

Here we have called n a load factor rather than a factor of safety, though the two ideas are somewhat related. Any value of $n > 1$ in Eq. (8-23) ensures that the bolt stress is less than the proof strength.

Another means of ensuring a safe joint is to require that the external load be smaller than that needed to cause the joint to separate. If separation does occur, then the entire external load will be imposed on the bolt. Let P_0 be the value of the external load that would cause joint separation. At separation, $F_m = 0$ in Eq. (8-22), and so

$$(1 - C)P_0 - F_i = 0 \qquad (d)$$

The factor of safety guarding against joint separation is

$$n = \frac{P_0}{P} \qquad (e)$$

By substituting $P_0 = nP$ in Eq. (d), we find

$$n = \frac{F_i}{P(1 - C)} \qquad (8\text{-}24)$$

as a load factor guarding against joint separation.

Figure 8-15 is the stress-strain diagram of a good-quality bolt material. Notice that there is no clearly defined yield point and that the diagram progresses smoothly up to fracture, which corresponds to the tensile strength. This means that no matter how much preload is given the bolt, it will retain its load-carrying capacity. This is what keeps the bolt tight and determines the joint strength. The pre-tension is the "muscle" of the joint, and its magnitude is determined by the bolt strength. If the full bolt strength is not used in developing the pre-tension, then money is wasted and the joint is weaker.

Good-quality bolts can be preloaded into the plastic range to develop more strength. Some of the bolt torque used in tightening produces torsion, which increases the princi-

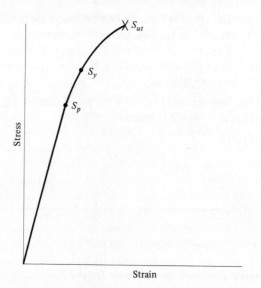

FIGURE 8-15

Typical stress-strain diagram for bolt materials showing proof strength S_p, yield strength S_y, and tensile strength S_{ut}.

pal tensile stress. However, this torsion is held only by the friction of the bolt head and nut; in time it relaxes and lowers the bolt tension slightly. Thus, as a rule, a bolt will either fracture during tightening, or not at all.

Above all, do not rely too much on wrench torque; it is not a good indicator of preload. Actual bolt elongation should be used whenever possible—especially with fatigue loading. In fact, if high reliability is a requirement of the design, then preload should always be determined by bolt elongation.

RB&W recommendations for preload are 60 kpsi for SAE grade 5 bolts for nonpermanent connections, and that A325 bolts (equivalent to SAE grade 5) used in structural applications be tightened to proof load or beyond (85 kpsi up to a diameter of 1 in).[*] Bowman[†] recommends a preload of 75 percent of proof load, which is about the same as the RB&W recommendations for reused bolts. In view of these guidelines, it is recommended for both static and fatigue loading that the following be used for preload:

$$F_i = \begin{cases} 0.75F_p & \text{for reused connections} \\ 0.90F_p & \text{for permanent connections} \end{cases} \tag{8-25}$$

where F_p is the proof load, obtained from the equation

$$F_p = A_t S_p \tag{8-26}$$

Here, S_p is the proof strength obtained from Tables 8-4 to 8-6. For other materials, an approximate value is $S_p = 0.85S_y$. But be very careful never to use a soft material as a threaded fastener.

EXAMPLE 8-2   Figure 8-16 is a cross section from a pressure cylinder. A total of N bolts are to be used to resist a separating force of 36 kip.

(a) Find the stiffnesses and the constant C.

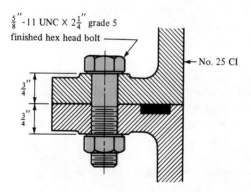

$\frac{5}{8}''$-11 UNC $\times$ $2\frac{1}{4}''$ grade 5 finished hex head bolt

No. 25 CI

$\frac{3}{4}''$

$\frac{3}{4}''$

FIGURE 8-16

---

[*]Russell, Burdsall & Ward Corp., *Helpful Hints for Fastener Design and Application*, Mentor, Ohio, 1976, p. 42.

[†]Loc. cit.

(b) Find the number of bolts required for a design factor of 2 and accounting for the fact that the bolts may be reused when the joint is taken apart.

Solution   (a) Using $L_T = 1.50$ in, $l_d = 0.75$ in, $l_T = 0.75$ in, $A_d = 0.3068$ iñ², $A_T = 0.226$ in², and $E = 30$ Mpsi, by Eq. (8-11) we find $k_b = 5.21$ Mlb/in.

Here the grip is $l = 1.5$ in. The modulus of elasticity of No. 25 cast iron is 12 Mpsi. Thus the stiffness of the members, from Eq. (8-16), is

$$k_m = \frac{0.577\pi Ed}{2\ln\left(5\dfrac{0.577l + 0.5d}{0.577l + 2.5d}\right)} = \frac{0.577\pi(12)(0.625)}{2\ln\left[5\dfrac{0.577(1.5) + 0.5(0.625)}{0.577(1.5) + 2.5(0.625)}\right]}$$

$$= 7.67 \text{ Mlb/in}$$

The constant C is now found to be

Answer $$C = \frac{k_b}{k_b + k_m} = \frac{5.21}{5.21 + 7.67} = 0.404$$

(b) From Tables 8-2 and 8-4 we get $A_t = 0.226$ in² and $S_p = 85$ kpsi. Then, using Eqs. (8-25) and (8-26), we find the recommended preload to be

$$F_i = 0.75A_tS_p = 0.75(0.226)(85) = 14.4 \text{ kip}$$

For N bolts, Eq. (8-23) can be written

$$n = \frac{S_pA_t - F_i}{C(F/N)} \tag{8-27}$$

or

$$N = \frac{CnF}{S_pA_t - F_i} \tag{8-28}$$

For this problem we find N to be

$$N = \frac{0.404(2)(36)}{85(0.226) - 14.4} = 6.05$$

and so we try seven bolts. Using this value for N in Eq. (8-27) gives

$$n = \frac{85(0.226) - 14.4}{0.404(36/7)} = 2.31$$

which is greater than the required value. Therefore we choose seven bolts and use the recommended preload in tightening.

8-10   GASKETED JOINTS

If a full gasket is present in the joint, the gasket pressure p is found by dividing the force in the member by the gasket area per bolt, or

$$p = -\frac{F_m}{A_g/N} \qquad (a)$$

For a load factor n, Eq. (8-23) can be written as

$$F_m = (1 - C)nP - F_i \qquad (b)$$

Substituting this in Eq. (a) gives the gasket pressure as

$$p = \frac{N}{A_g}[F_i - nP(1 - C)] \qquad (8\text{-}29)$$

In full-gasketed joints the uniformity of pressure is important. To maintain uniformity, bolts should not be spaced more than six bolt diameters apart. But to maintain wrench clearance, bolts should be spaced at least three diameters apart. So a rough rule for bolt spacing when the bolts are arranged circularly is

$$3 \le \frac{\pi D_b}{Nd} \le 6 \qquad (8\text{-}30)$$

where D_b is the diameter of the bolt circle and N is the number of bolts.

8-11   FATIGUE LOADING

Tension-loaded bolted joints subjected to fatigue action can be analyzed directly by the methods of Chap. 7. Table 8-11 lists average fatigue-strength-reduction factors for the fillet under the bolt head and also at the beginning of the threads on the bolt shank. These are already corrected for notch sensitivity and for surface finish. Designers should be aware that situations may arise in which it would be advisable to investigate these factors more closely, since they are only average values. In fact, Peterson[*] observes that the distribution of typical bolt failures is about 15 percent under the head, 20 percent at the end of the thread, and 65 percent in the thread at the nut face.

In using Table 8-11, it is usually safe to assume that the fasteners have rolled

---

[*]R. E. Peterson, *Stress Concentration Factors,* Wiley, New York, 1974, p. 253.

TABLE 8-11
Fatigue-Strength-Reduction
Factors K_f for Threaded
Elements

SAE GRADE	METRIC GRADE	ROLLED THREADS	CUT THREADS	FILLET
0 to 2	3.6 to 5.8	2.2	2.8	2.1
4 to 8	6.6 to 10.9	3.0	3.8	2.3

threads, unless specific information is available. Also, in computing endurance limit, use a machined finish for the body of the bolt if nothing is stated.

Most of the time, the type of fatigue loading encountered in the analysis of bolted joints is one in which the externally applied load fluctuates between zero and some maximum force P. This would be the situation in a pressure cylinder, for example, where a pressure either exists or does not exist. In order to determine the mean and alternating bolt stresses for such a situation, let us employ the notation of Sec. 8-7. Then $F_{\max} = F_b$ and $F_{\min} = F_i$. Therefore, the alternating component of bolt stress is, from Eq. (8-17),

$$\sigma_a = \frac{F_b - F_i}{2A_t} = \frac{k_b}{k_b + k_m}\frac{P}{2A_t} = \frac{CP}{2A_t}$$

(8-31)

Then, since the mean stress is equal to the alternating component plus the minimum stress, we have

$$\sigma_m = \sigma_a + \frac{F_i}{A_t} = \frac{CP}{2A_t} + \frac{F_i}{A_t}$$

(8-32)

We have already learned of the importance of getting a high preload in bolted joints. This is especially important in fatigue loading, because it makes the first term of Eq. (8-32) relatively small compared with the second term, which is the preload stress. Examination of Eq. (8-32) shows that it is made up of a constant term F_i/A_t and the alternating stress σ_a. Thus the load line is a unit slope line beginning at F_i/A_t on the σ_m axis, as shown in Fig. 8-17. The distance AC represents failure, and the distance AB, safety. Therefore AC divided by AB is the factor of safety according to the Goodman criterion. Thus

$$n = \frac{S_a}{\sigma_a}$$

(8-33)

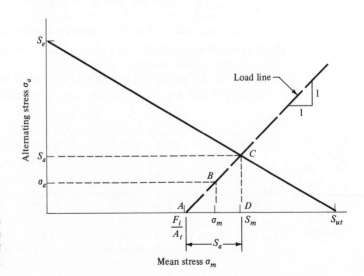

FIGURE 8-17

Goodman fatigue diagram showing how the load line is used to define failure and safety in bolted joints loaded in fatigue. Point B represents safety; point C, failure.

Noting that the distance AD is equal to S_a, we have

$$S_a = S_m - \frac{F_i}{A_t} \qquad (8\text{-}34)$$

Equation (7-35) for the modified Goodman line can be written in the form

$$S_m = S_{ut}\left(1 - \frac{S_a}{S_e}\right) \qquad (a)$$

Solving Eqs. (8-34) and (a) simultaneously gives

$$S_a = \frac{S_{ut} - F_i/A_t}{1 + S_{ut}/S_e} \qquad (8\text{-}35)$$

In using the relations of this section be sure to use K_f as a fatigue-strength-reduction factor in accordance with Eq. (7-24); otherwise, the slope of the load line will not remain 1 to 1.

After solving Eq. (8-33) for the factor of safety guarding against a fatigue failure, you should also check the possibility of yielding. Use the relation

$$n = \frac{S_y}{\sigma_{max}} = \frac{S_y}{\sigma_m + \sigma_a} \qquad (8\text{-}36)$$

It is easy to obtain fully corrected endurance limits using the Marin equation [Eq. (7-13)]. This has been done for the SAE and ISO bolts most useful in resisting fatigue; the results are shown in Table 8-12. The values shown include the effect of K_f on the strength.

EXAMPLE 8-3

Figure 8-18 shows a connection using cap screws. The joint is subjected to a fluctuating force whose maximum value is 5 kip per screw. The required data are: cap screw, $\frac{5}{8}''$-11 NC, SAE 5; hardened-steel washer $\frac{1}{16}$ in thick; steel cover plate, $t_1 = \frac{5}{8}$ in, $E = 30$ Mpsi; cast-iron base, $t_2 = \frac{5}{8}$ in, $E = 16$ Mpsi.

(a) Find k_b, k_m, and C using the assumption in the caption of Fig. 8-18.

(b) Find all factors of safety and explain what they mean.

Solution

(a) Using the symbols of Figs. 8-14 and 8-18, we find $l = 1$ in, $D_1 = 1.51$ in (not

TABLE 8-12

Fully Corrected Endurance Limits for Bolts and Screws with Rolled Threads

GRADE OR CLASS	SIZE RANGE	ENDURANCE LIMIT
SAE 5	$\frac{1}{4}$–1 in	18.6 kpsi
	$1\frac{1}{8}$–$1\frac{1}{2}$ in	16.3 kpsi
SAE 7	$\frac{1}{4}$–$1\frac{1}{2}$ in	20.6 kpsi
SAE 8	$\frac{1}{4}$–$1\frac{1}{2}$ in	23.2 kpsi
ISO 8.8	M16–M36	129 MPa
ISO 9.8	M1.6–M16	140 MPa
ISO 10.9	M5–M36	162 MPa
ISO 12.9	M1.6–M36	190 MPa

FIGURE 8-18

Assumed pressure-cone frusta for a joint using a cap screw. For this model the significant sizes are:

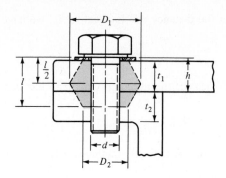

$$l = \begin{cases} h + \dfrac{t_2}{2} & t_2 < d \\ h + \dfrac{d}{2} & t_2 \geq d \end{cases}$$

$D_1 = d_w + l \tan \alpha = 1.5d + 0.577l$
$D_2 = d_w = 1.5d$

where l = effective grip. The solutions are for $\alpha = 30°$ and $d_w = 1.5d$.

required), and $D_2 = 0.9375$ in. The joint is composed of three frusta; the upper one has a thickness of $l/2$, and the lower one has a thickness of $d/2$. The two upper frusta are of steel in this example.

The upper frustum has $t = l/2 = 0.5$ in, $D = 0.9375$ in, and $E = 30$ Mpsi. Using these values in Eq. (8-14) gives $k_1 = 46.46$ Mlb/in.

The middle frustum has $t = 0.1875$ in and $D = 1.293$ in. With these and $E = 30$ Mpsi, Eq. (8-14) gives $k_2 = 197.60$ Mlb/in.

The lower frustum has $D = 0.9375$ in, $t = 0.3125$ in, and $E = 16$ Mpsi. The same equation yields $k_3 = 32.40$ Mlb/in.

Putting these three stiffnesses together using Eq. (8-12) gives $k_m = 17.4$ Mlb/in. The cap screw is short and threaded all the way. Using $l = 1$ in as the grip, we find the stiffness to be $k_b = 6.78$ Mlb/in. Thus the joint constant is

Answer
$$C = \frac{k_b}{k_b + k_m} = \frac{6.78}{6.78 + 17.4} = 0.280$$

(*b*) Equation (8-25) gives the preload as

$$F_i = 0.75F_p = 0.75A_tS_p = 0.75(0.226)(85) = 14.4 \text{ kip}$$

where $A_t = 0.226$ in², from Table 8-2, and $S_p = 85$ kpsi for an SAE grade 5 cap screw, from Table 8-4. Using Eq. (8-23), we obtain the load factor as

Answer
$$n = \frac{S_pA_t - F_i}{CP} = \frac{85(0.226) - 14.4}{0.280(5)} = 3.44$$

This factor prevents the bolt stress from becoming equal to the proof strength. Next, using Eq. (8-24), we have

Answer
$$n = \frac{F_i}{P(1 - C)} = \frac{14.4}{5(1 - 0.280)} = 4.00$$

If the force F gets too large, the joint will separate and the bolt will take the entire load. This factor guards against that event.

For the remaining factors, refer to Fig. 8-19. This diagram contains the modified Goodman line, the Gerber line, the proof-strength line, and the load line. The intersection of the load line L with the respective failure lines at points C, D, and E defines a set of strengths S_a and S_m at each intersection. Point B represents the

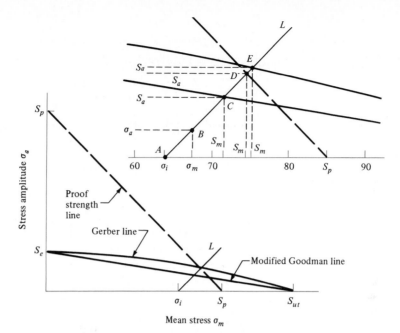

stress state σ_a, σ_m. Point A is the preload stress σ_i. Therefore the load line begins at A and makes an angle having a unit slope. This angle is 45° only when both axes have the same scale.

The factors of safety are found by dividing the distances AC, AD, and AE by the distance AB. Note that this is the same as dividing S_a, for each theory, by σ_a. Alternatively, this can be accomplished by dividing the quantity $S_m - \sigma_i$, for each case, by the quantity $\sigma_m - \sigma_i$.

The required distances can be obtained from a scaled drawing using direct measurement. Such drawings are easy to make, the results are obtained quickly, and there is small chance for error. Engineers taking the P.E. exam or students taking tests will find the graphical approach the safest and best method. But, if the Gerber theory is to be used, it is probably better to compute the distances to avoid plotting the Gerber curve.

The quantities shown in the caption of Fig. 8-19 are obtained as follows:

Point A

$$\sigma_i = \frac{F_i}{A_t} = \frac{14.4}{0.226} = 63.72 \text{ kpsi}$$

Point B

$$\sigma_a = \frac{CP}{2A_t} = \frac{0.280(5)}{2(0.226)} = 3.10 \text{ kpsi}$$

$$\sigma_m = \frac{CP}{2A_t} + \frac{F_i}{A_t} = 3.10 + 63.72 = 66.82 \text{ kpsi}$$

Point C

This is the modified Goodman theory. From Table 8-12, we find $S_e = 18.6$ kpsi. Then, using Eq. (8-35), we get

$$S_a = \frac{S_{ut} - (F_i/A_t)}{1 + (S_{ut}/S_e)} = \frac{120 - (14.4/0.226)}{1 + (120/18.6)} = 7.55 \text{ kpsi}$$

The mean strength is obtained by rearranging Eq. (7-35):

$$S_m = S_{ut}\left(1 - \frac{S_a}{S_e}\right) = 120\left(1 - \frac{7.55}{18.6}\right) = 71.29 \text{ kpsi}$$

But, using the triangle whose hypotenuse is AC in Fig. 8-19, we could have obtained S_m by noting that $S_m - \sigma_i = S_a$. The factor of safety is now found to be

Answer $$n = \frac{S_a}{\sigma_a} = \frac{7.55}{3.10} = 2.44$$

Point D

In Fig. 8-19, observe the right triangle whose hypotenuse is the line distance from D to S_p. The legs of this triangle are equal, and so

$$S_p - S_m = S_a \tag{1}$$

In addition, the right triangle whose hypotenuse is AD provides the relation

$$S_m = \sigma_i + S_a \tag{2}$$

Solving Eqs. (1) and (2) simultaneously gives

$$S_a = \frac{S_p - \sigma_i}{2} = \frac{85 - 63.72}{2} = 10.64 \text{ kpsi}$$

The factor that results is

Answer $$n = \frac{S_a}{\sigma_a} = \frac{10.64}{3.10} = 3.43$$

which, of course, is identical with the result previously obtained using Eq. (8-23). We have noted before that a fully loaded bolt is one in which the maximum bolt stress equals the proof strength.

A similar analysis of a fatigue diagram could have been done using yield strength instead of proof strength. Though the two strengths are somewhat related, proof strength is a much better and more positive indicator of a fully loaded bolt than is the yield strength. It is also worth remembering that proof-strength values are specified in design codes; yield strengths are not.

We found $n = 2.44$ on the basis of fatigue and the modified Goodman line, and $n = 3.43$ on the basis of proof strength. Thus the danger of failure is by fatigue, not by over-proof loading. These two factors should always be compared to determine where the greatest danger lies.

Point E

For the Gerber theory, we make use of Eqs. (2) and (7-36) to get

$$S_m - \sigma_i = S_e\left[1 - \left(\frac{S_m}{S_{ut}}\right)^2\right] \tag{3}$$

Solving this equation for S_m gives

$$S_m = -\frac{S_{ut}^2}{2S_e} \pm \frac{S_{ut}}{2}\left[\left(\frac{S_{ut}}{S_e}\right)^2 + \frac{4(\sigma_i + S_e)}{S_e}\right]^{1/2} \tag{8-37}$$

Substituting $S_{ut} = 120$ kpsi, $S_e = 18.6$ kpsi, and $\sigma_i = 63.72$ kpsi gives $S_m = 75.04$ kpsi. Then Eq. (2) gives $S_a = 11.32$ kpsi. Therefore

Answer $$n = \frac{S_a}{\sigma_a} = \frac{11.32}{3.10} = 3.65$$

which is the factor of safety guarding against fatigue failure according to Gerber.

8-12   STOCHASTIC CONSIDERATIONS

The data available to analyze the statistical performance of bolted joints are widely scattered, and so little exist that a reliable analysis really cannot be made. The methods of Parts 1 and 2 of this book can be used to perform the analysis when data are available.

One of the random variables needed is the tensile strength S_{ut}. The codes specify only the minimum value. Figure 8-20 reveals the data for one series of tests which might be used as a guide. For these tests the probability that S_{ut} will equal or exceed the minimum specification works out as 99.2 percent.

Another random variable that would be desirable is the proof strength S_p. The data

FIGURE 8-20

Histogram of bolt tensile strength for 539 tests of ISO 9.8 bolts having a minimum specified tensile strength of 120 kpsi (830 MPa). The report did not explain why U.S. customary units were used with ISO bolts. The statistical parameters are $\bar{S}_{ut} = 145.1$ kpsi and $\hat{\sigma}_{S_{ut}} = 10.3$ kpsi, giving $C_{S_{ut}} = 0.071$. *(Source: A. R. Breed and J. B. Bedford, "Mechanical and Material Requirements for Metric Threaded Fasteners," Trans. Tech. Conf. Metric Mechanical Fasteners, ANSI SR 17 ASTM STP 587, 1975, p. 117.)*

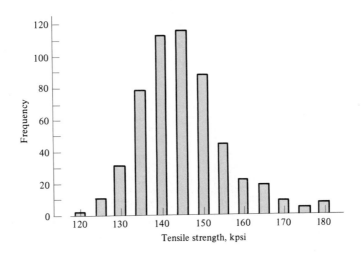

of Tables 8-8 and 8-9 do give some help on the preload variable $\mathbf{F}_i$. The coefficient of variation turns out to be $C_{Fi} = 0.143$ for unlubricated bolts and 0.084 for lubricated bolts.

The modified Goodman line will be of little value in analyzing the statistical performance of bolted joints, because it is always on the conservative side of failure points. Thus a nonlinear theory, such as the Gerber parabolic theory, is needed. Equation (8-37) reveals that the determination of the mean-strength variate $\mathbf{S}_m$ is quite complex for the Gerber theory. A computer solution might be the way to go.

8-13   BOLTED AND RIVETED JOINTS LOADED IN SHEAR*

Riveted and bolted joints loaded in shear are treated exactly alike in design and analysis.

In Fig. 8-21a is shown a riveted connection loaded in shear. Let us now study the various means by which this connection might fail.

Figure 8-21b shows a failure by bending of the rivet or of the riveted members. The

---

*The design of bolted and riveted connections for boilers, bridges, buildings, and other structures in which danger to human life is involved is strictly governed by various construction codes. When designing these structures, the engineer should refer to the *American Institute of Steel Construction Handbook,* the American Railway Engineering Association specifications, or the Boiler Construction Code of the American Society of Mechanical Engineers.

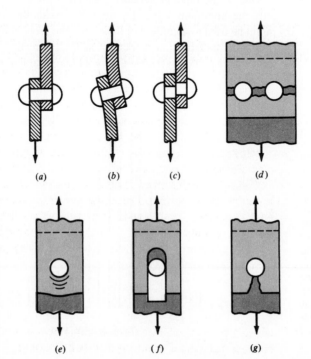

(a)   (b)   (c)   (d)

(e)   (f)   (g)

FIGURE 8-21

Modes of failure in shear loading of a bolted or riveted connection: (a) shear loading; (b) bending of rivet; (c) shear of rivet; (d) tensile failure of members; (e) bearing of rivet on members or bearing of members on rivet; (f) shear tear-out; (g) tensile tear-out.

bending moment is approximately $M = Ft/2$, where F is the shearing force and t is the grip of the rivet, that is, the total thickness of the connected parts. The bending stress in the members or in the rivet is, neglecting stress concentration,

$$\sigma = \frac{M}{I/c} \tag{8-38}$$

where I/c is the section modulus for the weakest member or for the rivet or rivets, depending upon which stress is to be found. The calculation of the bending stress in this manner is an assumption, because we do not know exactly how the load is distributed to the rivet or the relative deformations of the rivet and the members. Although this equation can be used to determine the bending stress, it is seldom used in design; instead its effect is compensated for by an increase in the factor of safety.

In Fig. 8-21c failure of the rivet by pure shear is shown; the stress in the rivet is

$$\tau = \frac{F}{A} \tag{8-39}$$

where A is the cross-sectional area of all the rivets in the group. It may be noted that it is standard practice in structural design to use the nominal diameter of the rivet rather than the diameter of the hole, even though a hot-driven rivet expands and nearly fills up the hole.

Rupture of one of the connected members or plates by pure tension is illustrated in Fig. 8-21d. The tensile stress is

$$\sigma = \frac{F}{A} \tag{8-40}$$

where A is the net area of the plate, that is, the area reduced by an amount equal to the area of all the rivet holes. For brittle materials and static loads and for either ductile or brittle materials loaded in fatigue, the stress-concentration effects must be included. It is true that the use of a bolt with an initial preload and, sometimes, a rivet will place the area around the hole in compression and thus tend to nullify the effects of stress concentration, but unless definite steps are taken to ensure that the preload does not relax, it is on the conservative side to design as if the full stress-concentration effect were present. The stress-concentration effects are not considered in structural design, because the loads are static and the materials ductile.

In calculating the area for Eq. (8-40), the designer should, of course, use the combination of rivet or bolt holes which gives the smallest area.

Figure 8-21e illustrates a failure by crushing of the rivet or plate. Calculation of this stress, which is usually called a *bearing stress,* is complicated by the distribution of the load on the cylindrical surface of the rivet. The exact values of the forces acting upon the rivet are unknown, and so it is customary to assume that the components of these forces are uniformly distributed over the projected contact area of the rivet. This gives for the stress

$$\sigma = \frac{F}{A} \tag{8-41}$$

where the projected area for a single rivet is $A = td$. Here, t is the thickness of the thinnest plate and d is the rivet or bolt diameter.

Shearing, or tearing, of the margin is shown in Fig. 8-21f and g. In structural practice this failure is avoided by spacing the rivet at least $1\frac{1}{2}$ diameters away from the margin. Bolted connections usually are spaced an even greater distance than this for satisfactory appearance, and hence this type of failure may usually be neglected.

In structural design it is customary to select in advance the number of rivets and their diameters and spacing. The strength is then determined for each method of failure. If the calculated strength is not satisfactory, a change is made in the diameter, spacing, or number of rivets used, to bring the strength in line with expected loading conditions. It is not usual, in structural practice, to consider the combined effects of the various methods of failure.

8-14 CENTROIDS OF BOLT GROUPS

In Fig. 8-22, let A_1 through A_5 be the respective cross-sectional areas of a group of five bolts. These bolts need not be of the same diameter. In order to determine the shear forces which act upon each bolt, it is necessary to know the location of the centroid of the bolt group. Using statics, we learn that the centroid G is located by the coordinates where x_i and y_i are the distances to the respective bolt centers. In many instances these

$$\bar{x} = \frac{A_1x_1 + A_2x_2 + A_3x_3 + A_4x_4 + A_5x_5}{A_1 + A_2 + A_3 + A_4 + A_5} = \frac{\sum_1^n A_i x_i}{\sum_1^n A_i}$$

$$(8\text{-}42)$$

$$\bar{y} = \frac{A_1y_1 + A_2y_2 + A_3y_3 + A_4y_4 + A_5y_5}{A_1 + A_2 + A_3 + A_4 + A_5} = \frac{\sum_1^n A_i y_i}{\sum_1^n A_i}$$

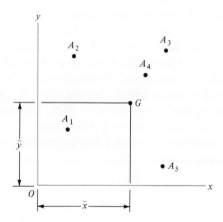

FIGURE 8-22

centroidal distances can be located by symmetry. Note that the Σ keys on your calculator are especially useful for finding these centroidal distances.

8-15 SHEAR OF BOLTS AND RIVETS DUE TO ECCENTRIC LOADING

An example of eccentric loading of fasteners is shown in Fig. 8-23. This is a portion of a machine frame containing a beam A subjected to the action of a bending load. In this case, the beam is fastened to vertical members at the ends with bolts. You will recognize the schematic representation in Fig. 8-23b as an indeterminate beam with both ends fixed and with the moment reaction M and the shear reaction V at the ends.

For convenience, the centers of the bolts at one end of the beam are drawn to a larger scale in Fig. 8-23c. Point O represents the centroid of the group, and it is assumed in this example that all the bolts are of the same diameter. The total load taken by each bolt will be calculated in three steps. In the first step the shear V is divided equally among the bolts so that each bolt takes $F' = V/n$, where n refers to the number of bolts in the group and the force F' is called the *direct load,* or *primary shear*.

It is noted that an equal distribution of the direct load to the bolts assumes an absolutely rigid member. The arrangement of the bolts or the shape and size of the members sometimes justify the use of another assumption as to the division of the load. The direct loads F' are shown as vectors on the loading diagram (Fig. 8-23c).

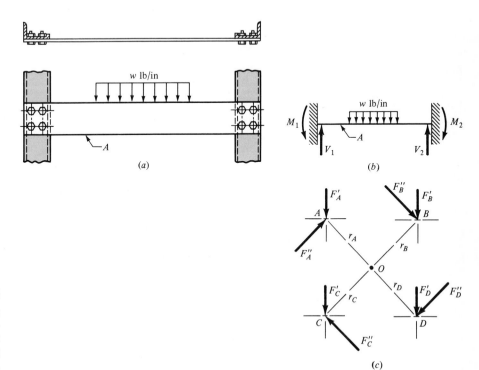

FIGURE 8-23

(a) Beam bolted at both ends with distributed load; (b) free-body diagram of beam; (c) enlarged view of bolt group showing primary and secondary shear forces.

The *moment load,* or *secondary shear,* is the additional load on each bolt due to the moment M. If r_A, r_B, r_C, etc., are the radial distances from the centroid to the center of each bolt, the moment and moment load are related as follows:

$$M = F''_A r_A + F''_B r_B + F''_C r_C + \cdots \tag{a}$$

where F'' is the moment load. The force taken by each bolt depends upon its radial distance from the centroid; that is, the bolt farthest from the centroid takes the greatest load, while the nearest bolt takes the smallest. We can therefore write

$$\frac{F''_A}{r_A} = \frac{F''_B}{r_B} = \frac{F''_C}{r_C} \tag{b}$$

Solving Eqs. (*a*) and (*b*) simultaneously, we obtain

$$F''_n = \frac{M r_n}{r_A^2 + r_B^2 + r_C^2 + \cdots} \tag{8-43}$$

where the subscript n refers to the particular bolt whose load is to be found. These moment loads are also shown as vectors on the loading diagram.

In the third step the direct and moment loads are added vectorially to obtain the resultant load on each bolt. Since all the bolts or rivets are usually the same size, only that bolt having the maximum load need be considered. When the maximum load is found, the strength may be determined, using the various methods already described.

EXAMPLE 8-4   Shown in Fig. 8-24 is a 15- by 200-mm rectangular steel bar cantilevered to a 250-mm steel channel using four bolts. On the basis of the external load of 16 kN, find:

(*a*) The resultant load on each bolt

(*b*) The maximum bolt shear stress

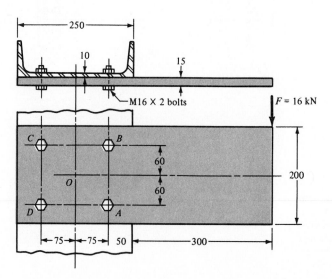

FIGURE 8-24

Dimensions in millimeters.

(c) The maximum bearing stress

(d) The critical bending stress in the bar

Solution   (a) Point O, the centroid of the bolt group in Fig. 8-24, is found by symmetry. If a free-body diagram of the beam were constructed, the shear reaction V would pass through O and the moment reaction M would be about O. These reactions are

$$V = 16 \text{ kN} \qquad M = 16(425) = 6800 \text{ N} \cdot \text{m}$$

In Fig. 8-25, the bolt group has been drawn to a larger scale and the reactions are shown. The distance from the centroid to the center of each bolt is

$$r = \sqrt{(60)^2 + (75)^2} = 96.0 \text{ mm}$$

The primary shear load per bolt is

$$F' = \frac{V}{n} = \frac{16}{4} = 4 \text{ kN}$$

Since the secondary shear forces are equal, Eq. (8-43) becomes

$$F'' = \frac{Mr}{4r^2} = \frac{M}{4r} = \frac{6800}{4(96.0)} = 17.7 \text{ kN}$$

The primary and secondary shear forces are plotted to scale in Fig. 8-25 and the resultants obtained by using the parallelogram rule. The magnitudes are found by measurement (or analysis) to be

Answer $$F_A = F_B = 21.0 \text{ kN}$$

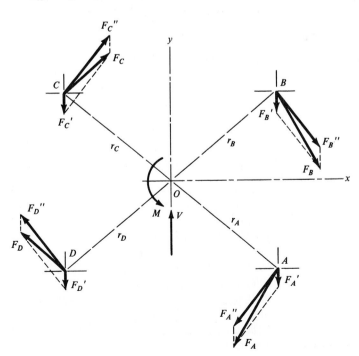

FIGURE 8-25

Answer $F_C = F_D = 13.8$ kN

(b) Bolts A and B are critical because they carry the largest shear load. Does this shear act on the threaded portion of the bolt, or on the unthreaded portion? The bolt length will be 25 mm plus the height of the nut plus about 2 mm for a washer. Table A-28 gives the nut height as 14.8 mm. This adds up to a length of 41.8 mm, and so a bolt 46 mm long will be needed. From Table 8-6 we compute the thread length as $L_T = 22$ mm. Thus the unthreaded portion of the bolt is $45 - 22 = 23$ mm long. This exceeds the 15 mm for the plate in Fig. 8-24, and so the bolt will tend to shear across its major diameter. Therefore the shear-stress area is $A_s = \pi d^2/4 = \pi(16)^2/4 = 201$ mm^2, and so the shear stress is

Answer $$\tau = \frac{F}{A_s} = \frac{21.0(10)^3}{201} = 104 \text{ MPa}$$

(c) The channel is thinner than the bar, and so the largest bearing stress is due to the pressing of the bolt against the channel web. The bearing area is $A_b = td = 10(16) = 160$ mm^2. Thus the bearing stress is

Answer $$\sigma = \frac{F}{A_b} = -\frac{21.0(10)^3}{160} = -131 \text{ MPa}$$

(d) The critical bending stress in the bar is assumed to occur in a section parallel to the y axis and through bolts A and B. At this section the bending moment is

$$M = 16(300 + 50) = 5600 \text{ N} \cdot \text{m}$$

The second moment of area through this section is obtained by the use of the transfer formula, as follows:

$$I = I_{\text{bar}} - 2(I_{\text{holes}} + d^2 A)$$

$$= \frac{15(200)^3}{12} - 2\left[\frac{15(16)^3}{12} + (60)^2(15)(16)\right] = 8.26(10)^6 \text{ mm}^4$$

Then

Answer $$\sigma = \frac{Mc}{I} = \frac{5600(100)}{8.26(10)^6} (10)^3 = 67.8 \text{ MPa}$$

8-16   SETSCREWS

Unlike bolts and cap screws, which depend upon tension to develop a clamping force, the setscrew depends upon compression to develop the clamping force. The resistance to axial motion or rotary motion of the collar or hub relative to the shaft is called the *holding power*. This holding power, which is really a force resistance, is due to the frictional resistance of the contacting portions of the collar and shaft as well as any slight penetration of the setscrew into the shaft.

Figure 8-26 shows the point styles available with socket setscrews. These are also manufactured with screwdriver slots and with square heads.

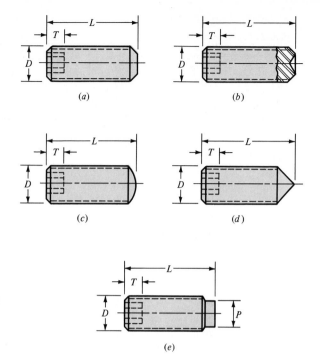

FIGURE 8-26

Socket setscrews: (a) flat point;
(b) cup point; (c) oval point;
(d) cone point; (e) half-dog point.

Table 8-13 lists values of the seating torque and the corresponding holding power for inch-series setscrews. The values listed apply to both axial holding power, for resisting thrust, and the tangential holding power, for resisting torsion. Typical factors of safety are 1.5 to 2.0 for static loads and 4 to 8 for various dynamic loads.

Setscrews should have a length of about half of the shaft diameter. Note that this practice also provides a rough rule for the radial thickness of a hub or collar.

8-17   KEYS AND PINS

Keys and pins are used on shafts to secure rotating elements, such as gears, pulleys, or other wheels. Keys are used to enable the transmission of torque from the shaft to the shaft-supported element. Pins are used for axial positioning and for the transfer of torque or thrust or both.

Figure 8-27 shows a variety of keys and pins. Pins are useful when the principal loading is shear and when both torsion and thrust are present. Taper pins are sized according to the diameter at the large end. Some of the most useful sizes of these are listed in Table 8-14. The diameter at the small end is

$$d = D - 0.0208L \qquad (8\text{-}44)$$

where d = diameter at small end, in
D = diameter at large end, in
L = length, in

TABLE 8-13

Typical Holding Power (Force) for Socket Setscrews*

SIZE, in	SEATING TORQUE, lb · in	HOLDING POWER, lb
#0	1.0	50
#1	1.8	65
#2	1.8	85
#3	5	120
#4	5	160
#5	10	200
#6	10	250
#8	20	385
#10	36	540
$\frac{1}{4}$	87	1000
$\frac{5}{16}$	165	1500
$\frac{3}{8}$	290	2000
$\frac{7}{16}$	430	2500
$\frac{1}{2}$	620	3000
$\frac{9}{16}$	620	3500
$\frac{5}{8}$	1325	4000
$\frac{3}{4}$	2400	5000
$\frac{7}{8}$	5200	6000
1	7200	7000

*Based on alloy-steel screw against steel shaft, class 3A coarse or fine threads in class 2B holes, and cup-point socket setscrews.

Source: Unbrako Division, SPS Technologies, Jenkintown, Pa.

FIGURE 8-27

(a) Square key; (b) round key; (c and d) round pins; (e) taper pin; (f) split tubular spring pin. The pins in parts (e) and (f) are shown longer than necessary, to illustrate the chamfer on the ends, but their lengths should be kept smaller than the hub diameters to prevent injuries due to projections on rotating parts.

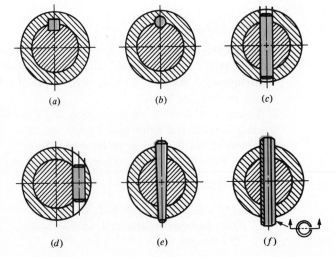

TABLE 8-14

Dimensions at Large End of Some Standard Taper Pins— Inch Series

SIZE	COMMERCIAL		PRECISION	
	MAXIMUM	MINIMUM	MAXIMUM	MINIMUM
4/0	0.1103	0.1083	0.1100	0.1090
2/0	0.1423	0.1403	0.1420	0.1410
0	0.1573	0.1553	0.1570	0.1560
2	0.1943	0.1923	0.1940	0.1930
4	0.2513	0.2493	0.2510	0.2500
6	0.3423	0.3403	0.3420	0.3410
8	0.4933	0.4913	0.4930	0.4920

For less important applications, a dowel pin or a drive pin can be used. A large variety of these are listed in manufacturers' catalogs.*

The square key, shown in Fig. 8-27a, is also available in rectangular sizes. Standard sizes of these, together with the range of applicable shaft diameters, are listed in Table 8-15. The length of the key is based on the hub length and the torsional load to be transferred.

The gib-head key, in Fig. 8-28, is tapered so that, when firmly driven, it acts to prevent relative axial motion. This also gives the advantage that the hub position can be

*See also Joseph E. Shigley, "Unthreaded Fasteners," Chap. 22 in Joseph E. Shigley and Charles R. Mischke (eds.), *Standard Handbook of Machine Design,* McGraw-Hill, New York, 1986.

TABLE 8-15

Inch Dimensions for Some Standard Square- and Rectangular-Key Applications

SHAFT DIAMETER		KEY SIZE		KEYWAY
OVER	TO (INCL.)	w	h	DEPTH
$\frac{5}{16}$	$\frac{7}{16}$	$\frac{3}{32}$	$\frac{3}{32}$	$\frac{3}{64}$
$\frac{7}{16}$	$\frac{9}{16}$	$\frac{1}{8}$	$\frac{3}{32}$	$\frac{3}{64}$
		$\frac{1}{8}$	$\frac{1}{8}$	$\frac{1}{16}$
$\frac{9}{16}$	$\frac{7}{8}$	$\frac{3}{16}$	$\frac{1}{8}$	$\frac{1}{16}$
		$\frac{3}{16}$	$\frac{3}{16}$	$\frac{3}{32}$
$\frac{7}{8}$	$1\frac{1}{4}$	$\frac{1}{4}$	$\frac{3}{16}$	$\frac{3}{32}$
		$\frac{1}{4}$	$\frac{1}{4}$	$\frac{1}{8}$
$1\frac{1}{4}$	$1\frac{3}{8}$	$\frac{5}{16}$	$\frac{1}{4}$	$\frac{1}{8}$
		$\frac{5}{16}$	$\frac{5}{16}$	$\frac{5}{32}$
$1\frac{3}{8}$	$1\frac{3}{4}$	$\frac{3}{8}$	$\frac{1}{4}$	$\frac{1}{8}$
		$\frac{3}{8}$	$\frac{3}{8}$	$\frac{3}{16}$
$1\frac{3}{4}$	$2\frac{1}{4}$	$\frac{1}{2}$	$\frac{3}{8}$	$\frac{3}{16}$
		$\frac{1}{2}$	$\frac{1}{2}$	$\frac{1}{4}$
$2\frac{1}{4}$	$2\frac{3}{4}$	$\frac{5}{8}$	$\frac{7}{16}$	$\frac{7}{32}$
		$\frac{5}{8}$	$\frac{5}{8}$	$\frac{5}{16}$
$2\frac{3}{4}$	$3\frac{1}{4}$	$\frac{3}{4}$	$\frac{1}{2}$	$\frac{1}{4}$
		$\frac{3}{4}$	$\frac{3}{4}$	$\frac{3}{8}$

Source: Joseph E. Shigley, "Unthreaded Fasteners," chap. 22 in Joseph E. Shigley and Charles R. Mischke (eds.), *Standard Handbook of Machine Design,* McGraw-Hill, New York, 1986.

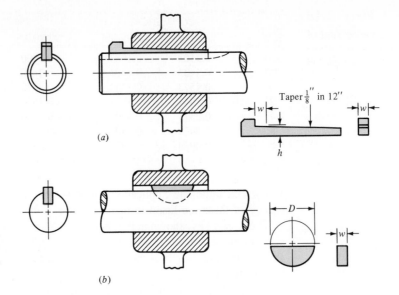

FIGURE 8-28

(a) Gib-head key; (b) Woodruff
key.

adjusted for the best axial location. The head makes removal possible without access to the other end, but the projection may be hazardous.

Also shown in Fig. 8-28a is the sled-runner keyway. This has less stress concentration than an end-milled keyway.

The Woodruff key, shown in Fig. 8-28b, is of general usefulness, especially when a wheel is to be positioned against a shaft shoulder, since the keyslot need not be machined into the shoulder stress-concentration region. The use of the Woodruff key also yields better concentricity after assembly of the wheel and shaft. This is especially important at high speeds, as, for example, with a turbine wheel and shaft. Dimensions for some standard Woodruff key sizes can be found in Table 8-16, and Table 8-17 gives the shaft diameters for which the different keyseat widths are suitable.

Stress-concentration factors for keyways depend for their values upon the fillet radius at the bottom and ends of the keyway, according to Peterson.* For fillets cut by standard milling-machine cutters, Peterson's charts give $K_t = 2.14$ for bending and $K_{ts} = 2.62$ for torsion. These are for end-milled keyseats, which permit definite key positioning longitudinally. In bending, the sled-runner seat will have less stress concentration. The values of K_t for the sled-runner keyway are based on the full values of I and J for shafting; that is, the second moments are not reduced by the slot dimensions. Other results reported are 1.79 for an end-milled keyway and 1.38 for the sled-runner type.

The use of shrink fits eliminates the use of keys to transmit torque. These were discussed in Sec. 4-8, and corresponding data are listed in Tables A-11 to A-14. Peterson reports $K_f = 2.0$ for bending at each end of a press-fitted collar. Tests seem to indicate that the interface pressure has little effect on K_t. However, fretting corrosion introduces a complication.

A retaining ring is frequently used instead of a shaft shoulder or a sleeve to axially

---

*R. E. Peterson, *Stress Concentration Factors*, Wiley, New York, 1974, pp. 245, 266, 267.

TABLE 8-16

Dimensions of Woodruff Keys—
Inch Series

KEY SIZE		HEIGHT	OFFSET	KEYSEAT DEPTH	
w	D	b	e	SHAFT	HUB
$\frac{1}{16}$	$\frac{1}{4}$	0.109	$\frac{1}{64}$	0.0728	0.0372
$\frac{1}{16}$	$\frac{3}{8}$	0.172	$\frac{1}{64}$	0.1358	0.0372
$\frac{3}{32}$	$\frac{3}{8}$	0.172	$\frac{1}{64}$	0.1202	0.0529
$\frac{3}{32}$	$\frac{1}{2}$	0.203	$\frac{3}{64}$	0.1511	0.0529
$\frac{3}{32}$	$\frac{5}{8}$	0.250	$\frac{1}{16}$	0.1981	0.0529
$\frac{1}{8}$	$\frac{1}{2}$	0.203	$\frac{3}{64}$	0.1355	0.0685
$\frac{1}{8}$	$\frac{5}{8}$	0.250	$\frac{1}{16}$	0.1825	0.0685
$\frac{1}{8}$	$\frac{3}{4}$	0.313	$\frac{1}{16}$	0.2455	0.0685
$\frac{5}{32}$	$\frac{5}{8}$	0.250	$\frac{1}{16}$	0.1669	0.0841
$\frac{5}{32}$	$\frac{3}{4}$	0.313	$\frac{1}{16}$	0.2299	0.0841
$\frac{5}{32}$	$\frac{7}{8}$	0.375	$\frac{1}{16}$	0.2919	0.0841
$\frac{3}{16}$	$\frac{3}{4}$	0.313	$\frac{1}{16}$	0.2143	0.0997
$\frac{3}{16}$	$\frac{7}{8}$	0.375	$\frac{1}{16}$	0.2763	0.0997
$\frac{3}{16}$	1	0.438	$\frac{1}{16}$	0.3393	0.0997
$\frac{1}{4}$	$\frac{7}{8}$	0.375	$\frac{1}{16}$	0.2450	0.1310
$\frac{1}{4}$	1	0.438	$\frac{1}{16}$	0.3080	0.1310
$\frac{1}{4}$	$1\frac{1}{4}$	0.547	$\frac{5}{64}$	0.4170	0.1310
$\frac{5}{16}$	1	0.438	$\frac{1}{16}$	0.2768	0.1622
$\frac{5}{16}$	$1\frac{1}{4}$	0.547	$\frac{5}{64}$	0.3858	0.1622
$\frac{5}{16}$	$1\frac{1}{2}$	0.641	$\frac{7}{64}$	0.4798	0.1622
$\frac{3}{8}$	$1\frac{1}{4}$	0.547	$\frac{5}{64}$	0.3545	0.1935
$\frac{3}{8}$	$1\frac{1}{2}$	0.641	$\frac{7}{64}$	0.4485	0.1935

position a component on a shaft or in a housing bore. As shown in Fig. 8-29, a groove is cut in the shaft or bore to receive the spring retainer. The tapered design of both the internal and external rings ensures uniform pressure against the bottom of the groove. For sizes, dimensions, and ratings, the manufacturers' catalogs should be consulted.

EXAMPLE 8-5

A UNS G10350 steel shaft, heat-treated to a minimum yield strength of 75 kpsi, has a diameter of $1\frac{7}{16}$ in. The shaft rotates at 600 rev/min and transmits 40 hp through a gear. Select an appropriate key for the gear.

TABLE 8-17

Sizes of Woodruff Keys
Suitable for Various Shaft
Diameters

KEYSEAT	SHAFT DIAMETER, in	
WIDTH, in	FROM	TO (INCLUSIVE)
$\frac{1}{16}$	$\frac{5}{16}$	$\frac{1}{2}$
$\frac{3}{32}$	$\frac{3}{8}$	$\frac{7}{8}$
$\frac{1}{8}$	$\frac{3}{8}$	$1\frac{1}{2}$
$\frac{5}{32}$	$\frac{1}{2}$	$1\frac{5}{8}$
$\frac{3}{16}$	$\frac{9}{16}$	2
$\frac{1}{4}$	$\frac{11}{16}$	$2\frac{1}{4}$
$\frac{5}{16}$	$\frac{3}{4}$	$2\frac{3}{8}$
$\frac{3}{8}$	1	$2\frac{5}{8}$

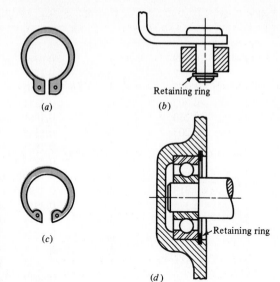

FIGURE 8-29

Typical uses for retaining rings. (a) External ring and (b) its application; (c) internal ring and (d) its application.

Solution   A $\frac{3}{8}$-in square key is selected, UNS G10200 cold-drawn steel being used. The design will be based on a yield strength of 65 kpsi. A factor of safety of 2.80 will be employed in the absence of exact information about the nature of the load.

The torque is obtained from the horsepower equation

$$T = \frac{63\ 000H}{n} = \frac{(63\ 000)\ (40)}{600} = 4200\ \text{lb} \cdot \text{in}$$

Referring to Fig. 8-30, the force F at the surface of the shaft is

$$F = \frac{T}{r} = \frac{4200}{0.719} = 5850\ \text{lb}$$

By the distortion-energy theory, the shear strength is

$$S_{sy} = 0.577S_y = (0.577)(65) = 37.5\ \text{kpsi}$$

Failure by shear across the area ab will create a stress of $\tau = F/tl$. Substituting the strength divided by the factor of safety for τ gives

$$\frac{S_{sy}}{n} = \frac{F}{tl} \quad \text{or} \quad \frac{37.5(10)^3}{2.80} = \frac{5850}{0.375l}$$

or $l = 1.16$ in. To resist crushing, the area of one-half the face of the key is used:

$$\frac{S_y}{n} = \frac{F}{tl/2} \quad \text{or} \quad \frac{65(10)^3}{2.80} = \frac{5850}{0.375l/2}$$

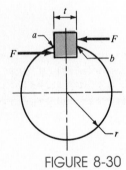

FIGURE 8-30

and $l = 1.34$ in. The hub length of a gear is usually greater than the shaft diameter, for stability. If the key, in this example, is made equal in length to the hub, it would therefore have ample strength, since it would probably be $1\frac{7}{16}$ in or longer.

PROBLEMS*

8-1 A power screw is 25 mm in diameter and has a thread pitch of 5 mm.
(a) Find the thread depth, the thread width, the mean and root diameters, and the lead, provided square threads are used.
(b) Repeat part (a) for Acme threads.

8-2 Using the information in the footnote of Table 8-1, show that the tensile-stress area is

$$A_t = \frac{\pi}{4}(d - 0.938\ 194p)^2$$

8-3 Show that for zero collar friction the efficiency of a square-thread screw is given by the equation

$$e = \tan \lambda \, \frac{1 - \mu \tan \lambda}{\tan \lambda + \mu}$$

Plot a curve of the efficiency for lead angles up to 45°. Use $\mu = 0.08$.

8-4 A single-threaded 25-mm power screw is 25 mm in diameter with a pitch of 5 mm. A vertical load on the screw reaches a maximum of 6 kN. The coefficients of friction are 0.05 for the collar and 0.08 for the threads. The frictional diameter of the collar is 40 mm. Find the overall efficiency and the torque to "raise" and "lower" the load.

8-5 The machine shown in the figure can be used for a tension test but not for a compression test. Why? Do both screws have the same hand?

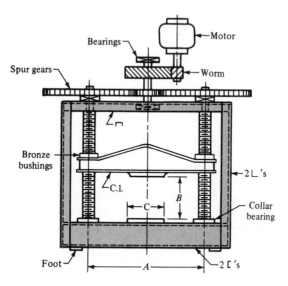

PROBLEM 8-5

8-6 The press shown for Prob. 8-5 has a rated load of 5000 lb. The twin screws have Acme threads, a diameter of 3 in, and a pitch of $\frac{1}{2}$ in. Coefficients of friction are 0.05 for the threads and 0.06 for the collar bearings. Collar diameters are 5 in. The gears have an efficiency of 95 percent and a speed ratio of 75:1. A slip clutch, on the motor shaft, prevents overloading. The full-load motor speed is 1720 rev/min.

---

*An asterisk indicates a design-type problem.

(a) When the motor is turned on, how fast will the press head move?

(b) What should be the horsepower rating of the motor?

8-7*   A screw clamp similar to the one shown in the figure has a handle with diameter $\frac{3}{16}$ in made of cold-drawn AISI 1006 steel. The overall length is 3 in. The screw is $\frac{7}{16}$"-14 UNC and is $5\frac{3}{4}$ in long, overall. Distance A is 2 in. The clamp will accommodate parts up to $4\frac{3}{16}$ in high.

(a) What screw torque will cause the handle to bend permanently?

(b) What clamping force will the answer to part (a) cause if the collar friction is neglected and if the thread friction is 0.075?

(c) What clamping force will cause the screw to buckle?

(d) Are there any other stresses or possible failures to be checked?

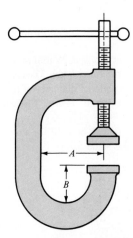

PROBLEM 8-7

8-8   The C clamp shown in the figure for Prob. 8-7 uses a $\frac{5}{8}$"-6 Acme thread. The frictional coefficients are 0.15 for the threads and for the collar. The collar, which in this case is the anvil striker's swivel joint, has a friction diameter of $\frac{7}{16}$ in. Calculations are to be based on a maximum force of 6 lb applied to the handle at a radius of $2\frac{3}{4}$ in from the screw centerline. Find the clamping force.

8-9   Find the power required to drive a 40-mm power screw having double square threads with a pitch of 6 mm. The nut is to move at a velocity of 48 mm/s and move a load of $F = 10$ kN. The frictional coefficients are 0.10 for the threads and 0.15 for the collar. The frictional diameter of the collar is 60 mm.

8-10   A single square-thread power screw has an input power of 3 kW at a speed of 1 rev/s. The screw has a diameter of 36 mm and a pitch of 6 mm. The frictional coefficients are 0.14 for the threads and 0.09 for the collar, with a collar friction radius of 45 mm. Find the axial resisting load F and the combined efficiency of the screw and collar.

8-11†   The figure illustrates the connection of a cylinder head to a pressure vessel using 10 bolts and a confined-gasket seal. The effective sealing diameter is 150 mm. Other dimensions are: $A = 100$, $B = 200$, $C = 300$, $D = 20$, and $E = 25$, all in millimeters. The cylinder is used to store gas at

---

†Unless specified otherwise, coarse threads are the rule in bolt and screw problems.

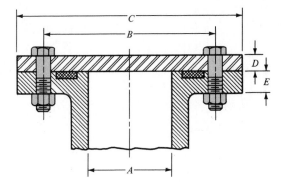

PROBLEM 8-11
Cylinder head is steel; cylinder is
No. 30 cast iron.

a static pressure of 6 MPa. ISO class 8.8 bolts with a diameter of 12 mm have been selected. This provides an acceptable bolt spacing. What load factor n results from this selection?

8-12   We wish to alter the figure for Prob. 8-11 by decreasing the inside diameter of the seal to the diameter $A = 100$ mm. This makes an effective sealing diameter of 120 mm. Then, by using cap screws instead of bolts, the bolt circle diameter B can be reduced as well as the outside diameter C. If the same bolt spacing and the same edge distance are used, then eight 12-mm cap screws can be used on a bolt circle with $B = 160$ mm and an outside diameter of 260 mm, a substantial savings. With these dimensions and all other data the same as in Prob. 8-11, find the load factor.

8-13   In the figure for Prob. 8-11, the bolts have a diameter of $\frac{1}{2}$ in and the cover plate is steel, with $D = \frac{1}{2}$ in. The cylinder is cast iron, with $E = \frac{5}{8}$ in and a modulus of elasticity of 18 Mpsi. The $\frac{1}{2}$-in SAE washer to be used under the nut has OD = 1.062 in and is 0.095 in thick. Find the stiffnesses of the bolt and the members and the joint constant C.

8-14   The same as Prob. 8-13, except that $\frac{1}{2}$-in cap screws are used with washers.

8-15   In addition to the data of Prob. 8-13, the dimensions of the cylinder are $A = 3.5$ in and an effective seal diameter of 4.25 in. The internal static pressure is 1500 psi. The outside diameter of the head is $C = 8$ in. The diameter of the bolt circle is 6 in, and so a bolt spacing in the range of 3 to 5 bolt diameters would require from 8 to 13 bolts. Select ten SAE grade 5 bolts and find the resulting load factor n.

8-16   A $\frac{3}{8}$-in class 5 cap screw and steel washer are used to secure a cast-iron frame of a machine having a blind threaded hole. The washer is 0.065 in thick. The cap has a modulus of elasticity of 14 Mpsi and is $\frac{1}{4}$ in thick. The screw is 1 in long. The material in the frame has a modulus of elasticity of 14 Mpsi. Find the stiffnesses k_b and k_m of the bolt and members.

8-17   Bolts distributed about a bolt circle are often called upon to resist an external bending moment as shown in the figure. The external moment is 12 kip · in and the bolt circle has a diameter of 8 in. The neutral axis for bending is a diameter of the bolt circle. What needs to be determined is the most severe external load seen by a bolt in the assembly.
        (a) View the effect of the bolts as placing a line load around the bolt circle whose intensity F_b', in pounds per inch, varies linearly with the distance from the neutral axis according to the relation $F_b' = F_{b,\max}' R \sin \theta$. The load on any particular bolt can be viewed as the effect of the line load over the arc associated with the bolt. For example, there are 12 bolts shown in

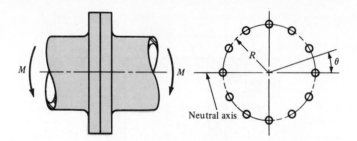

PROBLEM 8-17
Bolted connection subject to bending.

the figure. Thus each bolt load is assumed to be distributed on a 30° arc of the bolt circle. Under these conditions, what is the largest bolt load?

(b) View the largest load as the intensity $F'_{b,\text{max}}$ multiplied by the arc length associated with each bolt and find the largest bolt load.

(c) Express the load on any bolt as $F = F_{\text{max}} \sin \theta$, sum the moments due to all the bolts, and estimate the largest bolt load. Compare the results of these three approaches to decide how to attack such problems in the future.

8-18 A $\frac{3}{8}''$-24 UNF SAE grade 5 bolt is lubricated and then tightened to a torque-wrench specification of 450 lb · in. This corresponds to a preload of 90 percent of proof strength. [See Eq. (8-25).] Data in the footnote of Table 8-9 show that the coefficient of variation of the preload is $C_{Fi} = 2.88/34.18 = 0.0843$.

(a) Find the mean and the standard deviation of the bolt tension.

(b) What is the probability that the proof strength will be exceeded?

8-19† The figure shows a cast-iron bearing block which is to be bolted to a steel ceiling joist and is to support a gravity load. Bolts used are M20 ISO 8.8 with coarse threads and with 3.4-mm-thick steel washers under the bolt head and nut. The joist flanges are 16 mm in thickness, and the dimension A, shown in the figure, is 20 mm. The modulus of elasticity of the bearing is 135 GPa.

(a) Find the wrench torque required if the fasteners are lubricated during assembly and the joint is to be permanent.

(b) Estimate values of the least and the greatest preloads taken by the four bolts. Use $C_F = 0.09$.

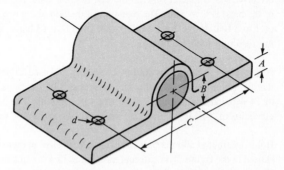

PROBLEM 8-19

8-20 The upside-down A frame shown in the figure is to be bolted to steel beams on the ceiling of a machine room using ISO grade 8.8 bolts. This frame is to support the 40-kN radial load as

---

†Don't be alarmed if you haven't used all the given information to solve the problem. This is typical of many real design problems. It also provides your instructor with an opportunity to pose problems of his or her own devising.

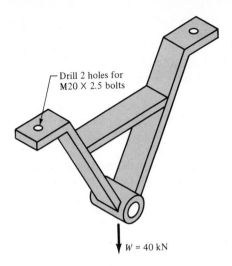

PROBLEM 8-20

illustrated. The total bolt grip is 48 mm, which includes the thickness of the steel beam, the A-frame feet, and the steel washers used. The bolts are size M20 × 2.5.

(*a*) What tightening torque should be used if the connection is permanent and the fasteners are lubricated?

(*b*) What portion of the external load is taken by the bolts? By the members?

8-21   In the figure for Prob. 8-11, let $A = 0.9$ m, $B = 1$ m, $C = 1.10$ m, $D = 20$ mm, and $E = 25$ mm. The cylinder is made of ASTM No. 30 cast iron ($E = 96$ GPa), and the head, of low-carbon steel. There are thirty-six M10 × 1.5 ISO 10.9 bolts tightened to 75 percent of proof load. During use, the cylinder pressure fluctuates between 0 and 550 kPa. Using the Gerber relation, find the factor of safety guarding against a fatigue failure of a bolt.

8-22   A 1-in-diameter hot-rolled AISI 1144 steel rod is hot-formed into an eyebolt similar to that shown in the figure for Prob. 2-59, with a 2-in-diameter eye. The threads are 1″-12 UNF and are die-cut.

(*a*) For a repeatedly applied load collinear with the thread axis, is fatigue failure more likely in the thread or in the eye?

(*b*) What can be done to strengthen the bolt at the weaker location?

(*c*) If the factor of safety guarding against a fatigue failure is $n_f = 2$, what repeatedly applied load can be applied to the eye?

8-23   The section of the sealed joint shown in the figure is loaded by a repeated force $P = 6$ kip. The members have $E = 16$ Mpsi. All bolts have been carefully preloaded to $F_i = 25$ kip each.

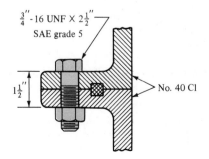

$\frac{3''}{4}$-16 UNF × $2\frac{1}{2}''$
SAE grade 5

$1\frac{1}{2}''$

No. 40 CI

PROBLEM 8-23

(a) If hardened-steel washers 0.134 in thick are to be used under the head and nut, what length of bolts should be used?

(b) Find k_b, k_m, and C.

(c) Using the modified Goodman line, find the factor of safety guarding against a fatigue failure.

(d) Find the load factor guarding against over-proof loading.

8-24   Suppose the welded steel bracket shown in the figure is bolted underneath a structural-steel ceiling beam to support a fluctuating tensile load imposed on it by a pin and yoke. The bolts are $\frac{1}{2}$ in, coarse threads, SAE grade 5, tightened to recommended preload. The stiffnesses have already been computed and are $k_b = 4.94$ Mlb/in and $k_m = 15.97$ Mlb/in.

(a) Assuming that the bolts, rather than the welds, govern the strength of this design, determine the safe repeated load P that can be imposed on this assembly using the modified Goodman criterion and a fatigue design factor of 2.

(b) Compute the load factors based on the load found in part (a).

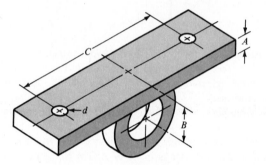

PROBLEM 8-24

8-25   Using the modified Goodman line and a fatigue-design factor of 2, determine the external repeated load P that a $1\frac{1}{4}$-in SAE grade 5 coarse-thread bolt can take compared with that for a fine-thread bolt. The joint constants are $C = 0.30$ for coarse- and 0.32 for fine-thread bolts.

8-26   An M30 × 3.5 ISO 8.8 bolt is used in a joint at recommended preload, and the joint is subject to a repeated tensile fatigue load of $P = 80$ kN per bolt. The joint constant is $C = 0.33$. Find the load factors and the factor of safety guarding against a fatigue failure based on the Goodman theory.

8-27   The figure shows a fluid-pressure linear actuator (hydraulic cylinder) in which $D = 4$ in, $t = \frac{3}{8}$ in, $L = 12$ in, and $w = \frac{3}{4}$ in. Both brackets as well as the cylinder are of steel. The actuator has been designed for a working pressure of 2000 psi. Six $\frac{3}{8}$-in SAE grade 5 coarse-thread bolts are used, tightened to 75 percent of proof load.

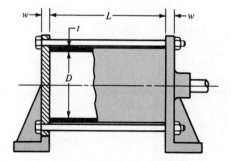

PROBLEM 8-27

(a) Find the stiffnesses of the bolts and members, assuming that the entire cylinder is compressed uniformly and that the end brackets are perfectly rigid.

(b) Using the Goodman criterion, find the factor of safety guarding against a fatigue failure.

(c) What pressure would be required to cause total joint separation?

8-28   The figure shows a bolted lap joint that uses SAE grade 8 bolts. The members are made of cold-drawn AISI 1040 steel. Find the safe tensile shear load F that can be applied to this connection if the following factors of safety are specified: shear of bolts 3, bearing on bolts 2, bearing on members 2.5, and tension of members 3.

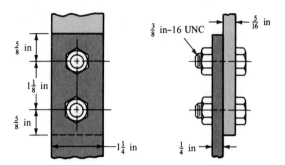

PROBLEM 8-28

8-29   The bolted connection shown in the figure uses SAE grade 5 bolts. The members are hot-rolled AISI 1018 steel. A tensile shear load $F = 4000$ lb is applied to the connection. Find the factor of safety for all possible modes of failure.

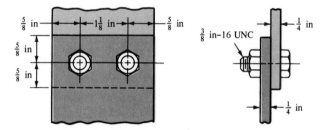

PROBLEM 8-29

8-30   A bolted lap joint using SAE grade 5 bolts and members made of cold-drawn SAE 1040 steel is shown in the figure. Find the tensile shear load F that can be applied to this connection if the following factors of safety are specified: shear of bolts 1.8, bearing on bolts 2.2, bearing on members 2.4, and tension of members 2.6.

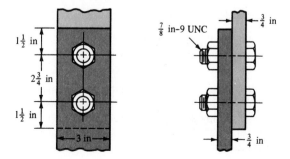

PROBLEM 8-30

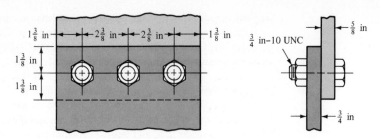

PROBLEM 8-31

8-31 The bolted connection shown in the figure is subjected to a tensile shear load of 20 kip. The bolts are SAE grade 5 and the material is cold-drawn AISI 1015 steel. Find the factor of safety of the connection for all possible modes of failure.

8-32 The figure shows a connection which employs three SAE grade 5 bolts. The tensile shear load on the joint is 5400 lb. The members are cold-drawn bars of AISI 1020 steel. Find the factor of safety for each possible mode of failure.

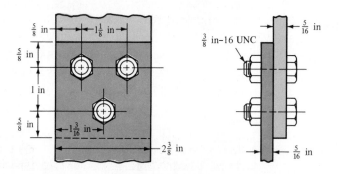

PROBLEM 8-32

8-33 A beam is made up by bolting together two cold-drawn bars of AISI 1018 steel as a lap joint, as shown in the figure. The bolts used are ISO 5.8. Ignoring any twisting, determine the factor of safety of the connection.

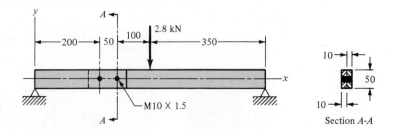

PROBLEM 8-33
Dimensions in millimeters.

8-34 Standard design practice, as exhibited by the solutions to Probs. 8-28 to 8-32, is to assume that the bolts, or rivets, share the shear equally. For many situations, such an assumption may lead to an unsafe design. Consider the yoke bracket of Prob. 8-24, for example. Suppose this bracket is bolted to a wide-flange column with the centerline through the two bolts in the vertical direction. A vertical load through the yoke-pin hole at distance B from the column flange would place a shear load on the bolts as well as a tensile load. The tensile load comes about because the bracket tends to pry itself about the bottom corner, much like a claw hammer, exerting a large tensile load on the upper bolt. In addition, it is almost certain that both the spacing of the bolt holes and their diameters will be slightly different on the column flange from what they are on the yoke bracket. Thus, unless yielding occurs, only one of the bolts will take the shear load. The designer has no way of knowing which bolt this will be.

In this problem the bracket is 8 in long, $A = \frac{1}{2}$ in, $B = 3$ in, $C = 6$ in, and the column flange is $\frac{1}{2}$ in thick. The bolts are $\frac{1}{2}$-in UNC SAE 5. Steel washers 0.095 in thick are used under the nuts. The nuts are tightened to 75 percent of proof load. The vertical yoke-pin load is 3000 lb. If the upper bolt takes all the shear load as well as the tensile load, how closely does the bolt stress approach the proof strength?

8-35   The bearing of Prob. 8-19 is bolted to a vertical surface and supports a horizontal shaft. The bolts used have coarse threads and are M20 ISO 5.8. The joint constant is $C = 0.30$, and the dimensions are $A = 20$ mm, $B = 50$ mm, and $C = 160$ mm. The bearing base is 240 mm long. The bearing load is 12 kN. If the bolts are tightened to 75 percent of proof load, will the bolt stress exceed the proof strength? Use worst-case loading, as discussed in Prob. 8-34.

8-36   A split-ring clamp-type shaft collar such as is described in Prob. 6-23 must resist an axial load of 1000 lb. Using a design factor of $n = 3$ and a coefficient of friction of 0.12, specify an SAE Grade 5 cap screw using fine threads. What wrench torque should be used if a lubricated screw is used?

8-37   The figure shows a welded steel bracket which is to support a force F as shown. The bracket is to be bolted to a vertical face by means of four SAE grade 5 bolts with diameter $\frac{3}{8}$ in and with coarse threads. The joint constant is $C = 0.173$, and the bolts are preloaded to 90 percent of the proof load.
(a) Estimate the external load carried by the upper two bolts if they resist the moment.
(b) What is the load factor for the upper pair of bolts if they carry no shear load?
(c) Suppose the shear load is carried by friction between the bracket and the supporting face. If $\mu = 0.25$, what is the factor of safety?
(d) What is the factor of safety for the lower pair of bolts if they carry the entire shear load and none of the moment load?

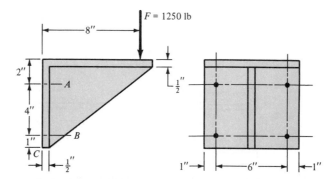

PROBLEM 8-37

8-38   A vertical channel 152×76 (see Table A-7) has a cantilever bolted to it as shown. The channel is hot-rolled AISI 1006 steel. The bar is of hot-rolled AISI 1015 steel. The bolts are M12 $\times$ 1.75 ISO 5.8. For a design factor of 2.8, find the safe force F that can be applied to the cantilever.

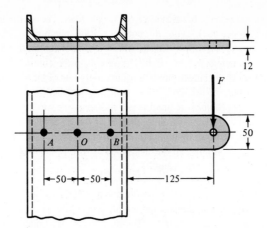

PROBLEM 8-38
Dimensions in millimeters.

8-39   Find the total shear load on each of the three bolts for the connection shown in the figure and compute the significant bolt shear stress and bearing stress. Find the second moment of area of the 8-mm plate on a section through the three bolt holes, and find the maximum bending stress in the plate.

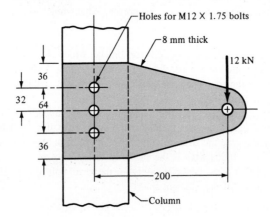

PROBLEM 8-39
Dimensions in millimeters.

8-40   A $\frac{3}{8}$- × 2-in AISI 1018 cold-drawn steel bar is cantilevered to support a static load of 300 lb as illustrated. The bar is secured to the support using two $\frac{1}{2}"$-13 UNC SAE 5 bolts. Find the factor of safety for the following modes of failure: shear of bolt, bearing on bolt, bearing on member, and strength of member.

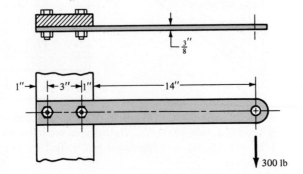

PROBLEM 8-40

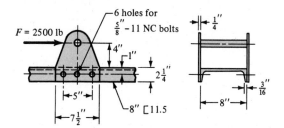

PROBLEM 8-41

8-41   The figure shows a welded fitting which has been tentatively designed to be bolted to a channel so as to transfer the 2500-lb load into the channel. The channel is made of hot-rolled low-carbon steel having a minimum yield strength of 46 kpsi; the two fitting plates are of hot-rolled stock having a minimum S_y of 45.5 kpsi. The fitting is to be bolted using six standard SAE grade 2 bolts. Check the strength of the design by computing the factor of safety for all possible modes of failure.

8-42   Tests with carefully torqued bolts lubricated with machine oil have shown the initial tension predicted by Eq. (8-20) to be approximately normally distributed with a coefficient of variation of 0.09. What fraction of proof load should be the target if a designer wishes to run a risk of 1 chance in 100 of tightening a bolt over the recommended proof load? Apply this result to a $\frac{5}{8}''$-11 UNC grade 5 bolt tightening torque.

8-43   Use the decision of Example 8-2 and estimate the probability of tightening any of the seven bolts too much during a maximum load of nP if the bolts are torqued to 60 percent of recommended value.

8-44   What should be the torque-wrench setting for the bolts of Example 8-2 if the chance of getting too much proof load is set at 1 in 100? The lubricant is machine oil.

ANSWERS   8-4 $e = 0.294$; 16.2 N · m, 6.62 N · m
8-6 (b) 0.579 hp
8-10 $F = 65.0$ kN, $e = 0.130$
8-13 $k_b = 3.94$ Mlb/in, $k_m = 12.1$ Mlb/in
8-17 (a) 494 lb; (b) 500 lb; (c) $F_{max} = 2M/RN$
8-21 4.86
8-24 (a) 4.54 kip; (b) 2.82, 2.61
8-26 3.19 based on proof strength; 4.71 based on joint separation; 2.17 based on fatigue
8-30 $F = 35.4$ kip
8-33 $n = 2.72$
8-35 $\sigma' = 289$ MPa; no
8-38 $F = 1.80$ kN

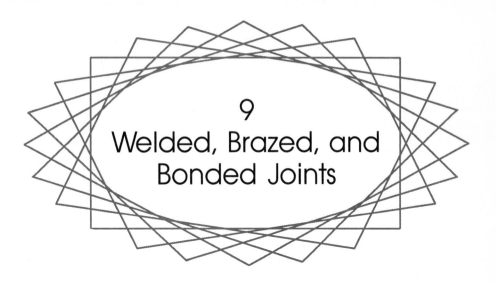

9
Welded, Brazed, and Bonded Joints

Processes such as welding, brazing, soldering, cementing, and gluing are used extensively in manufacturing today. Whenever parts have to be assembled or fabricated, there is usually good cause for considering one of these processes in preliminary design work. Especially when the sections to be joined are thin, one of these fastening methods may lead to significant savings. The elimination of individual fasteners and the adaptability of the method to rapid machine assembly are among the advantages.

9-1 WELDING SYMBOLS

A weldment is fabricated by welding together a collection of metal shapes, cut to particular configurations. During welding, the several parts are held securely together, often by clamping or jigging. The welds must be precisely specified on working drawings, and this is done using the welding symbol, shown in Fig. 9-1, as standardized by the American Welding Society (AWS). The arrow of this symbol points to the joint to be welded. The body of the symbol contains as many of the following elements as are deemed necessary:

- Reference line
- Arrow
- Basic weld symbols as in Fig. 9-2
- Dimensions and other data
- Supplementary symbols
- Finish symbols
- Tail
- Specification or process

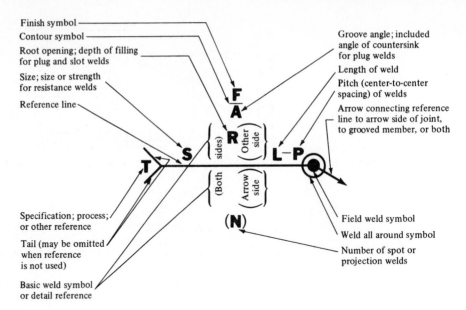

FIGURE 9-1

The AWS standard welding symbol showing the location of the symbol elements.

		Plug or slot	Type of weld				
				Groove			
Bead	Fillet		Square	V	Bevel	U	J

FIGURE 9-2

Arc- and gas-weld symbols.

The *arrow side* of a joint is the line, side, area, or near member to which the arrow points. The side opposite the arrow side is the *other side*.

Figures 9-3 to 9-6 illustrate the types of welds used most frequently by designers. For general machine elements most welds are fillet welds, though butt welds are used a great deal in designing pressure vessels. Of course, the parts to be joined must be arranged so that there is sufficient clearance for the welding operation. If unusual joints are required because of insufficient clearance or because of the section shape, the design may be a poor one and the designer should begin again and endeavor to synthesize another solution.

Since heat is used in the welding operation, there is a possibility of metallurgical

FIGURE 9-3

Fillet welds. (*a*) The number indicates the leg size; the arrow should point to only one weld when both sides are the same. (*b*) The symbol indicates that the welds are intermittent and staggered 60 mm long on 200-mm centers.

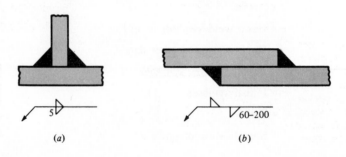

(*a*)   (*b*)

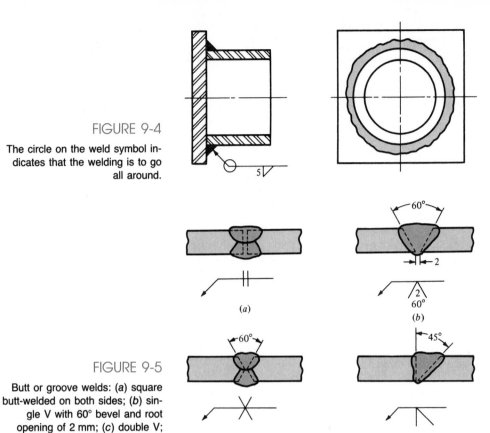

FIGURE 9-4

The circle on the weld symbol indicates that the welding is to go all around.

FIGURE 9-5

Butt or groove welds: (*a*) square butt-welded on both sides; (*b*) single V with 60° bevel and root opening of 2 mm; (*c*) double V; (*d*) single bevel.

changes in the parent metal in the vicinity of the weld. Also, residual stresses may be introduced because of clamping or holding or, sometimes, because of the order of welding. Usually these residual stresses are not severe enough to cause concern; in some cases a light heat treatment after welding has been found helpful in relieving

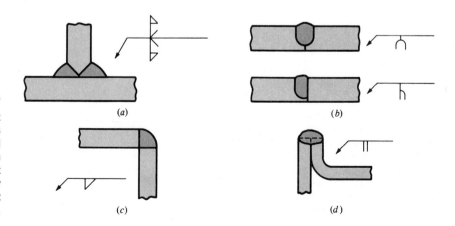

FIGURE 9-6

Special groove welds: (*a*) T joint for thick plates; (*b*) U and J welds for thick plates; (*c*) corner weld (may also have a bead weld on inside for greater strength but should not be used for heavy loads); (*d*) edge weld for sheet metal and light loads.

them. When the parts to be welded are thick, a preheating will also be of benefit. If the reliability of the component is to be quite high, a testing program should be established to learn what changes or additions to the operations are necessary to ensure the best quality.

9-2   BUTT AND FILLET WELDS

Figure 9-7 shows a single V-groove weld loaded by the tensile force F. For either tension or compression loading, the average normal stress is

$$\sigma = \frac{F}{hl} \tag{9-1}$$

where h is the weld throat and l is the length of the weld, as shown in the figure. Note that the value of h does not include the reinforcement. The reinforcement is desirable in order to compensate for flaws, but it varies somewhat and does produce stress concentration at point A in the figure. If fatigue loads exist, it is good practice to grind or machine off the reinforcement.

The average stress in a butt weld due to shear loading is

$$\tau = \frac{F}{hl} \tag{9-2}$$

Figure 9-8 illustrates a typical transverse fillet weld. Attempts to solve for the stress distribution in such welds, using the methods of theory of elasticity, have not been very successful. Conventional practice in welding engineering design has always been to base the size of the weld upon the magnitude of the stress on the throat area DB.

In Fig. 9-9a a portion of the weld has been selected from Fig. 9-8 so as to treat the weld throat as a problem in free-body analysis. The throat area is $A = hl \cos 45° = 0.707hl$, where l is the length of the weld. Thus the stress σ_x is

$$\sigma_x = \frac{F}{A} = \frac{F}{0.707hl} \tag{a}$$

This stress can be divided into two components, a shear stress τ and a normal stress σ. These are

$$\tau = \sigma_x \cos 45° = \frac{F}{hl} \qquad \sigma = \sigma_x \cos 45° = \frac{F}{hl} \tag{b}$$

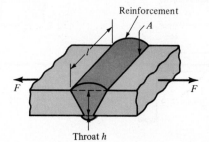

Reinforcement

FIGURE 9-7

A typical butt joint.

Throat h

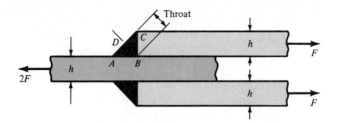

FIGURE 9-8

A transverse fillet weld.

In Fig. 9-9*b* these are entered into a Mohr's circle diagram. The largest principal stress is seen to be

$$\sigma_1 = \frac{F}{2hl} + \sqrt{\left(\frac{F}{2hl}\right)^2 + \left(\frac{F}{hl}\right)^2} = 1.618 \frac{F}{hl} \tag{c}$$

The maximum shear stress is

$$\tau_{max} = \sqrt{\left(\frac{F}{2hl}\right)^2 + \left(\frac{F}{hl}\right)^2} = 1.118 \frac{F}{hl} \tag{d}$$

However, for design purposes it is customary to base the shear stress on the throat area and to neglect the normal stress altogether. Thus the equation for *average stress* is

$$\tau = \frac{F}{0.707hl} = \frac{1.414F}{hl} \tag{9-3}$$

and is normally used in designing joints having fillet welds. Note that this gives a shear stress $1.414/1.118 = 1.26$ times greater than that given by Eq. (*d*).

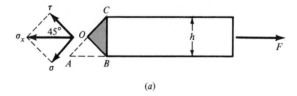

(*a*)

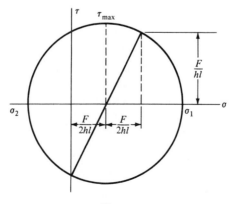

(*b*)

FIGURE 9-9

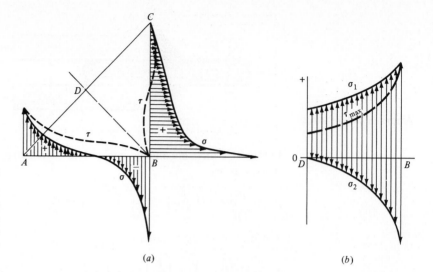

FIGURE 9-10

Stress distribution in fillet welds:
(a) Stress distribution on the legs
as reported by Norris; (b) distribution of principal stresses and maximum shear stress as reported by
Salakian.

(a)   (b)

There are some experimental and analytical results that are helpful in evaluating Eq.
(9-3). A model of the transverse fillet weld of Fig. 9-8 is easily constructed for
photoelastic purposes and has the advantage of a balanced loading condition. Norris
constructed such a model and reported the stress distribution along the sides AB and BC
of the weld.* An approximate graph of the results he obtained is shown as Fig. 9-10a.
Note that stress concentration exists at A and B on the horizontal leg and at B on the
vertical leg. Norris states that he could not determine the stresses at A and B with any
certainty.

Salakian† presents data for the stress distribution across the throat of a fillet weld
(Fig. 9-10b). This graph is of particular interest because we have just learned that it is
the throat stresses that are used in design. Again, the figure shows stress concentration
at point B. Note that Fig. 9-10a applies either to the weld metal or to the parent metal,
and that Fig. 9-10b applies only to the weld metal.

Suppose the double-filleted lap joint of Fig. 9-3b is loaded by tensile forces exerted
to the right and to the left. The throat area is $0.707hl$ for each weld. Since there are two
of them, the average stress is

$$\tau = \frac{F}{1.414hl} \tag{9-4}$$

In the case of tension-loaded parallel fillet welds, as in Fig. 9-11, it is probable that
the stress distribution along the length of the welds is not uniform. Still it is customary
to assume a uniform shear stress along the throat. Thus the average shear stress for Fig.
9-11 is also given by Eq. (9-4).

The results of this section are summarized in Table 9-1.

*C. H. Norris, "Photoelastic Investigation of Stress Distribution in Transverse Fillet Welds," *Welding J.,*
vol. 24, 1945, p. 557s.

†A. G. Salakian and G. E. Claussen, "Stress Distribution in Fillet Welds; A Review of the Literature,"
Welding J., vol. 16, May 1937, pp. 1–24.

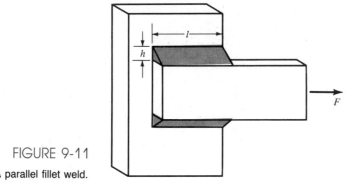

FIGURE 9-11

A parallel fillet weld.

TABLE 9-1

Transverse and Parallel Loading of Fillet Welds for Both Types of Loading. $\tau = P/(0.707h\ \Delta x)$*

KIND OF LOADING	INDUCED STRESS	STRESS MAGNITUDE		K_f
1. Transverse loading	Throat loading includes induced normal and shear stresses of equal magnitude.	Throat stresses as function of load: $\tau = P/(h\ \Delta x)$ $\sigma = P/(h\ \Delta x)$ $\tau_{max} = 1.118\ P/(h\ \Delta x)$ $= P/(0.894h\ \Delta x)$	Throat stresses as function of leg stresses: $\sigma = \tau = \tau_0 = \sigma_0$	$K_f = 1.5$ $K_f = 2$
2. Parallel loading	Throat loading includes shear stress only	Throat stress as function of load: $\tau = \sqrt{2}\ P/(h\ \Delta x)$ $= P/(0.707h\ \Delta x)$	Throat stress as function of leg stress: $\tau = \sqrt{2}\ \tau_0$	$K_f = 2.7$

*The approach to fillet-weld analysis and design is to provide for carrying the external loading as a pure shear stress on the throat area.

9-3   TORSION IN WELDED JOINTS

Figure 9-12 illustrates a cantilever of length l welded to a column by two fillet welds. The reaction at the support of a cantilever always consists of a shear force V and a

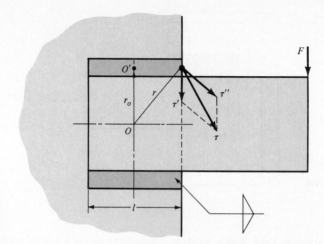

FIGURE 9-12

This is a *moment connection;* such a connection produces *torsion* in the welds.

moment M. The shear force produces a *primary shear* in the welds of magnitude

$$\tau' = \frac{V}{A} \tag{9-5}$$

where A is the throat area of all the welds.

The moment at the support produces *secondary shear* or *torsion* of the welds, and this stress is given by the equation

$$\tau'' = \frac{Mr}{J} \tag{9-6}$$

where r is the distance from the centroid of the weld group to the point in the weld of interest and J is the second polar moment of area of the weld group about the centroid of the group. When the sizes of the welds are known, these equations can be solved and the results combined to obtain the maximum shear stress. Note that r is usually the farthest distance from the centroid of the weld group.

Figure 9-13 shows two welds in a group. The rectangles represent the throat areas of the welds. Weld 1 has a throat width $b_1 = 0.707h_1$; and weld 2 has a throat width $d_2 = 0.707h_2$. Note that h_1 and h_2 are the respective weld sizes. The throat area of both welds together is

$$A = A_1 + A_2 = b_1d_1 + b_2d_2 \tag{a}$$

This is the area that is to be used in Eq. (9-5).

The x axis in Fig. 9-13 passes through the centroid G_1 of weld 1. The second moment of area about this axis is

$$I_x = \frac{b_1d_1^3}{12}$$

Similarly, the second moment of area about an axis through G_1 parallel to the y axis is

$$I_y = \frac{d_1b_1^3}{12}$$

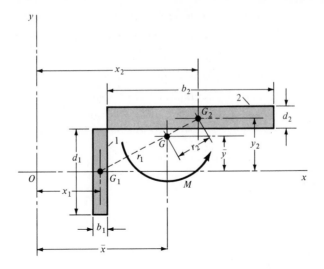

FIGURE 9-13

Thus the second polar moment of area of weld 1 about its own centroid is

$$J_{G1} = I_x + I_y = \frac{b_1 d_1^3}{12} + \frac{d_1 b_1^3}{12} \qquad (b)$$

In a similar manner, the second polar moment of area of weld 2 about its centroid is

$$J_{G2} = \frac{b_2 d_2^3}{12} + \frac{d_2 b_2^3}{12} \qquad (c)$$

The centroid G of the weld group is located at

$$\bar{x} = \frac{A_1 x_1 + A_2 x_2}{A} \qquad \bar{y} = \frac{A_1 y_1 + A_2 y_2}{A}$$

Using Fig. 9-13 again, we see that the distances r_1 and r_2 from G_1 and G_2 to G, respectively, are

$$r = [(\bar{x} - x_1)^2 + \bar{y}^2]^{1/2} \qquad r_2 = [(y_2 - \bar{y})^2 + (x_2 - \bar{x})^2]^{1/2}$$

Now, using the parallel-axis theorem, we find the second polar moment of area of the weld group to be

$$J = (J_{G1} + A_1 r_1^2) + (J_{G2} + A_2 r_2^2) \qquad (d)$$

This is the quantity to be used in Eq. (9-6). The distance r must be measured from G and the moment M computed about G.

The reverse procedure is that in which the allowable shear stress is given and we wish to find the weld size. The usual procedure is to estimate a probable weld size and then to use iteration.

Observe in Eq. (b) that the second term contains the quantity b_1^3, which is the cube of the weld width, and that the quantity d_2^3 in the first term of Eq. (c) is also the cube of the weld width. Both of these quantities can be set equal to unity. This leads to the idea of treating each fillet weld as a line. The resulting second moment of area is then a *unit second polar moment of area.* The advantage of treating the weld size as a line is that the value of J_u is the same regardless of the weld size. Since the throat width of a fillet weld is $0.707h$, the relationship between J and the unit value is

$$J = 0.707hJ_u \qquad (9\text{-}7)$$

in which J_u is found by conventional methods for an area having unit width. The transfer formula for J_u must be employed when the welds occur in groups, as in Fig. 9-12. Table 9-2 lists the throat areas and the unit second polar moments of area for the

TABLE 9-2

Torsional Properties of Fillet Welds*

WELD	THROAT AREA	LOCATION OF G	UNIT SECOND POLAR MOMENT OF AREA
	$A = 0.707hd$	$\bar{x} = 0$ $\bar{y} = d/2$	$J_u = d^3/12$
	$A = 1.414hd$	$\bar{x} = b/2$ $\bar{y} = d/2$	$J_u = \dfrac{d(3b^2 + d^2)}{6}$
	$A = 0.707h(2b + d)$	$\bar{x} = \dfrac{b^2}{2(b + d)}$ $\bar{y} = \dfrac{d^2}{2(b + d)}$	$J_u = \dfrac{(b + d)^4 - 6b^2 d^2}{12(b + d)}$
	$A = 0.707h(2b + d)$	$\bar{x} = \dfrac{b^2}{2b + d}$ $\bar{y} = d/2$	$J_u = \dfrac{8b^3 + 6bd^2 + d^3}{12} - \dfrac{b^4}{2b + d}$
	$A = 1.414h(b + d)$	$\bar{x} = b/2$ $\bar{y} = d/2$	$J_u = \dfrac{(b + d)^3}{6}$
	$A = 1.414\pi hr$		$J_u = 2\pi r^3$

*G is centroid of weld group; h is weld size; plane of torque couple is in the plane of the paper; all welds are of unit width.

most common fillet welds encountered. The example that follows is typical of the calculations normally made.

EXAMPLE 9-1   A 50-kN load is transferred from a welded fitting into a 200-mm steel channel as illustrated in Fig. 9-14. Compute the maximum stress in the weld.

Solution*   (*a*) Label the ends and corners of each weld by letter. Sometimes it is desirable to label each weld of a set by number. See Fig. 9-15.

(*b*) Compute the primary shear stress τ'. As shown in Fig. 9-14, each plate is welded to the channel by means of three 6-mm fillet welds. Figure 9-15 shows that we have divided the load in half and are considering only a single plate. From case 4 of Table 9-2 we find the throat area as

$$A = 0.707(6)\ [2(56) + 190] = 1280 \text{ mm}^2$$

Then the primary shear stress is

$$\tau' = \frac{V}{A} = \frac{25(10)^3}{1280} = 19.5 \text{ MPa}$$

(*c*) Draw the τ' stress, to scale, at each lettered corner or end. See Fig. 9-16.

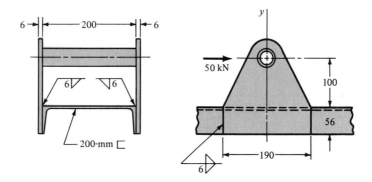

FIGURE 9-14

Dimensions in millimeters.

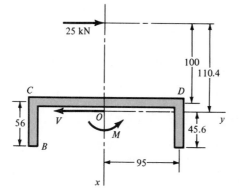

FIGURE 9-15

Diagram showing the weld geometry; all dimensions in millimeters. Note that *V* and *M* represent loads applied by the welds *to the plate.*

---

*We are indebted to Professor George Piotrowski of the University of Florida for the detailed steps, presented here, of his method of weld analysis. J.E.S., C.R.M.

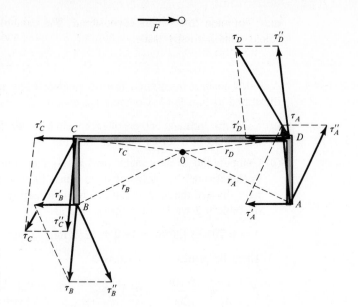

FIGURE 9-16

Free-body diagram of one of the side plates.

(*d*) Locate the centroid of the weld pattern. Using case 4 of Table 9-2, we find

$$\bar{x} = \frac{(56)^2}{2(56) + 190} = 10.4 \text{ mm}$$

This is shown as point O on Figs. 9-15 and 9-16.

(*e*) Find the distances r_i (see Fig. 9-16):

$$r_A = r_B = [(190/2)^2 + (56 - 10.4)^2]^{1/2} = 105 \text{ mm}$$
$$r_C = r_D = [(190/2)^2 + (10.4)^2]^{1/2} = 95.6 \text{ mm}$$

These distances can also be scaled from the drawing.

(*f*) Find J. Using case 4 of Table 9-2 again, we get

$$J = 0.707(6)\left[\frac{8(56)^3 + 6(56)(190)^2 + (190)^3}{12} - \frac{(56)^4}{2(56) + 190}\right]$$

$$= 7.07(10)^6 \text{ mm}^4$$

(*g*) Find M:

$$M = Fl = 25(100 + 10.4) = 2760 \text{ N} \cdot \text{m}$$

(*h*) Compute the secondary shear stresses τ'' at each lettered end or corner:

$$\tau_A'' = \tau_B'' = \frac{Mr}{J} = \frac{2760(10)^3(95.6)}{7.07(10)^6} = 37.3 \text{ MPa}$$

$$\tau_C'' = \tau_D'' = \frac{2760(10)^3(105)}{7.07(10)^6} = 41.0 \text{ MPa}$$

(i) Draw the τ' stress, to scale, at each corner and end. See Fig. 9-16. Note that this is a free-body diagram of one of the side plates, and therefore the τ' and τ'' stresses represent what the channel is doing to the plate (through the welds) to hold the plate in equilibrium.

(j) At each letter, combine the two stress components as vectors. This gives

$$\tau_A = \tau_B = 37 \text{ MPa}$$

$$\tau_C = \tau_D = 44 \text{ MPa}$$

(k) Identify the most highly stressed point:

Answer $\tau_{max} = \tau_C = \tau_D = 44 \text{ MPa}$

9-4 BENDING IN WELDED JOINTS

Figure 9-17a shows a cantilever welded to a support by fillet welds at top and bottom. A free-body diagram of the beam would show a shear-force reaction V and a moment reaction M. The shear force produces a primary shear in the welds of magnitude

$$\tau' = \frac{V}{A} \tag{a}$$

where A is the total throat area.

The moment M produces a normal bending stress σ in the welds. Though not rigorous, it is customary in the stress analysis of welds to assume that this stress acts normal to the throat area.* By treating the two welds in Fig. 9-17b as lines, we find the unit second moment of area to be

$$I_u = \frac{bd^2}{2} \tag{b}$$

Then the second moment of area based on the weld throat is

$$I = 0.707h \, \frac{bd^2}{2} \tag{c}$$

---

*In transverse loading of fillet welds, the load is resisted at the throat by a combination of shear and normal stresses of equal magnitude. Equation (d) of Sec. 9-2 shows that the shear stress alone approximates the maximum shear stress. In resisting bending by a transverse fillet weld, the same situation prevails, and Eq. (d) of this section predicts the throat shear stress as well as the normal stress.

FIGURE 9-17

A rectangular cantilever welded to a support at the top and bottom edges.

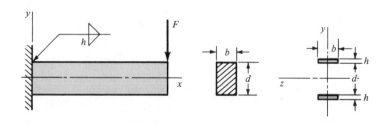

The normal stress is now found to be

$$\tau = \sigma = \frac{Mc}{I} = \frac{M(d/2)}{0.707bd^2h/2} = \frac{1.414M}{bdh} \tag{d}$$

The second moment of area in Eq. (d) is based on the distance d between the two welds. If this moment is found by treating the two welds as rectangles instead, the distance between the weld centroids would be $(d + h)$. This would produce a slightly larger moment and result in a smaller value of the stress σ. Thus the method of treating welds as lines produces more conservative results. Perhaps the added safety is appropriate in view of the stress distributions of Fig. 9-10.

Once the stress components σ and τ have been found for welds subjected to bending, they may be combined by using a Mohr's circle diagram to find the principal stresses or the maximum shear stress. Then an appropriate failure theory is applied to determine the likelihood of failure or safety.

Table 9-3 lists the bending properties most likely to be encountered in the analysis of welded beams.

TABLE 9-3

Bending Properties of Fillet Welds*

WELD	THROAT AREA	LOCATION OF G	UNIT SECOND MOMENT OF AREA
	$A = 0.707hd$	$\bar{x} = 0$ $\bar{y} = d/2$	$I_u = \dfrac{d^3}{12}$
	$A = 1.414hd$	$\bar{x} = b/2$ $\bar{y} = d/2$	$I_u = \dfrac{d^3}{6}$
	$A = 1.414hb$	$\bar{x} = b/2$ $\bar{y} = d/2$	$I_u = \dfrac{bd^2}{2}$
	$A = 0.707h(2b + d)$	$\bar{x} = \dfrac{b^2}{2b + d}$ $\bar{y} = d/2$	$I_u = \dfrac{d^2}{12}(6b + d)$

TABLE 9-3

(Continued)

WELD	THROAT AREA	LOCATION OF G	UNIT SECOND MOMENT OF AREA
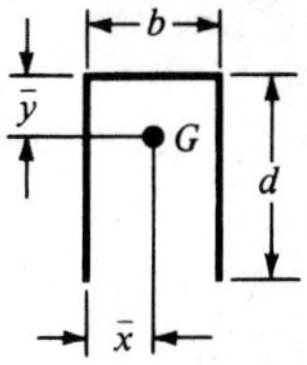	$A = 0.707h(b + 2d)$	$\bar{x} = b/2$ $\bar{y} = \dfrac{d^2}{b + 2d}$	$I_u = \dfrac{2d^3}{3} - 2d^2\bar{y} + (b + 2d)\bar{y}^2$
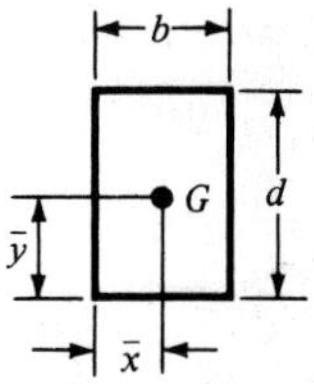	$A = 1.414h(b + d)$	$\bar{x} = b/2$ $\bar{y} = d/2$	$I_u = \dfrac{d^2}{6}(3b + d)$
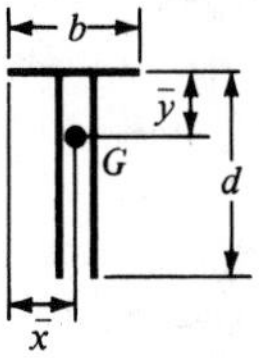	$A = 0.707h(b + 2d)$	$\bar{x} = b/2$ $\bar{y} = \dfrac{d^2}{b + 2d}$	$I_u = \dfrac{2d^3}{3} - 2d^2\bar{y} + (b + 2d)\bar{y}^2$
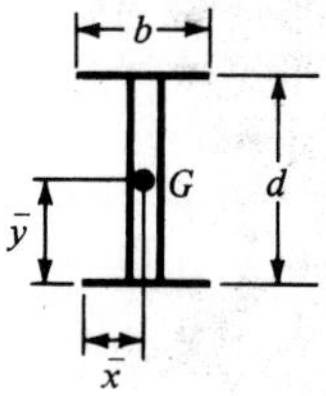	$A = 1.414h(b + d)$	$\bar{x} = b/2$ $\bar{y} = d/2$	$I_u = \dfrac{d^2}{6}(3b + d)$
	$A = 1.414\pi hr$		$I_u = \pi r^3$

*I_u, unit second moment of area, is taken about a horizontal axis through G, the centroid of the weld group; h is weld size; the plane of the bending couple is normal to the plane of the paper and parallel to the y axis; all welds are of the same size.

9-5   THE STRENGTH OF WELDED JOINTS

The matching of the electrode properties with those of the parent metal is usually not so important as speed, operator appeal, and the appearance of the completed joint. The properties of electrodes vary considerably, but Table 9-4 lists the minimum properties for some electrode classes.

TABLE 9-4
Minimum Weld-Metal
Properties

AWS ELECTRODE NUMBER*	TENSILE STRENGTH, kpsi (MPa)	YIELD STRENGTH, kpsi (MPa)	PERCENT ELONGATION
E60xx	62 (427)	50 (345)	17–25
E70xx	70 (482)	57 (393)	22
E80xx	80 (551)	67 (462)	19
E90xx	90 (620)	77 (531)	14–17
E100xx	100 (689)	87 (600)	13–16
E120xx	120 (827)	107 (737)	14

*The American Welding Society (AWS) specification code numbering system for electrodes. This system uses an E prefixed to a four- or five-digit numbering system in which the first two or three digits designate the approximate tensile strength. The last digit indicates variables in the welding technique, such as current supply. The next-to-last digit indicates the welding position, as, for example, flat, or vertical, or overhead. The complete set of specifications may be obtained from the AWS upon request.

It is preferable, in designing welded components, to select a steel that will result in a fast, economical weld even though this may require a sacrifice of other qualities such as machinability. Under the proper conditions, all steels can be welded, but best results will be obtained if steels having a UNS specification between G10140 and G10230 are chosen. All these steels have a tensile strength in the hot-rolled condition in the range of 60 to 70 kpsi.

The designer can choose factors of safety or permissible working stresses with more confidence if he or she is aware of the values of those used by others. One of the best standards to use is the American Institute of Steel Construction (AISC) code for building construction.* The permissible stresses are now based on the yield strength of the material instead of the ultimate strength, and the code permits the use of a variety of ASTM structural steels having yield strengths varying from 33 to 50 kpsi. Provided the loading is the same, the code permits the same stress in the weld metal as in the parent metal. For these ASTM steels, $S_y = 0.5S_u$. Table 9-5 lists the formulas specified by the code for calculating these permissible stresses for various loading conditions. The factors of safety implied by this code are easily calculated. For tension, $n = 1/0.60 = 1.67$. For shear, $n = 0.577/0.40 = 1.44$, if we accept the distortion-energy theory as the criterion of failure.

It is important to observe that the electrode material is often the strongest material present. If a bar of AISI 1010 steel is welded to one of 1018 steel, the weld metal is actually a mixture of the electrode material and the 1010 and 1018 steels. Furthermore,

*For a copy, write the AISC, 400 N. Michigan Ave., Chicago, IL 60611.

TABLE 9-5
Stresses Permitted by the
AISC Code for Weld Metal

TYPE OF LOADING	TYPE OF WELD	PERMISSIBLE STRESS	n^*
Tension	Butt	$0.60S_y$	1.67
Bearing	Butt	$0.90S_y$	1.11
Bending	Butt	0.60–$0.66S_y$	1.52–1.67
Simple compression	Butt	$0.60S_y$	1.67
Shear	Butt or fillet	$0.40S_y$	1.44

*The factor of safety n has been computed using the distortion-energy theory.

TABLE 9-6

Fatigue-Strength Reduction
Factors

TYPE OF WELD	K_f
Reinforced butt weld	1.2
Toe of transverse fillet weld	1.5
End of parallel fillet weld	2.7
T-butt joint with sharp corners	2.0

a welded cold-drawn bar has its cold-drawn properties replaced with the hot-rolled properties in the vicinity of the weld. Finally, remembering that the weld metal is usually the strongest, do check the stresses in the parent metals.

The AISC code, as well as the AWS code, for bridges includes permissible stresses when fatigue loading is present. The designer will have no difficulty in using these codes, but their empirical nature tends to obscure the fact that they have been established by means of the same knowledge of fatigue failure already discussed in Chap. 7. Of course, for structures covered by these codes, the actual stresses *cannot* exceed the permissible stresses; otherwise the designer is legally liable. But in general, codes tend to conceal the actual margin of safety involved.

The fatigue-strength reduction factors listed in Table 9-6 are suggested for use. These factors should be used for the parent metal as well as for the weld metal.

EXAMPLE 9-2   Brackets, such as the one of Fig. 9-18, are used in mooring small watercraft. Failure of such brackets is usually caused by the bearing pressure of the mooring-line clip against the side of the hole. To get an idea of the static margins of safety involved, we use a bracket $\frac{1}{4}$ in thick and made of hot-rolled AISI 1018 steel. We then assume that wave action on the boat will create no greater force F than 1200 lb. Under these conditions, determine the factors of safety guarding against a static failure.

Solution   Figure 9-19 shows a free-body diagram of the bracket with the external force F acting through the center of the hole. The centroid of the weld group and of the bottom of the bracket is G. The force F_G is the force of the weld group acting on the bracket. Since F_G and F have different lines of action, we also have a moment M. The dimensions shown are obtained from the trigonometry of the diagram.

Note that the forces and moment of the bracket acting on the welds are equal and opposite to those shown. Thus,

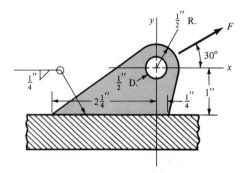

FIGURE 9-18

Welded mooring bracket.

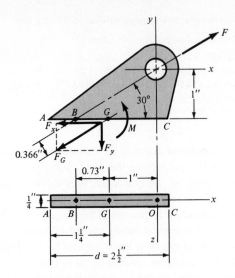

FIGURE 9-19

- The moment M produces a bending stress in the welds with tension at A and compression at C.
- The force component F_y produces tension throughout the weld.
- The force component F_x produces shear throughout the weld.

These effects are

$$M = 1200(0.366) = 439 \text{ lb} \cdot \text{in} \qquad F_x = 1200 \cos 30° = 1039 \text{ lb}$$
$$F_y = 1200 \sin 30° = 600 \text{ lb}$$

From Table 9-3, we find

$$A = 1.414h(b + d) = 1.414(0.25)(0.25 + 2.5) = 0.972 \text{ in}^2$$
$$I_u = \frac{d^2}{6}(3b + d) = \frac{(2.5)^2}{6}[3(0.25) + 2.5] = 3.39 \text{ in}^4$$

Then the second moment of area about an axis through G parallel to z is

$$I = 0.707hI_u = 0.707(0.25)(3.39) = 0.599 \text{ in}^4$$

At end A, the shear stress due to the moment and that due to F_y are additive. For the weld metal, the total shear stress is

$$\tau_1 = \frac{F_y}{A} + \frac{Mc}{I} = \frac{600}{0.972} + \frac{439(1.25)}{0.599} = 1530 \text{ psi} \tag{1}$$

The shear stress due to F_x is

$$\tau_2 = \frac{F_x}{A} = \frac{1039}{0.972} = 1069 \text{ psi} \tag{2}$$

From Table A-20, we find the minimum mechanical properties for the bracket material to be $S_u = 58$ kpsi and $S_y = 32$ kpsi. We shall use these same properties for the weld metal even though stronger electrodes could be used. The stresses τ_1 and τ_2 are perpendicular to each other, and their resultant is

$$\tau = (\tau_1^2 + \tau_2^2)^{1/2} = [(1530)^2 + (1069)^2]^{1/2} = 1866 \text{ psi}$$

The factor of safety guarding against static yielding in the weldment is

Answer $n = \dfrac{S_{sy}}{\tau} = \dfrac{0.577(32)(10^3)}{1866} = 9.89$

Next, we proceed to calculate the stresses in the parent metal. The area subject to shear is

$$A = bd = 0.25(2.5) = 0.625 \text{ in}^2$$

Thus the shear stress in the parent metal is

$$\tau_{yx} = \frac{F_x}{A} = \frac{1039}{0.625} = 1662 \text{ psi}$$

The section modulus of the bracket at the weld interface is

$$\frac{I}{c} = \frac{bd^2}{6} = \frac{0.25(2.5)^2}{6} = 0.260 \text{ in}^3$$

Thus the tensile stress at A in the parent metal is

$$\sigma_y = \frac{F_y}{A} + \frac{M}{I/c} = \frac{600}{0.625} + \frac{439}{0.260} = 2650 \text{ psi}$$

Using Eq. (6-18), we find

$$\sigma' = [\sigma_y^2 + 3\tau_{yx}^2]^{1/2} = [(2650)^2 + 3(1662)^2]^{1/2} = 3910 \text{ psi}$$

Then the factor of safety guarding against a static failure in the parent metal, at the weld, is

Answer $n = \dfrac{S_y}{\sigma'} = \dfrac{32(10^3)}{3910} = 8.18$

The clip on the mooring line bears against the side of the $\frac{1}{2}$-in hole. We have no idea of the size of this clip. But if we assume that the clip fills the hole, then the average bearing stress is

$$\sigma = \frac{F}{td} = \frac{-1200}{0.25(0.50)} = -9600 \text{ psi}$$

Thus the factor of safety is

Answer $n = \dfrac{S_y}{|\sigma|} = \dfrac{32(10^3)}{9600} = 3.33$

EXAMPLE 9-3   Find the factors of safety against a fatigue failure in the weld metal of the mooring bracket of the preceding example.

Solution   From Eq. (7-4), we find the rotating-beam endurance limit to be

$$S'_e = 0.504S_{ut} = 0.504(58) = 29.2 \text{ kpsi}$$

An as-forged surface should always be used for weldments unless a superior finish is specified and obtained. From Eq. (7-14) we obtain

$$k_a = aS_{ut}^b = 39.9(58)^{-0.995} = 0.702$$

For the size factor, we first employ Eq. (7-19) to get an equivalent size. This gives

$$d_e = 0.808(0.707hb)^{1/2} = 0.808[0.707(2.5)(0.25)]^{1/2} = 0.537 \text{ in}$$

Equation (7-15) is used to obtain the size factor. The result is

$$k_b = \left(\frac{d_e}{0.3}\right)^{-0.1133} = \left(\frac{0.537}{0.3}\right)^{-0.1133} = 0.936$$

For torsion, the load factor is $k_c = 0.577$; and $k_d = 1$. Also,

$$k_e = \frac{1}{K_f} = \frac{1}{2.70} = 0.370$$

Thus the fully corrected endurance limit of the weld metal in shear is

$$S_{se} = k_a k_b k_c k_e S'_e = 0.702(0.936)(0.577)(0.370)(29.2) = 4.10 \text{ kpsi}$$

The shear stresses in the weld metal are repeated, and so

$$\tau_a = \frac{1866}{2} = 933 \text{ psi} \qquad \tau_m = 933 \text{ psi}$$

Here we shall select the Goodman theory of failure. From Eq. (7-43), $S_{su} = 0.67S_{ut} = 0.67(58) = 38.9$ kpsi. For shear, the Goodman line is expressed as

$$n = \frac{1}{\dfrac{\tau_a}{S_{se}} + \dfrac{\tau_m}{S_{su}}} = \frac{S_{se}S_{su}}{\tau_a S_{su} + \tau_m S_{se}}$$

Thus, based on fatigue of the weld metal, the factor of safety is

Answer $$n = \frac{4100[38.9(10^3)]}{933[38.9(10^3)] + 933(4100)} = 3.98$$

It will be necessary to use the distortion-energy theory to analyze the parent metal. The procedure outlined in Sec. 7-15 is recommended.

9-6   RESISTANCE WELDING

The heating and consequent welding that occur when an electric current is passed through several parts which are pressed together is called resistance welding. *Spot welding* and *seam welding* are the forms of resistance welding most often used. The

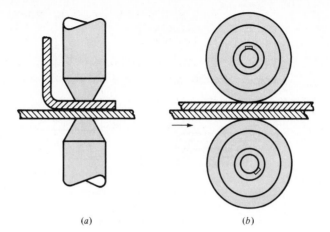

FIGURE 9-20

(a) Spot welding; (b) seam welding.

(a)   (b)

advantages of resistance welding over other forms are the speed, the accurate regulation of time and heat, the uniformity of the weld and the mechanical properties which result, the elimination of filler rods or fluxes, and the fact that the process is easy to automate.

The spot- and seam-welding processes are illustrated schematically in Fig. 9-20. Seam welding is actually a series of overlapping spot welds, since the current is applied in pulses as the work moves between the rotating electrodes.

Failure of a resistance weld is either by shearing of the weld or by tearing of the metal around the weld. Because of the possibility of tearing, it is good practice to avoid loading a resistance-welded joint in tension. Thus, for the most part, design so that the spot or seam is loaded in pure shear. The shear stress is then simply the load divided by the area of the spot. Because the thinner sheet of the pair being welded may tear, the strength of spot welds is often specified by stating the load per spot based on the thickness of the thinnest sheet. Such strengths are best obtained by experiment.

Somewhat larger factors of safety should be used when parts are fastened by spot welding rather than by bolts or rivets, to account for the metallurgical changes in the materials due to the welding.

9-7   BONDED JOINTS

When two parts or materials are connected together by a third material unlike the base materials, the process is called *bonding*. Thus *brazing, soldering,* and *cementing,* or the use of adhesives, are all means of bonding parts together. Figure 9-21 shows some examples of bonded joints that represent good joint design.

Brazing

Parts that are joined by heating them to more than 800°F to allow the filler metal to flow into the clearance space by capillary action are said to be brazed. Parts, like that in Fig. 9-21b, which are self-jigging can be furnace-brazed. Another method of brazing is

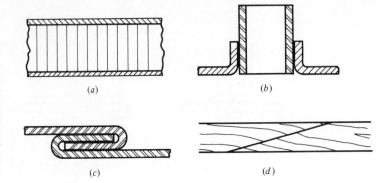

FIGURE 9-21

(a) Airplane wing section fabricated by bonding aluminum honeycomb to the skins, using resin bonding under heat and pressure; (b) tubing joined to sheet-metal section by brazing metal; (c) sheet-metal parts joined by soldering; (d) wood parts joined by gluing.

by torch, and this may be either automatic or manual. Some of the advantages offered by brazing are:

1 The ability to join materials of different thicknesses

2 The lack of radical change in the mechanical properties of the base materials after they are joined

3 The ability to join both cast and rolled materials as well as unlike materials

4 The fact that the completed joint requires no additional finishing procedures

A well-designed brazed joint is one that permits the base materials rather than the filler material to carry the load. Thus, if a stress analysis appears to be needed, it is good practice to review the joint design to see if a more effective geometry can be obtained.

Soldering

A soldered joint is much like a brazed joint, except that the filler metal is softer and the process is carried out at temperatures less than 800°F. Soldering is used to join sheet metals, such as tin cans and duct work, and for numerous electrical applications. Connections to be joined by soldering should always be designed so that the base materials carry the entire external load.

Cementing

In many instances, parts that must be connected together can be joined using adhesives, creating a significant cost advantage over the use of screws, rivets, or other mechanical fasteners. The stresses in an adhesive joint are also much more uniform than in, say, a riveted joint, where the load is shared by each rivet. Adhesive joints should be carefully designed so as to carry only compression or shear. Figure 9-22 shows that even a shear-loaded joint will have stress concentration at the ends.

Though adhesives have been around for a long time, the development of ''superglues'' has added significantly to their applications. For example, adhesive retaining compounds can now be used to assemble cylindrical parts formerly requiring press or

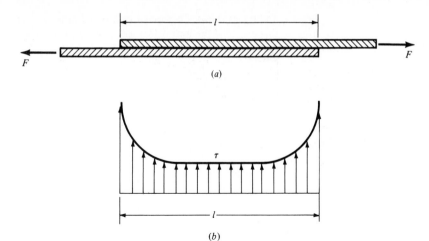

FIGURE 9-22

(a) Simple lap joint, adhesively bonded, and loaded in shear. (b) Approximate stress distribution showing stress concentration at the ends.

shrink fits; adhesives in such cases eliminate press-fit stresses and reduce machining costs.

Adhesives remain inert until a cure signal is given. Some of the most common curing methods are:

- *Anaerobic curing.* Here curing takes place when the adhesive is deprived of oxygen. Time required varies from $\frac{1}{2}$ to 24 h.

- *Heating.* Curing time varies from a few minutes to hours, depending upon the assembly size.

- *Interaction.* Here the cure begins when the components are first mixed. Longer curing times are generally required.

- *Solvent evaporation.* Used mostly for nonrigid materials; insulation to metal, for example.

- *Surface activation.* Adhesives are compounded to react to certain metals or coated surfaces.

- *Ultraviolet light.* Materials sensitive to ultraviolet light can be added to adhesives to cause nearly an instant cure when the adhesives are exposed to it.

Some of the advantages of adhesive joints are:

- They involve low cost because fewer parts and fewer steps are required for assembly.

- Unlike materials can be joined, since adhesives are flexible enough to absorb stresses due to differing coefficients of expansion.

- Joints are sealed against pressure or entry of moisture or other unwanted elements that might cause deterioration.

- Adhesives do absorb shock and vibration.

- Adhesives fill gaps and require less than precision tolerances in the assembly of parts.

PROBLEMS

9-1 The figure shows a horizontal steel bar, $\frac{3}{8}$ in thick, loaded in tension and welded to a vertical support. Find the load F that will cause a shear stress in the welds of 20 kpsi.

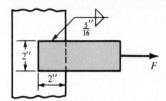

PROBLEM 9-1

9-2 Five of the weldments shown in the figure for Prob. 9-1 were tested for weld strength. The results were 17.1, 20.4, 21.2, 19.5, and 20.1 kip. Find the mean of the population with 90 percent confidence.

9-3 Twenty weldments of the parts shown in the figure for Prob. 9-1 have been made, and it is known that the mean load F required to stress the welds to 20 kpsi is 18.0 kip, with a standard deviation of 1.6 kip. Estimate the value of the least load F for the group.

9-4 A $\frac{5}{16}$-in steel bar is welded to a vertical support as shown. What is the shear stress in the welds if the force is $F = 32$ kip?

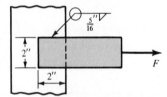

PROBLEM 9-4

9-5 A $\frac{3}{8}$-in-thick steel bar, to be used as a beam, is welded to a vertical support using two fillet welds as illustrated. Find the safe bending force F if the permissible shear stress in the welds is 20 kpsi.

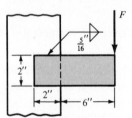

PROBLEM 9-5

9-6 The figure shows a weldment just like that of Prob. 9-5 except that there are four fillet welds instead of two. Show that this weldment is twice as strong as that of Prob. 9-5.

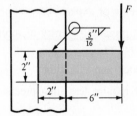

PROBLEM 9-6

9-7 The weldment shown in the figure is subjected to an alternating force F. The hot-rolled steel bar is 10 mm thick and is of AISI 1010 steel. Find the fatigue load F the bar will carry if three 6-mm fillet welds are used, as shown. Use a design factor of 3.

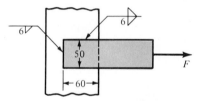

PROBLEM 9-7

9-8 A repeated force F acts on the welded tension member as illustrated. The member is 10-mm-thick AISI 1010 hot-rolled steel, and is welded to the support using two 6-mm parallel fillet welds. If the corrected endurance limit of the bar and welds is 52 MPa, estimate the safe force F based on a design factor of 2.8.

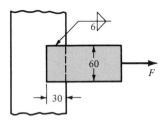

PROBLEM 9-8

9-9 The beam shown in the figure is a bar of AISI 1010 hot-rolled steel, 10 mm thick, and is welded to a support using three 6-mm fillet welds. The beam is loaded by a completely reversed force $F_a = 2$ kN. Estimate the factors of safety.

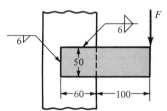

PROBLEM 9-9

9-10 The permissible shear stress for the weldment illustrated is 140 MPa. Estimate the bending load F that will cause this stress.

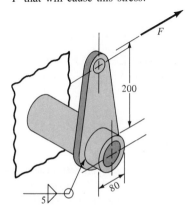

PROBLEM 9-10

9-11   A torque $T = 20(10^3)$ lb · in is applied to the weldment shown. Find the maximum shear stress in the welds.

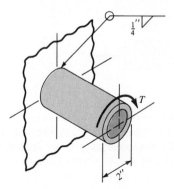

PROBLEM 9-11

9-12   Find the maximum shear stress in the weld metal of the bracket shown in the figure.

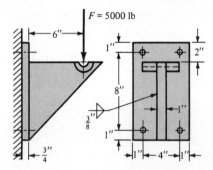

PROBLEM 9-12

9-13   The figure shows a welded steel bracket loaded by the static force F. What factor of safety results if the allowable shear stress in the weld metal is 120 MPa?

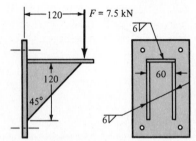

PROBLEM 9-13

9-14   The figure shows a formed sheet-steel bracket. Instead of securing it to the support with machine screws, welding has been proposed. If the combined stress in the weld metal is limited to 900 psi, what total load W will the bracket support?

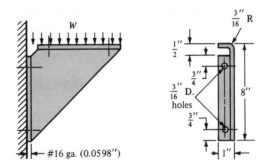

PROBLEM 9-14   #16 ga. (0.0598″)

9-15   Without bracing, a machinist can exert only about 100 lb on a wrench or tool handle. The lever shown in the figure has $t = \frac{1}{2}$ in and $w = 2$ in. We wish to specify the fillet weld size to secure the lever to the tubular part at A. Both parts are of steel, and the shear stress in the weld should not exceed 3000 psi. Find a safe weld size.

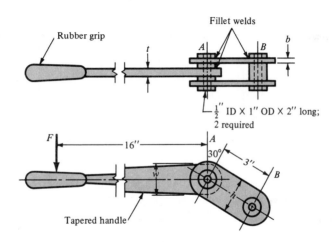

PROBLEM 9-15

9-16*   A fuel tanker has an elliptic tank of length 6 m. The cross section is shown in the figure and defined by the equation

$$4x^2 + 16y^2 = 9$$

The tank is mounted to the flat deck of the trailer at five stations, A through E. Each joint is welded to the tank and bolted to the trailer upper deck as shown. The free surface of the fuel is at a height of 1 m from the tank bottom. The specific gravity of the fuel is 0.9. Based on a

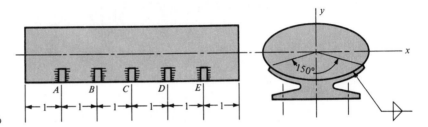

PROBLEM 9-16

*We are grateful to Dr. Ahmed F. Abdel Azim of Zagazig University, Cairo, for Probs. 9-16 to 9-18.

permissible shear stress of 140 MPa, determine the weld thickness at stations *A* and *B* for the following two cases:

(*a*) The trailer is decelerated from 60 km/h to a complete stop within 200 m.

(*b*) The trailer is driven in a turn of radius 200 m at a constant speed of 60 km/h.

9-17   A 10-in 15.3-lb structural-steel channel, 10 ft long, as shown in the figure, is used as a truss beam. As a result of solar energy during the daylight hours, the beam expands through an incremental elongation of 0.267 in. At night it shrinks back to its original length. If such a member is welded at both ends to other members so that no elongation is allowed, plot the force-time curve for the weld. Determine the weld size needed for 10^5 cycles of life using an E100xx electrode, a design factor of 1.8, and a stress-concentration factor of 1.5.

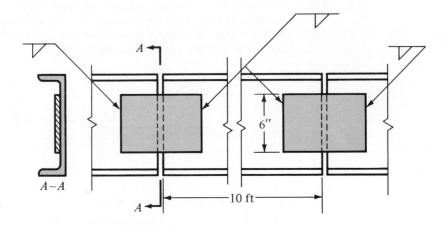

PROBLEM 9-17

9-18   Find the safe static load *F* for the weldment shown in the figure if an E6010 electrode is used and the factor of safety is to be 2 using the maximum-shear-stress theory.

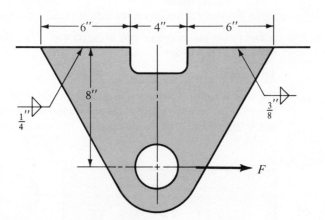

PROBLEM 9-18

9-19   The bar depicted in the figure is made of AISI 1018 hot-rolled steel and welded to a frame of the same material using E6010 electrodes. Estimate the safe amount of force that can be repeatedly applied using a design factor of 2.

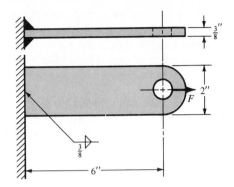

PROBLEM 9-19

9-20   A bar as shown in the figure is welded with E6010 electrodes. The bar is cut from AISI 1018 hot-rolled steel. The load is 500 lb completely reversed. Estimate the median cycles to failure.

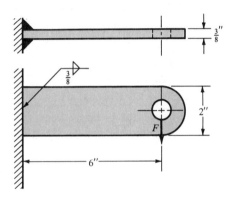

PROBLEM 9-20

9-21   An AISI 1010 $\frac{3}{8}$-in plate is T-butt-welded to an AISI 1015 beam with an E6010 electrode as depicted in the figure. For a design factor $n_d = 2$, what repeatedly applied load can be carried by the weldment indefinitely?

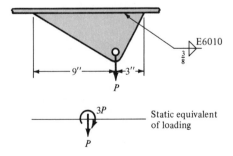

PROBLEM 9-21

9-22   A cantilevered bar such as is depicted in Fig. 9-17 can be welded as shown or with two vertical weld beads along the sides. If the cost of welding is $9.00 per inch of leg per inch of bead length, which welding scheme has the smaller welding deposit cost?

9-23   The stem portion of the fillet-welded joint shown in Fig. 9-3a carries a load of $2F$ directed vertically upward, and the cap of the tee carries a corresponding load directed downward. The length of the weld is l, and the leg size is h. Let θ be the abscissa angle of a plane cutting the right fillet and its mirror image on the left weld fillet.

(a) At what angle θ is the shear stress on the plane described by θ a maximum? What multiple of F/hl is its magnidue?

(b) How does this compare with τ_{max} of Table 9-1?

(c) Is the conventional design approach in fillet-weld analysis to provide for carrying the external load as a pure shear at the throat conservative or not in this case?

9-24   Consider the stem of the tee of Prob. 9-23 subjected to a load of $2F$ into the plane of the paper and to an equal load out of the plane of the paper applied to the cap of the tee.

(a) At what angle θ is the shear stress on the plane described by θ a maximum? What is its magnitude?

(b) How does this compare with τ_{max} of Table 9-1?

9-25   Examine the patterns of welds in torsion as listed in Table 9-2. In the interests of minimizing the amount of weldment (length) the strength per linear inch of weldment should be considered. If the pattern is restricted to a square $a \times a$, it is clear instinctively that the square weld pattern is best. Rank the others in order of decreasing effectiveness in static loading.

ANSWERS   **9-1**   17.7 kip
9-4   18.1 kpsi
9-7   7.21 kN
9-9   4.71 for static loads; 1.24 for fatigue loads
9-14 $W = 124$ lb
9-19 $F = 3610$ lb
9-21 $P_{max} = 7430$ lb

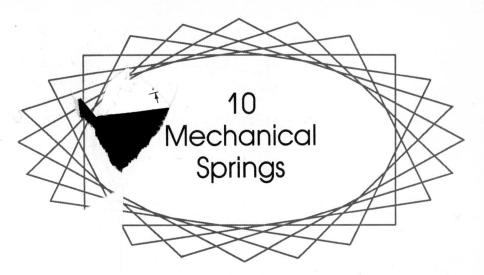

10
Mechanical Springs

Mechanic   are used in machines to exert force, to provide flexibility, and to store or absorb energy. In general, springs may be classified as either wire springs, flat springs, or special-shaped springs, and there are variations within these divisions. Wire springs include helical springs of round or square wire and are made to resist tensile, compressive, or torsional loads. Under flat springs are included the cantilever and elliptical types, the wound motor- or clock-type power springs, and the flat spring washers, usually called Belleville springs.

10-1 STRESSES IN HELICAL SPRINGS

Figure 10-1a shows a round-wire helical compression spring loaded by the axial force F. We designate D as the *mean spring diameter* and d as the *wire diameter*. Now imagine that the spring is cut at some point (Fig. 10-1b), a portion of it removed, and the effect of the removed portion replaced by the internal forces. Then, as shown in the figure, the cut portion would exert a direct shear force F and a torsion T on the remaining part of the spring.

To visualize the torsion, picture a coiled garden hose. Now pull one end of the hose in a straight line perpendicular to the plane of the coil. As each turn of hose is pulled off the coil, the hose twists or turns about its own axis. The flexing of a helical spring creates a torsion in the wire in a similar manner.

The maximum stress in the wire may be computed by superposition of Eqs. (2-15) and (2-41). The result is

$$\tau_{\max} = \pm \frac{Tr}{J} + \frac{F}{A} \tag{a}$$

where the term Tr/J is the torsion formula and F/A is the direct (not flexural) shear stress. Replacing the terms by $T = FD/2$, $r = d/2$, $J = \pi d^4/32$, and $A = \pi d^2/4$ gives

$$\tau = \frac{8FD}{\pi d^3} + \frac{4F}{\pi d^2} \tag{10-1}$$

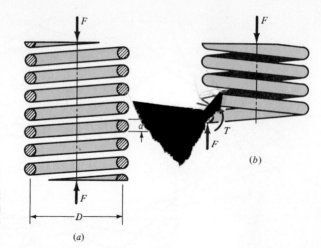

FIGURE 10-1

(a) Axially loaded helical spring;
(b) free-body diagram showing
that the wire is subjected to a direct shear and a torsional shear.

In this equation the subscript indicating maximum shear stress has been omitted as unnecessary. The positive signs of Eq. (*a*) have been retained, and hence Eq. (10-1) gives the shear stress at the *inside* fiber of the spring.

Now we define the *spring index*

$$C = \frac{D}{d} \tag{10-2}$$

as a measure of coil curvature. With this relation, Eq. (10-1) can be rearranged to give

$$\tau = K_s \frac{8FD}{\pi d^3} \tag{10-3}$$

where K_s is a *shear-stress correction factor* and is defined by the equation

$$K_s = \frac{2C + 1}{2C} \tag{10-4}$$

For most springs, C will range from about 6 to 12. Equation (10-3) is quite general and applies for both static and dynamic loads.

The use of square or rectangular wire is not recommended for springs unless space limitations make it necessary. Springs of special wire shapes are not made in large quantities, as are those of round wire; they have not had the benefit of refining development and hence may not be as strong as springs made from round wire. When space is severely limited, the use of nested round-wire springs should always be considered. They may have an economical advantage over the special-section springs, as well as a strength advantage.

10-2   THE CURVATURE EFFECT

An effect very similar to stress concentration occurs at the inside surface of a helical spring. The curvature of the wire increases the stress on the inside of the spring but decreases it only slightly on the outside. This curvature stress is so highly localized that it is important only when fatigue is present. For static loading, this stress can be neglected, because it will be relieved by local yielding with the first application of a load.

Unfortunately, it is necessary to find the curvature factor in a roundabout way. The reason for this is that the published equations include the effect of the direct shear stress too. Suppose K_s in Eq. (10-3) is replaced by another K factor which corrects for both curvature and direct shear. Then this factor is given by either of the equations

$$K_W = \frac{4C - 1}{4C - 4} + \frac{0.615}{C} \tag{10-5}$$

$$K_B = \frac{4C + 2}{4C - 3} \tag{10-6}$$

The first of these is called the Wahl factor, and the second, the Bergsträsser factor.* Since the results of these two equations differ by less than 1 percent, Eq. (10-6) is preferred. The curvature correction factor can now be obtained by canceling out the effect of the direct shear. Thus, using Eq. (10-6) with Eq. (10-4), the curvature correction factor is found to be

$$K_c = \frac{K_B}{K_s} = \frac{2C(4C + 2)}{(4C - 3)(2C + 1)} \tag{10-7}$$

When fatigue failure is likely or when the spring material must be considered as a brittle material, K_c is treated as a stress-concentration factor and used according to earlier recommendations.

10-3   DEFLECTION OF HELICAL SPRINGS

The deflection-force relations are quite easily obtained using Castigliano's theorem. The total strain energy for a helical spring is composed of a torsional component and a shear component. From Eqs. (3-30) and (3-31), the strain energy is

$$U = \frac{T^2 l}{2GJ} + \frac{F^2 l}{2AG} \tag{a}$$

Substituting $T = FD/2$, $l = \pi DN$, $J = \pi d^4/32$, and $A = \pi d^2/4$ results in

$$U = \frac{4F^2 D^3 N}{d^4 G} + \frac{F^2 DN}{d^2 G} \tag{b}$$

where $N = N_a$ = number of active coils. Then, using Castigliano's theorem,

$$y = \frac{\partial U}{\partial F} = \frac{8FD^3 N}{d^4 G} + \frac{4FDN}{d^2 G} \tag{c}$$

---

*Cyril Samónov, "Some Aspects of Design of Helical Compression Springs," *Int. Symp. Design and Synthesis,* Tokyo, 1984.

Since $C = D/d$, Eq. (c) can be rearranged to yield

$$y = \frac{8FD^3N}{d^4G}\left(1 + \frac{1}{2C^2}\right) \approx \frac{8FD^3N}{d^4G} \tag{10-8}$$

The spring rate is $k = F/y$, and so

$$k = \frac{d^4G}{8D^3N} \tag{10-9}$$

10-4   EXTENSION SPRINGS

Extension springs necessarily must have some means of transferring the load from the support to the body of the spring. Although this can be done with a threaded plug or a swivel hook, both of these add to the cost of the finished product, and so one of the methods shown in Fig. 10-2 is usually employed. In designing a spring with a hook end, the stress-concentration effect must be considered.

In Fig. 10-3a and b is shown a much-used method of designing the end. Stress concentration due to the sharp bend makes it impossible to design the hook as strong as the body. Tests as well as analysis show that the stress-concentration factor is given approximately by

$$K = \frac{r_m}{r_i} \tag{10-10}$$

which holds for bending stress and occurs when the hook is offset, and for torsional stress. Figure 10-3c and d shows an improved design due to a reduced coil diameter, not to elimination of stress concentration. The reduced coil diameter results in a lower stress because of the shorter moment arm. No stress-concentration factor is needed for the axial component of the load.

When extension springs are made with coils in contact with one another, they are said to be *close-wound*. Spring manufacturers prefer some initial tension in close-wound springs in order to hold the free length more accurately. Table 10-1 gives the range of torsional stress due to the pre-tension as preferred by spring manufacturers. Stresses below the given range cause difficulty in controlling the free length. The customary tolerance on preload is ± 10 percent, though lesser values can be obtained.

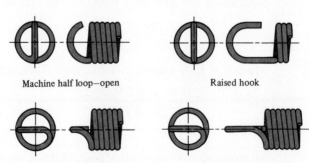

Machine half loop—open   Raised hook

FIGURE 10-2

Types of ends used on extension springs. *(Courtesy of Associated Spring Corporation.)*

Short twisted loop   Full twisted loop

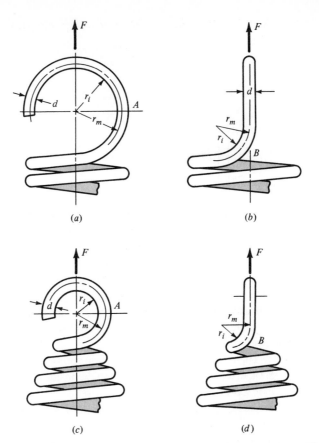

FIGURE 10-3

Ends for extension springs.
(a) Usual design; stress at A is
due to combined axial and bending forces. (b) Side view of part a;
stress is mostly torsion at B.
(c) Improved design; stress at A is
due to combined axial and bending forces. (d) Side view of part c;
stress at B is mostly torsion.

The initial tension is created in the winding process by twisting the wire as it is wound onto the mandrel. When the spring is completed and removed from the mandrel, the initial tension is locked in because the spring cannot get any shorter.

The direction of the stresses can be visualized by reference to Fig. 10-4. In Fig. 10-4a, block A simulates the effect of the stacked coils, and the free length of the spring is the length L_0 with no external force applied. In Fig. 10-4b, an external force F

TABLE 10-1

Preferred Range of Torsional
Stresses Due to Initial Tension
for Steel Helical Extension
Springs

INDEX	STRESS RANGE	
C	MPa	kpsi
4	115–183	16.7–26.6
6	95–160	13.8–23.2
8	82–127	11.9–18.4
10	60–106	8.71–15.4
12	48–86	6.97–12.5
14	37–60	5.37–8.71
16	25–50	3.63–7.26

Source: Associated Spring–Barnes Group, *Design Handbook*, Bristol, Conn., 1981, p. 50.

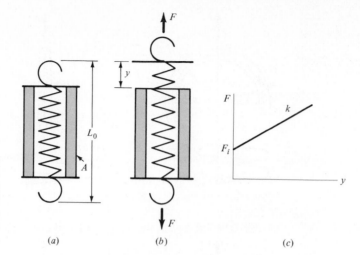

FIGURE 10-4

Simulation of an extension spring with initial tension. (*a*) No external force; spring compresses block *A* with initial force F_i. The free length is L_0. (*b*) Spring extended a distance *y* by external force *F*. (*c*) Force-deflection relation.

(*a*)   (*b*)   (*c*)

has been applied, causing the spring to elongate through the distance *y*. Note, very particularly, that the stresses in the spring are in the *same direction* in Fig. 10-4*a* and *b*.

Figure 10-4*c* shows the relation between the external force and the spring elongation. Here we see that *F* must exceed the initial tension F_i before a deflection *y* is experienced.

The free length L_0 of an extension spring is equal to the body length plus 2 times the hook distance, and is measured on the inside surface of the hooks. The body length L_B is given by the equation

$$L_B = d(N_a + 1)$$

where N_a is the number of active coils.

10-5   COMPRESSION SPRINGS

The four types of ends generally used for compression springs are illustrated in Fig. 10-5. A spring with *plain ends* has a noninterrupted helicoid; the ends are the same as if a long spring had been cut into sections. A spring with plain ends that are *squared* or

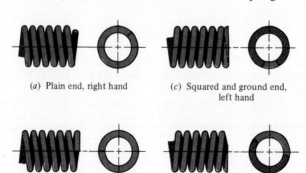

(*a*) Plain end, right hand

(*c*) Squared and ground end, left hand

FIGURE 10-5

Types of ends for compression springs: (*a*) both ends plain; (*b*) both ends squared; (*c*) both ends squared and ground; (*d*) both ends plain and ground.

(*b*) Squared or closed end, right hand

(*d*) Plain end, ground, left hand

TABLE 10-2

Formulas for Compression-Spring Dimensions. (N_a = Number of Active Coils)

TERM	PLAIN	PLAIN AND GROUND	SQUARED OR CLOSED	SQUARED AND GROUND
		TYPE OF SPRING ENDS		
End coils, N_e	0	1	2	2
Total coils, N_t	N_a	$N_a + 1$	$N_a + 2$	$N_a + 2$
Free length, L_0	$pN_a + d$	$p(N_a + 1)$	$pN_a + 3d$	$pN_a + 2d$
Solid length, L_s	$d(N_t + 1)$	dN_t	$d(N_t + 1)$	dN_t
Pitch, p	$(L_0 - d)/N_a$	$L_0/(N_a + 1)$	$(L_0 - 3d)/N_a$	$(L_0 - 2d)/N_a$

Source: Associated Spring–Barnes Group, *Design Handbook,* Bristol, Conn., 1981, p. 32.

closed is obtained by deforming the ends to a zero-degree helix angle. Springs should always be both squared and ground for important applications, because a better transfer of the load is obtained.

Table 10-2 shows how the type of end used affects the number of coils and the spring length.*

Set removal or *presetting* is a process used in the manufacture of compression springs to induce useful residual stresses. It is done by making the spring longer than needed and then compressing it to its solid height. This operation *sets* the spring to the required final free length and, since the torsional yield strength has been exceeded, induces residual stresses opposite in direction to those induced in service. Springs to be preset should be designed so that 10 to 30 percent of the initial free length is removed during the operation. If the stress at the solid height is greater than 1.3 times the torsional yield strength, distortion may occur. If this stress is much less than 1.1 times, it is difficult to control the resulting free length.

Set removal increases the strength of the spring and so is especially useful when the spring is used for energy-storage purposes. However, set removal should not be used when springs are subject to fatigue.

10-6 STABILITY

In Chap. 3 we learned that a column will buckle when the load becomes too large. Similarly, compression coil springs will buckle when the deflection becomes too large. The critical deflection is given by the equation

$$y_{cr} = L_0 C_1 \left[1 - \left(1 - \frac{C_2}{\lambda_{eff}^2} \right)^{1/2} \right] \tag{10-11}$$

where y_{cr} is the deflection corresponding to the onset of instability. Samónov† states that this equation is cited by Wahl‡ and verified experimentally by Haringx.§The

---

*For a thorough discussion and development of these relations, see Cyril Samónov, ''Computer-Aided Design of Helical Compression Springs,'' ASME paper no. 80-DET-69, 1980.

†''Computer-Aided Design.''

‡A. M. Wahl, *Mechanical Springs,* 2d ed., McGraw-Hill, New York, 1963.

§J. A. Haringx, ''On Highly Compressible Helical Springs and Rubber Rods and Their Application for Vibration-Free Mountings,'' I and II, *Philips Res. Rep.,* vol. 3, December 1948, pp. 401–449, and vol. 4, February 1949, pp. 49–80.

END CONDITION	CONSTANT α
Spring supported between flat parallel surfaces (fixed ends)	0.5
One end supported by flat surface perpendicular to spring axis (fixed); other end pivoted (hinged)	0.707
Both ends pivoted (hinged)	1
One end clamped; other end free	2

*Ends supported by flat surfaces must be squared and ground.

quantity λ_{eff} in Eq. (10-11) is the *effective slenderness ratio* and is given by the equation

$$\lambda_{\text{eff}} = \frac{\alpha L_0}{D} \tag{10-12}$$

C_1 and C_2 are the elastic constants and are defined by the equations

$$C_1 = \frac{E}{2(E - G)} \tag{10-13}$$

$$C_2 = \frac{2\pi^2(E - G)}{2G + E} \tag{10-14}$$

Equation (10-12) contains the *end-condition constant* α. This depends upon how the ends of the spring are supported. Table 10-3 gives values of α for usual end conditions. Note how closely these resemble the end conditions for columns.

Absolute stability occurs when, in Eq. (10-11), the term $C_2/\lambda_{\text{eff}}^2$ is less than unity. This means that the condition for absolute stability is that

$$L_0 < \frac{\pi D}{\alpha}\left[\frac{2(E - G)}{2G + E}\right]^{1/2} \tag{10-15}$$

For steels, this turns out to be

$$L_0 < 2.63\frac{D}{\alpha} \tag{10-16}$$

10-7   SPRING MATERIALS

Springs are manufactured either by hot- or cold-working processes, depending upon the size of the material, the spring index, and the properties desired. In general, prehardened wire should not be used if $D/d < 4$ or if $d > \frac{1}{4}$ in. Winding of the spring induces residual stresses through bending, but these are normal to the direction of the torsional working stresses in a coil spring. Quite frequently in spring manufacture, they are relieved, after winding, by a mild thermal treatment.

A great variety of spring materials are available to the designer, including plain carbon steels, alloy steels, and corrosion-resisting steels, as well as nonferrous materials such as phosphor bronze, spring brass, beryllium copper, and various nickel alloys. Descriptions of the most commonly used steels will be found in Table 10-4. The UNS

TABLE 10-4

High-Carbon and Alloy Spring Steels

NAME OF MATERIAL	SIMILAR SPECIFICATIONS	DESCRIPTION
Music wire, 0.80–0.95*C*	UNS G10850 AISI 1085 ASTM A228-51	This is the best, toughest, and most widely used of all spring materials for small springs. It has the highest tensile strength and can withstand higher stresses under repeated loading than any other spring material. Available in diameters 0.12 to 3 mm (0.005 to 0.125 in). Do not use above 120°C (250°F) or at subzero temperatures
Oil-tempered wire, 0.60–0.70*C*	UNS G10650 AISI 1065 ASTM 229-41	This general-purpose spring steel is used for many types of coil springs where the cost of music wire is prohibitive and in sizes larger than available in music wire. Not for shock or impact loading. Available in diameters 3 to 12 mm (0.125 to 0.5000 in), but larger and smaller sizes may be obtained. Not for use above 180°C (350°F) or at subzero temperatures
Hard-drawn wire, 0.60–0.70*C*	UNS G10660 AISI 1066 ASTM A227-47	This is the cheapest general-purpose spring steel and should be used only where life, accuracy, and deflection are not too important. Available in diameters 0.8 to 12 mm (0.031 to 0.500 in). Not for use above 120°C (250°F) or at subzero temperatures
Chrome vanadium	UNS G61500 AISI 6150 ASTM 231-41	This is the most popular alloy spring steel for conditions involving higher stresses than can be used with the high-carbon steels and for use where fatigue resistance and long endurance are needed. Also good for shock and impact loads. Widely used for aircraft-engine valve springs and for temperatures to 220°C (425°F). Available in annealed or pretempered sizes 0.8 to 12 mm (0.031 to 0.500 in) in diameter
Chrome silicon	UNS G92540 AISI 9254	This alloy is an excellent material for highly stressed springs that require long life and are subjected to shock loading. Rockwell hardnesses of C50 to C53 are quite common, and the material may be used up to 250°C (475°F). Available from 0.8 to 12 mm (0.031 to 0.500 in) in diameter

Source: By permission from Harold C. R. Carlson, ''Selection and Application of Spring Materials,'' *Mech. Eng.*, vol. 78, 1956, pp. 331–334.

TABLE 10-5

Constants for Computing
Minimum Tensile Strengths of
Common Spring Steels

MATERIAL	ASTM NO.	EXPONENT m	INTERCEPT	
			A, kpsi	A, MPa
Music wire[a]	A228	0.163	186	2060
Oil-tempered wire[b]	A229	0.193	146	1610
Hard-drawn wire[c]	A227	0.201	137	1510
Chrome vanadium[d]	A232	0.155	173	1790
Chrome silicon[e]	A401	0.091	218	1960

[a]Surface is smooth, free from defects, and has a bright lustrous finish.

[b]Has a slight heat-treating scale which must be removed before plating.

[c]Surface is smooth and bright with no visible marks.

[d]Aircraft-quality tempered wire; can also be obtained annealed.

[e]Tempered to Rockwell C49, but may be obtained untempered.

Source: Associated Spring–Barnes Group, *Design Handbook,* Bristol, Conn., 1981, p. 19.

steels listed in the Appendix should be used in designing hot-worked, heavy-coil springs, as well as flat springs, leaf springs, and torsion bars.

Spring materials may be compared by an examination of their tensile strengths; these vary so much with wire size that they cannot be specified until the wire size is known. The material and its processing also, of course, have an effect on tensile strength. It turns out that the graph of tensile strength versus wire diameter is almost a straight line for some materials when plotted on log-log paper. Writing the equation of this line as

$$S_{ut} = \frac{A}{d^m} \tag{10-17}$$

furnishes a good means of estimating minimum tensile strengths when the intercept A and the slope m of the line are known. Values of these constants have been worked out from recent data and are given for strengths in units of kpsi and MPa in Table 10-5.

Equation (10-17) is valid only for the materials listed in Table 10-5. Figure 10-6 has

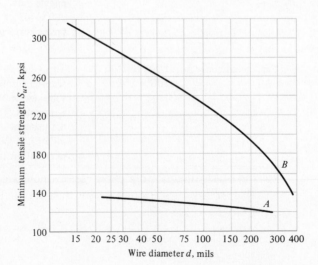

FIGURE 10-6

Minimum tensile strengths; *A*, hard phosphor bronze wire, *E* = 15 Mpsi, *G* = 6.3 Mpsi; *B*, ASTM A313 stainless steel wire (type 302), *E* = 28 Mpsi, *G* = 10 Mpsi.

been prepared to use in obtaining the strength of stainless steel wire (type 302) and of hard phosphor bronze wire. Note that phosphor bronze is very nearly a straight line on the semilog plot.

Although the torsional yield strength is needed to design the spring and to analyze the performance, spring materials customarily are tested only for tensile strength—perhaps because it is such an easy and economical test to make. A very rough estimate of the torsional yield strength can be obtained by assuming that the tensile yield strength is between 60 and 90 percent of the tensile strength. Then the distortion-energy theory can be employed to obtain the torsional yield strength. This approach results in the range

$$0.35S_{ut} \leq S_{sy} \leq 0.52S_{ut} \tag{10-18}$$

for steels.

Instead of the torsional yield strength, Joerres* uses the *maximum allowable torsional stress* for static applications. This is the best and most reliable information available and, for springs, can be used instead of S_{sy}. Joerres' values are

$$S_{sy} = \tau_{\text{all}} = \begin{cases} 0.45S_{ut} & \text{cold-drawn carbon steel} \\ 0.50S_{ut} & \text{hardened and tempered carbon and} \\ & \quad \text{low-alloy steel} \\ 0.35S_{ut} & \text{austenitic stainless steel and} \\ & \quad \text{nonferrous alloys} \end{cases} \tag{10-19}$$

Samónov† discusses this problem and shows that

$$S_{sy} = \tau_{\text{all}} \approx 0.56S_{ut} \tag{10-20}$$

for high-tensile spring steels. He points out that this value of allowable stress is specified by Draft Standard DIN 2089 of the German Federal Republic when Eq. (10-3) is used without the stress correction factor.

EXAMPLE 10-1   A helical compression spring is made of No. 16 (0.037-in) music wire. The outside diameter of the spring is $\frac{7}{16}$ in. The ends are squared and there are $12\frac{1}{2}$ total turns.

(a) Estimate the torsional yield strength of the wire.

(b) Find the static load corresponding to the yield strength.

(c) What is the scale of the spring?

(d) Compute the deflection that would be caused by the load in part (b).

(e) Compute the solid length of the spring.

(f) What length should the spring be so that when it is compressed solid and then released there will be no permanent change in the free length?

---

*Robert E. Joerres, "Springs," chap. 24 in Joseph E. Shigley and Charles R. Mischke (eds.), *Standard Handbook of Machine Design*, McGraw-Hill, New York, 1986, p. 24.19.

†"Computer-Aided Design."

(g) Given the length found in part (f), is buckling a possibility?

(h) What is the pitch of the spring?

Solution   (a) Using Eq. (10-17) and Table 10-5, we find $A = 186$ kpsi and $m = 0.163$. Therefore

$$S_{ut} = \frac{A}{d^m} = \frac{186}{(0.037)^{0.163}} = 318 \text{ kpsi}$$

Then, from Eq. (10-19),

Answer $S_{sy} = 0.45S_{ut} = 0.45(318) = 143 \text{ kpsi}$

(b) The mean spring diameter is $D = \frac{7}{16} - 0.037 = 0.400$ in, and so the spring index is $C = 0.400/0.037 = 10.8$. Then, from Eq. (10-4),

$$K_s = \frac{2C + 1}{2C} = \frac{2(10.8) + 1}{2(10.8)} = 1.046$$

Now rearrange Eq. (10-3) and solve for F_s using the torsional yield strength instead of the shear stress. This gives

Answer $$F_s = \frac{S_{sy}\pi d^3}{8K_s D} = \frac{\pi(143)(10^3)(0.037)^3}{8(1.046)(0.400)} = 6.80 \text{ lb}$$

(c) From Table 10-2, $N_a = 12.5 - 2 = 10.5$ turns. Using $G = 11.5$ Mpsi, the scale of the spring is found from Eq. (10-9):

Answer $$k = \frac{d^4 G}{8D^3 N_a} = \frac{(0.037)^4(11.5)(10^6)}{8(0.400)^3(10.5)} = 4.01 \text{ lb/in}$$

Answer   (d) $y_s = F_s/k = 6.80/4.01 = 1.70$ in

(e) From Table 10-2,

Answer $L_s = d(N_t + 1) = 0.037(12.5 + 1) = 0.50$ in

(f) To avoid yielding, the spring can be no longer than the solid length plus the deflection caused by a load just short of the amount required to initiate yielding. Summing the results of parts (d) and (e) gives this length as

Answer $L_0 = 1.70 + 0.50 = 2.20$ in

(g) Table 10-3 gives $\alpha = 0.5$. Then, from Eq. (10-16), we find

$$2.63\frac{D}{\alpha} = \frac{2.63(0.400)}{0.5} = 2.104 \text{ in}$$

Since $L_0 = 2.20$ in, buckling will occur.

(h) The pitch is found from Table 10-2. Based on $L_0 = 2.20$ in, the result is found to be

Answer $$p = \frac{L_0}{N_a + 1} = \frac{2.20}{10.5 + 1} = 0.191 \text{ in}$$

10-8   DESIGN OF HELICAL SPRINGS

The design of a new spring involves the following considerations:

- Space into which the spring must fit and operate
- Values of working forces and deflections
- Accuracy and reliability needed
- Tolerances and permissible variations in specifications
- Environmental conditions such as temperature and presence of a corrosive atmosphere
- Cost and quantities needed

The designer uses these factors to select a material and specify suitable values for the wire size, the number of turns, the diameter and free length, the type of ends, and the spring rate needed to satisfy the working force-deflection requirements. Samónov* states that, for compression springs, the primary design constraints are that the wire size be commercially available and that the stress at the solid length be no greater than the torsional yield strength. His goal, in a complete computer solution, is to fully utilize the material.

Charts and nomographs have been used by many to simplify the spring design problem.†

There are almost as many ways to create a spring-design program as there are programmers; and there is nothing unusual about the program presented here. It works. The program, which is for the design of compression springs, can be used as a starting point for the creation of other programs. It consists of seven separate subroutines, all of which utilize the same memory locations. The subroutines, which should be used in the order in which they are presented, are:

1 Enter and display (or print) the outside diameter.

2 Enter and display the total number of coils. Enter and display the number of dead coils. Compute and display the number of active coils.

3 Select a material and enter and display the exponent and coefficient.

4 Enter and display the wire diameter. Compute and display the torsional yield strength.

5 Enter and display the maximum torsional stress desired when the spring is closed solid. Compute and display the solid height, the free length, and the force required to compress the spring solid.

---

*"Computer-Aided Design."

†See M. Massoud and L. Hubert, "Brief Survey of Spring Design Nomographs," ASME paper no. 76-DET/77, 1977. See also M. Dao-Thien and M. Massoud, "Design Nomographs of Compression Helical Springs for Predetermined Reliability Levels," ASME paper no. 80-C2/DET 85, 1980; N. P. Chironis, *Spring Design and Application,* McGraw-Hill, New York, 1961; F. J. Camm, *Newnes Engineer's Reference Book,* 8th ed., Newnes-Butterworth, London, 1958; T. K. Tsai, "Speedy Design of Helical Compression Springs by Nomography Method," *Journal of Engineering for Industry,* February 1975, pp. 373–374.

6 Compute and display the spring constant.

7 Enter and display any desired operating force F. Compute and display the corresponding values of the torsional stress and the spring deflection.

Separate subroutines are used in this program in order to avoid reentering all the data when only a single parameter is to be changed. In this way it is easy to see the effect of the single change. For example, having run through the program once, you may wish to try a different wire size. This can be done by entering the new wire size in subroutine 4 and proceeding from that point.

EXAMPLE 10-2

Indexing is used in machine operations when a circular part being manufactured must be divided into a certain number of segments. Figure 10-7 shows a portion of an indexing fixture used to successively position a part for the operation. When the knob is pulled up, part 6, which holds the workpiece, is rotated to the next position and locked in place by releasing the index pin. In this example we wish to design the spring to exert a force of about 3 lb and to fit in the space defined in the figure caption.

Solution

Since the fixture is not a high-production item, a stock spring will be selected. These are available in music wire. In one catalog there are 76 stock springs available having an outside diameter of 0.480 in and designed to work in a $\frac{1}{2}$-in hole. These are made in seven different wire sizes, ranging from 0.038 up to 0.063 in, and in free lengths from $\frac{1}{2}$ to $2\frac{1}{2}$ in, depending upon the wire size.

Since the pull knob must be raised $\frac{3}{4}$ in for indexing and the space for the spring is $1\frac{3}{8}$ in long when the pin is down, the solid length cannot be more than $\frac{5}{8}$ in.

Let us begin by selecting a spring having an outside diameter of 0.480 in, a wire size of 0.051 in, a free length of $1\frac{3}{4}$ in, $11\frac{1}{2}$ total turns, and plain ends. Then $m = 0.163$ and $A = 186$ kpsi for music wire. Then

FIGURE 10-7

Part 1, pull knob; part 2, tapered retaining pin; part 3, hardened bushing with press fit; part 4, body of fixture; part 5, indexing pin; part 6, workpiece holder. Space for the spring is $\frac{3}{8}$ in OD by $\frac{1}{4}$ in ID and $1\frac{3}{8}$ in long, with the pin down as shown. The pull knob must be raised $\frac{3}{4}$ in to permit indexing.

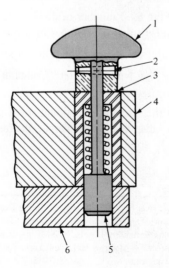

$$S_{ut} = \frac{A}{d^m} = \frac{186}{(0.051)^{0.163}} = 302 \text{ kpsi}$$

and, from Eq. (10-19),

$$S_{sy} = 0.45 S_{ut} = 0.45(302) = 136 \text{ kpsi}$$

From Table 10-2, $N_a = N_t = 11.5$ turns for plain ends. Also, $D = D_o - d = 0.480 - 0.051 = 0.429$ in. Then, from Eq. (10-9), the spring rate is

$$k = \frac{d^4 G}{8D^3 N_a} = \frac{(0.051)^4(11.5)(10^6)}{8(0.429)^3(11.5)} = 10.7 \text{ lb/in}$$

Also, from Table 10-2, the solid length is

$$L_s = d(N_t + 1) = 0.051(11.5 + 1) = 0.6375 \text{ in}$$

The spring force when the pin is down is

$$F_{min} = ky = 10.7(1.75 - 1.375) = 4.01 \text{ lb}$$

When the spring is pulled up, the spring force is

$$F_{max} = k(L_0 - L_s) = 10.7(1.75 - 0.6375) = 11.9 \text{ lb}$$

Using $C = D/d = 0.429/0.051 = 8.41$ and $K_s = (2C + 1)/2C = 1.059$, the stress at the solid length is

$$\tau = K_s \frac{8FD}{\pi d^3} = 1.059 \frac{8(11.9)(0.429)}{\pi(0.051)^3}(10^{-3}) = 104 \text{ kpsi}$$

But this spring is too strong and the solid length is too long; therefore we must go to a smaller wire size.

After several trials, a spring is found which yields $F_{min} = 2.34$ lb, $F_{max} = 7.40$ lb, and $\tau_{max} = 94.6$ kpsi. Buckling does not occur. The specifications and other results for this spring are found to be:

Material, music wire

Ends, plain

Wire diameter $d = 0.045$ in

Outside diameter $D_o = 0.480$ in

Mean diameter $D = 0.435$ in

Total number of coils $N_t = 11.5$

Free length $L_0 = 1.75$ in

Solid length $L_s = 0.5625$ in

Spring index $C = 9.67$

Spring rate $k = 6.23$ lb/in

Torsional yield strength $S_{sy} = 139$ kpsi

10-9

STOCHASTIC CONSIDERATIONS

In this section we shall consider the wire diameter **d**, the coil diameter **D**, the free length L_0, the shear modulus **G**, and, for extension springs, the pre-tension F_i, as random variables. When these variates have been specified, together with the number of active turns N_a, they encode the designer's target of a spring rate or a force corresponding to a specified end deflection. This section will assist our understanding of the limits of the springmaker's ability to achieve the designer's target.

If, say, a kilometer of wire is sized by drawing it through a die, the wire diameter slowly increases as a result of die wear. If no effort is made to identify the location of the spooled wire in the draw as the wire is spooled and cut, then the wire diameter changes from predictable to stochastic.

The coil diameter is toleranced and hence is a random variable; this tolerance reflects the springmaker's ability to wind to a chosen diameter over a mandrel.

The free length is also a random variable; it is random because it depends upon the two variates **D** and **d** and also because it depends upon the winding operation together with the pretwisting this may require.

It is well known that commercial tolerances are natural tolerances and represent 3 standard deviations. Commercial tolerances are the most economic; tighter tolerances increase the spring cost, sometimes quite dramatically. Tables 10-6, 10-7, and 10-8 list the commercial tolerances for wire diameter and free length. Tables 10-9 and 10-10 list the standard deviations of the outside coil diameter; these values were deduced from listings of commercial tolerances.

SIZE, in	TOLERANCE, in	C_d	SIZE, mm	TOLERANCE, mm	C_d	
0.020	±0.0004	0.0067	0.50	±0.010	0.0067	
0.024	±0.0004	0.0056	0.55	±0.010	0.0061	
0.028	±0.0004	0.0048	0.60	±0.010	0.0056	
0.035	±0.0006	0.0057	0.70	±0.010	0.0048	
0.042	±0.0006	0.0048	0.80	±0.015	0.0063	
0.048	±0.0006	0.0042	0.90	±0.015	0.0056	
0.055	±0.0006	0.0036	1.0	±0.015	0.0050	
0.063	±0.0006	0.0032	1.2	±0.015	0.0042	
0.072	±0.0006	0.0028	1.4	±0.0015	0.0036	
0.081	±0.0008	0.0033	1.6	±0.0015	0.0031	
0.092	±0.0008	0.0029	2.0	±0.020	0.0033	
0.105	±0.0008	0.0025	2.5	±0.020	0.0027	
0.112	±0.0008	0.0024	3.0	±0.020	0.0022	
0.125	±0.0012	0.0032	3.5	±0.030	0.0029	
0.135	±0.0012	0.0030	4.0	±0.030	0.0025	
0.148	±0.0012	0.0027	5.0	±0.030	0.0020	
0.162	±0.0012	0.0025	6.0	±0.030	0.0017	
0.177	±0.0012	0.0023	8.0	±0.050	0.0021	
0.192	±0.0012	0.0021	10.0	±0.070	0.0023	
0.207	±0.0012	0.0019	12.0	±0.070	0.0019	
0.225	±0.0012	0.0018	14.0	±0.070	0.0017	
0.250	±0.0020	0.0027	16.0	±0.070	0.0015	

TABLE 10-6

Tolerances and Coefficients of Variation C_d for Selected Sizes of Spring Steel Wire

TABLE 10-7

Unit Free-Length Tolerances T/L_0 of Squared and Ground Helical Compression Springs.* (Units are in $\pm$mm/mm ($\pm$in/in))

UNIT NO. COILS N_a/L_0, 1/mm (1/in)	SPRING INDEX $C = D/d$						
	4	6	8	10	12	14	16
0.02 (0.5)	0.010	0.011	0.012	0.013	0.015	0.016	0.016
0.04 (1)	0.011	0.013	0.015	0.016	0.017	0.018	0.019
0.08 (2)	0.013	0.015	0.017	0.019	0.020	0.022	0.023
0.20 (4)	0.016	0.018	0.021	0.023	0.024	0.026	0.027
0.30 (8)	0.019	0.022	0.024	0.026	0.028	0.030	0.032
0.50 (12)	0.021	0.024	0.027	0.030	0.032	0.034	0.036
0.60 (16)	0.022	0.026	0.029	0.032	0.034	0.036	0.038
0.80 (20)	0.023	0.027	0.031	0.034	0.036	0.038	0.040

*For springs having free lengths less than 12.7 mm (0.500 in), use tolerances for 12.7 mm (0.500 in). For springs having closed ends that are not ground, multiply tabulated values by 1.7.

Source: Associated Spring–Barnes Group, *Design Handbook*, Bristol, Conn., 1981, p. 42.

TABLE 10-8

Free-Length Tolerances for Helical Extension Springs with Initial Tension

FREE LENGTH INSIDE HOOKS, mm (in)	TOLERANCE, $\pm$mm ($\pm$in)
Up to 12.7 (0.500)	0.51 (0.020)
Over 12.7 to 25.4 (0.500 to 1.00)	0.76 (0.030)
Over 25.4 to 50.8 (1.00 to 2.00)	1.0 (0.040)
Over 50.8 to 102 (2.00 to 4.00)	1.5 (0.060)
Over 102 to 203 (4.00 to 8.00)	2.4 (0.093)
Over 203 to 406 (8.00 to 16.0)	4.0 (0.156)
Over 406 to 610 (16.0 to 24.0)	5.5 (0.218)

Source: Associated Spring–Barnes Group, *Design Handbook*, Bristol, Conn., 1981, p. 53.

TABLE 10-9

Standard Deviations in Millimeters of the Outside Coil Diameter of Helical Compression and Extension Springs

WIRE DIAMETER d, mm	SPRING INDEX C						
	4	6	8	10	12	14	16
0.5	0.0167	0.0227	0.0307	0.0433	0.0533	0.0600	0.0740
0.6	0.0167	0.0270	0.0343	0.0507	0.0610	0.0683	0.0850
1.0	0.0193	0.0360	0.0527	0.0663	0.0830	0.102	0.119
1.4	0.0277	0.0460	0.0637	0.0877	0.106	0.132	0.149
2.0	0.0347	0.0613	0.0853	0.113	0.140	0.164	0.182
4.0	0.0627	0.0943	0.135	0.184	0.223	0.264	0.305
6.0	0.0887	0.122	0.171	0.229	0.289	0.343	0.398
8.0	0.116	0.149	0.199	0.276	0.347	0.409	0.481
10.0	0.143	0.182	0.238	0.344	0.433	0.514	0.620

Table 10-6 gives the tolerances for wire diameter; the coefficient of variation C_d has been determined and tabulated as just described. Thus the diameter of a 1.2-mm wire, as a stochastic variable, is

$$\mathbf{d} = 1.2(1, 0.0042) \quad \text{mm}$$

Tables 10-7 and 10-8 are used to obtain the tolerance on free length. Suppose a compression spring has 14 active coils, a nominal free length of $1\frac{1}{2}$ in, and a spring index of $C = 8$. We enter Table 10-7 with $N_a/L_0 = 14/1.5 = 9.33$ and $C = 8$. Using

TABLE 10-10

Standard Deviations in Inches
of the Outside Coil Diameter
of Helical Compression and
Extension Springs

WIRE DIAMETER	SPRING INDEX C						
d, in	4	6	8	10	12	14	16
0.020	0.000 667	0.000 873	0.001 21	0.001 75	0.002 08	0.002 42	0.002 96
0.028	0.000 667	0.001 14	0.001 61	0.002 14	0.002 61	0.003 08	0.003 73
0.042	0.000 813	0.001 48	0.002 15	0.002 77	0.003 43	0.004 27	0.004 93
0.063	0.001 16	0.001 99	0.002 81	0.003 80	0.004 63	0.005 63	0.006 30
0.081	0.001 42	0.002 42	0.003 47	0.004 57	0.005 57	0.006 60	0.007 37
0.105	0.001 84	0.002 84	0.004 10	0.005 60	0.006 60	0.007 87	0.009 03
0.125	0.002 13	0.003 19	0.004 60	0.006 33	0.007 47	0.008 83	0.010 2
0.162	0.002 56	0.003 83	0.005 47	0.007 40	0.008 97	0.010 6	0.012 2
0.192	0.002 93	0.004 27	0.006 03	0.008 10	0.009 97	0.011 8	0.013 6
0.250	0.003 67	0.005 00	0.007 00	0.009 33	0.011 6	0.014 0	0.016 3

linear interpolation, we have

X	Y
8	0.024
9.33	?
12	0.027

and find $Y_2 = T/L_0 = 0.025$. Thus the tolerance is

$$T = \pm 0.025 L_0 = \pm 0.025(1.5) = \pm 0.0375 \text{ in}$$

The standard deviation is $\hat{\sigma}_L = 0.0375/3 = 0.0125$ in. Also, $C_L = \hat{\sigma}_L/L_0 = 0.0125/1.5 = 0.0083$, and so the free-length variable is

$$\mathbf{L}_0 = 1.50(1, 0.0083) \quad \text{in}$$

EXAMPLE 10-3   An extension spring is to be manufactured to the following specifications:

Material: ASTM A227 hard-drawn wire

Wire diameter: 0.9 mm

Number of active coils: 13.2

Outside coil diameter: 6.3 mm

Pre-tension: moderate

Ends: full loop (Fig. 10-3a)

Assume that $C_G = 0.008$ and find the following:

(a) Spring index and mean of the mean spring diameter $\mathbf{D}$

(b) Free length

(c) Preload

(d) Spring rate

(e) Force required to deflect the spring 7 mm

(f) Deflection of the spring caused by a force of 25 N

Solution   (a) The spring index C can be taken as deterministic except when it is used to replace D/d in Eq. (10-3) or (10-9). Using D_o as the outside coil diameter, the spring index is found to be

Answer
$$C = \frac{D_o - d}{d} = \frac{6.3 - 0.9}{0.9} = \frac{5.4}{0.9} = 6$$

From Table 10-6, find $\mathbf{d} = 0.9(1, 0.0056) = (0.9, 0.0050)$ mm. Using Table 10-9 with linear interpolation gives $\hat{\sigma}_{Do} = 0.033\ 75$ mm. Therefore

$$\mathbf{D}_o = (6.3, 0.033\ 75) = 6.3(1, 0.005\ 36) \qquad \text{mm}$$

Then the mean of the mean spring diameter is

$$\mathbf{D} = \mathbf{D}_o - \mathbf{d}$$

Here $\overline{D} = 6.3 - 0.9 = 5.4$ mm, and

$$\hat{\sigma}_D = (\hat{\sigma}_{Do}^2 + \hat{\sigma}_d^2)^{1/2} = [(0.033\ 75)^2 + (0.005\ 36)^2]^{1/2} = 0.0342 \text{ mm}$$

Therefore

Answer
$$\mathbf{D} = (5.4, 0.0342) = 5.4(1, 0.006\ 33) \qquad \text{mm}$$

(b) The free length is the body length plus 2 times the hook distance measured on the insides of the hooks. Thus, for Fig. 10-3a, it is given by the equation

$$L_0 = L_B + 2D_i = d(N_a + 1) + 2D_i \tag{10-21}$$

Substituting $D_i = 6.3 - 2(0.9) = 4.5$ mm gives

Answer
$$L_0 = 0.9(13.2 + 1) + 2(4.5) = 21.78 \text{ mm}$$

Table 10-8 gives the tolerance as ± 0.76 mm. Thus $\hat{\sigma}_L = 0.76/3 = 0.253$ mm. Therefore the free-length variate is

Answer
$$\mathbf{L}_0 = (21.78, 0.253) = 21.79(1, 0.0116) \qquad \text{mm}$$

(c) Using the midrange for the preload stress in Table 10-1 gives $\tau_i = 127$ MPa. The preload, or pre-tension, corresponding to this stress is found using Eqs. (10-3) and (10-4). Thus

$$K_s = \frac{2(6) + 1}{2(6)} = 1.083$$

$$F_i = \frac{\pi d^3 \tau_i}{8 K_s D} = \frac{\pi (0.9)^3 (127)}{8(1.083)(5.4)} = 6.22 \text{ N}$$

The commercial tolerance on preload is ± 10 percent, as indicated in Sec. 10-4. Therefore the standard deviation is

$$\hat{\sigma}_{Fi} = \frac{0.10(6.22)}{3} = 0.207 \text{ N}$$

and so the preload variate is

Answer $F_i = (6.22, 0.207) = 6.22(1, 0.033)$ N

(*d*) Equation (10-9), for spring rate, is written in the form

$$k = \frac{d^4 G}{8 D^3 N_a}$$

Substituting values in the form $\mathbf{x} = x(1, C_x)$ gives

$$k = \frac{[0.9(1, 0.0056)]^4[79.3(1, 0.008)]}{8[5.4(1, 0.006\ 33)]^3(13.2)}\left[\frac{(10^{-3})^4(10^9)}{(10^{-3})^3}\right]$$

Using Table 4-4, we find the coefficient of variation of the spring rate to be

$$C_k = \{[4(0.0056)]^2 + (0.008)^2 + [3(0.006\ 33)]^2\}^{1/2} = 0.0304$$

The mean is found to be $\bar{k} = 3129$ N/m. Therefore

Answer $k = 3129(1, 0.0304) = (3129, 95.2)$ N/m

With a natural spread of $3\hat{\sigma}_k$, the bilateral tolerance on the spring rate turns out to be about 9 percent.

(*e*) The force required to deflect this spring is given by the stochastic equation

$$F = F_i + ky$$

provided y is deterministic. Using $y = 7$ mm gives

$$F = (6.22, 0.207) + 7(3129, 95.2)(10^{-3})$$

Using Table 4-4 again, we obtain the standard deviation as

$$\hat{\sigma}_F = \{(0.207)^2 + [7(95.2)(10^{-3})]^2\}^{1/2} = 0.698 \text{ N}$$

The mean is

Answer $\bar{F} = 6.22 + 7(3129)(10^{-3}) = 28.1$ N

and so the variate is

Answer $F = (28.1, 0.698) = 28.1(1, 0.025)$ N

(*f*) Extension springs are often designed into machines and calibrated by built-in adjustments to provide a fixed or deterministic force. Under these conditions the deflection is stochastic and is given by the relation

$$y = \frac{F - F_i}{k}$$

For $F = 25$ N, this equation gives

$$y = \frac{25 - (6.22, 0.207)}{(3129, 95.2)}(10^3)$$

Using Table 4-4, this becomes

$$\mathbf{y} = \frac{(18.78, \, 0.207)}{(3129, \, 95.2)} \, (10^3) = \frac{18.78(1, \, 0.011)}{3129(1, \, 0.0304)} \, (10^3)$$

Thus $\bar{y} = 6.00$ mm and

$$C_y = [(0.011)^2 + (0.0304)^2]^{1/2} = 0.0323$$

so that

Answer $\mathbf{y} = 6.00(1, \, 0.0323) = (6.00, \, 0.194)$ mm

10-10 CRITICAL FREQUENCY OF HELICAL SPRINGS

If a wave is created by a disturbance at one end of a swimming pool, this wave will travel down the length of the pool, be reflected back at the far end, and continue in this back-and-forth motion until it is finally damped out. The same effect occurs in helical springs, and it is called *spring surge*. If one end of a compression spring is held against a flat surface and the other end is disturbed, a compression wave is created that travels back and forth from one end to the other exactly like the swimming-pool wave.

Spring manufacturers have taken slow-motion movies of automotive valve-spring surge. These pictures show a very violent surging, with the spring actually jumping out of contact with the end plates. Figure 10-8 is a photograph of such a failure.

When helical springs are used in applications requiring a rapid reciprocating mo-

FIGURE 10-8

Valve-spring failure in an over-revved engine. Fracture is along the 45° line of maximum principal stress associated with pure shear.

tion, the designer must be certain that the physical dimensions of the spring are not such as to create a natural vibratory frequency close to the frequency of the applied force; otherwise, resonance may occur, resulting in damaging stresses, since the internal damping of spring materials is quite low.

The governing equation for a spring placed between two flat and parallel plates is the wave equation

$$\frac{\partial^2 u}{\partial y^2} = \frac{W}{kgl^2} \frac{\partial^2 u}{\partial t^2} \tag{10-22}$$

where k = spring rate
$\quad\quad g$ = acceleration due to gravity
$\quad\quad l$ = length of spring between plates
$\quad\quad W$ = weight of spring
$\quad\quad y$ = coordinate along length of spring
$\quad\quad u$ = motion of any particle at distance y

The solutions to this equation are obtained using well-known methods. Here we are interested only in the natural frequencies; in radians per second, these are

$$\omega = m\pi \sqrt{\frac{kg}{W}}$$

where the fundamental frequency is found for $m = 1$, the second harmonic for $m = 2$, and so on. We are usually interested in the frequency in cycles per second; since $\omega = 2\pi f$, we have, for the fundamental frequency,

$$f = \frac{1}{2} \sqrt{\frac{kg}{W}} \tag{10-23}$$

Wolford and Smith* show that the frequency is

$$f = \frac{1}{4} \sqrt{\frac{kg}{W}} \tag{10-24}$$

where the spring has one end against a flat plate and the other end free. They also point out that Eq. (10-23) applies when one end is against a flat plate and the other end is driven with a sine-wave motion.

The weight of the active part of a helical spring is

$$W = AL\rho = \frac{\pi d^2}{4}(\pi DN_a)(\rho) = \frac{\pi^2 d^2 DN_a \rho}{4} \tag{10-25}$$

where ρ is the weight (not mass) per unit volume.

---

*J. C. Wolford and G. M. Smith, "Surge of Helical Springs," *Mech. Eng. News,* vol. 13, no. 1, February 1976, pp. 4–9.

The fundamental critical frequency should be from 15 to 20 times the frequency of the force or motion of the spring in order to avoid resonance with the harmonics. If the frequency is not high enough, the spring should be redesigned to increase k or decrease W.

10-11   FATIGUE LOADING

Springs are made to be used, and consequently they are almost always subject to fatigue loading. In many instances the number of cycles of required life may be small, say, several thousand for a padlock spring or a toggle-switch spring. But the valve spring of an automotive engine must sustain millions of cycles of operation without failure; so it must be designed for infinite life.

In the case of shafts and many other machine members, fatigue loading in the form of completely reversed stresses is quite ordinary. Helical springs, on the other hand, are never used as both compression and extension springs. In fact, they are usually assembled with a preload so that the working load is additional. Thus the stress-time diagram of Fig. 7-12d expresses the usual condition for helical springs. The worst condition, then, would occur when there is no preload, that is, when $\tau_{\min} = 0$.

Now, we define

$$F_a = \frac{F_{\max} - F_{\min}}{2} \tag{10-26}$$

$$F_m = \frac{F_{\max} + F_{\min}}{2} \tag{10-27}$$

where the subscripts have the same meaning as those of Fig. 7-12d when applied to the axial spring force F. Then the stress amplitude is

$$\tau_a = K_B \frac{8F_a D}{\pi d^3} \tag{10-28}$$

where K_B is the Bergsträsser factor, obtained from Eq. (10-6), and corrects for both direct shear and the curvature effect. As noted in Sec. 10-2, the Wahl factor K_W can be used instead, if desired.

The usual method of analyzing for fatigue loading would be to correct K_B to K_f because of notch sensitivity. But the notch sensitivity of high-hardness steels is so near unity that there is little value in making such a correction. Consequently, the full value of K_B is used in Eq. (10-28).

The mean stress is given by the equation

$$\tau_m = K_s \frac{8F_m D}{\pi d^3} \tag{10-29}$$

The best data on the torsional endurance limits of spring steels are those reported by Zimmerli.* He discovered the surprising fact that size, material, and tensile strength

---

*F. P. Zimmerli, ''Human Failures in Spring Applications,'' *The Mainspring,* no. 17, Associated Spring Corporation, Bristol, Conn., August–September 1957.

have no effect on the endurance limits (infinite life only) of spring steels in sizes under $\frac{3}{8}$ in (10 mm). We have already observed that endurance limits tend to level out at high tensile strengths (Fig. 7-7), but the reason for this is not clear. Zimmerli suggests that it may be because the original surfaces are alike or because plastic flow during testing makes them the same.

Interpreted in terms of the nomenclature of this book, Zimmerli's results are

must be corrected, look below

$$S_{se} = k_a k_b k_c S'_e = 45.0 \text{ kpsi (310 MPa)} \qquad \text{for unpeened springs}$$
$$S_{se} = k_a k_b k_c S'_e = 67.5 \text{ kpsi (465 MPa)} \qquad \text{for peened springs}$$

These results are valid for all the spring steels in Table 10-4. They are already corrected, as indicated, for surface finish, size, and loading, but not for temperature or miscellaneous effects.

In constructing the Goodman diagram or the S-N diagram, the torsional modulus of rupture is needed. Here we shall continue to employ Eq. (7-43), which is

$$S_{su} = 0.67 S_{ut} \tag{10-30}$$

EXAMPLE 10-4

A helical compression spring, made of music wire, has a wire size of 0.092 in, an outside diameter of $\frac{9}{16}$ in, a free length of $4\frac{1}{8}$ in, 21 active coils, and both ends squared and ground. The spring is to be assembled with a preload of 5 lb and will operate to a maximum load of 35 lb during use.

(a) Find the factor of safety guarding against a fatigue failure.

(b) Find the critical operating frequency.

Solution

(a) The mean diameter is $D = 0.5625 - 0.092 = 0.4705$ in. Then the spring index is $C = D/d = 0.4705/0.092 = 5.11$. Equations (10-4) and (10-6) give

$$K_s = \frac{2C + 1}{2C} = \frac{2(5.11) + 1}{2(5.11)} = 1.098$$

$$K_B = \frac{4C + 2}{4C - 3} = \frac{4(5.11) + 2}{4(5.11) - 3} = 1.287$$

Using Eqs. (10-26) and (10-27) next, we find

$$F_a = \frac{35 - 5}{2} = 15 \text{ lb}$$

$$F_m = \frac{35 + 5}{2} = 20 \text{ lb}$$

The alternating shear-stress component is found from Eq. (10-28) to be

$$\tau_a = K_B \frac{8 F_a D}{\pi d^3} = 1.287 \frac{8(15)(0.4705)}{\pi (0.092)^3} (10^{-3}) = 29.7 \text{ kpsi}$$

Equation (10-29) gives the mean shear-stress component as

$$\tau_m = K_s \frac{8 F_m D}{\pi d^3} = 1.098 \frac{8(20)(0.4705)}{\pi (0.092)^3} (10^{-3}) = 33.8 \text{ kpsi}$$

From Table 10-5, we find $A = 186$ kpsi and $m = 0.163$ for music wire. Then, using Eq. (10-17), we obtain

$$S_{ut} = \frac{A}{d^m} = \frac{186}{(0.092)^{0.163}} = 274 \text{ kpsi}$$

With Eq. (10-30), we find next that

$$S_{su} = 0.67 S_{ut} = 0.67(274) = 184 \text{ kpsi}$$

The spring is unpeened, and so the endurance limit is $S_{se} = 45.0$ kpsi. If we employ the Goodman criterion, then, for shear, Eq. (7-39) is written

$$\frac{\tau_a}{S_{se}} + \frac{\tau_m}{S_{su}} = \frac{1}{n} \quad \text{or} \quad n = \frac{S_{se}S_{su}}{\tau_a S_{su} + \tau_m S_{se}} \tag{10-31}$$

From this equation we find the factor of safety guarding against a fatigue failure to be

Answer

$$n = \frac{45(184)}{29.7(184) + 33.8(45)} = 1.19$$

(b) Using Eq. (10-9), we find the spring rate to be

$$k = \frac{d^4 G}{8D^3 N_a} = \frac{(0.092)^4 (11.5)(10^6)}{8(0.4705)^3(21)} = 47.1 \text{ lb/in}$$

Using $\rho = 0.282$ lb/in^3 and Eq. (10-25), we find the weight of the active part of the spring to be

$$W = \frac{\pi^2 d^2 D N_a \rho}{4} = \frac{\pi^2 (0.092)^2 (0.4705)(21)(0.282)}{4} = 0.0582 \text{ lb}$$

The critical frequency is found from Eq. (10-23) to be

$$f = \frac{1}{2}\left(\frac{kg}{W}\right)^{1/2} = \frac{1}{2}\left[\frac{47.1(386)}{0.0582}\right]^{1/2} = 279 \text{ Hz}$$

Thus if the operational frequency is much more than $279/20 \approx 14$ Hz, the spring may need to be redesigned.

10-12   HELICAL TORSION SPRINGS

The torsion springs illustrated in Fig. 10-9 are used in door hinges and automobile starters and, in fact, for any application where torque is required. They are wound in the same manner as extension or compression springs but have the ends shaped to transmit torque.

A torsion spring is subjected to the action of a bending moment $M = Fr$, producing a normal stress in the wire. Note that this is in contrast to a compression or an extension helical spring, in which the load produces a torsional stress in the wire. This means that

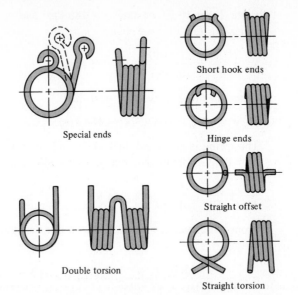

Short hook ends

Hinge ends

Special ends

Straight offset

FIGURE 10-9

Torsion springs. *(Courtesy of Associated Spring Corporation.)*

Double torsion

Straight torsion

the residual stresses built in during winding are in the same direction as but of opposite sign to the working stresses which occur during use. These residual stresses are useful in making the spring stronger by opposing the working stress, *provided* the load is always applied so as to cause the spring to wind up. Because the residual stress opposes the working stress, torsional springs can be designed to operate at stress levels which equal or even exceed the yield strength of the wire.

The bending stress can be obtained by using curved-beam theory as explained in Sec. 2-20. It is convenient to write the expression in the form

$$\sigma = K\frac{Mc}{I} \tag{a}$$

where K is a stress-concentration factor and, in this case, is treated as such, rather than as a strength-reduction factor. The value of K depends upon the shape of the wire and upon whether or not the stress is desired on the inner fiber of the coil or on the outer fiber. Wahl has analytically determined the following values for K for round wire:

$$K_i = \frac{4C^2 - C - 1}{4C(C - 1)} \qquad K_o = \frac{4C^2 + C - 1}{4C(C + 1)} \tag{10-32}$$

where C is the spring index and the subscripts i and o refer to the inner and outer fibers, respectively. In view of the fact that K_o is always less than unity, we shall use only K_i to determine stresses. When the bending moment $M = Fr$ and the section modulus $I/c = \pi d^3/32$ are substituted in Eq. (a), we obtain

$$\sigma = K_i\frac{32Fr}{\pi d^3} \tag{10-33}$$

which gives the bending stress for a round-wire torsion spring.

Deflection

The strain energy in bending is, from Eq. (3-32),

$$U = \int \frac{M^2 \, dx}{2EI} \tag{b}$$

For the torsion spring, $M = Fr$, and integration must be accomplished over the length of the wire. The force F will deflect through the distance $r\theta$, where θ is the total angular deflection of the spring. Applying Castigliano's theorem,

$$r\theta = \frac{\partial U}{\partial F} = \int_0^{\pi DN} \frac{\partial}{\partial F} \left(\frac{F^2 r^2 \, dx}{2EI} \right) = \int_0^{\pi DN} \frac{Fr^2 \, dx}{EI} \tag{c}$$

Substituting $I = \pi d^4/64$ for round wire and solving Eq. (c) for θ gives

$$\theta = \frac{64FrDN}{d^4 E} \tag{10-34}$$

where θ is the angular deflection of the spring in radians. The spring rate is therefore

$$k = \frac{Fr}{\theta} = \frac{d^4 E}{64DN} \tag{10-35}$$

The spring rate may also be expressed as the torque required to wind up the spring one turn. This is obtained by multiplying Eq. (10-35) by 2π. Thus

$$k' = \frac{d^4 E}{10.2DN} \tag{10-36}$$

These deflection equations have been developed without taking into account the curvature of the wire. Actual tests show that the constant 10.2 should be increased slightly. Thus the equation

$$k' = \frac{d^4 E}{10.8DN} \tag{10-37}$$

will give better results. Corresponding corrections may be made to Eqs. (10-34) and (10-35) if desired.

Torsion springs are frequently used over a round bar or pin. When a load is applied to a torsion spring, the spring winds up, causing a decrease in the inside diameter. Therefore it is necessary to ensure that the inside diameter of the spring never becomes equal to the diameter of the bar or pin; otherwise, a spring failure will occur. The inside diameter of a loaded torsion spring can be found from the equation

$$D_i' = \frac{N}{N'} D_i \tag{10-38}$$

where N = number of coils at no load
$\quad\quad D_i$ = inside diameter at no load
$\quad\quad N'$ = number of coils when loaded
$\quad\quad D_i'$ = inside diameter when loaded

Allowable tensile stresses for torsion springs can be estimated using Eq. (10-19). Using the distortion-energy theory, we divide the factor in each part of Eq. (10-19) by the quantity 0.577. This results in

$$S_y = \sigma_{all} = \begin{cases} 0.78S_{ut} & \text{cold-drawn carbon steel} \\ 0.87S_{ut} & \text{hardened and tempered carbon and} \\ & \quad \text{low-alloy steel} \\ 0.61S_{ut} & \text{austenitic stainless steel and} \\ & \quad \text{nonferrous alloys} \end{cases} \qquad (10\text{-}39)$$

The endurance limit for torsion springs can be found in a similar manner by using the value of S_{se} in Sec. 10-11. The result is

$$S_e = k_a k_b k_c S'_e = 78.0 \text{ kpsi (537 MPa)} \qquad (10\text{-}40)$$

Note that we have, in effect, substituted $k_c = 1$ for $k_c = 0.577$, which is the implied load factor in the result of Sec. 10-11. Of course, this result is for unpeened springs and has not been corrected for temperature or miscellaneous effects.

EXAMPLE 10-5

A stock torsion spring is shown in Fig. 10-10. It is made of 0.072-in music wire and has $4\frac{1}{4}$ total turns.

(a) Find the maximum operating torque and the corresponding angular rotation.

(b) What is the inside coil diameter when subject to the torque found in part (a)?

(c) Find the maximum operating torque and the angular rotation if failure is not to occur in an infinite number of cycles of operation.

Solution

(a) For music wire, we find from Table 10-5 that $A = 186$ kpsi and $m = 0.163$. Therefore, from Eq. (10-17),

$$S_{ut} = \frac{A}{d^m} = \frac{186}{(0.072)^{0.163}} = 286 \text{ kpsi}$$

Using Eq. (10-39), we estimate the yield strength as

$$S_y = 0.78S_{ut} = 0.78(286) = 223 \text{ kpsi}$$

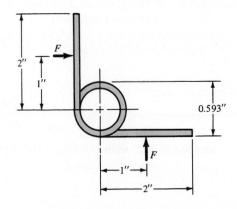

FIGURE 10-10

The mean coil diameter is $D = 0.593 - 0.072 = 0.521$ in. Then $C = D/d = 0.521/0.072 = 7.24$. The stress-concentration factor is found from the first part of Eq. (10-32) as

$$K_i = \frac{4(7.24)^2 - 7.24 - 1}{4(7.24)(7.24 - 1)} = 1.115$$

Now rearrange Eq. (10-33), substitute S_y for σ, and solve for the maximum torque Fr. This gives

Answer $$Fr = \frac{\pi d^3 S_y}{32 K_i} = \frac{\pi (0.072)^3 (223)(10^3)}{32(1.115)} = 7.33 \text{ lb} \cdot \text{in}$$

Note that no factor of safety has been used. One reason is that the value of S_y used is a permissible or allowable value. In addition, the design is considered safe because the assistance provided by the built-in winding stresses is significant.

Next, from Eq. (10-37), we estimate the spring rate to be

$$k' = \frac{d^4 E}{10.8 D N} = \frac{(0.072)^4 (30)(10^6)}{10.8(0.521)(4.25)} = 33.7 \text{ lb} \cdot \text{in/turn}$$

Thus a torque of $Fr = 7.33$ lb $\cdot$ in will wind the spring

Answer $$n = \frac{Fr}{k'} = \frac{7.33}{33.7} = 0.218 \text{ turns}$$

The corresponding angular deflection is

Answer $$\theta = 0.218(360°) = 78.5°$$

(b) With no load, the inside diameter of the spring is $D_i = 0.593 - 2(0.072) = 0.449$ in. At full torque, Eq. (10-38) gives the reduced inside diameter as

Answer $$D_i' = \frac{N}{N'} D_i = \frac{4.25}{4.25 + 0.218}(0.449) = 0.427 \text{ in}$$

(c) For the fatigue analysis, we assume that the torque $(Fr)_{\min}$ is zero. Then

$$(Fr)_a = (Fr)_m = \frac{(Fr)_{\max}}{2}$$

where the subscripts a and m refer to the alternating and mean components. We then have

$$\sigma_a = \sigma_m = S_a = S_m = K_i \frac{32(Fr)_{\max}}{2\pi d^3} = 1.115 \frac{32(Fr)_{\max}}{2\pi(0.072)^3}$$

$$= 15.2(10^3)(Fr)_{\max} \quad \text{psi}$$

Using the Goodman analysis with $S_e = 78.0$ kpsi [Eq. (10-40)], we have

$$\frac{S_a}{S_e} + \frac{S_m}{S_{ut}} = 1$$

or

$$15.2(Fr)_{\max}\left(\frac{1}{78.0} + \frac{1}{286}\right) = 1$$

Solving for the torque gives

Answer $(Fr)_{\max} = 4.03 \text{ lb} \cdot \text{in}$

This is the estimated spring torque corresponding to an infinite life. The corresponding maximum deflection is

Answer $\theta = \dfrac{4.03}{33.7}(360°) = 43.1°$

10-13   BELLEVILLE SPRINGS

The inset of Fig. 10-11 shows a coned-disk spring, commonly called a *Belleville spring*. Although the mathematical treatment is beyond the purposes of this book, you should at least be familiar with the remarkable characteristics of these springs.

Aside from the obvious advantage that a Belleville spring occupies only a small space, variation in the *h/t* ratio will produce a wide variety of load-deflection curve shapes, as illustrated in Fig. 10-11. For example, using an *h/t* ratio of 2.83 or larger gives an S curve which might be useful for snap-acting mechanisms. A reduction of the ratio to a value between 1.41 and 2.1 causes the central portion of the curve to become horizontal, which means that the load is constant over a considerable deflection range.

A higher load for a given deflection may be obtained by nesting, that is, by stacking

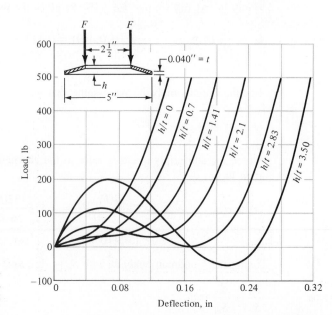

FIGURE 10-11

Load-deflection curves for Belleville springs. *(Courtesy of Associated Spring Corporation.)*

the springs in parallel. On the other hand, stacking in series provides a larger deflection for the same load, but in this case there is danger of instability.

10-14   MISCELLANEOUS SPRINGS

The extension spring shown in Fig. 10-12 is made of slightly curved strip steel, not flat, so that the force required to uncoil it remains constant; thus it is called a *constant-force spring*. This is equivalent to a zero spring rate. Such springs can also be manufactured having either a positive or a negative spring rate.

A *volute spring* is a wide, thin strip, or "flat," of steel wound on the flat so that the coils fit inside one another. Since the coils do not stack, the solid height of the spring is the width of the strip. A variable-spring scale, in a compression volute spring, is obtained by permitting the coils to contact the support. Thus, as the deflection increases, the number of active coils decreases. The volute spring, shown in Fig. 10-13a, has another important advantage which cannot be obtained with round-wire springs: if the coils are wound so as to contact or slide on one another during action, the sliding friction will serve to damp out vibrations or other unwanted transient disturbances.

A *conical spring,* as the name implies, is a coil spring wound in the shape of a cone. Most conical springs are compression springs and are wound with round wire. But a

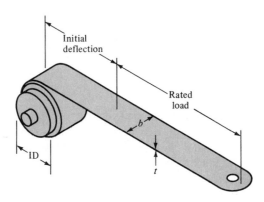

FIGURE 10-12

Constant-force spring. *(Courtesy of Vulcan Spring & Mfg. Co., Huntingdon Valley, Pa.)*

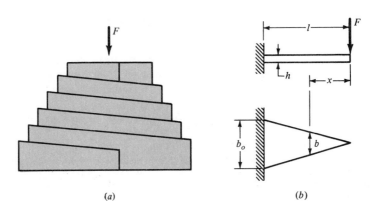

FIGURE 10-13

(a) A volute spring; (b) a flat triangular spring.

(a)   (b)

volute spring is a conical spring too. Probably the principal advantage of this type of spring is that it can be wound so that the solid height is only a single wire diameter.

Flat stock is used for a great variety of springs, such as clock springs, power springs, torsion springs, cantilever springs, and hair springs; frequently it is specially shaped to create certain spring actions for fuse clips, relay springs, spring washers, snap rings, and retainers.

In designing many springs of flat stock or strip material, it is often economical and of value to proportion the material so as to obtain a constant stress throughout the spring material. A uniform-section cantilever spring has a stress

$$\sigma = \frac{M}{I/c} = \frac{Fx}{I/c} \qquad\qquad (a)$$

which is proportional to the distance x if I/c is a constant. But there is no reason why I/c need be a constant. For example, one might design such a spring as that shown in Fig. 10-13b, in which the thickness h is constant but the width b is permitted to vary. Since, for a rectangular section, $I/c = bh^2/6$, we have, from Eq. (a),

$$\frac{bh^2}{6} = \frac{Fx}{\sigma}$$

or

$$b = \frac{6Fx}{h^2\sigma} \qquad\qquad (b)$$

Since b is linearly related to x, the width b_σ at the base of the spring is

$$b_\sigma = \frac{6Fl}{h^2\sigma} \qquad\qquad (10\text{-}41)$$

But the deflection of this triangular flat spring is more difficult to obtain, because the second moment of area is now a variable. Probably the quickest solution could be obtained by using singularity functions or the method of numerical integration.

The methods of stress and deflection analysis illustrated in previous sections of this chapter have served to illustrate that springs may be analyzed and designed by using the fundamentals discussed in the earlier chapters of this book. This is also true for most of the miscellaneous springs mentioned in this section, and you should now experience no difficulty in reading and understanding the literature of such springs.

PROBLEMS*

10-1   Make a mechanical drawing using two views, or a good freehand sketch, of a helical compression spring closed to its solid length and having a wire diameter of $\frac{1}{2}$ in, an outside diameter of 4 in, and one active coil. The spring is to have plain ends.

10-2   The same as Prob. 10-1, except that the ends are plain and ground.

10-3   A helical compression spring is wound using 0.105-in-diameter music wire. The spring has an outside diameter of 1.225 in with plain ground ends, and 12 total coils.

---

*An asterisk indicates a design-type problem.

(a) What should be the free length of the spring for which, when the spring is compressed solid, the stress is not greater than the yield strength?

(b) What force is needed to compress this spring to its solid length?

(c) Compute the spring rate.

(d) Is there any possibility that this spring might buckle in service?

10-4   A helical compression spring is made of hard-drawn spring steel wire of 2-mm diameter and has an outside diameter of 22 mm. The ends are plain and ground and there are $8\frac{1}{2}$ total coils.

(a) The spring is wound with a free length such that, when the spring is compressed solid, the stress will not exceed the torsional yield strength. Find the free length.

(b) What is the pitch of this spring?

(c) What force is needed to compress the spring to its solid length?

(d) What is the spring rate?

(e) Will the spring buckle in service?

10-5   The helical compression spring shown in the figure has the ends squared but not ground.

(a) Find the pitch, the solid length, and the number of active coils.

(b) What is the spring rate?

(c) If the spring is made of steel, what force is required to close it to its solid length?

(d) What is the stress in the spring due to the force found in part (c)?

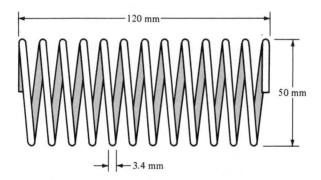

PROBLEM 10-5

10-6   A stock-size music-wire helical compression spring with ends squared and ground has an outside diameter of 0.850 in and a wire diameter of 0.074 in. The free length is 2 in and the solid length 0.618 in. Find the approximate load at the solid length and the spring rate.

10-7   A stock-size helical compression spring, with ends squared and ground, has a wire diameter of 1.40 mm, an outside diameter of 12.19 mm, and a solid length of 14.35 mm.

(a) What is the free length if the stress is never to be more than 90 percent of the torsional yield strength?

(b) What is the load corresponding to the solid length?

10-8*   Design a compression coil spring of music wire having squared and ground ends. The spring is to be assembled with a preload of 10 N and exert a force of 50 N when it is compressed an additional 140 mm. The force on the spring at the solid length should be about 25 percent more than the working force. Use only the sizes listed in Table 10-6.

10-9*   This problem is illustrative of many real design situations. Since it is also quite short, solving it will provide you with almost as much insight into the nature of design as would a lengthy design project. The problem provides you with the need to make decisions and with the opportunity, if

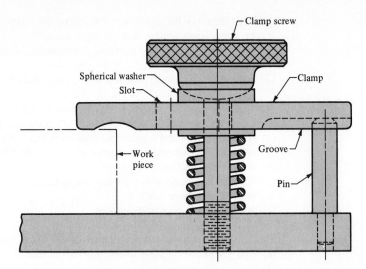

PROBLEM 10-9
Clamping fixture.

you wish to take it, of optimizing your solution on the basis of cost, wire size, full utilization of strength, or whatever.

Design a compression spring for the clamping fixture shown in the figure. The spring should exert a force of about 10 lb for a workpiece $1\frac{1}{2}$ in thick. The clamp screw has a diameter of $\frac{1}{2}$ in.

10-10*   Design a short helical compression spring to support a static load of 1200 lb. During assembly, impact produces an equivalent load of 2400 lb, and the spring should not bottom out in response to this load.

10-11   An extension spring has the following specifications:

Material, hard-drawn wire, $S_{yt} = 0.75S_{ut}$

Mean diameter $D = 10$ mm

Wire diameter $d = 2.0$ mm

Number of active coils $N_a = 120$

Ends, raised hook as in Fig. 10-2

Hook radius (Fig. 10-3a) $r_m = 6.0$ mm

Bend radius (Fig. 10-3b) $r_m = 3.0$ mm

Pre-tension $F_i = 30$ N

Distance between hooks (measured inside), 264 mm

(*a*) Estimate the tensile and torsional yield strengths of the wire.
(*b*) Compute the initial torsional stress in the wire.
(*c*) What is the spring rate?
(*d*) What force is required to cause the body of the spring to be stressed to the yield strength?
(*e*) What force is required to cause the torsional stress in the hook ends to reach the yield strength?
(*f*) What force is required to cause the normal stress in the hook ends to reach the tensile yield strength?
(*g*) What is the distance between the hook ends if the smallest of the three forces found in parts (*d*), (*e*), and (*f*) is applied?

10-12   Solve part (f) of Prob. 10-11 using curved-beam theory.

10-13   The extension spring shown in the figure has full twisted loop ends. The material is AISI 1065 oil-tempered wire. The spring has 84 coils and is close-wound with a preload of 16 lb.
(a) Find the closed length of the spring.
(b) Find the torsional stress in the spring corresponding to the preload.
(c) What is the estimated spring rate?
(d) What load would cause a permanent deformation?
(e) What is the spring deflection corresponding to the load found in part (d)?

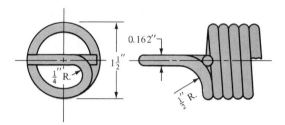

PROBLEM 10-13

10-14   A helical compression spring has a total of 14 coils, both ends squared and ground, a free length of $1\frac{1}{4}$ in, and an outside diameter of $\frac{7}{16}$ in; the material is hard-drawn wire of diameter 0.042-in. The operating forces are $F_{min} = 1.5$ lb and $F_{max} = 3.5$ lb.
(a) Compute the spring rate, the solid length, and the stress when the spring is compressed to the solid length.
(b) What is the factor of safety based on a possible fatigue failure?

10-15   An extension coil spring is made of 0.60-mm music wire and has an outside diameter of 5 mm. The spring is wound with a pre-tension of 1.0 N, and the load fluctuates from this value up to 6.0 N. Based on stresses in the body of the spring, find the factor of safety guarding against a fatigue failure.

10-16   A helical compression spring is wound with 3.0-mm-diameter music wire and has squared and ground ends. The spring has 9 total coils, an outside diameter of 28 mm, and a free length of 60 mm.
(a) Estimate the spring rate in newtons per millimeter.
(b) What factor of safety guards against a fatigue failure if the maximum force never exceeds 60 N?

10-17   Using a spread of 3 standard deviations, estimate the range of spring rates for the spring specified in Prob. 10-16. Use $C_G = 0.012$.

10-18   Estimate the commercial tolerance for the free length of the spring of Prob. 10-16.

10-19   A compression coil spring with squared and ground ends is wound using 0.162-in-diameter music wire. The spring has a total number of coils of 12.5, a nominal free-length specification of $4\frac{1}{2}$ in, and an outside diameter specified as 1.46 in. The coefficient of variation of the modulus of rigidity is $C_G = 0.010$.
(a) Estimate the free-length variate $\mathbf{L}_0$.
(b) Estimate the spring-rate variate $\mathbf{k}$.
(c) Estimate the deflection variate $\mathbf{y}$ due to a force $F = 50$ lb.
(d) Find the stress variate $\boldsymbol{\tau}$ caused by the force in part (c).

10-20   A compression coil spring is made of 2-mm music wire with both ends squared and ground and an outside diameter of 12.5 mm. The spring is subject to fatigue loads in the range of 10 to 130 N. Estimate the reliability based on infinite life. Use a coefficient of variation of $C = 0.05$ for both S_{se} and S_{su}, and assume that the quantity $S_{se} = 310$ MPa is the 99 percentile in a normal distribution.

10-21   The rat trap shown in the figure uses two opposite-image torsion springs. The wire has a diameter of 0.081 in, and the outside diameter of the spring in the position shown is $\frac{1}{2}$ in. Each spring has 11 turns. Use of a fish scale revealed a force of about 8 lb needed to set the trap.
(*a*) Find the probable configuration of the spring prior to assembly.
(*b*) Find the maximum stress in the spring when the trap is set.

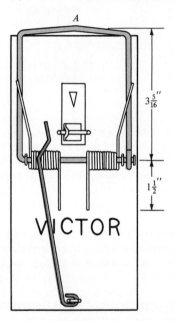

PROBLEM 10-21

10-22   A stock torsion spring is made of 0.055-in music wire and has 6 coils and straight ends 2 in long and 180° apart. The outside diameter is 0.654 in.
(*a*) What value of the torque would cause a maximum stress equal to the yield strength of the wire?
(*b*) If the torque found in part (*a*) is used as the maximum working torque, what is the smallest value of the inside diameter?
(*c*) Compute the angle of rotation corresponding to the torque found in part (*a*).

10-23   Based on commercial tolerances, what is the spread of the spring rate of the spring of Prob. 10-22? Use a coefficient of variation, for torsion springs, of 80 percent of that used for compression springs for coil diameter. For the modulus of elasticity, use $C_E = 0.02$.

10-24   The spring of Prob. 10-22 is to be used in an application in which the minimum torque is always 20 percent of the maximum. Estimate the maximum torque for an infinite life.

10-25   The figure shows a finger exerciser used by law-enforcement officers and athletes to strengthen their grip. It is formed by winding cold-drawn steel wire around a mandrel so as to obtain $2\frac{1}{2}$ turns when the grip is in the closed position. After winding, the wire is cut so as to leave the two

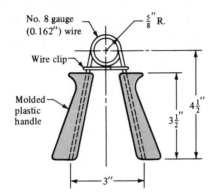

No. 8 gauge
(0.162") wire

$\frac{5}{8}''$ R.

Wire clip

Molded
plastic
handle

$4\frac{1}{2}''$

$3\frac{1}{2}''$

3"

PROBLEM 10-25

legs as handles. The plastic handles are then molded on, the grip squeezed together, and a wire clip placed around the legs to obtain initial "tension" and to space the handles for the best initial gripping position. The clip is formed like a figure eight to prevent it from coming off. The wire material is hard-drawn 0.60 carbon steel, plated. When the grip is in the closed position, the stress in the spring should not exceed the permissible stress. Determine

(*a*) The configuration of the spring before the grip is assembled.

(*b*) The force necessary to close the grip.

10-26   Specify the commercial tolerance on wire size, outside coil diameter, and free length if a spring has been specified as having a wire diameter of 0.060 in, an outside coil diameter of 0.600 in, squared and ground ends with five total turns, and a free length of 5.00 in using hard-drawn spring wire.

10-27   It is sometimes necessary to specify the free length and its tolerance and accept load variation without calling out a load tolerance in spring specifications. If the spring of Prob. 10-26 is specified in part as $d = 0.060 \pm 0.0006$ in, squared and ground ends, $D_o = 0.600 \pm 0.009$ in, hard-drawn AISI 1066 wire, $L_o = 5.000 \pm 0.066$ in with five total turns, and with the expectation that $C_G = 0.02$, then what variation in preload at a deflection of 0.6 in can be expected?

ANSWERS   **10-4** $L_0 = 50.5$ mm, $p = 5.94$ mm, $F_s = 88.4$ N; spring will not buckle.

10-6 $F_s = 20.08$ lb, $k = 14.53$ lb/in

10-10 One solution is $D_o = 4\frac{1}{2}$ in, $d = \frac{3}{4}$ in, $L_0 = 4.79$ in, $L_s = 3.75$ in, $N_a = 3$ coils.

10-13 (*a*) 16.45 in; (*b*) 13.6 kpsi; (*c*) 4.92 lb/in; (*d*) 93.6 lb; (*e*) 15.8 in

10-16 (*a*) 7.34 N/mm; (*b*) 3.02

10-17 $k = 7.34 \pm 0.45$ N/mm

10-22 (*a*) 3.53 lb · in; (*b*) 0.502 in; (*c*) 179.6°

10-24 2.13 lb · in

11
Rolling-Contact
Bearings

The terms *rolling-contact bearing, antifriction bearing,* and *rolling bearing* are all used to describe that class of bearing in which the main load is transferred through elements in rolling contact rather than in sliding contact. In a rolling bearing the starting friction is about twice the running friction, but still it is negligible in comparison with the starting friction of a sleeve bearing. Load, speed, and the operating viscosity of the lubricant do affect the frictional characteristics of a rolling bearing. It is probably a mistake to describe a rolling bearing as "antifriction," but the term is used generally throughout the industry.

From the mechanical designer's standpoint, the study of antifriction bearings differs in several respects when compared with the study of other topics because the bearings they specify have already been designed. The specialist in antifriction-bearing design is confronted with the problem of designing a group of elements which compose a rolling bearing; these elements must be designed to fit into a space whose dimensions are specified; they must be designed to receive a load having certain characteristics; and finally, these elements must be designed to have a satisfactory life when operated under the specified conditions. Bearing specialists must therefore consider such matters as fatigue loading, friction, heat, corrosion resistance, kinematic problems, material properties, lubrication, machining tolerances, assembly, use, and cost. From a consideration of all these factors, bearing specialists arrive at a compromise which, in their judgment, is a good solution to the problem as stated.

11-1 BEARING TYPES

Bearings are manufactured to take pure radial loads, pure thrust loads, or a combination of the two kinds of loads. The nomenclature of a ball bearing is illustrated in Fig. 11-1, which also shows the four essential parts of a bearing. These are the outer ring, the inner ring, the balls or rolling elements, and the separator. In low-priced bearings,

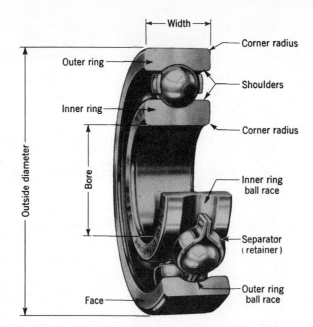

FIGURE 11-1

Nomenclature of a ball bearing.
*(Courtesy of New Departure–
Hyatt Division, General Motors
Corporation.)*

the separator is sometimes omitted, but it has the important function of separating the elements so that rubbing contact will not occur.

In this section we include a selection from the many types of standardized bearings which are manufactured. Most bearing manufacturers provide engineering manuals and brochures containing lavish descriptions of the various types available. In the small space available here, only a meager outline of some of the most common types can be given. So you should include a survey of bearing manufacturers' literature in your studies of this section.

Some of the various types of standardized bearings which are manufactured are shown in Fig. 11-2. The single-row deep-groove bearing will take radial load as well as some thrust load. The balls are inserted into the grooves by moving the inner ring to an eccentric position. The balls are separated after loading, and the separator is then inserted. The use of a filling notch (Fig. 11-2*b*) in the inner and outer rings enables a greater number of balls to be inserted, thus increasing the load capacity. The thrust capacity is decreased, however, because of the bumping of the balls against the edge of the notch when thrust loads are present. The angular-contact bearing (Fig. 11-2*c*) provides a greater thrust capacity.

All these bearings may be obtained with shields on one or both sides. The shields are not a complete closure but do offer a measure of protection against dirt. A variety of bearings are manufactured with seals on one or both sides. When the seals are on both sides, the bearings are lubricated at the factory. Although a sealed bearing is supposed to be lubricated for life, a method of relubrication is sometimes provided.

Single-row bearings will withstand a small amount of shaft misalignment or deflection, but where this is severe, self-aligning bearings may be used. Double-row bearings are made in a variety of types and sizes to carry heavier radial and thrust loads. Sometimes two single-row bearings are used together for the same reason, although a

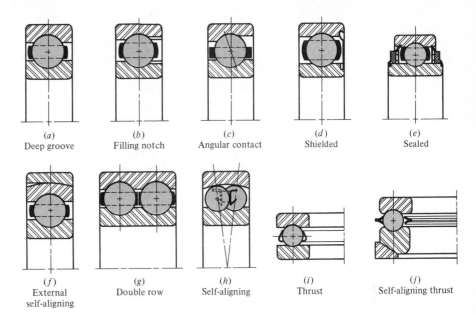

(a)
Deep groove

(b)
Filling notch

(c)
Angular contact

(d)
Shielded

(e)
Sealed

(f)
External
self-aligning

(g)
Double row

(h)
Self-aligning

(i)
Thrust

(j)
Self-aligning thrust

FIGURE 11-2

Various types of ball bearings.

double-row bearing will generally require fewer parts and occupy less space. The one-way ball thrust bearings (Fig. 11-2i) are made in many types and sizes.

Some of the large variety of standard roller bearings available are illustrated in Fig. 11-3. Straight roller bearings (Fig. 11-3a) will carry a greater load than ball bearings of the same size because of the greater contact area. However, they have the

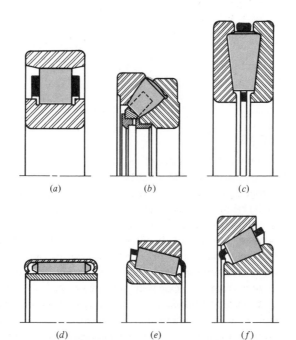

(a)

(b)

(c)

FIGURE 11-3

Types of roller bearings: (a) straight roller; (b) spherical roller thrust; (c) tapered roller thrust; (d) needle; (e) tapered roller; (f) steep-angle tapered roller. *(Courtesy of The Timken Company.)*

(d)

(e)

(f)

disadvantage of requiring almost perfect geometry of the raceways and rollers. A slight misalignment will cause the rollers to skew and get out of line. For this reason, the retainer must be heavy. Straight roller bearings will not, of course, take thrust loads.

Helical rollers are made by winding rectangular material into rollers, after which they are hardened and ground. Because of the inherent flexibility, they will take considerable misalignment. If necessary, the shaft and housing can be used for raceways instead of separate inner and outer races. This is especially important if radial space is limited.

The spherical-roller thrust bearing (Fig. 11-3b) is useful where heavy loads and misalignment occur. The spherical elements have the advantage of increasing their contact area as the load is increased.

Needle bearings (Fig. 11-3d) are very useful where radial space is limited. They have a high load capacity when separators are used, but may be obtained without separators. They are furnished both with and without races.

Tapered roller bearings (Fig. 11-3e, f) combine the advantages of ball and straight roller bearings, since they can take either radial or thrust loads or any combination of the two, and in addition, they have the high load-carrying capacity of straight roller bearings. The tapered roller bearing is designed so that all elements in the roller surface and the raceways intersect at a common point on the bearing axis.

The bearings described here represent only a small portion of the many available for selection. Many special-purpose bearings are manufactured, and bearings are also made for particular classes of machinery. Typical of these are:

- Instrument bearings, which are high-precision and are available in stainless steel and high-temperature materials
- Nonprecision bearings, usually made with no separator and sometimes having split or stamped sheet-metal races
- Ball bushings, which permit either rotation or sliding motion or both
- Bearings with flexible rollers

11-2   BEARING LIFE

When the ball or roller of an antifriction bearing rolls into the loading zone, Hertzian stresses occur on the inner ring, the rolling element, and the outer ring. Because the curvature of the contacting elements is different in the axial direction from what it is in the radial direction, the formulas for these stresses are much more complicated than the Hertzian equations presented in Sec. 2-21. If a bearing is clean and properly lubricated, is mounted and sealed against the entrance of dust or dirt, is maintained in this condition, and is operated at reasonable temperatures, then metal fatigue will be the only cause of failure. Since this implies many millions of stress applications, the term *bearing life* is in very general use.

The *life* of an *individual bearing* is defined as the total number of revolutions, or the number of hours at a given constant speed, of bearing operation required for the failure criteria to develop. Under ideal conditions the fatigue failure will consist of a spalling of the load-carrying surfaces. The Anti-Friction Bearing Manufacturers Association (AFBMA) standard states that the failure criterion is the first evidence of fatigue. It is noted, however, that the *useful life* is often used as the definition of fatigue life. The failure criterion used by the Timken Company laboratories is the spalling or pitting of an area of 0.01 in^2. But Timken observes that the useful life may extend considerably beyond this point.

Rating life is a term sanctioned by the AFBMA and used by most bearing manufacturers. The rating life of a group of apparently identical ball or roller bearings is defined as the number of revolutions, or hours at some given constant speed, that 90 percent of a group of bearings will complete or exceed before the failure criterion develops. The terms *minimum life*, L_{10} *life*, and B_{10} *life* are also used to denote rating life.

The terms *average life* and *median life* are both used quite generally in discussing the longevity of bearings. Both terms are intended to have the same import. When groups consisting of large numbers of bearings are tested to failure, the median lives of the groups are averaged. Thus, these terms are really intended to denote the *average median life*. In this book we shall use the term *median life* to signify the average of these medians.

In testing groups of bearings, the objective is to determine the median life and the L_{10} life, or rated life. When many groups of bearings are tested, it is found that the median life is somewhere between 4 and 5 times the L_{10} life. The graph of Fig. 11-4 shows approximately how the failures are distributed. This curve is only approximate; it must not be used for analytical or prediction purposes.

11-3   BEARING LOAD

Experiments show that two groups of identical bearings tested under different loads F_1 and F_2 will have respective lives L_1 and L_2 according to the relation

$$\frac{L_1}{L_2} = \left(\frac{F_2}{F_1}\right)^a \tag{11-1}$$

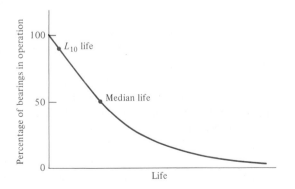

FIGURE 11-4

Typical curve of bearing life expectancy.

where $L = \begin{cases} \text{life, millions of revolutions or} \\ \text{life, hours at a given constant speed } n, \text{ in rev/min} \end{cases}$

$a = \begin{cases} 3 & \text{for ball bearings} \\ \frac{10}{3} & \text{for roller bearings} \end{cases}$

The AFBMA has established a standard load rating for bearings in which speed is not a consideration. This rating is called the basic load rating C. The *basic load rating is defined as the constant radial load which a group of apparently identical bearings can endure for a rating life of one million revolutions of the inner ring* (stationary load and stationary outer ring). The rating life of one million revolutions is a base value selected for ease of computation. The corresponding load rating is so high that plastic deformation of the contacting surfaces would occur were it actually applied. Consequently the basic load rating is purely a reference figure; such a large load would probably never be applied.

Using Eq. (11-1), we find the life of a bearing subjected to any other load F to be

$$L = \left(\frac{C}{F}\right)^a \tag{11-2}$$

where L from this equation is in millions of revolutions. The equation is more useful in the form

$$C = FL^{1/a} \tag{11-3}$$

Instead of tabulating the basic load rating for various sizes of bearings, most manufacturers prefer to publish ratings for their bearings corresponding to a certain number of hours of life at a specified speed. The Timken Company, for example, rates the bearings it manufactures for 3000 h at a speed of 500 rev/min. To illustrate the use of the equations above, suppose that a certain Timken bearing has a rated radial load of 2140 lb. Then the corresponding L_{10} life is

$$L_{10} = (3000 \text{ h})\left(\frac{60 \text{ min}}{\text{h}}\right)\left(\frac{500 \text{ rev}}{\text{min}}\right) = 90(10^6) \text{ rev}$$

Designating the corresponding rated load of 2140 lb as F_R means that the basic load rating is

$$C = F_R L^{1/a} = 2140(90)^{3/10} = 8263 \text{ lb}$$

Now, let

F_R = catalog radial rating, lb (kN)

L_R = catalog rated life, h

n_R = catalog rated speed, rev/min

F_D = required radial design load, lb (kN)

L_D = required design life, h

n_D = required design speed, rev/min

The designer's problem is: Given F_D, L_D, and n_D, what value of F_R should be used to enter the catalog to find an appropriate bearing? This problem can be solved using Eq. (11-1). The total number of design revolutions is

$$N_D = 60L_D n_D \qquad \text{rev}$$

and the total number of revolutions of the catalog bearing is

$$N_R = 60L_R n_R$$

Substituting in Eq. (11-1) gives

$$\frac{N_D}{N_R} = \left(\frac{F_R}{F_D}\right)^a$$

or

$$F_R = F_D \left(\frac{N_D}{N_R}\right)^{1/a} = F_D \left(\frac{L_D n_D}{L_R n_R}\right)^{1/a} \tag{11-4}$$

This means that the bearing selected to satisfy this design must have a catalog radial load rating equal to or greater than F_R.

11-4   BEARING SURVIVAL

The importance of knowing the probable survival of a group of bearings can be examined. Assume that the probability of failure for any single bearing is independent of that for the others in the same machine. If the machine is assembled with a total of N bearings, each having the same reliability R, then the reliability of the group must be

$$R_N = R^N$$

Suppose we have a gear-reduction unit consisting of six bearings, all loaded so that the L_{10} lives are equal. If the reliability of each bearing is 90 percent, the reliability of all the bearings in the assembly is

$$R_6 = (0.90)^6 = 0.531$$

This points up the need to select bearings having reliabilities greater than 90 percent.

The distribution of bearing failures at constant load can be best approximated by the Weibull distribution. When Eq. (4-22) is used, we have

$$R = \exp\left[-\left(\frac{x - x_0}{\theta - x_0}\right)^b\right] \tag{11-5}$$

where x = life measure
$\quad x_0$ = guaranteed value of life measure
$\quad \theta$ = Weibull characteristic life measure
$\quad b$ = Weibull shape parameter

To evaluate these parameters, we utilize the test data of Fig. 11-5, which were obtained from more than 2500 bearings tested. Table 11-1 has been obtained from Fig. 11-5. The three Weibull parameters can be obtained from these data using the method de-

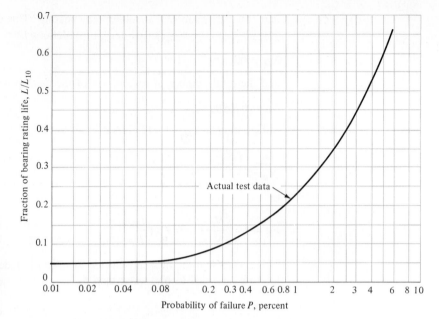

FIGURE 11-5

Reduction in life for reliabilities greater than 90 percent. Note that the abscissa is the probability of failure $P = 100 - R$, in percent. *(By permission from Tedric A. Harris, "Predicting Bearing Reliability," Machine Design, vol. 35, no. 1, Jan. 3, 1963, pp. 129–132.)*

scribed in Sec. 4-12. Using $x = L/L_{10}$, the results are found to be $x_0 = 0.02$, $\theta = 4.459$, and $b = 1.483$. Thus Eq. (11-5) gives the reliability as

$$R = \exp\left[-\left(\frac{L/L_{10} - 0.02}{4.439}\right)^{1.483}\right] \tag{11-6}$$

For example, if $L/L_{10} = 0.5$, this equation gives $R = 0.9637$.

The survival relation for tapered roller bearings utilizes the two-parameter Weibull distribution. Using Eq. (4-23), the result is*

$$R = \exp\left[-\left(\frac{x}{\theta}\right)^{b}\right] = \exp\left[-\left(\frac{L/L_{10}}{4.48}\right)^{1.5}\right] \tag{11-7}$$

11-5   THE RELIABILITY GOAL

As indicated in Sec. 11-3, the reliability is 90 percent when the design load F_D equals the catalog rated load F_R. In this section we wish to learn how to select bearings for any desired reliability.

---

*See Charles R. Mischke, "Rolling-Contact Bearings," chap. 27 in Joseph E. Shigley and Charles R. Mischke (eds.), *Standard Handbook of Machine Design,* McGraw-Hill, New York, 1986, p. 27.12. Timken uses the simpler $b = 1.5$ and $\theta = 4.48$ in its 1986 handbook.

TABLE 11-1

Coordinates of Points from Fig. 11-5

RELIABILITY R	LIFE MEASURE L/L_{10}	RELIABILITY R	LIFE MEASURE L/L_{10}
0.94	0.67	0.994	0.17
0.95	0.60	0.995	0.15
0.96	0.52	0.996	0.13
0.97	0.435	0.997	0.11
0.975	0.395	0.9975	0.095
0.98	0.35	0.998	0.08
0.985	0.29	0.9985	0.07
0.99	0.23	0.999	0.06
0.992	0.20	0.9995	0.05

We begin by noting that the reciprocal of Eq. (11-6) is

$$\frac{1}{R} = \exp\left[\left(\frac{L/L_{10} - 0.02}{4.439}\right)^{1.483}\right] \tag{a}$$

where L is the desired life and R is the desired reliability. The problem is to find the value of the L_{10} life that will satisfy these two requirements. So we solve Eq. (a) for L_{10} and get

$$L_{10} = \frac{L}{0.02 + 4.439[\ln{(1/R)}]^{1/1.483}} \tag{11-8}$$

To determine the catalog load rating, we incorporate Eq. (11-8) into Eq. (11-4) and get

$$F_R = F_D\left\{\frac{(L_D n_D/L_R n_R)}{0.02 + 4.439[\ln{(1/R)}]^{1/1.483}}\right\}^{1/a} \tag{11-9}$$

where F_R is the catalog radial load rating corresponding to L_R hours of life at the rated speed of n_R rev/min, and where F_D is the design radial load corresponding to the required life of L_D hours at a design speed of n_D rev/min and a reliability R.

A similar expression can be derived for tapered roller bearings. Beginning with Eq. (11-7), the result is found to be

$$F_R = F_D\left\{\frac{(L_D n_D/L_R n_R)}{4.48[\ln{(1/R)}]^{1/1.5}}\right\}^{3/10} \tag{11-10}$$

11-6   SELECTION OF BALL AND STRAIGHT ROLLER BEARINGS

Except for pure thrust bearings, as in Fig. 11-2i, ball bearings are usually operated with some combination of radial and thrust load. Since catalog ratings are based only on radial load, it is convenient to define an *equivalent radial load* F_e that will have the same effect on bearing life as do the applied loads. The AFBMA equation for equivalent radial load for ball bearings is the maximum of the two values

$$F_e = VF_r \tag{11-11}$$

$$F_e = XVF_r + YF_a \tag{11-12}$$

where F_e = equivalent radial load
F_r = applied radial load
F_a = applied thrust load
V = a rotation factor
X = a radial factor
Y = a thrust factor

In using these equations, the rotation factor V is to correct for the various rotating-ring conditions. For a rotating inner ring, $V = 1$. For a rotating outer ring, $V = 1.2$. The factor of 1.2 for outer-ring rotation is simply an acknowledgment that the fatigue life is reduced under these conditions. Self-aligning bearings are an exception; they have $V = 1$ for rotation of either ring.

The X and Y factors in Eq. (11-12) depend upon the geometry of the bearing, including the number of balls and the ball diameter. The AFBMA recommendations are based on the ratio of the thrust component F_a to the *basic static load rating* C_0 and a variable reference value e. The static load rating C_0 is tabulated, along with the basic dynamic load rating C, in many of the bearing manufacturers' publications; see Table 11-2, for example.

Since straight or cylindrical roller bearings will take no axial load, or very little, the Y factor is always zero.

The AFBMA has established standard boundary dimensions for bearings which define the bearing bore, the outside diameter (OD), the width, and the fillet sizes on the shaft and housing shoulders. The basic plan covers all ball and straight roller bearings in the metric sizes. The plan is quite flexible in that, for a given bore, there are an assortment of widths and outside diameters. Furthermore, the outside diameters se-

TABLE 11-2

Equivalent Radial-Load
Factors for Ball Bearings

F_a/C_0	e	$F_a/F_r \leq e$		$F_a/F_r > e$	
		X_1	Y_1	X_2	Y_2
0.014*	0.19	1.00	0	0.56	2.30
0.021	0.21	1.00	0	0.56	2.15
0.028	0.22	1.00	0	0.56	1.99
0.042	0.24	1.00	0	0.56	1.85
0.056	0.26	1.00	0	0.56	1.71
0.070	0.27	1.00	0	0.56	1.63
0.084	0.28	1.00	0	0.56	1.55
0.110	0.30	1.00	0	0.56	1.45
0.17	0.34	1.00	0	0.56	1.31
0.28	0.38	1.00	0	0.56	1.15
0.42	0.42	1.00	0	0.56	1.04
0.56	0.44	1.00	0	0.56	1.00

*Use 0.014 if $F_a/C_0 < 0.014$.

FIGURE 11-6

The basic AFBMA plan for boundary dimensions. These apply to ball bearings, straight roller bearings, and spherical roller bearings, but not to tapered roller bearings or to inch-series ball bearings. The contour of the corner is not specified; it may be rounded or chamfered, but it must be small enough to clear the fillet radius specified in the standards.

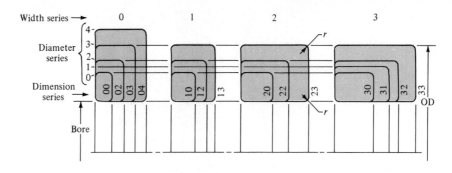

lected are such that, for a particular outside diameter, one can usually find a variety of bearings having different bores and widths.

This basic AFBMA plan is illustrated in Fig. 11-6. The bearings are identified by a two-digit number called the *dimension-series code*. The first number in the code is from the *width series,* 0, 1, 2, 3, 4, 5, and 6. The second number is from the *diameter series* (outside), 8, 9, 0, 1, 2, 3, and 4. Figure 11-6 shows the variety of bearings which may be obtained with a particular bore. Since the dimension-series code does not reveal the dimensions directly, it is necessary to resort to tabulations. The 02 series are used here as an example of what is available. See Table 11-3.

TABLE 11-3

Dimensions and Load Ratings for Single-Row 02-Series Deep-Groove and Angular-Contact Ball Bearings

BORE, mm	OD, mm	WIDTH, mm	FILLET RADIUS, mm	SHOULDER DIAMETER, mm d_S	d_H	LOAD RATINGS, kN DEEP GROOVE C	C_0	ANGULAR CONTACT C	C_0
10	30	9	0.6	12.5	27	5.07	2.24	4.94	2.12
12	32	10	0.6	14.5	28	6.89	3.10	7.02	3.05
15	35	11	0.6	17.5	31	7.80	3.55	8.06	3.65
17	40	12	0.6	19.5	34	9.56	4.50	9.95	4.75
20	47	14	1.0	25	41	12.7	6.20	13.3	6.55
25	52	15	1.0	30	47	14.0	6.95	14.8	7.65
30	62	16	1.0	35	55	19.5	10.0	20.3	11.0
35	72	17	1.0	41	65	25.5	13.7	27.0	15.0
40	80	18	1.0	46	72	30.7	16.6	31.9	18.6
45	85	19	1.0	52	77	33.2	18.6	35.8	21.2
50	90	20	1.0	56	82	35.1	19.6	37.7	22.8
55	100	21	1.5	63	90	43.6	25.0	46.2	28.5
60	110	22	1.5	70	99	47.5	28.0	55.9	35.5
65	120	23	1.5	74	109	55.9	34.0	63.7	41.5
70	125	24	1.5	79	114	61.8	37.5	68.9	45.5
75	130	25	1.5	86	119	66.3	40.5	71.5	49.0
80	140	26	2.0	93	127	70.2	45.0	80.6	55.0
85	150	28	2.0	99	136	83.2	53.0	90.4	63.0
90	160	30	2.0	104	146	95.6	62.0	106	73.5
95	170	32	2.0	110	156	108	69.5	121	85.0

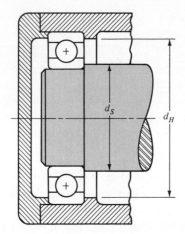

FIGURE 11-7

Shaft and housing shoulder diameters d_S and d_H should be adequate to ensure good bearing support.

The housing and shaft shoulder diameters listed in the tables should be used whenever possible to secure adequate support for the bearing and to resist the maximum thrust loads (Fig. 11-7). Table 11-4 lists the dimensions and load ratings of some straight roller bearings.

TABLE 11-4

Dimensions and Basic Load Ratings for Cylindrical Roller Bearings

	02 SERIES			03 SERIES		
BORE, mm	OD, mm	WIDTH, mm	LOAD RATING, kN	OD, mm	WIDTH, mm	LOAD RATING, kN
25	32	15	16.8	62	17	28.6
30	62	16	22.4	72	19	36.9
35	72	17	31.9	80	21	44.6
40	80	18	41.8	90	23	56.1
45	85	19	44.0	100	25	72.1
50	90	20	45.7	110	27	88.0
55	100	21	56.1	120	29	102
60	110	22	64.4	130	31	123
65	120	23	76.5	140	33	138
70	125	24	79.2	150	35	151
75	130	25	91.3	160	37	183
80	140	26	106	170	39	190
85	150	28	119	180	41	212
90	160	30	142	190	43	242
95	170	32	165	200	45	264
100	180	34	183	215	47	303
110	200	38	229	240	50	391
120	215	40	260	260	55	457
130	230	40	270	280	58	539
140	250	42	391	300	62	682
150	270	45	446	320	65	781

To assist the designer in the selection of bearings, most of the manufacturers' handbooks contain data on bearing life for many classes of machinery, as well as information on load-application factors. Such information has been accumulated the hard way, that is, by experience, and the beginner designer should utilize this information until he or she gains enough experience to know when deviations are possible. Table 11-5 contains recommendations on bearing life for some classes of machinery. The load-application factors in Table 11-6 serve the same purpose as factors of safety; use them to increase the equivalent load before selecting a bearing.

11-7   SELECTION OF TAPERED ROLLER BEARINGS

The nomenclature for a tapered roller bearing differs in some respects from that of ball and straight roller bearings. The inner ring is called the cone, and the outer ring is called the cup, as shown in Fig. 11-8. It can also be seen that a tapered roller bearing is separable in that the cup can be removed from the cone-and-roller assembly.

A tapered roller bearing can carry both radial and thrust (axial) loads or any combination of the two. However, even when an external thrust load is not present, the radial load will induce a thrust reaction within the bearing because of the taper. To avoid separation of the races and rollers, this thrust must be resisted by an equal and opposite force. One way of generating this force is to always use at least two tapered roller bearings on a shaft. These can be mounted with the cone backs facing each other, in the

TABLE 11-5		
Bearing-Life Recommendations for Various Classes of Machinery		

TYPE OF APPLICATION	LIFE, kh
Instruments and apparatus for infrequent use	Up to 0.5
Aircraft engines	0.5–2
Machines for short or intermittent operation where service interruption is of minor importance	4–8
Machines for intermittent service where reliable operation is of great importance	8–14
Machines for 8-h service which are not always fully utilized	14–20
Machines for 8-h service which are fully utilized	20–30
Machines for continuous 24-h service	50–60
Machines for continuous 24-h service where reliability is of extreme importance	100–200

TABLE 11-6	
Load-Application Factors	

TYPE OF APPLICATION	LOAD FACTOR
Precision gearing	1.0–1.1
Commercial gearing	1.1–1.3
Applications with poor bearing seals	1.2
Machinery with no impact	1.0–1.2
Machinery with light impact	1.2–1.5
Machinery with moderate impact	1.5–3.0

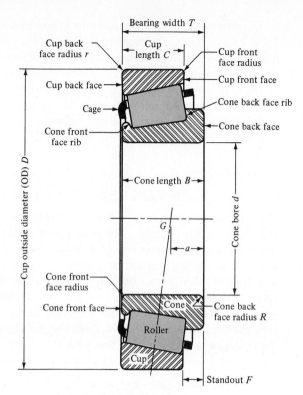

FIGURE 11-8

Nomenclature of a tapered roller bearing. Point *G* is the effective load center; use this point to calculate the radial bearing load. *(Courtesy of the Timken Company.)*

configuration called *direct mounting;* or with the cone fronts facing each other, in what is called *indirect mounting.*

The thrust component F_a produced by a pure radial load F_r is specified by the Timken Company as

$$F_a = \frac{0.47 F_r}{K} \tag{11-13}$$

where K is the ratio of the radial rating of the bearing to the thrust rating. The constant 0.47 is derived from a summation of the thrust components from the individual rollers supporting the load. The value of K is approximately 1.5 for radial bearings and 0.75 for steep-angle bearings. These values may be used for a preliminary bearing selection, after which the exact values may be obtained from the *Timken Engineering Journal* in order to verify the selection.

Figure 11-9 shows a typical bearing mounting subjected to an external thrust load T_e. The radial reactions F_{rA} and F_{rB} are computed by taking moments about the effective load centers G. The distance a (Fig. 11-8) is obtained from the catalog rating sheets. The equivalent radial loads are computed using an equation similar to Eq. (11-12), except that a rotation factor is not used with tapered roller bearings. We shall use subscripts A and B to designate each of the two bearings in Fig. 11-9. The equiva-

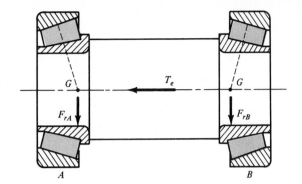

FIGURE 11-9

Schematic drawing showing a pair of tapered roller bearings assembled on a single shaft with direct mounting. The radial bearing forces are F_{rA} and F_{rB}. T_e is the external thrust.

lent radial load on bearing A is

$$F_{eA} = 0.4F_{rA} + K_A\left(\frac{0.47F_{rB}}{K_B} + T_e\right) \tag{11-14}$$

For bearing B, we have

$$F_{eB} = 0.4F_{rB} + K_B\left(\frac{0.47F_{rA}}{K_A} - T_e\right) \tag{11-15}$$

If the actual radial load on either bearing should happen to be larger than the corresponding value of F_e, then use the actual radial load instead of F_e for that bearing.

Figure 11-10 is a reproduction of a portion of a typical catalog page from the Timken Company *Bearing Selection Handbook*.

EXAMPLE 11-1   The gear-reduction unit shown in Fig. 11-11 is arranged to rotate the cup while the cone is stationary. Bearing A takes the thrust load of 250 lb and, in addition, has a radial load of 875 lb. Bearing B is subjected to a pure radial load of 625 lb. The speed is 150 rev/min. The desired L_{10} life is 90 kh. Note that this is the same as saying that a life of 90 kh is desired at a reliability of 90 percent. The desired shaft diameters are $1\frac{1}{8}$ in at A and 1 in at B. Select suitable tapered roller bearings, using an application factor of unity.

Solution   Since B carries only radial load, the thrust on A is augmented by the induced thrust due to B. Equation (11-14) applies. Using a trial value of 1.5 for both K values gives

$$F_{eA} = 0.4F_{rA} + K_A\left(\frac{0.47F_{rB}}{K_B} + T_e\right)$$

$$= 0.4(875) + 1.5\left[\frac{0.47(625)}{1.5} + 250\right] = 1020 \text{ lb} \tag{1}$$

Thus $F_{eA} > F_{rA}$, and so we use 1020 lb as the equivalent radial load to select bearing A. We next use Eq. (11-4) to obtain the L_{10} rating. As shown in Fig. 11-10 (columns 4 and 5), the load rating of Timken bearings is 300 h of L_{10} life at 500 rev/min. There-

FIGURE 11-10

Page 110 from The Timken Company *Bearing Selection Handbook*, revised 1986.

SINGLE-ROW STRAIGHT BORE

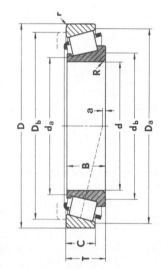

bore	outside diameter	width	rating at 500 rpm for 3000 hours L₁₀ one-row radial	thrust	factor	eff. load center	part numbers		cone				cup			
d	D	T	N lbf	N lbf	K	a②	cone	cup	max shaft fillet radius R①	width B	backing shoulder diameters db	da	max housing fillet radius r①	width C	backing shoulder diameters Db	Da
25.400 1.0000	65.088 2.5625	22.225 0.8750	13100 2950	16400 3690	0.80	-2.3 -0.09	23100	23256	1.5 0.06	21.463 0.8450	39.0 1.54	34.5 1.36	1.5 0.06	15.875 0.6250	53.0 2.09	63.0 2.48
25.400 1.0000	66.421 2.6150	23.812 0.9375	18400 4140	8000 1800	2.30	-9.4 -0.37	2687	2631	1.3 0.05	25.433 1.0013	33.5 1.32	31.5 1.24	1.3 0.05	19.050 0.7500	58.0 2.28	60.0 2.36
25.400 1.0000	68.262 2.6875	22.225 0.8750	15300 3440	10900 2450	1.40	-5.1 -0.20	02473	02420	0.8 0.03	22.225 0.8750	34.5 1.36	33.5 1.32	1.5 0.06	17.462 0.6875	59.0 2.32	63.0 2.48
25.400 1.0000	72.233 2.8438	25.400 1.0000	18400 4140	17200 3870	1.07	-4.6 -0.18	HM88630	HM88610	0.8 0.03	25.400 1.0000	39.5 1.56	39.5 1.56	2.3 0.09	19.842 0.7812	60.0 2.36	69.0 2.72
25.400 1.0000	72.626 2.8593	30.162 1.1875	22700 5110	13000 2910	1.76	-10.2 -0.40	3189	3120	0.8 0.03	29.997 1.1810	35.5 1.40	35.0 1.38	3.3 0.13	23.812 0.9375	61.0 2.40	67.0 2.64
26.157 1.0298	62.000 2.4409	19.050 0.7500	12100 2730	7280 1640	1.67	-5.8 -0.23	15103	15245	0.8 0.03	20.638 0.8125	33.0 1.30	32.5 1.28	1.3 0.05	14.288 0.5625	55.0 2.17	58.0 2.28

26.162 / 1.0300	63.100 / 2.4843	23.812 / 0.9375	18400 / 4140	8000 / 1800	2.30	−9.4 / −0.37	2682	2630	1.5 / 0.06	25.433 / 1.0013	34.5 / 1.36	32.0 / 1.26	0.8 / 0.03	19.050 / 0.7500	57.0 / 2.24	59.0 / 2.32
26.162 / 1.0300	66.421 / 2.6150	23.812 / 0.9375	18400 / 4140	8000 / 1800	2.30	−9.4 / −0.37	2682	2631	1.5 / 0.06	25.433 / 1.0013	34.5 / 1.36	32.0 / 1.26	1.3 / 0.05	19.050 / 0.7500	58.0 / 2.28	60.0 / 2.36
26.975 / 1.0620	58.738 / 2.3125	19.050 / 0.7500	11600 / 2610	6560 / 1470	1.77	−5.8 / −0.23	1987	1932	0.8 / 0.03	19.355 / 0.7620	32.5 / 1.28	31.5 / 1.24	1.3 / 0.05	15.080 / 0.5937	52.0 / 2.05	54.0 / 2.13
†26.988 / †1.0625	50.292 / 1.9800	14.224 / 0.5600	7210 / 1620	4620 / 1040	1.56	−3.3 / −0.13	L44649	L44610	3.5 / 0.14	14.732 / 0.5800	37.5 / 1.48	31.0 / 1.22	1.3 / 0.05	10.668 / 0.4200	44.5 / 1.75	47.0 / 1.85
†26.988 / †1.0625	60.325 / 2.3750	19.842 / 0.7812	11000 / 2480	6550 / 1470	1.69	−5.1 / −0.20	15580	15523	3.5 / 0.14	17.462 / 0.6875	38.5 / 1.52	32.0 / 1.26	1.5 / 0.06	15.875 / 0.6250	51.0 / 2.01	54.0 / 2.13
†26.988 / †1.0625	62.000 / 2.4409	19.050 / 0.7500	12100 / 2730	7280 / 1640	1.67	−5.8 / −0.23	15106	15245	0.8 / 0.03	20.638 / 0.8125	33.5 / 1.32	33.0 / 1.30	1.3 / 0.05	14.288 / 0.5625	55.0 / 2.17	58.0 / 2.28
†26.988 / †1.0625	66.421 / 2.6150	23.812 / 0.9375	18400 / 4140	8000 / 1800	2.30	−9.4 / −0.37	2688	2631	1.5 / 0.06	25.433 / 1.0013	35.0 / 1.38	33.0 / 1.30	1.3 / 0.05	19.050 / 0.7500	58.0 / 2.28	60.0 / 2.36
28.575 / 1.1250	56.896 / 2.2400	19.845 / 0.7813	11600 / 2610	6560 / 1470	1.77	−5.8 / −0.23	1985	1930	0.8 / 0.03	19.355 / 0.7620	34.0 / 1.34	33.5 / 1.32	0.8 / 0.03	15.875 / 0.6250	51.0 / 2.01	54.0 / 2.11
28.575 / 1.1250	57.150 / 2.2500	17.462 / 0.6875	11000 / 2480	6550 / 1470	1.69	−5.1 / −0.20	15590	15520	3.5 / 0.14	17.462 / 0.6875	39.5 / 1.56	33.5 / 1.32	1.5 / 0.06	13.495 / 0.5313	51.0 / 2.01	53.0 / 2.09
28.575 / 1.1250	58.738 / 2.3125	19.050 / 0.7500	11600 / 2610	6560 / 1470	1.77	−5.8 / −0.23	1985	1932	0.8 / 0.03	19.355 / 0.7620	34.0 / 1.34	33.5 / 1.32	1.3 / 0.05	15.080 / 0.5937	52.0 / 2.05	54.0 / 2.13
28.575 / 1.1250	58.738 / 2.3125	19.050 / 0.7500	11600 / 2610	6560 / 1470	1.77	−5.8 / −0.23	1988	1932	3.5 / 0.14	19.355 / 0.7620	39.5 / 1.56	33.5 / 1.32	1.3 / 0.05	15.080 / 0.5937	52.0 / 2.05	54.0 / 2.13
28.575 / 1.1250	60.325 / 2.3750	19.842 / 0.7812	11000 / 2480	6550 / 1470	1.69	−5.1 / −0.20	15590	15523	3.5 / 0.14	17.462 / 0.6875	39.5 / 1.56	33.5 / 1.32	1.5 / 0.06	15.875 / 0.6250	51.0 / 2.01	54.0 / 2.13
28.575 / 1.1250	60.325 / 2.3750	19.845 / 0.7813	11600 / 2610	6560 / 1470	1.77	−5.8 / −0.23	1985	1931	0.8 / 0.03	19.355 / 0.7620	34.0 / 1.34	33.5 / 1.32	1.3 / 0.05	15.875 / 0.6250	52.0 / 2.05	55.0 / 2.17

① These maximum fillet radii will be cleared by the bearing corners.
② Minus value indicates center is inside cone backface.
† For standard class ONLY, the maximum metric size is a whole millimetre value.
• For "J" part tolerances—see metric tolerances, page 73, and fitting practice, page 65.
ISO cone and cup combinations are designated with a common part number and should be purchased as an assembly.
◆ For ISO bearing tolerances—see metric tolerances, page 73, and fitting practice, page 65.

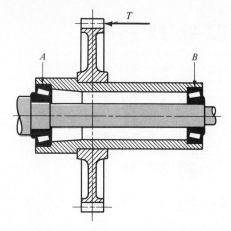

FIGURE 11-11

Tapered roller bearings applied to
a gear-reduction unit. *(Courtesy of
The Timken Company.)*

fore Eq. (11-4) gives the rating desired as

$$F_R = F_D \left(\frac{L_D n_D}{L_R n_R} \right)^{1/a} = 1020 \left[\frac{90(150)}{3(500)} \right]^{3/10} = 1970 \text{ lb} \tag{2}$$

Using this figure and a bore of $1\frac{1}{8}$ in, we enter the catalog sheets (Fig. 11-10 is typical) and select a 15590 cone and a 15520 cup. This selection has $K_A = 1.69$. The difference from the assumed value of 1.5 is small, and we will not recalculate F_{eA} at this time.

For bearing B, Eq. (11-14) applies. Thus

$$F_{eB} = 0.4 F_{rB} + K_B \left(\frac{0.47 F_{rA}}{K_A} - T_e \right)$$

$$= 0.4(625) + 1.5 \left[\frac{0.47(875)}{1.69} - 250 \right] = 240 \text{ lb}$$

Note that the actual value of K_A was used but K_B was assumed to be 1.5 as before. In this case, $F_{eB} < F_{rB}$, and so we use F_{rB} as the effective load. Using Eq. (11-4) again, we find the L_{10} desired rating as

$$F_R = 625 \left[\frac{90(150)}{3(500)} \right]^{3/10} = 1210 \text{ lb}$$

This bearing is to have a bore of 1 in. If we use Fig. 11-10 for the selection, there are five bearings from which to choose. The one at the top of the list has the smallest rating, the smallest OD, and the narrowest width. Thus we select the 23100 cone and 23256 cup. The L_{10} rating is 2950 lb with $K = 0.80$. The force rating is indeed ample.

If we now employ the correct K values and recalculate Eqs. (1) and (2), we find $F_R = 2690$ lb. But the bearing we selected had a rating of only 2480 lb. Since most of this increase seems to have been caused by the low value of K_B (0.80), we shall select the third bearing down on the list; this one has cone 02473, cup 02420, and $K = 1.40$. Solving Eqs. (1) and (2) for the third time yields $F_R = 2177$ lb, which is satisfactory.

11-8   LOAD-CYCLE ANALYSIS

Bearings are frequently subjected to cyclic loading as, for example, in the cycle *start, load, advance, unload, retract, stop*. Each phase of such a cycle can be expected to have its own set of operating characteristics. Let these be

F_{ei} = equivalent radial load for the ith event

T_i = time period of the ith event

n_i = speed of the ith event

Then

$$F_e = \left[\frac{\sum\limits_{i=1}^{i=j} T_i n_i (F_{ei})^a}{\sum\limits_{i=1}^{i=j} T_i n_i} \right]^{1/a} \tag{11-16}$$

When the character of loading changes from event to event, the appropriate application factors $(AF)_i$ can be prefixed to F_{ei}.

There are examples of variability of load within a revolution, as in cams, or in reciprocating compressors or engines. In such cases, Eq. (11-16) becomes an integral of the form

$$F_e = \left(\frac{1}{\phi} \int_0^\phi F^a \, d\theta \right)^{1/a} \tag{11-17}$$

where θ is the bearing rotation angle in radians and ϕ is the wavelength of the repetition in radians. Numerical integration procedures such as Simpson's rule can be useful here.

11-9   LUBRICATION

The contacting surfaces in rolling bearings have a relative motion that is both rolling and sliding, and so it is difficult to understand exactly what happens. If the relative velocity of the sliding surfaces is high enough, then the lubricant action is hydrodynamic (see Chap. 12). *Elastohydrodynamic lubrication* (EHD) is the phenomenon that occurs when a lubricant is introduced between surfaces that are in pure rolling contact. The contact of gear teeth and that found in rolling bearings and in cam-and-follower surfaces are typical examples. When a lubricant is trapped between two surfaces in rolling contact, a tremendous increase in the pressure within the lubricant film occurs. But viscosity is exponentially related to pressure, and so a very large increase in viscosity occurs in the lubricant that is trapped between the surfaces. Leibensperger* observes that the change in viscosity in and out of contact pressure is equivalent to the difference between cold asphalt and light sewing machine oil.

The purposes of an antifriction-bearing lubricant may be summarized as follows:

---

*R. L. Leibensperger, "When Selecting a Bearing," *Machine Design,* vol. 47, no. 8, April 3, 1975, pp. 142–147.

1 To provide a film of lubricant between the sliding and rolling surfaces

2 To help distribute and dissipate heat

3 To prevent corrosion of the bearing surfaces

4 To protect the parts from the entrance of foreign matter

Either oil or grease may be employed as a lubricant. The following rules may help in deciding between them.

USE GREASE WHEN	USE OIL WHEN
1. The temperature is not over 200°F.	1. Speeds are high.
2. The speed is low.	2. Temperatures are high.
3. Unusual protection is required from the entrance of foreign matter.	3. Oiltight seals are readily employed.
4. Simple bearing enclosures are desired.	4. Bearing type is not suitable for grease lubrication.
5. Operation for long periods without attention is desired.	5. The bearing is lubricated from a central supply which is also used for other machine parts.

11-10 MOUNTING AND ENCLOSURE

There are so many methods of mounting antifriction bearings that each new design is a real challenge to the ingenuity of the designer. The housing bore and shaft outside diameter must be held to very close limits, which of course is expensive. There are usually one or more counterboring operations, several facing operations, and drilling, tapping, and threading operations, all of which must be performed on the shaft, housing, or cover plate. Each of these operations contributes to the cost of production, so that the designer, in ferreting out a trouble-free and low-cost mounting, is faced with a difficult and important problem. The various bearing manufacturers' handbooks give many mounting details in almost every design area. In a text of this nature, however, it is possible to give only the barest details.

The most frequently encountered mounting problem is that which requires one bearing at each end of a shaft. Such a design might use one ball bearing at each end, one tapered roller bearing at each end, or a ball bearing at one end and a straight roller bearing at the other. One of the bearings usually has the added function of positioning or axially locating the shaft. Figure 11-12 shows a very common solution to this problem. The inner rings are backed up against the shaft shoulders and are held in position by round nuts threaded onto the shaft. The outer ring of the left-hand bearing is backed up against a housing shoulder and is held in position by a device which is not shown. The outer ring of the right-hand bearing floats in the housing.

There are many variations possible on the method shown in Fig. 11-12. For example, the function of the shaft shoulder may be performed by retaining rings, by the hub of a gear or pulley, or by spacing tubes or rings. The round nuts may be replaced by retaining rings or by washers locked in position by screws, cotters, or taper pins. The

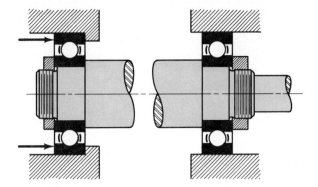

FIGURE 11-12

A common bearing mounting.

housing shoulder may be replaced by a retaining ring; the outer ring of the bearing may be grooved for a retaining ring, or a flanged outer ring may be used. The force against the outer ring of the left-hand bearing is usually applied by the cover plate, but if no thrust is present, the ring may be held in place by retaining rings.

Figure 11-13 shows an alternative method of mounting in which the inner races are backed up against the shaft shoulders as before but no retaining devices are required. With this method the outer races are completely retained. This eliminates the grooves or threads, which cause stress concentration on the overhanging end, but it requires accurate dimensions in an axial direction or the employment of adjusting means. This method has the disadvantage that if the distance between the bearings is great, the temperature rise during operation may expand the shaft enough to wreck the bearings.

It is frequently necessary to use two or more bearings at one end of a shaft. For example, two bearings could be used to obtain additional rigidity or increased load capacity or to cantilever a shaft. Several two-bearing mountings are shown in Fig. 11-14. These may be used with tapered roller bearings, as shown, or with ball bearings. In either case it should be noted that the effect of the mounting is to preload the bearings in an axial direction.

Figure 11-15 shows another two-bearing mounting. Note the use of washers against the cone backs.

When maximum stiffness and resistance to shaft misalignment is desired, pairs of

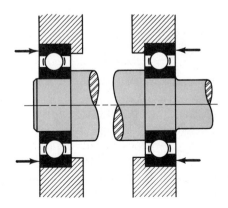

FIGURE 11-13

An alternative bearing mounting.

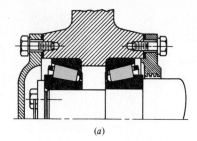
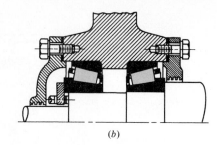

(a)   *(b)*

FIGURE 11-14

Two-bearing mountings. *(Courtesy of The Timken Company.)*

angular-contact ball bearings (Fig. 11-2) are often used in an arrangement called *duplexing*. Bearings manufactured for duplex mounting have their rings ground with an offset, so that when a pair of bearings is tightly clamped together, a preload is automatically established. As shown in Fig. 11-16, three mounting arrangements are used. The face-to-face mounting, called DF, will take heavy radial loads and thrust loads from either direction. The DB mounting (back to back) has the greatest aligning stiffness and is also good for heavy radial loads and thrust loads from either direction. The tandem arrangement, called the DT mounting, is used where the thrust is always in the same direction; since the two bearings have their thrust functions in the same direction, a preload, if required, must be obtained in some other manner.

Bearings are usually mounted with the rotating ring a press fit, whether it be the inner or outer ring. The stationary ring is then mounted with a push fit. This permits the stationary ring to creep in its mounting slightly, bringing new portions of the ring into the load-bearing zone to equalize wear.

Preloading

The object of preloading is to remove the internal clearance usually found in bearings, to increase the fatigue life, and to decrease the shaft slope at the bearing. Figure 11-17 shows a typical bearing in which the clearance is exaggerated for clarity.

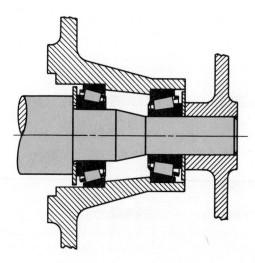

FIGURE 11-15

Mounting for a washing-machine spindle. *(Courtesy of The Timken Company.)*

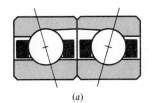
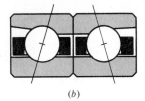
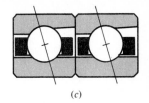

(a)   (b)   (c)

Preloading of straight roller bearings may be obtained by:

1 Mounting the bearing on a tapered shaft or sleeve to expand the inner ring

2 Using an interference fit for the outer ring

3 Purchasing a bearing with the outer ring preshrunk over the rollers

Ball bearings are usually preloaded by the axial load built in during assembly. However, the bearings of Fig. 11-16a and b are preloaded in assembly because of the differences in widths of the inner and outer rings.

It is always good practice to follow manufacturers' recommendations in determining preload, since too much will lead to early failure.

Alignment

Based on the general experience with rolling bearings as expressed in manufacturers' catalogs, the permissible misalignment in cylindrical and tapered roller bearings is limited to 0.001 rad. For spherical ball bearings, the misalignment should not exceed 0.0087 rad. But for deep-groove ball bearings, the allowable range of misalignment is 0.0035 to 0.0047 rad.

The life of the bearing decreases significantly when the misalignment exceeds the allowable limits. Figure 11-18 shows that there is about a 20 percent loss in life for every 0.001 rad of neutral-axis slope beyond 0.001 rad.

Additional protection against misalignment is obtained by providing the full shoulders (see Fig. 11-7) recommended by the manufacturer. Also, if there is any misalignment at all, it is good practice to provide a safety factor of around 2 to account for possible increases during assembly.

Clearance

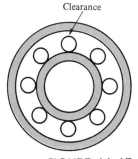

FIGURE 11-17

Clearance in an off-the-shelf bearing; exaggerated for clarity.

Enclosures

To exclude dirt and foreign matter and to retain the lubricant, the bearing mountings must include a seal. The three principal methods of sealings are the felt seal, the commercial seal, and the labyrinth seal (Fig. 11-19).

Felt seals may be used with grease lubrication when the speeds are low. The rubbing surfaces should have a high polish. Felt seals should be protected from dirt by placing them in machined grooves or by using metal stampings as shields.

The *commercial seal* is an assembly consisting of the rubbing element and, generally, a spring backing, which are retained in a sheet-metal jacket. These seals are usually made by press-fitting them into a counterbored hole in the bearing cover. Since

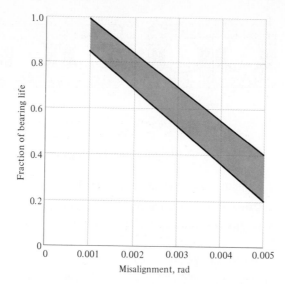

FIGURE 11-18

Effect of misalignment on bearing
life for line-contact bearings.

FIGURE 11-19

Typical sealing methods. *(Cour-
tesy of New Departure–Hyatt
Division, General Motors
Corporation.)*

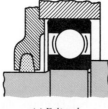  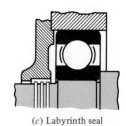

(*a*) Felt seal   (*b*) Commercial seal   (*c*) Labyrinth seal

they obtain the sealing action by rubbing, they should not be used for high speeds.

The *labyrinth seal* is especially effective for high-speed installations and may be used with either oil or grease. It is sometimes used with flingers. At least three grooves should be used, and they may be cut on either the bore or the outside diameter. The clearance may vary from 0.010 to 0.040 in, depending upon the speed and temperature.

PROBLEMS

11-1 A certain application requires a bearing to last for 1800 h with a reliability of 90 percent. What should be the rated life of the bearing?

11-2 A ball bearing is to be selected to withstand a radial load of 4 kN and have an L_{10} life of 1200 h at a speed of 600 rev/min. The bearing maker's catalog rating sheets are based on an L_{10} life of 3800 h at 500 rev/min. What load should be used to enter the catalog?

11-3 Suppose the bearing selected for Prob. 11-2 has a catalog load rating of 3.8 kN. What is the reliability of this application?

11-4   A certain application requires a bearing to last for 1800 h with a reliability of 96 percent. What should be the rated life for this application?

11-5   A certain ball-bearing manufacturer's catalog ratings are based not on L_{10} life, but on average life. A certain bearing in this catalog has a rated load of 1570 lb at a speed of 1800 rev/min, and an average life of 3800 h. To what basic load rating does this correspond?

11-6   An 02-series ball bearing is to be selected to carry a radial load of 8 kN and a thrust load of 4 kN. The L_{10} life is to be 5000 h with inner-ring rotation of 900 rev/min. What basic load rating should be used in selecting the bearing?

11-7   The bearing of Prob. 11-6 is to be sized to have a reliability of 99 percent. What basic load rating should be used in selecting the bearing?

11-8   A straight roller bearing is subjected to a radial load of 12 kN. The life is to be 4000 h at a speed of 750 rev/min. What load rating should be used to enter the bearing catalog?

11-9   Shown in the figure is a gear-driven squeeze roll which mates with an idler roll, not shown. The roll is designed to exert a normal force of 30 lb/in of roll length and a pull of 24 lb/in on the material being processed. The roll speed is 300 rev/min, and an L_{10} life of 30 kh is desired. Use an application factor of 1.2 and select a pair of radial-contact 02-series ball bearings to be mounted at O and A. Use same-size bearings.

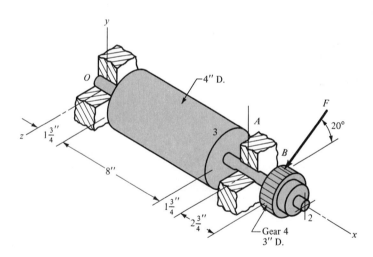

PROBLEM 11-9

11-10   The figure shows a geared countershaft with an overhanging pinion at C. Select a plain radial-contact ball bearing for mounting at O and a straight roller bearing for mounting at B. The force on gear A is $F_A = 600$ lb, and the shaft is to run at a speed of 480 rev/min. Solution of the statics problem gives the force of the bearings against the shaft at O as $\mathbf{R}_O = 388\mathbf{j} + 471\mathbf{k}$ lb and at B as $\mathbf{R}_B = 317\mathbf{j} - 1620\mathbf{k}$ lb. Find the size of bearings required, using an application factor of 1.4 and an L_{10} life of 50 kh.

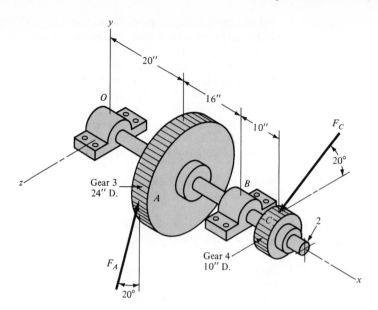

PROBLEM 11-10

11-11   The figure is a schematic drawing of a countershaft that supports two V-belt pulleys. The countershaft runs at 1200 rev/min and the bearings are to have a life of 60 kh at a reliability of 0.999. The belt tension on the loose side of pulley A is 15 percent of the tension on the tight side. Find the bearing reactions at O and E due to the belt pulls; assume the belt pulls are parallel. Select ball bearings for use at O and E, each to have a 25-mm bore, with an application factor of unity.

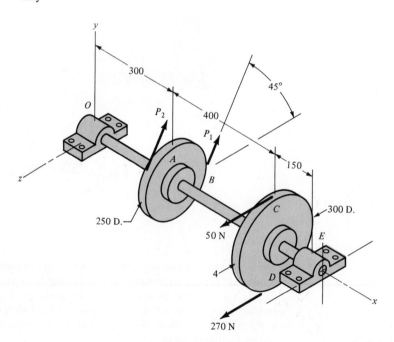

PROBLEM 11-11
Dimensions in millimeters.

11-12   A countershaft is supported by two tapered roller bearings using indirect mounting. The radial bearing loads are 1120 lb for the left-hand bearing and 2190 lb for the right-hand bearing. The shaft rotates at 400 rev/min and is to have an L_{10} life of 40 kh and an application factor of 1.4. Assume $K = 1.5$, and find the required radial rating for each bearing.

11-13   A gear-reduction unit uses the countershaft shown in the figure. Find the two bearing reactions. These bearings are to be plain radial ball bearings, selected for an L_{10} life of 40 kh corresponding to a shaft speed of 400 rev/min. Use 1.2 for the application factor and specify the bearings selected.

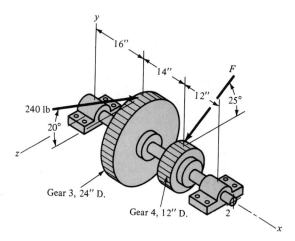

PROBLEM 11-13

11-14   The worm shaft shown in part *a* of the figure transmits 1.35 hp at 600 rev/min. A static force analysis gave the results shown in part *b* of the figure. Bearing *A* is to be an angular-contact ball

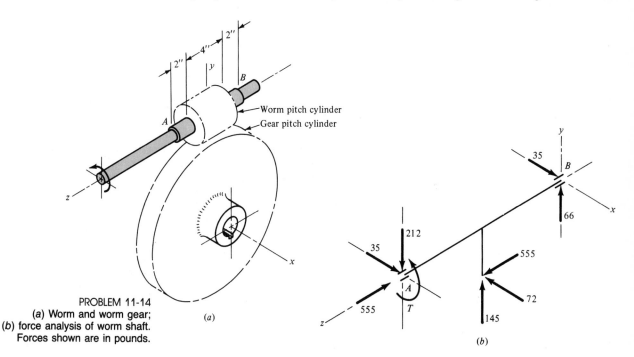

PROBLEM 11-14
(a) Worm and worm gear;
(b) force analysis of worm shaft.
Forces shown are in pounds.

bearing mounted to take the 555-lb thrust load. The bearing at B is to take only radial load, and so a straight roller bearing will be employed. Use an application factor of 1.3 and a life of 25 kh corresponding to a reliability of 99 percent, and specify each bearing.

11-15   The ratio of median life x to rating life $\tilde{x}_{10}$ can be expected to vary from manufacturer to manufacturer because of geometric and metallurgical differences. For a Timken roller bearing where $R = \exp[-(x/4.48)^{1.5}]$, what ratio $\tilde{x}/x_{10}$ has been the Timken experience as revealed by their two-parameter Weibull equation? Examine $\bar{x}/x_{10}$ as well.

11-16   The Torrington experience with their self-aligning spherical roller bearing has led to the survival equation

$$R = \exp\left[-\left(\frac{x - 0.0403}{6.58}\right)^{1.165}\right]$$

What ratio $\tilde{x}/x_{10}$ is revealed by this three-parameter Weibull survival equation?

11-17   The SKF experience with their ball bearings has led to the Weibull survival equation

$$R = \exp\left[-\left(\frac{x - 0.02}{4.439}\right)^{1.483}\right]$$

What ratio of median to rating life $\tilde{x}/x_{10}$ is revealed by the survival equation?

11-18   In a bearing tested at 2000 rev/min with a steady radial load of 18 kN, a set of bearings showed a B_{10} life of 115 h and a B_{80} life of 600 h. The basic load rating for this bearing is 39.6 kN. Estimate the Weibull shape factor b and the characteristic life θ for a two-parameter model. This manufacturer rates ball bearings at 1 million revolutions.

11-19   Different bearing metallurgy affects bearing life. A manufacturer reports that a particular heat treatment increases bearing life at least threefold. A bearing identical to that of Prob. 11-18, except for the heat treatment, loaded to 18 kN and run at 2000 rev/min revealed a B_{10} life of 360 h and a B_{80} life of 2000 h. Do you agree with the manufacturer's assertion concerning increased life?

11-20   Estimate the remaining life in revolutions of an 02-series 30-mm angular-contact ball bearing already subjected to 200 000 revolutions with a radial load of 18 kN and which is now to be subjected to a radial load of 30 kN.

11-21   The same 02-30 angular-contact ball bearing as in the previous problem is to be subjected to a two-step loading cycle of 4 min with a loading of 18 kN and of 6 min with a loading of 30 kN. This cycle is to be repeated until failure. Estimate the total life in revolutions.

ANSWERS   **11-2**   2.89 kN
               11-4   3373 h
               11-6   68.5 kN
               11-8   57.0 kN
               11-10   At O, $F_R = 42.9$ kN; use 55-mm-bore 02 series.
               11-13   Bearing C: 30-mm bore, 02 series; $F_R = 15.7$ kN
               11-18 $b = 1.65$; $\theta = 5.07$

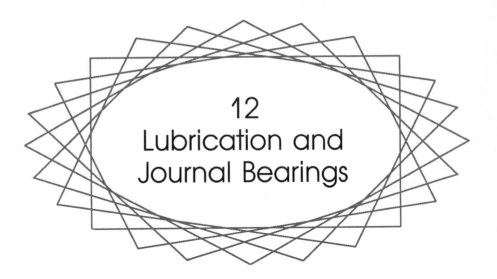

12
Lubrication and
Journal Bearings

The object of lubrication is to reduce friction, wear, and heating of machine parts which move relative to each other. A lubricant is any substance which, when inserted between the moving surfaces, accomplishes these purposes. In a sleeve bearing, a shaft, or *journal,* rotates or oscillates within a sleeve, or *bearing,* and the relative motion is sliding. In an antifriction bearing, the main relative motion is rolling. A follower may either roll or slide on the cam. Gear teeth mate with each other by a combination of rolling and sliding. Pistons slide within their cylinders. All these applications require lubrication to reduce friction, wear, and heating.

The field of application for journal bearings is immense. The crankshaft and connecting-rod bearings of an automotive engine must operate for thousands of miles at high temperatures and under varying load conditions. The journal bearings used in the steam turbines of power-generating stations are said to have reliabilities approaching 100 percent. At the other extreme there are thousands of applications in which the loads are light and the service relatively unimportant; a simple, easily installed bearing is required, using little or no lubrication. In such cases an antifriction bearing might be a poor answer because of the cost, the elaborate enclosures, the close tolerances, the radial space required, the high speeds, or the increased inertial effects. Instead, a nylon bearing requiring no lubrication, a powder-metallurgy bearing with the lubrication ''built in,'' or a bronze bearing with ring oiling, wick-feeding, or solid-lubricant film or grease lubrication might be a very satisfactory solution. Recent metallurgy developments in bearing materials, combined with increased knowledge of the lubrication process, now make it possible to design journal bearings with satisfactory lives and very good reliabilities.

Much of the material we have studied thus far in this book has been based on fundamental engineering studies, such as statics, dynamics, the mechanics of solids, metal processing, mathematics, and metallurgy. In the study of lubrication and journal bearings, additional fundamental studies, such as chemistry, fluid mechanics, thermodynamics, and heat transfer, must be utilized in developing the material. While we shall not utilize all of them in the material to be included here, you can now begin to appreciate better how the study of mechanical engineering design is really an integration of most of your previous studies and a directing of this total background toward the resolution of a single objective.

12-1   TYPES OF LUBRICATION

Five distinct forms of lubrication may be identified:

1 Hydrodynamic

2 Hydrostatic

3 Elastohydrodynamic

4 Boundary

5 Solid-film

Hydrodynamic lubrication means that the load-carrying surfaces of the bearing are separated by a relatively thick film of lubricant, so as to prevent metal-to-metal contact, and that the stability thus obtained can be explained by the laws of fluid mechanics. Hydrodynamic lubrication does not depend upon the introduction of the lubricant under pressure, though that may occur; but it does require the existence of an adequate supply at all times. The film pressure is created by the moving surface itself pulling the lubricant into a wedge-shaped zone at a velocity sufficiently high to create the pressure necessary to separate the surfaces against the load on the bearing. Hydrodynamic lubrication is also called *full-film,* or *fluid, lubrication.*

Hydrostatic lubrication is obtained by introducing the lubricant, which is sometimes air or water, into the load-bearing area at a pressure high enough to separate the surfaces with a relatively thick film of lubricant. So, unlike hydrodynamic lubrication, this kind of lubrication does not require motion of one surface relative to another. We shall not deal with hydrostatic lubrication in this book, but the subject should be considered in designing bearings where the velocities are small or zero and where the frictional resistance is to be an absolute minimum.

Elastohydrodynamic lubrication is the phenomenon that occurs when a lubricant is introduced between surfaces which are in rolling contact, such as mating gears or rolling bearings. The mathematical explanation requires the Hertzian theory of contact stress and fluid mechanics.

Insufficient surface area, a drop in the velocity of the moving surface, a lessening in the quantity of lubricant delivered to a bearing, an increase in the bearing load, or an increase in lubricant temperature resulting in a decrease in viscosity—any one of these—may prevent the buildup of a film thick enough for full-film lubrication. When this happens, the highest asperities may be separated by lubricant films only several molecular dimensions in thickness. This is called *boundary lubrication.* The change from hydrodynamic to boundary lubrication is not at all a sudden or abrupt one. It is probable that a mixed hydrodynamic- and boundary-type lubrication occurs first, and as the surfaces move closer together, the boundary-type lubrication becomes predominant. The viscosity of the lubricant is not of as much importance with boundary lubrication as is the chemical composition.

When bearings must be operated at extreme temperatures, a *solid-film lubricant* such as graphite or molybdenum disulfide must be used because the ordinary mineral oils are not satisfactory. Much research is currently being carried out in an effort, too, to find composite bearing materials with low wear rates as well as small frictional coefficients.

12-2   VISCOSITY

In Fig. 12-1 let a plate A be moving with a velocity U on a film of lubricant of thickness h. We imagine the film as composed of a series of horizontal layers and the force F causing these layers to deform or slide on one another just like a deck of cards. The layers in contact with the moving plate are assumed to have a velocity U; those in contact with the stationary surface are assumed to have a zero velocity. Intermediate layers have velocities which depend upon their distances y from the stationary surface. Newton's law of viscous flow states that the shear stress in the fluid is proportional to the rate of change of velocity with respect to y. Thus

$$\tau = \frac{F}{A} = \mu \frac{du}{dy} \tag{12-1}$$

where μ is the constant of proportionality and defines *absolute viscosity,* also called *dynamic viscosity*. The derivative du/dy is the rate of change of velocity with distance and may be called the rate of shear, or the velocity gradient. The viscosity μ is thus a measure of the internal frictional resistance of the fluid. If the assumption is made that the rate of shear is a constant, then $du/dy = U/h$, and from Eq. (12-1),

$$\tau = \frac{F}{A} = \mu \frac{U}{h} \tag{12-2}$$

The unit of viscosity in the ips system is seen to be the pound-force-second per square inch; this is the same as stress or pressure multiplied by time. The ips unit is called the *reyn*, in honor of Sir Osborne Reynolds.

The absolute viscosity is measured by the pascal-second (Pa · s) in SI; this is the same as a newton-second per square meter. The conversion from ips units to SI is the same as for stress. For example, multiply the absolute viscosity in reyns by 6890 to convert to units of Pa · s.

The American Society of Mechanical Engineers (ASME) has published a list of cgs

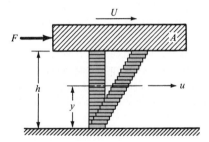

FIGURE 12-1

units which are not to be used in ASME documents.* This list results from a recommendation by the International Committee of Weights and Measures (CIPM) that the use of cgs units with special names be discouraged. Included in this list is a unit of force called the *dyne* (dyn), a unit of dynamic viscosity called the *poise* (P), and a unit of kinematic viscosity called the *stoke* (St). All of these units have been, and still are, used extensively in lubrication studies.

The poise is the cgs unit of dynamic or absolute viscosity, and its unit is the dyne-second per square centimeter (dyn · s/cm^2). It has been customary to use the centipoise (cP) in analysis, because its value is more convenient. When the viscosity is expressed in centipoises, it is designated by Z. The conversion from cgs units to SI and ips units is as follows:

$$\mu \ (\text{Pa} \cdot \text{s}) = (10)^{-3} Z \ (\text{cP})$$

$$\mu \ (\text{reyn}) = \frac{Z \ (\text{cP})}{6.89(10)^6}$$

The ASTM standard method for determining viscosity uses an instrument called the Saybolt Universal Viscosimeter. The method consists of measuring the time in seconds for 60 ml of lubricant at a specified temperature to run through a tube 17.6 mm in diameter and 12.25 mm long. The result is called the *kinematic viscosity,* and in the past the unit of the square centimeter per second has been used. One square centimeter per second is defined as a stoke. By the use of the *Hagen-Poiseuille law,* the kinematic viscosity based upon seconds Saybolt, also called *Saybolt Universal viscosity* (SUV) in seconds, is

$$Z_k = \left(0.22t - \frac{180}{t} \right) \tag{12-3}$$

where Z_k is in centistokes (cSt) and t is the number of seconds Saybolt.

In SI, the kinematic viscosity ν has the unit of the square meter per second (m^2/s), and the conversion is

$$\nu \ (\text{m}^2/\text{s}) = 10^{-6} Z_k \ (\text{cSt})$$

Thus, Eq. (12-3) becomes

$$\nu = \left(0.22t - \frac{180}{t} \right) (10^{-6}) \tag{12-4}$$

To convert to dynamic viscosity, we multiply ν by the density in SI units. Designating the density as ρ with the unit of the kilogram per cubic meter, we have

$$\mu = \rho \left(0.22t - \frac{180}{t} \right) (10^{-6}) \tag{12-5}$$

where μ is in pascal-seconds.

Figure 12-2 shows the absolute viscosity in the ips system of a number of fluids often used for lubrication purposes and their variation with temperature.

---

*ASME Orientation and Guide for Use of Metric Units, 2d ed., American Society of Mechanical Engineers, 1972, p. 13.

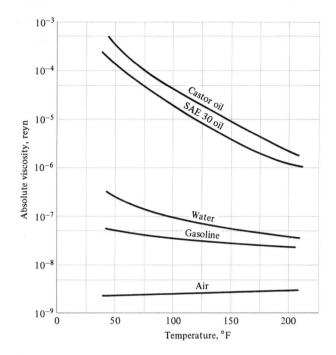

FIGURE 12-2

A comparison of the viscosities of various fluids.

12-3 PETROFF'S LAW

The phenomenon of bearing friction was first explained by Petroff using the assumption that the shaft is concentric. Though we shall seldom make use of Petroff's method of analysis in the material to follow, it is important because it defines groups of dimensionless parameters and because the coefficient of friction predicted by this law turns out to be quite good even when the shaft is not concentric.

Let us now consider a vertical shaft rotating in a guide bearing. It is assumed that the bearing carries a very small load, that the clearance space c is completely filled with oil, and that leakage is negligible (Fig. 12-3). We denote the radius of the shaft by r, the radial clearance by c, and the length of the bearing by l, all dimensions being in inches. If the shaft rotates at N rev/s, then its surface velocity is $U = 2\pi rN$ in/s. Since the shearing stress in the lubricant is equal to the velocity gradient times the viscosity, from Eq. (12-2) we have

$$\tau = \mu\frac{U}{h} = \frac{2\pi r\mu N}{c} \qquad (a)$$

where the radial clearance c has been substituted for the distance h. The force required to shear the film is the stress times the area. The torque is the force times the lever arm. Thus

$$T = (\tau A)(r) = \left(\frac{2\pi r\mu N}{c}\right)(2\pi rl)(r) = \frac{4\pi^2 r^3 l\mu N}{c} \qquad (b)$$

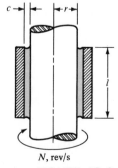

N, rev/s

FIGURE 12-3

If we now designate a small force on the bearing by W, in pounds-force, then the pressure P, in pounds-force per square inch of projected area, is $P = W/2rl$. The frictional force is fW, where f is the coefficient of friction, and so the frictional torque is

$$T = fWr = (f)(2rlP)(r) = 2r^2flP \qquad (c)$$

Substituting the value of the torque from Eq. (c) in Eq. (b) and solving for the coefficient of friction, we find

$$f = 2\pi^2 \frac{\mu N}{P} \frac{r}{c} \qquad (12\text{-}6)$$

Equation (12-6) is called *Petroff's law* and was first published in 1883. The two quantities $\mu N/P$ and r/c are very important parameters in lubrication. Substitution of the appropriate dimensions in each parameter will show that they are dimensionless.

The *bearing characteristic number,* or the *Sommerfeld number,* is defined by the equation

$$S = \left(\frac{r}{c}\right)^2 \frac{\mu N}{P} \qquad (12\text{-}7)$$

where S = bearing characteristic number
      r = journal radius, in
      c = radial clearance, in
      μ = absolute viscosity, reyn
      N = significant speed, rev/s
      P = load per unit of projected bearing area, psi

The Sommerfeld number is very important in lubrication analysis because it contains all the variables usually specified by the designer. Note that it is also dimensionless. The quantity r/c is called the *clearance ratio*. If we multiply both sides of Eq. (12-6) by this ratio, we obtain the interesting relation

$$f\frac{r}{c} = 2\pi^2 \frac{\mu N}{P}\left(\frac{r}{c}\right)^2 = 2\pi^2 S \qquad (12\text{-}8)$$

12-4   STABLE LUBRICATION

The difference between boundary and hydrodynamic lubrication can be explained by reference to Fig. 12-4. This plot of the change in the coefficient of friction versus the bearing characteristic $\mu N/P$ was obtained by the McKee brothers in an actual test of friction.* The plot is important because it defines stability of lubrication and helps us to understand hydrodynamic and boundary, or thin-film, lubrication.

Suppose we are operating to the right of line BA and something happens, say, an increase in lubricant temperature. This results in a lower viscosity and hence a smaller

---

*S. A. McKee and T. R. McKee, "Journal Bearing Friction in the Region of Thin Film Lubrication," *SAE J.,* vol. 31, 1932, pp. (T)371-377.

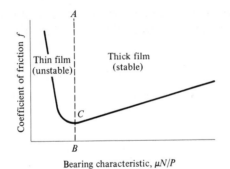

FIGURE 12-4

Variation of the coefficient of friction with $\mu N/P$.

value of $\mu N/P$. The coefficient of friction decreases, not as much heat is generated in shearing the lubricant, and consequently the lubricant temperature drops. Thus the region to the right of line BA defines *stable lubrication* because variations are self-correcting.

To the left of line BA, a decrease in viscosity would increase the friction. A temperature rise would ensue, and the viscosity would be reduced still more. The result would be compounded. Thus the region to the left of line BA represents *unstable lubrication*.

It is also helpful to see that a small viscosity, and hence a small $\mu N/P$, means that the lubricant film is very thin and that there will be a greater possibility of some metal-to-metal contact, and hence of more friction. Thus, point C represents what is probably the beginning of metal-to-metal contact as $\mu N/P$ becomes smaller.

12-5   THICK-FILM LUBRICATION

Let us now examine the formation of a lubricant film in a journal bearing. Figure 12-5*a* shows a journal which is just beginning to rotate in a clockwise direction. Under starting conditions, the bearing will be dry, or at least partly dry, and hence the journal will climb or roll up the right side of the bearing as shown in Fig. 12-5*a*. Under the conditions of a dry bearing, equilibrium will be obtained when the friction force is balanced by the tangential component of the bearing load.

Now suppose a lubricant is introduced into the top of the bearing as shown in Fig.

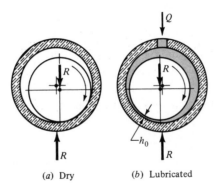

FIGURE 12-5

Formation of a film.

(a) Dry   (b) Lubricated

12-5*b*. The action of the rotating journal is to pump the lubricant around the bearing in a clockwise direction. The lubricant is pumped into a wedge-shaped space and forces the journal over to the other side. A *minimum film thickness* h_0 occurs, not at the bottom of the journal, but displaced clockwise from the bottom as in Fig. 12-5*b*. This is explained by the fact that a film pressure in the converging half of the film reaches a maximum somewhere to the left of the bearing center.

Figure 12-5 shows how to decide whether the journal, under hydrodynamic lubrication, is eccentrically located on the right or on the left side of the bearing. Visualize the journal beginning to rotate. Find the side of the bearing upon which the journal tends to roll. Then, if the lubrication is hydrodynamic, mentally place the journal on the opposite side.

The nomenclature of a journal bearing is shown in Fig. 12-6. The dimension c is the *radial clearance* and is the difference in the radii of the bearing and journal. In Fig. 12-6 the center of the journal is at O and the center of the bearing at O'. The distance between these centers is the *eccentricity* and is denoted by e. The *minimum film thickness* is designated by h_0, and it occurs at the line of centers. The film thickness at any other point is designated by h. We also define an *eccentricity ratio* ϵ as

$$\epsilon = \frac{e}{c}$$

The bearing shown in the figure is known as a *partial bearing*. If the radius of the bearing is the same as the radius of the journal, it is known as a *fitted bearing*. If the bearing encloses the journal, as indicated by the dashed lines, it becomes a *full bearing*. The angle β describes the angular length of a partial bearing. For example, a 120° partial bearing has the angle β equal to 120°.

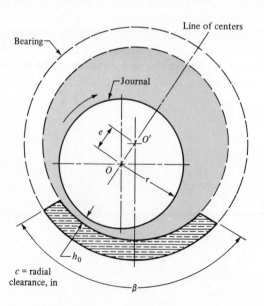

FIGURE 12-6

Nomenclature of a journal bearing.

c = radial clearance, in

12-6   HYDRODYNAMIC THEORY

The present theory of hydrodynamic lubrication originated in the laboratory of Beauchamp Tower in the early 1880s in England. Tower had been employed to study the friction in railroad journal bearings and learn the best methods of lubricating them. It was an accident or error, during the course of this investigation, that prompted Tower to look at the problem in more detail and that resulted in a discovery that eventually led to the development of the theory.

Figure 12-7 is a schematic drawing of the journal bearing which Tower investigated. It is a partial bearing, having a diameter of 4 in, a length of 6 in, and a bearing arc of 157°, and having bath-type lubrication, as shown. The coefficients of friction obtained by Tower in his investigations on this bearing were quite low, which is not now surprising. After testing this bearing, Tower later drilled a $\frac{1}{2}$-in-diameter lubricator hole through the top. But when the apparatus was set in motion, oil flowed out of this hole. In an effort to prevent this, a cork stopper was used, but this popped out, and so it was necessary to drive a wooden plug into the hole. When the wooden plug was pushed out too, Tower, at this point, undoubtedly realized that he was on the verge of discovery. A pressure gauge connected to the hole indicated a pressure of more than twice the unit bearing load. Finally, he investigated the bearing film pressures in detail throughout the bearing width and length and reported a distribution similar to that of Fig. 12-8.*

The results obtained by Tower had such regularity that Osborne Reynolds concluded that there must be a definite law relating the friction, the pressure, and the velocity. The present mathematical theory of lubrication is based upon Reynolds' work follow-ing the experiment by Tower.† The original differential equation, developed by Rey-nolds, was used by him to explain Tower's results. The solution is a challenging

---

*Beauchamp Tower, "First Report on Friction Experiments," *Proc. Inst. Mech. Eng.*, November 1883, pp. 632–666; "Second Report," ibid., 1885, pp. 58–70; "Third Report," ibid., 1888, pp. 173–205; "Fourth Report," ibid., 1891, pp. 111–140.

†Osborne Reynolds, "Theory of Lubrication, Part I," *Phil. Trans. Roy. Soc. London*, 1886.

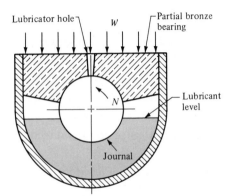

FIGURE 12-7

Schematic representation of the partial bearing used by Tower.

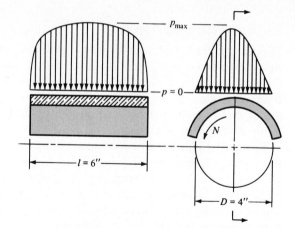

FIGURE 12-8

Approximate pressure-distribution
curves obtained by Tower.

problem which has interested many investigators ever since then, and it is still the
starting point for lubrication studies.

Reynolds pictured the lubricant as adhering to both surfaces and being pulled by the
moving surface into a narrowing, wedge-shaped space so as to create a fluid or film
pressure of sufficient intensity to support the bearing load. One of the important simplifying assumptions resulted from Reynolds' realization that the fluid films were so thin
in comparison with the bearing radius that the curvature could be neglected. This
enabled him to replace the curved partial bearing with a flat bearing, called a *plane
slider bearing*. Other assumptions made were:

1 The lubricant obeys Newton's law of viscous flow.

2 The forces due to the inertia of the lubricant are neglected.

3 The lubricant is assumed to be incompressible.

4 The viscosity is assumed to be constant throughout the film.

5 The pressure does not vary in the axial direction.

Figure 12-9a shows a journal rotating in the clockwise direction supported by a film
of lubricant of variable thickness h on a partial bearing which is fixed. We specify that
the journal has a constant surface velocity U. Using Reynolds' assumption that curvature can be neglected, we fix a right-handed xyz reference system to the stationary
bearing. We now make the following additional assumptions:

6 The bearing and journal extend infinitely in the z direction; this means there can be
no lubricant flow in the z direction.

7 The film pressure is constant in the y direction. Thus the pressure depends only on
the coordinate x.

8 The velocity of any particle of lubricant in the film depends only on the coordinates x
and y.

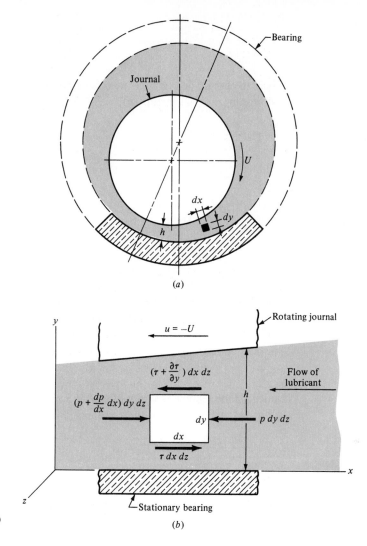

FIGURE 12-9

We now select an element of lubricant in the film (Fig. 12-9a) of dimensions dx, dy, and dz, and compute the forces which act on the sides of this element. As shown in Fig. 12-9b, normal forces, due to the pressure, act upon the right and left sides of the element, and shear forces, due to the viscosity and to the velocity, act upon the top and bottom sides. Summing the forces gives

$$\sum F = \left(p + \frac{dp}{dx}\, dx\right) dy\ dz + \tau\ dx\ dz - \left(\tau + \frac{\partial \tau}{\partial y}\, dy\right) dx\ dz - p\ dy\ dz = 0 \qquad (a)$$

This reduces to

$$\frac{dp}{dx} = \frac{\partial \tau}{\partial y} \qquad\qquad (b)$$

From Eq. (12-1), we have

$$\tau = \mu \frac{\partial u}{\partial y} \qquad (c)$$

where the partial derivative is used because the velocity u depends upon both x and y. Substituting Eq. (c) in Eq. (b), we obtain

$$\frac{dp}{dx} = \mu \frac{\partial^2 u}{\partial y^2} \qquad (d)$$

Holding x constant, we now integrate this expression twice with respect to y. This gives

$$\frac{\partial u}{\partial y} = \frac{1}{\mu} \frac{dp}{dx} y + C_1$$

$$u = \frac{1}{2\mu} \frac{dp}{dx} y^2 + C_1 y + C_2 \qquad (e)$$

Note that the act of holding x constant means that C_1 and C_2 can be functions of x. We now assume that there is no slip between the lubricant and the boundary surfaces. This gives two sets of boundary conditions for evaluating the constants C_1 and C_2:

$$\begin{array}{ll} y = 0 & y = h \\ u = 0 & u = -U \end{array} \qquad (f)$$

Notice, in the second condition, that h is a function of x. Substituting these conditions in Eq. (e) and solving for the constants gives

$$C_1 = -\frac{U}{h} - \frac{h}{2\mu} \frac{dp}{dx} \qquad C_2 = 0$$

or

$$u = \frac{1}{2\mu} \frac{dp}{dx} (y^2 - hy) - \frac{U}{h} y \qquad (12\text{-}9)$$

This equation gives the velocity distribution of the lubricant in the film as a function of the coordinate y and the pressure gradient dp/dx. The equation shows that the velocity distribution across the film (from $y = 0$ to $y = h$) is obtained by superposing a parabolic distribution (the first term) onto a linear distribution (the second term). Figure 12-10 shows the superposition of these two terms to obtain the velocity for particular values of x and dp/dx. In general, the parabolic term may be additive or subtractive to the linear term, depending upon the sign of the pressure gradient. When the pressure is maximum, $dp/dx = 0$ and the velocity is

$$u = -\frac{U}{h} y \qquad (g)$$

which is a linear relation.

We next define Q as the volume of lubricant flowing in the x direction per unit time.

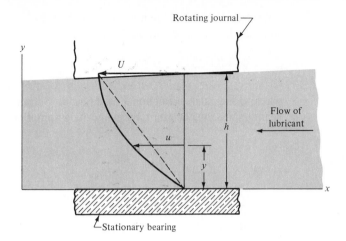

FIGURE 12-10

Velocity of the lubricant.

By using a width of unity in the z direction, the volume may be obtained by the expression

$$Q = \int_0^h u \, dy \tag{h}$$

Substituting the value of u from Eq. (12-9) and integrating gives

$$Q = -\frac{Uh}{2} - \frac{h^3}{12\mu} \frac{dp}{dx} \tag{i}$$

The next step uses the assumption of an incompressible lubricant and states that the flow is the same for any cross section. Thus

$$\frac{dQ}{dx} = 0$$

From Eq. (i),

$$\frac{dQ}{dx} = -\frac{U}{2} \frac{dh}{dx} - \frac{d}{dx}\left(\frac{h^3}{12\mu} \frac{dp}{dx}\right) = 0$$

or

$$\frac{d}{dx}\left(\frac{h^3}{\mu} \frac{dp}{dx}\right) = -6U \frac{dh}{dx} \tag{12-10}$$

which is the classical Reynolds equation for one-dimensional flow. It neglects side leakage, that is, flow in the z direction. A similar development is used when side leakage is not neglected. The resulting equation is

$$\frac{\partial}{\partial x}\left(\frac{h^3}{\mu} \frac{\partial p}{\partial x}\right) - \frac{\partial}{\partial z}\left(\frac{h^3}{\mu} \frac{\partial p}{\partial z}\right) = -6U\frac{\partial h}{\partial x} \tag{12-11}$$

There is no general solution to Eq. (12-11); approximate solutions have been obtained by using electrical analogies, mathematical summations, relaxation methods, and numerical and graphical methods. One of the important solutions is due to Sommerfeld[*] and may be expressed in the form.

$$\frac{r}{c}f = \phi\left[\left(\frac{r}{c}\right)^2 \frac{\mu N}{P}\right]$$

(12-12)

where ϕ indicates a functional relationship. Sommerfeld found the functions for half-bearings and full bearings by using the assumption of no side leakage.

12-7   DESIGN CONSIDERATIONS

We may distinguish between two groups of variables in the design of sliding bearings. In the first group are those whose values either are given or are under the control of the designer. These are:

1 The viscosity μ

2 The load per unit of projected bearing area, P

3 The speed N

4 The bearing dimensions r, c, β, and l

Of these four variables, the designer usually has no control over the speed, because it is specified by the overall design of the machine. Sometimes the viscosity is specified in advance, as, for example, when the oil is stored in a sump and is used for lubricating and cooling a variety of bearings. The remaining variables, and sometimes the viscosity, may be controlled by the designer and are therefore the *decisions* he or she makes. In other words, when these four variables are defined, the design is complete.

In the second group are the dependent variables. The designer cannot control these except indirectly by changing one or more of the first group. These are:

1 The coefficient of friction f

2 The temperature rise ΔT

3 The flow of oil Q

4 The minimum film thickness h_0

This group of variables tells us how well the bearing is performing, and hence we may regard them as *performance factors*. Certain limitations on their values must be imposed by the designer to ensure satisfactory performance. These limitations are specified by the characteristics of the bearing materials and of the lubricant. The fundamental problem in bearing design, therefore, is to define satisfactory limits for the second group of variables and then to decide upon values for the first group such that these limitations are not exceeded.

---

[*]A. Sommerfeld, "Zur Hydrodynamischen Theorie der Schmiermittel-Reibung" ("On the Hydrodynamic Theory of Lubrication"), *Z. Math. Physik*, vol. 50, 1904, pp. 97–155.

Significant Angular Velocity

Especial care must be used in computing the bearing speed N when both the bearing and the journal rotate. This situation occurs, for example, in epicyclic bearings. Let

ω_j = absolute angular velocity of the journal
ω_b = absolute angular velocity of the bearing
ω_f = absolute angular velocity of the load vector

Then the velocity of the journal relative to the load vector is

$$\omega_{jf} = \omega_j - \omega_f \tag{a}$$

The velocity of the bearing relative to the load vector is

$$\omega_{bf} = \omega_b - \omega_f \tag{b}$$

The significant angular velocity is simply the sum of Eqs. (a) and (b) and is

$$\omega^* = \omega_{jf} + \omega_{bf} = \omega_j + \omega_b - 2\omega_f \tag{12-13}$$

This is the angular velocity that should be used to compute the speed N as used in this chapter.

12-8 THE RELATIONS OF THE VARIABLES

Before proceeding to the problem of design, it is necessary to establish the relationships between the variables. Albert A. Raimondi and John Boyd, of Westinghouse Research Laboratories, used an iteration technique to solve Reynolds' equation on the digital computer.* This is the first time such extensive data have been available for use by designers, and consequently we shall employ them in this book.†

The Raimondi and Boyd papers were published in three parts and contain 45 detailed charts and 6 tables of numerical information. In all three parts, charts are used to define the variables for length-diameter (l/d) ratios of 1:4, 1:2, and 1 and for beta angles of 60 to 360°. Under certain conditions the solution to the Reynolds equation gives negative pressures in the diverging portion of the oil film. Since a lubricant cannot usually support a tensile stress, Part III of the Raimondi-Boyd papers assumes that the oil film is ruptured when the film pressure becomes zero. Part III also contains data for the infinitely long bearing; since it has no ends, this means that there is no side leakage. The charts appearing in this book are from Part III of the papers, and are for full journal bearings ($\beta = 360°$) only. Space does not permit the inclusion of charts for partial bearings. This means that you must refer to the charts in the original papers when beta angles of less than 360° are desired. The notation is very nearly the same as in this book, and so no problems should arise.

---

*A. A. Raimondi and John Boyd, "A Solution for the Finite Journal Bearing and Its Application to Analysis and Design, Parts I, II, and III," *Trans. ASLE,* vol. 1, no. 1, in *Lubrication Science and Technology,* Pergamon, New York, 1958, pp. 159–209.

†See also the earlier companion paper, John Boyd and Albert A. Raimondi, "Applying Bearing Theory to the Analysis and Design of Journal Bearings, Parts I and II," *J. Appl. Mechanics,* vol. 73, 1951, pp. 298–316.

Viscosity Charts (Figs. 12-11 to 12-13)

One of the most important assumptions made in the Raimondi-Boyd analysis is that *viscosity of the lubricant is constant as it passes through the bearing*. But since work is done on the lubricant during this flow, the temperature of the oil is higher when it leaves the loading zone than it was on entry. And the viscosity charts clearly indicate that the viscosity drops off significantly with a rise in temperature. Since the analysis is based on a constant viscosity, our problem now is to determine the value of viscosity to be used in the analysis.

Some of the lubricant that enters the bearing emerges as a side flow, which carries away some of the heat. The balance of the lubricant flows through the load-bearing zone and carries away the balance of the heat generated. In determining the viscosity to be used we shall employ a temperature that is the average of the inlet and outlet temperatures, or

$$T_{av} = T_1 + \frac{\Delta T}{2} \tag{12-14}$$

where T_1 is the inlet temperature and ΔT is the temperature rise of the lubricant from inlet to outlet. Of course, the viscosity used in the analysis must correspond to T_{av}.

One of the objectives of lubrication analysis is to determine the oil outlet temperature when the oil and its inlet temperature are specified. This is a trial-and-error type of problem.

To illustrate, suppose we have decided to use SAE 30 oil in an application in which the oil inlet temperature is $T_1 = 180°F$. We begin by estimating that the temperature rise will be $\Delta T = 30°F$. Then, from Eq. (12-14),

$$T_{av} = T_1 + \frac{\Delta T}{2} = 180 = \frac{30}{2} = 195°F$$

From Fig. 12-11 we follow the SAE 30 line and find that $\mu = 1.40$ μreyn at 195°F. So we use this viscosity (in an analysis to be explained in detail in due time) and find that the temperature rise is actually $\Delta T = 54°F$. Thus Eq. (12-14) gives

$$T_{av} = 180 + \tfrac{54}{2} = 207°F$$

This corresponds to point A on Fig. 12-11, which is above the SAE 30 line and indicates that the viscosity used in the analysis was too high.

For a second guess, try $\mu = 1.00$ μreyn. Again we run through an analysis and this time find that $\Delta T = 30°F$. This gives an average temperature of

$$T_{av} = 180 + \tfrac{30}{2} = 195°F$$

and locates point B on Fig. 12-11.

If points A and B are fairly close to each other and on opposite sides of the SAE 30 line, a straight line can be drawn between them with the intersection locating the correct values of viscosity and average temperature to be used in the analysis. For this illustration, we see from the viscosity chart that they are $T_{av} = 203°F$ and $\mu = 1.26$ μreyn.

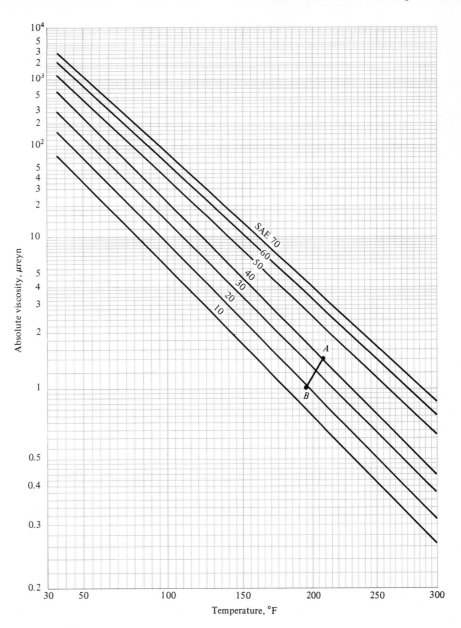

FIGURE 12-11

Viscosity-temperature chart in U.S. customary units. *(Boyd and Raimondi.)*

Minimum Film Thickness (Figs. 12-14 and 12-15)

Let us specify the following quantities for a full journal bearing:

$\mu = 4\ \mu$reyn
$N = 30$ rev/s
$W = 500$ lb (bearing load)
$r = 0.75$ in
$c = 0.0015$ in
$l = 1.50$ in

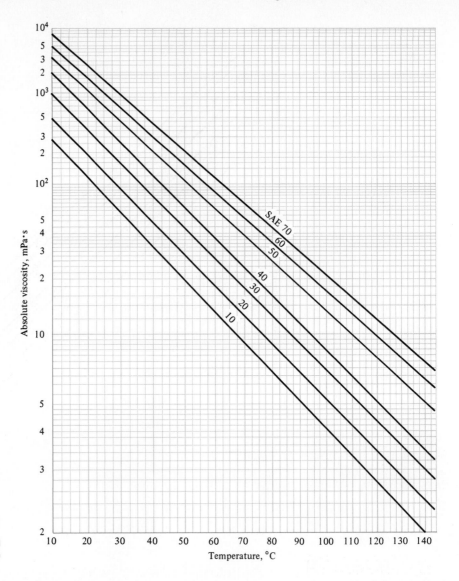

The unit load is

$$P = \frac{W}{2rl} = \frac{500}{2(0.75)(1.50)} = 222 \text{ psi}$$

The bearing characteristic number, from Eq. (12-7), is

$$S = \left(\frac{r}{c}\right)^2 \left(\frac{\mu N}{P}\right) = \left(\frac{0.75}{0.0015}\right)^2 \left[\frac{4(10)^{-6}(30)}{222}\right] = 0.135$$

Also, $l/d = 1.50/(2)(0.75) = 1$.

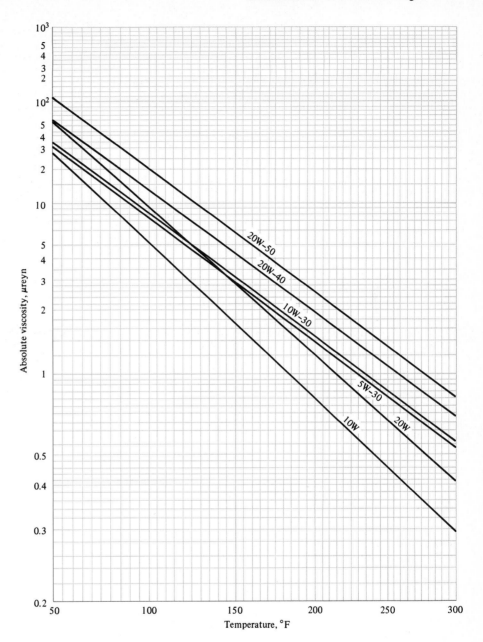

FIGURE 12-13

Chart for multiviscosity lubricants. This chart was derived from known viscosities at two points, 100 and 210°F, and the results are believed to be correct for other temperatures.

Entering Fig. 12-14 with $S = 0.135$ and $l/d = 1$ gives

$$\frac{h_0}{c} = 0.42 \qquad \epsilon = 0.58$$

The quantity h_0/c is called the *minimum-film-thickness variable*. Since $c = 0.0015$ in,

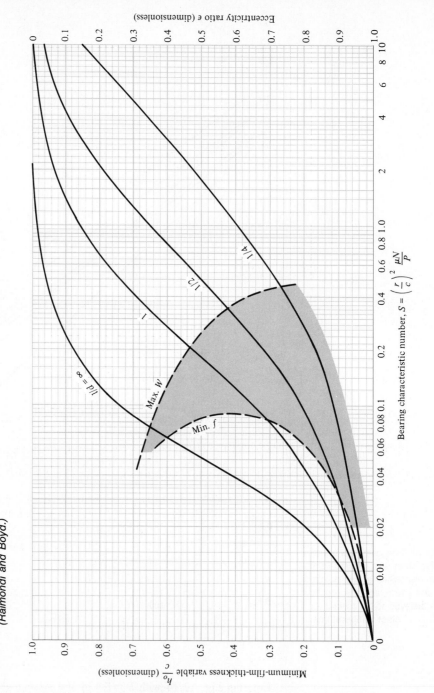

FIGURE 12-14

Chart for minimum-film-thickness variable and eccentricity ratio. The left boundary of the shaded zone defines the optimum h_0 for minimum friction; the right boundary is the optimum h_0 for maximum load. (*Raimondi and Boyd.*)

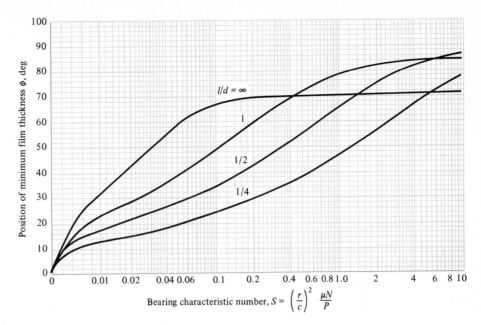

FIGURE 12-15

Chart for determining the position of the minimum film thickness h_0. For location of origin, see Fig. 12-16. *(Raimondi and Boyd.)*

the minimum film thickness is

$$h_0 = 0.42(0.0015) = 0.000\ 63\ \text{in}$$

This is shown in Fig. 12-16. We can find the angular location ϕ of the minimum film thickness from the chart of Fig. 12-15. Entering with $S = 0.135$ and $l/d = 1$ gives $\phi = 53°$.

The eccentricity ratio is $\epsilon = e/c = 0.58$. This means that the eccentricity is

$$e = 0.58(0.0015) = 0.000\ 87\ \text{in}$$

This is also shown in Fig. 12-16. Note that if the bearing is centered, $e = 0$ and $h_0 = c$; this corresponds to a very light or zero load, and since $e = 0$, the eccentricity ratio is zero. As the load is increased, the journal is forced downward; the limiting position is reached when $h_0 = 0$ and $e = c$, that is, when the journal is touching the bearing. For this condition the eccentricity ratio is unity.

Since

$$h_0 = c - e \tag{12-15}$$

we have, by dividing both sides by c,

$$\frac{h_0}{c} = 1 - \epsilon \tag{12-16}$$

Design optima frequently used are *maximum load,* which is a load-carrying characteristic, and *minimum power loss,* which is a function of the departure from thick-film relationships. Dashed lines for both these conditions have been constructed in Fig. 12-14 so that the optimum values of h_0 or ϵ can readily be found. The shaded zone

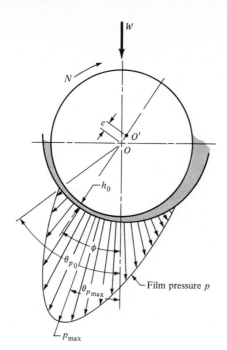

FIGURE 12-16

Polar diagram of film-pressure distribution showing the notation used. *(Raimondi and Boyd.)*

between the boundaries defined by these two optima may therefore be considered a recommended operating zone.

From this discussion and by examination of Fig. 12-14, you should have concluded that lightly loaded bearings operate with a large Sommerfeld number, while heavily loaded bearings will operate at a small number.

Coefficient of Friction (Fig. 12-17)

The friction chart has the *friction variable* $(r/c) f$ plotted against S for various values of the l/d ratio. Using the same data as before, we enter Fig. 12-17 with $S = 0.135$ and $l/d = 1$. We then find the friction variable to be

$$\frac{r}{c} f = 3.50$$

Therefore the coefficient of friction is

$$f = 3.50 \frac{c}{r} = 3.50\left(\frac{0.0015}{0.75}\right) = 0.007$$

With this known, other things can be learned about the bearing performance. For example, the torque required to overcome friction is

$$T = fWr = 0.007(500)(0.75) = 2.62 \text{ lb} \cdot \text{in}$$

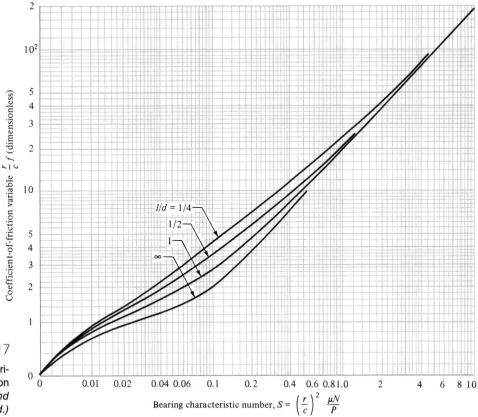

FIGURE 12-17

Chart for coefficient-of-friction variable; note that Petroff's equation is asymptotic. *(Raimondi and Boyd.)*

The power lost in the bearing, in horsepower, is

$$H = \frac{TN}{1050} = \frac{2.62(30)}{1050} = 0.0748 \text{ hp}$$

or, expressed in Btu, we have

$$H = \frac{2\pi TN}{778(12)} = \frac{2\pi(2.62)(30)}{778(12)} = 0.0529 \text{ Btu/s}$$

Lubricant Flow (Figs. 12-18 and 12-19)

The *flow variable* $Q/rcNl$, found from the chart of Fig. 12-18, is used to find the volume of lubricant Q which is pumped into the converging space by the rotating journal. This chart is based on the *assumption of atmospheric pressure and the absence of oil grooves or holes in the bearing*. The amount of oil supplied to the bearing must, at least, be equal to Q if the bearing is to perform according to the charts.

Of the amount of oil Q pumped by the rotating journal, an amount Q_s flows out the ends, and hence is called the *side leakage*. This side leakage can be computed from the *flow ratio* Q_s/Q of Fig. 12-19.

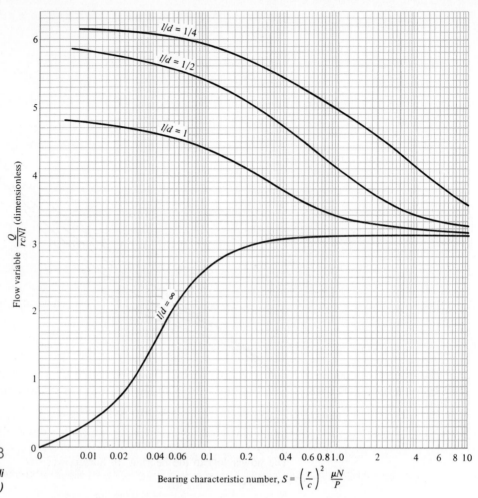

FIGURE 12-18

Chart for flow variable. *(Raimondi and Boyd.)*

Using the same data as before, we enter Fig. 12-18 with $S = 0.135$ and $l/d = 1$. From the chart, we find

$$\frac{Q}{rcNl} = 4.28$$

Therefore, the total flow is

$$Q = 4.28rcNl = 4.28(0.75)(0.0015)(30)(1.5) = 0.216 \text{ in}^3/\text{s}$$

From Fig. 12-19, we find the flow ratio to be

$$\frac{Q_s}{Q} = 0.655$$

Therefore the side leakage is

$$Q_s = 0.655Q = 0.655(0.216) = 0.142 \text{ in}^3/\text{s}$$

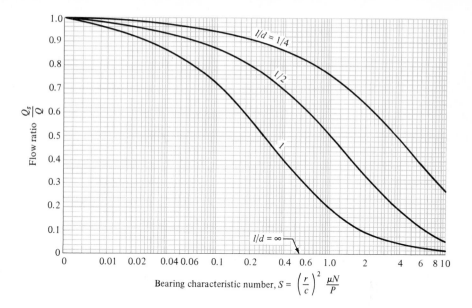

FIGURE 12-19

Chart for determining the ratio of side flow to total flow. *(Raimondi and Boyd.)*

Bearing characteristic number, $S = \left(\dfrac{r}{c}\right)^2 \dfrac{\mu N}{P}$

Film Pressure (Figs. 12-20 and 12-21)

The maximum film pressure developed in the film can be obtained by finding the *pressure ratio* P/p_{max} from the chart of Fig. 12-20. Using the same data as before, we enter this chart with $S = 0.135$ and $l/d = 1$. The maximum-film-pressure ratio is found to be

$$\frac{P}{p_{max}} = 0.42$$

Since $P = 222$ psi, the maximum pressure is found to be

$$p_{max} = \frac{P}{0.42} = \frac{222}{0.42} = 529 \text{ psi}$$

Figure 12-16 shows that the angular location of this point of maximum pressure is given by the angle $\theta_{p_{max}}$. Entering Fig. 12-21 with $S = 0.135$ and $l/d = 1$ gives $\theta_{p_{max}} = 18.5°$.

The terminating position of the oil film is θ_{p_0}, according to Fig. 12-16. Entering Fig. 12-21 again, we find this angle to be $\theta_{p_0} = 75°$.

Temperature Rise

Since the journal does work on the lubricant, heat is produced, as we have seen. This heat must be dissipated by conduction, convection, and radiation and carried away by the flow of oil. It is very difficult to calculate the rate of heat flow by each method with any accuracy. Later we shall examine this problem in more detail; but for the present we make the assumption that the oil flow carries away all the heat generated. Then, as far as oil temperature is concerned, we shall be on the conservative side.

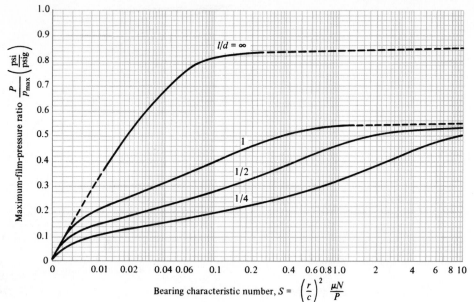

FIGURE 12-20

Chart for determining the maximum film pressure. *(Raimondi and Boyd.)*

The Raimondi-Boyd papers contain temperature-rise charts based on assumptions similar to these. Instead of presenting these charts in this book, we present an analytical approach based on information already obtained.

Let us use the following additional notation:

J = mechanical equivalent of heat, 9336 lbf-in per Btu

C_H = specific heat of lubricant, 0.42 Btu per lbf per °F being an average value for use

γ = weight per unit volume of the lubricant; at an average specific gravity of 0.86, $\gamma = (0.86)(62.4)/1728 = 0.0311$ lbf per in^3

ΔT_F = temperature rise, °F

$X = (r/c)\,f$ = friction variable

$Y = Q/rcNl$ = flow variable

The heat generated is

$$H = \frac{2\pi TN}{J} = \frac{2\pi fWrN}{J} \tag{a}$$

Substituting $(c/r)\,X$ for f gives

$$H = \left(\frac{2\pi WNc}{J}\right)X \tag{b}$$

We now assume that an oil flow Q is to carry away all the heat. Then the temperature rise of the oil will be

$$\Delta T_F = \frac{H}{\gamma C_H Q} \tag{c}$$

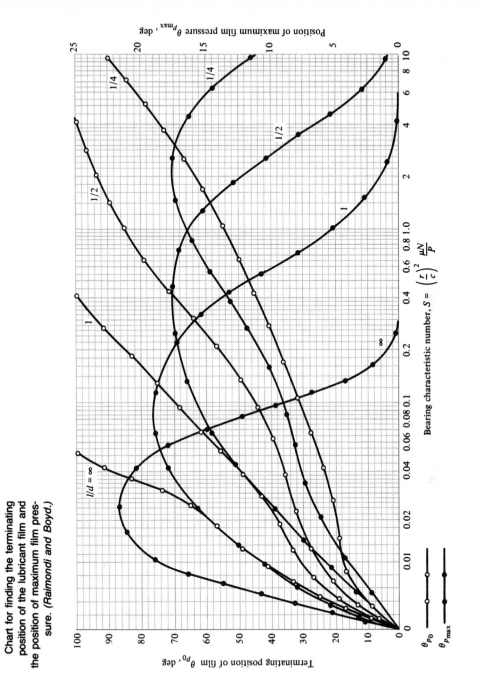

FIGURE 12-21

Chart for finding the terminating position of the lubricant film and the position of maximum film pressure. *(Raimondi and Boyd.)*

If for Q we substitute $(rcNl)Y$, then

$$\Delta T_F = \frac{H}{(\gamma C_H rcNl)Y} \tag{d}$$

We now multiply the numerator and denominator of Eq. (d) by the unit pressure P, noting that $P = W/2rl$, and substitute the value of H from Eq. (b). After canceling terms, this gives

$$\Delta T_F = \frac{4\pi P}{J\gamma C_H} \frac{X}{Y} \tag{e}$$

If then we assume average lubrication conditions and substitute the values of J, γ, and C_H, we finally obtain

$$\Delta T_F = 0.103P \frac{(r/c)f}{Q/rcNl} \tag{12-17}$$

where ΔT is in degrees Fahrenheit. But it is necessary to account for the oil flow out of the sides of the bearing before the hydrodynamic film is terminated. If we assume that the temperature of the side flow is the mean of the inlet and outlet temperatures, then it can be shown that

$$\Delta T_F = \frac{0.103P}{[1 - \frac{1}{2}(Q_s/Q)]} \frac{(r/c)f}{Q/rcNl} \tag{12-18}$$

In this equation the pressure P is in ips units and ΔT_F in degrees Fahrenheit. The corresponding equation using SI is

$$\Delta T_C = \frac{8.30P}{[1 - \frac{1}{2}(Q_s/Q)]} \frac{(r/c)f}{Q/rcNl} \tag{12-19}$$

where P is in MPa and ΔT_C in degrees Celsius.

Using the same data as before, Eq. (12-18) gives a temperature rise of

$$\Delta T_F = \frac{(0.103)(222)}{1 - (0.5)(0.655)} \frac{3.50}{4.28} = 27.8°F$$

Interpolation

According to Raimondi and Boyd, interpolation of the chart data for other l/d ratios can be done by using the equation

$$y = \frac{1}{(l/d)^3} \left[-\frac{1}{8}\left(1 - \frac{l}{d}\right)\left(1 - 2\frac{l}{d}\right)\left(1 - 4\frac{l}{d}\right)y_\infty + \frac{1}{3}\left(1 - 2\frac{l}{d}\right)\left(1 - 4\frac{l}{d}\right)y_1 \right.$$

$$\left. -\frac{1}{4}\left(1 - \frac{l}{d}\right)\left(1 - 4\frac{l}{d}\right)y_{1/2} + \frac{1}{24}\left(1 - \frac{l}{d}\right)\left(1 - 2\frac{l}{d}\right)y_{1/4} \right] \tag{12-20}$$

where y is the desired variable within the interval $\infty > l/d > \frac{1}{4}$ and y_∞, y_1, $y_{1/2}$, and $y_{1/4}$ are the variables corresponding to l/d ratios of ∞, 1, $\frac{1}{2}$, and $\frac{1}{4}$, respectively.

12-9   TEMPERATURE AND VISCOSITY CONSIDERATIONS

In a *self-contained bearing* there is no method of circulating or cooling the lubricant; it passes through the bearing, heats up, and is stored in a sump. Heat is removed by convection, conduction, and radiation, and eventually the system reaches an equilibrium temperature.

In a *force-feed lubricating system,* cool, clean lubricant is fed to the bearing from an external source.

For most problems it is possible to specify the inlet temperature, but, since the viscosity used in the analysis should correspond to the mean of the inlet and outlet temperatures, this does not yield a value of viscosity to use in the analysis. As we have learned, a solution to this problem, when the lubricant grade is specified, is to assume two trial values of viscosity. To repeat, one of these should be somewhat lower than expected, the other higher. By using each of these viscosities, the temperature rise is computed and the average temperature determined from Eq. (12-14). When these results are plotted on Fig. 12-11, a straight line, such as *AB,* can be drawn between them, and the intersection of this line with the SAE grade of oil gives the correct viscosity to use in the analysis. It should be noted that a series of trial viscosities will yield a curved line instead of a straight line if their values differ considerably; therefore the viscosities chosen should not be too different from one another. The following example illustrates this procedure.

EXAMPLE 12-1   If an SAE 20 oil at an inlet temperature of 100°F is the lubricant for the example of the preceding section, what viscosity should be used in the analysis?

Solution   We have already determined that a viscosity of 4 μreyn gives a temperature rise of 27.8°F. The mean temperature is

$$T_{av} = T_1 + \frac{\Delta T_F}{2} = 100 + \frac{27.8}{2} = 113.9°F$$

This yields one point on the viscosity-temperature chart, and when plotted, it is found to be below the SAE 20 line. Therefore we choose $\mu = 6$ μreyn for the second trial value. Computing the S number gives

$$S = \left(\frac{r}{c}\right)^2 \frac{\mu N}{P} = \left(\frac{0.75}{0.0015}\right)^2 \frac{(6)(10)^{-6}(30)}{222} = 0.202$$

Then, using Figs. 12-17, 12-18, and 12-19 gives $(r/c)f = 4.7$, $Q/rcNl = 4.1$, and $Q_s/Q = 0.56$. Equation (12-18) gives

$$\Delta T_F = \frac{0.103P}{1 - \frac{1}{2}(Q_s/Q)} \frac{(r/c)f}{Q/rcNl} = \frac{(0.103)(222)}{1 - (0.5)(0.56)} \frac{4.7}{4.1} = 36.4°F$$

Therefore the average temperature is

$$T_{av} = 100 + \frac{36.4}{2} = 118.2°F$$

When these points are plotted on Fig. 12-11 and joined, they cross the SAE 20 line at $\mu = 5.5$ μreyn and $T_{av} = 117°F$. Therefore this is the correct viscosity to use in completing the analysis. The temperature rise is twice 17, or 34°F.

12-10 CLEARANCE

In designing a journal bearing for thick-film lubrication, the engineer must select the grade of oil to be used, together with suitable values for P, N, r, c, and l. A poor selection of these or inadequate control of them during manufacture or in use may result in a film that is too thin, so that the oil flow is insufficient, causing the bearing to overheat and, eventually, fail. Furthermore, the radial clearance c is difficult to hold accurate during manufacture, and it may increase because of wear. What is the effect of an entire range of radial clearances, expected in manufacture, and what will happen to the bearing performance if c increases because of wear? Most of these questions can be answered and the design optimized by plotting curves of the performance as functions of the quantities over which the designer has control.

Figure 12-22 shows the results obtained when the performance of a particular bearing is calculated for a whole range of radial clearances and is plotted with clearance as the independent variable. The bearing used for this graph is the one of Example 12-1, with SAE 20 oil at an inlet temperature of 100°F. The graph shows that if the clearance is too tight, the temperature will be too high and the minimum film thickness too low. High temperatures may cause the bearing to fail by fatigue. If the oil film is too thin, dirt particles may be unable to pass without scoring or may embed themselves in the bearing. In either event, there will be excessive wear and friction, resulting in high temperatures and possible seizing.

To investigate the problem in more detail, Table 12-1 was prepared using the two types of preferred running fits which seem to be most useful for journal-bearing design (see Table 4-5). The results shown in Table 12-1 were obtained using Eqs. (4-16) and

FIGURE 12-22

A plot of some performance characteristics of the bearing of Example 12-1 for radial clearances of 0.0005 to 0.003 in. The bearing outlet temperature is designated T_2. New bearings should be designed for the shaded zone, because wear will move the operating point to the right.

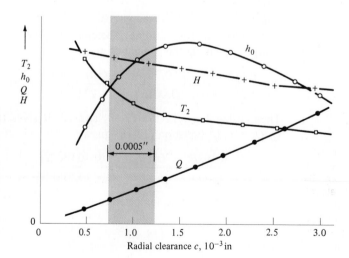

Radial clearance c, 10^{-3} in

TABLE 12-1

Maximum, Minimum, and Average Clearances for 1.5-in-Diameter Journal Bearings Based on Type of Fit

TYPE OF FIT	SYMBOL	CLEARANCE c, in		
		MAXIMUM	AVERAGE	MINIMUM
Close-running	H8/f7	0.001 75	0.001 125	0.000 5
Free-running	H9/d9	0.003 95	0.002 75	0.001 55

TABLE 12-2

Performance of 1.5-in-Diameter Journal Bearing with Various Clearances. (SAE 20 Lubricant, $T_1 = 100°F$, $N = 30$ rev/s, $W = 500$ lb, $L = 1.5$ in)

c, in	T_2, °F	h_0, in	f	Q, m³/s	H, Btu/s
0.000 5	226	0.000 38	0.011 3	0.061	0.086
0.001 125	142	0.000 65	0.009 0	0.153	0.068
0.001 55	133	0.000 77	0.008 7	0.218	0.066
0.001 75	128	0.000 76	0.008 4	0.252	0.064
0.002 75	118	0.000 73	0.007 9	0.419	0.060
0.003 95	113	0.000 69	0.007 7	0.617	0.059

(4-17) of Sec. 4-8. Notice that there is a slight overlap, but the range of clearances for the free-running fit is about twice that of the close-running fit.

The six clearances of Table 12-1 were used in a computer program to obtain the numerical results shown in Table 12-2. These conform to the results of Fig. 12-22, too. Both the table and the figure show that a tight clearance results in a high temperature. Figure 12-23 can be used to estimate an upper temperature limit when the characteristics of the application are known.

It would seem that a large clearance will permit the dirt particles to pass through and also will permit a large flow of oil, as indicated in Table 12-2. This lowers the temperature and increases the life of the bearing. However, if the clearance becomes too large, the bearing becomes noisy and the minimum film thickness begins to decrease again.

In between these two limitations there exist a rather large range of clearances that will result in satisfactory bearing performance.

When both the production tolerance and the future wear on the bearing are considered, it is seen, from Fig. 12-22, that the best compromise is a clearance range slightly to the left of the top of the minimum-film-thickness curve. In this way, future wear will

FIGURE 12-23

Temperature limits for mineral oils. The lower limit is for oils containing antioxidants and applies when oxygen supply is unlimited. The upper limit applies when insignificant oxygen is present. The life in the colored zone depends on the amount of oxygen and catalysts present. [Source: M. J. Neale (ed.), Tribology Handbook, Section B1, Newnes-Butterworth, London, 1975.]

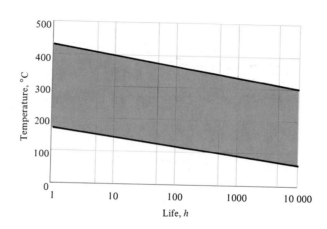

move the operating point to the right and increase the film thickness and decrease the operating temperature.

12-11   PRESSURE-FED BEARINGS

When so much heat is generated by hydrodynamic action that the normal lubricant flow is insufficient to carry it away, an additional supply of lubricant must be furnished under pressure. To force a greater flow through the bearing and thus obtain an increased cooling effect, a common practice is to use a circumferential groove at the center of the bearing, with an oil-supply hole located opposite the load-bearing zone. Such a bearing is shown in Fig. 12-24. The effect of the groove is to create two half-bearings, each having a smaller l/d ratio than the original. The groove divides the pressure-distribution curve into two lobes and reduces the minimum film thickness, but it has wide acceptance among lubrication engineers because such bearings carry more load without overheating.

To set up a method of solution for oil flow, we shall assume a groove ample enough that the pressure drop in the groove itself is small. Initially we will neglect eccentricity and then apply a correction factor for this condition. The oil flow, then, is the amount which flows out of the two halves of the bearing in the direction of the concentric shaft. If we neglect the rotation of the shaft, we obtain the force situation shown in Fig. 12-25. Here we designate the supply pressure by p_s and the pressure at any point by p. Laminar flow is assumed, and we are interested in the static equilibrium of an element of width dx, thickness $2y$, and unit depth. Note particularly that the origin of the reference system has been chosen at the midpoint of the clearance space. The pressure is $p + dp$ on the left face and p on the right face, and the upper and lower surfaces are acted upon by the shear stresses τ. The equilibrium equation is

$$2y(p + dp) - 2yp - 2\tau\, dx = 0 \qquad\qquad (a)$$

Expanding and canceling terms, we find that

$$\tau = y\,\frac{dp}{dx} \qquad\qquad (b)$$

Newton's law for viscous flow [Eq. (12-1)] is

$$\tau = \mu\,\frac{du}{dy}$$

However, in this case we have taken τ in a negative direction. Also, du/dy is negative

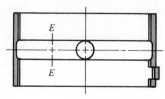

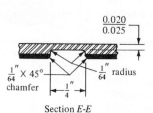

$\frac{1''}{64} \times 45°$ chamfer

$\frac{1''}{4}$

$\frac{0.020}{0.025}$

$\frac{1''}{64}$ radius

Section *E-E*

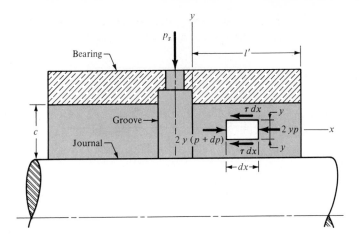

FIGURE 12-25

Flow of lubricant from a pressure-fed bearing having a central groove.

because u decreases as y increases. We therefore write Newton's law in the form

$$-\tau = \mu\left(-\frac{du}{dy}\right) \tag{c}$$

Now eliminating τ from Eqs. (b) and (c) gives

$$\frac{du}{dy} = \frac{1}{\mu}\frac{dp}{dx}\,y \tag{d}$$

Treating dp/dx as a constant and integrating with respect to y gives

$$u = \frac{1}{2\mu}\frac{dp}{dx}\,y^2 + C_1 \tag{e}$$

At the boundaries, where $y = \pm c/2$, the velocity u is zero. Using one of these conditions in Eq. (e) gives

$$0 = \frac{1}{2\mu}\frac{dp}{dx}\left(\frac{c}{2}\right)^2 + C_1$$

or

$$C_1 = -\frac{c^2}{8\mu}\frac{dp}{dx}$$

Substituting this constant in Eq. (e) yields

$$u = \frac{1}{8\mu}\frac{dp}{dx}\,(4y^2 - c^2) \tag{f}$$

Next let us assume that the oil pressure varies linearly from the center to the end of the bearing, as shown in Fig. 12-26. Since the equation of a straight line may be written

$$p = Ax + B$$

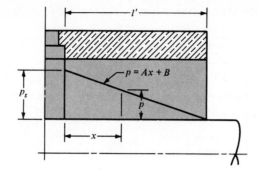

FIGURE 12-26

A linear oil-pressure distribution is assumed.

with $p = p_s$ at $x = 0$ and $p = 0$ at $x = l'$, substituting these end conditions gives

$$A = \frac{p_s}{l'} \qquad B = p_s$$

or

$$p = -\frac{p_s}{l'}x + p_s \qquad (g)$$

and therefore

$$\frac{dp}{dx} = -\frac{p_s}{l'} \qquad (h)$$

We can now substitute Eq. (h) in Eq. (f) to get the relationship between the oil velocity and the coordinate y:

$$u = \frac{p_s}{8\mu l'}(c^2 - 4y^2) \qquad (12\text{-}21)$$

Figure 12-27 shows a graph of this relation fitted into the clearance space c so that you can see how the velocity of the lubricant varies from the journal surface to the bearing surface. The distribution is parabolic, as shown, with the maximum velocity occurring

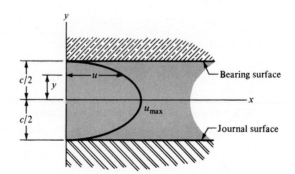

FIGURE 12-27

Parabolic distribution of the lubricant velocity.

at the center, where $y = 0$. The magnitude is, from Eq. (12-20),

$$u_{\max} = \frac{p_s c^2}{8 \mu l'} \tag{i}$$

The average ordinate of a parabola is two-thirds of the maximum; if we also generalize by substituting h for c in Eq. (i), then the average velocity at any angular position θ (Fig. 12-28) is

$$u_{av} = \frac{2}{3} \frac{p_s h^2}{8 \mu l'} = \frac{p_s}{12 \mu l'} (c - e \cos \theta)^2 \tag{j}$$

We still have a little further to go in this analysis; so be patient. Now that we have an expression for the lubricant velocity, we can compute the amount of lubricant that flows out both ends; the elemental side flow at any position θ (Fig. 12-28) is

$$dQ_s = 2u_{av} \, dA = 2u_{av}(rh \, d\theta) \tag{k}$$

where dA is the elemental area. Substituting u_{av} from Eq. (j) and h from Fig. 12-28 gives

$$dQ_s = \frac{p_s r}{6 \mu l'} (c - e \cos \theta)^3 \, d\theta \tag{l}$$

Integrating around the bearing gives the total side flow as

$$\begin{aligned} Q_s = \int dQ_s &= \frac{p_s r}{6 \mu l'} \int_0^{2\pi} (c - e \cos \theta)^3 \, d\theta \\ &= \frac{\pi p_s r c^3}{3 \mu l'} (1 + 1.5\epsilon^2) \end{aligned} \tag{12-22}$$

In analyzing the performance of pressure-fed bearings, the bearing length should be

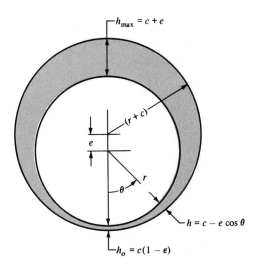

FIGURE 12-28

taken as l', as defined in Fig. 12-25. Thus the unit load is

$$P = \frac{W/2}{2rl'} = \frac{W}{4rl'} \tag{12-23}$$

because each half of the bearing carries half of the load.

The charts of Figs. 12-18 and 12-19 for flow variable and flow ratio, of course, do not apply to pressure-fed bearings. Also, to the maximum film pressure, given by Fig. 12-20, must be added the supply pressure p_s, to obtain the total film pressure.

Since the oil flow has been increased by forced feed, Eq. (12-18) will give a temperature rise that is too high because the side flow carries away all the heat generated. So we shall modify Eq. (c) of Sec. 12-8 to

$$\Delta T_F = \frac{2H}{\gamma C_H Q_s} \tag{m}$$

and the heat loss is

$$H = \frac{2\pi f W r N}{J} \tag{n}$$

Substituting Eq. (12-22) and Eq. (n) in Eq. (m) and canceling terms gives

$$\Delta T_F = \frac{12\mu l' f W N}{(1 + 1.5\epsilon^2)J\gamma C_H p_s c^3} \tag{12-24}$$

For average lubrication conditions, $\gamma = 0.0311$ lbf per in^3, $C_H = 0.42$ Btu per lbf per °F; also $J = 9336$ lbf · in per Btu, and so Eq. (12-24) can be written

$$\Delta T_F = \frac{0.0492\mu l' f W N}{(1 + 1.5\epsilon^2)p_s c^3} \tag{o}$$

Now multiply and divide Eq. (o) by the Sommerfeld number

$$S = \left(\frac{r}{c}\right)^2 \frac{\mu N}{P} = \left(\frac{r}{c}\right)^2 \frac{4rl'\mu N}{W} \tag{p}$$

Upon rearranging the terms, we find

$$\Delta T_F = \frac{0.0246}{1 + 1.5\epsilon^2} \frac{[(r/c)f]SW^2}{p_s r^4} \tag{12-25}$$

which is easier to solve than Eq. (o) because the number S must be computed anyway. Equation (12-25) is, of course, in the customary ips units. The corresponding equation in SI is

$$\Delta T_C = \frac{1956(10)^6}{1 + 1.5\epsilon^2} \frac{[(r/c)f]SW^2}{p_s r^4} \tag{12-26}$$

where ΔT_C = temperature rise, °C
$\quad\quad\quad W$ = bearing load, kN
$\quad\quad\quad p_s$ = supply pressure, kPa
$\quad\quad\quad r$ = radius, mm

12-12   HEAT BALANCE

The case in which the lubricant carries away all the generated heat has already been discussed. We shall now investigate self-contained bearings, in which the lubricant is stored in the bearing housing itself. These bearings find many applications in industrial machinery; are often described as pedestal, or pillow-block, bearings; and are used on fans, blowers, pumps, motors, and the like. The problem is to balance the heat-dissipation capacity of the housing with the heat generated in the bearing itself.

The heat given up by the bearing housing may be approximated by the equation

$$H = CA(T_H - T_A) \tag{12-27}$$

where H = heat dissipated, Btu/h
C = combined coefficient of radiation and convection, Btu/(h)(ft^2)(°F)
A = surface area of housing, ft^2
T_H = surface temperature of housing, °F
T_A = temperature of surrounding air, °F

The coefficient C depends upon the material, color, geometry, and roughness of the housing, the temperature difference between the housing and the surrounding objects, and the temperature and velocity of the air. Equation (12-27) should be used only when "ballpark" answers are sufficient. Exact results can be obtained by experimentation under actual, not simulated, operating conditions and environment. With these limitations, assume C is a constant having the values

$$C = \begin{cases} 2 \text{ Btu/(h)(ft}^2\text{)(°F)} & \text{for still air} \\ 2.7 \text{ Btu/(h)(ft}^2\text{)(°F)} & \text{for average design practice} \\ 5.9 \text{ Btu/(h)(ft}^2\text{)(°F)} & \text{for air moving at 500 fpm} \end{cases}$$

An expression quite similar to Eq. (12-27) can be written for the temperature difference $T_L - T_H$ between the lubricant film and the housing. Because the type of lubricating system and the quality of the lubricant circulation affect this relationship, the resulting expression is even more approximate than that of Eq. (12-27). An *oil-bath lubrication system*, in which a part of the journal is actually immersed in the lubricant, would provide good circulation. A *ring-oiled bearing*, in which oil rings ride on top of the journal, dip into the oil sump, and hence carry a moderate amount of lubricant into the load-bearing zone, would provide satisfactory circulation for many purposes. On the other hand, if the lubricant is supplied by *wick-feeding* methods, the circulation is so inadequate that it is doubtful if any heat at all can be carried away by the lubricant. No matter what type of self-contained lubrication system is used, a great deal of engineering judgment is necessary in computing the heat balance. On the basis of these limitations, the equation

$$T_L - T_H = B(T_H - T_A) \tag{a}$$

where T_L is the *average* film temperature and B is a constant which depends upon the lubrication system, may be used to get a rough estimate of the bearing temperature. Table 12-3 provides some guidance in deciding on a suitable value for B.

TABLE 12-3

LUBRICATION SYSTEM	CONDITIONS	RANGE OF B
Oil ring	Moving air	1–2
	Still air	$\frac{1}{2}$–1
Oil bath	Moving air	$\frac{1}{2}$–1
	Still air	$\frac{1}{5}$–$\frac{2}{5}$

Since T_L and T_A are usually known, Eqs. (12-27) and (a) can be combined to give

$$H = \frac{CA}{B+1}\,(T_L - T_A) \tag{12-28}$$

In beginning a heat-balance computation, the film temperature, and hence the viscosity of the lubricant, in a self-contained bearing is unknown. Thus, finding the equilibrium temperatures is an iterative procedure which starts with an estimate of the film temperature and ends with verification or nonverification of this estimate. Since the computations are lengthy, a computer should be used to make them if at all possible.

12-13   LOADS AND MATERIALS

Some help in choosing unit loads and bearing materials is afforded by Tables 12-4 and 12-5. Since the diameter and length of a bearing depend upon the unit load, these tables will help the designer to establish a starting point in the design.

The length-diameter ratio l/d of a bearing depends upon whether it is expected to run under thin-film-lubrication conditions. A long bearing (large l/d ratio) reduces the coefficient of friction and the side flow of oil and therefore is desirable where thin-film or boundary-value lubrication is present. On the other hand, where forced-feed or

TABLE 12-4

Range of Unit Loads in Current Use for Sleeve Bearings

APPLICATION	UNIT LOAD	
	psi	MPa
Diesel engines:		
Main bearings	900–1700	6–12
Crankpin	1150–2300	8–15
Wristpin	2000–2300	14–15
Electric motors	120–250	0.8–1.5
Steam turbines	120–250	0.8–1.5
Gear reducers	120–250	0.8–1.5
Automotive engines:		
Main bearings	600–750	4–5
Crankpin	1700–2300	10–15
Air compressors:		
Main bearings	140–280	1–2
Crankpin	280–500	2–4
Centrifugal pumps	100–180	0.6–1.2

TABLE 12-5

Some Characteristics of
Bearing Alloys

ALLOY NAME	THICKNESS, in	SAE NUMBER	CLEARANCE RATIO r/c	LOAD CAPACITY	CORROSION RESISTANCE
Tin-base babbitt	0.022	12	600–1000	1.0	Excellent
Lead-base babbitt	0.022	15	600–1000	1.2	Very good
Tin-base babbitt	0.004	12	600–1000	1.5	Excellent
Lead-base babbitt	0.004	15	600–1000	1.5	Very good
Leaded bronze	Solid	792	500–1000	3.3	Very good
Copper-lead	0.022	480	500–1000	1.9	Good
Aluminum alloy	Solid		400–500	3.0	Excellent
Silver plus overlay	0.013	17P	600–1000	4.1	Excellent
Cadmium (1.5% Ni)	0.022	18	400–500	1.3	Good
Trimetal 88*				4.1	Excellent
Trimetal 77†				4.1	Very good

*This is a 0.008-in layer of copper-lead on a steel back plus 0.001 in of tin-base babbitt.

†This is a 0.013-in layer of copper-lead on a steel back plus 0.001 in of lead-base babbitt.

positive lubrication is present, the l/d ratio should be relatively small. The short bearing length results in a greater flow of oil out of the ends, thus keeping the bearing cooler. Current practice is to use an l/d ratio of about unity, in general, and then to increase this ratio if thin-film lubrication is likely to occur and to decrease it for thick-film lubrication or high temperatures. If shaft deflection is likely to be severe, a short bearing should be used to prevent metal-to-metal contact at the ends of the bearings.

You should always consider the use of a partial bearing if high temperatures are a problem, because relieving the non-load-bearing area of a bearing can very substantially reduce the heat generated.

The two conflicting requirements of a good bearing material are that it must have a satisfactory compressive and fatigue strength to resist the externally applied loads and that it must be soft and have a low melting point and a low modulus of elasticity. The second set of requirements is necessary to permit the material to wear or break in, since the material can then conform to slight irregularities and absorb and release foreign particles. The resistance to wear and the coefficient of friction are also important because all bearings must operate, at least for part of the time, with thin-film lubrication.

Additional considerations in the selection of a good bearing material are its ability to resist corrosion and, of course, the cost of producing the bearing. Some of the commonly used materials are listed in Table 12-5, together with their composition and characteristics.

Bearing life can be increased very substantially by depositing a layer of babbitt, or other white metal, in thicknesses from 0.001 to 0.014 in over steel backup material. In fact, a copper-lead on steel to provide strength, with a babbitt overlay to provide surface and corrosion characteristics, makes an excellent bearing.

Small bushings and thrust collars are often expected to run with thin-film lubrication. When this is the case, improvements over a solid bearing material can be made to add significantly to the life. A powder-metallurgy bushing is porous and permits the oil to penetrate into the bushing material. Sometimes such a bushing may be enclosed by oil-soaked material to provide additional storage space. Bearings are frequently ball-indented to provide small basins for the storage of lubricant while the journal is at rest. This supplies some lubrication during starting. Another method of reducing friction is to indent the bearing wall and to fill the indentations with graphite.

With all these tentative decisions made, a lubricant can be selected and the hydrodynamic analysis made as already presented. The values of the various performance parameters, if plotted as in Fig. 12-22, for example, will then indicate whether a satisfactory design has been achieved or additional iterations are necessary.

12-14   BEARING TYPES

A bearing may be as simple as a hole machined into a cast-iron machine member. It may still be simple yet require detailed design procedures, as, for example, the two-piece grooved pressure-fed connecting-rod bearing in an automotive engine. Or it may be as elaborate as the large water-cooled, ring-oiled bearings with built-in reservoirs used on heavy machinery.

Figure 12-29 shows two types of bearings which are often called bushings. The solid bushing is made by casting, by drawing and machining, or by using a powder-metallurgy process. The lined bushing is usually a split type. In one method of manufacture the molten lining material is cast continuously on thin strip steel. The babbitted strip is then processed through presses, shavers, and broaches, resulting in a lined bushing. Any type of grooving may be cut into the bushings. Bushings are assembled as a press fit and finished by boring, reaming, or burnishing.

Flanged and straight two-piece bearings are shown in Fig. 12-30. These are available in many sizes in both thick- and thin-wall types, with or without lining material. A locking lug positions the bearing and effectively prevents axial or rotational movement of the bearing in the housing.

Some typical groove patterns are shown in Fig. 12-31. In general, the lubricant may be brought in from the end of the bushing, through the shaft, or through the bushing. The flow may be intermittent or continuous. The preferred practice is to bring the oil in at the center of the bushing so that it will flow out both ends, thus increasing the flow and cooling action.

12-15   THRUST BEARINGS

This chapter is devoted to the study of the mechanics of lubrication and its application to the design and analysis of journal bearings. The design and analysis of thrust bear-

FIGURE 12-29

Sleeve bearings.

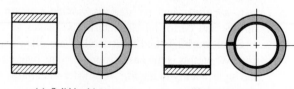

(*a*) Solid bushing   (*b*) Lined bushing

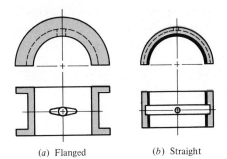

FIGURE 12-30

Two-piece bearings.

(a) Flanged   (b) Straight

ings is an important application of lubrication theory, too. A detailed study of thrust bearings is not included here, because it would not contribute anything significantly different and because of space limitations. Having studied this chapter, you should experience no difficulty in reading the literature on thrust bearings and applying that knowledge to actual design situations.*

Figure 12-32 shows a fixed-pad thrust bearing consisting essentially of a runner sliding over a fixed pad. The lubricant is brought into the radial grooves and pumped

---

*Harry C. Rippel, *Cast Bronze Thrust Bearing Design Manual,* International Copper Research Association, Inc., 825 Third Ave., New York, NY 10022, 1967. CBBI, 14600 Detroit Ave., Cleveland, OH, 44107, 1967.

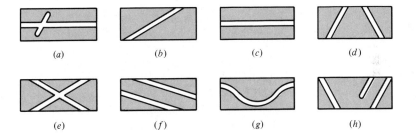

FIGURE 12-31

Developed views of typical groove patterns. *(Courtesy of the Cleveland Graphite Bronze Company, Division of Clevite Corporation.)*

(a)   (b)   (c)   (d)

(e)   (f)   (g)   (h)

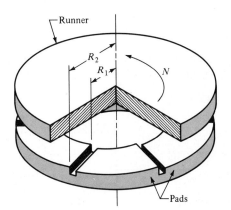

FIGURE 12-32

Fixed-pad thrust bearing. *(Courtesy of The Westinghouse Corporation Research Laboratories.)*

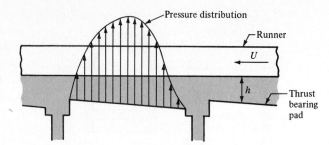

FIGURE 12-33

Pressure distribution of lubricant in a thrust bearing. *(Courtesy of Copper Research Corporation.)*

into the wedge-shaped space by the motion of the runner. Full-film, or hydrodynamic, lubrication is obtained if the speed of the runner is continuous and sufficiently high, if the lubricant has the correct viscosity, and if it is supplied in sufficient quantity. Figure 12-33 provides a picture of the pressure distribution under conditions of full-film lubrication.

We should note that bearings are frequently made with a flange, as shown in Fig. 12-34. The flange positions the bearing in the housing and also takes a thrust load. Even when it is grooved, however, and has adequate lubrication, such an arrangement is not a hydrodynamically lubricated thrust bearing. The reason for this is that the clearance space is not wedge-shaped but has a uniform thickness. Similar reasoning would apply to various designs of thrust washers.

12-16   BOUNDARY-LUBRICATED BEARINGS

When two surfaces slide relative to each other with only a partial lubricant film between them, *boundary lubrication* is said to exist. Boundary- or thin-film lubrication occurs in hydrodynamically lubricated bearings when they are starting or stopping, when the load increases, when the supply of lubricant decreases, or whenever other operating changes happen to occur. There are, of course, a very large number of cases in design in which boundary-lubricated bearings must be used because of the type of application or the competitive situation.

The coefficient of friction for boundary-lubricated surfaces may be greatly decreased by the use of animal or vegetable oils mixed with the mineral oil or grease. Fatty acids, such as stearic acid, palmitic acid, or oleic acid, or several of these, which

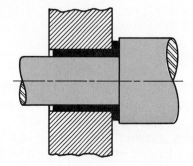

FIGURE 12-34

Flanged sleeve bearing takes both radial and thrust loads.

occur in animal and vegetable fats, are called *oiliness agents*. These acids appear to reduce friction, either because of their strong affinity for certain metallic surfaces or because they form a soap film which binds itself to the metallic surfaces by a chemical reaction. Thus the fatty-acid molecules bind themselves to the journal and bearing surfaces with such great strength that the metallic asperities of the rubbing metals do not weld or shear.

Fatty acids will break down at temperatures of 250°F or more, causing increased friction and wear in thin-film-lubricated bearings. In such cases the *extreme-pressure,* or EP, lubricants may be mixed with the fatty-acid lubricant. These are composed of chemicals such as chlorinated esters or tricresyl phosphate, which form an organic film between the rubbing surfaces. Though the EP lubricants make it possible to operate at higher temperatures, there is the added possibility of excessive chemical corrosion of the sliding surfaces.

When a bearing operates partly under hydrodynamic conditions and partly under dry or thin-film conditions, a *mixed-film lubrication* exists. If the lubricant is supplied by hand oiling, by drop or mechanical feed, or by wick feed, for example, the bearing is operating under mixed-film conditions. In addition to occurring with a scarcity of lubricant, mixed-film conditions may be present when

- The viscosity is too low.
- The bearing speed is too low.
- The bearing is overloaded.
- The clearance is too tight.
- Journal and bearing are not properly aligned.

The range of operating conditions encountered is so great that it is virtually impossible to devise reliable design procedures. The best approach is that of establishing design guidelines or targets and then designing to meet or exceed these standards.

One method of design is based on the ability of the bearing to dissipate heat; after all, a cool bearing is likely to have a long life. In this approach a PV value is computed using the equation

$$PV = \frac{k(T_B - T_A)}{f_M} \tag{12-29}$$

where P = load per unit of projected bearing area, psi
$\quad\quad V$ = surface velocity of journal relative to bearing surface, fpm
$\quad\quad T_A$ = ambient air temperature, °F
$\quad\quad T_B$ = bearing bore temperature, °F
$\quad\quad f_M$ = coefficient of mixed-film friction

Table 12-6 shows some of the materials commonly used when dry or mixed-film conditions are present. Note that all quantities listed are maximum values. However, they cannot all be maximum at the same time.

The coefficient of friction used depends upon whether there is any lubrication at all. Figure 12-35 is a graph of suggested frictional coefficients plotted against the percent-

TABLE 12-6

Some Materials for Boundary-Lubricated Bearings and Their Operating Limits

MATERIAL	MAXIMUM LOAD, psi	MAXIMUM TEMPERATURE, °F	MAXIMUM SPEED, fpm	MAXIMUM PV VALUE*
Cast bronze	4 500	325	1 500	50 000
Porous bronze	4 500	150	1 500	50 000
Porous iron	8 000	150	800	50 000
Phenolics	6 000	200	2 500	15 000
Nylon	1 000	200	1 000	3 000
Teflon	500	500	100	1 000
Reinforced Teflon	2 500	500	1 000	10 000
Teflon fabric	60 000	500	50	25 000
Delrin	1 000	180	1 000	3 000
Carbon-graphite	600	750	2 500	15 000
Rubber	50	150	4 000	
Wood	2 000	150	2 000	15 000

*P = load, psi; V = speed, fpm.

age of mixed lubrication. Except for the graph at $f_B = 0.20$, these are values recommended by Rippel.*

The constant k in Eq. (12-29) depends upon the ability of the bearing to dissipate heat. The best way to evaluate k is to analyze the bearing performance in known

*Harry C. Rippel, *Cast Bronze Bearing Design Manual*, 2d ed., International Copper Research Assoc., New York, 1965.

FIGURE 12-35

Coefficient of friction corresponding to various percentages of mixed-film lubrication. Coefficient of dry friction is f_B. Coefficient of mixed-film friction is f_M. The lower 40 percent is the approximate range of thin-film lubrication.

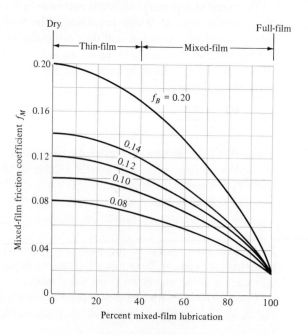

situations or to obtain the value from previous designs known to be satisfactory. Here we can estimate a value of k by using minimum f_M and maximum PV and $(T_B - T_A)$. Using values for cast bronze from Table 12-6, $f_M = 0.02$ from Fig. 12-35, and $T_A = 75°F$ gives

$$k = \frac{f_M(PV)_{max}}{(T_B - T_A)_{max}} = \frac{0.02(50\ 000)}{325 - 75} = 4.0$$

Thus the equation

$$PV = \frac{4(T_B - T_A)}{f_M} \tag{12-30}$$

can be used as a guideline in designing cast-bronze bearings for mixed- or dry-film conditions

PROBLEMS*

12-1 A full journal bearing has a diameter of 1 in and an l/d ratio of unity. The bearing load is 250 lb, and the journal runs at 1100 rev/min. Use a radial clearance of 0.000 75 in and an average viscosity of 8 μreyn, and find the minimum film thickness, the power loss, and the side flow.

12-2* Purchase a quart of your favorite engine oil and determine the viscosity in your laboratory according to ASTM standards. Plot the absolute viscosity on the chart of Fig. 12-11 or Fig. 12-12 for later use.

12-3 A full journal bearing has a diameter of 1.25 in and is 2.5 in long. The bearing load is 400 lb, and the journal runs at a speed of 1150 rev/min. Use a radial clearance of 0.001 in and find the minimum oil-film thickness, the total oil flow, and the maximum film pressure. Use $\mu = 10$ μreyn.

12-4 A journal bearing has a diameter of 3 in and is $1\frac{1}{2}$ in long; it supports a load of 800 lb. The journal speed is 600 rev/min and the radial clearance is 0.0025 in. Find the minimum oil-film thickness and the maximum film pressure for both SAE 10 and SAE 40 lubricants if the operating temperature is 150°F.

12-5 A journal bearing has a diameter of 3 in, is 3 in long, and supports a load of 600 lb. The journal speed is 750 rev/min, and the radial clearance is 0.003 in. Find the minimum oil-film thickness and the maximum film pressure for both a 10W and a 20W-40 lubricant. Use an operating temperature of 140°F.

12-6 A full journal bearing has a diameter of 2 in, is 1 in long, and supports a load of 600 lb. The speed is 800 rev/min, and the radial clearance is 0.0012 in. Find the minimum film thickness, the power loss, and the total flow if the operating temperature is 130°F and SAE 30 lubricant is used.

12-7 A full journal bearing is 25 mm in diameter and has an l/d ratio of unity. The bearing load is 1.25 kN, and the journal rotates at 1200 rev/min. Use a radial clearance of 0.02 mm and an average viscosity of 50 mPa · s, and find the minimum oil-film thickness, the power loss, and the percentage of side flow.

---

*An asterisk indicates a design-type problem or a problem that may not have a unique answer.

12-8   A full journal bearing is 30 mm in diameter and 50 mm in length. The bearing load is 2.75 kN, and the journal rotates at 1120 rev/min. Use a radial clearance of 0.025 mm, and find the minimum film thickness, the friction coefficient, and the total oil flow if $\mu = 60$ mPa $\cdot$ s.

12-9   A journal bearing is 75 mm in diameter and 36 mm long and supports a load of 2 kN. The journal speed is 720 rev/min, and the radial clearance is 0.05 mm. Find the minimum film thickness, the heat loss, and the maximum film pressure for SAE 20 and SAE 40 lubricants if the operating temperature is 60°C.

12-10   A full journal bearing is 25 mm long and 50 mm in diameter and supports a load of 2000 N. The speed is 840 rev/min, and the radial clearance is 0.025 mm. Find the minimum oil-film thickness, the power loss, and the side flow if the operating temperature is 55°C and SAE 30 oil is used.

12-11   A $1\frac{1}{4} \times 1\frac{1}{4}$-in sleeve bearing supports a load of 700 lb and has a journal speed of 3600 rev/min. An SAE 10 oil used is assumed to have an average operating temperature of 160°F. Using Fig. 12-14, determine the radial clearance for minimum f and for maximum W. The difference between these two clearances is called the clearance range. Is the resulting range attainable in manufacture?

12-12   A full journal bearing has a diameter of 80 mm and an l/d ratio of unity and runs at a journal speed of 8 rev/s. The oil supply is SAE 30 at an inlet temperature of 60°C. The radial load is 3000 N, and the radial clearance is 0.040 mm. Estimate the temperature rise, the minimum oil-film thickness, the heat loss, and the side flow.

12-13   An SAE 50 oil at 25°C inlet temperature is used to lubricate a sleeve bearing 150 mm in length and 50 mm in diameter. The bearing load is 8 kN, and the journal speed is 160 rev/min. Using a clearance ratio of $r/c = 600$, find the temperature rise, the eccentricity ratio, and the maximum film pressure.

12-14   A 10- $\times$ 10-mm sleeve bearing is lubricated with SAE 10 oil at an inlet temperature of 50°C. The radial clearance is 0.010 mm, the load 65 N, and the journal speed 3200 rev/min. Find the temperature rise and the minimum oil-film thickness together with its angular location.

12-15   A sleeve bearing is 40 mm in diameter and 40 mm in length and has a journal speed of 1200 rev/min. The r/c ratio is 1000, the load is 2500 N, and the bearing is lubricated with an SAE 40 oil at an inlet temperature of 40°C. Find the oil outlet temperature, the minimum oil-film thickness, and the magnitude and angular location of the maximum film pressure.

12-16*   A set of sleeve bearings are 38 mm in diameter, with $l/d = 1$. The bearings are to be rated at 2500 N radial load corresponding to a journal speed of 20 rev/s. SAE 40 lubricant is specified, with an inlet temperature not to exceed 35°C. Specify the clearance limits if the tolerance is 0.01 mm.

12-17   A set of sleeve bearings have diameter $1\frac{1}{4}$ in and length $1\frac{1}{4}$ in and are planned for radial loads not more than 250 lb at rotational speeds up to 1750 rev/min. The lubricant recommended is SAE 10 at an inlet temperature not to exceed 120°F. The mean clearance is 0.001 in, with a tolerance of 0.0005 in. Find the outlet temperature corresponding to the extremes of the clearance.

12-18   A $2\frac{1}{2}$- $\times$ $2\frac{1}{2}$-in sleeve bearing uses SAE 20 oil at an inlet temperature of 110°F. The journal speed is 1120 rev/min, and the radial load is 1200 lb. Based on a radial clearance of 0.002 in, find:

(*a*) The magnitude and location of the minimum oil-film thickness

(*b*) The eccentricity

(*c*) The coefficient of friction

(*d*) The power loss

(*e*) Both the total and side oil flows

(*f*) The maximum oil-film pressure and its angular location

(*g*) The terminating position of the oil film

(*h*) The average temperature of the side flow

(*i*) The oil temperature at the terminating position of the oil film

12-19   A sleeve bearing is 1 in long and has a diameter of 2 in and a radial clearance of 0.002 in. If the load is 500 lb corresponding to a speed of 1500 rev/min and SAE 30 oil at 90°F is used, find:

(*a*) The oil outlet temperature

(*b*) The minimum oil-film thickness

(*c*) The total oil flow

(*d*) The side flow

12-20   A $\frac{1}{2}$- $\times$ $\frac{1}{2}$-in sleeve bearing has a radial clearance of 0.0005 in and uses SAE 20 oil at an inlet temperature of 120°F. The journal runs at 3000 rev/min and supports a load of 24 lb. Find the temperature rise of the lubricant.

12-21   A $1\frac{3}{4}$-in-diameter bearing is 2 in long and has a central annular oil groove $\frac{1}{4}$ in wide which is fed by SAE 10 oil at 120°F and 30 psi supply pressure. The radial clearance is 0.0015 in. The journal rotates at 3000 rev/min, and the average load is 600 psi of projected area. Find the temperature rise, the minimum film thickness, and the maximum film pressure.

12-22   An eight-cylinder diesel engine has a front main bearing with diameter $3\frac{1}{2}$ in and length 2 in. The bearing has a central annular oil groove 0.250 in wide. It is pressure-lubricated with SAE 30 oil at an inlet temperature of 180°F and at a supply pressure of 50 psi. Corresponding to a radial clearance of 0.0025 in, a speed of 2800 rev/min, and a radial load of 4600 lb, find the temperature rise and the minimum oil-film thickness.

12-23   A 50-mm-diameter bearing is 55 mm long and has a central annular oil groove 5 mm wide which is fed by SAE 30 oil at 55°C and 200-kPa supply pressure. The radial clearance is 42 μm. The journal speed is 48 rev/s corresponding to a bearing load of 10 kN. Find the temperature rise of the lubricant, the total oil flow, and the minimum film thickness.

ANSWERS

12-1 $h_0 = 446$ μin, $H = 0.0172$ Btu/s, $Q_s = 0.0137$ in^3/s

12-5 For 10W, $h_0 = 0.001$ 02 in, $p_{max} = 169$ psi; for 20W-40, $h_0 = 0.001$ 71 in, $p_{max} = 142$ psi

12-9 SAE 20: $h_0 = 0.0145$ mm, $H = 38.5$ W, $p_{max} = 2.35$ MPa

12-12 $\Delta T = 10$°C, $h_0 = 0.0268$ mm, $H = 45.2$ W, $Q_s = 1628$ mm^3/s

12-17 138°F $\leq T_2 \leq$ 153°F

12-21 $\Delta T = 78$°F, $h_0 = 0.000$ 18 in, $p_{max} = 2820$ psi

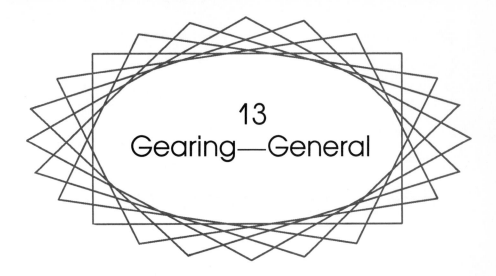

13
Gearing—General

This chapter deals with the geometry, the kinematic relations, and the force analysis of the four principal types of gears. The two chapters that follow deal with other design considerations, such as stress, strength, safety, and reliability.

13-1 TYPES OF GEARS

Spur gears, illustrated in Fig. 13-1, have teeth parallel to the axis of rotation and are used to transmit motion from one shaft to another, parallel, shaft. Of all types, the spur gear is the simplest and, for this reason, will be used to develop the primary kinematic relationships of the tooth form.

Helical gears, shown in Fig. 13-2, have teeth inclined to the axis of rotation. Helical gears can be used for the same applications as spur gears and, when so used, are not as noisy, because of the more gradual engagement of the teeth during meshing. The inclined tooth also develops thrust loads and bending couples, which are not present with spur gearing. Sometimes helical gears are used to transmit motion between nonparallel shafts.

Bevel gears, shown in Fig. 13-3, have teeth formed on conical surfaces and are used mostly for transmitting motion between intersecting shafts. The figure actually illustrates *straight-tooth bevel gears. Spiral bevel gears* are cut so that the tooth is no longer straight, but forms a circular arc. *Hypoid gears* are quite similar to spiral bevel gears except that the shafts are offset and nonintersecting.

Shown in Fig. 13-4 is the fourth basic gear type, the *worm* and *worm gear.* As shown, the worm resembles a screw. The direction of rotation of the worm gear, also called the worm wheel, depends upon the direction of rotation of the worm and upon whether the worm teeth are cut right-hand or left-hand. Worm-gear sets are also made so that the teeth of one or both wrap partly around the other. Such sets are called *single-enveloping* and *double-enveloping* worm-gear sets. Worm-gear sets are mostly used when the speed ratios of the two shafts are quite high, say, 3 or more.

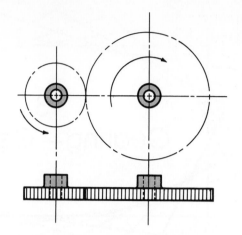

FIGURE 13-1

Spur gears are used to transmit
rotary motion between parallel
shafts.

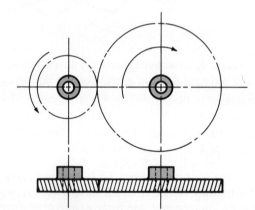

FIGURE 13-2

Helical gears are used to transmit
motion between parallel or nonparallel shafts.

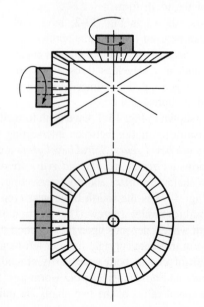

FIGURE 13-3

Bevel gears are used to transmit
motion between intersecting
shafts.

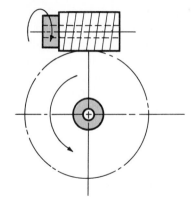

FIGURE 13-4

Worm gears are used to transmit motion between nonparallel nonintersecting shafts.

13-2   NOMENCLATURE

The terminology of spur-gear teeth is illustrated in Fig. 13-5. The *pitch circle* is a theoretical circle upon which all calculations are usually based; its diameter is the *pitch diameter*. The pitch circles of a pair of mating gears are tangent to each other. A *pinion* is the smaller of two mating gears. The larger is often called the *gear*.

The *circular pitch p* is the distance, measured on the pitch circle, from a point on one tooth to a corresponding point on an adjacent tooth. Thus the circular pitch is equal to the sum of the *tooth thickness* and the *width of space*.

The *module m* is the ratio of the pitch diameter to the number of teeth. The customary unit of length used is the millimeter. The module is the index of tooth size in SI.

The *diametral pitch P* is the ratio of the number of teeth on the gear to the pitch diameter. Thus, it is the reciprocal of the module. Since diametral pitch is used only with U.S. units, it is expressed as teeth per inch.

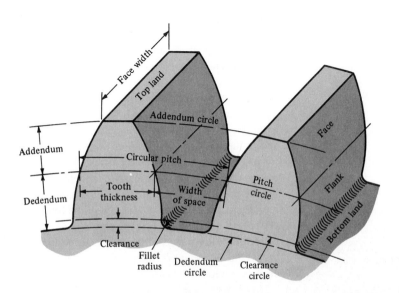

FIGURE 13-5

Nomenclature of spur-gear teeth.

The *addendum a* is the radial distance between the *top land* and the pitch circle. The *dedendum b* is the radial distance from the *bottom land* to the pitch circle. The *whole depth h_t* is the sum of the addendum and dedendum.

The *clearance circle* is a circle that is tangent to the addendum circle of the mating gear. The *clearance c* is the amount by which the dedendum in a given gear exceeds the addendum of its mating gear. The *backlash* is the amount by which the width of a tooth space exceeds the thickness of the engaging tooth measured on the pitch circles.

You should prove for yourself the validity of the following useful relations:

$$P = \frac{N}{d} \tag{13-1}$$

where P = diametral pitch, teeth per inch
$\qquad N$ = number of teeth
$\qquad d$ = pitch diameter, in

$$m = \frac{d}{N} \tag{13-2}$$

where m = module, mm
$\qquad d$ = pitch diameter, mm

$$p = \frac{\pi d}{N} = \pi m \tag{13-3}$$

where p = circular pitch

$$pP = \pi \tag{13-4}$$

13-3   CONJUGATE ACTION

The following discussion assumes the teeth to be perfectly formed, perfectly smooth, and absolutely rigid. Such an assumption is, of course, unrealistic, because the application of forces will cause deflections.

Mating gear teeth acting against each other to produce rotary motion are similar to cams. When the tooth profiles, or cams, are designed so as to produce a constant angular-velocity ratio during meshing, they are said to have *conjugate action*. In theory, at least, it is possible arbitrarily to select any profile for one tooth and then to find a profile for the meshing tooth which will give conjugate action. One of these solutions is the *involute profile,* which, with few exceptions, is in universal use for gear teeth and is the only one with which we shall be concerned.

When one curved surface pushes against another (Fig. 13-6), the point of contact occurs where the two surfaces are tangent to each other (point c), and the forces at any instant are directed along the common normal ab to the two curves. The line ab, representing the direction of action of the forces, is called the *line of action*. The line of action will intersect the line of centers O-O at some point P. The angular-velocity ratio between the two arms is inversely proportional to their radii to the point P. Circles drawn through point P from each center are called *pitch circles,* and the radius of each circle is called the *pitch radius.* Point P is called the *pitch point.*

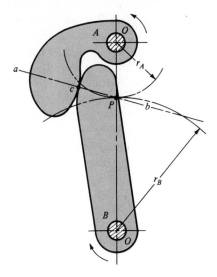

FIGURE 13-6

To transmit motion at a constant angular-velocity ratio, the pitch point must remain fixed; that is, all the lines of action for every instantaneous point of contact must pass through the same point P. In the case of the involute profile, it will be shown that all points of contact occur on the same straight line ab, that all normals to the tooth profiles at the point of contact coincide with the line ab, and, thus, that these profiles transmit uniform rotary motion.

13-4   INVOLUTE PROPERTIES

An involute curve may be generated as shown in Fig. 13-7a. A partial flange B is attached to the cylinder A, around which is wrapped a cord def which is held tight.

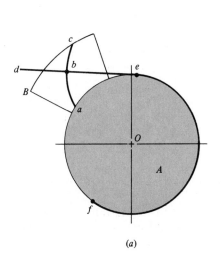

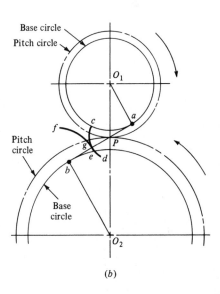

FIGURE 13-7

(a) Generation of an involute; (b) involute action.

(a)   (b)

Point b on the cord represents the tracing point, and as the cord is wrapped and unwrapped about the cylinder, point b will trace out the involute curve ac. The radius of curvature of the involute varies continuously, being zero at point a and a maximum at point c. At point b the radius is equal to the distance be, since point b is instantaneously rotating about point e. Thus the generating line de is normal to the involute at all points of intersection and, at the same time, is always tangent to the cylinder A. The circle on which the involute is generated is called the *base circle*.

Let us now examine the involute profile to see how it satisfies the requirement for the transmission of uniform motion. In Fig. 13-7b, two gear blanks with fixed centers at O_1 and O_2 are shown having base circles whose respective radii are O_1a and O_2b. We now imagine that a cord is wound clockwise around the base circle of gear 1, pulled tight between points a and b, and wound counterclockwise around the base circle of gear 2. If, now, the base circles are rotated in different directions so as to keep the cord tight, a point g on the cord will trace out the involutes cd on gear 1 and ef on gear 2. The involutes are thus generated simultaneously by the tracing point. The tracing point, therefore, represents the point of contact, while the portion of the cord ab is the generating line. The point of contact moves along the generating line; the generating line does not change position, because it is always tangent to the base circles; and since the generating line is always normal to the involutes at the point of contact, the requirement for uniform motion is satisfied.

13-5   FUNDAMENTALS

Among other things, it is necessary that you actually be able to draw the teeth on a pair of meshing gears. You should understand, however, that you are not doing this for manufacturing or shop purposes. Rather, we make drawings of gear teeth to obtain an understanding of the problems involved in the meshing of the mating teeth.

First, it is necessary to learn how to construct an involute curve. As shown in Fig. 13-8, divide the base circle into a number of equal parts and construct radial lines OA_0, OA_1, OA_2, etc. Beginning at A_1, construct perpendiculars A_1B_1, A_2B_2, A_3B_3, etc. Then along A_1B_1 lay off the distance A_1A_0, along A_2B_2 lay off twice the distance A_1A_0, etc., producing points through which the involute curve can be constructed.

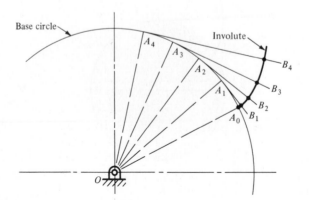

FIGURE 13-8

Construction of an involute curve.

To investigate the fundamentals of tooth action, let us proceed step by step through the process of constructing the teeth on a pair of gears.

When two gears are in mesh, their pitch circles roll on one another without slipping. Designate the pitch radii as r_1 and r_2 and the angular velocities as ω_1 and ω_2, respectively. Then the pitch-line velocity is

$$V = |r_1\omega_1| = |r_2\omega_2|$$

Thus the relation between the radii and the angular velocities is

$$\left|\frac{\omega_1}{\omega_2}\right| = \frac{r_2}{r_1} \tag{13-5}$$

Suppose now we wish to design a speed reducer such that the input speed is 1800 rev/min and the output speed is 1200 rev/min. This is a ratio of 3:2; the pitch diameters would be in the same ratio, for example, a 4-in pinion driving a 6-in gear. The various dimensions found in gearing are always based on the pitch circles.

We next specify that an 18-tooth pinion is to mesh with a 30-tooth gear and that the diametral pitch of the gearset is to be 2 teeth per inch. Then, from Eq. (13-1), the pitch diameters of the pinion and gear are, respectively,

$$d_1 = \frac{N_1}{P} = \frac{18}{2} = 9 \text{ in} \qquad d_2 = \frac{N_2}{P} = \frac{30}{2} = 15 \text{ in}$$

The first step in drawing teeth on a pair of mating gears is shown in Fig. 13-9. The center distance is the sum of the pitch radii, in this case 12 in. So locate the pinion and gear centers O_1 and O_2, 12 in apart. Then construct the pitch circles of radii r_1 and r_2. These are tangent at P, the *pitch point*. Next draw line ab, the common tangent, through the pitch point. We now designate gear 1 as the driver, and since it is rotating counterclockwise, we draw a line cd through point P at an angle ϕ to the common tangent ab. The line cd has three names, all of which are in general use. It is called the *pressure line*, the *generating line*, and the *line of action*. It represents the direction in which the resultant force acts between the gears. The angle ϕ is called the *pressure angle*, and it usually has values of 20 or 25°, though $14\frac{1}{2}°$ was once used.

Next, on each gear draw a circle tangent to the pressure line. These circles are the

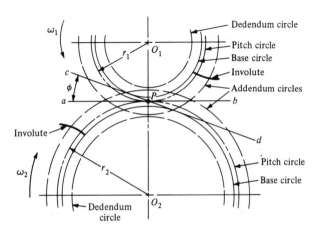

FIGURE 13-9

Gear layout.

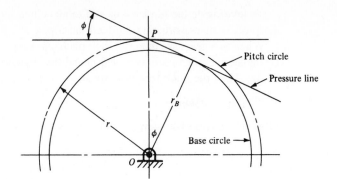

FIGURE 13-10

base circles. Since they are tangent to the pressure line, the pressure angle determines their size. As shown in Fig. 13-10, the radius of the base circle is

$$r_b = r \cos \phi \tag{13-6}$$

where r is the pitch radius.

Now generate an involute on each base circle as previously described and as shown in Fig. 13-9. This involute is to be used for one side of a gear tooth. It is not necessary to draw another curve in the reverse direction for the other side of the tooth, because we are going to use a template which can be turned over to obtain the other side.

The addendum and dedendum distances for standard interchangeable teeth are, as we shall learn later, $1/P$ and $1.25/P$, respectively. Therefore, for the pair of gears we are constructing,

$$a = \frac{1}{P} = \frac{1}{2} = 0.500 \text{ in} \qquad b = \frac{1.25}{P} = \frac{1.25}{2} = 0.625 \text{ in}$$

Using these distances, draw the addendum and dedendum circles on the pinion and on the gear as shown in Fig. 13-9.

Next, using heavy drawing paper, or preferably, a sheet of 0.015- to 0.020-in clear plastic, cut a template for each involute, being careful to locate the gear centers properly with respect to each involute. Figure 13-11 is a reproduction of the template used to create some of the illustrations for this book. Note that only one side of the tooth profile is formed on the template. To get the other side, turn the template over. For some problems you might wish to construct a template for the entire tooth.

To draw a tooth, we must know the tooth thickness. From Eq. (13-4), the circular pitch is

$$p = \frac{\pi}{P} = \frac{\pi}{2} = 1.57 \text{ in}$$

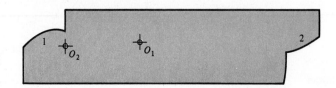

FIGURE 13-11

Template for drawing gear teeth.

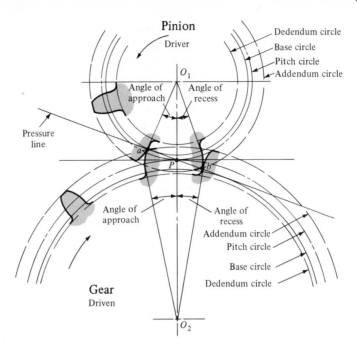

FIGURE 13-12

Tooth action.

Therefore the tooth thickness is

$$t = \frac{p}{2} = \frac{1.57}{2} = 0.785 \text{ in}$$

measured on the pitch circle. Using this distance for the tooth thickness as well as the tooth space, draw as many teeth as are desired, using the template, after the points have been marked on the pitch circle. In Fig. 13-12 only one tooth has been drawn on each gear. You may run into trouble in drawing these teeth if one of the base circles happens to be larger than the dedendum circle. The reason for this is that the involute begins at the base circle and is undefined below this circle. So, in drawing gear teeth, we usually draw a radial line for the profile below the base circle. The actual shape, however, will depend upon the kind of machine tool used to form the teeth in manufacture, that is, how the profile is generated.

The portion of the tooth between the clearance circle and the dedendum circle includes the fillet. In this instance the clearance is

$$c = b - a = 0.625 - 0.500 = 0.125 \text{ in}$$

The construction is finished when these fillets have been drawn.

Referring again to Fig. 13-12, the pinion with center at O_1 is the driver and turns counterclockwise. The pressure, or generating, line is the same as the cord used in Fig. 13-7a to generate the involute, and contact occurs along this line. The initial contact will take place when the flank of the driver comes into contact with the tip of the driven tooth. This occurs at point a in Fig. 13-12, where the addendum circle of the driven gear crosses the pressure line. If we now construct tooth profiles through point a and draw radial lines from the intersections of these profiles with the pitch circles to the gear centers, we obtain the *angle of approach* for each gear.

As the teeth go into mesh, the point of contact will slide up the side of the driving tooth so that the tip of the driver will be in contact just before contact ends. The final point of contact will therefore be where the addendum circle of the driver crosses the pressure line. This is point *b* in Fig. 13-12. By drawing another set of tooth profiles through *b*, we obtain the *angle of recess* for each gear in a manner similar to that of finding the angles of approach. The sum of the angle of approach and the angle of recess for either gear is called the *angle of action*. The line *ab* is called the *line of action*.

We may imagine a *rack* as a spur gear having an infinitely large pitch diameter. Therefore the rack has an infinite number of teeth and a base circle which is an infinite distance from the pitch point. The sides of involute teeth on a rack are straight lines making an angle to the line of centers equal to the pressure angle. Figure 13-13 shows an involute rack in mesh with a pinion.

Corresponding sides of involute teeth are parallel curves; the *base pitch* is the constant and fundamental distance between them along a common normal as shown in Fig. 13-13. The base pitch is related to the circular pitch by the equation

$$p_b = p_c \cos \phi \tag{13-7}$$

where p_b is the base pitch.

Figure 13-14 shows a pinion in mesh with an *internal, or annular, gear*. Note that both of the gears now have their centers of rotation on the same side of the pitch point. Thus the positions of the addendum and dedendum circles with respect to the pitch circle are reversed; the addendum circle of the internal gear lies *inside* the pitch circle. Note, too, from Fig. 13-14, that the base circle of the internal gear lies inside the pitch circle near the addendum circle.

Another interesting observation concerns the fact that the operating diameters of the pitch circles of a pair of meshing gears need not be the same as the respective design pitch diameters of the gears, though this is the way they have been constructed in Fig. 13-12. If we increase the center distance, we create two new operating pitch circles having larger diameters because they must be tangent to each other at the pitch point. Thus the pitch circles of gears really do not come into existence until a pair of gears are brought into mesh.

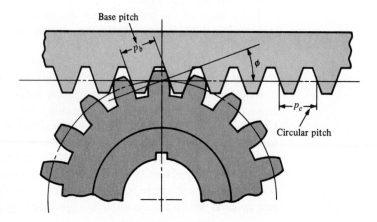

FIGURE 13-13

Involute pinion and rack.

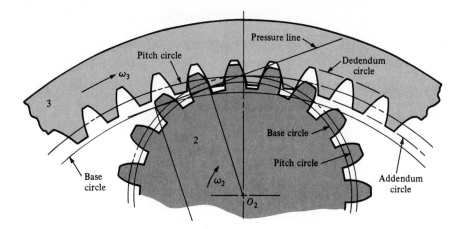

FIGURE 13-14

Internal gear and pinion.

Changing the center distance has no effect on the base circles, because these were used to generate the tooth profiles. Thus the base circle is basic to a gear. Increasing the center distance increases the pressure angle and decreases the length of the line of action, but the teeth are still conjugate, the requirement for uniform motion transmission is still satisfied, and the angular-velocity ratio has not changed.

EXAMPLE 13-1

A gearset consists of a 16-tooth pinion driving a 40-tooth gear. The diametral pitch is 2, and the addendum and dedendum are $1/P$ and $1.25/P$, respectively. The gears are cut using a pressure angle of 20°.

(a) Compute the circular pitch, the center distance, and the radii of the base circles.

(b) In mounting these gears, the center distance was incorrectly made $\frac{1}{4}$ in larger. Compute the new values of the pressure angle and the pitch-circle diameters.

Solution

Answer   (a) $p = \dfrac{\pi}{P} = \dfrac{\pi}{2} = 1.57$ in

The pitch diameters of the pinion and gear are, respectively,

$$d_P = \tfrac{16}{2} = 8 \text{ in} \qquad d_G = \tfrac{40}{2} = 20 \text{ in}$$

Therefore the center distance is

$$\frac{d_P + d_G}{2} = \frac{8 + 20}{2} = 14 \text{ in}$$

Since the teeth were cut on the 20° pressure angle, the base-circle radii are found to be, using $r_b = r \cos \phi$,

Answer r_b (pinion) $= \tfrac{8}{2} \cos 20° = 3.76$ in

Answer r_b (gear) $= \tfrac{20}{2} \cos 20° = 9.40$ in

(b) Designating d'_P and d'_G as the new pitch-circle diameters, the $\frac{1}{4}$-in increase in the center distance requires that

$$\frac{d'_P + d'_G}{2} = 14.250 \tag{1}$$

Also, the velocity ratio does not change, and hence

$$\frac{d'_P}{d'_G} = \frac{16}{40} \tag{2}$$

Solving Eqs. (1) and (2) simultaneously yields

Answer $d'_P = 8.143$ in $d'_G = 20.357$ in

Since $r_b = r \cos \phi$, the new pressure angle is

Answer $\phi' = \cos^{-1} \dfrac{r_b \text{ (pinion)}}{d'_P/2} = \cos^{-1} \dfrac{3.76}{8.143/2} = 22.56°$

13-6   CONTACT RATIO

The zone of action of meshing gear teeth is shown in Fig. 13-15. We recall that tooth contact begins and ends at the intersections of the two addendum circles with the pressure line. In Fig. 13-15 initial contact occurs at *a* and final contact at *b*. Tooth profiles drawn through these points intersect the pitch circle at *A* and *B*, respectively. As shown, the distance *AP* is called the *arc of approach* q_a, and the distance *PB*, the *arc of recess* q_r. The sum of these is the *arc of action* q_t.

Now, consider a situation in which the arc of action is exactly equal to the circular pitch, that is, $q_t = p$. This means that one tooth and its space will occupy the entire arc *AB*. In other words, when a tooth is just beginning contact at *a*, the previous tooth is simultaneously ending its contact at *b*. Therefore, during the tooth action from *a* to *b*, there will be exactly one pair of teeth in contact.

Next, consider a situation in which the arc of action is greater than the circular pitch, but not very much greater, say, $q_t \approx 1.2p$. This means that when one pair of teeth is just entering contact at *a*, another pair, already in contact, will not yet have

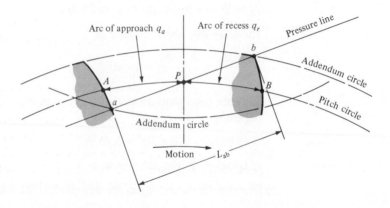

FIGURE 13-15

Definition of contact ratio.

reached b. Thus, for a short period of time, there will be two pairs of teeth in contact, one in the vicinity of A and another near B. As the meshing proceeds, the pair near B must cease contact, leaving only a single pair of contacting teeth, until the procedure repeats itself.

Because of the nature of this tooth action, either one or two pairs of teeth in contact, it is convenient to define the term *contact ratio* m_c as

$$m_c = \frac{q_t}{p} \qquad\qquad (13\text{-}8)$$

a number which indicates the average number of pairs of teeth in contact. Note that this ratio is also equal to the length of the path of contact divided by the base pitch. Gears should not generally be designed having contact ratios less than about 1.20, because inaccuracies in mounting might reduce the contact ratio even more, increasing the possibility of impact between the teeth as well as an increase in the noise level.

An easier way to obtain the contact ratio is to measure the line of action ab instead of the arc distance AB. Since ab in Fig. 13-15 is tangent to the base circle when extended, the base pitch p_b must be used to calculate m_c instead of the circular pitch as in Eq. (13-8). Designating the length of the line of action as L_{ab}, the contact ratio is

$$m_c = \frac{L_{ab}}{p \cos \phi} \qquad\qquad (13\text{-}9)$$

in which Eq. (13-7) was used for the base pitch.

13-7   INTERFERENCE

The contact of portions of tooth profiles which are not conjugate is called *interference*. Consider Fig. 13-16. Illustrated are two 16-tooth gears which have been cut using the now obsolete $14\frac{1}{2}°$ pressure angle. The driver, gear 2, turns clockwise. The initial and final points of contact are designated A and B, respectively, and are located on the pressure line. Now notice that the points of tangency of the pressure line with the base circles C and D are located *inside* of points A and B. Interference is present.

The interference is explained as follows. Contact begins when the tip of the driven tooth contacts the flank of the driving tooth. In this case the flank of the driving tooth first makes contact with the driven tooth at point A, and this occurs *before* the involute portion of the driving tooth comes within range. In other words, contact is occurring below the base circle of gear 2 on the *noninvolute* portion of the flank. The actual effect is that the involute tip or face of the driven gear tends to dig out the noninvolute flank of the driver.

In this example the same effect occurs again as the teeth leave contact. Contact should end at point D or before. Since it does not end until point B, the effect is for the tip of the driving tooth to dig out, or interfere with, the flank of the driven tooth.

When gear teeth are produced by a generation process, interference is automatically eliminated because the cutting tool removes the interfering portion of the flank. This effect is called *undercutting;* if undercutting is at all pronounced, the undercut tooth is considerably weakened. Thus the effect of eliminating interference by a generation process is merely to substitute another problem for the original one.

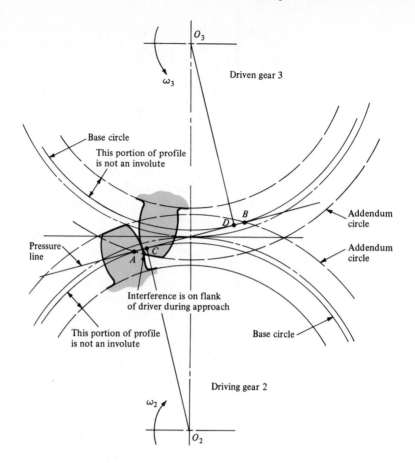

FIGURE 13-16

Interference in the action of gear teeth. (This is actually a rather poor figure; I drew the tooth shape using circle arcs, which is incorrect, many years ago. *J.E.S.*)

TABLE 13-1

Minimum Tooth Numbers to Avoid Interference Numbers are Based on a Normal Pressure Angle of $\phi_n = 20°$ and Full-Depth Teeth. For Spur Gears, $\psi = 0$

NUMBER OF PINION TEETH, N_P	NUMBER OF GEAR TEETH, N_G							
	HELIX ANGLE ψ, deg							
	0	5	10	15	20	25	30	35
8								12
9							12	34
10						12	26	∞
11					13	23	93	
12			12	16	24	57	∞	
13	16	17	20	27	50	1409		
14	26	27	34	53	207	∞		
15	45	49	69	181	∞			
16	101	121	287	∞				
17	1309	∞	∞					

Source: R. Lipp, "Avoiding Tooth Interference in Gears," *Machine Design*, vol. 54, no. 1, 1982, p. 122.

The importance of the problem of teeth which have been weakened by undercutting cannot be overemphasized. Of course, interference can be eliminated by using more teeth on the gears. However, if the gears are to transmit a given amount of power, more teeth can be used only by increasing the pitch diameter. This makes the gears larger, which is seldom desirable, and it also increases the pitch-line velocity. This increased pitch-line velocity makes the gears noisier and reduces the power transmission somewhat, although not in direct ratio. In general, however, the use of more teeth to eliminate interference or undercutting is seldom an acceptable solution.

Interference can also be reduced by using a larger pressure angle. This results in a smaller base circle, so that more of the tooth profile becomes involute. The demand for smaller pinions with fewer teeth thus favors the use of a 25° pressure angle even though the frictional forces and bearing loads are increased and the contact ratio decreased.

See Table 13-1 for minimum tooth numbers to avoid interference problems.

13-8 THE FORMING OF GEAR TEETH

There are a large number of ways of forming the teeth of gears, such as *sand casting, shell molding, investment casting, permanent-mold casting, die casting,* and *centrifugal casting*. Teeth can be formed by using the *powder-metallurgy process;* or, by using *extrusion,* a single bar of aluminum may be formed and then sliced into gears. Gears which carry large loads in comparison with their size are usually made of steel and are cut with either *form cutters* or *generating cutters*. In form cutting, the tooth space takes the exact shape of the cutter. In generating, a tool having a shape different from the tooth profile is moved relative to the gear blank so as to obtain the proper tooth shape. One of the newest and most promising of the methods of forming teeth is called *cold forming,* or *cold rolling,* in which dies are rolled against steel blanks to form the teeth. The mechanical properties of the metal are greatly improved by the rolling process, and a high-quality generated profile is obtained at the same time.

Gear teeth may be machined by milling, shaping, or hobbing. They may be finished by shaving, burnishing, grinding, or lapping.

Milling

Gear teeth may be cut with a form milling cutter shaped to conform to the tooth space. With this method it is theoretically necessary to use a different cutter for each gear, because a gear having 25 teeth, for example, will have a different-shaped tooth space from one having, say, 24 teeth. Actually, the change in space is not too great, and it has been found that eight cutters may be used to cut with reasonable accuracy any gear in the range of 12 teeth to a rack. A separate set of cutters is, of course, required for each pitch.

Shaping

Teeth may be generated with either a pinion cutter or a rack cutter. The pinion cutter (Fig. 13-17) reciprocates along the vertical axis and is slowly fed into the gear blank to

FIGURE 13-17

Generating a spur gear with a pinion cutter. *(Courtesy of Boston
Gear Works, Inc.)*

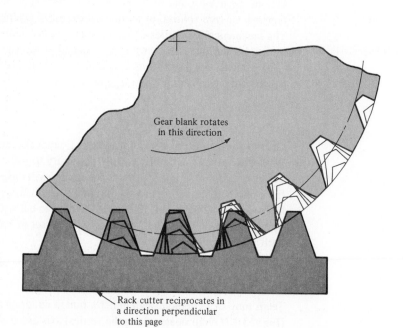

Gear blank rotates
in this direction

Rack cutter reciprocates in
a direction perpendicular
to this page

FIGURE 13-18

Shaping teeth with a rack cutter.
(This is a drawing-board figure
that I executed about 25 years
ago in response to a question
from a student at the University of
Michigan. *J.E.S.*)

the required depth. When the pitch circles are tangent, both the cutter and blank rotate slightly after each cutting stroke. Since each tooth of the cutter is a cutting tool, the teeth are all cut after the blank has completed one revolution.

The sides of an involute rack tooth are straight. For this reason, a rack-generating tool provides an accurate method of cutting gear teeth. This is also a shaping operation and is illustrated by the drawing of Fig. 13-18. In operation, the cutter reciprocates and is first fed into the gear blank until the pitch circles are tangent. Then, after each cutting stroke, the gear blank and cutter roll slightly on their pitch circles. When the blank and cutter have rolled a distance equal to the circular pitch, the cutter is returned to the starting point, and the process is continued until all the teeth have been cut.

Hobbing

The hobbing process is illustrated in Fig. 13-19. The hob is simply a cutting tool which is shaped like a worm. The teeth have straight sides, as in a rack, but the hob axis must be turned through the lead angle in order to cut spur-gear teeth. For this reason, the teeth generated by a hob have a slightly different shape from those generated by a rack cutter. Both the hob and the blank must be rotated at the proper angular-velocity ratio. The hob is then fed slowly across the face of the blank until all the teeth have been cut.

Finishing

Gears which run at high speeds and transmit large forces may be subjected to additional dynamic forces if there are errors in tooth profiles. Errors may be diminished some-

FIGURE 13-19

Hobbing a worm gear. *(Courtesy of Boston Gear Works, Inc.)*

what by finishing the tooth profiles. The teeth may be finished, after cutting, by either shaving or burnishing. Several shaving machines are available which cut off a minute amount of metal, bringing the accuracy of the tooth profile within the limits of 250 μin.

Burnishing, like shaving, is used with gears which have been cut but not heat-treated. In burnishing, hardened gears with slightly oversize teeth are run in mesh with the gear until the surfaces become smooth.

Grinding and lapping are used for hardened gear teeth after heat treatment. The grinding operation employs the generating principle and produces very accurate teeth. In lapping, the teeth of the gear and lap slide axially so that the whole surface of the teeth is abraded equally.

13-9   STRAIGHT BEVEL GEARS

When gears are to be used to transmit motion between intersecting shafts, some form of bevel gear is required. A bevel gearset is shown in Fig. 13-20. Although bevel gears are usually made for a shaft angle of 90°, they may be produced for almost any angle. The teeth may be cast, milled, or generated. Only the generated teeth may be classed as accurate.

The terminology of bevel gears is illustrated in Fig. 13-20. The pitch of bevel gears is measured at the large end of the tooth, and both the circular pitch and the pitch diameter are calculated in the same manner as for spur gears. It should be noted that the clearance is uniform. The pitch angles are defined by the pitch cones meeting at the apex, as shown in the figure. They are related to the tooth numbers as follows:

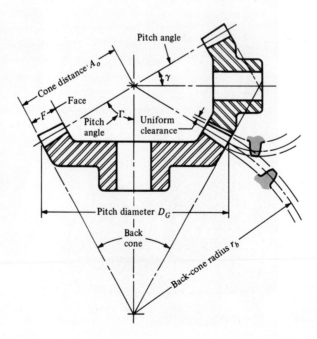

FIGURE 13-20

Terminology of bevel gears.

$$\tan \gamma = \frac{N_P}{N_G} \qquad \tan \Gamma = \frac{N_G}{N_P} \tag{13-10}$$

where the subscripts P and G refer to the pinion and gear, respectively, and where γ and Γ are, respectively, the pitch angles of the pinion and gear.

Figure 13-20 shows that the shape of the teeth, when projected on the back cone, is the same as in a spur gear having a radius equal to the back-cone distance r_b. This is called Tredgold's approximation. The number of teeth in this imaginary gear is

$$N' = \frac{2\pi r_b}{p} \tag{13-11}$$

where N' is the *virtual number of teeth* and p is the circular pitch measured at the large end of the teeth.

Standard straight-tooth bevel gears are cut by using a 20° pressure angle, unequal addenda and dedenda, and full-depth teeth. This increases the contact ratio, avoids undercut, and increases the strength of the pinion.

13-10 PARALLEL HELICAL GEARS

Helical gears, used to transmit motion between parallel shafts, are shown in Fig. 13-2. The helix angle is the same on each gear, but one gear must have a right-hand helix and the other a left-hand helix. The shape of the tooth is an involute helicoid and is illustrated in Fig. 13-21. If a piece of paper cut in the shape of a parallelogram is wrapped around a cylinder, the angular edge of the paper becomes a helix. If we unwind this paper, each point on the angular edge generates an involute curve. This surface obtained when every point on the edge generates an involute is called an *involute helicoid*.

The initial contact of spur-gear teeth is a line extending all the way across the face of the tooth. The initial contact of helical-gear teeth is a point which extends into a line as the teeth come into more engagement. In spur gears the line of contact is parallel to the axis of rotation; in helical gears the line is diagonal across the face of the tooth. It is this gradual engagement of the teeth and the smooth transfer of load from one tooth to another which give helical gears the ability to transmit heavy loads at high speeds. Because of the nature of contact between helical gears, the contact ratio is of only

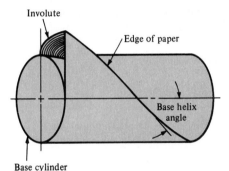

FIGURE 13-21

An involute helicoid.

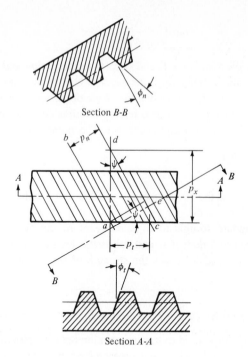

FIGURE 13-22

Nomenclature of helical gears.

minor importance, and it is the contact area, which is proportional to the face width of the gear, that becomes significant.

Helical gears subject the shaft bearings to both radial and thrust loads. When the thrust loads become high or are objectionable for other reasons, it may be desirable to use double helical gears. A double helical gear (herringbone) is equivalent to two helical gears of opposite hand, mounted side by side on the same shaft. They develop opposite thrust reactions and thus cancel out the thrust load.

When two or more single helical gears are mounted on the same shaft, the hand of the gears should be selected so as to produce the minimum thrust load.

Figure 13-22 represents a portion of the top view of a helical rack. Lines *ab* and *cd* are the centerlines of two adjacent helical teeth taken on the pitch plane. The angle ψ is the *helix angle*. The distance *ac* is the *transverse circular pitch* p_t in the plane of rotation (usually called the *circular pitch*). The distance *ae* is the *normal circular pitch* p_n and is related to the transverse circular pitch as follows:

$$p_n = p_t \cos \psi \tag{13-12}$$

The distance *ad* is called the *axial pitch* p_x and is related by the expression

$$p_x = \frac{p_t}{\tan \psi} \tag{13-13}$$

Since $p_n P_n = \pi$, the *normal diametral pitch* is

$$P_n = \frac{P_t}{\cos \psi} \tag{13-14}$$

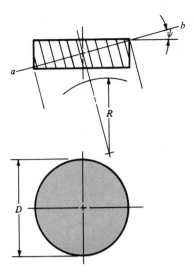

FIGURE 13-23

A cylinder cut by an oblique
plane.

The pressure angle ϕ_n in the normal direction is different from the pressure angle ϕ_t in the direction of rotation, because of the angularity of the teeth. These angles are related by the equation

$$\cos \psi = \frac{\tan \phi_n}{\tan \phi_t} \qquad (13\text{-}15)$$

Figure 13-23 illustrates a cylinder cut by an oblique plane ab at an angle ψ to a right section. The oblique plane cuts out an arc having a radius of curvature of R. For the condition that $\psi = 0$, the radius of curvature is $R = D/2$. If we imagine the angle ψ to be slowly increased from zero to $90°$, we see that R begins at a value of $D/2$ and increases until, when $\psi = 90°$, $R = \infty$. The radius R is the apparent pitch radius of a helical-gear tooth when viewed in the direction of the tooth elements. A gear of the same pitch and with the radius R will have a greater number of teeth, because of the increased radius. In helical-gear terminology this is called the *virtual number of teeth*. It can be shown by analytical geometry that the virtual number of teeth is related to the actual number by the equation

$$N' = \frac{N}{\cos^3 \psi} \qquad (13\text{-}16)$$

where N' is the virtual number of teeth and N is the actual number of teeth. It is necessary to know the virtual number of teeth in design for strength and also, sometimes, in cutting helical teeth. This apparently larger radius of curvature means that fewer teeth may be used on helical gears, because there will be less undercutting.

EXAMPLE 13-2   A stock helical gear has a normal pressure angle of $14\frac{1}{2}°$, a helix angle of $45°$, and a transverse diametral pitch of 6 teeth/in, and has 18 teeth. Find:

(a) The pitch diameter

(b) The transverse, the normal, and the axial circular pitches

(c) The normal diametral pitch

(d) The transverse pressure angle

Solution

Answer (a) $d = \dfrac{N}{P_t} = \dfrac{18}{6} = 3$ in

Answer (b) $p_t = \dfrac{\pi}{P_t} = \dfrac{\pi}{6} = 0.5236$ in

Answer $p_n = p_t \cos \psi = 0.5236 \cos 45° = 0.3702$ in

Answer $p_x = \dfrac{p_t}{\tan \psi} = \dfrac{0.5236}{\tan 45°} = 0.5236$ in

Answer (c) $P_n = \dfrac{P_t}{\cos \psi} = \dfrac{6}{\cos 45°} = 8.485$ teeth/in

Answer (d) $\phi_t = \tan^{-1}\left(\dfrac{\tan \phi_n}{\cos \psi}\right) = \tan^{-1}\left(\dfrac{\tan 14.5°}{\cos 45°}\right) = 20.09°$

13-11 WORM GEARS

The nomenclature of a worm and worm gear is shown in Fig. 13-24. The worm and worm gear of a set have the same hand of helix as for crossed helical gears, but the

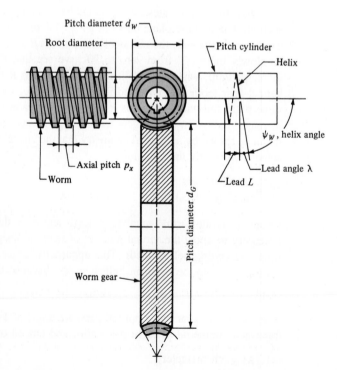

FIGURE 13-24

Nomenclature of a single-enveloping worm gearset.

helix angles are usually quite different. The helix angle on the worm is generally quite large, and that on the gear very small. Because of this, it is usual to specify the lead angle λ on the worm and helix angle ψ_G on the gear; the two angles are equal for a 90° shaft angle. The worm lead angle is the complement of the worm helix angle, as shown in Fig. 13-24.

In specifying the pitch of worm gearsets, it is customary to state the *axial pitch* p_x of the worm and the *transverse circular pitch* p_t, often simply called the circular pitch, of the mating gear. These are equal if the shaft angle is 90°. The pitch diameter of the gear is the diameter measured on a plane containing the worm axis, as shown in Fig. 13-24; it is the same as for spur gears and is

$$d_G = \frac{N_G p_t}{\pi} \tag{13-17}$$

Since it is not related to the number of teeth, the worm may have any pitch diameter; this diameter should, however, be the same as the pitch diameter of the hob used to cut the worm-gear teeth. Generally, the pitch diameter of the worm should be selected so as to fall into the range

$$\frac{C^{0.875}}{3.0} \le d_W \le \frac{C^{0.875}}{1.7} \tag{13-18}$$

where C is the center distance. These proportions appear to result in optimum horsepower capacity of the gearset.

The *lead L* and the *lead angle* λ of the worm have the following relations:

$$L = p_x N_W \tag{13-19}$$

$$\tan \lambda = \frac{L}{\pi d_W} \tag{13-20}$$

13-12   TOOTH SYSTEMS*

A *tooth system* is a standard which specifies the relationships involving addendum, dedendum, working depth, tooth thickness, and pressure angle. The standards were originally planned to attain interchangeability of gears of all tooth numbers, but of the same pressure angle and pitch.

Table 13-2 contains the standards most used for spur gears. A $14\frac{1}{2}°$ pressure angle was once used for these but is now obsolete; the resulting gears had to be comparatively larger to avoid interference problems.

Table 13-3 is particularly useful in selecting the pitch or module of a gear. Cutters are generally available for the sizes shown in this table.

Table 13-4 lists the standard tooth proportions for straight bevel gears. These sizes apply to the large end of the teeth. The nomenclature is defined in Fig. 13-20b.

---

*Standardized by the American Gear Manufacturers Association (AGMA). Write AGMA for a complete list of standards, because changes are made from time to time. The address is: 1500 King Street, Suite 201, Alexandria, VA 22314.

TABLE 13-2

Standard and Commonly Used Tooth Systems for Spur Gears

TOOTH SYSTEM	PRESSURE ANGLE ϕ, deg	ADDENDUM a	DEDENDUM b
Full depth	20	$1/P_d$ or $1m$	$1.25/P_d$ or $1.25m$ $1.35/P_d$ or $1.35m$
	$22\frac{1}{2}$	$1/P_d$ or $1m$	$1.25/P_d$ or $1.25m$ $1.35/P_d$ or $1.35m$
	25	$1/P_d$ or $1m$	$1.25/P_d$ or $1.25m$ $1.35/P_d$ or $1.35m$
Stub	20	$0.8/P_d$ or $0.8m$	$1/P_d$ or $1m$

TABLE 13-3

Tooth Sizes in General Use

DIAMETRAL PITCHES	
Coarse	2, $2\frac{1}{4}$, $2\frac{1}{2}$, 3, 4, 6, 8, 10, 12, 16
Fine	20, 24, 32, 40, 48, 64, 80, 96, 120, 150, 200

MODULES	
Preferred	1, 1.25, 1.5, 2, 2.5, 3, 4, 5, 6, 8, 10, 12, 16, 20, 25, 32, 40, 50
Next choice	1.125, 1.375, 1.75, 2.25, 2.75, 3.5, 4.5, 5.5, 7, 9, 11, 14, 18, 22, 28, 36, 45

TABLE 13-4

Tooth Proportions for 20° Straight Bevel-Gear Teeth

ITEM	FORMULA
Working depth	$h_k = 2.0/P$
Clearance	$c = (0.188/P) + 0.002$ in
Addendum of gear	$a_G = \dfrac{0.54}{P} + \dfrac{0.460}{P(m_{90})^2}$
Gear ratio	$m_G = N_G/N_P$
Equivalent 90° ratio	$m_{90} = m_G$ when $\Sigma = 90°$
	$m_{90} = \sqrt{m_G \dfrac{\cos \gamma}{\cos \Gamma}}$ when $\Sigma \neq 90°$
Face width	$F = \dfrac{A_o}{3}$ or $F = \dfrac{10}{P}$, whichever is smaller
Minimum number of teeth	Pinion | 16 15 14 13 Gear | 16 17 20 30

TABLE 13-5

Standard Tooth Proportions for
Helical Gears

QUANTITY*	FORMULA	QUANTITY*	FORMULA
Addendum	$\dfrac{1.00}{P_n}$	External gears:	
Dedendum	$\dfrac{1.25}{P_n}$	Standard center distance	$\dfrac{D + d}{2}$
Pinion pitch diameter	$\dfrac{N_P}{P_n \cos \psi}$	Gear outside diameter	$D + 2a$
Gear pitch diameter	$\dfrac{N_G}{P_n \cos \psi}$	Pinion outside diameter	$d + 2a$
Normal arc tooth thickness	$\dfrac{\pi}{P_n} - \dfrac{B_n}{2}$	Gear root diameter	$D - 2b$
Pinion base diameter	$d \cos \phi_t$	Pinion root diameter	$d - 2b$
		Internal gears:	
Gear base diameter	$D \cos \phi_t$	Center distance	$\dfrac{D - d}{2}$
Base helix angle	$\tan^{-1}(\tan \psi \cos \phi_t)$	Inside diameter	$d - 2a$
		Root diameter	$D + 2b$

*All dimensions are in inches, and angles are in degrees.

TABLE 13-6

Recommended Pressure
Angles and Tooth Depths for
Worm Gearing

LEAD ANGLE λ, deg	PRESSURE ANGLE ϕ_n, deg	ADDENDUM a	DEDENDUM b_G
0–15	$14\frac{1}{2}$	$0.3683p_x$	$0.3683p_x$
15–30	20	$0.3683p_x$	$0.3683p_x$
30–35	25	$0.2865p_x$	$0.3314p_x$
35–40	25	$0.2546p_x$	$0.2947p_x$
40–45	30	$0.2228p_x$	$0.2578p_x$

Standard tooth proportions for helical gears are listed in Table 13-5. Tooth proportions are based on the normal pressure angle; these angles are standardized the same as for spur gears. Though there will be exceptions, the face width of helical gears should be at least 2 times the axial pitch to obtain good helical-gear action.

Tooth forms for worm gearing have not been highly standardized, perhaps because there has been less need for it. The pressure angles used depend upon the lead angles and must be large enough to avoid undercutting of the worm-gear tooth on the side at which contact ends. A satisfactory tooth depth, which remains in about the right proportion to the lead angle, may be obtained by making the depth a proportion of the axial circular pitch. Table 13-6 summarizes what may be regarded as good practice for pressure angle and tooth depth.

The *face width* F_G of the worm gear should be made equal to the length of a tangent to the worm pitch circle between its points of intersection with the addendum circle, as shown in Fig. 13-25.

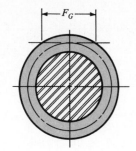

FIGURE 13-25

13-13   GEAR TRAINS

Consider a pinion 2 driving a gear 3. The speed of the driven gear is

$$n_3 = \left| \frac{N_2}{N_3} n_2 \right| = \left| \frac{d_2}{d_3} n_2 \right| \tag{13-21}$$

where n = revolutions or rev/min
   N = number of teeth
   d = pitch diameter

Equation (13-21) applies to any gearset no matter whether the gears are spur, helical, bevel, or worm. The absolute-value signs are used to permit complete freedom in choosing positive and negative directions. In the case of spur and parallel helical gears, the directions ordinarily correspond to the right-hand rule and are positive for counter-clockwise rotation.

Rotational directions are somewhat more difficult to deduce for worm and crossed helical gearsets. Figure 13-26 will be of help in these situations.

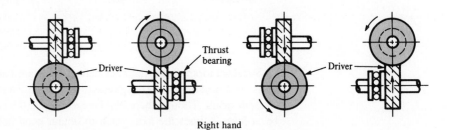

Right hand

FIGURE 13-26

Thrust, rotation, and hand relations for crossed helical gears. Note that each pair of drawings refers to a single gearset. These relations also apply to worm gears. *(Source: Boston Gear Works, Inc.)*

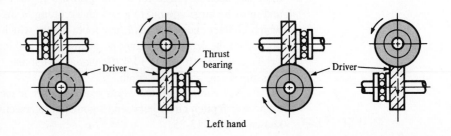

Left hand

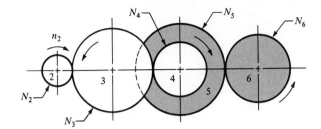

FIGURE 13-27

The gear train shown in Fig. 13-27 is made up of five gears. The speed of gear 6 is

$$n_6 = -\frac{N_2}{N_3}\frac{N_3}{N_4}\frac{N_5}{N_6}n_2 \qquad\qquad (a)$$

Here we notice that gear 3 is an idler, that its tooth numbers cancel in Eq. (a), and hence that it affects only the direction of rotation of gear 6. We notice, furthermore, that gears 2, 3, and 5 are drivers, while 3, 4, and 6 are driven members. We define *train value e* as

$$e = \frac{\text{product of driving tooth numbers}}{\text{product of driven tooth numbers}} \qquad\qquad (13\text{-}22)$$

Note that pitch diameters can be used in Eq. (13-22) as well. When Eq. (13-22) is used for spur gears, e is positive if the last gear rotates in the same sense as the first, and negative if the last rotates in the opposite sense.

Now we can write

$$n_L = en_F \qquad\qquad (13\text{-}23)$$

where n_L is the speed of the last gear in the train and n_F is the speed of the first.

Unusual effects can be obtained in a gear train by permitting some of the gear axes to rotate about others. Such trains are called *planetary*, or *epicyclic*, *gear trains*. Planetary trains always include a *sun gear*, a *planet carrier* or *arm*, and one or more *planet gears*, as shown in Fig. 13-28. Planetary gear trains are unusual mechanisms

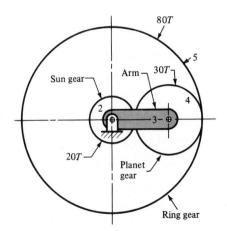

FIGURE 13-28

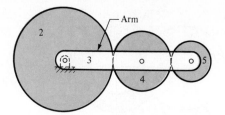

FIGURE 13-29

because they have two degrees of freedom; that is, for constrained motion, a planetary train must have two inputs. For example, in Fig. 13-28 these two inputs could be the motion of any two of the elements of the train. We might, in Fig. 13-28, say, specify that the sun gear rotates at 100 rev/min clockwise and that the ring gear rotates at 50 rev/min counterclockwise; these are the inputs. The output would be the motion of the arm. In most planetary trains one of the elements is attached to the frame and has no motion.

Figure 13-29 shows a planetary train composed of a sun gear 2, an arm or carrier 3, and planet gears 4 and 5. The angular velocity of gear 2 relative to the arm in rev/min is

$$n_{23} = n_2 - n_3 \tag{b}$$

Also, the velocity of gear 5 relative to the arm is

$$n_{53} = n_5 - n_3 \tag{c}$$

Dividing Eq. (c) by Eq. (b) gives

$$\frac{n_{53}}{n_{23}} = \frac{n_5 - n_3}{n_2 - n_3} \tag{d}$$

Equation (d) expresses the ratio of the relative velocity of gear 5 to that of gear 2, and both velocities are taken relative to the arm. Now this ratio is the same and is proportional to the tooth numbers, whether the arm is rotating or not. It is the train value. Therefore we may write

$$e = \frac{n_5 - n_3}{n_2 - n_3} \tag{e}$$

This equation can be used to solve for the output motion of any planetary train. It is more conveniently written in the form

$$e = \frac{n_L - n_A}{n_F - n_A} \tag{13-24}$$

where n_F = rev/min of first gear in planetary train
n_L = rev/min of last gear in planetary train
n_A = rev/min of arm

EXAMPLE 13-3   In Fig. 13-28 the sun gear is the input, and it is driven clockwise at 100 rev/min. The ring gear is held stationary by being fastened to the frame. Find the rev/min and direction of rotation of the arm.

Solution   Designate $n_F = n_2 = -100$ rev/min, and $n_L = n_5 = 0$. Unlocking gear 5 and holding the arm stationary, in our imagination, we find

$$e = -(\tfrac{20}{30})(\tfrac{30}{80}) = -0.25$$

Substituting in Eq. (13-24),

$$-0.25 = \frac{0 - n_A}{(-100) - n_A}$$

or

Answer $n_A = -20$ rev/min

To obtain the speed of gear 4, we follow the procedure outlined by Eqs. (*b*), (*c*), and (*d*). Thus

$$n_{43} = n_4 - n_3 \qquad n_{23} = n_2 - n_3$$

and so

$$\frac{n_{43}}{n_{23}} = \frac{n_4 - n_3}{n_2 - n_3} \tag{1}$$

But

$$\frac{n_{43}}{n_{23}} = -\frac{20}{30} = -\frac{2}{3} \tag{2}$$

Substituting the known values in Eq. (1) gives

$$-\frac{2}{3} = \frac{n_4 - (-20)}{(-100) - (-20)}$$

Solving gives

Answer $n_4 = 33\tfrac{1}{3}$ rev/min

EXAMPLE 13-4   Figure 13-30 shows a gear train consisting of a pair of *miter gears* (same-size bevel gears) having 16 teeth each, a 4-tooth right-hand worm, and a 40-tooth worm gear. The

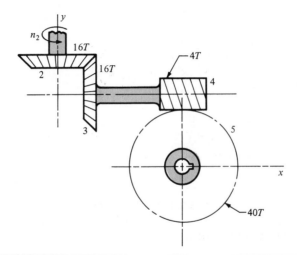

FIGURE 13-30

speed of gear 2 is given as $n_2 = +200$ rev/min, which corresponds to counterclockwise about the y axis. What is the speed and direction of rotation of the worm gear?

Solution

Answer $n_5 = -(\frac{16}{16})(\frac{4}{40})(200) = -20$ rev/min

Gear 5 rotates clockwise (negative) 20 rev/min about the z axis in a right-handed coordinate system.

13-14   FORCE ANALYSIS OF SPUR GEARS

Before beginning the force analysis of gear trains, let us agree on the notation to be used. Beginning with the numeral 1 for the frame of the machine, we shall designate the input gear as gear 2, and then number the gears successively 3, 4, etc., until we arrive at the last gear in the train. Next, there may be several shafts involved, and usually one or two gears are mounted on each shaft as well as other elements. We shall designate the shafts, using lowercase letters of the alphabet, a, b, c, etc.

With this notation we can now speak of the force exerted by gear 2 against gear 3 as F_{23}. The force of gear 2 against shaft a is F_{2a}. We can also write F_{a2} to mean the force of shaft a against gear 2. Unfortunately, it is also necessary to use superscripts to indicate directions. The coordinate directions will usually be indicated by the x, y, and z coordinates, and the radial and tangential directions by superscripts r and t. With this notation, F_{43}^t is the tangential component of the force of gear 4 acting against gear 3.

Figure 13-31a shows a pinion mounted on shaft a rotating clockwise at n_2 rev/min and driving a gear on shaft b at n_3 rev/min. The reactions between the mating teeth occur along the pressure line. In Fig. 13-31b the pinion has been separated from the gear and from the shaft, and their effects have been replaced by forces. F_{a2} and T_{a2} are the force and torque, respectively, exerted by shaft a against pinion 2. F_{32} is the force

FIGURE 13-31

Free-body diagrams of the forces acting upon two gears in a simple gear train.

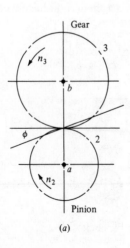

(a)

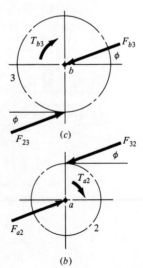

(b)

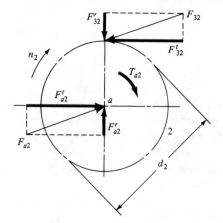

FIGURE 13-32

Resolution of gear forces.

exerted by gear 3 against the pinion. Using a similar approach, we obtain the free-body diagram of the gear shown in Fig. 13-31c.

In Fig. 13-32, the free-body diagram of the pinion has been redrawn and the forces have been resolved into tangential and radial components. We now define

$$W_t = F'_{32} \tag{a}$$

as the *transmitted load*. This tangential load is really the useful component, because the radial component F^r_{32} serves no useful purpose. It does not transmit power. The applied torque and the transmitted load are seen to be related by the equation

$$T = \frac{d}{2} W_t \tag{13-25}$$

where we have used $T = T_{a2}$ and $d = d_2$ to obtain a general relation.

If next we designate the pitch-line velocity by V, where $V = \pi dn/12$ and is in feet per minute, the tangential load may be obtained from the equation

$$H = \frac{W_t V}{33\ 000} \tag{13-26}$$

The corresponding equation in SI is

$$W_t = \frac{60(10)^3 H}{\pi dn} \tag{13-27}$$

where W_t = transmitted load, kN
$\quad\quad H$ = power, kW
$\quad\quad d$ = gear diameter, mm
$\quad\quad n$ = speed, rev/min

EXAMPLE 13-5   Pinion 2 in Fig. 13-33a runs at 1750 rev/min and transmits 2.5 kW to idler gear 3. The teeth are cut on the 20° full-depth system and have a module of $m = 2.5$ mm. Draw a free-body diagram of gear 3 and show all the forces which act upon it.

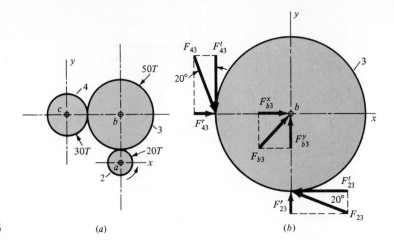

FIGURE 13-33   (a)   (b)

Solution   The pitch diameters of gears 2 and 3 are

$$d_2 = N_2 m = 20(2.5) = 50 \text{ mm}$$

$$d_3 = N_3 m = 50(2.5) = 125 \text{ mm}$$

From Eq. (13-27) we find the transmitted load to be

$$W_t = \frac{60(10)^3 H}{\pi d_2 n} = \frac{60(10)^3(2.5)}{\pi(50)(1750)} = 0.546 \text{ kN}$$

Thus, the tangential force of gear 2 on gear 3 is $F_{23}^t = 0.546$ kN, as shown in Fig. 13-33b. Therefore

$$F_{23}^r = F_{23}^t \tan 20° = (0.546) \tan 20° = 0.199 \text{ kN}$$

and so

$$F_{23} = \frac{F_{23}^t}{\cos 20°} = \frac{0.546}{\cos 20°} = 0.581 \text{ kN}$$

Since gear 3 is an idler, it transmits no power (torque) to its shaft, and so the tangential reaction of gear 4 on gear 3 is also equal to W_t. Therefore

$$F_{43}^t = 0.546 \text{ kN} \qquad F_{43}^r = 0.199 \text{ kN} \qquad F_{43} = 0.581 \text{ kN}$$

and the directions are as shown in Fig. 13-33b.

The shaft reactions in the x and y directions are

$$F_{b3}^x = -(F_{23}^t + F_{43}^r) = -(-0.546 + 0.199) = 0.347 \text{ kN}$$

$$F_{b3}^y = -(F_{23}^r + F_{43}^t) = -(0.199 - 0.546) = 0.347 \text{ kN}$$

The resultant shaft reaction is

$$F_{b3} = \sqrt{(0.347)^2 + (0.347)^2} = 0.491 \text{ kN}$$

These are shown on the figure.

13-15

BEVEL GEARS—FORCE ANALYSIS

In determining shaft and bearing loads for bevel-gear applications, the usual practice is to use the tangential or transmitted load which would occur if all the forces were concentrated at the midpoint of the tooth. While the actual resultant occurs somewhere between the midpoint and the large end of the tooth, there is only a small error in making this assumption. For the transmitted load, this gives

$$W_t = \frac{T}{r_{av}} \tag{13-28}$$

where T is the torque and r_{av} is the pitch radius at the midpoint of the tooth for the gear under consideration.

The forces acting at the center of the tooth are shown in Fig. 13-34. The resultant force W has three components: a tangential force W_t, a radial force W_r, and an axial force W_a. From the trigonometry of the figure,

$$W_r = W_t \tan \phi \cos \gamma \tag{13-29}$$

$$W_a = W_t \tan \phi \sin \gamma \tag{13-30}$$

The three forces W_t, W_r, and W_a are at right angles to each other and can be used to determine the bearing loads by using the methods of statics.

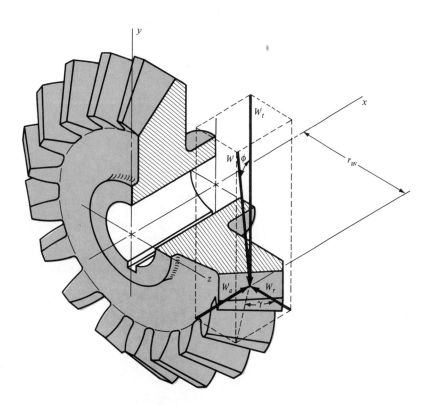

FIGURE 13-34

Bevel-gear tooth forces.

EXAMPLE 13-6   The bevel pinion in Fig. 13-35 rotates at 600 rev/min in the direction shown and transmits 5 hp to the gear. The mounting distances, the location of all bearings, and the average pitch radii of the pinion and gear are shown in the figure. For simplicity, the teeth have been replaced by the pitch cones. Bearings A and C should take the thrust loads. Find the bearing forces on the gearshaft.

Solution   The pitch angles are

$$\gamma = \tan^{-1} \left(\tfrac{3}{9}\right) = 18.4° \qquad \Gamma = \tan^{-1} \left(\tfrac{9}{3}\right) = 71.6°$$

The pitch-line velocity corresponding to the average pitch radius is

$$V = \frac{2\pi r_P n}{12} = \frac{2\pi(1.293)(600)}{12} = 406 \text{ ft/min}$$

Therefore the transmitted load is

$$W_t = \frac{33\,000H}{V} = \frac{(33\,000)(5)}{406} = 406 \text{ lb}$$

which acts in the positive z direction, as shown in Fig. 13-36. We next have

$$W_r = W_t \tan \phi \cos \Gamma = 406 \tan 20° \cos 71.6° = 46.6 \text{ lb}$$

$$W_a = W_t \tan \phi \sin \Gamma = 406 \tan 20° \sin 71.6° = 140 \text{ lb}$$

where W_r is in the $-x$ direction and W_a in the $-y$ direction, as illustrated in the isometric sketch of Fig. 13-36.

In preparing to take a sum of the moments about bearing D, define the position

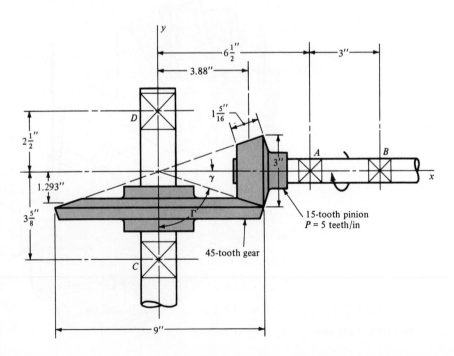

FIGURE 13-35

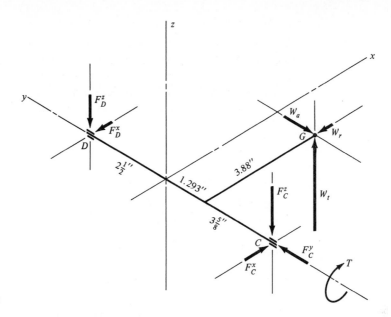

FIGURE 13-36

vector from D to G as

$$\mathbf{R}_G = 3.88\mathbf{i} - (2.5 + 1.293)\mathbf{j} = 3.88\mathbf{i} - 3.793\mathbf{j}$$

We shall also require a vector from D to C:

$$\mathbf{R}_C = -(2.5 + 3.625)\mathbf{j} = -6.125\mathbf{j}$$

Then, summing moments about D gives

$$\mathbf{R}_G \times \mathbf{W} + \mathbf{R}_C \times \mathbf{F}_C + \mathbf{T} = \mathbf{0} \tag{1}$$

When we place the details in Eq. (1), we get

$$(3.88\mathbf{i} - 3.793\mathbf{j}) \times (-46.6\mathbf{i} - 140\mathbf{j} + 406\mathbf{k})$$
$$+ (6.125\mathbf{j}) \times (F_C^x\mathbf{i} + F_C^y\mathbf{j} + F_C^z\mathbf{k}) + T\mathbf{j} = \mathbf{0} \tag{2}$$

After the two cross products are taken, the equation becomes

$$(-1504\mathbf{i} - 1580\mathbf{j} - 721\mathbf{k}) + (-6.125F_C^z\mathbf{i} + 6.125F_C^x\mathbf{k}) + T\mathbf{j} = \mathbf{0}$$

from which

$$\mathbf{T} = 1580\mathbf{j}\ \text{lb}\cdot\text{in} \qquad F_C^x = 118\ \text{lb} \qquad F_C^z = -246\ \text{lb} \tag{3}$$

Now sum the forces to zero. Thus

$$\mathbf{F}_D + \mathbf{F}_C + \mathbf{W} = \mathbf{0} \tag{4}$$

When the details are inserted, Eq. (4) becomes

$$(F_D^x\mathbf{i} + F_D^z\mathbf{k}) + (118\mathbf{i} + F_C^y\mathbf{j} - 246\mathbf{k}) + (-46.6\mathbf{i} - 140\mathbf{j} + 406\mathbf{k}) = \mathbf{0} \tag{5}$$

First we see that $F_C^y = 140$ lb, and so

Answer $$\mathbf{F}_C = 118\mathbf{i} + 140\mathbf{j} - 246\mathbf{k}\ \text{lb}$$

Then, from Eq. (5),

Answer $\mathbf{F}_D = -71\mathbf{i} - 160\mathbf{k}$ lb

These are all shown in Fig. 13-36 in the proper directions. The analysis for the pinion shaft is quite similar.

13-16   HELICAL GEARS—FORCE ANALYSIS

Figure 13-37 is a three-dimensional view of the forces acting against a helical-gear tooth. The point of application of the forces is in the pitch plane and in the center of the gear face. From the geometry of the figure, the three components of the total (normal) tooth force W are

$$W_r = W \sin \phi_n$$
$$W_t = W \cos \phi_n \cos \psi \qquad\qquad (13\text{-}31)$$
$$W_a = W \cos \phi_n \sin \psi$$

where W = total force
W_r = radial component
W_t = tangential component; also called transmitted load
W_a = axial component; also called thrust load

Usually W_t is given and the other forces are desired. In this case, it is not difficult to discover that

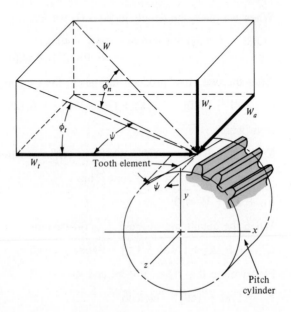

FIGURE 13-37

Tooth forces acting on a right-hand helical gear.

$$W_r = W_t \tan \phi_t$$

$$W_a = W_t \tan \psi \tag{13-32}$$

$$W = \frac{W_t}{\cos \phi_n \cos \psi}$$

EXAMPLE 13-7

In Fig. 13-38 a 1-hp electric motor runs at 1800 rev/min in the clockwise direction, as viewed from the positive x axis. Keyed to the motor shaft is an 18-tooth helical pinion having a normal pressure angle of 20°, a helix angle of 30°, and a normal diametral pitch of 12 teeth/in. The hand of the helix is shown in the figure. Make a three-dimensional sketch of the motor shaft and pinion, and show the forces acting on the pinion and the bearing reactions at A and B. The thrust should be taken out at A.

Solution

From Eq. (13-15) we find

$$\phi_t = \tan^{-1} \frac{\tan \phi_n}{\cos \psi} = \tan^{-1} \frac{\tan 20°}{\cos 30°} = 22.8°$$

Also, $P_t = P_n \cos \psi = 12 \cos 30° = 10.4$ teeth/in. Therefore the pitch diameter of the pinion is $d_p = 18/10.4 = 1.73$ in. The pitch-line velocity is

$$V = \frac{\pi d n}{12} = \frac{\pi(1.73)(1800)}{12} = 815 \text{ ft/min}$$

The transmitted load is

$$W_t = \frac{33\,000 H}{V} = \frac{(33\,000)(1)}{815} = 40.5 \text{ lb}$$

From Eq. (13-32) we find

$$W_r = W_t \tan \phi_t = (40.5)(0.420) = 17.0 \text{ lb}$$

$$W_a = W_t \tan \psi = (40.5)(0.577) = 23.4 \text{ lb}$$

$$W = \frac{W_t}{\cos \phi_n \cos \psi} = \frac{40.5}{(0.940)(0.866)} = 49.8 \text{ lb}$$

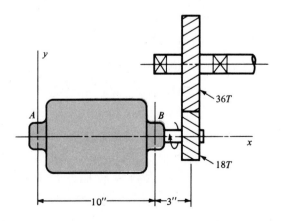

FIGURE 13-38

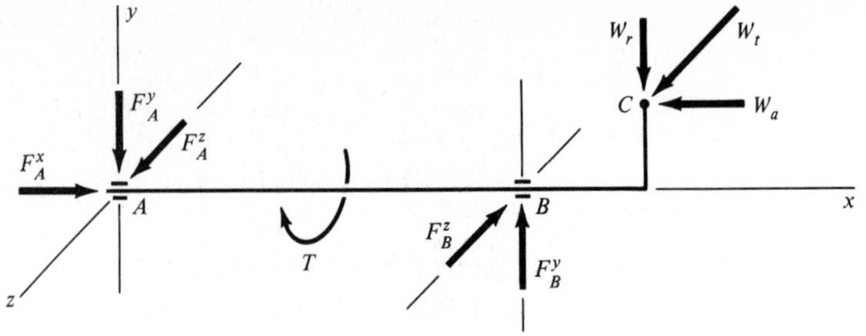

FIGURE 13-39

These three forces, W_r in the $-y$ direction, W_a in the $-x$ direction, and W_t in the $+z$ direction, are shown acting at point C in Fig. 13-39. We assume bearing reactions at A and B as shown. Then $F_a^x = W_a = 23.4$ lb. Taking moments about the z axis,

$$-(17.0)(13) + (23.4)\left(\frac{1.73}{2}\right) + 10F_B^y = 0$$

or $F_B^y = 20$ lb. Summing forces in the y direction then gives $F_A^y = 3.0$ lb. Taking moments about the y axis, next,

$$10F_B^z - (40.5)(13) = 0$$

or $F_B^z = 52.7$ lb. Summing forces in the z direction and solving gives $F_A^z = 12.2$ lb. Also, the torque is $T = W_t d_p/2 = (40.5)(1.73/2) = 35$ lb · in.

EXAMPLE 13-8   Solve Example 13-7 using vectors.

Solution   The force at C is

$$\mathbf{W} = -23.4\mathbf{i} - 17.1\mathbf{j} + 40.5\mathbf{k}$$

Position vectors to B and C from origin A are

$$\mathbf{R}_B = 10\mathbf{i} \qquad \mathbf{R}_C = 13\mathbf{i} + 0.865\mathbf{j}$$

Taking moments about A, we have

$$\mathbf{R}_B \times \mathbf{F}_B + \mathbf{T} + \mathbf{R}_C \times \mathbf{W} = \mathbf{0}$$

Using the directions assumed in Fig. 13-39 and substituting values gives

$$10\mathbf{i} \times (F_B^y\mathbf{j} - F_B^z\mathbf{k}) - T\mathbf{i} + (13\mathbf{i} + 0.865\mathbf{j}) \times (-23.4\mathbf{i} - 17.0\mathbf{j} + 40.5\mathbf{k}) = \mathbf{0}$$

When the cross products are formed, we get

$$(10F_B^y\mathbf{k} + 10F_B^z\mathbf{j}) - T\mathbf{i} + (35\mathbf{i} - 527\mathbf{j} - 200\mathbf{k}) = \mathbf{0}$$

whence $T = 35$ lb · in, $F_B^y = 20$ lb, and $F_B^z = 52.7$ lb.
    Next, $\mathbf{F}_A = -\mathbf{F}_B - \mathbf{W}$, and so $\mathbf{F}_A = 23.4\mathbf{i} - 2.9\mathbf{j} + 12.1\mathbf{k}$ lb.

13-17   WORM GEARING—FORCE ANALYSIS

If friction is neglected, then the only force exerted by the gear will be the force W, shown in Fig. 13-40, having the three orthogonal components W^x, W^y, and W^z. From the geometry of the figure, we see that

$$W^x = W \cos \phi_n \sin \lambda$$
$$W^y = W \sin \phi_n \tag{13-33}$$
$$W^z = W \cos \phi_n \cos \lambda$$

We now use the subscripts W and G to indicate forces acting against the worm and gear, respectively. We note that W^y is the separating, or radial, force for both the worm and the gear. The tangential force on the worm is W^x and is W^z on the gear, assuming a 90° shaft angle. The axial force on the worm is W^z, and on the gear, W^x. Since the gear forces are opposite to the worm forces, we can summarize these relations by writing

$$W_{Wt} = -W_{Ga} = W^x$$
$$W_{Wr} = -W_{Gr} = W^y \tag{13-34}$$
$$W_{Wa} = -W_{Gt} = W^z$$

It is helpful in using Eq. (13-33) and also Eq. (13-34) to observe that *the gear axis is parallel to the x direction and the worm axis is parallel to the z direction* and that we are employing a right-handed coordinate system.

In our study of spur-gear teeth we have learned that the motion of one tooth relative to the mating tooth is primarily a rolling motion; in fact, when contact occurs at the pitch point, the motion is pure rolling. In contrast, the relative motion between worm and worm-gear teeth is pure sliding, and so we must expect that friction plays an important role in the performance of worm gearing. By introducing a coefficient of friction μ, we can develop another set of relations similar to those of Eq. (13-33). In Fig. 13-40 we see that the force W acting normal to the worm-tooth profile produces a

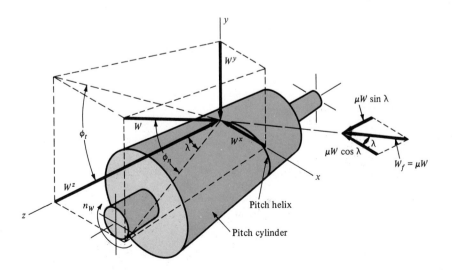

FIGURE 13-40

Drawing of the pitch cylinder of a worm, showing the forces exerted upon it by the worm gear.

frictional force $W_f = \mu W$, having a component $\mu W \cos \lambda$ in the negative x direction and another component $\mu W \sin \lambda$ in the positive z direction. Equation (13-33) therefore becomes

$$W^x = W(\cos \phi_n \sin \lambda + \mu \cos \lambda)$$
$$W^y = W \sin \phi_n \qquad\qquad (13\text{-}35)$$
$$W^z = W(\cos \phi_n \cos \lambda - \mu \sin \lambda)$$

Equation (13-34), of course, still applies.

If we substitute W^z in the third part of Eq. (13-34) and multiply both sides by μ, we find the frictional force to be

$$W_f = \mu W = \frac{\mu W_{Gt}}{\mu \sin \lambda - \cos \phi_n \cos \lambda} \qquad (13\text{-}36)$$

Another useful relation can be obtained by solving the first and third parts of Eq. (13-34) simultaneously to get a relation between the two tangential forces. The result is

$$W_{Wt} = W_{Gt} \frac{\cos \phi_n \sin \lambda + \mu \cos \lambda}{\mu \sin \lambda - \cos \phi_n \cos \lambda} \qquad (13\text{-}37)$$

Efficiency η can be defined by using the equation

$$\eta = \frac{W_{Wt} \text{ (without friction)}}{W_{Wt} \text{ (with friction)}} \qquad (a)$$

Substitute Eq. (13-37) with $\mu = 0$ in the numerator of Eq. (*a*) and the same equation in the denominator. After some rearranging, you will find the efficiency to be

$$\eta = \frac{\cos \phi_n - \mu \tan \lambda}{\cos \phi_n + \mu \cot \lambda} \qquad (13\text{-}38)$$

Selecting a typical value of the coefficient of friction, say $\mu = 0.05$, and the pressure angles shown in Table 13-6, we can use Eq. (13-38) to get some useful design information. Solving this equation for helix angles from 1 to 30° gives the interesting results shown in Table 13-7.

Many experiments have shown that the coefficient of friction is dependent on the

TABLE 13-7

Efficiency of Worm Gearsets for $\mu = 0.05$

HELIX ANGLE ψ, deg	EFFICIENCY η, percent
1.0	25.2
2.5	46.8
5.0	62.6
7.5	71.2
10.0	76.8
15.0	82.7
20.0	86.0
25.0	88.0
30.0	89.2

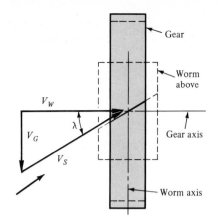

FIGURE 13-41

Velocity components in worm gearing.

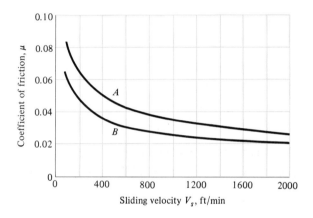

FIGURE 13-42

Representative values of the coefficient of friction for worm gearing. These values are based on good lubrication. Use curve *B* for high-quality materials, such as a case-hardened worm mating with a phosphor-bronze gear. Use curve *A* when more friction is expected, as, for example, with a cast-iron worm and worm gear.

relative or sliding velocity. In Fig. 13-41, V_G is the pitch-line velocity of the gear and V_W the pitch-line velocity of the worm. Vectorially, $\mathbf{V}_W = \mathbf{V}_G + \mathbf{V}_S$; consequently,

$$V_S = \frac{V_W}{\cos \lambda} \tag{13-39}$$

Published values of the coefficient of friction vary as much as 20 percent, undoubtedly because of the differences in surface finish, materials, and lubrication. The values on the chart of Fig. 13-42 are representative and indicate the general trend.

EXAMPLE 13-9   A 2-tooth right-hand worm transmits 1 hp at 1200 rev/min to a 30-tooth worm gear. The gear has a transverse diametral pitch of 6 teeth/in and a face width of 1 in. The worm has a pitch diameter of 2 in and a face width of $2\frac{1}{2}$ in. The normal pressure angle is $14\frac{1}{2}°$. The materials and quality of work needed are such that curve *B* of Fig. 13-42 should be used to obtain the coefficient of friction.

(*a*) Find the axial pitch, the center distance, the lead, and the lead angle.

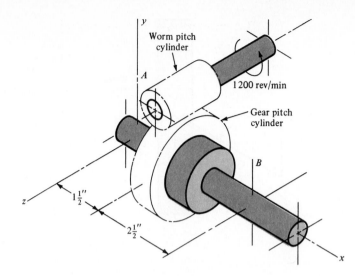

Worm pitch
cylinder

A

1200 rev/min

Gear pitch
cylinder

B

z

$1\frac{1}{2}''$

$2\frac{1}{2}''$

x

y

FIGURE 13-43

(b) Figure 13-43 is a drawing of the worm gear oriented with respect to the coordinate system described earlier in this section; the gear is supported by bearings A and B. Find the forces exerted by the bearings against the worm-gear shaft, and the output torque.

Solution   (a) The axial pitch is the same as the transverse circular pitch of the gear, which is

Answer
$$p_t = \frac{\pi}{P} = \frac{\pi}{6} = 0.5236 \text{ in}$$

The pitch diameter of the gear is $d_G = N_G/P = 30/6 = 5$ in. Therefore the center distance is

Answer
$$C = \frac{d_W + d_G}{2} = \frac{2 + 5}{2} = 3.5 \text{ in}$$

From Eq. (13-19), the lead is

$$L = p_x N_W = (0.5236)(2) = 1.0472 \text{ in}$$

Answer   Also, using Eq. (13-20), find

Answer
$$\lambda = \tan^{-1} \frac{L}{\pi d_W} = \tan^{-1} \frac{1.0472}{\pi(2)} = 9.47°$$

(b) Using the right-hand rule for the rotation of the worm, you will see that your thumb points in the positive z direction. Now use the bolt-and-nut analogy (the worm is right-handed, as is the screw thread of a bolt), and turn the bolt clockwise with the right hand while preventing nut rotation with the left. The nut will move axially along the bolt toward your right hand. Therefore the surface of the gear (Fig. 13-43) in contact with the worm will move in the negative z direction. Thus the gear rotates clockwise about x, with your right thumb pointing in the negative x direction.

The pitch-line velocity of the worm is

$$V_W = \frac{\pi d_W n_W}{12} = \frac{\pi(2)(1200)}{12} = 628 \text{ ft/min}$$

The speed of the gear is $n_G = (\frac{2}{30})(1200) = 80$ rev/min. Therefore the pitch-line velocity is

$$V_G = \frac{\pi d_G n_G}{12} = \frac{\pi(5)(80)}{12} = 105 \text{ ft/min}$$

Then, using Eq. (13-39), the sliding velocity V_S is found to be

$$V_S = \frac{V_W}{\cos \lambda} = \frac{628}{\cos 9.47°} = 637 \text{ ft/min}$$

Getting to the forces now, we begin with the horsepower formula

$$W_{Wt} = \frac{33\,000H}{V_W} = \frac{33\,000(1)}{628} = 52.5 \text{ lb}$$

This force acts in the negative z direction, the same as in Fig. 13-40. Using Fig. 13-42, we find $\mu = 0.03$. Then, the first equation of group (13-35) gives

$$W = \frac{W^x}{\cos \phi_n \sin \lambda + \mu \cos \lambda}$$

$$= \frac{52.5}{\cos 14.5° \sin 9.47° + 0.03 \cos 9.47°} = 278 \text{ lb}$$

Also, from Eq. (13-35),

$$W^y = W \sin \phi_n = 278 \sin 14.5° = 69.6 \text{ lb}$$

$$W^z = W(\cos \phi_n \cos \lambda - \mu \sin \lambda)$$

$$= 278(\cos 14.5° \cos 9.47° - 0.03 \sin 9.47°) = 264 \text{ lb}$$

We now identify the components acting on the gear as

$$W_{Ga} = -W^x = 52.5 \text{ lb}$$

$$W_{Gr} = -W^y = 69.6 \text{ lb}$$

$$W_{Gt} = -W^z = -264 \text{ lb}$$

At this point a three-dimensional line drawing should be made in order to simplify the work to follow. An isometric sketch, such as the one of Fig. 13-44, is easy to make and will help you to avoid errors. Note that the y axis is vertical, while the x and z axes make angles of 30° with the horizontal. The illusion of depth is enhanced by sketching lines parallel to each of the coordinate axes through every point of interest.

We shall make B a thrust bearing in order to place the gearshaft in compression. Thus, summing forces in the x direction gives

Answer $$F_B^x = -52.5 \text{ lb}$$

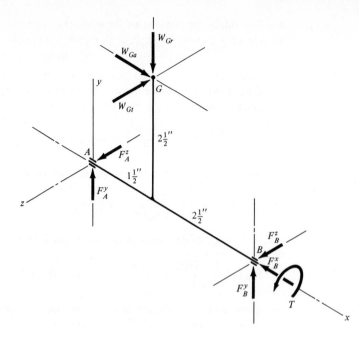

FIGURE 13-44

Taking moments about the z axis, we have

Answer $-(52.5)(2.5) - (69.6)(1.5) + 4F_B^y = 0 \qquad F_B^y = 58.9$ lb

Taking moments about the y axis,

Answer $(264)(1.5) - 4F_B^z = 0 \qquad F_B^z = 99$ lb

These three components are now inserted on the sketch as shown at B in Fig. 13-44. Summing forces in the y direction,

Answer $-69.6 + 58.9 + F_A^y = 0 \qquad F_A^y = 10.7$ lb

Similarly, summing forces in the z direction,

Answer $-264 + 99 + F_A^z = 0 \qquad F_A^z = 165$ lb

These two components can now be placed at A on the sketch. We still have one more equation to write. Summing moments about x,

Answer $-(264)(2.5) + T = 0 \qquad T = 660$ lb · in

It is because of the frictional loss that this output torque is less than the product of the gear ratio and the input torque.

PROBLEMS

13-1   A 17-tooth spur pinion has a diametral pitch of 8, runs at 1120 rev/min, and drives a gear at a speed of 544 rev/min. Find the number of teeth on the gear and the theoretical center distance.

13-2   A 15-tooth spur pinion has a module of 3 mm and runs at a speed of 1600 rev/min. The driven gear has 60 teeth. Find the speed of the driven gear, the circular pitch, and the theoretical center distance.

13-3   A spur gearset has a module of 4 mm, and a velocity ratio of 2.80. The pinion has 20 teeth. Find the number of teeth on the driven gear, the pitch diameters, and the theoretical center distance.

13-4   A 21-tooth spur pinion mates with a 28-tooth gear. The diametral pitch is 3 teeth/in and the pressure angle is 20°. Make a drawing of the gears showing one tooth on each gear. Find and tabulate the following results: the addendum, dedendum, clearance, circular pitch, tooth thickness, and base-circle diameters; the lengths of the arcs of approach, recess, and action; and the base pitch and contact ratio.

13-5   A 17-tooth spur pinion paired with a 50-tooth gear has a diametral pitch of $2\frac{1}{2}$ teeth/in and a 20° pressure angle. Make a drawing of the gears showing one tooth on each gear. Find the arcs of approach, recess, and action and the contact ratio.

13-6   Draw a 26-tooth spur pinion in mesh with a rack having a diametral pitch of 2 teeth/in and a pressure angle of 20°.

(*a*) Find the arcs of approach, recess, and action and the contact ratio.

(*b*) Draw a second rack in mesh with the same gear but offset $\frac{1}{8}$ in away from the pinion center. Determine the new contact ratio. Has the pressure angle changed?

13-7   A 15-tooth spur pinion having a 25° pressure angle and a diametral pitch of 3 teeth/in is to drive an 18-tooth gear. Without drawing the teeth, make a full-scale drawing and show the pitch circles, base circles, addendum circles, dedendum circles, and pressure line. Locate both interference points and show the amount of interference, if it exists. Locate the initial and final points of contact and label them. Compute the base pitch and find the contact ratio.

13-8   By employing a pressure angle larger than standard, it is possible to use fewer pinion teeth, and hence obtain smaller gears, without undercutting during tooth generation. If the gears are spur gears, what is the smallest possible pressure angle that can be obtained without undercutting for a 9-tooth pinion to mesh with a rack?

13-9   A 20° straight-tooth bevel pinion having 14 teeth and a diametral pitch of 6 teeth/in drives a 32-tooth gear. The two shafts are at right angles and in the same plane. Find:

(*a*) The cone distance

(*b*) The pitch angles

(*c*) The pitch diameters

(*d*) The face width

13-10 A parallel helical gearset uses a 17-tooth pinion driving a 34-tooth gear. The pinion has a right-hand helix angle of 30°, a normal pressure angle of 20°, and a normal diametral pitch of 5 teeth/in. Find:

(*a*) The normal, transverse, and axial circular pitches

(*b*) The normal base circular pitch

(*c*) The transverse diametral pitch and the transverse pressure angle

(*d*) The addendum, dedendum, and pitch diameter of each gear

13-11   A parallel helical gearset consists of a 19-tooth pinion driving a 57-tooth gear. The pinion has a left-hand helix angle of 20°, a normal pressure angle of $14\frac{1}{2}°$, and a normal diametral pitch of 10 teeth/in. Find:

(*a*) The normal, transverse, and axial circular pitches

(*b*) The transverse diametral pitch and the transverse pressure angle

(*c*) The addendum, dedendum, and pitch diameter of each gear

13-12   A parallel-shaft gearset consists of an 18-tooth helical pinion driving a 32-tooth gear. The pinion has a left-hand helix angle of 25°, a normal pressure angle of 20°, and a normal module of 3 mm. Find:

(*a*) The normal, transverse, and axial circular pitches

(*b*) The transverse module and the transverse pressure angle

(*c*) The pitch diameters of the two gears

13-13   The double-reduction helical gearset shown in the figure is driven through shaft *a* at a speed of 900 rev/min. Gears 2 and 3 have a normal diametral pitch of 10 teeth/in, a 30° helix angle, and a normal pressure angle of 20°. The second pair of gears in the train, gears 4 and 5, have a normal diametral pitch of 6 teeth/in, a 25° helix angle, and a normal pressure angle of 20°. The tooth numbers are: $N_2 = 14$, $N_3 = 54$, $N_4 = 16$, $N_5 = 36$. Find:

(*a*) The direction of the thrust force exerted by each gear upon its shaft

(*b*) The speed and direction of rotation of shaft *c*

(*c*) The center distance between shafts

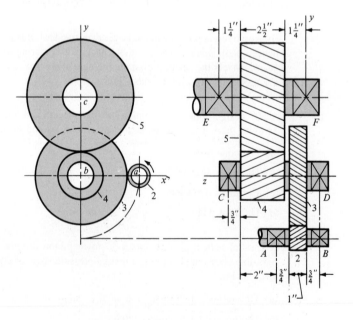

PROBLEM 13-13

13-14   Shaft *a* in the figure rotates at 600 rev/min in the direction shown. Find the speed and direction of rotation of shaft *d*.

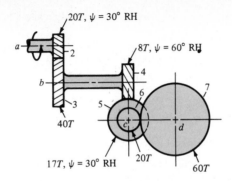

20T, $\psi = 30°$ RH

8T, $\psi = 60°$ RH

a

2

b

4

6

7

3

5

40T

c

d

17T, $\psi = 30°$ RH

20T

60T

PROBLEM 13-14

13-15 The mechanism train shown consists of an assortment of gears and pulleys to drive gear 9. Pulley 2 rotates at 1200 rev/min in the direction shown. Determine the speed and direction of rotation of gear 9.

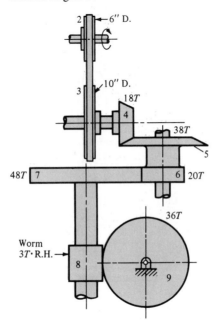

2   6" D.

3   10" D.

18T

4

38T

5

48T 7   6 20T

36T

Worm
3T·R.H.

8   9

PROBLEM 13-15

13-16 The figure shows a gear train consisting of a pair of helical gears and a pair of miter gears. The

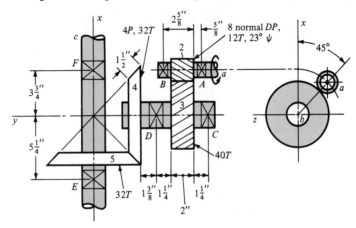

x

c

4P, 32T

$2\frac{5}{8}$

$\frac{5}{8}$ 8 normal DP,
12T, 23° ψ

x

45°

F

$1\frac{1}{2}$

2

B   A

a

a

$3\frac{3}{4}$

4

3

y

z

b

$5\frac{1}{4}$

D   C

40T

5

E

32T $1\frac{3}{8}$ $1\frac{1}{4}$

$1\frac{1}{4}$

2"

PROBLEM 13-16

573

helical gears have a $17\frac{1}{2}°$ normal pressure angle and a helix angle as shown. Find:

(a) The speed of shaft c

(b) The distance between shafts a and b

(c) The diameter of the miter gears

13-17 The tooth numbers for the automotive differential shown in the figure are $N_2 = 17$, $N_3 = 54$, $N_4 = 11$, $N_5 = N_6 = 16$. The drive shaft turns at 1200 rev/min.

(a) What are the wheel speeds if the car is traveling in a straight line on a good road surface?

(b) Suppose the right wheel is jacked up and the left wheel is resting on the road surface. What is the speed of the right wheel?

(c) Suppose, with a rear-wheel drive vehicle, the auto is parked with the right wheel resting on a wet icy surface. Does the answer to part (b) give you any hint as to what would happen if you started the car and attempted to drive on?

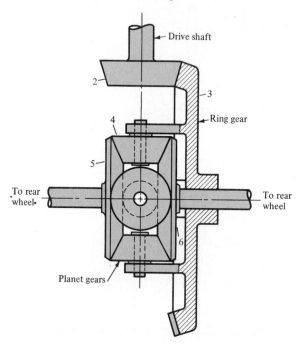

PROBLEM 13-17

13-18 The figure illustrates an all-wheel drive concept using three differentials, one for the front axle, another for the rear, and the third connected to the drive shaft.

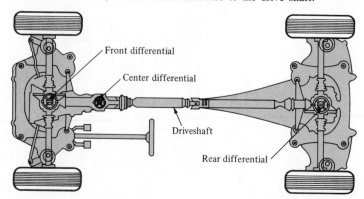

PROBLEM 13-18
The Audi "Quattro concept,"
showing the three differentials
which provide permanent all-wheel
drive. *(By permission, Audi of
America, Inc., Troy, Michigan.)*

574

(*a*) Explain why this concept may allow greater acceleration.

(*b*) Suppose either the center or the rear differential, or both, can be locked for certain road conditions. Would either or both of these actions provide greater traction? Why?

13-19   In the reverted planetary train illustrated, find the speed and direction of rotation of the arm if gear 2 is unable to rotate and gear 6 is driven at 12 rev/min in the clockwise direction.

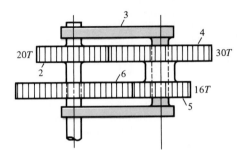

PROBLEM 13-19

13-20   In the train of Prob. 13-19, let gear 2 be driven at 180 rev/min ccw while gear 6 is held stationary. What is the speed of the arm?

13-21   Tooth numbers for the gear train shown in the figure are $N_2 = 12$, $N_3 = 16$, and $N_4 = 12$. How many teeth must internal gear 5 have? Suppose gear 5 is fixed. What is the speed of the arm if shaft *a* rotates counterclockwise at 320 rev/min?

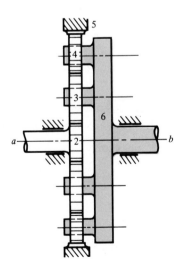

PROBLEM 13-21

13-22   The tooth numbers for the gear train illustrated are $N_2 = 24$, $N_3 = 18$, $N_4 = 30$, $N_6 = 36$, and $N_7 = 54$. Gear 7 is fixed. If shaft *b* is turned through 5 revolutions, how many turns will shaft *a* make?

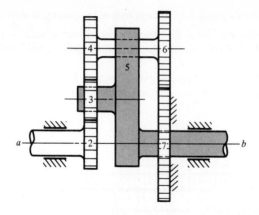

PROBLEM 13-22

13-23   By using nonstandard gears, it is possible to mate a 99-T gear with a 100-T gear at the same distance between centers as would be required to mate a 100-T gear with a 101-T gear. The planetary train shown in the figure is based on this idea.

(*a*) Find the ratio of the speed of the output shaft to the speed of the input shaft.

(*b*) The housing for this planetary train is cylindrical, with the axis of the cylinder coincident with the axes of the input and output shafts. If the pitch of gears 4 and 5 is 10 teeth/in of pitch diameter, and if these gears have standard addenda, what should be the inside diameter of this housing?

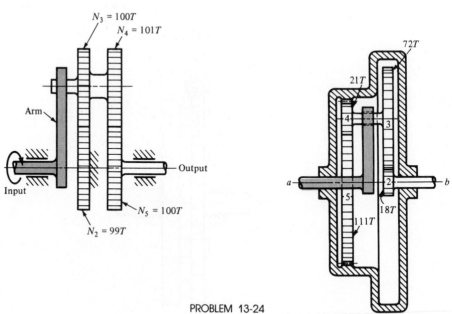

PROBLEM 13-23

PROBLEM 13-24

13-24   The epicyclic train shown in the figure has the arm attached to shaft a, and sun gear 2 to shaft b. Gear 5, with 111 teeth, is an internal gear and is part of the frame. The two planets, gears 3 and 4, are both fixed to the same planet shaft. If this train is used as an in-line speed reducer, which is the input shaft, a or b? Will both shafts then rotate in the same or in the opposite directions?

13-25   The figure shows a speed reducer in which the input shaft a is in line with output shaft b. The tooth numbers are $N_2 = 24$, $N_3 = 18$, $N_5 = 22$, and $N_6 = 64$. Find the ratio of the output speed to the input speed. Will both shafts rotate in the same direction? Note that gear 6 is a fixed internal gear.

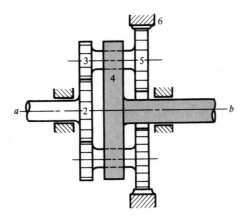

PROBLEM 13-25

13-26   The speed reducer shown in the figure has pinion 2 fixed. The planets are gears 3 and 4, both fixed to the planet shaft. Sun gear 5 is attached to the output shaft. Input shaft a drives the arm. Find the overall speed ratio of this reducer, and the direction of rotation of the output shaft.

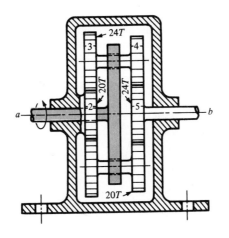

PROBLEM 13-26

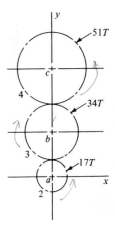

PROBLEM 13-27

13-27   Shaft a in the figure has a power input of 75 kW at a speed of 1000 rev/min in the counterclockwise direction. The gears have a module of 5 mm and a 20° pressure angle. Gear 3 is an idler.

(a) Find the force F_{3b} that gear 3 exerts against shaft b.

(b) Find the torque T_{4c} that gear 4 exerts on shaft c.

13-28   The 24-T 6-pitch 20° pinion 2 shown in the figure rotates clockwise at 1000 rev/min and is driven at a power of 25 hp. Gears 4, 5, and 6 have 24, 36, and 144 teeth, respectively. What torque can arm 3 deliver to its output shaft? Draw free-body diagrams of the arm and of each gear and show all forces which act upon them.

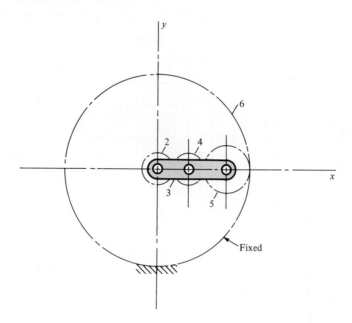

PROBLEM 13-28

13-29   The gears shown in the figure have a diametral pitch of 2 teeth per inch and a 20° pressure angle. The pinion rotates at 1800 rev/min clockwise and transmits 200 hp through the idler pair to gear 5 on shaft c. What forces do gears 3 and 4 transmit to the idler shaft?

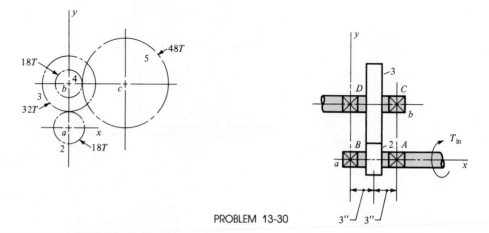

PROBLEM 13-29

PROBLEM 13-30

13-30   The figure shows a pair of shaft-mounted spur gears having a diametral pitch of 5 teeth/in with an 18-tooth 20° pinion driving a 45-tooth gear. The horsepower input is 32 maximum at 1800 rev/min. Find the direction and magnitude of the maximum forces acting on bearings A, B, C, and D.

13-31 The figure shows the electric-motor frame dimensions for a 30-hp 900 rev/min motor. The frame is bolted to its support using four $\frac{3}{4}$-in bolts spaced $11\frac{1}{4}$ in apart in the view shown and 14 in apart when viewed from the end of the motor. A 4-diametral pitch 20° spur pinion having 20 teeth and a face width of 2 in is keyed to the motor shaft. This pinion drives another gear whose axis is in the same xz plane. Determine the maximum shear and tensile forces on the mounting bolts based on 200 percent overload torque. Does the direction of rotation matter?

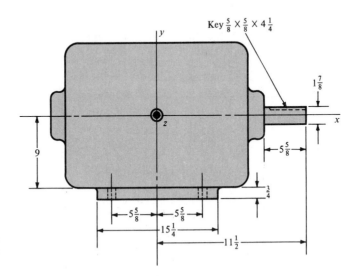

PROBLEM 13-31
NEMA No. 364 frame; dimensions in inches. The z axis is directed out of the paper.

13-32 The figure shows a 16-T 20° straight bevel pinion driving a 32-T gear, and the location of the bearing centerlines. Pinion shaft a receives 2.5 hp at 240 rev/min. Determine the bearing reactions at A and B if A is to take both radial and thrust loads.

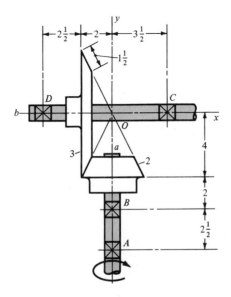

PROBLEM 13-32
Dimensions in inches.

13-33   The figure shows a 10-diametral pitch 15-tooth 20° straight bevel pinion driving a 25-tooth gear. The transmitted load is 30 lb. Find the bearing reactions at C and D on the output shaft if D is to take both radial and thrust loads.

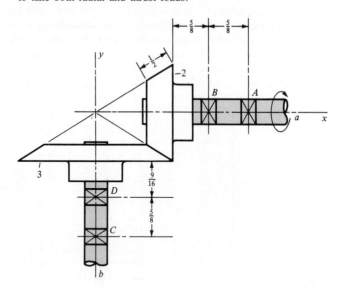

PROBLEM 13-33
Dimensions in inches.

13-34   The gears in the two trains shown in the figure have a normal diametral pitch of 4, a normal pressure angle of 20°, and a 30° helix angle. For both gear trains the transmitted load is 800 lb. In part a the pinion rotates counterclockwise about the y axis. Find the force exerted by each gear in part a on its shaft.

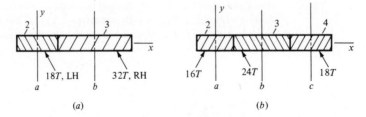

PROBLEM 13-34   (a)   (b)

13-35   This is a continuation of Prob. 13-34. Here, you are asked to find the forces exerted by gears 2 and 3 on their shafts as shown in part b. Gear 2 rotates clockwise about the y axis. Gear 3 is an idler.

13-36   A gear train is composed of four helical gears with the three shaft axes in a single plane, as shown in the figure. The gears have a normal pressure angle of 20° and a 30° helix angle. Shaft b

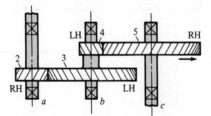

PROBLEM 13-36

is an idler and the transmitted load acting on gear 3 is 500 lb. The gears on shaft b both have a normal diametral pitch of 7 teeth/in and have 54 and 14 teeth, respectively. Find the forces exerted by gears 3 and 4 on shaft b.

13-37 In the figure for Prob. 13-30, pinion 2 is to be a right-hand helical gear having a helix angle of 30°, a normal pressure angle of 20°, 16 teeth, and a normal diametral pitch of 6 teeth/in. A 25-hp motor drives shaft a at a speed of 1720 rev/min clockwise about the x axis. Gear 3 has 42 teeth. Find the reaction exerted by bearings C and D on shaft b. One of these bearings is to take both radial and thrust loads. This bearing should be selected so as to place the shaft in compression.

13-38 Gear 2, in the figure, has 16 teeth, a 20° transverse angle, a 15° helix angle, and a normal diametral pitch of 8 teeth/in. Gear 2 drives the idler on shaft b, which has 36 teeth. The driven gear on shaft c has 28 teeth. If the driver rotates at 1720 rev/min and transmits $7\frac{1}{2}$ hp, find the radial and thrust load on each shaft.

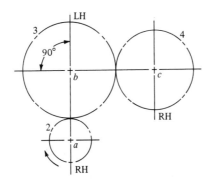

PROBLEM 13-38

13-39 The figure shows a double-reduction helical gearset. Pinion 2 is the driver, and it receives a torque of 1200 lb · in from its shaft in the direction shown. Pinion 2 has a normal diametral pitch of 8 teeth/in, 14 teeth, and a normal pressure angle of 20° and is cut right-handed with a helix

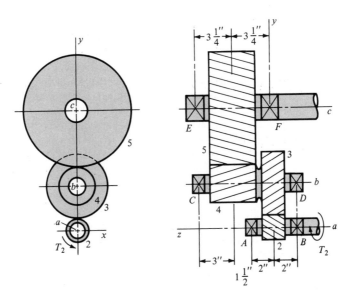

PROBLEM 13-39

angle of 30°. The mating gear 3 on shaft *b* has 36 teeth. Gear 4, which is the driver for the second pair of gears in the train, has a normal diametral pitch of 5 teeth/in, 15 teeth, and a normal pressure angle of 20° and is cut left-handed with a helix angle of 15°. Mating gear 5 has 45 teeth. Find the magnitude and direction of the force exerted by the bearings at *C* and *D* on shaft *b* if bearing *C* can take only radial load while bearing *D* is mounted to take both radial and thrust load.

13-40 A right-hand single-tooth hardened-steel (hardness not specified) worm has a catalog rating of 2000 W at 600 rev/min when meshed with a 48-tooth cast-iron gear. The axial pitch of the worm is 25 mm, the normal pressure angle is $14\frac{1}{2}°$, the pitch diameter of the worm is 100 mm, and the face widths of the worm and gear are, respectively, 100 mm and 50 mm. The figure shows bearings *A* and *B* on the worm shaft symmetrically located with respect to the worm and 200 mm apart. Determine which should be the thrust bearing, and find the magnitudes and directions of the forces exerted by both bearings.

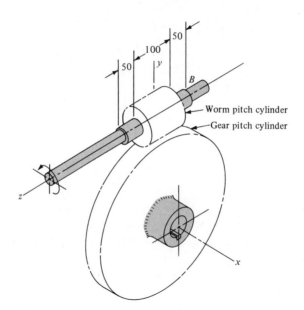

PROBLEM 13-40
Dimensions in millimeters.

13-41 The hub diameter and projection for the gear of Prob. 13-40 are 100 and 37.5 mm, respectively. The face width of the gear is 50 mm. Locate bearings *C* and *D* on opposite sides, spacing *C* 10 mm from the gear on the hidden face (see figure) and *D* 10 mm from the hub face. Find the output torque and the magnitudes and directions of the forces exerted by the bearings on the gearshaft.

13-42 A 2-tooth left-hand worm transmits $\frac{1}{2}$ hp at 900 rev/min to a 36-tooth gear having a transverse diametral pitch of 10 teeth/in. The worm has a normal pressure angle of $14\frac{1}{2}°$, a pitch diameter of $1\frac{1}{4}$ in, and a face width of $1\frac{1}{2}$ in. Use a coefficient of friction of 0.05 and find the force exerted by the gear on the worm and the torque input. For the same geometry as shown for Prob. 13-40, the worm velocity is clockwise about the *z* axis.

13-43 The diametral pitch of the teeth in all the spur gears of the epicyclic gear train of Prob. 13-23 is 48 teeth per inch. The pressure angle is 20°. If the 100:101 gear mesh is set to the proper center-to-center distance, what is the operative pressure angle of the 99:100 gear mesh? Show that the pressure angle ϕ' is given by the equation

$$\phi' = \cos^{-1}\left[\frac{(1 + m_G)N_P \cos \phi}{2PC'}\right]$$

13-2 400 rev/min, 9.425 mm, 112.5 mm

13-6 (a) $d = 13$ in, $p = 1.57$ in, $t = 0.785$ in, $a = 0.5$ in, $b = 0.625$ in, $q_a = 1.54$ in, $q_r = 1.52$ in, $q_t = 3.06$ in, $m_c = 1.95$; (b) $m_c = 1.55$, no

13-8 28.13°

13-11 (a) $P_n = 0.3142$ in, $p_t = 0.3343$ in, $p_x = 0.9185$ in; (b) $P_t = 9.397$ teeth/in, $\phi_t = 15.39°$; (c) $a = 0.1$ in, $b = 0.125$ in, $d_P = 2.022$ in, $d_G = 6.066$ in

13-14 11.82 rev/min ccw

13-16 (a) 162 rev/min; (b) 3.531 in; (c) 8 in

13-19 −17.49 rev/min

13-22 −14.375 turns

13-26 11/36 in same direction

13-28 7875 lb · in

13-32 $\mathbf{F}_A = -42.72\mathbf{i} + 106.8\mathbf{j} + 262.4\mathbf{k}$ lb

13-35 The shaft forces are two couples tending to rotate the shaft end over end; $\mathbf{T}_b = 5542\mathbf{j} + 3201\mathbf{k}$ lb · in.

13-39 $\mathbf{F}_C = 1564\mathbf{i} + 674\mathbf{j}$ lb, $\mathbf{F}_D = 1610\mathbf{i} - 426\mathbf{j} + 154\mathbf{k}$ lb

13-42 $\mathbf{W} = 46.7\mathbf{i} + 65.8\mathbf{j} + 251\mathbf{k}$ lb, $T = 35$ lb · in

14
Spur and
Helical Gears

This chapter is devoted primarily to the analysis and design of spur and helical gears to resist bending failure of the teeth and to resist pitting or failure of the tooth surfaces. Failure by bending will occur when the significant tooth stress equals or exceeds either the yield strength or the endurance limit. A surface failure occurs when the significant contact stress equals or exceeds the surface endurance strength.

The American Gear Manufacturers Association* (AGMA) has for many years been the responsible authority for the dissemination of knowledge pertaining to the design and analysis of gearing. The methods this organization presents are in universal use in this country when strength and wear are the primary considerations. In view of this fact it is important that the AGMA approach to the subject be presented here without changes.

The general AGMA approach requires a great many charts and graphs—too many for a single chapter in this book. We have omitted many of these here by choosing a single pressure angle and by using only full-depth teeth. This simplification reduces the complexity but does not prevent the development of a basic understanding of the approach. Furthermore, the simplification makes possible a better development of the fundamentals and hence should constitute an ideal introduction to the use of the general AGMA method.† Sections 14-1 and 14-2 are elementary and serve as an examination of the foundations of the AGMA method.

---

*1500 King Street, Suite 201, Alexandria, VA 22314.

†The standard used in this chapter is *AGMA Standard for Rating Pitting Resistance and Bending Strength of Spur and Helical Involute Gear Teeth,* AGMA 218.01, December 1982. The following is quoted from the Foreword:

> This AGMA Standard and related publications are based on typical or average data, conditions, or applications. The standards are subject to continued improvement, revision, or withdrawal as dictated by increased experience. Any person who refers to AGMA technical publications should be sure that he has the latest information available from the Association on the subject matter.
>
> Tables or other self-supporting sections may be quoted or extracted in their entirety. Credit line should read: "Extracted from AGMA Standard for Rating Pitting Resistance and Bending Strength of Spur and Helical Involute Gear Teeth, AGMA 218.01, with the permission of the publisher, American Gear Manufacturers Association."

Table 14-1 is a listing of the symbols used for gearing and their units.

TABLE 14-1

Symbols, Their Names, and Location of Values

SYMBOL	NAME	WHERE FOUND
C_a	Application factor	Sec. 14-8
C_f	Surface-condition factor	Sec. 14-9
C_H	Hardness-ratio factor	Eq. (14-30)
C_L	Life factor	Fig. 14-8
C_m	Load-distribution factor	Table 14-6
C_p	Elastic coefficient	Eq. (14-13), Table 14-5
C_R	Reliability factor	Table 14-7
C_s	Size factor	Sec. 14-10
C_T	Temperature factor	Sec. 14-15
C_V	Velocity factor [for use in Eq. (14-14)]	Eqs. (14-5), (14-6)
C_v	AGMA dynamic factor	Eq. (14-27)
d	Pitch diameter	Given
d_G	Pitch diameter of gear	Given
d_P	Pitch diameter of pinion	Given
E	Modulus of elasticity	Given
F	Face width	Given
H	Power	Given
H_{BG}	Brinell hardness of gear tooth	Given
H_{BP}	Brinell hardness of pinion tooth	Given
I	Geometry factor	Eq. (14-23)
J	Geometry factor	Figs. 14-4, 14-5, 14-6
K_a	Application factor	Sec. 14-8
K_f	Fatigue stress-concentration factor	Eq. (14-20)
K_L	Life factor	Fig. 14-9
K_m	Load-distribution factor	Table 14-6
K_R	Reliability factor	Table 14-7
K_s	Size factor	Sec. 14-10
K_T	Temperature factor	Sec. 14-15
K_V	Velocity factor (for use in Lewis equation only)	Eqs. (14-5), (14-6)
K_v	AGMA dynamic factor	Eq. (14-27)
m	Metric module	Given
m_F	Face-contact ratio	Eq. (14-19)
m_G	Speed ratio	Eq. (14-22)
m_N	Load-sharing ratio	Eq. (14-21)
m_p	Transverse-contact ratio	Eqs. (13-8), (13-9)
n	Speed	Given
P	Diametral pitch	Given
P_d	Diametral pitch in plane of rotation	Given
p	Circular pitch	Given
p_n	Normal circular pitch	Eq. (13-12)
p_x	Axial pitch	Eq. (13-13)
Q_v	Transmission accuracy-level number	Sec. 14-7
r	Radius of curvature	Eq. (14-12)
r_f	Tooth fillet radius	Given

TABLE 14-1
(Continued)

SYMBOL	NAME	WHERE FOUND
S_C	Surface endurance strength (not AGMA)	Eq. (7-60)
S_c	AGMA surface fatigue strength	Fig. 14-3, Table 14-4
S_t	AGMA bending strength	Fig. 14-2, Table 14-3
t	Tooth thickness	Fig. 14-1
V	Pitch-line velocity	Given
W_t	Transmitted (tangential) load	Eqs. (13-27), (13-31)
Y	Lewis form factor	Table 14-2
Y	AGMA form factor	Eq. (14-20)
Z	Length of line of action	Fig. 13-15
ν	Poisson's ratio	Given
σ	Tooth bending stress	Eqs. (14-7), (14-8), (14-15)
σ_{all}	Allowable bending stress	Eq. (14-17)
σ_C	Surface compressive stress	Eqs. (14-11), (14-14)
σ_c	Contact stress (AGMA formula)	Eq. (14-16)
$\sigma_{c,\text{all}}$	Allowable contact stress	Eq. (14-18)
ϕ	Pressure angle	Given
ϕ_t	Transverse pressure angle	Fig. 13-22

14-1   THE LEWIS FORMULA

Wilfred Lewis was the first to present a formula for computing the bending stress in gear teeth in which the tooth form entered into the equation. The formula was announced in 1892, and it still remains the basis for most gear design today.

To derive the basic Lewis equation, refer to Fig. 14-1a, which shows a cantilever of cross-sectional dimensions F and t, having a length l and a load W_t uniformly distributed across the distance F. The section modulus is $I/c = Ft^2/6$, and therefore the bending stress is

$$\sigma = \frac{M}{I/c} = \frac{6W_t l}{Ft^2} \qquad (a)$$

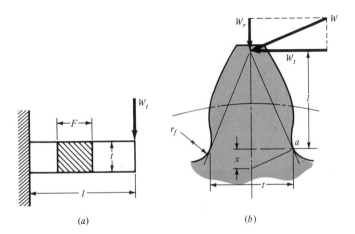

FIGURE 14-1   (a)   (b)

Referring now to Fig. 14-1b, we assume that the maximum stress in a gear tooth occurs at point a. By similar triangles, you can write

$$\frac{t/2}{x} = \frac{l}{t/2} \qquad \text{or} \qquad x = \frac{t^2}{4l} \tag{b}$$

By rearranging Eq. (a),

$$\sigma = \frac{6W_t l}{Ft^2} = \frac{W_t}{F} \frac{1}{t^2/6l} = \frac{W_t}{F} \frac{1}{t^2/4l} \frac{1}{\frac{4}{6}} \tag{c}$$

If we now substitute the value of x from Eq. (b) in Eq. (c) and multiply the numerator and denominator by the circular pitch p, we find

$$\sigma = \frac{W_t p}{F(\frac{2}{3})xp} \tag{d}$$

Letting $y = 2x/3p$, we have

$$\sigma = \frac{W_t}{Fpy} \tag{14-1}$$

This completes the development of the original Lewis equation. The factor y is called the *Lewis form factor,* and it may be obtained by a graphical layout of the gear tooth or by digital computation.

In using this equation, most engineers prefer to employ the diametral pitch in determining the stresses. This is done by substituting $P = \pi/p$ and $Y = \pi y$ in Eq. (14-1). This gives

$$\sigma = \frac{W_t P}{FY} \tag{14-2}$$

where

$$Y = \frac{2xP}{3} \tag{14-3}$$

The use of this equation for Y means that only the bending of the tooth is considered and that the compression due to the radial component of the force is neglected. Values of Y obtained from this equation are tabulated in Table 14-2.

The use of Eq. (14-3) also implies that the teeth do not share the load and that the greatest force is exerted at the tip of the tooth. But we have already learned that the contact ratio should be somewhat greater than unity, say about 1.5, to achieve a quality gearset. If, in fact, the gears are cut with sufficient accuracy, the tip-load condition is not the worst, because another pair of teeth will be in contact when this condition occurs. Examination of run-in teeth will show that the heaviest loads occur near the middle of the tooth. Therefore the maximum stress probably occurs while a single pair of teeth is carrying the full load, at a point where another pair of teeth is just on the verge of coming into contact.

TABLE 14-2

Values of the Lewis Form Factor Y. (These Values Are for a Normal Pressure Angle of 20°, Full-Depth Teeth, and a Diametral Pitch of Unity in the Plane of Rotation)

NUMBER OF TEETH	Y	NUMBER OF TEETH	Y
12	0.245	28	0.353
13	0.261	30	0.359
14	0.277	34	0.371
15	0.290	38	0.384
16	0.296	43	0.397
17	0.303	50	0.409
18	0.309	60	0.422
19	0.314	75	0.435
20	0.322	100	0.447
21	0.328	150	0.460
22	0.331	300	0.472
24	0.337	400	0.480
26	0.346	Rack	0.485

Dynamic Effects

When a pair of gears is driven at moderate or high speeds and noise is generated, it is certain that dynamic effects are present. One of the earliest efforts to account for an increase in the load due to velocity employed a number of gears of the same size, material, and strength. Several of these gears were tested to destruction by meshing and loading them at zero velocity. The remaining gears were tested to destruction at various pitch-line velocities. Then, for example, if a pair of gears failed at 500 lb tangential load at zero velocity, and at 250 lb at a velocity V_1, then a *velocity factor,* designated K_V, of 0.5 was specified for the gears at velocity V_1. Then, another, identical pair of gears running at a pitch-line velocity V_1 could be assumed to have a dynamic load equal to twice the tangential or transmitted load.

In the nineteenth century, Carl G. Barth first expressed the velocity factor by the equation

$$K_V = \frac{600}{600 + V} \tag{14-4}$$

where V is the pitch-line velocity in feet per minute. This equation is called the *Barth equation,* and it is known to be based on tests of cast-iron gears with cast teeth. It is also quite probable, because of the date, that the tests were made on teeth having a cycloidal profile, instead of an involute; cycloidal teeth were in quite general use in the nineteenth century because they were easier to cast than involute teeth.

The Barth equation is often modified to

$$K_V = \frac{1200}{1200 + V} \tag{14-5}$$

which is then used for cut or milled teeth.

In SI units, the above equations become, respectively,

$$K_V = \frac{3.05}{3.05 + V} \quad \text{and} \quad K_V = \frac{6.1}{6.1 + V} \tag{14-6}$$

where V is now in meters per second (m/s).

Introducing the velocity factor into Eq. (14-2) gives

$$\sigma = \frac{W_t P}{K_V F Y} \tag{14-7}$$

The metric version of this equation is

$$\sigma = \frac{W_t}{K_V F m Y} \tag{14-8}$$

where the face width F and the module m are both in millimeters (mm). Specifying the tangential load W_t in newtons (N) then results in the stress in units of megapascals (MPa).

As a general rule, spur gears should be designed with a face width between 3 and 5 times the circular pitch.

Equations (14-7) and (14-8) are important because they form the basis for the AGMA approach to the bending strength of gear teeth. They are in general use for estimating the capacity of gear drives when life and reliability are not important considerations. The equations are also quite useful in obtaining a preliminary estimate of gear sizes needed for various applications.

EXAMPLE 14-1

A stock spur gear is available having a diametral pitch of 8 teeth/in, a $1\frac{1}{2}$-in face, 16 teeth, and a pressure angle of 20° with full-depth teeth. The material is AISI 1020 steel as rolled. Use a factor of safety of 3 to rate the horsepower output of the gear corresponding to a speed of 1200 rev/min and moderate applications.

Solution

The term *moderate applications* seems to imply that the gear can be rated using the yield strength as a criterion of failure. From Table A-20, we find $S_{ut} = 55$ kpsi and $S_y = 30$ kpsi. Using a factor of safety of 3 means that the allowable bending stress is $30/3 = 10$ kpsi. The pitch diameter is $N/P = 16/8 = 2$ in, and so the pitch-line velocity is

$$V = \frac{\pi d n}{12} = \frac{\pi (2)(1200)}{12} = 628 \text{ ft/min}$$

The velocity factor is found from Eq. (14-5) to be

$$K_V = \frac{1200}{1200 + V} = \frac{1200}{1200 + 628} = 0.656$$

Table 14-2 gives the form factor as $Y = 0.296$. We now arrange and substitute in Eq. (14-7) as follows:

$$W_t = \frac{K_V F Y \sigma_{all}}{P} = \frac{(0.656)(1.5)(0.296)(10)(10^3)}{8} = 364 \text{ lb}$$

and so the horsepower output is

Answer $H = \dfrac{W_t V}{33\ 000} = \dfrac{364(628)}{33\ 000} = 6.93$

It is important to emphasize that this is a very rough estimate and that this approach must not be used for important applications. The example is intended to help you to understand the fundamentals that will enter into the AGMA approach.

EXAMPLE 14-2   Estimate the horsepower rating of the gear of the previous example based on obtaining an infinite life in bending.

Solution   The rotating-beam endurance limit is first estimated using Eq. (7-4). The result is

$S'_e = 0.504 S_{ut} = 0.504(55) = 27.7$ kpsi

To obtain the surface-finish factor k_a we refer to Table 7-4 and find $a = 14.4$ and $b = -0.718$ for hot-rolled steel. Then Eq. (7-14) gives the surface factor as

$k_a = a S_{ut}^b = 14.4(55)^{-0.718} = 0.811$

The next problem is that of finding a size factor. The height of the tooth is the addendum plus the dedendum which is the distance l in Fig. 14-1b. Thus, from Table 13-2 we have

$l = \dfrac{1}{P} + \dfrac{1.25}{P} = \dfrac{1}{8} + \dfrac{1.25}{8} = 0.281$ in

The tooth thickness t in Fig. 14-1b is given in Sec. 14-1 [Eq. (b)] as $t = (4lx)^{1/2}$, where $x = 3Y/2P$ from Eq. (14-3). Therefore

$x = \dfrac{3Y}{2P} = \dfrac{3(0.296)}{2(8)} = 0.0555$

and

$t = (4lx)^{1/2} = [4(0.281)(0.0555)]^{1/2} = 0.250$ in

We have recognized the tooth as a beam of rectangular cross section; so the effective rotating-beam diameter must be obtained from Eq. (7-19). Thus

$d_e = 0.808(hb)^{1/2} = 0.808(Ft)^{1/2} = 0.808[1.5(0.250)]^{1/2} = 0.495$ in

Then Eq. (7-15) gives the size factor k_b as

$k_b = \left(\dfrac{d}{0.3}\right)^{-0.1133} = \left(\dfrac{0.495}{0.3}\right)^{-0.1133} = 0.945$

The load factor k_c, from Eq. (7-22), is unity. So is the temperature factor k_d.

Two effects are used to evaluate the miscellaneous-effects factor k_e. The first of these is the effect of one-way bending. In general, a gear tooth is subject to only one-way bending. Exceptions to this rule occur when the gear is used as an idler or if it is to be used in a reversing mechanism.

For one-way bending, the mean and alternating stress components are

$$\sigma_a = \sigma_m = \frac{\sigma}{2} \qquad (1)$$

where σ is the tooth bending stress as given by Eq. (14-7). Equation (7-35), for the modified Goodman relation, is

$$\frac{S_a}{S_e'} + \frac{S_m}{S_{ut}} = 1 \qquad (2)$$

Since S_a and S_m are equal for one-way bending, we solve Eq. (2) and get

$$S_a = \frac{S_e' S_{ut}}{S_e' + S_{ut}} \qquad (3)$$

Now, from Eq. (1), replace S_a with $\sigma/2$; and in the denominator, replace S_e' with $0.504S_{ut}$. Solving gives

$$\sigma = \frac{2S_e' S_{ut}}{0.504S_{ut} + S_{ut}} = 1.33S_e' \qquad (4)$$

This shows that a miscellaneous-effects factor of $k_e = 1.33$ can be used in the Marin equation to account for one-way bending for this material.

The second effect to be accounted for in using the miscellaneous-effects factor is that of stress concentration. AGMA uses a rather elaborate approach to this problem requiring a good drawing of the generated tooth profile or a computer program. Since our purpose here is to learn fundamental concepts, we shall bypass the complexity by again considering the tooth to be a simple cantilever connected to a fixed support and having a fillet radius at the root of the tooth of r_f. For 20° full-depth teeth, this radius is

$$r_f = \frac{0.300}{P} = \frac{0.300}{8} = 0.0375 \text{ in}$$

Refer now to Appendix Fig. A-15-6 and compute

$$\frac{r}{d} = \frac{r_f}{t} = \frac{0.0375}{0.250} = 0.15$$

Since $D/d = \infty$, we find $K_t = 1.68$.

We are still not finished. Next, use Fig. 5-16 with $S_{ut} = 55$ kpsi and find the notch sensitivity $q = 0.78$ at the end of the chart. Then the fatigue stress-concentration factor is

$$K_f = 1 + q(K_t - 1) = 1 + 0.78(1.68 - 1) = 1.53$$

The miscellaneous-effects factor for stress concentration alone is

$$k_e = \frac{1}{K_f} = \frac{1}{1.53} = 0.653$$

The final value of k_e is obtained by multiplying these two factors. So we have, finally, that

$$k_e = 1.33(0.653) = 0.869$$

The last step is to combine all these in the Marin equation. The fully corrected endurance limit is

$$S_e = k_a k_b k_c k_d k_e S'_e = 0.811(0.945)(1)(1)(0.869)(27.7) = 18.4 \text{ kpsi}$$

Using the same factor of safety as in the previous example gives an allowable stress as

$$\sigma_{\text{all}} = \frac{S_e}{n} = \frac{18.4}{3} = 6.15 \text{ kpsi}$$

Using the same calculations as before, we now find $W_t = 224$ lb and $H = 4.26$ hp.

Again it is emphasized that these results should be accepted only as preliminary estimates.

14-2   SURFACE DURABILITY

In this section we are interested in the failure of the surfaces of gear teeth, which is generally called *wear*. *Pitting,* as explained in Sec. 7-18, is a surface fatigue failure due to many repetitions of high contact stresses. Other surface failures are *scoring,* which is a lubrication failure, and *abrasion,* which is wear due to the presence of foreign material.

To obtain an expression for the surface-contact stress, we shall employ the Hertz theory. In Eq. (2-94) it was shown that the contact stress between two cylinders may be computed from the equation

$$p_{\text{max}} = \frac{2F}{\pi b l} \tag{14-9}$$

where p_{max} = largest surface pressure
   F = force pressing the two cylinders together
   l = length of cylinders

and b is obtained from the equation

$$b = \left\{ \frac{2F}{\pi l} \frac{[(1 - \nu_1^2)/E_1] + [(1 - \nu_2^2)/E_2]}{(1/d_1) + (1/d_2)} \right\}^{1/2} \tag{14-10}$$

[Eq. (2-93)] where ν_1, ν_2, E_1, and E_2 are the elastic constants and d_1 and d_2 are the diameters, respectively, of the two contacting cylinders.

To adapt these relations to the notation used in gearing, we replace F by $W_t/\cos \phi$, d by $2r$, and l by the face width F. With these changes, we can substitute the value of b as given by Eq. (14-10) in Eq. (14-9). Replacing p_{max} by σ_C, the *surface compressive stress (Hertzian stress)* is found from the equation

$$\sigma_C^2 = \frac{W_t}{\pi F \cos \phi} \frac{(1/r_1) + (1/r_2)}{[(1 - \nu_1^2)/E_1] + [(1 - \nu_2^2)/E_2]} \tag{14-11}$$

where r_1 and r_2 are the instantaneous values of the radii of curvature on the pinion- and gear-tooth profiles, respectively, at the point of contact. By accounting for load sharing in the value of W_t used, Eq. (14-11) can be solved for the Hertzian stress for any or all points from the beginning to the end of tooth contact. Of course, pure rolling exists only at the pitch point. Elsewhere the motion is a mixture of rolling and sliding. Equation (14-11) does not account for any sliding action in the evaluation of stress.

We note that AGMA uses μ for Poisson's ratio instead of ν as is used here.

We have already noted that the first evidence of wear occurs near the pitch line. The radii of curvature of the tooth profiles at the pitch point are

$$r_1 = \frac{d_P \sin \phi}{2} \qquad r_2 = \frac{d_G \sin \phi}{2} \tag{14-12}$$

where ϕ is the pressure angle and d_P and d_G are the pitch diameters of the pinion and gear, respectively.

Note, in Eq. (14-11), that the denominator of the second group of terms contains four elastic constants, two for the pinion and two for the gear. As a simple means of combining and tabulating the results for various combinations of pinion and gear materials, AGMA defines an *elastic coefficient* C_p by the equation

$$C_p = \left[\frac{1}{\pi \left(\dfrac{1 - \nu_P^2}{E_P} + \dfrac{1 - \nu_G^2}{E_G} \right)} \right]^{1/2} \tag{14-13}$$

With this simplification, and the addition of a velocity factor C_V, Eq. (14-11) can be written as

$$\sigma_C = -C_P \left[\frac{W_t}{C_V F \cos \phi} \left(\frac{1}{r_1} + \frac{1}{r_2} \right) \right]^{1/2} \tag{14-14}$$

where the sign is negative because σ_C is a compressive stress. The velocity factor C_V has the same values as K_V. AGMA practice is to distinguish modification factors by using C's for surface durability and K's for bending.

EXAMPLE 14-3

The pinion of Examples 14-1 and 14-2 is to be mated with a 50-tooth gear manufactured of ASTM No. 30 cast iron. Using the tangential load of 270 lb found in Example 14-2, estimate the factor of safety of the drive based on the possibility of a surface fatigue failure.

Solution

From Table A-5 we find the elastic constants to be $E_P = 30$ Mpsi, $\nu_P = 0.292$, $E_G = 14.5$ Mpsi, $\nu_G = 0.211$. We substitute these in Eq. (14-13) to get the elastic coefficient as

$$C_P = \left\{ \frac{1}{\pi \left[\dfrac{1 - (0.292)^2}{30(10^6)} + \dfrac{1 - (0.211)^2}{14.5(10^6)} \right]} \right\}^{1/2} = 1817$$

From Example 14-1, the pinion pitch diameter is $d_P = 2$ in. The value for the gear is $d_G = 50/8 = 6.25$ in. Then Eq. (14-12) is used to obtain the radii of curvature at the pitch points. Thus

$$r_1 = \frac{2 \sin 20°}{2} = 0.342 \text{ in} \qquad r_2 = \frac{6.25 \sin 20°}{2} = 1.069 \text{ in}$$

The face width is given as $F = 1.5$ in. Use $C_V = K_V = 0.656$ from Example 14-1. Substituting all these values directly in Eq. (14-14) gives the contact stress as

$$\sigma_C = -1817 \left[\frac{270}{(0.656)(1.5)(\cos 20°)} \left(\frac{1}{0.342} + \frac{1}{1.069} \right) \right]^{1/2}$$

$$= -61\ 000 \text{ psi}$$

The surface endurance strength of cast iron can be estimated from the formula

$$S_C \approx 0.32 H_B$$

for 10^8 cycles, where S_C is in kpsi.* Table A-24 gives $H_B = 201$ for ASTM No. 30 cast iron. Therefore $S_C = 0.32(201) = 64.3$ kpsi. If the factor of safety is defined as the ratio of the strength to the stress, then

$$n = \frac{S_C}{\sigma_C} = \frac{64.3}{|61.0|} = 1.05$$

However, σ_C is not directly proportional to W_t, and so we substitute S_C for $|\sigma_C|$ in Eq. (14-14) and solve for W_t. This gives 300 lb. Since the actual value is 270 lb, the factor of safety based on the ratio of the loads is $300/270 = 1.11$.

14-3   THE AGMA STRESS FORMULAS

Two fundamental formulas are used in the AGMA approach, one for bending stress and one for pitting resistance. The results obtained from these two formulas are called *stress numbers* in the standard and are designated using a lowercase letter s, instead of the Greek letter σ we have used in this book (and shall continue to use).

The fundamental formulas for bending stress are

$$\sigma = \begin{cases} \dfrac{W_t K_a}{K_v} \dfrac{P_d}{F} \dfrac{K_s K_m}{J} \\[2ex] \dfrac{W_t K_a}{K_v} \dfrac{1.0}{Fm} \dfrac{K_s K_m}{J} \end{cases} \tag{14-15}$$

where the first uses U.S. customary units and the second SI units. Also,

σ = bending stress
W_t = transmitted tangential load
K_a = application factor
K_v = dynamic factor
P_d = nominal diametral pitch in plane of rotation
m = nominal metric module in plane of rotation
F = face width
K_s = size factor
K_m = load-distribution factor
J = geometry factor

---

*But see footnote in Sec. 14-4.

Notice that there are three groups of terms. The first refers to the loading characteristics, the second to the gear geometry, and the third to the tooth form. These terms will be explained and evaluated in the sections to follow.

The fundamental AGMA formula for pitting resistance is

$$\sigma_c = C_P \left(\frac{W_t C_a}{C_v} \frac{C_s}{Fd} \frac{C_m C_f}{I} \right)^{1/2} \tag{14-16}$$

where σ_c = absolute value of contact stress
$\quad C_p$ = elastic coefficient
$\quad C_a$ = application factor
$\quad C_v$ = dynamic factor
$\quad C_s$ = size factor
$\quad d$ = pitch diameter of pinion
$\quad C_m$ = load-distribution factor
$\quad C_f$ = surface-condition factor
$\quad I$ = geometry factor

The evaluation of all these factors is explained in the sections that follow. The development of Eq. (14-16) is clarified in the second part of Sec. 14-5.

14-4   THE AGMA STRENGTH FORMULAS

Instead of using the term *strength*, AGMA uses data termed *allowable stress numbers* and designates these by the symbol s_a. It will be less confusing here if we continue the practice in this book of using the capital letter S to denote strength and the Greek letters σ and τ for stress. To make it perfectly clear, we shall use the term *AGMA strength* as a replacement for the phrase *allowable stress numbers* as used by AGMA.

Following this convention, values for *AGMA bending strengths*, designated here as S_t, are to be found in Fig. 14-2 and Table 14-3.

Similarly, *AGMA surface fatigue strengths* S_c are given in Fig. 14-3 and Table 14-4.*

Since AGMA strengths are not identified with other strengths, such as S_{ut}, S_e, or S_y, as used elsewhere in this book, their use should be restricted to the analysis of gear problems.

In the AGMA approach, the strengths are modified by various factors which produce limiting values of the bending stress and the contact stress. Using the same notation as in Eq. (1-1), we shall term the resulting modifications the *allowable bending stress* σ_{all} and the *allowable contact stress* $\sigma_{c,all}$. The formula for the allowable

---

*As an incidental, for cast iron in grades 30 to 40,

$S_t = 0.18H_B - 23.0$ kpsi   bending

$S_c = 0.4H_B - 5$ kpsi   contact

These are not AGMA recommendations, but they are valid for 10^7 cycles at 0.99 quantile.

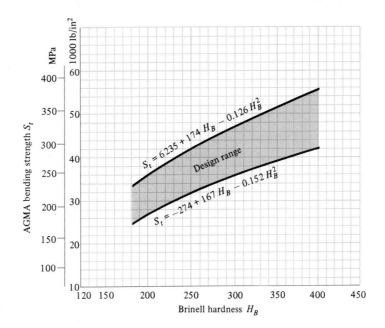

FIGURE 14-2

AGMA bending strength S_t for steel gears. *(Source: AGMA 218.01.)*

bending stress is

$$\sigma_{\text{all}} = \frac{S_t K_L}{K_T K_R} \qquad (14\text{-}17)$$

where K_L = life factor
K_T = temperature factor
K_R = reliability factor

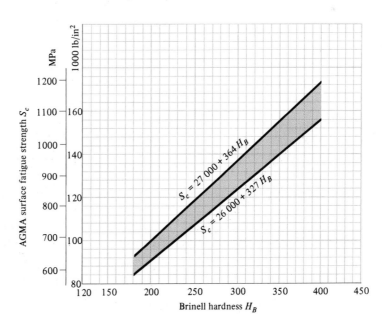

FIGURE 14-3

AGMA surface fatigue strength S_c for steel gears. *(Source: AGMA 218.01.)*

TABLE 14-3

AGMA Bending Strength S_t

MATERIAL	AGMA CLASS	COMMERCIAL DESIGNATION	HEAT TREATMENT	MINIMUM HARDNESS AT SURFACE	CORE	S_t psi	S_t MPa
Steel	A-1 through A-5	—	Through-hardened and tempered	180 BHN	—	25–33 000	(170–230)
				240 BHN	—	31–41 000	(210–280)
				300 BHN	—	36–47 000	(250–320)
				360 BHN	—	40–52 000	(280–360)
				400 BHN	—	42–56 000	(290–390)
			Flame- or induction-hardened with type A pattern	50–54 HRC	—	45–55 000	(310–380)
			Flame- or induction-hardened with type B pattern		—	22 000	(150)
			Carburized and case-hardened	55 HRC	—	55–65 000	(380–450)
				60 HRC	—	55–70 000	(380–480)
		AISI 4140	Nitrided*	48 HRC	300 BHN	34–45 000	(230–310)
		AISI 4340	Nitrided*	46 HRC	300 BHN	36–47 000	(250–325)
		Nitralloy 135M	Nitrided*	60 HRC	300 BHN	38–48 000	(260–330)
		2½% chrome	Nitrided*	54–60 HRC	350 BHN	55–65 000	(380–450)
Cast iron	20		As cast	—	—	5000	(35)
	30		As cast	175 BHN	—	8500	(69)
	40		As cast	200 BHN	—	13 000	(90)
Nodular (ductile) iron	A-7-a	60-40-18	Annealed, quenched, and tempered	140 BHN	—	90–100% of S_t for steel of same hardness
	A-7-c	80-55-06		180 BHN	—	
	A-7-d	100-70-03		230 BHN	—	
	A-7-e	120-90-02		270 BHN	—	
Malleable iron (pearlitic)	A-8-c	45007	—	165 BHN	—	10 000	(70)
	A-8-e	50005	—	180 BHN	—	13 000	(90)
	A-8-f	53007	—	195 BHN	—	16 000	(110)
	A-8-i	80002	—	240 BHN	—	21 000	(145)
Bronze	Bronze 2	AGMA 2C	Sand-cast Sand-cast	Tensile strength minimum 40 000 lb/in^2 (275 MPa)		5 700	(40)

TABLE 14-3

(Continued)

MATERIAL	AGMA CLASS	COMMERCIAL DESIGNATION	HEAT TREATMENT	MINIMUM HARDNESS AT SURFACE	CORE	S_t	
						psi	MPa
Bronze	Al/Br 3	ASTM B-148-52 alloy 9C	Heat-treated	Tensile strength minimum 90 000 lb/in² (620 MPa)		23 600	(160)

*The overload capacity of nitrided gears is low, since the shape of the effective *S-N* curve is flat. The sensitivity to shock should be investigated before proceeding with the design.

Source: AGMA 218.01

The formula for the allowable contact stress is

$$\sigma_{c,\text{all}} = \frac{S_c C_L C_H}{C_T C_R} \qquad (14\text{-}18)$$

where C_L = life factor
C_H = hardness-ratio factor
C_T = temperature factor
C_R = reliability factor

Note the use of K for bending factors and C for the contact-strength factors. It happens that the two factors are equal for some effects. But the use of separate symbols could be useful in the future if research reveals differing values.

Note also that the strength-modification factors in this section are subscripted with capital letters, while the stress modification factors in the previous section are subscripted with lowercase letters.

The factors in this section, too, will be evaluated in the sections to follow.

14-5 GEOMETRY FACTORS I AND J

We have seen how the factor Y is used in the Lewis equation to introduce the effect of tooth form into the stress equation. The AGMA factors I and J are intended to accomplish the same purpose in a more involved manner.

The determination of I and J depends upon the *face-contact ratio* m_F. This is defined as

$$m_F = \frac{F}{p_x} \qquad (14\text{-}19)$$

where p_x is the axial pitch and F is the face width. For spur gears, $m_F = 0$.

Low-contact-ratio (LCR) helical gears having a small helix angle or a thin face width, or both, have face-contact ratios less than unity ($m_F \leq 1$), and will not be

TABLE 14-4

AGMA Surface Fatigue Strength S_c

MATERIAL	AGMA CLASS	COMMERCIAL DESIGNATION	HEAT TREATMENT	MINIMUM HARDNESS AT SURFACE	S_c psi	S_c MPa
Steel	A-1 through A-5	—	Through-hardened and tempered	180 BHN and less	85–95 000	(590–660)
				240 BHN	105–115 000	(720–790)
				300 BHN	120–135 000	(830–930)
				360 BHN	145–160 000	(1000–1100)
				400 BHN	155–170 000	(1100–1200)
			Flame- or induction-hardened	50 HRC	170–190 000	(1200–1300)
				54 HRC	175–195 000	(1200–1300)
			Carburized and case-hardened	55 HRC	180–200 000	(1250–1400)
				60 HRC	200–225 000	(1400–1550)
		AISI 4140	Nitrided	48 HRC	155–180 000	(1100–1250)
		AISI 4340	Nitrided	46 HRC	150–175 000	(1050–1200)
		Nitralloy 135M	Nitrided	60 HRC	170–195 000	(1170–1350)
		2½% chrome	Nitrided	54 HRC	155–172 000	(1100–1200)
		2½% chrome	Nitrided	60 HRC	192–216 000	(1300–1500)
Cast iron	20		As cast	—	50–60 000	(340–410)
	30		As cast	175 BHN	65–75 000	(450–520)
	40		As cast	200 BHN	75–85 000	(520–590)
Nodular (ductile) iron	A-7-a	60-14-18	Annealed, quenched, and tempered	140 BHN	90–100% of S_c value of steel with same hardness	
	A-7-c	80-55-06		180 BHN		
	A-7-d	100-70-03		230 BHN		
	A-7-e	120-90-02		270 BHN		
Malleable iron (pearlitic)	A-8-c	45007	—	165 BHN	72 000	(500)
	A-8-e	50005	—	180 BHN	78 000	(540)
	A-8-f	53007	—	195 BHN	83 000	(570)
	A-8-i	80002	—	240 BHN	94 000	(650)
Bronze	Bronze 2	AGMA 2C	Sand-cast	Tensile strength minimum 40 000 lb/in^2 (275 MPa)	30 000	(205)
	Al/Br 3	ASTM B-148-52 alloy 9C	Heat-treated	Tensile strength minimum 90 000 lb/in^2 (620 MPa)	65 000	(450)

Source: AGMA 218.01

considered here. Such gears have a noise level not too different from that for spur gears. Consequently we shall consider here only spur gears with $m_F = 0$ and conventional helical gears with $m_F > 1$.

Bending-Strength Geometry Factor J

The AGMA factor J employs a modified value of the Lewis form factor, also denoted by Y; a *fatigue stress-concentration factor* K_f; and a tooth *load-sharing ratio* m_N. The resulting equation for J is

$$J = \frac{Y}{K_f m_N} \tag{14-20}$$

It is important to note that the form factor Y in Eq. (14-20) *is not* the Lewis factor at all. The value of Y here is obtained from a generated layout of the tooth profile in the normal plane and is based on the highest point of single-tooth contact.

The factor K_f in Eq. (14-20) is called a *stress correction factor* by AGMA. It is based on a formula deduced from a photoelastic investigation of stress concentration in gear teeth over 40 years ago.

The load-sharing ratio m_N is equal to the face width divided by the minimum total length of the lines of contact. This factor depends on the transverse contact ratio m_p, the face-contact ratio m_F, the effects of any profile modifications, and the tooth deflection. For spur gears, $m_N = 1.0$. For helical gears having a face-contact ratio $m_F > 2.0$, a conservative approximation is given by the equation

$$m_N = \frac{p_N}{0.95Z} \tag{14-21}$$

where p_N is the normal base pitch and Z is the length of the line of action in the transverse plane (distance L_{ab} in Fig. 13-15).

Use Fig. 14-4 to obtain the geometry factor J for spur gears having a $20°$ pressure angle and full-depth teeth. Use Figs. 14-5 and 14-6 for helical gears having a $20°$ normal pressure angle and face-contact ratios of $m_F = 2$ or greater. For other gears, consult the AGMA standard.

Surface-Strength Geometry Factor I

The factor I is also called the *pitting-resistance geometry factor* by AGMA. We begin by noting that the sum of the reciprocals of Eq. (14-12) can be expressed as

$$\frac{1}{r_1} + \frac{1}{r_2} = \frac{2}{\sin \phi_t} \left(\frac{1}{d_P} + \frac{1}{d_G} \right) \tag{a}$$

where we have replaced ϕ by ϕ_t, the transverse pressure angle, so that the relation will apply to helical gears too. Now define *speed ratio* m_G as

$$m_G = \frac{N_G}{N_P} = \frac{d_G}{d_P} \tag{14-22}$$

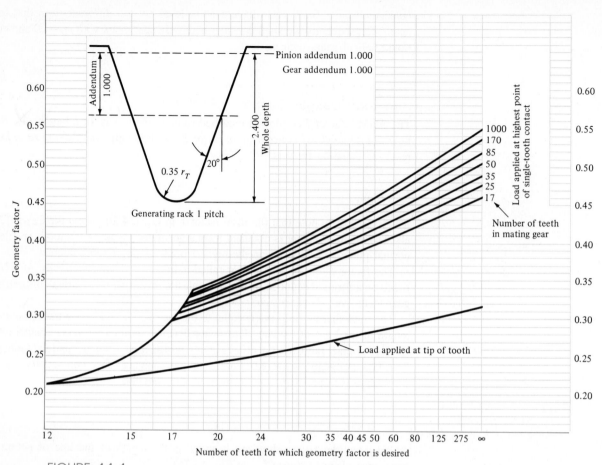

FIGURE 14-4

Spur-gear geometry factors J.
(Source: AGMA 218.01.)

Equation (a) can now be written

$$\frac{1}{r_1} + \frac{1}{r_2} = \frac{1}{d_P} \frac{1}{\sin \phi_t} \frac{m_G + 1}{m_G} \tag{b}$$

Now substitute Eq. (b) for the sum of the reciprocals in Eq. (14-14). The result is found to be

$$\sigma_c = -\sigma_C = C_P \left[\frac{W_t}{d_P F} \frac{1}{\dfrac{\cos \phi_t \sin \phi_t}{2} \dfrac{m_G}{m_G + 1}} \right]^{1/2} \tag{c}$$

The geometry factor I for external spur gears is the denominator of the second term in the brackets in Eq. (c). By adding the load-sharing ratio m_N, we obtain a factor valid

$$m_N = \frac{p_N}{0.95Z}$$

Value for Z is for an element of indicated numbers of teeth and a 75-tooth mate

Normal tooth thickness of pinion and gear tooth each reduced 0.024 in to provide 0.048 in total backlash for one normal diametral pitch

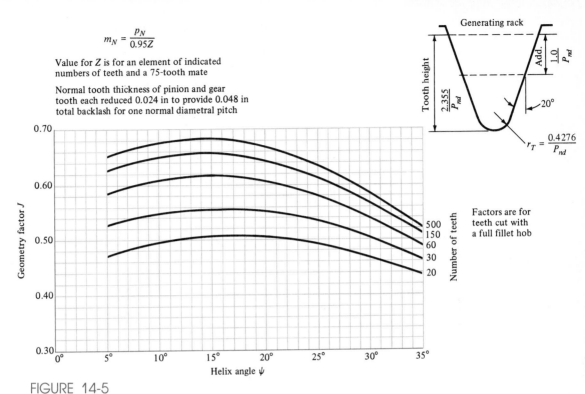

Factors are for teeth cut with a full fillet hob

FIGURE 14-5

Helical-gear geometry factors *J*.
(Source: AGMA 218.01.)

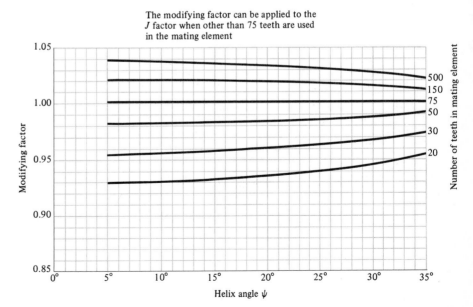

The modifying factor can be applied to the *J* factor when other than 75 teeth are used in the mating element

FIGURE 14-6

J-factor multipliers for use with Fig. 14-5. *(Source: AGMA 218.01.)*

for both spur and helical gears. The equation is then written as

$$
I = \begin{cases} \dfrac{\cos \phi_t \sin \phi_t}{2m_N} \dfrac{m_G}{m_G + 1} & \text{external gears} \\[4mm] \dfrac{\cos \phi_t \sin \phi_t}{2m_N} \dfrac{m_G}{m_G - 1} & \text{internal gears} \end{cases} \tag{14-23}
$$

where $m_N = 1$ for spur gears. In solving Eq. (14-21) for m_N, note that

$$
p_N = p_n \cos \phi_n \tag{14-24}
$$

where p_n is the normal circular pitch. If a layout of the gears is not made, the quantity Z, for use in Eq. (14-21), can be obtained from the equation

$$
Z = [(r_P + a)^2 - r_{bP}^2]^{1/2} + [(r_G + a)^2 - r_{bG}^2]^{1/2} - (r_P + r_G) \sin \phi_t \tag{14-25}
$$

where r_P and r_G are the pitch radii and r_{bP} and r_{bG} the base-circle radii.* The radius of the base circle is

$$
r_b = r \cos \phi_t \tag{14-26}
$$

Certain precautions must be taken in using Eq. (14-25). The tooth profiles are not conjugate below the base circle, and consequently, if either one or the other of the first two terms in brackets is larger than the third term, then it should be replaced by the third term. In addition, the effective outside radius is sometimes less than $r + a$, owing to removal of burrs or rounding of the tips of the teeth. When this is the case, always use the effective outside radius instead of $r + a$.

14-6 THE ELASTIC COEFFICIENT C_p

Values of C_p may be computed directly from Eq. (14-13) or obtained from Table 14-5.

14-7 DYNAMIC FACTORS C_V AND K_V

As noted earlier, dynamic factors are used to account for inaccuracies in the manufacture and meshing of gear teeth in action. *Transmission error* is defined as the departure from uniform angular velocity of the gear pair. Some of the effects which produce transmission error are:

- Inaccuracies produced in the generation of the tooth profile; these include errors in tooth spacing, profile lead, and runout
- Vibration of the tooth during meshing due to the tooth stiffness
- Magnitude of the pitch-line velocity

---

*For a development, see Joseph E. Shigley and John J. Uicker, Jr., *Theory of Machines and Mechanisms*, McGraw-Hill, New York, 1980, p. 262.

TABLE 14-5

Elastic Coefficient C_p

PINION MATERIAL	PINION MODULUS OF ELASTICITY E_P, lb/in² (MPa)*	GEAR MATERIAL AND MODULUS OF ELASTICITY E_G, lb/in² (MPa)*					
		STEEL 30×10^6 (2×10^5)	MALLEABLE IRON 25×10^6 (1.7×10^5)	NODULAR IRON 24×10^6 (1.7×10^5)	CAST IRON 22×10^6 (1.5×10^5)	ALUMINUM BRONZE 17.5×10^6 (1.2×10^5)	TIN BRONZE 16×10^6 (1.1×10^5)
Steel	30×10^6 (2×10^5)	2300 (191)	2180 (181)	2160 (179)	2100 (174)	1950 (162)	1900 (158)
Malleable iron	25×10^6 (1.7×10^5)	2180 (181)	2090 (174)	2070 (172)	2020 (168)	1900 (158)	1850 (154)
Nodular iron	24×10^6 (1.7×10^5)	2160 (179)	2070 (172)	2050 (170)	2000 (166)	1880 (156)	1830 (152)
Cast iron	22×10^6 (1.5×10^5)	2100 (174)	2020 (168)	2000 (166)	1960 (163)	1850 (154)	1800 (149)
Aluminum bronze	17.5×10^6 (1.2×10^5)	1950 (162)	1900 (158)	1880 (156)	1850 (154)	1750 (145)	1700 (141)
Tin bronze	16×10^6 (1.1×10^5)	1900 (158)	1850 (154)	1830 (152)	1800 (149)	1700 (141)	1650 (137)

Poisson's ratio = 0.30.

*When more exact values for modulus of elasticity are obtained from roller contact tests, they may be used.

Source: AGMA 218.01.

- Dynamic unbalance of the rotating members
- Wear and permanent deformation of contacting portions of the teeth
- Gearshaft misalignment and the linear and angular deflection of the shaft
- Tooth friction

In an attempt to gain some control over these effects, AGMA has defined a set of *quality-control numbers*.* These numbers define the tolerances for gears of various sizes manufactured to a specified quality class. Classes 3 to 7 will include most commercial-quality gears. Classes 8 to 12 are of precision quality. The AGMA *transmission accuracy-level number* Q_v can be taken as the same as the quality number. The following equations for the dynamic factor are based on these Q_v numbers:

$$C_v = K_v = \begin{cases} \left(\dfrac{A}{A + V^{1/2}} \right)^B & V \text{ in ft/min} \\ \left[\dfrac{A}{A + (200V)^{1/2}} \right]^B & V \text{ in m/s} \end{cases}$$

(14-27)

*AGMA 390.01.

where $A = 50 + 56(1 - B)$ (14-28)
            $B = (12 - Q_v)^{2/3}/4$ (14-29)

The values obtained from Eq. (14-27) are plotted in Fig. 14-7 for a range of useful Q_v values. The end of each curve corresponds to the maximum pitch-line velocity permitted for the given accuracy level.

14-8   APPLICATION FACTORS C_a AND K_a

The purpose of the application factor is to compensate for the fact that situations arise where the actual load exceeds the nominal tangential load W_t. See, for example, Table 1-2. Application factors are generally assigned on the basis of the opinion of the engineering designer.*

FIGURE 14-7

Dynamic factors C_v and K_v.
(Source: AGMA 218.01.)

---

*An extensive list of application factors is to be found in Howard B. Schwerdlin, ''Couplings,'' chap. 29 in Joseph E. Shigley and Charles R. Mischke (eds.), *Standard Handbook of Machine Design*, McGraw-Hill, New York, 1986.

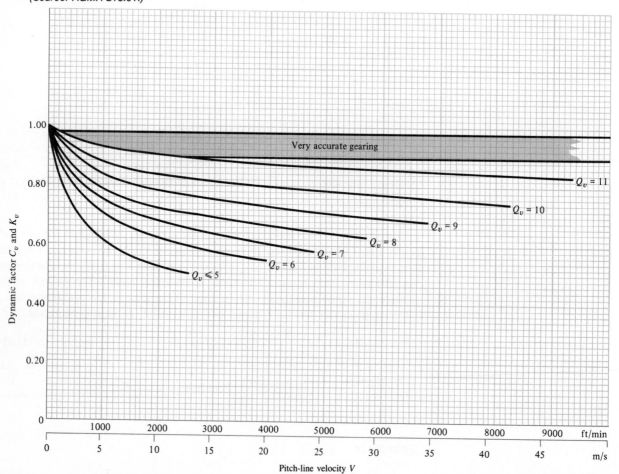

14-9   SURFACE-CONDITION FACTOR C_f

AGMA has not yet established values for the surface factor C_f, but suggests values greater than unity be used when obvious surface defects are present.

14-10   SIZE FACTORS C_s AND K_s

The AGMA recommendation is to use a size factor of unity "for most gears provided a proper choice of steel is made for the size of the part and the heat treatment and hardening process." The original intent of the size factor is to account for any nonuniformity of the material properties. The presence of the size factor in the formulas for stress is an important reminder that the effect should always be evaluated. When such effects are deemed to be present, then a factor greater than unity should be used.

14-11   LOAD-DISTRIBUTION FACTORS C_m AND K_m

The load-distribution factor is used to account for

• Misalignment of rotational axes for any reason

• Deviations in lead

• Load-caused elastic deflections of shafts, bearings, and/or housing

In the latest standard (218.01, December 1982), AGMA presents two methods, one empirical and the other analytical, of obtaining values for the load-distribution factor. Both of these are quite lengthy and are supported by many new definitions. AGMA acknowledges that the empirical method gives results similar to those obtained in previous standards. For this reason we shall use a simpler approach for values of C_m and K_m, that of Table 14-6, obtained from a previous standard.

TABLE 14-6

Load-Distribution Factors C_m and K_m for Spur and Helical Gears. (Helical-Gear Values are Shown in Brackets)

CONDITION OF SUPPORT	FACE WIDTH F, in (mm)			
	≤2(50)	6(150)	9(225)	≥16(400)
Accurate mounting, low bearing clearances, minimum deflections, precision gears	1.3 [1.2]	1.4 [1.3]	1.5 [1.4]	1.8 [1.7]
Less rigid mountings, less accurate gears, contact across full face	1.6 [1.5]	1.7 [1.6]	1.8 [1.7]	2.0 [2.0]
Accuracy and mounting such that less than full-face contact exists	>2.0 [>2.0]			

Source: AGMA 225.01, December 1967.

14-12

HARDNESS-RATIO FACTOR C_H

The pinion generally has a smaller number of teeth than the gear and consequently is subjected to more cycles of contact stress. If both the pinion and the gear are through-hardened, then a uniform surface strength can be obtained by making the pinion harder than the gear. A similar effect can be obtained when a surface-hardened pinion is mated with a through-hardened gear. The hardness-ratio factor C_H is used *only for the gear*. Its purpose is to adjust the surface strengths for this effect. The values of C_H are obtained from the equation

$$C_H = 1.0 + A(m_G - 1.0) \tag{14-30}$$

where

$$A = 8.98(10^{-3}) \left(\frac{H_{BP}}{H_{BG}} \right) - 8.29(10^{-3}) \tag{14-31}$$

The terms H_{BP} and H_{BG} are the Brinell hardnesses (10-mm ball at 3000-kg load) of the pinion and gear, respectively. The term m_G is the speed ratio and is given by Eq. (14-22). Equation (14-30) is valid only when $(H_{BP}/H_{BG}) \leq 1.70$.

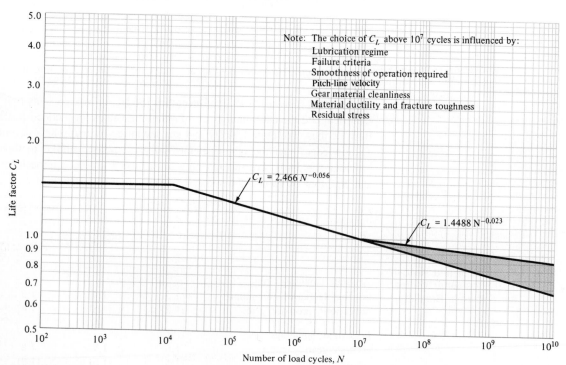

FIGURE 14-8

Pitting-resistance life factor C_L for steel gears. *(Source: AGMA 218.01.)*

14-13   LIFE FACTORS C_L AND K_L

The AGMA strengths as given in Figs. 14-2 and 14-3 and in Tables 14-3 and 14-4 are based on 10^7 tooth-load cycles, defined as "the number of mesh contacts under load." The purpose of the tooth-life factors is to modify the AGMA strengths for lives other than 10^7. Values for these factors are given in Figs. 14-8 and 14-9. Note that for 10^7 cycles, $C_L = K_L = 1.0$ on each graph.

14-14   RELIABILITY FACTORS C_R AND K_R

The AGMA strengths presented in this chapter are all based on a reliability $R = 0.99$ corresponding to 10^7 cycles of life. For other reliabilities, use Table 14-7. AGMA notes that values other than these may be used if supported by statistical data. Be aware that at $R = 0.90$, yielding of the tooth may occur rather than pitting. Note that Table 14-7 must be interpolated logarithmically.

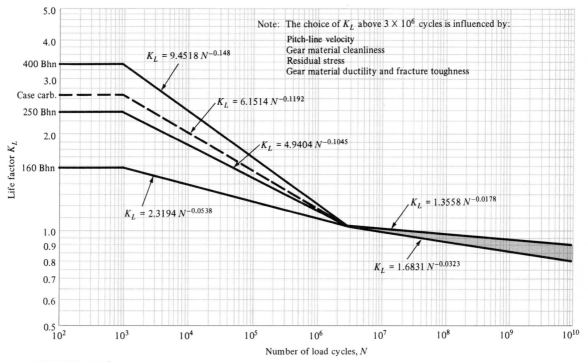

FIGURE 14-9

Bending-stength life factor K_L for steel gears. (Source: AGMA 218.01.)

TABLE 14-7

AGMA Reliability Factors
C_R and K_R

RELIABILITY	C_R, K_R
0.90	0.85
0.99	1.00
0.999	1.25
0.9999	1.50

$$C_R = \begin{cases} 0.7 - 0.15 \log (1 - R) & 0.9 \le R < 0.99 \\ 0.5 - 0.25 \log (1 - R) & 0.99 \le R < 0.9999 \end{cases}$$

14-15   TEMPERATURE FACTORS C_T AND K_T

For oil or gear-blank temperatures up to 250°F (120°C), use $C_T = K_T = 1.0$. For higher temperatures, these factors should be greater than unity.

PROBLEMS*

Section 14-1

14-1   A steel spur pinion has a pitch of 6 teeth/in, 22 full-depth teeth, and a 20° pressure angle. This pinion runs at a speed of 1200 rev/min and transmits 15 hp to a 60-tooth gear. If the face width is 2 in, estimate the bending stress.

14-2   A steel spur pinion has a diametral pitch of 12 teeth/in, 16 teeth cut full-depth with a 20° pressure angle, and a face width of $\frac{3}{4}$ in. This pinion is expected to transmit 1.5 hp at a speed of 700 rev/min. Determine the bending stress.

14-3   A steel spur pinion has a module of 1.25 mm, 18 teeth cut on the 20° full-depth system, and a face width of 12 mm. At a speed of 1800 rev/min, this pinion is expected to carry a steady load of 0.5 kW. Determine the resulting bending stress.

14-4   A spur pinion has 15 teeth cut on the 20° full-depth system with a module of 5 mm and a face width of 60 mm. The pinion rotates at 200 rev/min and transmits 5 kW to the mating gear. What is the resulting bending stress?

14-5   A spur pinion has a module of 1 mm and 16 teeth cut on the 20° full-depth system and is to carry 0.15 kW at 400 rev/min. Determine a suitable face width based on an allowable bending stress of 150 MPa.

14-6   A 20° full-depth spur pinion has 17 teeth and a module of 1.5 mm and is to transmit 0.25 kW at a speed of 400 rev/min. Find an appropriate face width if the bending stress is not to exceed 75 MPa.

14-7*   A 20° full-depth spur pinion is to transmit 1.5 kW at a speed of 900 rev/min. If the pinion has 18 teeth, determine suitable values for the module and face width. The bending stress should not exceed 75 MPa.

14-8*   A pinion gear is to be designed to transmit 3.5 kW at a speed of 1200 rev/min. Based on 20° full-depth teeth, 19 teeth, and an allowable bending stress of 70 MPa, find suitable values for the face width and module.

*An asterisk indicates a design-type problem.

14-9   Estimate the power rating in kW of a 20° full-depth spur pinion having a module of 4 mm, 20 teeth, and a face width of 50 mm if the speed is 1000 rev/min and the maximum allowable bending stress is 62.5 MPa.

14-10   A 20° full-depth spur pinion has a module of 6 mm, 21 teeth, and a face width of 75 mm, and a bending stress of 60 MPa is allowed. What is the power capacity if the speed is 800 rev/min?

14-11   A 20° full-depth spur pinion has a diametral pitch of 5 teeth/in and 24 teeth and transmits 6 hp at a speed of 50 rev/min. Find an appropriate face width if the allowable bending stress is 20 kpsi.

14-12   A spur pinion is to transmit 15 hp at a speed of 600 rev/min. The pinion is cut on the 20° full-depth system and has a diametral pitch of 5 teeth/in and 16 teeth. Find a suitable face width based on an allowable stress of 10 kpsi.

14-13* A spur pinion is to transmit 5 hp at a speed of 1000 rev/min with a bending stress not to exceed 8 kpsi. Find suitable values for the diametral pitch and face width, using 17 teeth.

14-14* A 20° full-depth spur pinion with 18 teeth is to transmit 2.5 hp at a speed of 600 rev/min. Determine appropriate values for the face width and diametral pitch based on an allowable bending stress of 10 kpsi.

14-15   Determine the horsepower rating of a 20° full-depth spur pinion having 18 teeth, a diametral pitch of 16 teeth/in, and a face width of 0.75 in. The speed is 1000 rev/min, and the allowable bending stress is 10 kpsi.

14-16   A 20° full-depth spur pinion has a diametral pitch of 20 teeth/in, 20 teeth, and a face width of 0.50 in and rotates at 900 rev/min. What is the horsepower capacity if the allowable bending stress is determined to be 10 400 psi?

Section 14-2   14-17   A speed reducer has 20° full-depth teeth and consists of a 22-tooth steel spur pinion driving a 60-tooth cast-iron gear. The horsepower transmitted is 15 at a pinion speed of 1200 rev/min. For a diametral pitch of 6 teeth/in and a face width of 2 in, find the contact stress.

14-18   A gear drive consists of a 16-tooth 20° spur pinion and a 48-tooth cast-iron gear having a pitch of 12 teeth/in. For a power input of 1.5 hp at a pinion speed of 700 rev/min, select a face width based on an allowable contact stress of 100 kpsi.

14-19   A gearset has a diametral pitch of 5 teeth/in, a 20° pressure angle, and a 24-tooth cast-iron spur pinion driving a 48-tooth cast-iron gear. The pinion is to rotate at 50 rev/min. What horsepower input can be used with this gearset if the contact stress is limited to 100 kpsi? and $F = 2.5$ in?

14-20   A 20° 20-tooth cast-iron spur pinion having a module of 4 mm drives a 32-tooth cast-iron gear. Find the contact stress if the pinion speed is 1000 rev/min, the face width is 50 mm, and 10 kW of power is transmitted.

14-21* A gearset is to consist of a 19-tooth steel spur pinion driving a 30-tooth cast-iron gear. The teeth are cut on the 20° full-depth system. Determine suitable values for the module and face width corresponding to a power input of 3.5 kW at a pinion speed of 1200 rev/min and a maximum contact stress of 600 MPa.

14-22   A gear drive consists of a 21-tooth cast-iron spur pinion rotating at 800 rev/min and driving a

44-tooth cast-iron gear. The drive has a pressure angle of 20°, a face width of 75 mm, and a module of 6 mm. For an allowable contact stress of 480 MPa, estimate the maximum power transmission capability in kilowatts.

Sections 14-3 to 14-15†

14-23   A gear drive consists of a 20° steel spur pinion having 16 teeth driving a 48-tooth cast-iron gear. The pinion speed is 300 rev/min, the face width 2 in, and the diametral pitch 6 teeth/in. The gears are made to No. 7 quality standards and are to be accurately and rigidly mounted. Determine the AGMA contact stress if 5 hp is to be transmitted.

14-24   An 18-tooth 20° steel spur pinion drives a 64-tooth cast-iron gear and transmits 4 hp at a pinion speed of 600 rev/min. This drive has a face width of 1.5 in and a diametral pitch of 8 teeth/in and is manufactured to a quality standard of 8. Find the AGMA contact stress based on a load-distribution factor of 1.6.

14-25   A 20°–pressure angle steel spur pinion with 20 teeth and a module of 1.5 mm transmits 120 W to a 36-tooth cast-iron gear. The pinion speed is 100 rev/min, and the gears are manufactured to a No. 6 quality standard and installed with less-than-perfect mounting rigidity, though contact is across the full face. Find the AGMA contact stress if the face width is 18 mm.

14-26   A 20° spur gearset having a module of 3 mm is to transmit 3 kW at a pinion speed of 1200 rev/min. Pinion and gear materials are of cast iron, with tooth numbers $N_P = 21$ and $N_G = 26$ cut to a quality standard of No. 7. For a face width of 36 mm and average mounting rigidity, find the AGMA contact stress.

14-27   A 17-tooth 20°–pressure angle spur pinion rotates at 1800 rev/min and transmits 4 hp to a 52-tooth gear. The diametral pitch is 10, the face width is 1.25 in, the quality standard is No. 6, and the mounting is with exceptional rigidity. The pinion is steel, hardened to Brinell 240, and the gear is of No. 30 cast iron. Find the factor of safety based on surface fatigue strength if the life is to be no more than 10^6 cycles corresponding to a reliability of 90 percent.

14-28   Determine the factor of safety based on contact stress for a life of 10^7 cycles and a reliability of 99 percent for a gearset having the following specifications:
Pinion. $n = 900$ rev/min, 15 teeth, 300 Bhn steel, $\phi = 20°$, $H = 15$ hp, $F = 2.5$ in, $P = 5$ teeth/in, $Q_v = 6$, and $C_m = 1.4$
Gear. 30 teeth, 240 Bhn steel.

14-29   Find the factor of safety based on surface fatigue failure for a life of 10^7 cycles and a reliability of 99.9 percent for the following 20° gearset:
Pinion. 500 rev/min, 19 teeth, 360 Bhn steel, $H = 6$ kW, $F = 50$ mm, $m = 4$ mm, $Q_v = 9$, and $C_m = 1.3$
Gear. 25 teeth, 300 Bhn steel

14-30   A 20°–pressure angle spur pinion rotates at 400 rev/min; has 22 teeth; is of AGMA No. 30 cast iron; has a face width of 32 mm, a module of 2.5 mm, and a quality level of 5 with average mounting conditions; and is to transmit 1 kW. The gear is of AGMA No. 30 cast iron and has 32 teeth. Find the factor of safety of this gearset based on contact stress if the life is to be 10^5 cycles at a reliability of 90 percent.

14-31   The 20°–pressure angle spur gears in a speed reducer are specified as follows:

---

†Unless specific information is supplied, the life, reliability, and temperature factors are taken as unity.

Pinion. 1200 rev/min, 16 teeth, 240 Bhn steel, $F = 0.5$ in, $H = 0.5$ hp, $P_d = 20$ teeth/in, $Q_v = 6$, $K_m = 1.3$, $K_a = K_s = 1$

Gear. 56 teeth, cast iron

Find the AGMA bending stress in the pinion and in the gear teeth.

14-32   A spur-gear drive consists of a 20-tooth 20° full-depth, cast-iron pinion driving a 72-tooth cast-iron gear having a face width of 1 in and a diametral pitch of 12 teeth/in. The power input is $\frac{3}{4}$ hp to the drive at a pinion speed of 200 rev/min. Find the AGMA bending stresses in the gear and pinion using $K_a = K_s = 1$, $K_m = 1.3$, and $Q_v = 7$.

14-33   An 18-tooth steel spur pinion rotates at 900 rev/min and transmits 25 kW to a 96-tooth cast-iron gear. The gears have a 20° pressure angle, a face width of 100 mm, and a module of 8 mm. Find the AGMA bending stress in both gears, using an application factor of unity, a size factor of 1.1, a load-distribution factor of 2.0, and a quality level of 7.

14-34   A speed reducer with a gear ratio of 4.42 has 19 teeth on the pinion, which runs at 600 rev/min and transmits 4 kW to the gear. The pinion has a pressure angle of 20°, a face width of 40 mm, and a module of 3 mm and is made of steel. The gear is of cast iron. Determine the AGMA bending stress in the pinion and gear based on $Q_v = 6$, $K_m = 1.3$, and $K_a = K_s = 1$.

14-35†   A 20° spur-gear drive is built to a quality standard of 6 and has a fairly rigid support for the shafts. Other specifications are:

Pinion. 17 teeth, 1400 rev/min, 240 Bhn steel, 12.5 hp, 2-in face, 6 teeth/in diametral pitch

Gear. 120 teeth, AGMA No. 30 cast iron

Based on bending strength, what is the factor of safety of this drive?

14-36   A 16-tooth 240 Bhn steel spur pinion having a pressure angle of 20° and rotating at 720 rev/min drives a 28-tooth 240 Bhn steel gear having a 2.5-in face and a diametral pitch of 5 teeth/in. Find the factor of safety of this drive based on bending strength if 17.5 hp is transmitted. Use $K_a = K_s = 1$, $K_m = 1.6$, and $Q_v = 7$.

14-37   A 19-tooth 300 Bhn steel spur pinion transmits 15 kW at a pinion speed of 360 rev/min to a 77-tooth AGMA No. 30 cast-iron gear. The face width is 75 mm, the pressure angle 20°, and the module 6 mm. Find the factor of safety of this drive based on bending if $K_a = 1$, $K_s = 1.1$, $K_m = 1.6$, and $Q_v = 6$.

14-38   A 20° spur-gear drive has a speed ratio of 1.1 with a 20-tooth pinion rotating at 1600 rev/min and transmitting 50 kW to the gear. Both gears are of No. 30 cast iron and have a module of 10 mm with a face width of 125 mm. With relatively poor mounting conditions, but a quality level of 7 and a size factor of 1.2, find the factor of safety of this drive based on the AGMA bending equations.

14-39   A 20-tooth helical pinion has a normal pressure angle of 20° and a helix angle of 25°, rotates at 1800 rev/min, and transmits 4 hp to a 50-tooth gear. The normal diametral pitch is 10, the face width 1.5 in, the quality standard 6, and the mounting with exceptional rigidity. Pinion and gear are both of steel, hardened to $H_B = 240$. Find the factor of safety based on surface fatigue strength if the life is to be no more than 10^6 cycles corresponding to a reliability of 90 percent.

---

†Unless otherwise specified, the K factors for bending strength should be taken as unity.

14-40   The same gears, operating conditions, and life requirements as in Prob. 14-39, find the factor of safety based on the AGMA bending strength.

14-41   A 20-tooth 240 Bhn steel helical pinion having a normal pressure angle of 20° and rotating at 720 rev/min drives a 30-tooth 180 Bhn steel gear having a 3-in face. The normal diametral pitch of this pair of gears is 5 teeth/in, and the helix angle is 30°. Find the factor of safety of this drive based on AGMA bending strength with a horsepower input of 17.5. Use $K_a = K_s = 1$, $K_m = 1.5$, and $Q_v = 7$.

14-42   For the same gears, conditions, and requirements as in Prob. 14-41, find the factor of safety based on the AGMA contact strength for $L = 10^7$ cycles and $R = 0.99$.

14-43   For an AISI 4340 steel, Poisson's ratio was found to be $v \sim N(0.287, 0.0072)$. Young's modulus for the same steel was found to be $E \sim N(29.0, 0.73)10$ Mpsi. Write a computer simulation program to find the mean and standard deviation of C_p from Eq. (14-13), and compare the mean to the entry in Table 14-5.

ANSWERS   **14-1** $d = 3.67$ in, $V = 1152$ ft/min, $W_t = 430$ lb, $K_V = 0.510$, $\sigma = 7.63$ kpsi
14-4 $d = 75$ mm, $V = 0.785$ m/s, $W_t = 6.37$ kN, $K_V = 0.886$, $\sigma = 82.6$ MPa
14-8 $m = 2.5$ mm, $F = 32$ mm
14-10 $H = 25$ kW
14-14 $F = 1.25$ in, $P = 10$ teeth/in
14-17 $d_P = 3.67$ in, $d_G = 10$ in, $V = 1152$ ft/min, $C_V = 0.510$, $W_t = 430$ lb, $\sigma_C = -65.5$ kpsi
14-22   25 kW
14-24   85 kpsi
14-27   1.24
14-29   Based on gear, $n = 0.976$
14-31   Pinion, $\sigma = 10.5$ kpsi; gear, $\sigma = 15.4$ kpsi
14-34   Pinion, $\sigma = 91.7$ MPa; gear, $\sigma = 69.8$ MPa
14-37   Pinion, $n = 2.36$; gear, $n = 0.82$ (failure)
14-39 $n = 3.25$
14-42 $n = 1.73$

15
Bevel and
Worm Gears

The American Gear Manufacturers Association (AGMA) has established standards for the analysis and design of the various kinds of bevel and worm gears.* Chapter 14 was an introduction to the AGMA methods for spur and helical gears. AGMA has established similar methods for other types of gearing, which all follow the same general approach.

15-1 BEVEL GEARING—GENERAL

Bevel gears may be classified as follows:

• Straight bevel gears

• Spiral bevel gears

• Zerol bevel gears

• Hypoid gears

• Spiroid gears

A straight bevel gear was illustrated in Fig. 13-34. These gears are usually used for pitch-line velocities up to 1000 ft/min (5 m/s) when the noise level is not an important consideration. They are available in many stock sizes and are less expensive to produce than other bevel gears, especially in small quantities.

A *spiral bevel* gear is shown in Fig. 15-1; the definition of the *spiral angle* is illustrated in Fig. 15-2. These gears are recommended for higher speeds and where the noise level is an important consideration. Spiral bevel gears are the bevel counterpart of the helical gear; it can be seen in Fig. 15-1 that the pitch surfaces and the nature of

---

*Write to AGMA, 1500 King Street, Suite 201, Alexandria, VA 22314 for a listing of current standards.

FIGURE 15-1

Spiral bevel gears. *(Courtesy of Gleason Works, Rochester, N.Y.)*

contact are the same as for straight bevel gears except for the differences brought about by the spiral-shaped teeth.

The *Zerol bevel gear* is a patented gear having curved teeth but with a zero spiral angle. The axial thrust loads permissible for Zerol bevel gears are not as large as those for the spiral bevel gear, and so they are often used instead of straight bevel gears. The Zerol bevel gear is generated using the same tool as for regular spiral bevel gears. For design purposes, use the same procedure as for straight-bevel gears and then simply substitute a Zerol bevel gear.

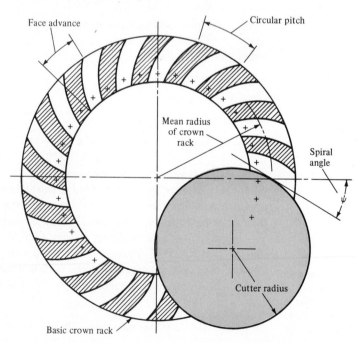

FIGURE 15-2

Cutting spiral-gear teeth on the basic crown rack.

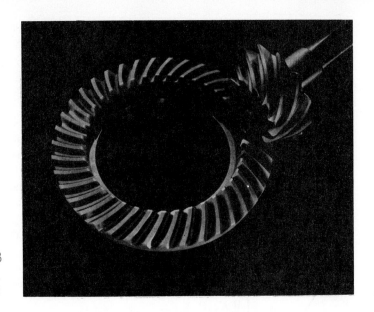

FIGURE 15-3

Hypoid gears. *(Courtesy of Glea-son Works, Rochester, N.Y.)*

It is frequently desirable, as in the case of automotive differential applications, to have gearing similar to bevel gears but with the shafts offset. Such gears are called *hypoid gears,* because their pitch surfaces are hyperboloids of revolution. The tooth action between such gears is a combination of rolling and sliding along a straight line and has much in common with that of worm gears. Figure 15-3 shows a pair of hypoid gears in mesh.

Figure 15-4 is included to assist in the classification of spiral bevel gearing. It is

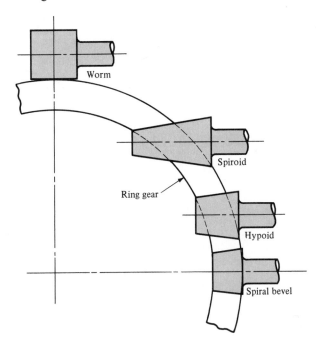

FIGURE 15-4

Comparison of intersecting- and offset-shaft bevel-type gearings. *(By permission, from Gear Handbook, McGraw-Hill, New York, 1962, p. 2-24.)*

seen that the hypoid gear has a relatively small shaft offset. For larger offsets, the pinion begins to resemble a tapered worm and the set is then called *spiroid gearing*.

15-2 BEVEL-GEAR STRESSES

In a typical bevel-gear mounting, Fig. 13-35, for example, one of the gears is often mounted outboard of the bearings. This means that the shaft deflections can be more pronounced and can have a greater effect on the nature of the tooth contact. Another difficulty which occurs in predicting the stress in bevel-gear teeth is the fact that the teeth are tapered. Thus, to achieve perfect line contact passing through the cone center, the teeth ought to bend more at the large end than at the small end. To obtain this condition requires that the load be proportionately greater at the large end. Because of this varying load across the face of the tooth, it is desirable to have a fairly short face width.

The AGMA equation for bending stress in spur and helical gears retains its form for bevel gears, too, and is repeated here for convenience:

$$\sigma = \begin{cases} \dfrac{W_t K_a}{K_v} \dfrac{P}{F} \dfrac{K_s K_m}{J} \\[2em] \dfrac{W_t K_a}{K_v} \dfrac{1}{Fm} \dfrac{K_s K_m}{J} \end{cases} \tag{15-1}$$

where the first version uses U.S. customary units and the second SI units.

Caution: The transmitted load W_t must be computed using the pitch radius at the

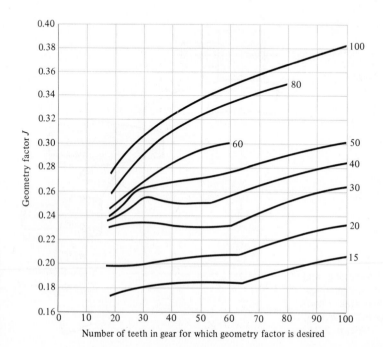

FIGURE 15-5

Geometry factors J for straight bevel gears; these are for a 90° shaft angle, a 20° pressure angle, and a clearance of $c = 0.240/P$ in. (AGMA 225.1.)

TABLE 15-1

Approximate Bevel-Gear Load-
Distribution Factors K_m
and C_m

APPLICATION	BOTH GEARS INBOARD	ONE GEAR OUTBOARD	BOTH GEARS OUTBOARD
General industrial	1.00–1.10	1.10–1.25	1.25–1.40
Automotive	1.00–1.10	1.10–1.25	
Aircraft	1.00–1.25	1.10–1.40	1.25–1.50

Source: AGMA Information Sheet 225.01, 1967, Table 4.

large end of the teeth in Eq. (15-1). Note that this is not the same transmitted load used in force analysis (Sec. 13-15), though the symbol is the same.

The geometry factor J is different for bevel gears, because the long-and-short addendum system is used and because the teeth are tapered. Use Fig. 15-5. The modification and correction factors used for spur and helical gears can also be used for bevel gears except for the load-distribution factor, which is given in Table 15-1. The AGMA strengths are also different. These are given in Table 15-2.

Equation (14-16), for AGMA contact stress in spur and helical gears, can be used for bevel gears too, with changes only in the elastic coefficient and the geometry factor. The equation is

$$\sigma_c = C_P \left(\frac{W_t C_a}{C_v} \frac{C_s}{Fd} \frac{C_m C_f}{I} \right)^{1/2} \tag{15-2}$$

where all values apply to the large end of the teeth. For bevel-gear teeth, the elastic coefficient C_p is obtained from a Hertzian stress analysis of contacting spheres. The resulting values are given in Table 15-3.

TABLE 15-2

AGMA Bending and Surface
Fatigue Strengths S_t and S_c
for Bevel Gears

MATERIAL	CONDITION	MINIMUM HARDNESS	BENDING STRENGTH S_t, kpsi (MPa)	SURFACE STRENGTH* S_c, kpsi (MPa)
Steel	Normalized	140 Bhn	11.0 (76)	
	Q&T	180 Bhn	14.0 (96)	85 (586)
	Q&T	300 Bhn	19.0 (131)	120 (827)
	Q&T	450 Bhn	25.0 (172)	145 (1000)
	Case carb.	55 R_c	27.5 (189)	180 (1240)
	Case carb.	60 R_C	30.0 (207)	200 (1380)
Cast iron	AGMA #20		2.7 (19)	50 (345)
	AGMA #30	175 Bhn	4.6 (32)	65 (448)
	AGMA #40	200 Bhn	7.0 (48)	75 (517)
Nodular iron:				
60-40-18	Annealed	165 Bhn	8.0 (55)	75 (517)
100-70-03	Normalized	210 Bhn	14.0 (96)	88 (606)
120-90-02	Q&T	255 Bhn	18.5 (127)	94 (648)
Bronze	10–12% tin	S_{ut} = 40 kpsi	3.0 (21)	30 (207)
Aluminum bronze	9C-H.T.	S_{ut} = 90 kpsi	12.0 (83)	65 (448)

*Minimum of a range of values.

Source: AGMA 215.01, 225.01.

TABLE 15-3

Values of the Elastic Coefficient C_p in Units of $\sqrt{\text{psi}}$ ($\sqrt{\text{MPa}}$) for Bevel Gears and Others with Localized Contact*

PINION MATERIAL	MODULUS OF ELASTICITY	GEAR MATERIAL			
		STEEL	CAST IRON	ALUMINUM BRONZE	TIN BRONZE
Steel	30 Mpsi	2800	2450	2400	2350
	(207 GPa)	(232)	(203)	(199)	(195)
Cast iron	19 Mpsi	2450	2250	2200	2150
	(131 GPa)	(203)	(187)	(183)	(178)
Aluminum bronze	17.5 Mpsi	2400	2200	2150	2100
	(121 GPa)	(199)	(183)	(178)	(174)
Tin bronze	16 Mpsi	2350	2150	2100	2050
	(110 GPa)	(195)	(178)	(174)	(170)

*$\nu = 0.30$.

Source: AGMA 212.02.

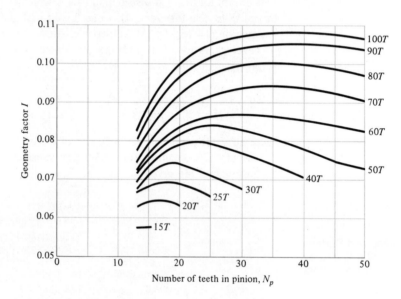

FIGURE 15-6

Geometry factors *I* for straight bevel gears of 20° pressure angle mounted at a 90° shaft angle. *(AGMA 212.02.)*

The geometry factor *I* for straight bevel gears is obtained from Fig. 15-6.*

EXAMPLE 15-1

A pair of miter gears listed in a catalog have a diametral pitch of 5, 25 teeth, a 1.10-in face width, and a 20° pressure angle and are made of a 20-point carbon steel, case-hardened to 55 R_C. The gears are intended for general industrial use and probably are made to a quality standard of $Q_v = 5$. It is quite likely that some applications will require outboard mounting.

*The stress analysis of other types of bevel gears may be found in Theodore J. Krenzer and Robert G. Hotchkiss, "Bevel and Hypoid Gears," chap. 34 in Joseph E. Shigley and Charles R. Mischke (eds.), *Standard Handbook of Machine Design,* McGraw-Hill, New York, 1986. This reference contains a new, comprehensive, and previously unpublished presentation on hypoid gears.

(a) For a speed of 600 rev/min, find the power rating in horsepower based on the AGMA bending strength.

(b) For the same speed as in part (a), find the rating based on the surface durability.

Solution (a) The pitch diameter at the large end of the teeth is $d = 25/5 = 5$ in. The corresponding pitch-line velocity is

$$V = \frac{\pi dn}{12} = \frac{\pi(5)(600)}{12} = 785 \text{ ft/min}$$

From Eqs. (14-28) and (14-29), we find

$$B = \frac{(12 - Q_v)^{2/3}}{4} = \frac{(12 - 5)^{2/3}}{4} = 0.9148$$

$$A = 50 + 56(1 - B) = 50 + 56(1 - 0.9148) = 54.77$$

Then, with Eq. (14-27), we find the dynamic factors to be

$$C_v = K_v = \left(\frac{A}{A + V^{1/2}}\right)^B = \left[\frac{54.77}{54.77 + (785)^{1/2}}\right]^{0.9148} = 0.685$$

From Fig. 15-5, we find the geometry factor to be $J = 0.21$. We also select $K_a = K_s = 1$ and choose $K_m = 1.40$ from Table 15-1. Equation (15-1) now gives

$$\sigma = \frac{W_t K_a}{K_v} \frac{P}{F} \frac{K_s K_m}{J} = \frac{W_t(1)}{0.685} \frac{5}{1.1} \frac{1(1.4)}{0.21} = 44.2 W_t \quad \text{psi} \tag{1}$$

The next step is to determine the allowable stress. For this we use Eq. (14-17), with $K_L = 1$, $K_T = 1$, and $K_R = 1$. From Table 15-2 we find $S_t = 14.0$ kpsi for 180 Bhn steel. Substituting in Eq. (14-17) gives

$$\sigma_{\text{all}} = \frac{S_t K_L}{K_T K_R} = \frac{14.0(1)}{1(1)} = 14.0 \text{ kpsi}$$

The rated transmitted load at the large end of the teeth, from Eq. (1), is, therefore,

$$W_t = \frac{14.0(10^3)}{44.2} = 317 \text{ lb}$$

and so the rated power based on bending is

Answer $$H = \frac{W_t V}{33\ 000} = \frac{317(785)}{33\ 000} = 7.54 \text{ hp}$$

(b) From Table 15-2, we find $S_c = 85$ kpsi. Using Eq. (14-18), with $C_L = C_H = C_T = C_R = 1$, gives the allowable contact stress as

$$\sigma_{c,\text{all}} = \frac{S_c C_L C_H}{C_T C_R} = \frac{85(1)(1)}{1(1)} = 85 \text{ kpsi} \tag{2}$$

From Fig. 15-6 we find the geometry factor as $I = 0.065$. Choosing $C_a = C_s = C_f = 1$, and $C_m = 1.4$ (as before), Eq. (15-2) yields

$$\sigma_c = C_P \left(\frac{W_t C_a}{C_v} \frac{C_s}{Fd} \frac{C_m C_f}{I} \right)^{1/2} = 2800 \left[\frac{W_t(1)}{0.685} \frac{1}{5(1.10)} \frac{1.4(1)}{0.065} \right]^{1/2}$$

$$= 6695\sqrt{W_t} \qquad \text{psi} \tag{3}$$

for the contact stress, where $C_p = 2800 \sqrt{\text{psi}}$. Equating Eqs. (2) and (3) and solving for W_t yields

$$W_t = \left[\frac{85(10^3)}{6695} \right]^2 = 161 \text{ lb}$$

Then

Answer $$H = \frac{W_t V}{33\ 000} = \frac{161(785)}{33\ 000} = 3.83 \text{ hp}$$

15-3   WORM GEARING

Since they are essentially nonenveloping worm gears, the *crossed helical* gears, shown in Fig. 15-7, can be considered with other worm gearing. Because the teeth of worm gears have *point contact* changing to *line contact* as the gears are used, worm gears are said to ''wear in,'' whereas other types ''wear out.''

Crossed helical gears, and worm gears too, usually have a 90° shaft angle, though this need not be so. The relation between the shaft and helix angles is

$$\Sigma = \psi_P \pm \psi_G \tag{15-3}$$

where Σ is the shaft angle. The plus sign is used when both helix angles are of the same hand, and the minus sign when they are of opposite hand. The subscript P in Eq. (15-3)

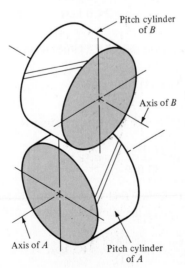

FIGURE 15-7

View of the pitch cylinders of a pair of crossed helical gears.

TABLE 15-4

Allowable Stresses for Use in Gear-Wheel Analysis*

MATERIAL	CONDITION OR GRADE	MINIMUM TENSILE STRENGTH, kpsi (MPa)	BRINELL HARDNESS, Bhn	ALLOWABLE STRESSES BENDING σ_B, kpsi (MPa)	ALLOWABLE STRESSES CONTACT σ_C, kpsi (MPa)
Phosphor	Sand-cast	12 (83)	70	7.00 (48.2)	1.50 (10.3)
bronze,	Chill-cast	15 (103)	82	8.50 (58.6)	1.80 (12.4)
BS 1400 PB2	Centrifugally cast	17 (117)	90	10.0 (68.9)	2.20 (15.2)
Cast iron,	Ordinary grade	12 (83)	150	6.00 (41.3)	1.00 (6.89)
BS 821	Medium grade	16 (110)	165	7.50 (51.7)	1.00 (6.89)
	High grade	22 (152)	180	10.0 (68.9)	1.00 (6.89)

*The pinion (worm) should be of steel harder than the wheel material.

Source: M. J. Neale, *Tribology Handbook,* Butterworth, London, 1973, sec. A-24.

refers to the pinion (worm); sometimes the subscript W is used for this same purpose. The subscript G refers to the *gear,* also called the *gear wheel* or simply the *wheel.*

For a change of pace, we shall analyze worm gearing using the methods employed in the United Kingdom.* For class 3 and 4 gears, the standards include equations which give the allowable wheelshaft torque for unlimited life for both crossed helical and worm gears.† Separate equations are used to predict the allowable torque for contact stress and for bending stress. When the proper units are used, the equations apply using either U.S. customary units or SI units.

The limiting value of the wheelshaft torque for unlimited life based on the allowable contact stress σ_C, given in Table 15-4, is

$$T_G = \begin{cases} \dfrac{\Omega \sigma_C C_v d_G^3 \sqrt{d_P/d_G}}{75} & \text{crossed helical gears} \\[3mm] \dfrac{\sigma_C C_v d_G^2 F_G}{30} & \text{worm gears} \end{cases} \tag{15-4}$$

and the limiting value based on the allowable bending stress σ_B, also given in Table 15-4, is

$$T_G = \begin{cases} \dfrac{\sigma_B N_G p_n^3}{20} & \text{crossed helical gears} \\[3mm] \dfrac{\sigma_B d_G^2 F_G \cos \psi_G}{1.5 N_G} & \text{worm gears} \end{cases} \tag{15-5}$$

*British Standards Institution (BSI), BS 721. For methods in general use in this country, see K. S. Edwards, "Worm Gearing," chap. 36 in Joseph E. Shigley and Charles R. Mischke (eds.), *Standard Handbook of Machine Design,* McGraw-Hill, New York, 1986. But see also Robert L. Mott, *Machine Elements in Mechanical Design,* Merrill, Columbus, Ohio, 1985, pp. 357–387; this book even contains a comprehensive computer program on worm gearing written in BASIC.

†See M. J. Neale (ed.), *Tribology Handbook,* Butterworth, London, 1973, sec. A-24.

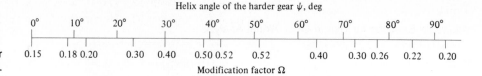

FIGURE 15-8

Contact-stress modification factor Ω for crossed helical gears.

The symbols and units for Eqs. (15-4) and (15-5) are designated as follows:

T_G = wheelshaft torque, lb · in (N · m)
Ω = dimensionless factors; use Fig. 15-8
σ_C = allowable contact stress, psi (GPa)
σ_B = allowable bending stress, psi (GPa)
C_V = velocity factor; use Eq. (15-6) or (15-7)
d_G = pitch diameter of gear wheel, in (mm)
d_p = pitch diameter of pinion (worm), in (mm)
F_G = face width of gear, in (mm)
N_G = number of teeth on gear wheel
p_n = normal pitch of generating cutter, in (mm)
ψ_G = helix angle of gear

The velocity factor is given by either of the equations

$$C_V = \begin{cases} \dfrac{55}{\sqrt{V_S}} & V_S > 3025 \text{ ft/min} \\[2ex] 1 & V_S \leq 3025 \text{ ft/min} \end{cases} \tag{15-6}$$

$$C_V = \begin{cases} \dfrac{237}{\sqrt{V_S}} & V_S > 15.4 \text{ m/s} \\[2ex] 1 & V_S \leq 15.4 \text{ m/s} \end{cases} \tag{15-7}$$

where V_S is the sliding velocity. From Eq. (13-39), this is

$$V_S = \frac{V_P}{\cos \lambda_P} \tag{15-8}$$

where V_P = pitch-line velocity of pinion (worm)
λ = lead angle of pinion (worm)

Note that the allowable stresses listed in Table 15-4 are in kpsi (MPa). These must be converted to psi or GPa for use in Eqs. (15-4) and (15-5).

The following precautions must be taken when using these equations for the limiting wheel torques:

1 For crossed-axis helical gears, the face width should be in the range

$2.8p_n \sin \psi \leq F \leq p_n (2 \sin \psi + 2.8 \cos \psi)$

2 Jet lubrication is necessary when $V_S \geq 2000$ ft/min (10 m/s).

3 The lubricant used should be approved by the supplier and, in any event, should have a Saybolt Universal viscosity in seconds* at the teeth of

$$SUV = 44 + \frac{160(10^3)}{V_S}$$

where V_S is in ft/min.

4 Both gears of a crossed helical gear pair may be of case-hardened steel. In this case use $\sigma_C = 700$ psi (4.80 MPa) for Eq. (15-4) and do not consider the bending stress. This requires that a lubricant with antiweld characteristics be used and that the teeth be formed with high precision and smoothness of the contacting surfaces.

Once it is ensured that the gears are strong enough to carry the power needed, they may still be destroyed by the heat generated. Temperatures over about 200°F will reduce the life of the lubricant and result in more heat generated, excessive wear, and eventual gear failure.

The power loss, due only to tooth friction, can be estimated from the equation

$$H = \begin{cases} \dfrac{V_S W_f}{33\,000} & \text{hp} \\ V_S W_f & \text{kW} \end{cases} \tag{15-9}$$

where V_S = sliding velocity, ft/min (m/s)
W_f = frictional force, lb (kN)

The friction force is found from Eq. (13-36) as follows:

$$W_f = \frac{\mu W_{Gt}}{\mu \sin \lambda - \cos \phi_n \cos \lambda} \tag{15-10}$$

where μ = coefficient of friction
W_{Gt} = gear-wheel transmitted load
λ = lead angle of pinion (worm)
ϕ_n = normal pressure angle

The coefficient of friction will seldom exceed 0.05 in worm gearing. Note that Eq. (15-9) does not account for the power losses due to bearing friction or to oil turbulence.

PROBLEMS

15-1 A straight bevel pinion has 20 teeth, a diametral pitch of 6 teeth/in, and a quality standard of $Q_v = 6$ and is made of steel hardened to 300 Bhn. The driven gear has 60 teeth and is made of No. 30 cast iron. The shaft angle is 90°, the pinion speed 900 rev/min, the face width 1.25 in, and the pressure angle 20°. The pinion is mounted outboard; the gear, straddle-mounted. Find the power rating based on the AGMA bending strength.

15-2 For the gears and conditions of Prob. 15-1, find the power rating based on the AGMA surface durability.

---

*Converted from a viscosity given in the standard in Redwood-seconds.

15-3   A straight bevel pinion has 22 teeth, a module of 4 mm, and a quality standard of $Q_v = 5$ and is made of steel having a Brinell hardness of 180. The gear has 24 teeth and is made of the same material. The shaft angle is 90°, the pinion speed 1800 rev/min, the face width 25 mm, and the pressure angle 20°. Both gears have outboard mounting. Find the power rating based on the AGMA pitting resistance.

15-4   For the same gears and conditions as in Prob. 15-3, find the power rating based on the AGMA bending strength.

15-5   A catalog of stock-size bevel gears lists a power rating of 5.2 hp at 1200 rev/min pinion speed for a straight bevel-gear set consisting of a 20-tooth pinion driving a 40-tooth gear. This gear pair has a 20° pressure angle, a face width of 0.71 in, a diametral pitch of 10 teeth/in, and hardened teeth. Assume the gears are for general industrial use, have a hardness of 300 Bhn, and are generated to a quality standard of $Q_v = 5$. Given these data, how do you feel about the catalog power rating?

15-6   A 4-diametral-pitch $14\frac{1}{4}°$–pressure angle single-thread hardened-steel worm is to mate with a 24-tooth gear. The worm has a lead of 0.7854 in, a lead angle of 4.767°, a face width of $4\frac{1}{2}$ in, and a pitch diameter of 3 in. The gear has a $1\frac{1}{2}$-in face width, and is made of sand-cast phosphor bronze, and the worm is of AISI 1040 steel. Estimate the maximum power output of this gearset for an unlimited life. Use a maximum worm speed of 1800 rev/min.

15-7   A 4-tooth worm has an axial pitch of 6 mm, a 20° pressure angle, and a pitch diameter of 15 mm and is made of hardened steel. The 40-tooth mating gear has a face width of 12 mm and is made of chill-cast phosphor bronze. The pinion (worm) speed is to be 1200 rev/min maximum.

(*a*) Find the maximum permissible gearshaft torque based on tooth breakage.

(*b*) Find the maximum permissible gearshaft torque based on surface durability.

15-8   A double-thread worm has an axial pitch of $\pi/8$ in, a $14\frac{1}{2}°$ pressure angle, and a pitch diameter of 1.5 in and is made of hardened steel. The 30-tooth mating gear has a face width of $\frac{3}{4}$ in and is made of medium-grade cast iron. The pinion speed is 900 rev/min. If it is to have an infinite life, what is the allowable power output of this drive?

15-9   A case-hardened steel helical pinion of a pair of crossed helical gears has a $14\frac{1}{2}°$ normal pressure angle, a 45° helix angle, 8 teeth, cut left hand, and a pitch diameter of 1 in. The mating 12-tooth gear is also case-hardened and has a helix angle of 45°. The face width of both is 1 in. Estimate the power that can be transmitted by this pair of crossed helical gears if the pinion speed is 600 rev/min.

15-10   A set of crossed-axis helical gears is composed of a 10-tooth 6-pitch (transverse) $14\frac{1}{2}°$–normal pressure angle hardened-steel pinion and a 30-tooth cast-bronze gear. The gear set has a 45° helix angle and a 1-in face width. The pinion speed is to be 120 rev/min.

(*a*) Find the maximum gearshaft torque based on the possibility of tooth breakage.

(*b*) Find the maximum gearshaft torque based on contact stress.

ANSWERS   **15-3** $H = 3.30$ kW, $d_P = 88$ mm, $d_G = 96$ mm, $V = 8.29$ m/s, $W_t = 0.398$ kN, $C_V = 0.601$
15-7 (*a*) 60.9 N · m; (*b*) 28.9 N · m

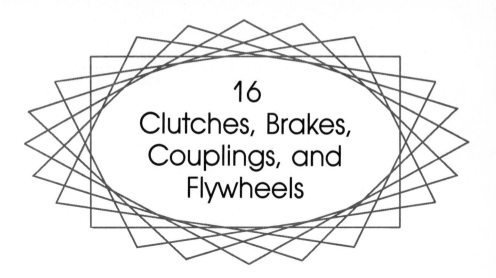

16
Clutches, Brakes, Couplings, and Flywheels

This chapter is concerned with a group of elements usually associated with rotation that have in common the function of storing and/or transferring rotating energy. Because of this similarity of function, clutches, brakes, couplings, and flywheels are treated together in this book.

A simplified dynamic representation of a friction clutch or brake is shown in Fig. 16-1a. Two inertias, I_1 and I_2, traveling at the respective angular velocities ω_1 and ω_2, one of which may be zero in the case of brakes, are to be brought to the same speed by engaging the clutch or brake. Slippage occurs because the two elements are running at different speeds and energy is dissipated during actuation, resulting in a temperature rise. In analyzing the performance of these devices we shall be interested in:

1 The actuating force

2 The torque transmitted

3 The energy loss

4 The temperature rise

The torque transmitted is related to the actuating force, the coefficient of friction, and the geometry of the clutch or brake. This is a problem in statics which will have to be studied separately for each geometric configuration. However, temperature rise is related to energy loss and can be studied without regard to the type of brake or clutch, because the geometry of interest is that of the heat-dissipating surfaces.

The various types of devices to be studied may be classified as follows:

1 Rim types with internal expanding shoes

2 Rim types with external contracting shoes

3 Band types

4 Disk or axial types

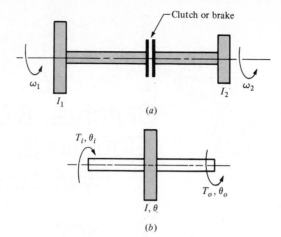

-Clutch or brake

ω_1

I_1

I_2 ω_2

(a)

T_i, θ_i

T_o, θ_o

I, θ

(b)

FIGURE 16-1

(a) Dynamic representation of a clutch or brake; (b) mathematical representation of a flywheel.

5 Cone types

6 Miscellaneous types

A flywheel is an inertial energy-storage device. It absorbs mechanical energy by increasing its angular velocity and delivers energy by decreasing its velocity. Fig. 16-1b is a mathematical representation of a flywheel. An input torque T_i, corresponding to a coordinate θ_i, will cause the flywheel speed to increase. And a load or output torque T_o, with coordinate θ_o, will absorb energy from the flywheel and cause it to slow down. We shall be interested in designing flywheels so as to obtain a specified amount of speed regulation.

16-1   STATICS

The analysis of all types of friction clutches and brakes uses the same general procedure. The following steps are necessary:

1 Estimate or determine the distribution of pressure on the frictional surfaces.

2 Find a relation between the maximum pressure and the pressure at any point.

3 Apply the conditions of static equilibrium to find the actuating force, the torque, and the support reactions.

Let us now apply these steps to the theoretical problem shown in Fig. 16-2. The figure shows a short shoe hinged at A, having an actuating force F, a normal force N pushing the surfaces together, and a frictional force fN, f being the coefficient of friction. The body is moving to the right, and the shoe is stationary. We designate the pressure at any point by p and the maximum pressure by p_a. The area of the shoe is designated by A.

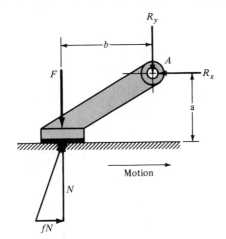

FIGURE 16-2

Forces acting upon a hinged friction shoe.

Step 1. Since the shoe is short, we assume the pressure is uniformly distributed over the frictional area.

Step 2. From step 1 it follows that

$$p = p_a \qquad (a)$$

Step 3. Since the pressure is uniformly distributed, we may replace the unit normal forces by an equivalent normal force. Thus

$$N = p_a A \qquad (b)$$

We now apply the conditions of static equilibrium by taking a summation of moments about the hinge pin. This gives

$$\sum M_A = Fb - Nb + fNa = 0 \qquad (c)$$

Substituting $p_a A$ for N and solving Eq. (c) for the actuating force, we get

$$F = \frac{p_a A(b - fa)}{b} \qquad (d)$$

Taking a summation of forces in the horizontal and vertical directions gives the hinge-pin reactions:

$$\sum F_x = 0 \qquad R_x = f p_a A \qquad (e)$$

$$\sum F_y = 0 \qquad R_y = p_a A - F \qquad (f)$$

This completes the analysis of the problem.

The preceding analysis is very useful when the dimensions of the clutch or brake are known and the characteristics of the friction material are specified. In design, however, we are interested more in synthesis than in analysis; that is, our purpose is to select a set of dimensions to obtain the best brake or clutch within the limitations of the frictional material we have specified.

In the preceding problem (Fig. 16-2) we make good use of the frictional material because the pressure is at an allowable maximum at all points of contact. Furthermore, note that the quantity $(b - fa)/b$ in Eq. (d) is always less than unity. This effect is called *self-energizing,* because friction is reducing the necessary actuating force. But note that a certain critical value of the coefficient of friction f will cause the term $(b - fa)$ to become zero. This is the condition for *self-locking.* It is the designer's responsibility to select values for the dimensions a and b to ensure that self-locking will never occur unless it is specifically desired. If the coefficient of friction f can be treated as a random variable, then the methods of Chap. 4 can be used to determine safe values for the dimensions a and b.

The terms *fail-safe* and *dead-man* are often encountered in studying the operation of brakes and clutches (as well as other mechanical elements). The terms are pretty well self-explanatory. *Fail-safe* means that the operating mechanism has been designed so that, if any element should fail to perform its function, then no accidents will occur in the machine or befall its operator. *Dead-man* is a term from the railway industry. It refers to the locomotive-control mechanism which causes the engine to come to a stop if the operator should suffer a blackout or die at the controls.

16-2   INTERNAL EXPANDING RIM CLUTCHES AND BRAKES

The internal-shoe rim clutch shown in Fig. 16-3 consists essentially of three elements: the mating frictional surface, the means of transmitting the torque to and from the surfaces, and the actuating mechanism. Depending upon the operating mechanism, such clutches are further classified as *expanding-ring, centrifugal, magnetic, hydraulic,* and *pneumatic.*

FIGURE 16-3

An internal expanding centrifugal-acting rim clutch. *(Courtesy of the Hilliard Corporation.)*

The expanding-ring clutch is often used in textile machinery, excavators, and machine tools where the clutch may be located within the driving pulley. Expanding-ring clutches benefit from centrifugal effects; transmit high torque, even at low speeds; and require both positive engagement and ample release force.

The centrifugal clutch is used mostly for automatic operation. If no spring is used, the torque transmitted is proportional to the square of the speed. This is particularly useful for electric-motor drives where, during starting, the driven machine comes up to speed without shock. Springs can also be used to prevent engagement until a certain motor speed has been reached, but some shock may occur.

Magnetic clutches are particularly useful for automatic and remote-control systems. Such clutches are also useful in drives subject to complex load cycles (see Sec. 11-8).

Hydraulic and pneumatic clutches are also useful in drives having complex loading cycles and in automatic machinery, or in robots. Here the fluid flow can be controlled remotely using solenoid valves. These clutches are also available as disk, cone, and multiple-plate clutches.

In braking systems, the *internal-shoe* or *drum* brake is used mostly for automotive applications.

To analyze an internal-shoe device, refer to Fig. 16-4, which shows a shoe pivoted at point A, with the actuating force acting at the other end of the shoe. Since the shoe is long, we cannot make the assumption that the distribution of normal forces is uniform. The mechanical arrangement permits no pressure to be applied at the heel, and we will therefore assume the pressure at this point to be zero.

It is the usual practice to omit the friction material for a short distance away from the heel (point A). This eliminates interference, and the material would contribute little to the performance anyway, as will be shown. In some designs the hinge pin is made movable to provide additional heel pressure. This gives the effect of a floating shoe. (Floating shoes will not be treated in this book, although their design follows the same general principles.)

Let us consider the unit pressure p acting upon an element of area of the frictional material located at an angle θ from the hinge pin (Fig. 16-4). We designate the maxi-

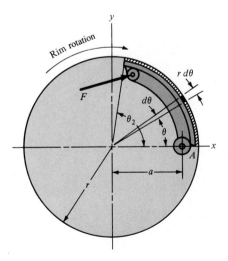

FIGURE 16-4

Internal friction shoe.

mum pressure by p_a located at the angle θ_a from the hinge pin. We now make the assumption (step 1) that the pressure at any point is proportional to the vertical distance from the hinge pin. This vertical distance is proportional to $\sin \theta$, and (step 2) the relation between the pressures is

$$\frac{p}{\sin \theta} = \frac{p_a}{\sin \theta_a} \tag{a}$$

Rearranging,

$$p = p_a \frac{\sin \theta}{\sin \theta_a} \tag{16-1}$$

From Eq. (16-1) we see that p will be a maximum when $\theta = 90°$, or if the toe angle θ_2 is less than $90°$, then p will be a maximum at the toe.

When $\theta = 0$, Eq. (16-2) shows that the pressure is zero. The frictional material located at the heel therefore contributes very little to the braking action and might as well be omitted. A good design would concentrate as much frictional material as possible in the neighborhood of the point of maximum pressure. Such a design is shown in Fig. 16-5. In this figure the frictional material begins at an angle θ_1, measured from the hinge pin A, and ends at an angle θ_2. Any arrangement such as this will give a good distribution of the frictional material.

Proceeding now to step 3 (Fig. 16-5), the hinge-pin reactions are R_x and R_y. The actuating force F has components F_x and F_y and operates at distance c from the hinge pin. At any angle θ from the hinge pin there acts a differential normal force dN whose magnitude is

$$dN = pbr\, d\theta \tag{b}$$

where b is the face width (perpendicular to the paper) of the friction material. Substitut-

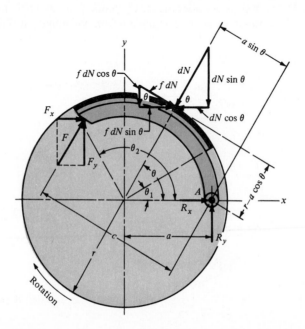

FIGURE 16-5

Forces on the shoe.

ing the value of the pressure from Eq. (16-1), the normal force is

$$dN = \frac{p_a b r \sin \theta \, d\theta}{\sin \theta_a} \tag{c}$$

The normal force dN has horizontal and vertical components $dN \cos \theta$ and $dN \sin \theta$, as shown in the figure. The frictional force $f \, dN$ has horizontal and vertical components whose magnitudes are $f \, dN \sin \theta$ and $f \, dN \cos \theta$, respectively. By applying the conditions of static equilibrium, we may find the actuating force F, the torque T, and the pin reactions R_x and R_y.

We shall find the actuating force F, using the condition that the summation of the moments about the hinge pin is zero. The frictional forces have a moment arm about the pin of $r - a \cos \theta$. The moment M_f of these frictional forces is

$$M_f = \int f \, dN \, (r - a \cos \theta) = \frac{f p_a b r}{\sin \theta_a} \int_{\theta_1}^{\theta_2} \sin \theta \, (r - a \cos \theta) \, d\theta \tag{16-2}$$

which is obtained by substituting the value of dN from Eq. (c). It is convenient to integrate Eq. (16-2) for each problem, and we shall therefore retain it in this form. The moment arm of the normal force dN about the pin is $a \sin \theta$. Designating the moment of the normal forces by M_N and summing these about the hinge pin give

$$M_N = \int dN \, (a \sin \theta) = \frac{p_a b r a}{\sin \theta_a} \int_{\theta_1}^{\theta_2} \sin^2 \theta \, d\theta \tag{16-3}$$

The actuating force F must balance these moments. Thus

$$F = \frac{M_N - M_f}{c} \tag{16-4}$$

We see here that a condition for zero actuating force exists. In other words, if we make $M_N = M_f$, self-locking is obtained, and no actuating force is required. This furnishes us with a method for obtaining the dimensions for some self-energizing action. Thus the dimension a in Fig. 16-5 must be such that

$$M_N > M_f \tag{16-5}$$

The torque T applied to the drum by the brake shoe is the sum of the frictional forces $f \, dN$ times the radius of the drum:

$$T = \int f r \, dN = \frac{f p_a b r^2}{\sin \theta_a} \int_{\theta_1}^{\theta_2} \sin \theta \, d\theta$$

$$= \frac{f p_a b r^2 (\cos \theta_1 - \cos \theta_2)}{\sin \theta_a} \tag{16-6}$$

The hinge-pin reactions are found by taking a summation of the horizontal and vertical forces. Thus, for R_x, we have

$$R_x = \int dN \cos \theta - \int f \, dN \sin \theta - F_x$$

$$= \frac{p_a b r}{\sin \theta_a} \left(\int_{\theta_1}^{\theta_2} \sin \theta \cos \theta \, d\theta - f \int_{\theta_1}^{\theta_2} \sin^2 \theta \, d\theta \right) - F_x \tag{d}$$

The vertical reaction is found in the same way:

$$R_y = \int dN \sin \theta + \int f \, dN \cos \theta - F_y$$

$$= \frac{p_a b r}{\sin \theta_a} \left(\int_{\theta_1}^{\theta_2} \sin^2 \theta \, d\theta + f \int_{\theta_1}^{\theta_2} \sin \theta \cos \theta \, d\theta \right) - F_y \tag{e}$$

The direction of the frictional forces is reversed if the rotation is reversed. Thus, for counterclockwise rotation the actuating force is

$$F = \frac{M_N + M_f}{c} \tag{16-7}$$

and since both moments have the same sense, the self-energizing effect is lost. Also, for counterclockwise rotation the signs of the frictional terms in the equations for the pin reactions change, and Eqs. (d) and (e) become

$$R_x = \frac{p_a b r}{\sin \theta_a} \left(\int_{\theta_1}^{\theta_2} \sin \theta \cos \theta \, d\theta + f \int_{\theta_1}^{\theta_2} \sin^2 \theta \, d\theta \right) - F_x \tag{f}$$

$$R_y = \frac{p_a b r}{\sin \theta_a} \left(\int_{\theta_1}^{\theta_2} \sin^2 \theta \, d\theta - f \int_{\theta_1}^{\theta_2} \sin \theta \cos \theta \, d\theta \right) - F_y \tag{g}$$

Equations (d), (e), (f), and (g) can be simplified to ease computations. Thus, let

$$A = \int_{\theta_1}^{\theta_2} \sin \theta \cos \theta \, d\theta = \left(\frac{1}{2} \sin^2 \theta \right)_{\theta_1}^{\theta_2}$$

$$B = \int_{\theta_1}^{\theta_2} \sin^2 \theta \, d\theta = \left(\frac{\theta}{2} - \frac{1}{4} \sin 2\theta \right)_{\theta_1}^{\theta_2} \tag{16-8}$$

Then, for clockwise rotation as shown in Fig. 16-5, the hinge-pin reactions are

$$R_x = \frac{p_a b r}{\sin \theta_a} (A - fB) - F_x$$

$$R_y = \frac{p_a b r}{\sin \theta_a} (B + fA) - F_y \tag{16-9}$$

For counterclockwise rotation, Eqs. (f) and (g) become

$$R_x = \frac{p_a b r}{\sin \theta_a} (A + fB) - F_x$$

$$R_y = \frac{p_a b r}{\sin \theta_a} (B - fA) - F_y \tag{16-10}$$

In using these equations, the reference system always has its origin at the center of the drum. The positive x axis is taken through the hinge pin. The positive y axis is always in the direction of the shoe, even if this should result in a left-handed system.

The following assumptions are implied by the preceding analysis:

1 The pressure at any point on the shoe is assumed to be proportional to the distance

from the hinge pin, being zero at the heel. This should be considered from the standpoint that pressures specified by manufacturers are averages rather than maxima.

2 The effect of centrifugal force has been neglected. In the case of brakes, the shoes are not rotating, and no centrifugal force exists. In clutch design, the effect of this force must be considered in writing the equations of static equilibrium.

3 The shoe is assumed to be rigid. Since this cannot be true, some deflection will occur, depending upon the load, pressure, and stiffness of the shoe. The resulting pressure distribution may be different from that which has been assumed.

4 The entire analysis has been based upon a coefficient of friction which does not vary with pressure. Actually, the coefficient may vary with a number of conditions, including temperature, wear, and environment.

EXAMPLE 16-1   The brake shown in Fig. 16-6 is 300 mm in diameter and is actuated by a mechanism that exerts the same force F on each shoe. The shoes are identical and have a face width of 32 mm. The lining is a molded asbestos having a coefficient of friction of 0.32 and a pressure limitation of 1000 kPa.

(a) Determine the actuating force F.

(b) Find the braking capacity.

(c) Calculate the hinge-pin reactions.

Solution   (a) The right-hand shoe is self-energizing, and so the force F is found on the basis that the maximum pressure will occur on this shoe. Here $\theta_1 = 0°$, $\theta_2 = 126°$, $\theta_a = 90°$, and $\sin \theta_a = 1$. Also,

$$a = \sqrt{(112)^2 + (50)^2} = 123 \text{ mm}$$

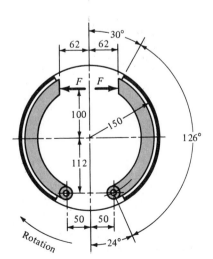

FIGURE 16-6

Brake with internal expanding shoes; dimensions in millimeters.

Integrating Eq. (16-2) from 0 to θ_2 yields

$$M_f = \frac{fp_a br}{\sin \theta_a} \left[\left(-r \cos \theta \right)_0^{\theta_2} - a\left(\frac{1}{2} \sin^2 \theta \right)_0^{\theta_2} \right]$$

$$= \frac{fp_a br}{\sin \theta_a} \left(r - r \cos \theta_2 - \frac{a}{2} \sin^2 \theta_2 \right)$$

Changing all lengths to meters, we have

$$M_f = (0.32)[1000(10)^3](0.032)(0.150)$$

$$\times \left[0.150 - 0.150 \cos 126° - \left(\frac{0.123}{2} \right) \sin^2 126° \right]$$

$$= 304 \text{ N} \cdot \text{m}$$

The moment of the normal forces is obtained from Eq. (16-3). Integrating from 0 to θ_2 gives

$$M_N = \frac{p_a bra}{\sin \theta_a} \left(\frac{\theta}{2} - \frac{1}{4} \sin 2\theta \right)_0^{\theta_2}$$

$$= \frac{p_a bra}{\sin \theta_a} \left(\frac{\theta_2}{2} - \frac{1}{4} \sin 2\theta_2 \right)$$

$$= [1000(10)^3](0.032)(0.150)(0.123) \left[\frac{\pi}{2} \frac{126}{180} - \frac{1}{4} \sin (2)(126°) \right]$$

$$= 790 \text{ N} \cdot \text{m}$$

From Eq. (16-4), the actuating force is

Answer
$$F = \frac{M_N - M_f}{c} = \frac{790 - 304}{100 + 112} = 2.29 \text{ kN}$$

(b) From Eq. (16-6), the torque applied by the right-hand shoe is

$$T_R = \frac{fp_a br^2(\cos \theta_1 - \cos \theta_2)}{\sin \theta_a}$$

$$= \frac{0.32[1000(10)^3](0.032)(0.150)^2(\cos 0° - \cos 126°)}{1} = 366 \text{ N} \cdot \text{m}$$

The torque contributed by the left-hand shoe cannot be obtained until we learn its maximum operating pressure. Equations (16-2) and (16-3) indicate that the frictional and normal moments are proportional to this pressure. Thus, for the left-hand shoe,

$$M_N = \frac{790p_a}{1000} \qquad M_f = \frac{304p_a}{1000}$$

Then, from Eq. (16-7),

$$F = \frac{M_N + M_f}{c}$$

or

$$2.29 = \frac{(790/1000)p_a + (304/1000)p_a}{100 + 112}$$

Solving gives $p_a = 444$ kPa. Then, from Eq. (16-6), the torque on the left-hand shoe is

$$T_L = \frac{f p_a b r^2 (\cos \theta_1 - \cos \theta_2)}{\sin \theta_a}$$

Since $\sin \theta_a = 1$, we have

$$T_L = 0.32[444(10)^3](0.032)(0.150)^2(\cos 0° - \cos 126°) = 162 \text{ N} \cdot \text{m}$$

The braking capacity is the total torque:

Answer $T = T_R + T_L = 366 + 162 = 528 \text{ N} \cdot \text{m}$

(c) In order to find the hinge-pin reactions, we note that $\sin \theta_a = 1$ and $\theta_1 = 0$. Then Eq. (16-8) gives

$$A = \tfrac{1}{2} \sin^2 \theta_2 = \tfrac{1}{2} \sin^2 126° = 0.3273$$

$$B = \frac{\theta_2}{2} - \frac{1}{4} \sin 2\theta_2 = \frac{\pi(126)}{2(180)} - \frac{1}{4} \sin (2)(126°) = 1.3373$$

Also, let

$$D = \frac{p_a b r}{\sin \theta_a} = \frac{1000(0.032)(0.150)}{1} = 4.8 \text{ kN}$$

where $p_a = 1000$ kPa for the right-hand shoe. Then, using Eq. (16-9), we have

$$R_x = D(A - fB) - F_x = 4.8[0.3273 - 0.32(1.3373)] - 2.29 \sin 24°$$
$$= -1.414 \text{ kN}$$

$$R_y = D(B + fA) - F_y = 4.8[1.3373 + 0.32(0.3273)] - 2.29 \cos 24°$$
$$= 4.830 \text{ kN}$$

The resultant on this hinge pin is

Answer $R = \sqrt{(1.414)^2 + (4.830)^2} = 5.03 \text{ kN}$

The reactions at the hinge pin of the left-hand shoe are found using Eqs. (16-10) for a pressure of 444 kPa. They are found to be $R_x = 0.678$ kN and $R_y = 0.535$ kN. The resultant is

Answer $R = \sqrt{(0.678)^2 + (0.535)^2} = 0.864 \text{ kN}$

The reactions for both hinge pins, together with their directions, are shown in Fig. 16-7.

This example dramatically shows the benefit to be gained by arranging the shoes to be self-energizing. If the left-hand shoe were turned over so as to place the hinge pin at the top, it could apply the same torque as the right-hand shoe. This would

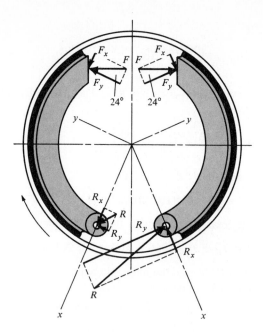

FIGURE 16-7

make the capacity of the brake $(2)(366) = 732$ N $\cdot$ m instead of the present 528 N $\cdot$ m, a 30 percent improvement. In addition, some of the friction material at the heel could be eliminated without seriously affecting the capacity, because of the low pressure in this area. This change might actually improve the overall design because the additional rim exposure would improve the heat-dissipation capacity.

16-3 EXTERNAL CONTRACTING RIM CLUTCHES AND BRAKES

The patented clutch-brake of Fig. 16-8 has external contracting friction elements, but the actuating mechanism is pneumatic. Here we shall study only pivoted external shoe brakes and clutches, though the methods presented can easily be adapted to the clutch-brake of Fig. 16-8.

Operating mechanisms can be classified as:

1 Solenoids

2 Levers, linkages, or toggle devices

3 Linkages with spring loading

4 Hydraulic and pneumatic devices

The static analysis required for these devices has already been covered in Sec. 2-7. The methods there apply to any mechanism system, including all those used in brakes and clutches. It is not necessary to repeat the material in Chap. 2 that applies directly to

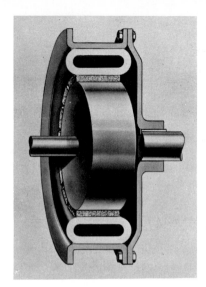

FIGURE 16-8

An external contracting clutch-brake that is engaged by expanding the flexible tube with compressed air. *(Courtesy of Twin Disc Clutch Company.)*

such mechanisms. Omitting the operating mechanisms from consideration allows us to concentrate on brake and clutch performance without the extraneous influences introduced by the need to analyze the statics of the control mechanisms.

The notation for external contracting shoes is shown in Fig. 16-9. The moments of the frictional and normal forces about the hinge pin are the same as for the internal

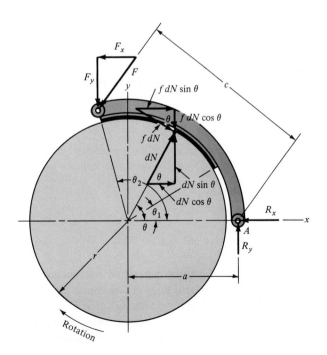

FIGURE 16-9

Notation for external contracting shoes.

expanding shoes. Equations (16-2) and (16-3) apply and are repeated here for convenience:

$$M_f = \frac{fp_a br}{\sin \theta_a} \int^{\theta_2}_{\theta_1} \sin \theta \, (r - a \cos \theta) \, d\theta \tag{16-2}$$

$$M_N = \frac{p_a bra}{\sin \theta_a} \int^{\theta_2}_{\theta_1} \sin^2 \theta \, d\theta \tag{16-3}$$

Both these equations give positive values for clockwise moments (Fig. 16-9) when used for external contracting shoes. The actuating force must be large enough to balance both moments:

$$F = \frac{M_N + M_f}{c} \tag{16-11}$$

The horizontal and vertical reactions at the hinge pin are found in the same manner as for internal expanding shoes. They are

$$R_x = \int dN \cos \theta + \int f \, dN \sin \theta - F_x \tag{a}$$

$$R_y = \int f \, dN \cos \theta - \int dN \sin \theta + F_y \tag{b}$$

By using Eq. (16-8), we have

$$R_x = \frac{p_a br}{\sin \theta_a} (A + fB) - F_x$$

$$R_y = \frac{p_a br}{\sin \theta_a} (fA - B) + F_y \tag{16-12}$$

If the rotation is counterclockwise, the sign of the frictional term in each equation is reversed. Thus Eq. (16-11) for the actuating force becomes

$$F = \frac{M_N - M_f}{c} \tag{16-13}$$

and self-energization exists for counterclockwise rotation. The horizontal and vertical reactions are found, in the same manner as before, to be

$$R_x = \frac{p_a br}{\sin \theta_a} (A - fB) - F_x$$

$$R_y = \frac{p_a br}{\sin \theta_a} (-fA - B) + F_y \tag{16-14}$$

It should be noted that, when external contracting designs are used as clutches, the effect of centrifugal force is to decrease the normal force. Thus, as the speed increases, a larger value of the actuating force F is required.

A special case arises when the pivot is symmetrically located and also placed so that the moment of the friction forces about the pivot is zero. The geometry of such a brake will be similar to that of Fig. 16-10a. To get a pressure-distribution relation, we assume that the lining wears so as always to retain its cylindrical shape. This means that the wear Δx in Fig. 16-10b is constant regardless of the angle θ. Thus the radial wear of the shoe is $\Delta r = \Delta x \cos \theta$. If, on any elementary area of the shoe, we assume that the energy or frictional loss is proportional to the radial pressure, and if we also assume that the wear is directly related to the frictional loss, then, by direct analogy,

$$p = p_a \cos \theta \tag{c}$$

and p is maximum at $\theta = 0°$.

Proceeding to the force analysis, we observe from Fig. 16-10a that

$$dN = pbr\, d\theta \tag{d}$$

or

$$dN = p_a br \cos \theta\, d\theta \tag{e}$$

The distance a to the pivot is to be chosen such that the moment of the frictional forces M_f is zero. Symmetry means that $\theta_1 = \theta_2$, and so

$$M_f = 2 \int_0^{\theta_2} (f\, dN)(a \cos \theta - r) = 0$$

Substituting Eq. (e) gives

$$2fp_a br \int_0^{\theta_2} (a \cos^2 \theta - r \cos \theta)\, d\theta = 0$$

from which

$$a = \frac{4r \sin \theta_2}{2\theta_2 + \sin 2\theta_2} \tag{16-15}$$

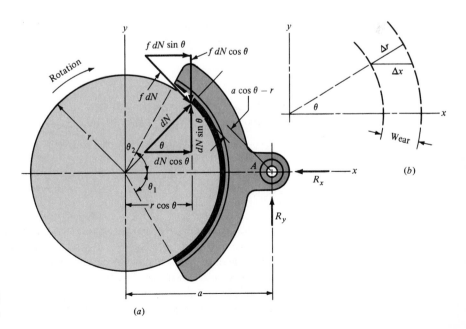

FIGURE 16-10

(a) Brake with symmetrical pivoted shoe; (b) wear of brake lining.

With the pivot located according to this equation, the moment about the pin is zero, and the horizontal and vertical reactions are

$$R_x = 2 \int_0^{\theta_2} dN \cos \theta = \frac{p_a b r}{2} (2\theta_2 + \sin 2\theta_2) \tag{16-16}$$

where, because of symmetry,

$$\int f \, dN \sin \theta = 0$$

Also,

$$R_y = 2 \int_0^{\theta_2} f \, dN \cos \theta = \frac{p_a b r f}{2} (2\theta_2 + \sin 2\theta_2) \tag{16-17}$$

where

$$\int dN \sin \theta = 0$$

also because of symmetry. Note, too, that $R_x = -N$ and $R_y = -fN$, as might be expected for the particular choice of the dimension a. Therefore the torque is

$$T = afN \tag{16-18}$$

16-4   BAND-TYPE CLUTCHES AND BRAKES

Flexible clutch and brake bands are used in power excavators and in hoisting and other machinery. The analysis follows the notation of Fig. 16-11.

Because of friction and the rotation of the drum, the actuating force P_2 is less than the pin reaction P_1. Any element of the band, of angular length $d\theta$, will be in equilibrium under the action of the forces shown in the figure. Summing these forces in the

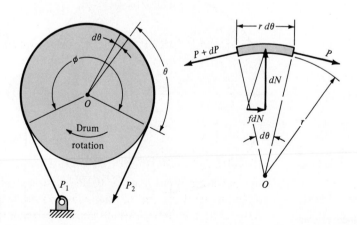

FIGURE 16-11

Forces on a brake band; $r \, d\theta$ is an arc distance.

vertical direction, we have

$$(P + dP) \sin \frac{d\theta}{2} + P \sin \frac{d\theta}{2} - dN = 0 \qquad (a)$$

$$dN = P \, d\theta \qquad (b)$$

since for small angles $\sin d\theta/2 = d\theta/2$. Summing the forces in the horizontal direction gives

$$(P + dP) \cos \frac{d\theta}{2} - P \cos \frac{d\theta}{2} - f \, dN = 0 \qquad (c)$$

$$dP - f \, dN = 0 \qquad (d)$$

Substituting the value of dN from Eq. (b) in (d) and integrating,

$$\int_{P_2}^{P_1} \frac{dP}{P} = f \int_0^\phi d\theta \qquad \ln \frac{P_1}{P_2} = f\phi$$

and

$$\frac{P_1}{P_2} = e^{f\phi} \qquad (16\text{-}19)$$

The torque may be obtained from the equation

$$T = (P_1 - P_2)\frac{D}{2} \qquad (16\text{-}20)$$

The normal force dN acting on an element of area of width b and length $r \, d\theta$ is

$$dN = pbr \, d\theta \qquad (e)$$

where p is the pressure. Substitution of the value of dN from Eq. (b) gives

$$P \, d\theta = pbr \, d\theta$$

Therefore

$$p = \frac{P}{br} = \frac{2P}{bD} \qquad (16\text{-}21)$$

The pressure is therefore proportional to the tension in the band. The maximum pressure p_a will occur at the toe and has the value

$$p_a = \frac{2P_1}{bD} \qquad (16\text{-}22)$$

16-5   FRICTIONAL-CONTACT AXIAL CLUTCHES

An axial clutch is one in which the mating frictional members are moved in a direction parallel to the shaft. One of the earliest of these is the cone clutch, which is simple in construction and quite powerful. However, except for relatively simple installations, it has been largely displaced by the disk clutch employing one or more disks as the operating members. Advantages of the disk clutch include the freedom from centrifugal effects, the large frictional area which can be installed in a small space, the more effective heat-dissipation surfaces, and the favorable pressure distribution. Figure 16-12 shows a single-plate disk clutch; a multiple-disk clutch-brake is shown in Fig. 16-13. Let us now determine the capacity of such a clutch or brake in terms of the material and geometry.

Figure 16-14 shows a friction disk having an outside diameter D and an inside diameter d. We are interested in obtaining the axial force F necessary to produce a certain torque T and pressure p. Two methods of solving the problem, depending upon the construction of the clutch, are in general use. If the disks are rigid, then the greatest amount of wear will at first occur in the outer areas, since the work of friction is greater in those areas. After a certain amount of wear has taken place, the pressure distribution will change so as to permit the wear to be uniform. This is the basis of the first method of solution.

Another method of construction employs springs to obtain a uniform pressure over the area. It is this assumption of uniform pressure that is used in the second method of solution.

Uniform Wear

After initial wear has taken place and the disks have worn down to the point where uniform wear becomes possible, the greatest pressure must occur at $r = d/2$ in order for the wear to be uniform. Denoting the maximum pressure by p_a, we can then write

$$pr = p_a \frac{d}{2} \qquad \text{or} \qquad p = p_a \frac{d}{2r} \tag{a}$$

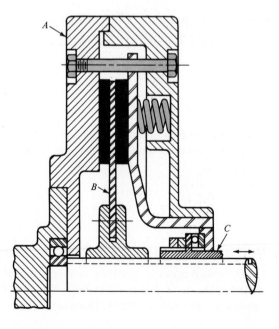

FIGURE 16-12

Cross-sectional view of a single-plate clutch; A, driver; B, driven plate (keyed to driven shaft); C, actuator.

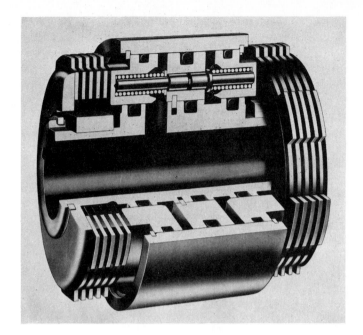

FIGURE 16-13

An oil-actuated multiple-disk clutch-brake for enclosed operation in an oil bath or spray. It is especially useful for rapid cycling. *(Courtesy of Twin Disc Clutch Company.)*

which is the condition for having the same amount of work done at radius r as is done at radius $d/2$. Referring to Fig. 16-14, we have an element of area of radius r and thickness dr. The area of this element is $2\pi r\, dr$, so that the normal force acting upon this element is $dF = 2\pi pr\, dr$. We can find the total normal force by letting r vary from $d/2$ to $D/2$ and integrating. Thus

$$F = \int_{d/2}^{D/2} 2\pi pr\, dr = \pi p_a d \int_{d/2}^{D/2} dr = \frac{\pi p_a d}{2}(D - d) \tag{16-23}$$

The torque is found by integrating the product of the frictional force and the radius:

$$T = \int_{d/2}^{D/2} 2\pi f p r^2\, dr = \pi f p_a d \int_{d/2}^{D/2} r\, dr = \frac{\pi f p_a d}{8}(D^2 - d^2) \tag{16-24}$$

By substituting the value of F from Eq. (16-23) we may obtain a more convenient

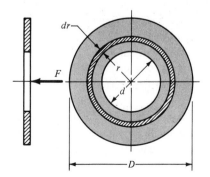

FIGURE 16-14

Disk friction member.

expression for the torque. Thus

$$T = \frac{Ff}{4}(D + d) \tag{16-25}$$

In use, Eq. (16-23) gives the actuating force for the selected maximum pressure p_a. This equation holds for any number of friction pairs or surfaces. Equation (16-25), however, gives the torque capacity for only a single friction surface.

Uniform Pressure

When uniform pressure can be assumed over the area of the disk, the actuating force F is simply the product of the pressure and the area. This gives

$$F = \frac{\pi p_a}{4}(D^2 - d^2) \tag{16-26}$$

As before, the torque is found by integrating the product of the frictional force and the radius:

$$T = 2\pi f p \int_{d/2}^{D/2} r^2 \, dr = \frac{2\pi f p}{24}(D^3 - d^3) \tag{16-27}$$

Since $p = p_a$, we can rewrite Eq. (16-27) as

$$T = \frac{Ff}{3}\frac{D^3 - d^3}{D^2 - d^2} \tag{16-28}$$

It should be noted for both equations that the torque is for a single pair of mating surfaces. This value must therefore be multiplied by the number of pairs of surfaces in contact.

16-6 DISK BRAKES

As indicated in Fig. 16-13, there is no fundamental difference between a disk clutch and a disk brake. The analysis of the preceding section applies to disk brakes too.

We have seen that rim or drum brakes can be designed for self-energization. While this feature is important in reducing the braking effort required, it also has a disadvantage. When drum brakes are used as vehicle brakes, only a slight change in the coefficient of friction will cause a large change in the pedal force required for braking. A not unusual 30 percent reduction in the coefficient of friction due to a temperature change or moisture, for example, can result in a 50 percent change in the pedal force required to obtain the same braking torque obtainable prior to the change. The disk brake has no self-energization, and hence is not so susceptible to changes in the coefficient of friction.

Another type of disk brake is the *floating caliper brake*, shown in Fig. 16-15. The caliper supports a single floating piston actuated by hydraulic pressure. The action is much like that of a screw clamp, with the piston replacing the function of the screw. The floating action also compensates for wear and ensures a fairly constant pressure

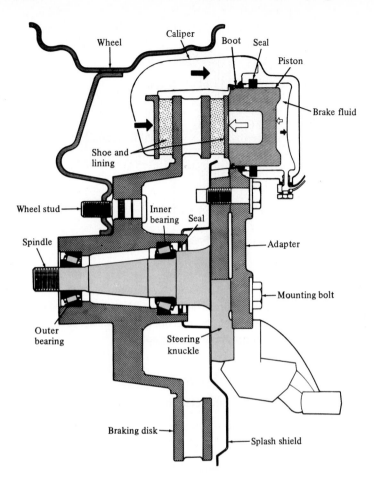

FIGURE 16-15

An automotive disk brake. *(Courtesy of Chrysler Corporation.)*

over the area of the friction pads. The seal and boot of Fig. 16-15 are designed to obtain clearance by backing off from the piston when the piston is released.

16-7 CONE CLUTCHES AND BRAKES

The drawing of a *cone clutch* in Fig. 16-16 shows that it consists of a *cup* keyed or splined to one of the shafts, a *cone* which must slide axially on splines or keys on the mating shaft, and a helical *spring* to hold the clutch in engagement. The clutch is disengaged by means of a fork which fits into the shifting groove on the friction cone. The *cone angle* α and the diameter and face width of the cone are the important geometric design parameters. If the cone angle is too small, say, less than about 8°, then the force required to disengage the clutch may be quite large. And the wedging effect lessens rapidly when larger cone angles are used. Depending upon the characteristics of the friction materials, a good compromise can usually be found using cone angles between 10 and 15°.

To find a relation between the operating force F and the torque transmitted, desig-

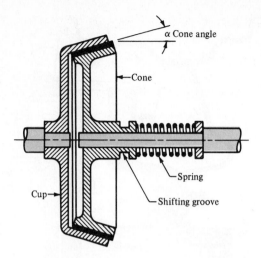

FIGURE 16-16

A cone clutch.

nate the dimensions of the friction cone as shown in Figure 16-17. As in the case of the axial clutch, we can obtain one set of relations for a uniform-wear and another set for a uniform-pressure assumption.

Uniform Wear

The pressure relation is the same as for the axial clutch:

$$p = p_a \frac{d}{2r} \tag{a}$$

Next, referring to Fig. 16-17, we see that we have an element of area dA of radius r and width $dr/\sin \alpha$. Thus $dA = (2\pi r \, dr)/\sin \alpha$. As shown in Fig. 16-17, the operating

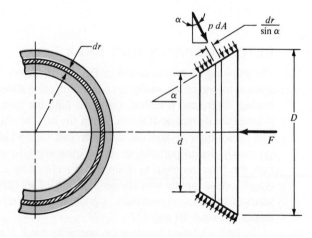

FIGURE 16-17

force will be the integral of the axial component of the differential force $p \, dA$. Thus

$$F = \int p \, dA \sin \alpha = \int_{d/2}^{D/2} \left(p_a \frac{d}{2r} \right) \left(\frac{2\pi r \, dr}{\sin \alpha} \right) (\sin \alpha)$$

$$= \pi p_a d \int_{d/2}^{D/2} dr = \frac{\pi p_a d}{2} (D - d) \qquad (16\text{-}29)$$

which is the same result as in Eq. (16-23).

The differential friction force is $fp \, dA$, and the torque is the integral of the product of this force with the radius. Thus

$$T = \int rfp \, dA = \int_{d/2}^{D/2} (rf) \left(p_a \frac{d}{2r} \right) \left(\frac{2\pi r \, dr}{\sin \alpha} \right)$$

$$= \frac{\pi f p_a d}{\sin \alpha} \int_{d/2}^{D/2} r \, dr = \frac{\pi f p_a d}{8 \sin \alpha} (D^2 - d^2) \qquad (16\text{-}30)$$

Note that Eq. (16-24) is a special case of Eq. (16-30), with $\alpha = 90°$. Using Eq. (16-29), we find that the torque can also be written

$$T = \frac{Ff}{4 \sin \alpha} (D + d) \qquad (16\text{-}31)$$

Uniform Pressure

Using $p = p_a$, the actuating force is found to be

$$F = \int p_a \, dA \sin \alpha = \int_{d/2}^{D/2} (p_a) \left(\frac{2\pi r \, dr}{\sin \alpha} \right) (\sin \alpha) = \frac{\pi p_a}{4} (D^2 - d^2) \qquad (16\text{-}32)$$

The torque is

$$T = \int rfp_a \, dA = \int_{d/2}^{D/2} (rfp_a) \left(\frac{2\pi r \, dr}{\sin \alpha} \right) = \frac{\pi f p_a}{12 \sin \alpha} (D^3 - d^3) \qquad (16\text{-}33)$$

or, using Eq. (16-32) in Eq. (16-33),

$$T = \frac{Ff}{3 \sin \alpha} \frac{D^3 - d^3}{D^2 - d^2} \qquad (16\text{-}34)$$

16-8   ENERGY CONSIDERATIONS

When the rotating members of a machine are caused to stop by means of a brake, the kinetic energy of rotation must be absorbed by the brake. This energy appears in the brake in the form of heat. In the same way, when the members of a machine which are initially at rest are brought up to speed, slipping must occur in the clutch until the driven members have the same speed as the driver. Kinetic energy is absorbed during slippage of either a clutch or a brake, and this energy appears as heat.

We have seen how the torque capacity of a clutch or brake depends upon the coefficient of friction of the material and upon a safe normal pressure. However, the character of the load may be such that, if this torque value is permitted, the clutch or brake may be destroyed by its own generated heat. The capacity of a clutch is therefore limited by two factors, the characteristics of the material and the ability of the clutch to dissipate heat. In this section we shall consider the amount of heat generated by a clutching or braking operation. If the heat is generated faster than it is dissipated, we have a temperature-rise problem; that is the subject of the next section.

To get a clear picture of what happens during a simple clutching or braking operation, refer to Fig. 16-1a, which is a mathematical model of a two-inertia system connected by a clutch. As shown, inertias I_1 and I_2 have initial angular velocities of ω_1 and ω_2, respectively. During the clutch operation both angular velocities change and eventually become equal. We assume that the two shafts are rigid and that the clutch torque is constant.

Writing the equation of motion for inertia 1 gives

$$I_1\ddot{\theta}_1 = -T \tag{a}$$

where $\ddot{\theta}_1$ is the angular acceleration of I_1 and T is the clutch torque. A similar equation for I_2 is

$$I_2\ddot{\theta}_2 = T \tag{b}$$

We can determine the instantaneous angular velocities $\dot{\theta}_1$ and $\dot{\theta}_2$ of I_1 and I_2 after any period of time t has elapsed by integrating Eqs. (a) and (b). The results are

$$\dot{\theta}_1 = -\frac{T}{I_1}t + \omega_1 \tag{c}$$

$$\dot{\theta}_2 = \frac{T}{I_2}t + \omega_2 \tag{d}$$

The difference in the velocities, sometimes called the relative velocity, is

$$\dot{\theta} = \dot{\theta}_1 - \dot{\theta}_2 = -\frac{T}{I_1}t + \omega_1 - \left(\frac{T}{I_2}t + \omega_2\right)$$

$$= \omega_1 - \omega_2 - T\left(\frac{I_1 + I_2}{I_1 I_2}\right)t \tag{16-35}$$

The clutching operation is completed at the instant in which the two angular velocities $\dot{\theta}_1$ and $\dot{\theta}_2$ become equal. Let the time required for the entire operation be t_1. Then $\dot{\theta} = 0$ when $\dot{\theta}_1 = \dot{\theta}_2$, and so Eq. (16-35) gives the time as

$$t_1 = \frac{I_1 I_2(\omega_1 - \omega_2)}{T(I_1 + I_2)} \tag{16-36}$$

This equation shows that the time required for the engagement operation is directly proportional to the velocity difference and inversely proportional to the torque.

We have assumed the clutch torque to be constant. Therefore, using Eq. (16-35), we find the rate of energy-dissipation during the clutching operation to be

$$u = T\dot{\theta} = T\left[\omega_1 - \omega_2 - T\left(\frac{I_1 + I_2}{I_1 I_2}\right)t\right] \tag{e}$$

This equation shows that the energy-dissipation rate is greatest at the start, when $t = 0$.

The total energy dissipated during the clutching operation or braking cycle is obtained by integrating Eq. (e) from $t = 0$ to $t = t_1$. The result is found to be

$$E = \int_0^{t_1} u \, dt = T \int_0^{t_1} \left[\omega_1 - \omega_2 - T\left(\frac{I_1 + I_2}{I_1 I_2}\right)t\right] dt$$

$$= \frac{I_1 I_2 (\omega_1 - \omega_2)^2}{2(I_1 + I_2)} \tag{16-37}$$

Note that the energy dissipated is proportional to the velocity difference squared and is independent of the clutch torque.

Note that E in Eq. (16-37) is the energy lost or dissipated; this is the energy that is absorbed by the clutch or brake. If the inertias are expressed in U.S. customary units (lb · s²/in), then the energy absorbed by the clutch assembly is in lb · in. Using these units, the heat generated in Btu is

$$H = \frac{E}{9336} \tag{16-38}$$

In SI, the inertias are expressed in kilogram-meter units, and the energy dissipated is expressed in joules.

16-9   TEMPERATURE RISE

The temperature rise of the clutch or brake assembly can be approximated by the classic expression

$$\Delta T = \frac{H}{CW} \tag{16-39}$$

where ΔT = temperature rise, °F
C = specific heat, Btu/(lb$_m$ · °F); use 0.12 for steel or cast iron
W = mass of clutch or brake parts, lb$_m$

A similar equation can be written using SI units. It is

$$\Delta T = \frac{E}{Cm} \tag{16-40}$$

where ΔT = temperature rise, °C
C = specific heat; use 500 J/kg · °C for steel or cast iron
m = mass of clutch or brake parts, kg

The temperature-rise equations above can be used to explain what happens when a

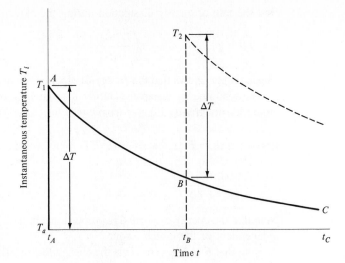

FIGURE 16-18

The effect of clutching or braking operations on temperature. T_a is the ambient temperature. Note that the temperature rise ΔT may be different for each operation.

clutch or brake is operated. However, there are so many variables involved that it would be most unlikely that such an analysis would even approximate experimental results. For this reason such analyses are most useful, for repetitive cycling, in pin-pointing those design parameters that have the greatest effect on performance.

An object heated to a temperature T_1 cools to an ambient temperature T_a according to the exponential relation

$$T_i - T_a = (T_1 - T_a)e^{-(AU/WC)t} \tag{16-41}$$

where T_1 = instantaneous temperature at time t, °F
A = heat-transfer area, ft^2
U = surface coefficient, Btu/(ft$^2 \cdot$ s $\cdot$ °F)

A similar expression can be written in SI units.

Figure 16-18 shows an application of Eq. (16-41). At time t_A a clutching or braking operation causes the temperature to rise to T_1 at A. Though the rise occurs in a finite time interval, it is assumed to occur instantaneously. The temperature then drops along the decay line ABC unless interrupted by another clutching operation. If a second operation occurs at time t_B, the temperature will rise along the dashed line to T_2 and then begin an exponential drop as before.

16-10   FRICTION MATERIALS

A brake or clutch friction material should have the following characteristics to a degree which is dependent upon the severity of the service:

1 A high and uniform coefficient of friction

2 Imperviousness to environmental conditions, such as moisture

3 The ability to withstand high temperatures, together with good heat conductivity

4 Good resiliency

5 High resistance to wear, scoring, and galling

The manufacture of friction materials is a highly specialized process, and it is advisable to consult manufacturers' catalogs and handbooks, as well as manufacturers directly, in selecting friction materials for specific applications. Selection involves a consideration of the many characteristics as well as the standard sizes available.

The *woven-cotton lining* is produced as a fabric belt which is impregnated with resins and polymerized. It is used mostly in heavy machinery and is usually supplied in rolls up to 50 ft in length. Thicknesses available range from $\frac{1}{8}$ to 1 in, in widths up to about 12 in.

A *woven-asbestos lining* is made in a similar manner to the cotton lining and may also contain metal particles. It is not quite as flexible as the cotton lining and comes in a smaller range of sizes. Along with the cotton lining, the asbestos lining is widely used as a brake material in heavy machinery.

Molded-asbestos linings contain asbestos fiber and friction modifiers; a thermoset polymer is used, with heat, to form a rigid or semirigid molding. The principal use is in drum brakes.

Molded-asbestos pads are similar to molded linings but have no flexibility; they are used for both clutches and brakes.

Sintered-metal pads are made of a mixture of copper and/or iron particles with friction modifiers, molded under high pressure and then heated to a high temperature to fuse the material. These pads are used in both brakes and clutches for heavy-duty applications.

Cermet pads are similar to the sintered-metal pads and have a substantial ceramic content.

Table 16-1 lists properties of typical brake linings. The linings may consist of a mixture of asbestos fibers to provide strength and ability to withstand high temperatures, various friction particles to obtain a degree of wear resistance as well as a higher coefficient of friction, and bonding materials.

Table 16-2 includes a wider variety of clutch friction materials, together with some

TABLE 16-1

Some Properties of Brake Linings

	WOVEN LINING	MOLDED LINING	RIGID BLOCK
Compressive strength, kpsi	10–15	10–18	10–15
Compressive strength, MPa	70–100	70–125	70–100
Tensile strength, kpsi	2.5–3	4–5	3–4
Tensile strength, MPa	17–21	27–35	21–27
Max. temperature, °F	400–500	500	750
Max. temperature, °C	200–260	260	400
Max. speed, ft/min	7500	5000	7500
Max. speed, m/s	38	25	38
Max. pressure, psi	50–100	100	150
Max. pressure, kPa	340–690	690	1000
Frictional coefficient, mean	0.45	0.47	0.40–45

TABLE 16-2

Friction Materials for Clutches

MATERIAL	FRICTION COEFFICIENT		MAX. TEMPERATURE		MAX. PRESSURE	
	WET	DRY	°F	°C	psi	kPa
Cast iron on cast iron	0.05	0.15–0.20	600	320	150–250	1000–1750
Powdered metal* on cast iron	0.05–0.1	0.1–0.4	1000	540	150	1000
Powdered metal* on hard steel	0.05–0.1	0.1–0.3	1000	540	300	2100
Wood on steel or cast iron	0.16	0.2–0.35	300	150	60–90	400–620
Leather on steel or cast iron	0.12	0.3–0.5	200	100	10–40	70–280
Cork on steel or cast iron	0.15–0.25	0.3–0.5	200	100	8–14	50–100
Felt on steel or cast iron	0.18	0.22	280	140	5–10	35–70
Woven asbestos* on steel or cast iron	0.1–0.2	0.3–0.6	350–500	175–260	50–100	350–700
Molded asbestos* on steel or cast iron	0.08–0.12	0.2–0.5	500	260	50–150	350–1000
Impregnated asbestos* on steel or cast iron	0.12	0.32	500–750	260–400	150	1000
Carbon graphite on steel	0.05–0.1	0.25	700–1000	370–540	300	2100

*The friction coefficient can be maintained within ±5 percent for specific materials in this group.

of their properties. Some of these materials may be run wet by allowing them to dip in oil or to be sprayed by oil. This reduces the coefficient of friction somewhat but carries away more heat and permits higher pressures to be used.

16-11   MISCELLANEOUS CLUTCHES AND COUPLINGS

The square-jaw clutch shown in Fig. 16-19a is one form of positive-contact clutch. These clutches have the following characteristics:

1 They do not slip.

2 No heat is generated.

3 They cannot be engaged at high speeds.

4 Sometimes they cannot be engaged when both shafts are at rest.

5 Engagement at any speed is accompanied by shock.

The greatest differences among the various types of positive clutches are concerned with the design of the jaws. To provide a longer period of time for shift action during engagement, the jaws may be ratchet-shaped, spiral-shaped, or gear-tooth-shaped. Sometimes a great many teeth or jaws are used, and they may be cut either circumferentially, so that they engage by cylindrical mating, or on the faces of the mating elements.

Although positive clutches are not used to the extent of the frictional-contact types,

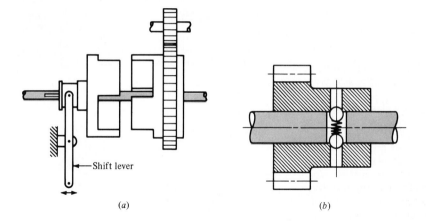

FIGURE 16-19

(a) Square-jaw clutch; (b) overload release clutch.

they do have important applications where synchronous operation is required, as, for example, in power presses or rolling-mill screw-downs.

Devices such as linear drives or motor-operated screwdrivers must run to a definite limit and then come to a stop. An overload-release type of clutch is required for these applications. Figure 16-19*b* is a schematic drawing illustrating the principle of operation of such a clutch. These clutches are usually spring-loaded so as to release at a predetermined torque. The clicking sound which is heard when the overload point is reached is considered to be a desirable signal.

Both fatigue and shock loads must be considered in obtaining the stresses and deflections of the various portions of positive clutches. In addition, wear must generally be considered. The application of the fundamentals discussed in Parts 1 and 2 is usually sufficient for the complete design of these devices.

An overrunning clutch or coupling permits the driven member of a machine to "freewheel" or "overrun" because the driver is stopped or because another source of power increases the speed of the driven mechanism. The construction uses rollers or balls mounted between an outer sleeve and an inner member having cam flats machined around the periphery. Driving action is obtained by wedging the rollers between the sleeve and the cam flats. This clutch is therefore equivalent to a pawl and ratchet with an infinite number of teeth.

There are many varieties of overrunning clutches available, and they are built in capacities up to hundreds of horsepower. Since no slippage is involved, the only power loss is that due to bearing friction and windage.

The shaft couplings shown in Fig. 16-20 are representative of the selection available in catalogs.

16-12   FLYWHEELS

The equation of motion for the flywheel represented in Fig. 16-1*b* is

$$\sum M = T_i(\theta_i, \dot{\theta}_i) - T_o(\theta_o, \dot{\theta}_o) - I\ddot{\theta} = 0$$

or

$$I\ddot{\theta} = T_i(\theta_i, \omega_i) - T_o(\theta_o, \omega_o) \qquad (a)$$

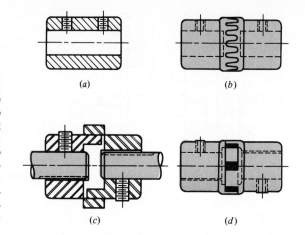

(a)

(b)

(c)

(d)

FIGURE 16-20

Shaft couplings. (a) Plain.
(b) Light-duty toothed coupling.
(c) BOST-FLEX® through-bore
design having elastomer insert to
transmit torque by compression;
insert permits 1° misalignment.
(d) Three-jaw coupling available
with bronze, rubber, or polyure-
thane insert to minimize vibration.
(Reproduced by permission, Boston Gear Division, Incom International, Inc., Quincy, Mass.)

where T_i is considered positive and T_o negative, and where $\dot\theta$ and $\ddot\theta$ are the first and second time derivatives of θ, respectively. Note that both T_i and T_o may depend for their values on the angular displacements θ_i and θ_o as well as their angular velocities ω_i and ω_o. In many cases the torque characteristic depends upon only one of these. Thus, the torque delivered by an induction motor depends upon the speed of the motor. In fact, motor manufacturers publish charts detailing the torque-speed characteristics of their various motors.

When the input and output torque functions are given, Eq. (a) can be solved for the motion of the flywheel using well-known techniques for solving linear and nonlinear differential equations. We can dispense with this here by assuming a rigid shaft, giving $\theta_i = \theta = \theta_o$. Thus, Eq. (a) becomes

$$I\ddot\theta = T_i(\theta, \omega) - T_o(\theta, \omega) \qquad (b)$$

When the two torque functions are known and the starting values of the displacement θ and velocity ω are given, Eq. (b) can be solved for ω and $\ddot\theta$ as functions of time. However, we are not really interested in the instantaneous values of these terms at all. Primarily we want to know the overall performance of the flywheel. What should its moment of inertia be? How do we match the power source to the load? And what are the resulting performance characteristics of the system that we have selected?

To gain insight into the problem, a hypothetical situation is diagrammed in Fig. 16-21. An input power source subjects a flywheel to a constant torque T_i while the

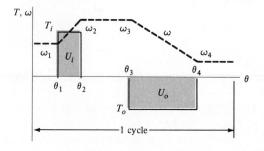

FIGURE 16-21

shaft rotates from θ_1 to θ_2. This is a positive torque and is plotted upward. Equation (b) indicates that a positive acceleration $\ddot{\theta}$ will be the result, and so the shaft velocity increases from ω_1 to ω_2. As shown, the shaft now rotates from θ_2 to θ_3 with zero torque and hence, from Eq. (b), with zero acceleration. Therefore $\omega_3 = \omega_2$. From θ_3 to θ_4 a load, or output torque, of constant magnitude is applied, causing the shaft to slow down from ω_3 to ω_4. Note that the output torque is plotted in the negative direction in accordance with Eq. (b).

The work input to the flywheel is the area of the rectangle between θ_1 and θ_2, or

$$U_i = T_i(\theta_2 - \theta_1) \tag{c}$$

The work output of the flywheel is the area of the rectangle from θ_3 to θ_4, or

$$U_o = T_o(\theta_4 - \theta_3) \tag{d}$$

If U_o is greater than U_i, the load uses more energy than has been delivered to the flywheel and so ω_4 will be less than ω_1. If $U_o = U_i$, ω_4 will be equal to ω_1 because the gains and losses are equal; we are assuming no friction losses. And finally, ω_4 will be greater than ω_1 if $U_i > U_o$.

We can also write these relations in terms of kinetic energy. At $\theta = \theta_1$ the flywheel has a velocity of ω_1 rad/s, and so its kinetic energy is

$$E_1 = \tfrac{1}{2}I\omega_1^2 \tag{e}$$

At $\theta = \theta_2$ the velocity is ω_2, and so

$$E_2 = \tfrac{1}{2}I\omega_2^2 \tag{f}$$

Thus the change in kinetic energy is

$$E_2 - E_1 = \tfrac{1}{2}I(\omega_2^2 - \omega_1^2) \tag{16-42}$$

Many of the torque displacement functions encountered in practical engineering situations are so complicated that they must be integrated by approximate methods. Figure 16-22, for example, is a typical plot of the engine torque for one cycle of motion of a single-cylinder internal combustion engine. Since a part of the torque curve is negative, the flywheel must return part of the energy back to the engine. Approximate integration of this curve for a cycle of 4π yields a mean torque T_m available to drive a load.

The simplest integration routine is Simpson's rule; this approximation can be handled on any computer and is short enough to use on the smallest programmable calculators. See Prob. 1-9. In fact, this routine is usually found as part of the library for most calculators and minicomputers. The equation used is

$$\int_{x_0}^{x_n} f(x)\, dx = \frac{h}{3}(f_0 + 4f_1 + 2f_2 + 4f_3 + 2f_4 + \cdots + 2f_{n-2} + 4f_{n-1} + f_n) \tag{16-43}$$

where

$$h = \frac{x_n - x_0}{n} \qquad x_n > x_0$$

and n is the number of subintervals used, 2, 4, 6, ... If memory is limited, solve Eq. (16-43) in two or more steps, say, from 0 to $n/2$ and then from $n/2$ to n.

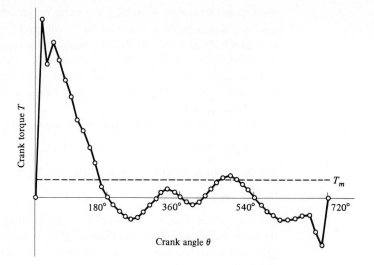

FIGURE 16-22

Relation between torque and crank angle for a one-cylinder four-cycle internal combustion engine.

It is convenient to define a *coefficient of speed fluctuation* as

$$C_s = \frac{\omega_2 - \omega_1}{\omega} \tag{16-44}$$

where ω is the nominal angular velocity, given by

$$\omega = \frac{\omega_2 + \omega_1}{2} \tag{16-45}$$

Equation (16-42) can be factored to give

$$E_2 - E_1 = \frac{I}{2}(\omega_2 - \omega_1)(\omega_2 + \omega_1)$$

Since $\omega_2 - \omega_1 = C_s\omega$ and $\omega_2 + \omega_1 = 2\omega$, we have

$$E_2 - E_1 = C_s I\omega^2 \tag{16-46}$$

Equation (16-46) can be used to obtain an appropriate flywheel inertia corresponding to the energy change $E_2 - E_1$.

EXAMPLE 16-2   Table 16-3 lists values of the torque used to plot Fig. 16-22. The nominal speed of the engine is to be 250 rad/s.

(a) Integrate the torque-displacement function for one cycle and find the energy that can be delivered to a load during the cycle.

(b) Determine the mean torque T_m (see Fig. 16-22).

(c) The greatest energy fluctuation is approximately between $\theta = 15°$ and $\theta = 150°$ on the torque diagram; see Fig. 16-22 and note that $T_o = -T_m$. Using a coefficient of speed fluctuation $C_s = 0.1$, find a suitable value for the flywheel inertia.

TABLE 16-3

θ, deg	T, lb · in	θ, deg	T, lb · in	θ, deg	T, lb · in	θ, deg	T, lb · in
0	0	180	0	360	0	540	0
15	2800	195	−107	375	−85	555	−107
30	2090	210	−206	390	−125	570	−206
45	2430	225	−280	405	−89	585	−292
60	2160	240	−323	420	8	600	−355
75	1840	255	−310	435	126	615	−371
90	1590	270	−242	450	242	630	−362
105	1210	285	−126	465	310	645	−312
120	1066	300	−8	480	323	660	−272
135	803	315	89	495	280	675	−274
150	532	330	125	510	206	690	−548
165	184	345	85	525	107	705	−760

(d) Find ω_2 and ω_1.

Solution (a) Using $n = 48$ and $h = 4\pi/48$, we enter the data of Table 16-3 into a computer program and get $E = 3490$ lb · in. This is the energy that can be delivered to the load.

Answer (b) $T_m = \dfrac{3490}{4\pi} = 278$ lb · in

(c) The largest positive loop on the torque-displacement diagram occurs between $\theta = 0°$ and $\theta = 180°$. We select this loop as yielding the largest speed change. Subtracting 278 lb · in from the values in Table 16-3 for this loop gives, respectively, −278, 2522, 1812, 2152, 1882, 1562, 1312, 932, 788, 525, 254, −94, and −278 lb · in. Entering Simpson's approximation again, using $n = 12$ and $h = 4\pi/48$, gives $E_2 - E_1 = 3660$ lb · in. We now solve Eq. (16-46) for I and substitute. This gives

Answer $I = \dfrac{E_2 - E_1}{C_s\omega^2} = \dfrac{3660}{0.1(250)^2} = 0.586$ lb · s^2 · in

(d) Equations (16-44) and (16-45) can be solved simultaneously for ω_2 and ω_1. Substituting appropriate values in these two equations yields

Answer $\omega_2 = \dfrac{\omega}{2}(2 + C_s) = \dfrac{250}{2}(2 + 0.1) = 262.5$ rad/s

Answer $\omega_1 = 2\omega - \omega_2 = 2(250) - 262.5 = 237.5$ rad/s

These two speeds occur at $\theta = 180°$ and $\theta = 0°$, respectively.

PROBLEMS*

16-1 The figure shows an internal rim-type brake having an inside rim diameter of 12 in and a dimension $R = 5$ in. The shoes have a face width of $1\frac{1}{2}$ in and are both actuated by a force $F = 500$ lb. The mean coefficient of friction is 0.28.

*An asterisk indicates a design-type problem.

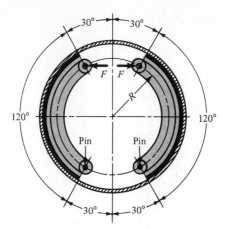

PROBLEM 16-1

(a) Find the maximum pressure and indicate the shoe on which it occurs.
(b) Estimate the braking torque effected by each shoe and find the total braking torque.
(c) Compute the resulting hinge-pin reactions.

16-2   Let the coefficient of friction for the brake of Prob. 16-1 be a random variable given as $\mathbf{f} =$ (0.28, 0.03). Find the maximum and minimum pressures within three standard deviations on each shoe assuming a deterministic actuating force.

16-3   In the figure for Prob. 16-1, the inside rim diameter is 280 mm and the dimension R is 90 mm. The shoes have a face width of 30 mm. Find the braking torque and the maximum pressure for each shoe if the actuating force is 1000 N, the drum rotation is counterclockwise, and $f = 0.30$.

16-4   The figure shows a 400-mm-diameter brake drum with four internally expanding shoes. Each of the hinge pins A and B supports a pair of shoes. The actuating mechanism is to be arranged to produce the same force F on each shoe. The face width of the shoes is 75 mm. The material used permits a coefficient of friction of 0.24 and a maximum pressure of 1000 kPa.
(a) Determine the actuating force.
(b) Calculate the braking capacity.
(c) Noting that rotation may be in either direction, compute the hinge-pin reactions.

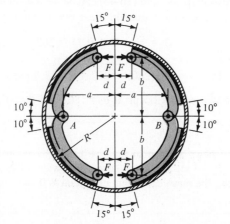

PROBLEM 16-4
The dimensions in millimeters are
$a = 150$, $b = 165$, $R = 200$, and
$d = 50$.

16-5 The block-type hand brake shown in the figure has a face width of 30 mm and a mean coefficient of friction of 0.25. For an estimated actuating force of 400 N, determine the maximum pressure on the shoe and the braking torque.

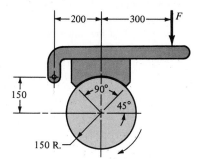

PROBLEM 16-5
Dimensions in millimeters.

16-6 Suppose the standard deviation of the coefficient of friction of Prob. 16-5 is $\hat{\sigma}_f = 0.025$, where the deviation from the mean is due entirely to environmental conditions. Find the braking torque corresponding to deviations of $3\hat{\sigma}$ above and below the mean.

16-7 The brake shown in the figure has a coefficient of friction of 0.30, a face width of 2 in, and a limiting shoe pressure of 150 psi. Find the limiting actuating force and the torque capacity.

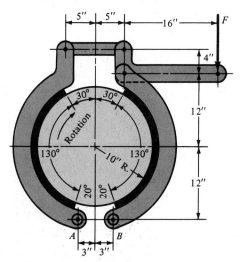

PROBLEM 16-7

16-8 The maximum lining pressure for the band brake shown is 90 psi. Use a 14-in-diameter drum, a

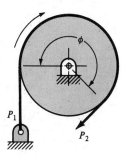

PROBLEM 16-8

band width of 4 in, a coefficient of friction of 0.25, and a wrap angle of 270°, and find the band tensions and the torque capacity.

16-9 The drum for the band brake in Prob. 16-8 is 300 mm in diameter. The band selected has a mean coefficient of friction of 0.28 and a width of 80 mm and can safely support a tension of 7.5 kN. If the angle of wrap is 270°, find the lining pressure and the torque capacity.

16-10 The brake shown in the figure has a coefficient of friction of 0.30 and is to operate using a maximum force $F = 400$ N. If the band width is 50 mm, find the band tensions and the braking torque.

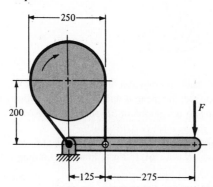

PROBLEM 16-10
Dimensions in millimeters.

16-11 A plate clutch has a single pair of mating friction surfaces 300 mm OD by 225 mm ID. The mean value of the coefficient of friction is 0.25, and the actuating force is 5 kN. Find the maximum pressure and the torque capacity, using the uniform-wear assumption.

16-12 What are the pressure and torque capacity of the clutch of Prob. 16-11 based on the assumption of uniform pressure?

16-13 Quality-control requirements dictate that the clutch of Prob. 16-11 deliver a torque of at least 150 N · m 99.9 percent of the time. If the standard deviation of the coefficient of friction is 0.02, estimate the least acceptable value of the mean coefficient of friction.

16-14 A plate clutch has a single pair of mating friction surfaces, is 200 mm OD by 100 mm ID, and has a coefficient of friction of 0.30. Using the uniform-wear method, find the actuating force and the clutch torque corresponding to a pressure limited to 250 kPa.

16-15 Solve Prob. 16-14 using the uniform-pressure assumption.

16-16 A plate clutch has two pairs of mating friction surfaces each 8 in OD by 6 in ID. The coefficient of friction is 0.24, and the pressure is not to exceed 120 psi. Using both methods of solution, estimate the actuating force and the resulting torques.

16-17 A disk clutch has four pairs of mating friction surfaces, is 5 in OD by 3 in ID, and has a mean coefficient of friction of 0.10. Using the uniform-pressure assumption, find the pressure and the torque corresponding to an actuating force of 1200 lb.

16-18 A cone clutch has $D = 330$ mm, $d = 306$ mm, a cone height of 60 mm, and a coefficient of friction of 0.26. A torque of 200 N · m is to be transmitted. For this requirement, estimate the actuating force and pressure by each method.

16-19*   A 12.5-hp one-cylinder two-cycle engine develops a maximum torque at a speed of 3400 rev/min. A cone clutch, with diameter not to exceed 4 in, is desired to couple the engine to its load. Is it feasible? If so, specify a suitable set of design values and find the resulting performance characteristics.

16-20   A two-jaw clutch has the dimensions shown in the figure and is made of ductile steel. The clutch has been designed to transmit 2 kW at 500 rev/min. Find the bearing and shear stress in the key and in the jaws.

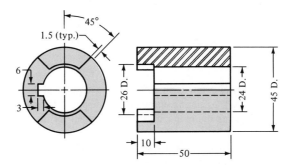

PROBLEM 16-20
Dimensions in millimeters.

16-21   A brake has a normal braking torque of 320 N · m and heat-dissipating surfaces whose mass is 18 kg. Suppose a load is brought to rest in 8.3 s from an initial velocity of 1800 rev/min using the normal braking torque; estimate the temperature rise of the heat-dissipating surfaces.

16-22   An inertial type of load is to be brought to rest in 10 s by a braking effort of 2800 lb · in of torque. The initial speed is 1600 rev/min, and the heat-dissipating surfaces weigh 40 lb. Estimate the temperature rise.

16-23   A flywheel is made from a steel disk 250 mm in diameter and 25 mm thick. A brake having a torque capacity of 7.5 N · m is used to bring the flywheel to a stop from a speed of 1000 rev/min. Estimate the energy absorbed by the brake, and the stopping time.

16-24   A cast-iron flywheel has a rim whose OD is 60 in and whose ID is 56 in. The flywheel weight is to be such that an energy fluctuation of 5000 lb · ft will cause the velocity to vary no more than 240 to 260 rev/min. Find the coefficient of speed fluctuation. If the weight of the spokes is neglected, what should be the thickness of the rim?

16-25   A single-geared blanking press has a stroke of 8 in and a rated capacity of 35 tons. A cam-driven ram is assumed to be capable of delivering the full press load at constant force during the last 15 percent of a constant-velocity stroke. The camshaft has an average speed of 90 rev/min and is geared to a flywheel shaft at a 6:1 ratio. The total work done is to include an allowance of 16 percent for friction.
(a) Estimate the maximum energy fluctuation.
(b) Find the rim weight for a mean rim diameter of 48 in and a coefficient of speed fluctuation of 0.10.

16-26   The load torque required by a 200-ton punch press is displayed in Table 16-4 for one revolution of the flywheel. The flywheel is to have a nominal speed of 240 rev/min and is to be designed for a coefficient of speed fluctuation of 0.075.

TABLE 16-4

θ, deg	T, lb · in	θ, deg	T, lb · in	θ, deg	T, lb · in	θ, deg	T, lb · in
0	857	90	7888	180	1801	270	857
10	857	100	8317	190	1629	280	857
20	857	110	8488	200	1458	290	857
30	857	120	8574	210	1372	300	857
40	857	130	8403	220	1115	310	857
50	1287	140	7717	230	1029	320	857
60	2572	150	3515	240	943	330	857
70	5144	160	2144	250	857	340	857
80	6859	170	1972	260	857	350	857

(*a*) Determine the mean motor torque required at the flywheel shaft and the motor horsepower needed, assuming constant-torque speed characteristics for the motor.

(*b*) Estimate the moment of inertia needed for the flywheel.

16-27   Using the data of Table 16-3, find the mean output torque and flywheel inertia required for a three-cylinder in-line engine corresponding to a nominal speed of 2400 rev/min. Use $C_s = 0.30$.

16-28   The disk of a clutch under uniform-wear condition has a geometry such as is shown in Fig. 16-14 and is faced with a friction material that can be subjected to a largest pressure p_a. Common proportions of such plate friction clutches lie in the range $0.45 \le d/D \le 0.80$. Is there a specific value of the ratio d/D, termed $(d/D)^*$, which permits the greatest allowable torque capacity to be built into a clutch plate with a fixed outer diameter D?

16-29   The diagram of Fig. 16-17 illustrates the force situation during cone-clutch slippage. If the axial force F continues to be exerted after the relative motion ceases, there is a friction force along the slant height of the cone. Under these circumstances, show that while the clutch is transmitting no torque,

$$F = \frac{p_a d(D - d)(1 + f \cot \alpha)}{2}$$

When the force F is released, the friction force reverses and the cone may or may not be expelled. Show that the critical angle is $\alpha_{cr} = \tan^{-1} f$.

ANSWERS   16-4 (*a*) 5.70 kN; (*b*) 1750 N · m; (*c*) 5.90 kN
16-7 $F = 360$ lb, $T = 23\ 400$ lb · in
16-12 $T = 165$ N · m
16-17 $T = 980$ lb · in
16-21 27.9°C
16-24 5.3 in
16-26 (*a*) 10.1 hp; (*b*) 150 lb · s^2 · in

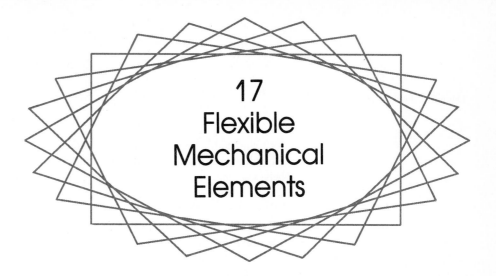

17
Flexible Mechanical Elements

Belts, ropes, chains, and other similar elastic or flexible machine elements are used in conveying systems and in the transmission of power over comparatively long distances. It often happens that these elements can be used as a replacement for gears, shafts, bearings, and other relatively rigid power-transmission devices. In many cases their use simplifies the design of a machine and substantially reduces the cost.

In addition, since these elements are elastic and usually quite long, they play an important part in absorbing shock loads and in damping out and isolating the effects of vibration. This is an important advantage as far as machine life is concerned.

Most flexible elements do not have an infinite life. When they are used, it is important to establish an inspection schedule to guard against wear, aging, and loss of elasticity. The elements should be replaced at the first sign of deterioration.

17-1 BELTS

The four principal types of belts are shown, with some of their characteristics, in Table 17-1. *Crowned pulleys* are used for flat belts, and *grooved pulleys*, or *sheaves,*

TABLE 17-1

Characteristics of Some Common Belt Types

BELT TYPE	FIGURE	JOINT	SIZE RANGE	CENTER DISTANCE
Flat		Yes	$t = \begin{cases} 0.03 \text{ to } 0.20 \text{ in} \\ 0.75 \text{ to } 5 \text{ mm} \end{cases}$	No upper limit
Round		Yes	$d = \frac{1}{8}$ to $\frac{3}{4}$ in	No upper limit
V		None	$b = \begin{cases} 0.31 \text{ to } 0.91 \text{ in} \\ 8 \text{ to } 19 \text{ mm} \end{cases}$	Limited
Timing		None	$p = 2$ mm and up	Limited

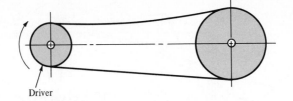

FIGURE 17-1

Open-belt drive; slack side should
be above.

Driver

for round and V belts. Timing belts require *toothed wheels,* or *sprockets*. In all cases,
the pulley axes must be separated by a certain minimum distance, depending upon the
belt type and size, to operate properly. Other characteristics of belts are:

- They may be used for long center distances.

- Except for timing belts, there is some slip and creep, and so the angular-velocity ratio
  between the driving and driven shafts is neither constant nor exactly equal to the ratio
  of the pulley diameters.

- In some cases an idler or tension pulley can be used to avoid adjustments in center
  distance that are ordinarily necessitated by age or the installation of new belts.

Figure 17-1 illustrates the usual open-belt drive. For a flat belt with this drive the
belt tension is such that the sag or droop is visible, as shown, when the belt is running.
Although the top is preferred for the loose side of the belt, for other belt types either the
top or the bottom may be used, because their installed tension is usually greater.

Two types of reversing drives are shown in Fig. 17-2. Notice that both sides of the
belt contact the pulleys in both figures, and so these drives cannot be used with V belts
or timing belts.

Figure 17-3 shows a flat-belt drive with out-of-plane pulleys. The shafts need not be
at right angles as in this case. Note the top view of the drive in Fig. 17-3. The pulleys
must be positioned so that the belt leaves each pulley in the midplane of the other

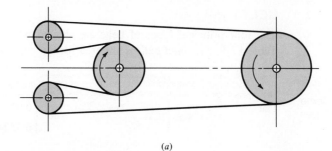

(*a*)

FIGURE 17-2

Reversing drives: (*a*) open;
(*b*) crossed. Crossed belts must
be separated to prevent rubbing if
high-friction materials are used.

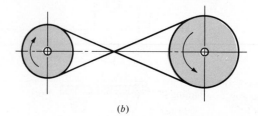

(*b*)

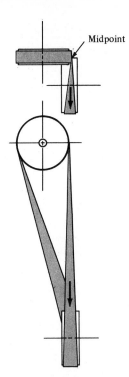

Midpoint

FIGURE 17-3

Quarter-twist belt drive; an idler
guide pulley must be used if motion is to be in both directions.

pulley face. Other arrangements may require guide pulleys to achieve this condition.

Another advantage of flat belts is shown in Fig. 17-4, where clutching action is obtained by shifting the belt from a loose to a tight or driven pulley.

Figure 17-5 shows two variable-speed drives. The drive in Fig. 17-5*a* is commonly used only for flat belts. The drive of Fig. 17-5*b* can also be used for V belts and round belts by using grooved sheaves.

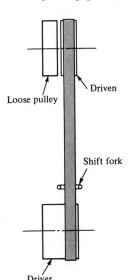

Driven

Loose pulley

Shift fork

FIGURE 17-4

This drive eliminates the need for
a clutch. Flat belt can be shifted
left or right by use of fork.

Driver

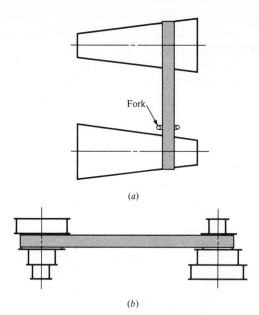

(a)

(b)

FIGURE 17-5

Variable-speed belt drives.

Flat belts are made of urethane and also of rubber-impregnated fabric reinforced with steel wire or nylon cords to take the tension load. One or both surfaces may have a friction surface coating. Flat belts are quiet, they are efficient at high speeds, and they can transmit large amounts of power over long center distances. Usually, flat belting is purchased by the roll and cut and the ends are joined using special kits furnished by the manufacturer. Two or more flat belts running side by side, instead of a single wide belt, are often used to form a conveying system.

A V belt is made of fabric and cord, usually cotton, rayon, or nylon, and impregnated with rubber. In contrast with flat belts, V belts are used with similar sheaves and at shorter center distances. V belts are slightly less efficient than flat belts, but a number of them can be used on a single sheave, thus making a multiple drive. V belts are made only in certain lengths and have no joints.

Timing belts are made of rubberized fabric and steel wire and have teeth which fit into grooves cut on the periphery of the sprockets. The timing belt does not stretch or slip and consequently transmits power at a constant angular-velocity ratio. The fact that the belt is toothed provides several advantages over ordinary belting. One of these is that no initial tension is necessary, so that fixed-center drives may be used. Another is the elimination of the restriction on speeds; the teeth make it possible to run at nearly any speed, slow or fast. Disadvantages are the first cost of the belt, the necessity of grooving the sprockets, and the attendant dynamic fluctuations caused at the belt-tooth meshing frequency.

17-2 FLAT- AND ROUND-BELT DRIVES

Modern flat-belt drives consist of a strong elastic core surrounded by an elastomer; these drives have distinct advantages over gear drives or V-belt drives. A flat-belt drive has an efficiency of about 98 percent, which is about the same as for a gear drive. On the other hand, the efficiency of a V-belt drive ranges from about 70 to 96 percent.* Flat-belt drives produce very little noise and absorb more torsional vibration from the system than either V-belt or gear drives.

When an open-belt drive (Fig. 17-1) is used, the contact angles are found to be

$$\theta_d = \pi - 2 \sin^{-1}\frac{D-d}{2C}$$

$$\theta_D = \pi + 2 \sin^{-1}\frac{D-d}{2C}$$

(17-1)

where D = diameter of large pulley
d = diameter of small pulley
C = center distance
θ = angle of contact

The length of the belt is found by summing the two arc lengths with twice the distance between the beginning and end of contact. The result is

$$L = [4C^2 - (D-d)^2]^{1/2} + \tfrac{1}{2}(D\theta_D + d\theta_d)$$

(17-2)

A similar set of equations can be derived for the crossed belts of Fig. 17-2b. For these the angle of wrap is the same for both pulleys and is

$$\theta = \pi + 2 \sin^{-1}\frac{D+d}{2C}$$

(17-3)

The belt length for crossed belts is found to be

$$L = [4C^2 - (D+d)^2]^{1/2} + \frac{\theta}{2}(D+d)$$

(17-4)

Firbank† explains flat-belt-drive theory in the following way. A change in belt tension due to friction forces between the belt and pulley will cause the belt to elongate or contract and move relative to the surface of the pulley. This motion is caused by *elastic creep* and is associated with sliding friction as opposed to static friction. The action at the driving pulley, through that portion of the angle of contact that is actually transmitting power, is such that the belt moves more slowly than the surface speed of the pulley because of the elastic creep. The angle of contact is made up of the *effective arc,* through which power is transmitted, and the *idle arc.* For the driving pulley the belt first contacts the pulley with a *tight-side tension* F_1 and a velocity V_1, which is the same as the surface velocity of the pulley. The belt then passes through the idle arc with no change in F_1 or V_1. Then creep or sliding contact begins, and the belt tension changes in accordance with the friction forces. At the end of the effective arc the belt leaves the pulley with a *loose-side tension* F_2 and a reduced speed V_2.

Firbank has used this theory to express the mechanics of flat-belt drives in mathematical form and has verified the results by experiment. His observations include the

*A. W. Wallin, "Efficiency of Synchronous Belts and V-Belts," *Proc. Nat. Conf. Power Transmission,* vol. 5, Illinois Institute of Technology, Chicago, Nov. 7–9, 1978, pp. 265–271.

†T. C. Firbank, *Mechanics of the Flat Belt Drive,* ASME paper no. 72-PTG-21.

fact that substantial amounts of power are transmitted by static friction as opposed to sliding friction. He also found that the coefficient of friction for a belt having a nylon core and leather surface was typically 0.7, but that it could be raised to 0.9 by employing special surface finishes.

Because of the space required to present the Firbank analysis, we present here a simplified approach instead; it is a conventional analysis that has been used for many years. We assume that the friction force on the belt is uniform throughout the entire arc of contact and that the centrifugal forces on the belt can be neglected. Then the relation between the tight-side tension F_1 and the slack-side tension F_2 is the same as for band brakes and is

$$\frac{F_1}{F_2} = e^{f\theta} \tag{17-5}$$

where f is the coefficient of friction and θ is the contact angle. The power transmitted is

$$P = (F_1 - F_2)V \tag{17-6}$$

In this equation the power P is in watts when the tension F is in newtons and the belt velocity V is in meters per second. The horsepower transmitted is

$$H = \frac{(F_1 - F_2)V}{33\ 000} \tag{17-7}$$

where the tensions F are in pounds and the velocity is in feet per minute.

The centrifugal force was neglected in writing Eq. (17-5). This force is given by the equation

$$F_c = mv^2 \tag{17-8}$$

where m is the mass of the belt per unit length and v is in units of length per second. When the centrifugal force is included, Eq. (17-5) becomes

$$\frac{F_1 - F_c}{F_2 - F_c} = e^{f\theta} \tag{17-9}$$

We note that the net tension ratio must be below $e^{f\theta}$, since this is the point of potential slippage of the belt-sheave interface. Now consider an added constraint. When the belt is installed, an initial tension F_i is set into the belt. Now think of each belt segment leaving the sheave as a spring under an initial tension F_i. As power is demanded, the sheave rotates, stretching the high-tension side and contracting the low-tension side. Thus

$$F_1 = F_i + \Delta F \tag{a}$$

$$F_2 = F_i - \Delta F \tag{b}$$

Solving for the initial tension gives

$$F_i = \frac{F_1 + F_2}{2} \tag{17-10}$$

The importance of Eq. (17-10) is that it really defines the maximum belt tension.

Consider this: When no power is being transmitted, the belt tensions on both sides are equal and so $F_1 = F_2 = F_i$. If now a slight load is added, some power is transmitted and F_1 increases by ΔF and F_2 decreases by the same amount. If the load is increased more and more, then eventually F_2 will become zero because the belt cannot support compression. At this point $F_1 = 2F_i$, which is the maximum belt tension. Thus, the only way to transmit more power is to increase the initial belt tension.

On the basis of the reasoning above, the drive is designed by limiting the maximum tension F_1 according to the allowable tension specified for the belt size and material. By making $F_2 = 0$ in Eq. (17-7) and substituting $2F_i$ for F_1, we get

$$H = \frac{F_i V}{16\ 500} \tag{17-11}$$

This is the basic design equation for flat- and round-belt drives. Certain modifications are needed, however, to account for the conditions of operation and the belt materials used.

It is unfortunate that many of the available data on belting are from sources in which they are presented in a very simplistic manner. These sources use a variety of charts, nomographs, and tables to enable someone who knows nothing about belting to apply them. Little, if any, computation is needed for such a person to obtain valid results. Since a basic understanding of the process, in many cases, is lacking, there is no way this person can vary the steps in the process to obtain a better design.

Incorporating the available belt-drive data into a form which provides a good understanding of belt mechanics involved certain adjustments in the data. Because of this, the results from the analysis presented here will not correspond exactly with those of the sources from which they were obtained.

A moderate variety of belt materials, with some of their properties, are listed in Table 17-2. These are sufficient for solving a large variety of design and analysis problems. The design equation to be used is, from Eq. (17-11),

$$H = \frac{C_P C_V F_a V}{33\ 000 K_S} \tag{17-12}$$

where H = horsepower transmitted
  C_P = pulley correction factor
  C_V = velocity correction factor
  F_a = allowable belt tension, lb
  V = belt speed, ft/min
  K_S = service factor

Minimum pulley sizes for the various belts are listed in Tables 17-2 and 17-3. The pulley correction factor accounts for the amount of bending or flexing of the belt and how this affects the life of the belt. For this reason it is dependent on the size and material of the belt used. Use Table 17-4. Use $C_P = 1.0$ for urethane belts.

Flat-belt pulleys should be crowned to keep the belts from running off the pulleys. If only one pulley is crowned, it should be the larger one. Both pulleys must be crowned whenever the pulley axes are not in a horizontal position. Use Table 17-5 for the crown height.

TABLE 17-2

Properties of Some Flat- and Round-Belt Materials. (Diameter = d, thickness = t, width = w)

MATERIAL	SPECIFICATION	SIZE, in	MINIMUM PULLEY DIAMETER, in	ALLOWABLE TENSION PER UNIT WIDTH AT 600 ft/min, lb/in	WEIGHT, lb/in³	COEFFICIENT OF FRICTION
Leather	1 ply	$t = \frac{11}{64}$	3	30	0.035–0.045	0.4
		$t = \frac{13}{64}$	$3\frac{1}{2}$	33	0.035–0.045	0.4
	2 ply	$t = \frac{18}{64}$	$4\frac{1}{2}$	41	0.035–0.045	0.4
		$t = \frac{20}{64}$	6^a	50	0.035–0.045	0.4
		$t = \frac{23}{64}$	9^a	60	0.035–0.045	0.4
Polyamide[b]	F–0[c]	$t = 0.03$	0.60	10	0.035	0.5
	F–1[c]	$t = 0.05$	1.0	35	0.035	0.5
	F–2[c]	$t = 0.07$	2.4	60	0.051	0.5
	A–2[c]	$t = 0.11$	2.4	60	0.037	0.8
	A–3[c]	$t = 0.13$	4.3	100	0.042	0.8
	A–4[c]	$t = 0.20$	9.5	175	0.039	0.8
	A–5[c]	$t = 0.25$	13.5	275	0.039	0.8
Urethane[d]	$w = 0.50$	$t = 0.062$	See	5.2^e	0.038–0.045	0.7
	$w = 0.75$	$t = 0.078$	Table	9.8^e	0.038–0.045	0.7
	$w = 1.25$	$t = 0.090$	17-3	18.9^e	0.038–0.045	0.7
	Round	$d = \frac{1}{4}$	See	8.3^e	0.038–0.045	0.7
		$d = \frac{3}{8}$	Table	18.6^e	0.038–0.045	0.7
		$d = \frac{1}{2}$	17-3	33.0^e	0.038–0.045	0.7
		$d = \frac{3}{4}$		74.3^e	0.038–0.045	0.7

[a]Add 2 in to pulley size for belts 8 in wide or more.

[b]Source: *Habasit Engineering Manual*, Habasit Belting, Inc., Chamblee (Atlanta), Ga.

[c]Friction cover of acrylonitrile-butadiene rubber on both sides.

[d]Source: Eagle Belting Co., Des Plaines, Ill.

[e]At 6% elongation; 12% is maximum allowable value.

TABLE 17-3

Minimum Pulley Sizes for Flat and Round Urethane Belts. (Listed are the Pulley Diameters in Inches)

BELT STYLE	BELT SIZE, in	RATIO OF PULLEY SPEED TO BELT LENGTH, rev/(ft · min)		
		UP TO 250	250 TO 499	500 TO 1000
Flat	0.50 × 0.062	0.38	0.44	0.50
	0.75 × 0.078	0.50	0.63	0.75
	1.25 × 0.090	0.50	0.63	0.75
Round	$\frac{1}{4}$	1.50	1.75	2.00
	$\frac{3}{8}$	2.25	2.62	3.00
	$\frac{1}{2}$	3.00	3.50	4.00
	$\frac{3}{4}$	5.00	6.00	7.00

Source: Eagle Belting Co., Des Plaines, Ill.

TABLE 17-4

Pulley Correction Factor C_P for Flat Belts*

MATERIAL	SMALL-PULLEY DIAMETER, in					
	1.6 TO 4	4.5 TO 8	9 TO 12.5	14, 16	18 TO 31.5	OVER 31.5
Leather	0.5	0.6	0.7	0.8	0.9	1.0
Polyamide, F–0	0.95	1.0	1.0	1.0	1.0	1.0
F–1	0.70	0.92	0.95	1.0	1.0	1.0
F–2	0.73	0.86	0.96	1.0	1.0	1.0
A–2	0.73	0.86	0.96	1.0	1.0	1.0
A–3	—	0.70	0.87	0.94	0.96	1.0
A–4	—	—	0.71	0.80	0.85	0.92
A–5	—	—	—	0.72	0.77	0.91

*Average values of C_P for the given ranges were approximated from curves in the *Habasit Engineering Manual*, Habasit Belting, Inc., Chamblee (Atlanta), Ga.

TABLE 17-5

Crown Height and ISO Pulley Diameters for Flat Belts*

ISO PULLEY DIAMETER, in	CROWN HEIGHT, in	ISO PULLEY DIAMETER, in	CROWN HEIGHT, in	
			$w \leq 10$ in	$w > 10$ in
1.6, 2, 2.5	0.012	12.5, 14	0.03	0.03
2.8, 3.15	0.012	12.5, 14	0.04	0.04
3.55, 4, 4.5	0.012	22.4, 25, 28	0.05	0.05
5, 5.6	0.016	31.5, 35.5	0.05	0.06
6.3, 7.1	0.020	40	0.05	0.06
8, 9	0.024	45, 50, 56	0.06	0.08
10, 11.2	0.030	63, 71, 80	0.07	0.10

*Crown should be rounded, not angled; maximum roughness is R_a = AA 63 μin.

The values given in Table 17-2 for the allowable belt tension are based on a belt speed of 600 ft/min. For higher speeds, use Fig. 17-6 to obtain C_V values for leather belts. For polyamide and urethane belts, use $C_V = 1.0$.

The service factors K_S for V-belt drives, given in Table 17-11 in Sec. 17-3, are also recommended here for flat- and round-belt drives.

17-3 V BELTS

The cross-sectional dimensions of V belts have been standardized by manufacturers, with each section designated by a letter of the alphabet for sizes in inch dimensions. Metric sizes are designated in numbers. Though these have not been included here, the procedure for analyzing and designing them is the same as presented here. Dimensions, minimum sheave diameters, and the horsepower range for each of the lettered sections are listed in Table 17-6.

To specify a V belt, give the belt-section letter, followed by the inside circumference in inches (standard circumferences are listed in Table 17-7). For example, B75 is a B-section belt having an inside circumference of 75 in.

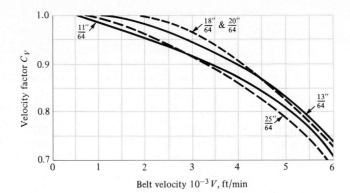

FIGURE 17-6

Velocity correction factor C_V for leather belts. *(Data source: Machinery's Handbook, 20th ed., Industrial Press, New York, 1976, p. 1047.)*

TABLE 17-6

Standard V-Belt Sections

BELT SECTION	WIDTH a, in	THICKNESS b, in	MINIMUM SHEAVE DIAMETER, in	hp RANGE, ONE OR MORE BELTS
A	$\frac{1}{2}$	$\frac{11}{32}$	3.0	$\frac{1}{4}$–10
B	$\frac{21}{32}$	$\frac{7}{16}$	5.4	1–25
C	$\frac{7}{8}$	$\frac{17}{32}$	9.0	15–100
D	$1\frac{1}{4}$	$\frac{3}{4}$	13.0	50–250
E	$1\frac{1}{2}$	1	21.6	100 and up

TABLE 17-7

Inside Circumferences of Standard V Belts

SECTION	CIRCUMFERENCE, in
A	26, 31, 33, 35, 38, 42, 46, 48, 51, 53, 55, 57, 60, 62, 64, 66, 68, 71, 75, 78, 80, 85, 90, 96, 105, 112, 120, 128
B	35, 38, 42, 46, 48, 51, 53, 55, 57, 60, 62, 64, 65, 66, 68, 71, 75, 78, 79, 81, 83, 85, 90, 93, 97, 100, 103, 105, 112, 120, 128, 131, 136, 144, 158, 173, 180, 195, 210, 240, 270, 300
C	51, 60, 68, 75, 81, 85, 90, 96, 105, 112, 120, 128, 136, 144, 158, 162, 173, 180, 195, 210, 240, 270, 300, 330, 360, 390, 420
D	120, 128, 144, 158, 162, 173, 180, 195, 210, 240, 270, 300, 330, 360, 390, 420, 480, 540, 600, 660
E	180, 195, 210, 240, 270, 300, 330, 360, 390, 420, 480, 540, 600, 660

Calculations involving the belt length are usually based on the pitch length. For any given belt section, the pitch length is obtained by adding a quantity to the inside circumference (Tables 17-7 and 17-8). For example, a B75 belt has a pitch length of 76.8 in. Similarly, calculations of the velocity ratios are made using the pitch diameters of the sheaves, and for this reason the stated diameters are usually understood to be the pitch diameters even though they are not always so specified.

TABLE 17-8

Length Conversion Dimensions. (Add the Listed Quantity to the Inside Circumference to Obtain the Pitch Length in Inches)

Belt section	A	B	C	D	E
Quantity to be added	1.3	1.8	2.9	3.3	4.5

The groove angle of a sheave is made somewhat less than the belt-section angle. This causes the belt to wedge itself into the groove, thus increasing the friction. The exact value of this angle depends upon the belt section, the sheave diameter, and the angle of contact. If it is made too much smaller than the belt, the force required to pull the belt out of the groove as the belt leaves the pulley will be excessive. Optimum values are given in the commercial literature.

The minimum sheave diameters have been listed in Table 17-6. For best results, a V belt should run quite fast; 4000 ft/min is a good speed. Trouble may be encountered if the belt runs much faster than 5000 ft/min or much slower than 1000 ft/min.

The *pitch,* or *effective length,* of a V belt is given by the equation

$$L_p = 2C + 1.57(D + d) + \frac{(D - d)^2}{4C} \tag{17-13}$$

where C = center distance
D = pitch diameter of large sheave
d = pitch diameter of small sheave
L_p = pitch length of belt

In the case of flat belts, there is virtually no limit to the center distance. Long center distances, however, are not recommended for V belts, because the excessive vibration of the slack side will shorten the belt life materially. In general, the center distance should not be greater than 3 times the sum of the sheave diameters or less than the diameter of the larger sheave. Link-type V belts have less vibration, because of better balance, and hence may be used with longer center distances.

The selection of V belts is based on obtaining a long and trouble-free life. Table 17-9 gives the horsepower capacity of standard single V belts for various sheave diameters and belt speeds corresponding to a satisfactory life. These ratings are based on a 180° contact angle. For smaller angles, this rating must be reduced. Figure 17-7 gives the values of the correction factor K_1, which is used to reduce the rated horsepower when the angle of contact is less than 180°.

For a given sheave speed, the hours of life of a short belt are less than those of a long belt, because the short belt is subjected to the action of the load a greater number of times. For this reason it is necessary to apply a second factor, K_2, which is called the *belt-length correction factor*. These factors are itemized in Table 17-10 for various belt sections and lengths. The rated belt horsepower must be multiplied by this factor to obtain the corrected horsepower.

The characteristics of both the driving and the driven machinery must be considered in selecting the belt. Manufacturers of V belts list these factors in great detail.

FIGURE 17-7

Correction factor K_1 for angle of contact. Multiply the rated horsepower per belt by this factor to obtain the corrected horsepower.

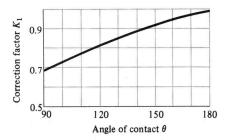

TABLE 17-9

Horsepower Ratings of
Standard V Belts

BELT SECTION	SHEAVE PITCH DIAMETER, in	BELT SPEED, ft/min				
		1000	2000	3000	4000	5000
A	2.6	0.47	0.62	0.53	0.15	
	3.0	0.66	1.01	1.12	0.93	0.38
	3.4	0.81	1.31	1.57	1.53	1.12
	3.8	0.93	1.55	1.92	2.00	1.71
	4.2	1.03	1.74	2.20	2.38	2.19
	4.6	1.11	1.89	2.44	2.69	2.58
	5.0 and up	1.17	2.03	2.64	2.96	2.89
B	4.2	1.07	1.58	1.68	1.26	0.22
	4.6	1.27	1.99	2.29	2.08	1.24
	5.0	1.44	2.33	2.80	2.76	2.10
	5.4	1.59	2.62	3.24	3.34	2.82
	5.8	1.72	2.87	3.61	3.85	3.45
	6.2	1.82	3.09	3.94	4.28	4.00
	6.6	1.92	3.29	4.23	4.67	4.48
	7.0 and up	2.01	3.46	4.49	5.01	4.90
C	6.0	1.84	2.66	2.72	1.87	
	7.0	2.48	3.94	4.64	4.44	3.12
	8.0	2.96	4.90	6.09	6.36	5.52
	9.0	3.34	5.65	7.21	7.86	7.39
	10.0	3.64	6.25	8.11	9.06	8.89
	11.0	3.88	6.74	8.84	10.0	10.1
	12.0 and up	4.09	7.15	9.46	10.9	11.1
D	10.0	4.14	6.13	6.55	5.09	1.35
	11.0	5.00	7.83	9.11	8.50	5.62
	12.0	5.71	9.26	11.2	11.4	9.18
	13.0	6.31	10.5	13.0	13.8	12.2
	14.0	6.82	11.5	14.6	15.8	14.8
	15.0	7.27	12.4	15.9	17.6	17.0
	16.0	7.66	13.2	17.1	19.2	19.0
	17.0 and up	8.01	13.9	18.1	20.6	20.7
E	16.0	8.68	14.0	17.5	18.1	15.3
	18.0	9.92	16.7	21.2	23.0	21.5
	20.0	10.9	18.7	24.2	26.9	26.4
	22.0	11.7	20.3	26.6	30.2	30.5
	24.0	12.4	21.6	28.6	32.9	33.8
	26.0	13.0	22.8	30.3	35.1	36.7
	28.0 and up	13.4	23.7	31.8	37.1	39.1

Table 17-11 can be used to obtain these factors when the characteristics of the driving and driven machinery have been identified.

EXAMPLE 17-1   A 10-hp split-phase motor running at 1750 rev/min is to be used to drive a rotary pump which operates 24 h per day. The pump should run at approximately 1175 rev/min.

TABLE 17-10

Belt-Length Correction Factor
K_2*

LENGTH FACTOR	NOMINAL BELT LENGTH, in				
	A BELTS	B BELTS	C BELTS	D BELTS	E BELTS
0.85	Up to 35	Up to 46	Up to 75	Up to 128	
0.90	38–46	48–60	81–96	144–162	Up to 195
0.95	48–55	62–75	105–120	173–210	210–240
1.00	60–75	78–97	128–158	240	270–300
1.05	78–90	105–120	162–195	270–330	330–390
1.10	96–112	128–144	210–240	360–420	420–480
1.15	120 and up	158–180	270–300	480	540–600
1.20		195 and up	330 and up	540 and up	660

*Multiply the rated horsepower per belt by this factor to obtain the corrected horsepower.

TABLE 17-11

Suggested Service Factors
K_S for V-Belt Drives

DRIVEN MACHINERY	SOURCE OF POWER	
	NORMAL TORQUE CHARACTERISTIC	HIGH OR NONUNIFORM TORQUE
Uniform	1.0 to 1.2	1.1 to 1.3
Light shock	1.1 to 1.3	1.2 to 1.4
Medium shock	1.2 to 1.4	1.4 to 1.6
Heavy shock	1.3 to 1.5	1.5 to 1.8

The center distance should not exceed 44 in. Space limits the diameter of the driven sheave to 11.5 in. Determine the sheave diameters, the belt size, and the number of belts required.

Solution

We first make the following decisions:

- An overload service factor of 1.2 corresponding to a 20 percent overload is selected from Table 17-11.

- From Table 17-6, a B-section belt is selected.

- Since the driven sheave should not exceed 11.5 in, the next smaller standard size of 11 in will be selected on a tentative basis.

- A center distance of 42 in is also selected tentatively.

The pump is to operate 24 h per day, and so 0.1 should be added to the service factor, making it 1.3. We must therefore design to obtain a horsepower of

$$H = 10(1.3) = 13 \text{ hp}$$

The diameter of the small sheave is

Answer

$$d = D\frac{n_1}{n_2} = 11\frac{1175}{1750} \approx 7.40 \text{ in}$$

This is a standard pitch diameter for B-section belts, though not listed in Table 17-9, and is also over the minimum diameter listed in Table 17-6. It will therefore be used.

Using Eq. (17-13), we find the pitch length to be

$$L_P = 2C + 1.57(D + d) + \frac{(D - d)^2}{4C}$$

$$= 2(42) + 1.57(11 + 7.4) + \frac{(11 - 7.4)^2}{4(42)} = 112.97 \text{ in}$$

This same result can also be obtained by using Eqs. (17-1) and (17-2). The nearest standard size, a B112, is selected from Table 17-7. This belt has a pitch length of 113.8 in.

The belt speed is

$$V = \frac{\pi dn}{12} = \frac{\pi(7.4)(1750)}{12} = 3390 \text{ ft/min}$$

Using Table 17-9 and interpolating, the rated horsepower per belt is 4.66. This must be corrected for the contact angle and the belt length. The contact angle for the small sheave turns out to be 3.056 rad, or 175°, obtainable from Eq. (17-1). From Fig. 17-7, the correction factor is 0.99. The belt-length correction factor is 1.05, from Table 17-10. Therefore the corrected horsepower per belt is

$$H = 0.99(1.05)(4.66) = 4.84 \text{ hp}$$

and so the number of belts required is

$$N = \frac{13}{4.85} = 2.69$$

Three B-section belts will therefore be specified.

17-4   TIMING BELTS

A timing belt is made of rubberized fabric with steel wire to take the tension load. It has teeth that fit into grooves cut on the periphery of the pulleys (Fig. 17-8); these are

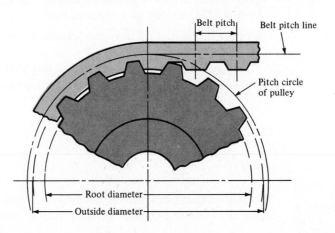

FIGURE 17-8

Timing-belt drive showing portions of the pulley and belt. Note that the pitch diameter of the pulley is greater than the diametral distance between the top lands of the teeth.

TABLE 17-12

Standard Pitches of Timing Belts

SERVICE	DESIGNATION	PITCH p, in
Extra light	XL	$\frac{1}{5}$
Light	L	$\frac{3}{8}$
Heavy	H	$\frac{1}{2}$
Extra heavy	XH	$\frac{7}{8}$
Double extra heavy	XXH	$1\frac{1}{4}$

coated with a nylon fabric. A timing belt does not stretch or slip and consequently transmits power at a constant angular-velocity ratio. No initial tension is needed. Such belts can operate over a very wide range of speeds, have efficiencies in the range of 97 to 99 percent, require no lubrication, and are quieter than chain drives. There is no chordal-speed variation, as in chain drives (see Sec. 17-5), and so they are an attractive solution for precision-drive requirements.

The steel-wire, or tension member, of a timing belt is located at the belt pitch line (Fig. 17-8). Thus the pitch length is the same regardless of the thickness of the backing.

The five standard inch-series pitches available are listed in Table 17-12 with their letter designations. Standard pitch lengths are available in sizes from 6 to 180 in. Pulleys come in sizes from 0.60 in pitch diameter up to 35.8 in and with groove numbers from 10 to 120.

The design and selection process for timing belts is so similar to that for V belts that the process will not be presented here. As in the case of other belt drives, the manufacturers will provide an ample supply of information and details on sizes and strengths.

17-5   ROLLER CHAINS

Basic features of chain drives include a constant ratio, since no slippage or creep is involved; long life; and the ability to drive a number of shafts from a single source of power.

Roller chains have been standardized as to sizes by the ANSI. Figure 17-9 shows

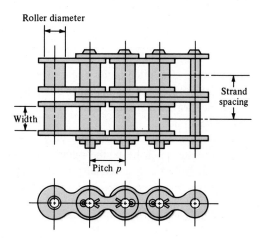

FIGURE 17-9

Portion of a double-strand roller chain.

TABLE 17-13

Dimensions of American
Standard Roller Chains—
Single Strand

ANSI CHAIN NUMBER	PITCH, in (mm)	WIDTH, in (mm)	MINIMUM TENSILE STRENGTH, lb (N)	AVERAGE WEIGHT, lb/ft (N/m)	ROLLER DIAMETER, in (mm)	MULTIPLE-STRAND SPACING, in (mm)
25	0.250 (6.35)	0.125 (3.18)	780 (3 470)	0.09 (1.31)	0.130 (3.30)	0.252 (6.40)
35	0.375 (9.52)	0.188 (4.76)	1 760 (7 830)	0.21 (3.06)	0.200 (5.08)	0.399 (10.13)
41	0.500 (12.70)	0.25 (6.35)	1 500 (6 670)	0.25 (3.65)	0.306 (7.77)	— —
40	0.500 (12.70)	0.312 (7.94)	3 130 (13 920)	0.42 (6.13)	0.312 (7.92)	0.566 (14.38)
50	0.625 (15.88)	0.375 (9.52)	4 880 (21 700)	0.69 (10.1)	0.400 (10.16)	0.713 (18.11)
60	0.750 (19.05)	0.500 (12.7)	7 030 (31 300)	1.00 (14.6)	0.469 (11.91)	0.897 (22.78)
80	1.000 (25.40)	0.625 (15.88)	12 500 (55 600)	1.71 (25.0)	0.625 (15.87)	1.153 (29.29)
100	1.250 (31.75)	0.750 (19.05)	19 500 (86 700)	2.58 (37.7)	0.750 (19.05)	1.409 (35.76)
120	1.500 (38.10)	1.000 (25.40)	28 000 (124 500)	3.87 (56.5)	0.875 (22.22)	1.789 (45.44)
140	1.750 (44.45)	1.000 (25.40)	38 000 (169 000)	4.95 (72.2)	1.000 (25.40)	1.924 (48.87)
160	2.000 (50.80)	1.250 (31.75)	50 000 (222 000)	6.61 (96.5)	1.125 (28.57)	2.305 (58.55)
180	2.250 (57.15)	1.406 (35.71)	63 000 (280 000)	9.06 (132.2)	1.406 (35.71)	2.592 (65.84)
200	2.500 (63.50)	1.500 (38.10)	78 000 (347 000)	10.96 (159.9)	1.562 (39.67)	2.817 (71.55)
240	3.00 (76.70)	1.875 (47.63)	112 000 (498 000)	16.4 (239)	1.875 (47.62)	3.458 (87.83)

Source: Compiled from ANSI B29.1-1975.

the nomenclature. The pitch is the linear distance between the centers of the rollers. The width is the space between the inner link plates. These chains are manufactured in single, double, triple, and quadruple strands. The dimensions of standard sizes are listed in Table 17-13.

In Fig. 17-10 is shown a sprocket driving a chain in a counterclockwise direction. Denoting the chain pitch by p, the pitch angle by γ, and the pitch diameter of the sprocket by D, from the trigonometry of the figure we see

$$\sin \frac{\gamma}{2} = \frac{p/2}{D/2} \quad \text{or} \quad D = \frac{p}{\sin (\gamma/2)} \tag{a}$$

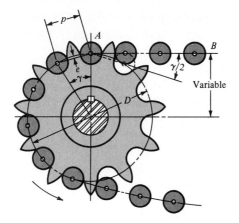

FIGURE 17-10

Engagement of a chain and sprocket.

Since $\gamma = 360°/N$, where N is the number of sprocket teeth, Eq. (*a*) can be written

$$D = \frac{p}{\sin (180°/N)} \tag{17-14}$$

The angle $\gamma/2$, through which the link swings as it enters contact, is called the *angle of articulation*. It can be seen that the magnitude of this angle is a function of the number of teeth. Rotation of the link through this angle causes impact between the rollers and the sprocket teeth and also wear in the chain joint. Since the life of a properly selected drive is a function of the wear and the surface fatigue strength of the rollers, it is important to reduce the angle of articulation as much as possible.

The number of sprocket teeth also affects the velocity ratio during the rotation through the pitch angle γ. At the position shown in Fig. 17-10, the chain AB is tangent to the pitch circle of the sprocket. However, when the sprocket has turned an angle of $\gamma/2$, the chain line AB moves closer to the center of rotation of the sprocket. This means that the chain line AB is moving up and down, and that the lever arm varies with rotation through the pitch angle, all resulting in an uneven chain exit velocity. You can think of the sprocket as a polygon in which the exit velocity of the chain depends upon whether the exit is from a corner, or from a flat of the polygon. Of course, the same effect occurs when the chain first enters into engagement with the sprocket.

The chain velocity V is defined as the number of feet coming off the sprocket in unit time. Thus the chain velocity in feet per minute is

$$V = \frac{Npn}{12} \tag{17-15}$$

where N = number of sprocket teeth
  p = chain pitch, in
  n = sprocket speed, rev/min

The maximum exit velocity of the chain is

$$v_{\max} = \frac{\pi Dn}{12} = \frac{\pi np}{12 \sin (\gamma/2)} \tag{b}$$

where Eq. (*a*) has been substituted for the pitch diameter *D*. The minimum exit velocity occurs at a diameter *d*, smaller than *D*. Using the geometry of Fig. 17-10, we find

$$d = D \cos \frac{\gamma}{2} \tag{c}$$

Thus the minimum exit velocity is

$$v_{\min} = \frac{\pi d n}{12} = \frac{\pi n p}{12} \frac{\cos (\gamma/2)}{\sin (\gamma/2)} \tag{d}$$

Now substituting $\gamma/2 = 180°/N$ and employing Eqs. (17-15), (*b*), and (*d*), we find the speed variation to be

$$\frac{\Delta V}{V} = \frac{v_{\max} - v_{\min}}{V} = \frac{\pi}{N} \left[\frac{1}{\sin (180°/N)} - \frac{1}{\tan (180°/N)} \right] \tag{17-16}$$

This is called the *chordal speed variation* and is plotted in Fig. 17-11. When chain drives are used to synchronize precision components or processes, due consideration must be given to these variations. For example, if a chain drive synchronized the cutting of photographic film with the forward drive of the film, the lengths of the cut sheets of film might vary too much because of this chordal speed variation. Such variations can also cause vibrations within the system.

Although a large number of teeth is considered desirable for the driving sprocket, in the usual case it is advantageous to obtain as small a sprocket as possible, and this requires one with a small number of teeth. For smooth operation at moderate and high speeds it is considered good practice to use a driving sprocket with at least 17 teeth; 19 or 21 will, of course, give a better life expectancy with less chain noise. Where space limitations are severe or for very slow speeds, smaller tooth numbers may be used by sacrificing the life expectancy of the chain.

Driven sprockets are not made in standard sizes over 120 teeth, because the pitch elongation will eventually cause the chain to "ride" high long before the chain is worn out. The most successful drives have velocity ratios up to 6:1, but higher ratios may be used at the sacrifice of chain life.

Roller chains seldom fail because they lack tensile strength; they more often fail because they have been subjected to a great many hours of service. Actual failure may be due either to wear of the rollers on the pins or to fatigue of the surfaces of the rollers. Roller-chain manufacturers have compiled tables which give the horsepower capacity corresponding to a life expectancy of 15 kh for various sprocket speeds. These capacities are tabulated in Tables 17-14 and 17-15 for 17-tooth sprockets.

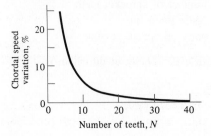

FIGURE 17-11

TABLE 17-14

Rated Horsepower Capacity of
Single-Strand Single-Pitch
Roller Chain for a 17-Tooth
Sprocket

SPROCKET SPEED, rev/min	ANSI CHAIN NUMBER					
	25	35	40	41	50	60
50	0.05	0.16	0.37	0.20	0.72	1.24
100	0.09	0.29	0.69	0.38	1.34	2.31
150	0.13*	0.41*	0.99*	0.55*	1.92*	3.32
200	0.16*	0.54*	1.29	0.71	2.50	4.30
300	0.23	0.78	1.85	1.02	3.61	6.20
400	0.30*	1.01*	2.40	1.32	4.67	8.03
500	0.37	1.24	2.93	1.61	5.71	9.81
600	0.44*	1.46*	3.45*	1.90*	6.72*	11.6
700	0.50	1.68	3.97	2.18	7.73	13.3
800	0.56*	1.89*	4.48*	2.46*	8.71*	15.0
900	0.62	2.10	4.98	2.74	9.69	16.7
1000	0.68*	2.31*	5.48	3.01	10.7	18.3
1200	0.81	2.73	6.45	3.29	12.6	21.6
1400	0.93*	3.13*	7.41	2.61	14.4	18.1
1600	1.05*	3.53*	8.36	2.14	12.8	14.8
1800	1.16	3.93	8.96	1.79	10.7	12.4
2000	1.27*	4.32*	7.72*	1.52*	9.23*	10.6
2500	1.56	5.28	5.51*	1.10*	6.58*	7.57
3000	1.84	5.64	4.17	0.83	4.98	5.76

Type A   Type B   Type C

*Estimated from ANSI tables by linear interpolation.
Note: Type A—manual or drip lubrication; type B—bath or disk lubrication; type C—oil-stream lubrication.
Source: Compiled from ANSI B29.1-1975 information only section, and from B29.9-1958.

The characteristics of the load are important considerations in the selection of roller chain. In general, extra chain capacity is required for any of the following conditions:

- The small sprocket has fewer than 9 teeth for low-speed drives or fewer than 16 teeth for high-speed drives.
- The sprockets are unusually large.
- Shock loading occurs, or there are frequent load reversals.
- There are three or more sprockets in the drive.
- The lubrication is poor.
- The chain must operate under dirty or dusty conditions.

To account for these and other conditions of operation, the ratings must be modified by two factors to get the corrected value for a single- or multiple-strand chain. These are

- The tooth correction factor K_1, which accounts for the fact that the driving sprocket may have more or fewer than 17 teeth (use Table 17-16)

TABLE 17-15

Rated Horsepower Capacity of Single-Strand Single-Pitch Roller Chain for a 17-Tooth Sprocket (Concluded)

SPROCKET SPEED, rev/min	ANSI CHAIN NUMBER							
	80	100	120	140	160	180	200	240
50 Type A	2.88	5.52	9.33	14.4	20.9	28.9	38.4	61.8
100	5.38	10.3	17.4	26.9	39.1	54.0	71.6	115
150	7.75	14.8	25.1	38.8	56.3	77.7	103	166
200	10.0	19.2	32.5	50.3	72.9	101	134	215
300	14.5	27.7	46.8	72.4	105	145	193	310
400	18.7	35.9	60.6	93.8	136	188	249	359
500	22.9	43.9	74.1	115	166	204	222	0
600	27.0	51.7	87.3	127	141	155	169	
700	31.0	59.4	89.0	101	112	123	0	
800	35.0	63.0	72.8	82.4	91.7	101		
900	39.9	52.8	61.0	69.1	76.8	84.4		
1000	37.7	45.0	52.1	59.0	65.6	72.1		
1200	28.7	34.3	39.6	44.9	49.9	0		
1400	22.7	27.2	31.5	35.6	0			
1600	18.6	22.3	25.8	0				
1800	15.6	18.7	21.6					
2000	13.3	15.9	0					
2500	9.56	0.40						
3000	7.25	0						

(Type B label appears in left margin, Type C and Type C′ below table)

Type C   Type C′

Note: Type A—manual or drip lubrication; type B—bath or disk lubrication; type C—oil-stream lubrication; type C′—type C, but this is a galling region; submit design to manufacturer for evaluation.

Source: Compiled from ANSI B29.1-1975 information only section, and from B29.9-1958.

TABLE 17-16

Tooth Correction Factors

NUMBER OF TEETH ON DRIVING SPROCKET	TOOTH CORRECTION FACTOR K_1	NUMBER OF TEETH ON DRIVING SPROCKET	TOOTH CORRECTION FACTOR K_1
11	0.53	22	1.29
12	0.62	23	1.35
13	0.70	24	1.41
14	0.78	25	1.46
15	0.85	30	1.73
16	0.92	35	1.95
17	1.00	40	2.15
18	1.05	45	2.37
19	1.11	50	2.51
20	1.18	55	2.66
21	1.26	60	2.80

The multiple-strand factor K_2, which accounts for the fact that the rating is not linearly related to the number of strands (see Table 17-17)

The corrected horsepower is obtained by applying these two factors to the rated horsepower as follows:

$$H'_r = K_1 K_2 H_r \tag{17-17}$$

TABLE 17-17

Multiple-Strand Factors K_2

NUMBER OF STRANDS	K_2
1	1.0
2	1.7
3	2.5
4	3.3

where H_r' is the fully corrected rating.

The service factor K_S in Table 17-11 can be used to account for variations in the driving and driven sources for roller chains too. Multiply the computed or given horsepower by K_S to get the design horsepower.

The length of a chain should be determined in pitches. It is preferable to have an even number of pitches; otherwise an offset link is required. The approximate length may be obtained from the following equation:

$$\frac{L}{p} = \frac{2C}{p} + \frac{N_1 + N_2}{2} + \frac{(N_2 - N_1)^2}{4\pi^2(C/p)} \tag{17-18}$$

where L = chain length
   p = chain pitch
   C = center distance
   N_1 = number of teeth on small sprocket
   N_2 = number of teeth on large sprocket

The length of chain for a multiple-sprocket drive is most easily obtained by making an accurate scale layout and determining the length by measurement.

Lubrication of roller chains is essential in order to obtain a long and trouble-free life. Either a drip feed or a shallow bath in the lubricant is satisfactory. A medium or light mineral oil, without additives, should be used. Except for unusual conditions, heavy oils and greases are not recommended, because they are too viscous to enter the small clearances in the chain parts.

EXAMPLE 17-2

A $7\frac{1}{2}$-hp speed reducer which runs at 300 rev/min is to drive a conveyor at 200 rev/min. The center distance is to be approximately 28 in. Select a suitable chain drive.

Solution

Although an odd number of sprocket teeth is preferred, sprockets of 20 and 30 teeth are tentatively chosen in order to obtain the proper velocity ratio. A 20-tooth sprocket will have a longer life and generate less noise than a 16- or an 18-tooth sprocket. It is chosen because space does not seem to be at a premium.

From Table 17-11 a service factor of $K_S = 1.3$ is chosen for operation with moderate shock. Thus, the design horsepower is

$$H = 1.3(7.5) = 9.75 \text{ hp}$$

Examination of Table 17-14 indicates that multiple strands of a No. 50 or 60 chain may be satisfactory. For a triple-strand No. 50, $H_r = 3.61$ hp. Tables 17-16 and 17-17 give $K_1 = 1.18$ and $K_2 = 2.5$. Thus, from Eq. (17-17),

$$H_r' = K_1 K_2 H_r = 1.18(2.5)(3.61) = 10.65 \text{ hp}$$

which is quite satisfactory. This is designated as a 50-3 chain.

For a 60-2 chain, we find $H_r = 6.20$ hp and $K_2 = 1.7$. Therefore

$$H_r' = 1.18(1.7)(6.2) = 12.44 \text{ hp}$$

which is also satisfactory.

The No. 60 chain would require larger sprockets, and hence it would run at a higher velocity, generate more noise, and have a shorter life. The No. 50 seems to be the better choice and is selected for this example. A comparison of the prices of the sprockets and chain for both cases might, however, make the No. 60 a better solution.

From Table 17-13, the pitch of No. 50 chain is $\frac{5}{8}$ in. Using a center distance in Eq. (17-18) of 28 in, the required length of triple-strand chain in pitches is

$$\frac{L}{p} = \frac{2C}{p} + \frac{N_1 + N_2}{2} + \frac{(N_2 - N_1)^2}{4\pi^2(C/p)}$$

$$= \frac{(2)(28)}{0.625} + \frac{20 + 30}{2} + \frac{(30 - 20)^2}{4\pi^2(28/0.625)}$$

$$= 114.7 \text{ pitches}$$

The nearest even number of pitches is 114, and this will be used. A slight adjustment in the center distance is required. Substituting $L/p = 114$ in Eq. (17-14) and solving for C gives approximately $27\frac{3}{4}$ in as the new center distance.

In general, the center distance should not exceed 80 pitches; 30 to 50 pitches is a better value. In this problem the center distance is $27.75/0.625 = 44.4$ pitches, which is satisfactory.

17-6 WIRE ROPE

Wire rope is made with two types of winding, as shown in Fig. 17-12. The *regular lay*, which is the accepted standard, has the wire twisted in one direction to form the strands, and the strands twisted in the opposite direction to form the rope. In the completed rope the visible wires are approximately parallel to the axis of the rope. Regular-lay ropes do not kink or untwist and are easy to handle.

Lang-lay ropes have the wires in the strand and the strands in the rope twisted in the same direction, and hence the outer wires run diagonally across the axis of the rope.

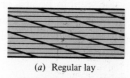

(a) Regular lay

(c) Section of 6 X 7 rope

FIGURE 17-12

Types of wire rope; both lays are available in either right or left hand.

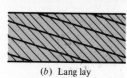

(b) Lang lay

Lang-lay ropes are more resistant to abrasive wear and failure due to fatigue than are regular-lay ropes, but they are more likely to kink and untwist.

Standard-ropes are made with a hemp core which supports and lubricates the strands. When the rope is subjected to heat, either a steel center or a wire-strand center must be used.

Wire rope is designated as, for example, a $1\frac{1}{8}$-in 6×7 haulage rope. The first figure is the diameter of the rope (Fig. 17-12c). The second and third figures are the number of strands and the number of wires in each strand, respectively. Table 17-18 lists some of the various ropes which are available, together with their characteristics and properties. The area of the metal in standard hoisting and haulage rope is $A_m = 0.38d^2$.

When a wire rope passes around a sheave, there is a certain amount of readjustment of the elements. Each of the wires and strands must slide on several others, and presumably some individual bending takes place. It is probable that in this complex action there exists some stress concentration. The stress in one of the wires of a rope passing around a sheave may be calculated as follows. From solid mechanics, we have

$$M = \frac{EI}{r} \quad \text{and} \quad M = \frac{\sigma I}{c} \tag{a}$$

where the quantities have their usual meaning. Eliminating M and solving for the stress

TABLE 17-18
Wire-Rope Data

ROPE	WEIGHT PER FOOT, lb	MINIMUM SHEAVE DIAMETER, in	STANDARD SIZES d, in	MATERIAL	SIZE OF OUTER WIRES	MODULUS OF ELASTICITY,* Mpsi	STRENGTH,† kpsi
6×7 haulage	$1.50d^2$	$42d$	$\frac{1}{4}-1\frac{1}{2}$	Monitor steel	$d/9$	14	100
				Plow steel	$d/9$	14	88
				Mild plow steel	$d/9$	14	76
6×19 standard hoisting	$1.60d^2$	$26d-34d$	$\frac{1}{4}-2\frac{3}{4}$	Monitor steel	$d/13-d/16$	12	106
				Plow steel	$d/13-d/16$	12	93
				Mild plow steel	$d/13-d/16$	12	80
6×37 special flexible	$1.55d^2$	$18d$	$\frac{1}{4}-3\frac{1}{2}$	Monitor steel	$d/22$	11	100
				Plow steel	$d/22$	11	88
8×19 extra flexible	$1.45d^2$	$21d-26d$	$\frac{1}{4}-1\frac{1}{2}$	Monitor steel	$d/15-d/19$	10	92
				Plow steel	$d/15-d/19$	10	80
7×7 aircraft	$1.70d^2$	—	$\frac{1}{16}-\frac{3}{8}$	Corrosion-resistant steel	—	—	124
				Carbon steel	—	—	124
7×9 aircraft	$1.75d^2$	—	$\frac{1}{8}-1\frac{3}{8}$	Corrosion-resistant steel	—	—	135
				Carbon steel	—	—	143
19-wire aircraft	$2.15d^2$	—	$\frac{1}{32}-\frac{5}{16}$	Corrosion-resistant steel	—	—	165
				Carbon steel	—	—	165

*The modulus of elasticity is only approximate; it is affected by the loads on the rope and, in general, increases with the life of the rope.

†The strength is based on the nominal area of the rope. The figures given are only approximate and are based on 1-in rope sizes and $\frac{1}{4}$-in aircraft-cable sizes.

Source: Compiled from *American Steel and Wire Company Handbook*.

gives

$$\sigma = \frac{Ec}{r} \qquad\qquad (b)$$

For the radius of curvature r, we can substitute the sheave radius $D/2$. Also, $c = d_w/2$, where d_w is the wire diameter. These substitutions give

$$\sigma = E\frac{d_w}{D} \qquad\qquad (17\text{-}19)$$

To understand this equation, observe that the individual wire makes a corkscrew figure in space and if you pull on it to determine E it will stretch or give more than its native E would suggest. Therefore E is still the modulus of elasticity of the *wire,* but in its peculiar configuration as part of the rope. A value for E equal to the modulus of elasticity of the *rope* gives an approximately correct value for the stress σ. For this reason we say that E in Eq. (17-19) is the modulus of elasticity of the rope, not the wire, recognizing that one can quibble over the name used.

Equation (17-19) gives the tensile stress σ in the outer wires. The sheave diameter is represented by D. This equation reveals the importance of using a large-diameter sheave. The suggested minimum sheave diameters in Table 17-18 are based on a D/d_w ratio of 400. If possible, the sheaves should be designed for a larger ratio. For elevators and mine hoists, D/d_w is usually taken from 800 to 1000. If the ratio is less than 200, heavy loads will often cause a permanent set in the rope.

A wire rope may fail because the static load exceeds the ultimate strength of the rope. Failure of this nature is generally not the fault of the designer, but rather that of the operator in permitting the rope to be subjected to loads for which it was not designed.

The first consideration in selecting a wire rope is to determine the static load. This load is composed of the following items:

* The known or dead weight

* Additional loads caused by sudden stops or starts

* Shock loads

* Sheave-bearing friction

When these loads are summed, the total can be compared with the ultimate strength of the rope to find a factor of safety. However, the ultimate strength used in this determination must be reduced by the strength loss that occurs when the rope passes over a curved surface such as a stationary sheave or a pin; see Fig. 17-13.

For an average operation, use a factor of safety of 5. Factors of safety up to 8 or 9 are used if there is danger to human life and for very critical situations. Table 17-19 lists minimum factors of safety for a variety of design situations. Here, the factor of safety is defined as

$$n = \frac{F_u}{F_t}$$

where F_u is the ultimate wire load and F_t is the largest working tension.

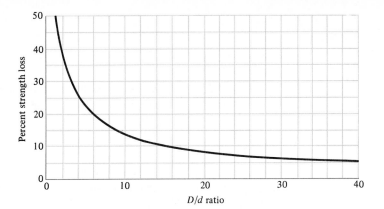

FIGURE 17-13

Percent strength loss due to different *D/d* ratios; derived from standard test data for 6 × 19 and 6 × 17 class ropes. *(From Wire Rope Users Manual; reproduced by permission from American Iron and Steel Institute.)*

Once you have made a tentative selection of a rope based upon static strength, the next consideration is to ensure that the wear life of the rope and the sheave or sheaves meets certain requirements. When a loaded rope is bent over a sheave, the rope stretches like a spring, rubs against the sheave, and causes wear of both the rope and the sheave. The amount of wear that occurs depends upon the pressure of the rope in the sheave groove. This pressure is called the *bearing pressure;* a good estimate of its magnitude is given by

$$p = \frac{2F}{dD} \tag{17-20}$$

where F = tensile force on rope
d = rope diameter
D = sheave diameter

TABLE 17-19

Minimum Factors of Safety for Wire Rope*

Track cables	3.2	Passenger elevators, ft/min:	
Guys	3.5	50	7.60
Mine shafts, ft:		300	9.20
Up to 500	8.0	800	11.25
1000–2000	7.0	1200	11.80
2000–3000	6.0	1500	11.90
Over 3000	5.0	Freight elevators, ft/min:	
Hoisting	5.0	50	6.65
Haulage	6.0	300	8.20
Cranes and derricks	6.0	800	10.00
Electric hoists	7.0	1200	10.50
Hand elevators	5.0	1500	10.55
Private elevators	7.5	Powered dumbwaiters, ft/min:	
Hand dumbwaiter	4.5	50	4.8
Grain elevators	7.5	300	6.6
		500	8.0

*Use of these factors does not preclude a fatigue failure.

Source: Compiled from a variety of sources, including ANSI A17.1-1978.

TABLE 17-20

Maximum Allowable Bearing
Pressures of Ropes on
Sheaves (in psi)

ROPE	WOOD[a]	CAST IRON[b]	CAST STEEL[c]	CHILLED CAST IRONS[d]	MANGANESE STEEL[e]
			MATERIAL		
Regular lay:					
6 × 7	150	300	550	650	1470
6 × 19	250	480	900	1100	2400
6 × 37	300	585	1075	1325	3000
8 × 19	350	680	1260	1550	3500
Lang lay:					
6 × 7	165	350	600	715	1650
6 × 19	275	550	1000	1210	2750
6 × 37	330	660	1180	1450	3300

[a]On end grain of beech, hickory, or gum.
[b]For H_B (min.) = 125.
[c]30–40 carbon; H_B (min.) = 160.
[d]Use only with uniform surface hardness.
[e]For high speeds with balanced sheaves having ground surfaces.
Source: Wire Rope Users Manual, AISI, 1979.

The allowable pressures given in Table 17-20 are to be used only as a rough guide; they may not prevent a fatigue failure or severe wear. They are presented here because they represent past practice and furnish a starting point in design.

A fatigue diagram not unlike an S-N diagram can be obtained for wire rope. Such a diagram is shown in Fig. 17-14. Here the ordinate is the pressure-strength ratio p/S_u, and S_u is the ultimate tensile strength of the wire. The abscissa is the number of bends which occur in the total life of the rope. The curve implies that a wire rope has a fatigue limit; but this is not true at all. A wire rope that is used over sheaves will eventually fail in fatigue or in wear. However, the graph does show that the rope will have a long life if the ratio p/S_u is less than 0.001. Substitution of this ratio in Eq. (17-20) gives

FIGURE 17-14

Experimentally determined relation
between the fatigue life of wire
rope and the sheave pressure.

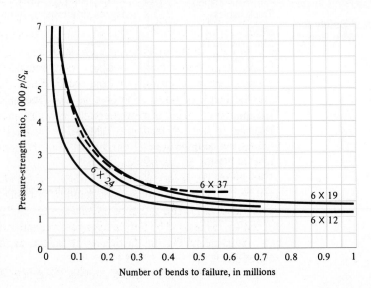

$$S_u = \frac{2000F}{dD} \tag{17-21}$$

where S_u is the ultimate strength of the *wire*, not the rope, and the units of S_u are related to the units of F. This interesting equation contains the wire strength, the load, the rope diameter, and the sheave diameter—all four variables in a single equation! Dividing both sides of Eq. (17-20) by the ultimate strength of the wires S_u and solving for F gives

$$F_f = \frac{(p/S_u)S_u dD}{2} \tag{17-22}$$

where F_f is interpreted as the fatigue tension allowable as the wire is flexed a number of times corresponding to p/S_u selected from Fig. 17-14 for a particular rope and life expectancy. The factor of safety can be defined in fatigue as

$$n = \frac{F_f}{F_t} \tag{17-23}$$

where F_f is the rope tension strength under flexing and F_t is the tension at the place where the rope is flexing. Unfortunately, the designer often has vendor information which tabulates ultimate rope tension and gives no ultimate-strength S_u information concerning the wires from which the rope is made. Some guidance in strength of individual wires is

IMPROVED PLOW STEEL (MONITOR):	$240 < S_u < 280$ kpsi
PLOW STEEL:	$210 < S_u < 240$ kpsi
MILD PLOW STEEL:	$180 < S_u < 210$ kpsi

In wire-rope usage, the factor of safety has been defined in static loading as $n = F_u/F_t$ or $n = (F_u - F_b)/F_t$, where F_b is the rope tension that would induce the same outer-wire stress as that given by Eq. (17-19). The factor of safety in fatigue loading can be defined as in Eq. (17-23), or by using a static analysis and compensating with a large factor of safety applicable to static loading, as in Table 17-19. When using factors of safety expressed in codes, standards, corporate design manuals, or wire-rope manufacturers' recommendations or from the literature, be sure to ascertain upon which basis the factor of safety is to be evaluated, and proceed accordingly.

If the rope is made of plow steel, the wires are probably hard-drawn AISI 1070 or 1080 carbon steel. Referring to Table 10-4, we see that this lies somewhere between hard-drawn spring wire and music wire. But the constants m and A needed to solve Eq. (10-17) for S_u are lacking.

Practicing engineers who desire to solve Eq. (17-21) should determine the wire strength S_u for the rope under consideration by unraveling enough wire to test for the Brinell hardness. Then S_u can be found using Eq. (5-20). Fatigue failure in wire rope is not sudden, as in solid bodies, but progressive, and shows as the breaking of an outside wire. This means that the beginning of fatigue can be detected by periodic routine inspection.

Figure 17-15 is another graph showing the gain in life to be obtained by using large

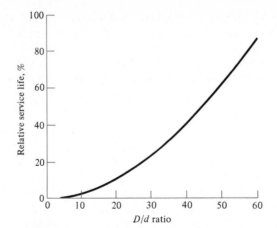

FIGURE 17-15

Service-life curve based on bending and tensile stresses only. This curve shows, for example, that the life corresponding to $D/d = 48$ is twice that of $D/d = 33$. *(From Wire Rope Users Manual; reproduced by permission from American Iron and Steel Institute.)*

D/d ratios. In view of the fact that the life of wire rope used over sheaves is only finite, it is extremely important that the designer specify and insist that periodic inspection, lubrication, and maintenance procedures be carried out during the life of the rope.

17-7 FLEXIBLE SHAFTS

One of the greatest limitations of the solid shaft is that it cannot transmit motion or power around corners. It is therefore necessary to resort to belts, chains, or gears, together with bearings and the supporting framework associated with them. The flexible shaft may often be an economical solution to the problem of transmitting motion around corners. In addition to the elimination of costly parts, its use may reduce noise considerably.

There are two main types of flexible shafts: the power-drive shaft for the transmit-

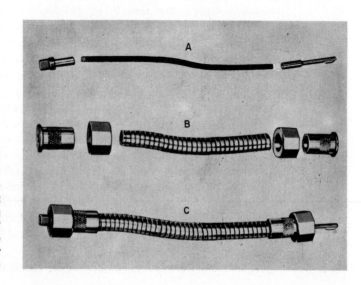

FIGURE 17-16

Construction of a flexible shaft. The wire-wound cable and end fittings are shown at *A*. These must be enclosed by the flexible casing and end fittings, shown at *B*. At *C* is shown the assembled flexible shaft. *(Courtesy of F. W. Stewart Corporation.)*

tion of power in a single direction, and the remote-control or manual-control shaft for the transmission of motion in either direction.

The construction of a flexible shaft is shown in Fig. 17-16. The cable is made by winding several layers of wire around a central core. For the power-drive shaft, rotation should be in a direction such that the outer layer is wound up. Remote-control cables have a different lay of the wires forming the cable, with more wires in each layer, so that the torsional deflection is approximately the same for either direction of rotation.

Flexible shafts are rated by specifying the torque corresponding to various radii of curvature of the casing. A 15-in radius of curvature, for example, will give from 2 to 5 times more torque capacity than a 7-in radius. When flexible shafts are used in a drive in which gears are also used, the gears should be placed so that the flexible shaft runs at as high a speed as possible. This permits the transmission of the maximum amount of horsepower.

PROBLEMS*

17-1 A flat belt is 6 in wide and $\frac{9}{32}$ in thick and transmits 15 hp. The pulley axes are parallel, in a horizontal plane, and 8 ft apart. The driving pulley has a diameter of 6 in and rotates at 1750 rev/min in such a way that the loose side of the belt is on top. The driven pulley has diameter 18 in. The weight of the belt material is 0.035 lb/in³.

(a) Determine the tension in the tight and slack sides of the belt if the coefficient of friction is 0.30.

(b) What belt tensions would result if adverse conditions caused the coefficient of friction to drop to 0.20? Would the belt slip?

(c) Calculate the length of the belt.

17-2 A nylon-core flat belt has an elastomer envelope, is 200 mm wide, and transmits 60 kW at a belt speed of 25 m/s. The belt has a mass of 2 kg/m of belt length. The belt is used in a crossed configuration to connect a 300-mm-diameter driving pulley to a 900-mm-diameter pulley at a shaft spacing of 6 m.

(a) Calculate the belt length and the angles of wrap.

(b) Compute the belt tensions based on a coefficient of friction of 0.38.

17-3* A flat-belt drive is to consist of two 4-ft cast-iron pulleys spaced 16 ft apart. Determine a satisfactory belt type and size to transmit 60 hp at a pulley speed of 380 rev/min. Use a service factor of 1.10.

17-4 A polyamide type A-3 flat belt is 10 in wide and connects a 16-in cast-iron driving pulley to a 36-in driven pulley in an open configuration and a 15-ft center distance. If the belt speed is 3600 ft/min, what maximum power can be transmitted? Use $K_S = 1.3$. What are the resulting belt tensions?

17-5* The line shaft illustrated in the figure is used to transmit power from an electric motor by means of flat-belt drives to various machines. Pulley A is driven by a vertical belt from the motor pulley. A belt from pulley B drives a machine tool at an angle of 70° from the vertical and at a center distance of 9 ft. Another belt from pulley C drives a grinder at a center distance of 11 ft. Pulley C has a double width to permit belt shifting as shown in Fig. 17-4. The belt from pulley D

---

*The asterisk indicates a design-type problem or one which may have no unique result.

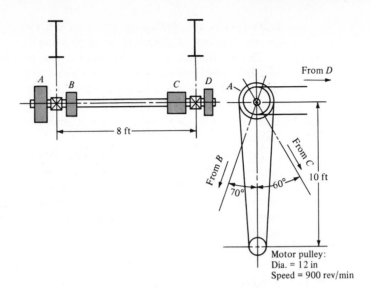

PROBLEM 17-5
(Courtesy of Dr. Ahmed F. Abdel Azim, Zagazig University, Cairo.)

drives a dust-extractor fan whose axis is located horizontally 8 ft from the axis of the line shaft. Additional data are:

MACHINE	SPEED, rev/min	POWER, hp	LINESHAFT PULLEY	DIAMETER, in
Machine tool	400	12.5	*B*	16
Grinder	300	4.5	*C*	14
Dust extractor	500	8.0	*D*	18

The power requirements, listed above, account for the overall efficiencies of the equipment. The two lineshaft bearings are mounted on hangers suspended from two overhead wide-flange beams (see Prob. 8-20, for example). Select the belt types and sizes for each of the four drives. Make provision for replacing belts from time to time because of wear or loss of elasticity.

17-6* Determine an appropriate belt size and length for pulley *D* in Prob. 17-5 if the belt is to have a crossed configuration.

17-7 A single V belt is to be selected to deliver engine power to the wheel-drive transmission of a riding tractor. A 5-hp single-cylinder engine is used. At most, 60 percent of this power is transmitted by the belt. The driving sheave has a diameter of 6.2 in, the driven, 12.0 in. The belt selected should be as close to a 92-in pitch length as possible. The engine speed is governor-controlled to a maximum of 3100 rev/min. An idler system is employed which causes the contact angles to be 180° for both pulleys. Select a satisfactory belt, and specify it using the standard designation.

17-8 A 60-hp four-cylinder internal combustion engine is used to drive brick-making machinery under a schedule of two shifts per day. The drive consists of two 26-in sheaves, spaced about 12 ft apart, with a sheave speed of 400 rev/min. Select a V-belt arrangement to handle this job.

17-9* A 2-hp electric motor running at 1720 rev/min is to drive a blower at a speed of 240 rev/min.

Select a V-belt drive for this application and specify standard V belts, sheave sizes, and the resulting center distance. The motor size limits the center distance to at least 22 in.

17-10 Two B85 V belts are used in a drive composed of a 5.4-in driving sheave, rotating at 1200 rev/min, and a 16-in driven sheave. Find the power capacity of the drive based on a service factor of 1.25, and the center distance.

17-11 A double-strand No. 60 roller chain is used to transmit power between a 13-tooth driving sprocket rotating at 300 rev/min and a 52-tooth driven sprocket.
(*a*) What is the rated horsepower of this drive?
(*b*) Determine the approximate center distance if the chain length is 82 pitches.

17-12 Calculate the torque and the bending force on the driving shaft produced by the chain of Prob. 17-11 if the actual horsepower transmitted is 30 percent less than the corrected rating.

17-13 A four-strand No. 40 roller chain transmits power from a 21-tooth driving sprocket which turns at 1200 rev/min. The velocity ratio is 4:1.
(*a*) Calculate the rated horsepower of this drive.
(*b*) Find the tension in the chain
(*c*) What is the factor of safety for the chain based upon the minimum tensile strength?
(*d*) What should be the chain length if the center distance is to be about 20 in?
(*e*) Estimate the maximum Hertzian shear stress in a roller; assume that the radius of curvature of the sprocket tooth at the point of contact is very large and that one tooth takes the full load.

17-14* A roller chain is to transmit 90 hp from a 17-tooth sprocket to a 34-tooth sprocket at a speed of 300 rev/min. The load characteristics are moderate shock with abnormal service conditions (poor lubrication, cold temperatures, and dirty surroundings). The equipment is to run 18 h/day. Specify the length and size of chain required for a center distance of about 25 pitches.

17-15* A 720 rev/min 25-hp squirrel-cage motor is to drive a two-cylinder reciprocating pump which is to be located out-of-doors under a shed. The pump is to run at full load for 24 h/day, and freedom from breakdowns is especially desired. The pump speed is 144 rev/min. Select suitable chain and sprocket sizes.

17-16* A mine hoist uses a 2-in 6 × 19 monitor-steel wire rope. The rope is used to haul loads up to 4 tons from a shaft 480 ft deep. The drum has a diameter of 6 ft; the sheaves are of good quality cast steel and the smallest is 3 ft in diameter.
(*a*) Using a maximum hoisting speed of 1200 ft/min and a maximum acceleration of 2 ft/s^2, determine the stresses in the rope.
(*b*) What are the factors of safety?

17-17* A temporary construction elevator is to be designed to carry workers and materials to a height of 90 ft. The maximum estimated load to be hoisted is 5000 lb at a velocity not exceeding 2 ft/s. Based on minimum sheave diameters, a minimum factor of safety, and an acceleration of 4 ft/s^2, find the number of ropes required. Use 1-in plow-steel 6 × 19 standard hoisting ropes.

17-18 A 2000-ft mine hoist operates with a 72-in drum using a 6 × 19 monitor-steel wire rope. The cage and load weigh 8000 lb, and the cage is subjected to an acceleration of 2 ft/s^2 when starting.
(*a*) For a single strand of wire rope, how does the factor of safety $n = F_f/F_t$ vary with the choice of rope diameter?

(*b*) For four strands of wire rope supporting the cage, how does the factor of safety vary with the choices of rope diameter?

17-19 Generalize the results of Prob. 17-18 by representing the factor of safety *n* as

$$n = \frac{ad}{(b/m) + cd^2}$$

where *m* is the number of ropes supporting the cage and *a*, *b*, and *c*, are constants. Show that the optimal diameter is $d^* = [b/(mc)]^{1/2}$ and that the corresponding maximum attainable factor of safety is $n^* = a[m/(bc)]^{1/2}/2$.

17-20 From your results in Prob. 17-19, show that to meet a fatigue factor of safety n_1, the optimal solution is

$$m = \frac{4bcn_1^2}{a^2} \quad \text{ropes}$$

having a diameter of

$$d = \frac{a}{2cn_1}$$

Solve Prob. 17-18 if a fatigue factor of safety of 2 is required. Show what to do in order to accommodate to the necessary discreteness in rope diameter *d* and number of ropes *m*.

17-21 For Prob. 17-16, estimate the elongation of the rope if a 9000-lb loaded mine cart is placed on the cage. The results of Prob. 3-16 may be useful.

17-1 (*a*) 349 lb, 169 lb; (*b*) 321 lb, 196 lb, but the belt will slip; (*c*) 230.1 in
17-7 Use a B90 belt.
17-10 3.44 hp
17-13 (*a*) 26.8 hp; (*b*) 842 lb; (*c*) 14.9 lb; (*d*) 135 pitches; (*e*) 45.3 kpsi
17-18 (*a*) The factor of safety *n* exhibits a stationary point maximum at about $d = 1\frac{5}{8}$ in.

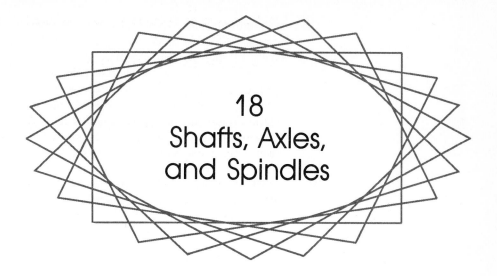

18
Shafts, Axles,
and Spindles

18-1   INTRODUCTION

A *shaft* is a rotating member, usually of circular cross section, used to transmit power or motion. It provides the axis of rotation, or oscillation, of elements such as gears, pulleys, flywheels, cranks, sprockets, and the like and controls the geometry of their motion. An *axle* is a nonrotating member which carries no torque and is used to support rotating wheels, pulleys, and the like. The automotive axle is not a true axle; the term is a carryover from the horse-and-buggy era, when the wheels rotated on nonrotating members. A *spindle* is a short shaft. Terms such as *lineshaft, headshaft, stub shaft, transmission shaft, countershaft,* and *flexible shaft* are names associated with special usage.

A shaft design really begins after much preliminary work. The design of the machine itself will dictate that certain gears, pulleys, bearings, and other elements will have at least been partially analyzed and their size and spacing tentatively determined. At this stage the design must be studied from the following points of view:

1 Deflection and rigidity
  (a) Bending deflection
  (b) Torsional deflection
  (c) Slope at bearings and shaft-supported elements
  (d) Shear deflection due to transverse loading of short shafts

2 Stress and strength
  (a) Static strength
  (b) Fatigue strength
  (c) Reliability

The geometry of a shaft is generally that of a stepped cylinder. Although a uniform-diameter cold-drawn round would require no finishing cuts, and appear to be inexpensive, it would be difficult to locate bearings, gears, pulleys, and other members on it in a positive manner. These elements must always be accurately positioned and provision made to accept thrust loads. The use of shaft shoulders is an excellent means of axially locating the shaft elements; these shoulders can be used to preload rolling bearings and to provide the necessary thrust reactions to the rotating elements. For these reasons our analysis will usually involve stepped shafts.

Figure 18-1 shows the stepped shaft supporting the gear of a worm-gear speed reducer. You should be able to determine the reason for each step of the shaft and for the sleeve. Note also how the lubricant is supplied and drained. Can this reducer be used for counterclockwise as well as clockwise rotation? Why?

In deciding on an approach to design, it is necessary to realize that a stress analysis at a specific point on a shaft can be made using only the shaft geometry in the vicinity of that point. Thus the geometry of the entire shaft is not needed. In design it is usually possible to locate the critical areas, size these to meet the strength requirements, and then size the rest of the shaft to meet the requirements of the shaft-supported elements.

Note that the deflection and slope analyses cannot be made until the geometry of the entire shaft has been defined. Thus deflection is a function of the geometry *everywhere*, whereas the stress at a section of interest is a function of *local geometry and moments*. For this reason shaft design allows a consideration of stress and strength first. Then, after tentative values for the shaft dimensions have been established, the determination of the deflection and slope can be made.

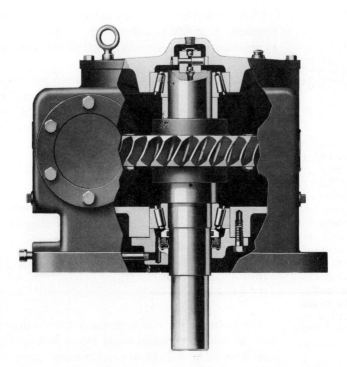

FIGURE 18-1

A vertical worm-gear speed reducer. *(Courtesy of Cleveland Worm and Gear Company.)*

18-2   DETERMINATION OF SHAFT GEOMETRY

The geometric configuration of the shaft to be designed is usually determined from past experience and, most often, is simply a revision of existing models in which a limited number of changes must be made. These changes can result for a variety of reasons, such as the use of a newly designed seal or coupling, a change in the power or speed, bearings of a different size, or the use of newly designed rotating components. Such changes are easy for the designer and need no further explanation.

If there is no existing design to use as a starter, then the determination of the shaft geometry may have many solutions. This problem is illustrated by the two examples of Fig. 18-2. In Fig. 18-2a a geared countershaft is to be supported by two bearings. In Fig. 18-2c a fanshaft is to be configured. A variety of solutions should occur to you for each of these problems. The solutions shown are not necessarily the best ones, but they do illustrate how the shaft-mounted devices are fixed and located in the axial direction, and how provision is made for torque transfer from one element to another. Note that the shaft in Fig. 18-2a is subject to bending, torsional, and axial loads. The shaft in Fig. 18-2c is subject only to bending and torsion.

There is no magic formula to determine the shaft geometry for any given design situation. The best approach is that of studying existing designs to learn how similar problems have been solved and then of combining the best of these to solve your own problem.

Many shaft-design situations include the problem of transmitting torque from one element to another on the shaft. Common *torque-transfer elements* are:

- Keys

- Splines

FIGURE 18-2

(*a*) Choose shaft configuration to support and locate the two gears and two bearings. (*b*) Solution uses integral pinion, three shaft shoulders, key and keyway, and sleeve. The housing locates the bearings on the outer rings and receives the thrust loads. (*c*) Choose fan-shaft configuration. (*d*) Solution uses sleeve bearings, a straight-through shaft, locating collars and set screws for collars, fan, and pulley. The sleeve bearings are supported by the fan housing.

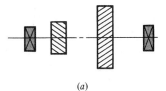

(*a*)

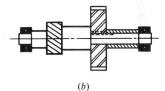

(*b*)

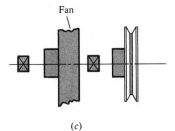

Fan

(*c*)

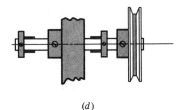

(*d*)

- Setscrews

- Pins

- Press or shrink fits

- Tapered fits

Keys, pins, and *setscrews* have already been described in Chap. 8. Pins for this purpose include not only straight pins, tapered pins, and cotters, but also a wide variety of patented pins many of which act as spring pins. Some of these are for locational purposes only—cotters, for example, should not be used to transmit very much torque—but others will serve as good torque transmitters. The use of these devices requires radial holes through the shaft, and hence stress concentration could be a problem, depending upon their location.

Shaft splines resemble gear teeth cut or forged into the shaft surface. They are used when large amounts of torque are to be transferred. When splines are used, stress concentration is generally quite moderate.

Press and shrink fits for securing hubs to shafts are used both for torque transfer and for preserving axial location. The resulting stress-concentration factor is usually quite small. A similar method is to use a split hub with screws to clamp the hub to the shaft. This method allows for disassembly and lateral adjustments. Another similar method uses a two-part hub consisting of a split inner member which fits into a tapered hole. The assembly is then tightened to the shaft using screws which force the inner part into the wheel and clamp the whole assembly against the shaft.

Plain tapered fits between the shaft and the shaft-mounted device are often used on the overhanging end of a shaft. Screw threads at the shaft end then permit the use of a nut to lock the wheel tightly to the shaft. This approach is useful because it can be disassembled, but it does not provide good axial location of the wheel on the shaft.

All these torque-transfer means solve the problem of securely anchoring the wheel or device to the shaft, but not all of them solve the problem of accurate axial location of the device. Some of the most-used *locational devices* include:

- Cotter and washer

- Nut and washer

- Sleeve

- Shaft shoulder

- Ring and groove

- Setscrew

- Split hub or tapered two-piece hub

- Collar and screw

- Pins

We have already discussed some of these items. The use of a ring fitted into a shaft groove is an economical solution to some problems. The grooves are quite shallow; many of the ring styles available do exert a spring force against the device to be anchored; and sometimes the grooves can be located where the stress-concentration effect is small or unimportant.

Figure 18-2*d* shows the use of a collar in which holding is obtained by tightening the setscrews against the shaft. An alternative and better arrangement uses a split collar and a cap screw to clamp the collar in place.

Figures 18-3 to 18-8 are intended as idea generators.* These successful designs have been developed and refined over a period of many years and represent very good design practice. Some of the figures in Chap. 11 will also be of help in the development of a good shaft design. Catalogues and commercial literature also will include various illustrations and recommendations leading to good design practice.

18-3   STATIC LOADING—GENERAL†

The determination of shaft dimensions is a far simpler problem when only static loads are present than it is when the loading is dynamic. And even with dynamic, that is, fatigue, loading, a preliminary estimate of dimensions is needed many times in order to get a good start on the problem. In that sense the material in this and in the following section will be useful in getting a first estimate of shaft dimensions for any kind of loading.

The stresses at a point on the surface of a solid round shaft of diameter d subject to bending, axial loading, and twisting are

$$\sigma_x = \frac{32M}{\pi d^3} + \frac{4F}{\pi d^2} \tag{a}$$

$$\tau_{xy} = \frac{16T}{\pi d^3} \tag{b}$$

---

*Figures 18-5 to 18-8 were redrawn from material furnished by New Departure-Hyatt Division, General Motors Corporation.

†We are grateful to Adjunct Professor Shankar Lal of the Naval Postgraduate School for his suggestions concerning much of the material in this and other sections. J.E.S., C.R.M.

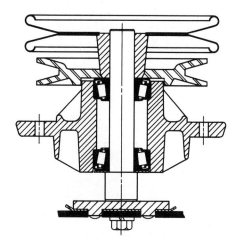

FIGURE 18-3

Tapered roller bearings used in a mowing-machine spindle. This design represents good practice for situations where one or more torque-transfer elements must be mounted outboard. (*Source: Redrawn from material furnished by The Timken Company.*)

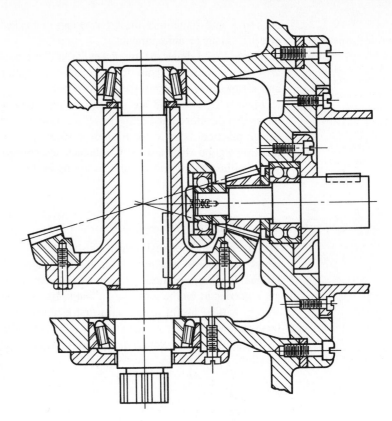

FIGURE 18-4

A bevel-gear drive in which both pinion and gear are straddle-mounted. *(Source: Redrawn from material furnished by Gleason Machine Division.)*

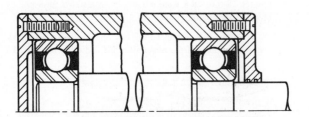

FIGURE 18-5

Arrangement showing bearing inner rings press-fitted to shaft while outer rings float in the housing. The axial clearance should be sufficient only to allow for machining variations. Note the labyrinth seal on the right.

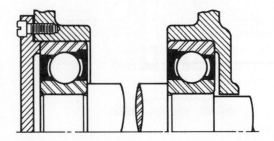

FIGURE 18-6

Similar to the arrangement of Fig. 18-5 except that the outer bearing rings are preloaded. Note the use of adjusting shims under the end cap.

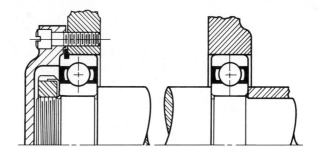

where the axial component of σ_x may be either additive or subtractive. We observe that the three loadings, M, F, and T, occur at the section containing the specific surface point.

By use of a Mohr's circle it can be shown that the two nonzero principal stresses are

$$\sigma_A,\ \sigma_B = \frac{\sigma_x}{2} \pm \left[\left(\frac{\sigma_x}{2}\right)^2 + \tau_{xy}^2\right]^{1/2} \tag{18-1}$$

These principal stresses can be combined to obtain the maximum shear stress τ_{max} and the von Mises stress σ'. The results are

$$\tau_{max} = \frac{\sigma_A - \sigma_B}{2} = \left[\left(\frac{\sigma_x}{2}\right)^2 + \tau_{xy}^2\right]^{1/2} \tag{18-2}$$

$$\sigma' = (\sigma_A^2 - \sigma_A\sigma_B + \sigma_B^2)^{1/2} = (\sigma_x^2 + 3\tau_{xy}^2)^{1/2} \tag{18-3}$$

By substituting Eqs. (a) and (b) in Eqs. (18-2) and (18-3) we obtain

$$\tau_{max} = \frac{2}{\pi d^3}\,[(8M + Fd)^2 + (8T)^2]^{1/2} \tag{18-4}$$

$$\sigma' = \frac{4}{\pi d^3}\,[(8M + Fd)^2 + 48T^2]^{1/2} \tag{18-5}$$

These equations enable us to determine τ_{max} or σ' when d is given, or to determine d when the allowable value of τ_{max} or σ' is given.

If the analysis, or design, is to be based on the maximum-shear-stress theory, then the allowable value of τ_{max} is

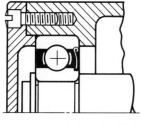

$$\tau_{all} = \frac{S_{sy}}{n} = \frac{S_y}{2n} \tag{18-6}$$

Equations (18-4) and (18-6) together can be used to determine the factor of safety n if the diameter d is given, or to find the diameter if the factor of safety is given.

A similar analysis can be carried out based on the distortion-energy theory of failure. In this case the allowable von Mises stress is

$$\sigma'_{all} = \frac{S_y}{n} \tag{18-7}$$

18-4

STATIC LOADING—BENDING AND TORSION

Under many conditions, the axial component F in Eqs. (18-4) and (18-5) is either zero or so small that it can be neglected. With $F = 0$, Eqs. (18-4) and (18-5) become

$$\tau_{max} = \frac{16}{\pi d^3} (M^2 + T^2)^{1/2} \tag{18-8}$$

$$\sigma' = \frac{16}{\pi d^3} (4M^2 + 3T^2)^{1/2} \tag{18-9}$$

It is easier to solve these equations for the diameter than Eqs. (18-4) and (18-5). By substituting the values of the allowable stresses from Eqs. (18-6) and (18-7), we find

$$d = \left[\frac{32n}{\pi S_y} (M^2 + T^2)^{1/2} \right]^{1/3} \tag{18-10}$$

using the maximum-shear-stress theory. Alternatively, if the diameter is known, the factor of safety is obtained from

$$\frac{1}{n} = \frac{32}{\pi d^3 S_y} (M^2 + T^2)^{1/2} \tag{18-11}$$

Similar relations can be obtained using the distortion-energy theory. The corresponding results are

$$d = \left[\frac{16n}{\pi S_y} (4M^2 + 3T^2)^{1/2} \right]^{1/3} \tag{18-12}$$

$$\frac{1}{n} = \frac{16}{\pi d^3 S_y} (4M^2 + 3T^2)^{1/2} \tag{18-13}$$

EXAMPLE 18-1   The integral pinion shaft shown in Fig. 18-9a is to be mounted in bearings at the locations shown and is to have a gear (not shown) mounted on the right-hand, or overhanging, end. The loading diagram (Fig. 18-9b) shows that the pinion force at A and the gear force at C are in the same xy plane. Equal and opposite torques T_A and T_C are assumed to be concentrated at A and C, as are the forces.

The bending-moment diagram of Fig. 18-9c shows a maximum moment at A, but the diameter of the pinion is quite large, and so the stress due to this moment can be neglected. On the other hand, another moment M_B, almost as large, occurs at the center of the right-hand bearing. Find the diameter d of the shaft at the right-hand bearing based upon a material having a yield strength of 66 kpsi and using a factor of safety of 1.80.

Solution   If we choose to design using the maximum-shear-stress theory, then Eq. (18-10) applies, giving

$$d = \left[\frac{32n}{\pi S_y} (M^2 + T^2)^{1/2} \right]^{1/3} = \left\{ \frac{32(1.80)}{\pi(66)(10^3)} [(-1925)^2 + (3300)^2]^{1/2} \right\}^{1/3}$$

Answer $$= 1.02 \text{ in}$$

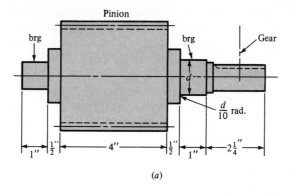

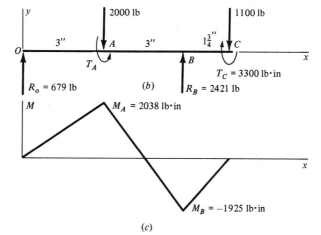

FIGURE 18-9

Left-bearing diameter is 1.000 in; left-bearing shoulder diameter = 2.000 in; dedendum diameter of gear is 3.43 in; diameter of right-hand bearing shoulder is 2.000 in; the overhang shoulder is $1\frac{1}{8}$ in by $\frac{1}{4}$ in long; diameter of overhanging gear seat is 1.000 in.

Equation (18-12) applies if the distortion-energy theory is chosen. This gives

$$d = \left[\frac{16n}{\pi S_y}(4M^2 + 3T^2)^{1/2}\right]^{1/3} = \left\{\frac{16(1.80)}{\pi(66)(10^3)}[4(-1925)^2 + 3(3300)^2]^{1/2}\right\}^{1/3}$$

Answer $= 0.986$ in

18-5 FATIGUE

Any rotating shaft loaded by stationary bending and torsional moments will be stressed by completely reversed bending stress, because of shaft rotation, but the torsional stress will remain steady. By using the subscript *a* for *alternating stress amplitude* and *m* for *midrange stress* or *steady stress*, as in Chap. 7, we can express Eqs. (*a*) and (*b*) of Sec. 18-3 as

$$\sigma_{xa} = \frac{32M_a}{\pi d^3} \qquad \tau_{xym} = \frac{16T_m}{\pi d^3}$$ (*a*)

These two stress components can be manipulated using separate Mohr's circles for each and using either the maximum-shear-stress theory or the distortion-energy theory to obtain equivalent values of the mean and alternating stresses. When these values have been obtained, one of the failure relations shown in the fatigue diagram of Fig. 18-10 can be selected for analysis or design. The intersection of the load line with the relation selected then establishes the limiting values of the stress components.

If the maximum-shear-stress theory is to be used, then the two Mohr's circles yield τ_a and τ_m. The components to be used in Fig. 18-10 are then

$$\sigma_a = 2\tau_a \qquad \sigma_m = 2\tau_m \tag{18-14}$$

If the distortion-energy theory is to be used, then the values are

$$\sigma_a = \sigma_{xa} \qquad \sigma_m = \sqrt{3}\tau_{xym} \tag{18-15}$$

Some of the relations shown in Fig. 18-10 have already been discussed in Chap. 7. Table 18-1 has been prepared to show the basic formula corresponding to each prediction.

The Soderberg approach (1936) uses data from the simple tension test and is quite conservative. This approach provides for the possibility of failure by yielding on the first half-cycle, and so Eq. (18-21) in Table 18-1 need not be used when this relation is selected.

The Goodman approach, which is really a modification of the original Goodman proposal (1894), is linear and also uses data from simple tension tests. However, it does not address the possibility of yielding on the first half-cycle. Therefore both Eqs. (18-17) and (18-21), in Table 18-1, should be used when the Goodman relation is selected. When so used, this combined approach is sometimes designated the *Goodman-Langer* approach.

The relation suggested by Gerber (1874) is based on tension-test data and is rather centrally located with respect to the actual test points. For this reason it is a favored approach when reliability analyses are to be made. The possibility of yielding must be checked when the Gerber method is used; the result is then called the *Gerber-Langer* method.

The ASME elliptic formula in Table 18-1 is included in Appendix A of the new

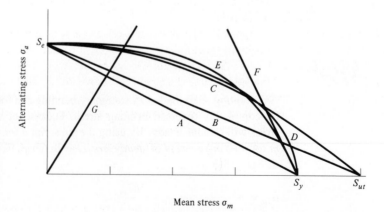

FIGURE 18-10

Fatigue diagram plotted to scale showing the relations between some well-known theories of fatigue failure. *A*, Soderberg line; *B*, Goodman line; *C*, Gerber curve; *D*, ASME elliptic curve; *E*, Bagci curve; *F*, yielding on first half-cycle (Langer line); *G*, load line.

THEORY	BASIC FORMULA	EQUATION NO.
Soderberg	$\dfrac{n\sigma_a}{S_e} + \dfrac{n\sigma_m}{S_y} = 1$	(18-16)
Goodman	$\dfrac{n\sigma_a}{S_e} + \dfrac{n\sigma_m}{S_{ut}} = 1$	(18-17)
Gerber	$\dfrac{n\sigma_a}{S_e} + \left(\dfrac{n\sigma_m}{S_{ut}}\right)^2 = 1$	(18-18)
ASME elliptic	$\left(\dfrac{n\sigma_a}{S_e}\right)^2 + \left(\dfrac{n\sigma_m}{S_y}\right)^2 = 1$	(18-19)
Bagci	$\dfrac{n\sigma_a}{S_e} + \left(\dfrac{n\sigma_m}{S_y}\right)^4 = 1$	(18-20)
Yielding (Langer)	$\dfrac{n}{S_y}(\sigma_a + \sigma_m) = 1$	(18-21)

TABLE 18-1

Fatigue-Strength Predictions for Shafts

ASME standard for shafting.* Despite the presence of the S_y term in Eq. (18-19), the possibility of yielding on the first half-cycle must be checked when the ASME formula is used. The reason for this can be seen in Fig. 18-10, where curve D is seen to cross the yield line F.

Bagci (1979) suggested a failure locus in quartic form so as to pass through the test points. This method uses the same intercepts as Soderberg and also includes protection against yielding on the first half-cycle.

Though not shown in Fig. 18-10, the line used to predict yielding has intercepts at S_y on both axes. This line makes a 45° angle with the axes *only* if both are plotted to the same scale.

18-6   AN EXAMPLE OF FATIGUE ANALYSIS

In this example of relating strength to shaft size, we opt for the maximum-shear-stress theory to predict the damaging stress and the modified Goodman line to predict the significant strength. The analysis is restricted to the case of reversed bending and steady torque.

Figure 18-11a shows a stress element on the surface of a solid round shaft whose rotational speed is ω, in radians per second. Now suppose that a plane PQ is passed through the upper right-hand corner of the element. Then, below plane PQ there will be a wedge-shaped element, as shown in Fig. 18-11b. The angle α shown in the figure is the angle between plane PQ and a horizontal plane. We shall consider all possible values of α to see whether or not we can decide what it will be for those planes on which failure occurs.

By taking an equilibrium equation for all forces in the direction of τ_α, we get

$$\tau_\alpha + \sigma_x \sin \alpha \cos \alpha + \tau_{xy} \sin^2 \alpha - \tau_{xy} \cos^2 \alpha = 0$$

Design of Transmission Shafting, ANSI/ASME B106.1M-1985.

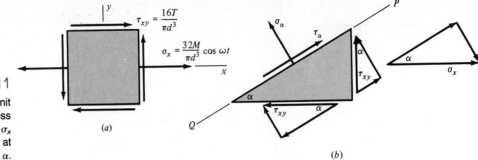

FIGURE 18-11

(a) Shaft stress element of unit depth has a steady shear stress τ_{xy} and an alternating stress σ_x due to rotation; (b) element cut at angle α.

or

$$\tau_\alpha = \tau_{xy}(\cos^2 \alpha - \sin^2 \alpha) - \sigma_x \sin \alpha \cos \alpha \qquad (a)$$

Substituting the values of τ_{xy} and σ_x in Eq. (a) and using several trigonometric identities yields

$$\tau_\alpha = \frac{16T_m}{\pi d^3} \cos 2\alpha - \frac{16M_a}{\pi d^3} \sin 2\alpha \cos \omega t \qquad (b)$$

For any plane making an angle α with the horizontal plane, Eq. (b) reveals that the shear stress has a steady component of

$$\tau_{\alpha m} = \frac{16T_m}{\pi d^3} \cos 2\alpha \qquad (c)$$

and an alternating component with amplitude

$$\tau_{\alpha a} = \frac{16M_a}{\pi d^3} \sin 2\alpha \qquad (d)$$

Shown in Fig. 18-12 is the modified Goodman diagram for shear. The shear endur-

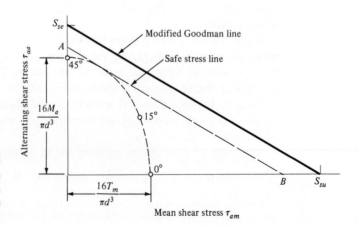

FIGURE 18-12

Fatigue diagram showing how the line of safe stress AB is drawn parallel to the modified Goodman line and tangent to the ellipse.

ance limit S_{se} is the fatigue strength of the shaft at a specified location after correction for size, surface finish, life, stress concentration, and the like.

To determine whether or not failure will occur on certain planes making an angle α with the horizontal, we plot a point in Fig. 18-12 for each value of α. The coordinates of the point are $(\tau_{\alpha m}, \tau_{\alpha a})$ as given by Eqs. (c) and (d). Notice that all such points fall on a quarter-ellipse in Fig. 18-12.

The safe-stress line in Fig. 18-12 is a line tangent to the ellipse and parallel to the Goodman line. The slopes of both lines are the same and are equal to $-S_{se}/S_{su}$. Differentiating Eqs. (c) and (d) with respect to α yields

$$\frac{d\tau_{\alpha m}}{d\alpha} = -\frac{32T_m}{\pi d^3} \sin 2\alpha \tag{e}$$

$$\frac{d\tau_{\alpha a}}{d\alpha} = \frac{32M_a}{\pi d^3} \cos 2\alpha \tag{f}$$

The slope of any tangent corresponding to the angle α is the ratio of Eq. (f) to Eq. (e). Since this is the same as the slope of the modified Goodman line, we have

$$-\frac{S_{se}}{S_{su}} = \frac{d\tau_{\alpha a}/d\alpha}{d\tau_{\alpha m}/d\alpha} = -\frac{M_a}{T_m \tan 2\alpha}$$

which gives

$$\tan 2\alpha = \frac{M_a}{T_m} \frac{S_{su}}{S_{se}} = \frac{M_a S_{ut}}{T_m S_e} \tag{18-22}$$

because, by the maximum-shear-stress theory, we take $S_{ut} = 2S_{su}$ and $S_e = 2S_{se}$. Then

$$\alpha = \frac{1}{2} \tan^{-1} \frac{M_a S_{ut}}{T_m S_e} \tag{18-23}$$

Also, because we are using the maximum-shear-stress theory, $\sigma_a = 2\tau_a$ and $\sigma_m = 2\tau_m$. Equations (c) and (d) should now be rewritten as

$$\sigma_a = \frac{32M_a}{\pi d^3} \sin 2\alpha$$

$$\sigma_m = \frac{32T_m}{\pi d^3} \cos 2\alpha \tag{18-24}$$

From this pair of equations and Eq. (18-22) we find the slope of the load line as

$$r = \frac{\sigma_a}{\sigma_m} = \frac{M_a \sin 2\alpha}{T_m \cos 2\alpha} = \frac{M_a^2 S_{ut}}{T_m^2 S_e} \tag{18-25}$$

Note that the load line is specific to material properties and to geometric effects in S_e and corresponds to the worst case, that is, the conditions at angle α.

We have elected to use the modified Goodman line, and from Eq. (18-17) in Table 18-1, this is

$$\frac{\sigma_a}{S_e} + \frac{\sigma_m}{S_{ut}} = \frac{1}{n}$$

By using the Pythagorean theorem with Eqs. (18-22) and (18-24) and with the modified Goodman relation, it can be shown that

$$d = \left\{ \frac{32n}{\pi} \left[\left(\frac{M_a}{S_e} \right)^2 + \left(\frac{T_m}{S_{ut}} \right)^2 \right]^{1/2} \right\}^{1/3} \tag{18-26}$$

Alternatively, the factor of safety is found from the equation

$$\frac{1}{n} = \frac{32}{\pi d^3} \left[\left(\frac{M_a}{S_e} \right)^2 + \left(\frac{T_m}{S_{ut}} \right)^2 \right]^{1/2} \tag{18-27}$$

As noted earlier, these two equations do not account for yielding on the first half-cycle.

In the usual case of a shaft subject to reversed bending and steady torsion, the critical bending stress will be located at a point of stress concentration. The stress-concentration effect should be included in Eq. (18-27). This can be done by refraining from correcting the value of S_e in the Marin equation [Eq. (7-13)] for stress concentration. Then Eqs. (18-26) and (18-27) are written in the forms

$$d = \left\{ \frac{32n}{\pi} \left[\left(\frac{K_f M_a}{S_e} \right)^2 + \left(\frac{T_m}{S_{ut}} \right)^2 \right]^{1/2} \right\}^{1/3} \tag{18-28}$$

$$\frac{1}{n} = \frac{32}{\pi d^3} \left[\left(\frac{K_f M_a}{S_e} \right)^2 + \left(\frac{T_m}{S_{ut}} \right)^2 \right]^{1/2} \tag{18-29}$$

EXAMPLE 18-2   Find the diameter of the shaft of Example 18-1 at the bearing shoulder based on the possibility of a fatigue failure. Use the methods derived in Sec. 18-6 and $S_{ut} = 80$ kpsi. The fatigue stress-concentration factor at the shoulder is $K_f = 1.90$. The endurance limit, corrected for all effects except stress concentration, is $S_e = 24$ kpsi. Use $n = 1.80$ as before, and find a safe diameter d, the slope r of the load line, and the angle α corresponding to the location of the tangent to the ellipse.

Solution   Using the loading diagram of Fig. 18-9b, we find the bending moment at the bearing shoulder to be

$$M = 679(5.5) - 2000(2.5) = -1265 \text{ lb} \cdot \text{in}$$

Substituting directly in Eq. (18-28) gives

$$d = \left(\frac{32(1.80)}{\pi} \left\{ \left[\frac{1.90(1265)}{24\ 000} \right]^2 + \left(\frac{3300}{80\ 000} \right)^2 \right\}^{1/2} \right)^{1/3} = 1.26 \text{ in}$$

In practice, a metric-size bearing having a bore of 35 mm (1.378 in) might be selected. However, we shall conclude the problem using the 1.26-in dimension.

Using Eq. (18-23), we find the angle corresponding to the tangent to be

Answer $$\alpha = \frac{1}{2} \tan^{-1} \frac{M_a S_{ut}}{T_m S_e} = \frac{1}{2} \tan^{-1} \frac{1.90(1265)(80\ 000)}{3300(24\ 000)} = 33.81°$$

Notice that the moment M_a has been augmented by the factor K_f in this solution.

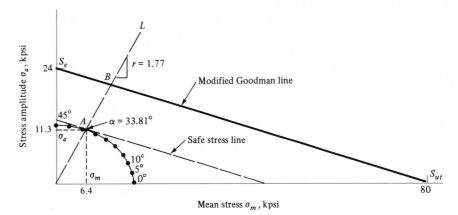

FIGURE 18-13

Graphical display of solution using the maximum-shear-stress theory combined with the modified Goodman relation.

Equation (18-25) is used to obtain the slope of the load line. The result is

Answer
$$r = \frac{M_a^2 S_{ut}}{T_m^2 S_e} = \frac{[1.90(1265)]^2(80\ 000)}{(3300)^2(24\ 000)} = 1.77$$

A graphic analysis of this problem is shown to true scale in Fig. 18-13. The Goodman line extends from $S_{ut} = 80$ kpsi to $S_e = 24$ kpsi. The quarter-ellipse shown in the figure is plotted using 5° increments for the angle α. The two equations used to plot the ellipse are the two parts of Eq. (18-24). The values used are

$$\sigma_a = \frac{32K_f M_a}{\pi d^3} = \frac{32(1.90)(1.265)}{\pi(1.26)^3}\ \sin 2\alpha = 12.2\ \sin 2\alpha \qquad \text{kpsi}$$

$$\sigma_m = \frac{32T_m}{\pi d^3}\ \cos 2\alpha = \frac{32(3.300)}{\pi(1.26)^3}\ \cos 2\alpha = 16.8\ \cos 2\alpha \qquad \text{kpsi}$$

As shown in Fig. 18-13, the safe-stress line is tangent to the ellipse at point A, where $\alpha = 33.81°$. At this point the two stresses are

$$\sigma_a = 11.3 \text{ kpsi} \qquad \sigma_m = 6.40 \text{ kpsi}$$

as shown in the figure. The load line is drawn from the origin through point A and intersecting the Goodman line at B, where it can be used, if desired, to define corresponding values of the strength.

Of course, the possibility of static failure must be checked when this approach is used. In this case, however, the results obtained from Example 18-1, though obtained at a different section of the shaft, indicate no likelihood of a static failure.

18-7   ADDITIONAL SOLUTIONS

In the previous section the problem solved was that of steady torsion combined with reversed bending using the modified Goodman relation for fatigue. If the Soderberg relation is used instead, then the corresponding equations are

$$d = \left\{ \frac{32n}{\pi} \left[\left(\frac{K_f M_a}{S_e} \right)^2 + \left(\frac{T_m}{S_y} \right)^2 \right]^{1/2} \right\}^{1/3}$$

(18-30)

$$\frac{1}{n} = \frac{32}{\pi d^3} \left[\left(\frac{K_f M_a}{S_e} \right)^2 + \left(\frac{T_m}{S_y} \right)^2 \right]^{1/2}$$

(18-31)

$$r = \frac{(K_f M_a)^2 S_y}{T_m^2 S_e}$$

(18-32)

$$\alpha = \frac{1}{2} \tan^{-1} \frac{K_f M_a S_y}{T_m S_e}$$

(18-33)

The solutions work out somewhat differently when the distortion energy theory is used. First designate the normal and shear stresses as

$$\sigma_{xa} = \frac{32 K_f M_a}{\pi d^3} \qquad \tau_{xym} = \frac{16 T_m}{\pi d^3}$$

(a)

Then the corresponding alternating and steady von Mises stresses are

$$\sigma_a' = (\sigma_{xa}^2 + 3\tau_{xya}^2)^{1/2} = \left[\left(\frac{32 K_f M_a}{\pi d^3} \right)^2 + 3(0)^2 \right]^{1/2} = \frac{32 K_f M_a}{\pi d^3}$$

(b)

$$\sigma_m' = (\sigma_{xm}^2 + 3\tau_{xym}^2)^{1/2} = \left[(0)^2 + 3 \left(\frac{16 T_m}{\pi d^3} \right)^2 \right]^{1/2} = \frac{16\sqrt{3} T_m}{\pi d^3}$$

(c)

If we select the modified Goodman approach, then we substitute Eqs. (b) and (c) in Eq. (18-17) from Table 18-1. We then have

$$\frac{32 K_f M_a}{\pi d^3 S_e} + \frac{16\sqrt{3} T_m}{\pi d^3 S_{ut}} = \frac{1}{n}$$

Solving for d and n yields

$$d = \left[\frac{32n}{\pi} \left(\frac{K_f M_a}{S_e} + \frac{\sqrt{3} T_m}{2 S_{ut}} \right) \right]^{1/3}$$

(18-34)

$$\frac{1}{n} = \frac{32}{\pi d^3} \left(\frac{K_f M_a}{S_e} + \frac{\sqrt{3} T_m}{2 S_{ut}} \right)$$

(18-35)

The slope of the load line is

$$r = \frac{\sigma_a'}{\sigma_m'} = \frac{2 K_f M_a}{\sqrt{3} T_m}$$

(18-36)

Equations (18-34) and (18-35) do not guard against yielding on the first half-cycle as do Eqs. (18-30) and (18-31).

Similar methods can be employed to obtain additional relations. If, for example, the distortion-energy theory is used for stress combined with the Gerber parabolic line for fatigue strength, then the corresponding equations turn out to be

$$d = \left(\frac{16nK_fM_a}{\pi S_e} \left\{ 1 + \left[1 + 3\left(\frac{T_mS_e}{K_fM_aS_{ut}} \right)^2 \right]^{1/2} \right\} \right)^{1/3} \tag{18-37}$$

$$\frac{1}{n} = \frac{16K_fM_a}{\pi d^3 S_e} \left\{ 1 + \left[1 + 3\left(\frac{T_mS_e}{K_fM_aS_{ut}} \right)^2 \right]^{1/2} \right\} \tag{18-38}$$

If we employ the ASME elliptic relation [Eq. (18-19)] for fatigue and distortion energy for stress, then the resulting relations are

$$d = \left\{ \frac{32n}{\pi} \left[\left(\frac{K_fM_a}{S_e} \right)^2 + \frac{3}{4}\left(\frac{T_m}{S_y} \right)^2 \right]^{1/2} \right\}^{1/3} \tag{18-39}$$

$$\frac{1}{n} = \frac{32}{\pi d^3} \left[\left(\frac{K_fM_a}{S_e} \right)^2 + \frac{3}{4}\left(\frac{T_m}{S_y} \right)^2 \right]^{1/2} \tag{18-40}$$

If Eq. (18-34), representing a combined use of the distortion energy theory and the modified Goodman relation, is used to solve for the diameter in Example 18-2, the result turns out to be $d = 1.36$ in. Try it. Note that this is a more conservative result than that obtained using the maximum-shear-stress theory. This is why many designers prefer to use the combination of the distortion-energy theory with the modified Goodman line for shaft design. In Example 18-2 we found $d = 1.26$ in, which is not as conservative.

The load lines are also different. In Example 18-2 the slope of the load line was found to be $r = 1.77$. But the slope of the load line using the distortion-energy theory is

$$r = \frac{2K_fM_a}{\sqrt{3}T_m} = \frac{2(1.90)(1265)}{\sqrt{3}(3300)} = 0.841$$

which is independent of the material properties; for this reason, a comparison with the load time for maximum-shear methods does not give any generalization.

It should be clear now that you can derive an equation for every combination of stress-failure theory and fatigue relation available. Here we present a general graphical method which makes these derivations, of which there could be many, unnecessary.

Suppose the stress state at a point on a shaft consists of a combination of reversed and steady moments and torques. This is quite general, but an even more general case would be one in which reversed and steady axial forces exist too; or one in which the reversals occur at differing frequencies.

For a shaft with both steady moments and torques and reversed moments and torques at the same frequency, the stresses are

$$\sigma_{xm} = \frac{32M_m}{\pi d^3} \qquad \tau_{xym} = \frac{16T_m}{\pi d^3} \tag{d}$$

$$\sigma_{xa} = K_f\frac{32M_a}{\pi d^3} \qquad \tau_{xya} = K_{fs}\frac{16T_a}{\pi d^3} \tag{e}$$

where we have assumed the use of a material in which no stress-concentration factor need be used for static loading.

Using the maximum-shear-stress theory combined with the Soderberg line, the following results can be found:

$$d = \left\{ \frac{32n}{\pi} \left[\left(\frac{M_m}{S_y} + \frac{K_f M_a}{S_e} \right)^2 + \left(\frac{T_m}{S_y} + \frac{K_{fs} T_a}{S_e} \right)^2 \right]^{1/2} \right\}^{1/3} \tag{18-41}$$

$$\frac{1}{n} = \frac{32}{\pi d^3} \left[\left(\frac{M_m}{S_y} + \frac{K_f M_a}{S_e} \right)^2 + \left(\frac{T_m}{S_y} + \frac{K_{fs} T_a}{S_e} \right)^2 \right]^{1/2} \tag{18-42}$$

Equation (18-41) is sometimes described as the *Westinghouse code formula*.

Though many similar relationships can be obtained, one of the most useful is the combination of the distortion-energy theory for stress and the modified Goodman line for fatigue strength. To develop this, use Eqs. (d) and (e) to obtain the von Mises stresses. Thus

$$\sigma'_a = \frac{16}{\pi d^3} [4(K_f M_a)^2 + 3(K_{fs} T_a)^2]^{1/2} \tag{f}$$

$$\sigma'_m = \frac{16}{\pi d^3} (4M_m^2 + 3T_m^2)^{1/2} \tag{g}$$

Now substitute these in the modified Goodman relation [Eq. (18-17)]. This gives

$$d = \left(\frac{32n}{\pi} \left\{ \left[\left(\frac{K_f M_a}{S_e} \right)^2 + \frac{3}{4} \left(\frac{K_{fs} T_a}{S_e} \right)^2 \right]^{1/2} + \left[\left(\frac{M_m}{S_{ut}} \right)^2 + \frac{3}{4} \left(\frac{T_m}{S_{ut}} \right)^2 \right]^{1/2} \right\} \right)^{1/3} \tag{18-43}$$

$$\frac{1}{n} = \frac{32}{\pi d^3} \left\{ \left[\left(\frac{K_f M_a}{S_e} \right)^2 + \frac{3}{4} \left(\frac{K_{fs} T_a}{S_e} \right)^2 \right]^{1/2} + \left[\left(\frac{M_m}{S_{ut}} \right)^2 + \frac{3}{4} \left(\frac{T_m}{S_{ut}} \right)^2 \right]^{1/2} \right\} \tag{18-44}$$

18-8   STIFFNESS CONSIDERATIONS

Many of the preceding sections have dealt with the problem of designing a shaft that will not be overly stressed. But a shaft so designed may still be unsatisfactory because it lacks rigidity. Insufficient rigidity, or stiffness, can result in poor performance of various shaft-mounted elements such as gears, clutches, bearings, and flywheels.

The angular deflection at bearings must be kept within the limits prescribed for bearings; see Chap. 11. The methods of Chap. 3 should be used to ensure meeting these limits.

If the linear deflection of geared shafts, for example, is too large, the life of the gears will be shortened because of additional impact forces during meshing and greater wear of the tooth surfaces. Gears mounted on shafts with inadequate stiffness will also be noisier. Use the methods of Chap. 3 to check the bending deflection at important points along the shaft. This should be done for other elements as well as gears.

The subject of shaft vibration, usually covered in dynamics studies, is not included in this book. Lack of shaft stiffness will result in both linear and torsional vibration, the effects of which can show up in many ways. Not only will the machine perform poorly, but this poor performance may affect the quality of products produced by the machine. If it is a metal-cutting machine, the effect will be seen in the resulting tolerance ranges.

If it is a pharmaceutical bottle-filling machine, vibration may cause variations in the amounts filled. And at the price of some drugs, this could result in very substantial losses. So both linear and torsional vibration must be controlled.

18-9   ESTIMATING RELIABILITY—A STOCHASTIC TASK

When any of the terms that combine to form the stress, the strength, the angular or linear distortion, or the rigidity of the shaft are random variables, it may be desirable to estimate the resulting reliability. This assessment is made from either of the two following cases:

1 A strength-limited case in which an interference analysis is obtained from the load-induced stress at a critical location with the corresponding strength.

2 A distortion-limited case in which the interference analysis is made from the load-induced distortion at a critical section with the limiting value of the distortion or deflection.

Interference was addressed in Sec. 6-14 for strength-limited designs involving arbitrary distributions, in Sec. 6-11 for normal distributions, and in Sec. 6-13 for lognormal distributions.

In Sec. 7-13, Example 7-6, it was pointed out that the factor of safety could be assessed in the following three ways:

$$n = \frac{S_a}{\sigma_a} = \frac{S_m}{\sigma_m} = \frac{S}{\sigma} \tag{18-45}$$

where S_a is the strength amplitude; σ_a is the uniaxial stress amplitude; S_m is the steady-amplitude component of strength; σ_m is the uniaxial steady component of stress on the fatigue diagram; S is the constructive strength, which is the magnitude of the vector sum of S_a and S_m; and σ is the constructive stress, which is the magnitude of the vector sum of σ_a and σ_m. In the stochastic viewpoint, each of these six quantities is a random variable with its own distribution, and interfering S_a with σ_a, S_m with σ_m, or S with σ will give an estimate of the reliability or probability of failure. If the reliability estimate is needed, then the designer's task is to formulate an estimate of one of these pairs of strength and stress distributions.

The load-induced stress distribution is sought first, since the load-line slope is needed for the strength distribution. In power-transmission shafting, the torque is given by

$$\mathbf{T} = \frac{63\ 000\mathbf{H}}{n}$$

where $\mathbf{T}$ = torque, lb · in; $\mathbf{H}$ = horsepower; and n = speed, rev/min. The variability in power is reflected to the torque. The shaft bending moment at the critical location is directly correlated with the torque. The coefficient of variation of the power is transmitted to both the torque $\mathbf{T}$ and the bending moment $\mathbf{M}$. If the coefficient of variation of the loading (in this case, power) is denoted C_L, then the von Mises alternating- and steady-stress components can be written, respectively, as

$$\sigma'_a = \frac{32\overline{M}_a(1,\,C_L)}{\pi d^3} \qquad \sigma'_m = \frac{16\sqrt{3}\,\overline{T}_m(1,\,C_L)}{\pi d^3}$$

which have means and standard deviations given by, respectively,

$$\overline{\sigma}'_a = \frac{32\overline{M}_a}{\pi d^3} \qquad \overline{\sigma}'_m = \frac{16\sqrt{3}\,\overline{T}_m}{\pi d^3} \tag{18-46}$$

$$\hat{\sigma}_{\sigma'a} = C_L\overline{\sigma}'_a \qquad \hat{\sigma}_{\sigma'm} = C_L\overline{\sigma}'_m \tag{18-47}$$

These forms imply that stress-concentration factor K_f, if present, will be applied to S'_e through $k_e = 1/K_f$. This is proper only for ductile materials, where the stress-concentration factor is applied only to the amplitude component. If one chooses to interfere σ'_a with S_a, then $\sigma'_a = \overline{\sigma}'_a(1,\,C_L)$. The slope of the load line is σ'_a/σ'_m, or

$$r = \frac{\sigma'_a}{\sigma'_m} = \frac{32\overline{M}_a(1,\,C_L)/\pi d^3}{16\sqrt{3}\,\overline{T}_m(1,\,C_L)/\pi d^3}$$

We have a quotient of two random variables which are directly correlated with $\rho = 1$ and with the same coefficients of variation. From Table 4-4 for the quotient x/y, using second-order terms, the standard deviation of the quotient is zero. This makes $r(1,\,0)$ a deterministic variable r. Thus

$$r = \frac{2\overline{M}_a}{\sqrt{3}\,\overline{T}_m} \tag{18-48}$$

The strength-distribution estimate is constructed as follows:

• Estimate S'_e from fatigue testing or from tensile testing using $S'_e = \overline{\phi}S_{ut}$, or from tabular information using $S'_e = \overline{\phi}S_{ut}$.

• Estimate S_e from $k_a k_b k_c k_d k_e S'_e$.

• Construct a stochastic fatigue diagram based on a Gerber locus and knowledge about intercepts S_e and S_{ut}, or based on the distortion-energy theory for stress combined with the ASME elliptic theory for fatigue and the intercepts S_e and S_y.

• Find S_a and S_m, or S, from the load-line slope and fatigue diagram. The median fatigue locus has the same equation as the deterministic form except that the parameters are the mean values of S_e and S_{ut}, or S_e and S_y.

The relation for $\overline{S}_a$ for a Gerber locus is

$$\overline{S}_a = \overline{S}_e - \frac{\overline{S}_e}{\overline{S}_{ut}^2}\,\overline{S}_m^2 = \overline{S}_e - \frac{\overline{S}_e}{\overline{S}_{ut}^2}\,\frac{\overline{S}_a^2}{r^2} \tag{18-49}$$

The solution to the quadratic in $\overline{S}_a$ in Eq. (18-49) is

$$\overline{S}_a = \frac{(r\overline{S}_{ut})^2}{2\overline{S}_e}\left\{-1 + \left[1 + \left(\frac{2\overline{S}_e}{r\overline{S}_{ut}}\right)^2\right]^{1/2}\right\} \tag{18-50}$$

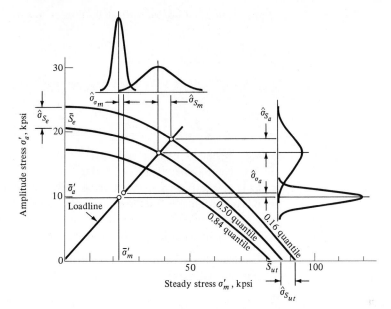

FIGURE 18-14

A stochastic fatigue diagram showing a Gerber fatigue-failure locus. The variability of the von Mises stress components is indicated. The two convenient strength-stress interferences are shown.

In Fig. 18-14, the 0.50 quantile is a plot of the left part of Eq. (18-49). Presuming a normal distribution of data about the regression line, or nearly so, then the 0.16 quantile will be about one standard deviation above the regression line. The ordinate intercept will be $\overline{S}_e + \hat{\sigma}_{Se}$, and the abscissa intercept will be $\overline{S}_{ut} + \hat{\sigma}_{Sut}$. If the values of $\overline{S}_e$ and $\overline{S}_{ut}$ in Eq. (18-50) are incremented by their respective standard deviations, then the left side of the equation is $\overline{S}_a + \hat{\sigma}_{Sa}$. It is easily shown that

$$\hat{\sigma}_{Sa} = \frac{r^2(\overline{S}_{ut} + \hat{\sigma}_{Sut})^2}{2(\overline{S}_e + \hat{\sigma}_{Se})}\left(-1 + \left\{1 + \left[\frac{2(\overline{S}_e + \hat{\sigma}_{Se})}{r(\overline{S}_{ut} + \hat{\sigma}_{Sut})}\right]^2\right\}^{1/2}\right) - \overline{S}_a \tag{18-51}$$

The corresponding values of $\overline{S}_m$ and $\hat{\sigma}_{Sm}$ can be found from

$$\overline{S}_m = \frac{\overline{S}_a}{r} \tag{18-52}$$

$$\hat{\sigma}_{Sm} = \frac{\hat{\sigma}_{Sa}}{r} \tag{18-53}$$

and for the constructive strength, $S = (S_a^2 + S_m^2)^{1/2}$. The mean of S is

$$\overline{S} = (\overline{S}_a^2 + \overline{S}_m^2)^{1/2} = \left[S_a^2 + \left(\frac{\overline{S}_a}{r}\right)^2\right] = \overline{S}_a\left(\frac{1 + r^2}{r^2}\right)^{1/2} \tag{18-54}$$

In Pythagorean relations with correlated ($\rho = 1$) elements, the standard deviation is

$$\hat{\sigma}_S = \hat{\sigma}_{Sa}\left(\frac{1 + r^2}{r^2}\right)^{1/2} \tag{18-55}$$

If one chooses to interfere S_a with σ_a', we are ready for the Gerber locus. If the ASME elliptic relation combined with distortion energy is preferred, it is easy to show that

$$\bar{S}_a = \left[\frac{(r\bar{S}_y\bar{S}_e)^2}{r^2\bar{S}_y^2 + \bar{S}_e^2} \right]^{1/2} \tag{18-56}$$

$$\hat{\sigma}_{Sa} = \left[\frac{r^2(\bar{S}_y + \sigma_{Sy})^2(\bar{S}_e + \hat{\sigma}_{Se})^2}{r^2(\bar{S}_y + \hat{\sigma}_{Sy})^2 + (\bar{S}_e + \hat{\sigma}_{Se})^2} \right]^{1/2} - \bar{S}_a \tag{18-57}$$

EXAMPLE 18-3

Suppose the integral pinion and shaft in Fig. 18-9a has a 1.65-in shoulder to the left of the right-hand journal that has a diameter of 1.100 in. The loading is depicted in Fig. 18-9b. The coefficient of variation of the power is 0.05, and the power variation is normally distributed. The torque is $\mathbf{T}_m = 3300(1, 0.05)$ lb · in, and the bending moment at the shoulder is taken as $\mathbf{M}_a = 1260(1, 0.05)$ lb · in. Apply the stress concentration to S_e through $\mathbf{k}_e = 1/K_f$. The material is AISI 1040 hot-rolled steel.

(a) Estimate the reliability using a stochastic Gerber locus combined with the distortion-energy theory.

(b) Estimate the reliability using a stochastic ASME elliptic locus combined with the distortion-energy theory.

Solution

(a) Here we choose to interfere S_a with σ_a'. The von Mises stress amplitude is

$$\sigma_a' = \frac{32\overline{M}_a(1, 0.05)}{\pi d^3} = \frac{32(1260)(1, 0.05)}{\pi(1.1)^3}$$

From Eq. (18-46),

$$\overline{\sigma}_a' = \frac{32\overline{M}_a}{\pi d^3} = \frac{32(1260)}{\pi(1.1)^3} = 9643 \text{ psi}$$

$$\hat{\sigma}_{\sigma'a} = C_L\overline{\sigma}_a' = 0.05(9643) = 482 \text{ psi}$$

Therefore $\sigma_a' = (9.643, 0.482)$ kpsi. The load-line slope is

$$r = \frac{2\overline{M}_a}{\sqrt{3}\overline{T}_m} = \frac{2(1260)}{\sqrt{3}(3300)} = 0.441$$

For this example, we estimate $\bar{S}_{ut} = 86$ kpsi for AISI 1040 hot-rolled steel. From Tables 7-4 and 7-6, and using a machined finish,

$$\mathbf{k}_a = 2.70(1, 0.06)\bar{S}_{ut}^{-0.265} = 2.70(1, 0.06)86^{-0.265} = 0.829(1, 0.06)$$

Using Eq. (7-15), we have

$$k_b = \left(\frac{1.1}{0.3} \right)^{-0.1133} = 0.863$$

We note that $k_c = k_d = 1$. Using Fig. A-15-9 in the Appendix, we get $r/d = (d/10)/10 = 0.10$, $D/d = 1.65/1.1 = 1.5$, from which $K_t = 1.67$. Also, $r = d/10 = 0.11$. From Fig. 5-16, $q = 0.80$, and so

$$\overline{K}_f = 1 + 0.80(1.67 - 1) = 1.54$$

Then, from Table 5-5,

$\mathbf{K}_f = 1.54(1, 0.08)$

and $\overline{k}_e = 1/\overline{K}_f = 1/1.54 = 0.649$. We therefore use $\mathbf{k}_e = 0.650(1, 0.08)$. We next estimate $\mathbf{S}_e$ as

$$\mathbf{S}_e = \mathbf{k}_a\mathbf{k}_b\mathbf{k}_c\mathbf{k}_d\mathbf{k}_e\phi\overline{S}_{ut}$$
$$= 0.829(1, 0.06)(0.863)(1)(1)(0.650)(1, 0.08)[0.504(1, 0.146)](86)$$

The mean and the coefficient of variation of $\mathbf{S}_e$ are

$$\overline{S}_e = 0.829(0.863)(0.650)(0.504)(86) = 20.2 \text{ kpsi}$$
$$C_{Se} = [(0.06)^2 + (0.08)^2 + (0.146)^2]^{1/2} = 0.177$$

Putting these two values together gives the estimate of the endurance strength at the shoulder as

$$\mathbf{S}_e = 20.2(1, 0.177) = (20.2, 3.58) \qquad \text{kpsi}$$

For this example we shall use $C_{Sut} = 0.06$. Then

$$\mathbf{S}_{ut} = 86(1, 0.06) = (86, 5.16) \qquad \text{kpsi}$$

The mean-strength-amplitude component to the Gerber curve from Eq. (18-50) is

$$\overline{S}_a = \frac{[0.441(86)]^2}{2(20.2)} \left(-1 + \left\{ 1 + \left[\frac{2(20.2)}{0.441(86)} \right]^2 \right\}^{1/2} \right) = 16.42 \text{ kpsi}$$

The standard deviation in $\mathbf{S}_a$ is from Eq. (18-51):

$$\hat{\sigma}_{Sa} = \frac{(0.441)^2(86 + 5.16)^2}{2(20.2 + 3.58)} \left(-1 + \left\{ 1 + \left[\frac{2(20.2 + 3.58)}{0.441(86 + 5.16)} \right]^2 \right\}^{1/2} \right) - 16.42$$
$$= 18.66 - 16.42 = 2.24 \text{ kpsi}$$

Thus we now have the estimate $\mathbf{S}_a = (16.42, 2.24)$ kpsi. Using Eq. (6-22) for normal-normal interference, we obtain

$$z = -\frac{16.42 - 9.643}{[(2.24)^2 + (0.482)^2]^{1/2}} = -2.96$$

as a quick answer. From Table A-10, the probability of failure is $p_f = 0.001\ 54$ with a corresponding reliability of 0.998. A numerical LN-N interference suggested by Table 18-2 gives $R = 0.9998$. Figure 18-14 represents this example.

(b) We begin by estimating the coefficient of variation of the yield strength as $C_{Sy} = 0.07$. Therefore

$$\hat{\sigma}_{Sy} = 0.07(54) = 3.78 \text{ kpsi}$$

From Eq. (18-57),

$$\overline{S}_a = \left\{ \frac{[0.441(54)(20.2)]^2}{(0.441)^2(54)^2 + (20.2)^2} \right\}^{1/2} = 15.40 \text{ kpsi}$$

TABLE 18-2

Interference Plan as a Function of Numbers of Stochastic Variables in Estimates of Strength and Stress*†

	STRENGTH	
STRESS	ONE OR TWO STOCHASTIC VARIABLES	THREE OR MORE STOCHASTIC VARIABLES
One or two stochastic variables	*N-N*	*LN-N*
Three or more stochastic variables	*N-LN*	*LN-LN*

*The symbols *N-N, N-LN, LN-N,* and *LN-LN* apply to the strength distribution and the stress distribution, respectively. For example, *LN-N* means lognormal strength interfered with normal stress.

†If in doubt, the normal strength–lognormal stress (*N-LN*) represents a worst-case interference.

From Eq. (18-57),

$$\hat{\sigma}_{Sa} = \left[\frac{(0.441)^2(54 + 3.78)^2(20.2 + 3.58)^2}{(0.441)^2(54 + 3.78)^2 + (20.2 + 3.58)^2} \right]^{1/2} - 15.40 = 1.99 \text{ kpsi}$$

Again using the coupling equation [Eq. (6-22)] for a quick answer, we find

$$z = - \frac{15.40 - 9.643}{[(1.99)^2 + (0.482)^2]^{1/2}} = -2.81$$

From Table A-10, the probability of failure is $p_f = 0.002\,48$ with a corresponding reliability of $R = 0.9975$. A numerical *LN-N* interference suggested by Table 18-2 gives $R = 0.9996$. The worst-case recommendation from Table 18-2 is an *N-LN* interference, which, carried out numerically, gives a reliability of 0.9984 in part (*a*) and 0.9975 in part (*b*) of this example. These are not materially different from our *N-N* estimate. The best answers to this problem are from the *LN-N* interference with a reliability of 0.9998 for the Gerber model and 0.9996 for the elliptic model.

PROBLEMS*

18-1 The geared industrial roll shown in the figure is driven at 300 rev/min by a force F acting on a gear of 3-in pitch diameter, as shown. The roll exerts a normal force of 30 lb per inch of roll length on the material being pulled through. The coefficient of friction is 0.40. A steel has been tentatively selected having $S_{ut} = 72$ kpsi and $S_y = 39$ kpsi. Based on static loading and a factor of safety of 3.5, find the theoretical shaft diameter at the critical points. Use both static-failure theories.

*An asterisk indicates a design-type problem or a problem that may have no unique answer.

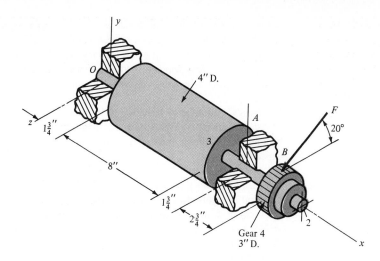

PROBLEM 18-1

18-2 The shaft for the industrial roll in Prob. 18-1 has been sized as shown in the figure, and all surfaces have been specified as machined. Note the shoulders used to position the shaft between the journal bearings. Find the factor of safety guarding against a fatigue failure. Use the modified Goodman relation for fatigue and the distortion-energy theory for stress.

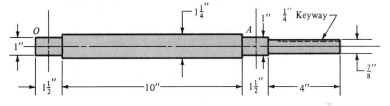

PROBLEM 18-2
All fillets are of $\frac{1}{16}$-in radius;
sledrunner keyway is $3\frac{1}{2}$ in long.

18-3 As shown in the figure, the axle of a railroad freight car is tapered, with the least diameter midway between the wheels. This problem will give some insight as to why. A freight-car axle is force-fitted to its wheels with the journals outboard of the wheels and with the journal centers about 80 in apart. The track gauge is $56\frac{1}{2}$ in between the rails, and so the span between the rail

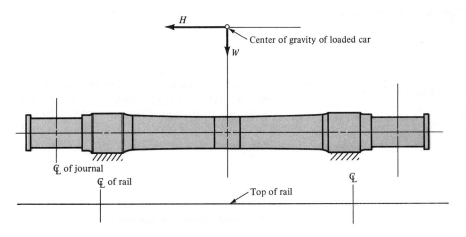

PROBLEM 18-3
A railroad freight-car axle.

centers is about $59\frac{1}{2}$ in. Freight-car wheels are usually of diameter 33 in for cars up to 70-ton capacity. Consider the worst-case location of the center of mass of the car and load to be 72 in above the rails, the worst-case vertical load to be 42 750 lb per axle, and the worst-case horizontal load due to cross winds and track curvature to be 17 100 lb per axle through the center of mass. Construct a bending-moment diagram for the axle.

18-4   In a class C American Association of Railroads standard axle, the diameter of the wheel seat is 7 in and the diameter of the axle at midspan is $5\frac{3}{8}$ in. Using the results of Prob. 18-3, compare the bending-stress level at the center of the wheel seat and at axle midspan. Is this difference to be expected?

18-5*   In order to obtain the rigidity needed for better roll performance, the shaft of Prob. 18-1 is designed as a cold-drawn steel tube made of AISI 1018 steel having an OD of $1\frac{1}{4}$ in and a wall thickness of $\frac{3}{16}$ in. The roll has been redesigned too, and is to be of a 4-in tube having flanges welded to each end with holes for straight pins through the ends, spaced 9 in apart; these secure the roll to the shaft. The gear is to be secured to the shaft using a straight pin through the hub.
(a) Estimate the size of pins required.
(b) What is the approximate value of the factor of safety based on the use of the maximum-shear-stress theory for stress and the Soderberg criterion for fatigue?

18-6*   Design a countershaft to support gears C and D shown in the figure. Gear C is a straight spur gear having 40 teeth and a diametral pitch of 2 teeth/in. It is driven at a speed of 250 rev/min in the direction shown. Gear D is also a straight-tooth spur gear; it has 18 teeth and a diametral pitch of 2 teeth/in, and it transmits 140 hp to a driven gear. Both gears have a pressure angle of 20° and hub diameters of 6 in. Specify the material and the factors of safety, and make a complete dimensioned drawing of the shaft showing shoulder diameters, fillet radii, keyway sizes, and any other details that apply.

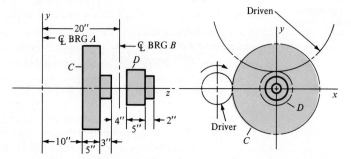

PROBLEM 18-6
Countershaft on z axis supports gears C and D with bearings at A and B. Note that gear D is overhung.

18-7   The figure shows a section of an AISI 1040 forged and heat-treated steel shaft having machined surfaces with dimensions $D = 1.5$ in and $d = 1$ in. The heat-treat process resulted in minimum tensile strengths of $S_u = 100$ kpsi and $S_y = 70$ kpsi. The shaft section at the shoulder is subject to a completely reversed bending moment of 800 lb · in and a steady torsion of 400 lb · in. Determine the factor of safety for infinite life based on the modified Goodman relation coupled with the maximum-shear-stress theory.

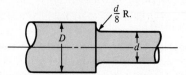

PROBLEM 18-7

18-8   Solve Prob. 18-7 using the Soderberg relation for fatigue.

18-9   Solve Prob. 18-7 using the distortion-energy theory for stress.

18-10   Solve Prob. 18-7 using the ASME elliptic relation for fatigue and the distortion-energy theory for stress.

18-11   Suppose the shaft section of Prob. 18-7 is subjected to a mean or steady moment of 800 lb · in and a completely reversed torsion of 600 lb · in. Estimate the resulting factor of safety based upon the maximum-shear-stress theory combined with the Soderberg line for fatigue.

18-12   The section of shaft shown in the figure is to be designed to approximate relative sizes of $d = 0.75D$ and $r = D/20$, with the diameter d conforming to that of standard metric rolling-bearing bore sizes. It is to be made of an SAE 2340 steel, heat-treated to obtain minimum tensile strengths in the shoulder area of $S_u = 1226$ MPa and $S_y = 1130$ MPa with a Brinell hardness not less than 368. At the shoulder, the shaft is subjected to a completely reversed bending moment of 70 N · m accompanied by a steady torsion of 45 N · m. Use a design factor of 2.5, and size the shaft for an infinite life. The results should be based on the distortion-energy theory for stress and the modified Goodman line for fatigue.

PROBLEM 18-12
Section of a shaft containing a grinding-relief groove. Unless otherwise specified, the diameter at the root of the groove is $d_R = d - 2r$, and though the section of diameter d is ground, the root of the groove is still a machined surface.

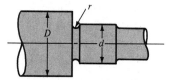

18-13   What factors of safety would result if a bearing having a 20-mm bore were to be used on the shaft of Prob. 18-12? What dimensions would you use for D and r in this case?

18-14   The portion of the shaft shown in the figure for Prob. 18-12 has $d = 25$ mm, $r = 1.8$ mm, and $D = 35$ mm. The shaft material is AISI 3140 steel, heat-treated to obtain minimum tensile strengths of $S_y = 1082$ MPa and $S_u = 1300$ MPa with a Brinell hardness of 378, all figures applying to the shoulder area. The shaft is subjected to a steady bending moment of 15 N · m and an alternating moment of 65 N · m. These effects are accompanied by a steady torsion of 50 N · m. Find the factor of safety of this section of the shaft using the Westinghouse code formula.

18-15   The section of shaft shown in the figure for Prob. 18-12 is to be designed to the approximate proportions of $d = 0.80D$ and $r = D/20$, with a diameter d selected in increments of $\frac{1}{8}$ in. An AISI 3140 steel has been selected and is to be heat-treated to obtain minimum tensile properties of $S_u = 188$ kpsi, $S_y = 157$ kpsi, and $H_B = 376$, all applying to the shoulder section. The shaft is subjected to a bending moment amplitude $M_a = 650$ lb · in and a steady torsion of 400 lb · in. The shaft is to be sized for an infinite life using a design factor of 2.6, the distortion-energy theory for stress, and the modified Goodman relation for fatigue. Report the values of D, d, and r selected, and the value of the factor of safety that results from these choices.

18-16   The section of shaft shown in the figure for Prob. 18-12 has $D = 1\frac{1}{8}$ in, $d = \frac{3}{4}$ in, and $r = \frac{1}{16}$ in. It is made of AISI 3120 steel heat-treated to give minimum properties in tension of $S_u = 112$ kpsi, $S_y = 91$ kpsi, and $H_B = 222$, all of which apply to the shoulder area. The loading on the shaft is a completely reversed torsion $T_a = 400$ lb · in. Find the factors of safety for an infinite life. Use the maximum-shear-stress theory for stress and the Soderberg line for fatigue strength.

18-17   A transverse drilled and reamed hole can be used in a solid shaft to hold a pin which locates and holds a mechanical element, such as the hub of a gear, in axial position and allows the transmission of torque. Since a small-diameter hole introduces high stress concentration and a large-diameter hole erodes the area resisting bending or torsion, investigate the existence of a pin diameter with minimum adverse effect on the shaft. Then formulate a design rule. (*Hint:* Use Table A-16.)

18-18* A slow-speed spur gear with a 1.75-in bore and a $1\frac{1}{2}$-in-long hub has a pitch diameter of 8 in and involute teeth of 20° pressure angle and transmits 4.5 hp at 112 rev/min. The shaft is to have a bearing span of 10 in, with the gear placed 3 in from the right-hand bearing. Deep-groove ball bearings are to be used. A 2-in overhang for a coupling having a 1-in-diameter seat is to be provided at the left of the left-hand bearing. The bearing life is to be 10 kh at a reliability of 0.99 for both bearings. Some a priori decisions are shown in the figure. Select appropriate bearings, and make a dimensioned drawing of the shaft.

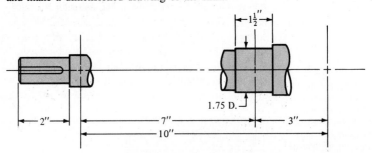

PROBLEM 18-18

18-19 The design of Prob. 18-18 has been completed, resulting in a shaft which is 13 in long overall. It is important to have assurance that the deflection is within reasonable limits. A testing machine which can accurately apply a compressive load is available, but the throat area is such that a specimen and fixture only up to about 8 in long can be handled. For an iconic scale model made half-size, what load should be applied to simulate the gear? How should a measured deflection, slope, or stress be interpreted if the model is made of the same material as the prototype?

18-20 An AISI 1020 cold-drawn steel shaft with the geometry shown in the figure carries a transverse load of 7 kN and a torque of 107 N · m. Examine the shaft for strength and deflection. If the largest allowable slope at the bearings is 0.001, what is the factor of safety guarding against damaging distortion? What is the factor of safety guarding against a fatigue failure? If the shaft turns out to be unsatisfactory, what would you recommend to correct the problem?

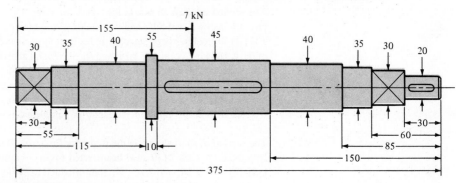

PROBLEM 18-20   All fillets 2 mm

ANSWERS   **18-1** At center of roll, $d = 0.895$ in; at A, $d = 0.727$ in
18-7 $n = 2.86$ (fatigue); $n = 8.11$ (yielding)
18-10 $n = 2.57$
18-12 $D = 32$ mm, $d = 25$ mm, $n = 2.66$
18-16 $n = 1.68$ (fatigue); $n = 5.45$ (yielding)

APPENDIX TABLES

TABLE A-1
Standard SI Prefixes*†

NAME	SYMBOL	FACTOR
exa	E	$1\ 000\ 000\ 000\ 000\ 000\ 000 = 10^{18}$
peta	P	$1\ 000\ 000\ 000\ 000\ 000 = 10^{15}$
tera	T	$1\ 000\ 000\ 000\ 000 = 10^{12}$
giga	G	$1\ 000\ 000\ 000 = 10^{9}$
mega	M	$1\ 000\ 000 = 10^{6}$
kilo	k	$1\ 000 = 10^{3}$
hecto‡	h	$100 = 10^{2}$
deka‡	da	$10 = 10^{1}$
deci‡	d	$0.1 = 10^{-1}$
centi‡	c	$0.01 = 10^{-2}$
milli	m	$0.001 = 10^{-3}$
micro	μ	$0.000\ 001 = 10^{-6}$
nano	n	$0.000\ 000\ 001 = 10^{-9}$
pico	p	$0.000\ 000\ 000\ 001 = 10^{-12}$
femto	f	$0.000\ 000\ 000\ 000\ 001 = 10^{-15}$
atto	a	$0.000\ 000\ 000\ 000\ 000\ 001 = 10^{-18}$

*If possible use multiple and submultiple prefixes in steps of 1000.

†Spaces are used in SI instead of commas to group numbers to avoid confusion with the practice in some European countries of using commas for decimal points.

‡Not recommended but sometimes encountered.

TABLE A-2

Conversion Factors A to Convert Input X to Output Y Using the Formula $Y = AX$*

MULTIPLY INPUT X	BY FACTOR A	TO GET OUTPUT Y	MULTIPLY INPUT X	BY FACTOR A	TO GET OUTPUT Y
British thermal unit, Btu	1055	joule, J	moment of inertia, lbm · ft^2	0.0421	kilogram-meter^2, kg · m^2
Btu/second, Btu/s	1.05	kilowatt, kW	moment of inertia, lbm · in^2	293	kilogram-millimeter^2, kg · mm^2
calorie	4.19	joule, J			
centimeter of mercury (0°C)	1.333	kilopascal, kPa	moment of section (second moment of area), in^4	41.6	centimeter^4, cm^4
centipoise, cP	0.001	pascal-second, Pa · s			
degree (angle)	0.0174	radian, rad	ounce-force, oz	0.278	newton, N
foot, ft	0.305	meter, m	ounce-mass	0.0311	kilogram, kg
foot^2, ft^2	0.0929	meter^2, m^2	pound, lb†	4.45	newton, N
foot/minute, ft/min	0.0051	meter/second, m/s	pound-foot, lb · ft	1.36	newton-meter, N · m
foot-pound, ft · lb	1.35	joule, J	pound/foot^2, lb/ft^2	47.9	pascal, Pa
foot-pound/ second, ft · lb/s	1.35	watt, W	pound-inch, lb · in	0.113	joule, J
			pound-inch, lb · in	0.113	newton-meter N · m
foot/second, ft/s	0.305	meter/second, m/s			
gallon (U.S.), gal	3.785	liter, l	pound/inch, lb/in	175	newton/meter, N/m
horsepower, hp	0.746	kilowatt, kW			
inch, in	0.0254	meter, m	pound/inch^2, psi (lb/in^2)	6.89	kilopascal, kPa
inch, in	25.4	millimeter, mm			
inch^2, in^2	645	millimeter^2, mm^2	pound-mass, lbm	0.454	kilogram, kg
inch of mercury (32°F)	3.386	kilopascal, kPa	pound-mass/ second, lbm/s	0.454	kilogram/second, kg/s
kilopound, kip	4.45	kilonewton, kN	quart (U.S. liquid), qt	946	milliliter, ml
kilopound/inch^2, kpsi (ksi)	6.89	megapascal, MPa (N/mm^2)	section modulus, in^3	16.4	centimeter^3, cm^3
mass, lb · s^2/in	175	kilogram, kg	slug	14.6	kilogram, kg
mile, mi	1.610	kilometer, km	ton (short 2000 lbm)	907	kilogram, kg
mile/hour, mi/h	1.61	kilometer/hour, km/h	yard, yd	0.914	meter, m
mile/hour, mi/h	0.447	meter/second, m/s			

*Approximate.

†The U.S. Customary System unit of the pound-force is often abbreviated as lbf to distinguish it from the pound-mass, which is abbreviated as lbm. In most places in this book the pound-force is usually written simply as the pound and abbreviated as lb.

TABLE A-3

Optional SI Units for Bending Stress $\sigma = Mc/I$, Torsion Stress $\tau = Tr/J$, Axial Stress $\sigma = F/A$, and Direct Shear Stress $\tau = F/A$

BENDING AND TORSION				AXIAL AND DIRECT SHEAR		
M, T	I, J	c, r	σ, τ	F	A	σ, τ
N · m*	m^4	m	Pa	N*	m^2	Pa
N · m	cm^4	cm	MPa (N/mm^2)	N†	mm^2	MPa (N/mm^2)
N · m†	mm^4	mm	GPa	kN	m^2	kPa
kN · m	cm^4	cm	GPa	kN†	mm^2	GPa
N · mm†	mm^4	mm	MPa (N/mm^2)			

*Basic relation.

†Often preferred.

TABLE A-4

Optional SI Units for Bending Deflection $y = f(F\ell^3/EI)$ or $y = f(w\ell^4/EI)$ and Torsional Deflection $\theta = T\ell/GJ$

BENDING DEFLECTION						TORSIONAL DEFLECTION			
$F, w\ell$	ℓ	I	E	y	T	ℓ	J	G	θ
N*	m	m^4	Pa	m	N · m*	m	m^4	Pa	rad
kN†	mm	mm^4	GPa	mm	N · m†	mm	mm^4	GPa	rad
kN	m	m^4	GPa	μm	N · mm	mm	mm^4	MPa (N/mm^2)	rad
N	mm	mm^4	KPa	m	N · m	cm	cm^4	MPa (N/mm^2)	rad

*Basic relation.

†Often preferred.

TABLE A-5

Physical Constants of Materials

MATERIAL	MODULUS OF ELASTICITY E		MODULUS OF RIGIDITY G		POISSON's RATIO ν	UNIT WEIGHT w		
	Mpsi	GPa	Mpsi	GPa		lb/in^3	lb/ft^3	kN/m^3
Aluminum (all alloys)	10.3	71.0	3.80	26.2	0.334	0.098	169	26.6
Beryllium copper	18.0	124.0	7.0	48.3	0.285	0.297	513	80.6
Brass	15.4	106.0	5.82	40.1	0.324	0.309	534	83.8
Carbon steel	30.0	207.0	11.5	79.3	0.292	0.282	487	76.5
Cast iron, gray	14.5	100.0	6.0	41.4	0.211	0.260	450	70.6
Copper	17.2	119.0	6.49	44.7	0.326	0.322	556	87.3
Douglas fir	1.6	11.0	0.6	4.1	0.33	0.016	28	4.3
Glass	6.7	46.2	2.7	18.6	0.245	0.094	162	25.4
Inconel	31.0	214.0	11.0	75.8	0.290	0.307	530	83.3
Lead	5.3	36.5	1.9	13.1	0.425	0.411	710	111.5
Magnesium	6.5	44.8	2.4	16.5	0.350	0.065	112	17.6
Molybdenum	48.0	331.0	17.0	117.0	0.307	0.368	636	100.0
Monel metal	26.0	179.0	9.5	65.5	0.320	0.319	551	86.6
Nickel silver	18.5	127.0	7.0	48.3	0.322	0.316	546	85.8
Nickel steel	30.0	207.0	11.5	79.3	0.291	0.280	484	76.0
Phosphor bronze	16.1	111.0	6.0	41.4	0.349	0.295	510	80.1
Stainless steel (18-8)	27.6	190.0	10.6	73.1	0.305	0.280	484	76.0

TABLE A-6

Properties of Structural-Steel Angles*·†

w = weight per foot, lb/ft
m = mass per meter, kg/m
A = area, in^2 (cm^2)
I = second moment of area, in^4 (cm^4)
k = radius of gyration, in (cm)
y = centroidal distance, in (cm)
Z = section modulus, in^3 (cm^3)

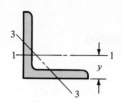

SIZE, in	w	A	$I_{1\text{-}1}$	$k_{1\text{-}1}$	$Z_{1\text{-}1}$	y	$k_{3\text{-}3}$
$1 \times 1 \times \frac{1}{8}$	0.80	0.234	0.021	0.298	0.029	0.290	0.191
$\times \frac{1}{4}$	1.49	0.437	0.036	0.287	0.054	0.336	0.193
$1\frac{1}{2} \times 1\frac{1}{2} \times \frac{1}{8}$	1.23	0.36	0.074	0.45	0.068	0.41	0.29
$\times \frac{1}{4}$	2.34	0.69	0.135	0.44	0.130	0.46	0.29
$2 \times 2 \times \frac{1}{8}$	1.65	0.484	0.190	0.626	0.131	0.546	0.398
$\times \frac{1}{4}$	3.19	0.938	0.348	0.609	0.247	0.592	0.391
$\times \frac{3}{8}$	4.7	1.36	0.479	0.594	0.351	0.636	0.389
$2\frac{1}{2} \times 2\frac{1}{2} \times \frac{1}{4}$	4.1	1.19	0.703	0.769	0.394	0.717	0.491
$\times \frac{3}{8}$	5.9	1.73	0.984	0.753	0.566	0.762	0.487
$3 \times 3 \times \frac{1}{4}$	4.9	1.44	1.24	0.930	0.577	0.842	0.592
$\times \frac{3}{8}$	7.2	2.11	1.76	0.913	0.833	0.888	0.587
$\times \frac{1}{2}$	9.4	2.75	2.22	0.898	1.07	0.932	0.584
$3\frac{1}{2} \times 3\frac{1}{2} \times \frac{1}{4}$	5.8	1.69	2.01	1.09	0.794	0.968	0.694
$\times \frac{3}{8}$	8.5	2.48	2.87	1.07	1.15	1.01	0.687
$\times \frac{1}{2}$	11.1	3.25	3.64	1.06	1.49	1.06	0.683
$4 \times 4 \times \frac{1}{4}$	6.6	1.94	3.04	1.25	1.05	1.09	0.795
$\times \frac{3}{8}$	9.8	2.86	4.36	1.23	1.52	1.14	0.788
$\times \frac{1}{2}$	12.8	3.75	5.56	1.22	1.97	1.18	0.782
$\times \frac{5}{8}$	15.7	4.61	6.66	1.20	2.40	1.23	0.779
$6 \times 6 \times \frac{3}{8}$	14.9	4.36	15.4	1.88	3.53	1.64	1.19
$\times \frac{1}{2}$	19.6	5.75	19.9	1.86	4.61	1.68	1.18
$\times \frac{5}{8}$	24.2	7.11	24.2	1.84	5.66	1.73	1.18
$\times \frac{3}{4}$	28.7	8.44	28.2	1.83	6.66	1.78	1.17

TABLE A-6

Properties of Structural-Steel Angles *(Continued)*

SIZE, mm	m	A	I_{1-1}	k_{1-1}	Z_{1-1}	y	k_{3-3}
25 × 25 × 3	1.11	1.42	0.80	0.75	0.45	0.72	0.48
× 4	1.45	1.85	1.01	0.74	0.58	0.76	0.48
× 5	1.77	2.26	1.20	0.73	0.71	0.80	0.48
40 × 40 × 4	2.42	3.08	4.47	1.21	1.55	1.12	0.78
× 5	2.97	3.79	5.43	1.20	1.91	1.16	0.77
× 6	3.52	4.48	6.31	1.19	2.26	1.20	0.77
50 × 50 × 5	3.77	4.80	11.0	1.51	3.05	1.40	0.97
× 6	4.47	5.59	12.8	1.50	3.61	1.45	0.97
× 8	5.82	7.41	16.3	1.48	4.68	1.52	0.96
60 × 60 × 5	4.57	5.82	19.4	1.82	4.45	1.64	1.17
× 6	5.42	6.91	22.8	1.82	5.29	1.69	1.17
× 8	7.09	9.03	29.2	1.80	6.89	1.77	1.16
× 10	8.69	11.1	34.9	1.78	8.41	1.85	1.16
80 × 80 × 6	7.34	9.35	55.8	2.44	9.57	2.17	1.57
× 8	9.63	12.3	72.2	2.43	12.6	2.26	1.56
× 10	11.9	15.1	87.5	2.41	15.4	2.34	1.55
100 × 100 × 8	12.2	15.5	145	3.06	19.9	2.74	1.96
× 12	17.8	22.7	207	3.02	29.1	2.90	1.94
× 15	21.9	27.9	249	2.98	35.6	3.02	1.93
150 × 150 × 10	23.0	29.3	624	4.62	56.9	4.03	2.97
× 12	27.3	34.8	737	4.60	67.7	4.12	2.95
× 15	33.8	43.0	898	4.57	83.5	4.25	2.93
× 18	40.1	51.0	1050	4.54	98.7	4.37	2.92

*Metric sizes also available in sizes of 45, 70, 90, 120, and 200 mm.

†These sizes are also available in aluminum alloy.

TABLE A-7

Properties of Structural-Steel Channels*

a, b = size, in (mm)
w = weight per foot, lb/ft
m = mass per meter, kg/m
t = web thickness, in (mm)
A = area, in^2 (cm^2)
I = second moment of area, in^4 (cm^4)
k = radius of gyration, in (cm)
x = centroidal distance, in (cm)
Z = section modulus, in^3 (cm^3)

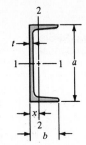

a, in	b, in	t	A	w	$I_{1\text{-}1}$	$k_{1\text{-}1}$	$Z_{1\text{-}1}$	$I_{2\text{-}2}$	$k_{2\text{-}2}$	$Z_{2\text{-}2}$	x
3	1.410	0.170	1.21	4.1	1.66	1.17	1.10	0.197	0.404	0.202	0.436
3	1.498	0.258	1.47	5.0	1.85	1.12	1.24	0.247	0.410	0.233	0.438
3	1.596	0.356	1.76	6.0	2.07	1.08	1.38	0.305	0.416	0.268	0.455
4	1.580	0.180	1.57	5.4	3.85	1.56	1.93	0.319	0.449	0.283	0.457
4	1.720	0.321	2.13	7.25	4.59	1.47	2.29	0.433	0.450	0.343	0.459
5	1.750	0.190	1.97	6.7	7.49	1.95	3.00	0.479	0.493	0.378	0.484
5	1.885	0.325	2.64	9.0	8.90	1.83	3.56	0.632	0.489	0.450	0.478
6	1.920	0.200	2.40	8.2	13.1	2.34	4.38	0.693	0.537	0.492	0.511
6	2.034	0.314	3.09	10.5	15.2	2.22	5.06	0.866	0.529	0.564	0.499
6	2.157	0.437	3.83	13.0	17.4	2.13	5.80	1.05	0.525	0.642	0.514
7	2.090	0.210	2.87	9.8	21.3	2.72	6.08	0.968	0.581	0.625	0.540
7	2.194	0.314	3.60	12.25	24.2	2.60	6.93	1.17	0.571	0.703	0.525
7	2.299	0.419	4.33	14.75	27.2	2.51	7.78	1.38	0.564	0.779	0.532
8	2.260	0.220	3.36	11.5	32.3	3.10	8.10	1.30	0.625	0.781	0.571
8	2.343	0.303	4.04	13.75	36.2	2.99	9.03	1.53	0.615	0.854	0.553
8	2.527	0.487	5.51	18.75	44.0	2.82	11.0	1.98	0.599	1.01	0.565
9	2.430	0.230	3.91	13.4	47.7	3.49	10.6	1.75	0.669	0.962	0.601
9	2.485	0.285	4.41	15.0	51.0	3.40	11.3	1.93	0.661	1.01	0.586
9	2.648	0.448	5.88	20.0	60.9	3.22	13.5	2.42	0.647	1.17	0.583
10	2.600	0.240	4.49	15.3	67.4	3.87	13.5	2.28	0.713	1.16	0.634
10	2.739	0.379	5.88	20.0	78.9	3.66	15.8	2.81	0.693	1.32	0.606
10	2.886	0.526	7.35	25.0	91.2	3.52	18.2	3.36	0.676	1.48	0.617
10	3.033	0.673	8.82	30.0	103	3.43	20.7	3.95	0.669	1.66	0.649
12	3.047	0.387	7.35	25.0	144	4.43	24.1	4.47	0.780	1.89	0.674
12	3.170	0.510	8.82	30.0	162	4.29	27.0	5.14	0.763	2.06	0.674

TABLE A-7

Properties of Structural-Steel Channels *(Continued)*

$a \times b$, mm	m	t	A	$I_{1\text{-}1}$	$k_{1\text{-}1}$	$Z_{1\text{-}1}$	$I_{2\text{-}2}$	$k_{2\text{-}2}$	$Z_{2\text{-}2}$	x
76 × 38	6.70	5.1	8.53	74.14	2.95	19.46	10.66	1.12	4.07	1.19
102 × 51	10.42	6.1	13.28	207.7	3.95	40.89	29.10	1.48	8.16	1.51
127 × 64	14.90	6.4	18.98	482.5	5.04	75.99	67.23	1.88	15.25	1.94
152 × 76	17.88	6.4	22.77	851.5	6.12	111.8	113.8	2.24	21.05	2.21
152 × 89	23.84	7.1	30.36	1166	6.20	153.0	215.1	2.66	35.70	2.86
178 × 76	20.84	6.6	26.54	1337	7.10	150.4	134.0	2.25	24.72	2.20
178 × 89	26.81	7.6	34.15	1753	7.16	197.2	241.0	2.66	39.29	2.76
203 × 76	23.82	7.1	30.34	1950	8.02	192.0	151.3	2.23	27.59	2.13
203 × 89	29.78	8.1	37.94	2491	8.10	245.2	264.4	2.64	42.34	2.65
229 × 76	26.06	7.6	33.20	2610	8.87	228.3	158.7	2.19	28.22	2.00
229 × 89	32.76	8.6	41.73	3387	9.01	296.4	285.0	2.61	44.82	2.53
254 × 76	28.29	8.1	36.03	3367	9.67	265.1	162.6	2.12	28.21	1.86
254 × 89	35.74	9.1	45.42	4448	9.88	350.2	302.4	2.58	46.70	2.42
305 × 89	41.69	10.2	53.11	7061	11.5	463.3	325.4	2.48	48.49	2.18
305 × 102	46.18	10.2	58.83	8214	11.8	539.0	499.5	2.91	66.59	2.66

*These sizes are also available in aluminum alloy.

TABLE A-8
Properties of Round Tubing

w_a = unit weight of aluminum tubing, lb/ft
w_s = unit weight of steel tubing, lb/ft
m = unit mass, kg/m
A = area, in^2 (cm^2)
I = second moment of area, in^4 (cm^4)
J = second polar moment of area, in^4 (cm^4)
k = radius of gyration, in (cm)
Z = section modulus, in^3 (cm^3)
d, t = size (OD) and thickness, in (mm)

SIZE, in	w_a	w_s	A	I	k	Z	J
$1 \times \frac{1}{8}$	0.416	1.128	0.344	0.034	0.313	0.067	0.067
$1 \times \frac{1}{4}$	0.713	2.003	0.589	0.046	0.280	0.092	0.092
$1\frac{1}{2} \times \frac{1}{8}$	0.653	1.769	0.540	0.129	0.488	0.172	0.257
$1\frac{1}{2} \times \frac{1}{4}$	1.188	3.338	0.982	0.199	0.451	0.266	0.399
$2 \times \frac{1}{8}$	0.891	2.670	0.736	0.325	0.664	0.325	0.650
$2 \times \frac{1}{4}$	1.663	4.673	1.374	0.537	0.625	0.537	1.074
$2\frac{1}{2} \times \frac{1}{8}$	1.129	3.050	0.933	0.660	0.841	0.528	1.319
$2\frac{1}{2} \times \frac{1}{4}$	2.138	6.008	1.767	1.132	0.800	0.906	2.276
$3 \times \frac{1}{4}$	2.614	7.343	2.160	2.059	0.976	1.373	4.117
$3 \times \frac{3}{8}$	3.742	10.51	3.093	2.718	0.938	1.812	5.436
$4 \times \frac{3}{16}$	2.717	7.654	2.246	4.090	1.350	2.045	8.180
$4 \times \frac{3}{8}$	5.167	14.52	4.271	7.090	1.289	3.544	14.180

SIZE, mm	m	A	I	k	Z	J
12×2	0.490	0.628	0.082	0.361	0.136	0.163
16×2	0.687	0.879	0.220	0.500	0.275	0.440
16×3	0.956	1.225	0.273	0.472	0.341	0.545
20×4	1.569	2.010	0.684	0.583	0.684	1.367
25×4	2.060	2.638	1.508	0.756	1.206	3.015
25×5	2.452	3.140	1.669	0.729	1.336	3.338
30×4	2.550	3.266	2.827	0.930	1.885	5.652
30×5	3.065	3.925	3.192	0.901	2.128	6.381
42×4	3.727	4.773	8.717	1.351	4.151	17.430
42×5	4.536	5.809	10.130	1.320	4.825	20.255
50×4	4.512	5.778	15.409	1.632	6.164	30.810
50×5	5.517	7.065	18.118	1.601	7.247	36.226

TABLE A-9

**Shear, Moment, and
Deflection of Beams**

1 Cantilever—end load

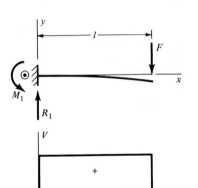

$$R_1 = V = F \qquad M_1 = -Fl$$

$$M = F(x - l)$$

$$y = \frac{Fx^2}{6EI}(x - 3l)$$

$$y_{max} = -\frac{Fl^3}{3EI}$$

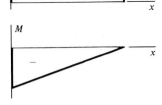

2 Cantilever—intermediate load

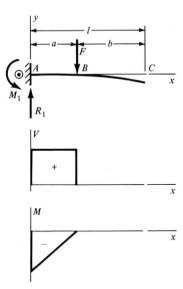

$$R_1 = V = F \qquad M_1 = -Fa$$

$$M_{AB} = F(x - a) \qquad M_{BC} = 0$$

$$y_{AB} = \frac{Fx^2}{6EI}(x - 3a)$$

$$y_{BC} = \frac{Fa^2}{6EI}(a - 3x)$$

$$y_{max} = \frac{Fa^2}{6EI}(a - 3l)$$

TABLE A-9

Shear, Moment, and Deflection of Beams (Continued)

3 Cantilever—uniform load

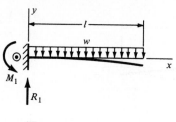

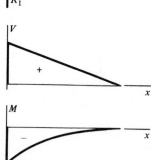

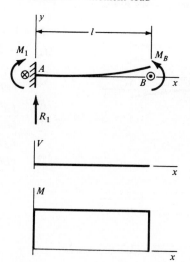

$$R_1 = wl \qquad M_1 = -\frac{wl^2}{2}$$

$$V = w(l - x) \qquad M = -\frac{w}{2}(l - x)^2$$

$$y = \frac{wx^2}{24EI}(4lx - x^2 - 6l^2)$$

$$y_{max} = -\frac{wl^4}{8EI}$$

4 Cantilever—moment load

$$R_1 = 0 \qquad M_1 = M_B \qquad M = M_B$$

$$y = \frac{M_B x^2}{2EI} \qquad y_{max} = \frac{M_B l^2}{2EI}$$

TABLE A-9

**Shear, Moment, and
Deflection of Beams
(Continued)**

5 Simple supports—center load

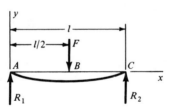

$$R_1 = R_2 = \frac{F}{2} \qquad V_{AB} = R_1$$

$$V_{AB} = R_1 \qquad V_{BC} = -R_2$$

$$M_{AB} = \frac{Fx}{2} \qquad M_{BC} = \frac{F}{2}(l - x)$$

$$y_{AB} = \frac{Fx}{48EI}(4x^2 - 3l^2)$$

$$y_{max} = -\frac{Fl^3}{48EI}$$

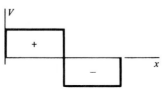

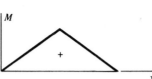

6 Simple supports—intermediate load, $a < b$

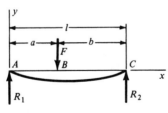

$$R_1 = \frac{Fb}{l} \qquad R_2 = \frac{Fa}{l}$$

$$V_{AB} = R_1 \qquad V_{BC} = -R_2$$

$$M_{AB} = \frac{Fbx}{l} \qquad M_{BC} = \frac{Fa}{l}(l - x)$$

$$y_{AB} = \frac{Fbx}{6EIl}(x^2 + b^2 - l^2)$$

$$y_{BC} = \frac{Fa(l - x)}{6EIl}(x^2 + a^2 - 2lx)$$

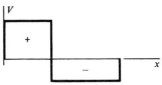

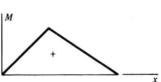

TABLE A-9

Shear, Moment, and
Deflection of Beams
(Continued)

7 Simple supports—uniform load

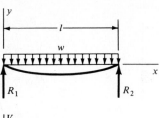

$$R_1 = R_2 = \frac{wl}{2} \qquad V = \frac{wl}{2} - wx$$

$$M = \frac{wx}{2}(l - x)$$

$$y = \frac{wx}{24EI}(2lx^2 - x^3 - l^3)$$

$$y_{max} = -\frac{5wl^4}{384EI}$$

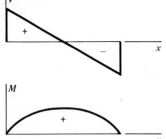

8 Simple supports—moment load

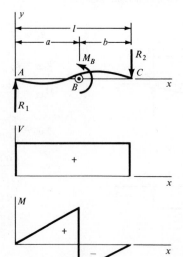

$$R_1 = -R_2 = \frac{M_B}{l} \qquad V = \frac{M_B}{l}$$

$$M_{AB} = \frac{M_B x}{l} \qquad M_{BC} = \frac{M_B}{l}(x - l)$$

$$y_{AB} = \frac{M_B x}{6EIl}(x^2 + 3a^2 - 6al + 2l^2)$$

$$y_{BC} = \frac{M_B}{6EIl}[x^3 - 3lx^2 + x(2l^2 + 3a^2) - 3a^2 l]$$

TABLE A-9

Shear, Moment, and
Deflection of Beams
(Continued)

9 Simple supports—twin loads

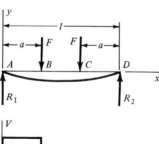

$$R_1 = R_2 = F \qquad V_{AB} = F \qquad V_{BC} = 0$$

$$V_{CD} = -F$$

$$M_{AB} = Fx \qquad M_{BC} = Fa \qquad M_{CD} = F(l - x)$$

$$y_{AB} = \frac{Fx}{6EI}(x^2 + 3a^2 - 3la)$$

$$y_{BC} = \frac{Fa}{6EI}(3x^2 + a^2 - 3lx)$$

$$y_{max} = \frac{Fa}{24EI}(4a^2 - 3l^2)$$

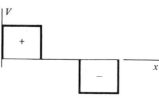

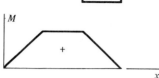

10 Simple supports—overhanging load

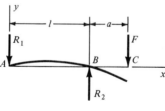

$$R_1 = -\frac{Fa}{l} \qquad R_2 = \frac{F}{l}(l + a)$$

$$V_{AB} = -\frac{Fa}{l} \qquad V_{BC} = F$$

$$M_{AB} = -\frac{Fax}{l} \qquad M_{BC} = F(x - l - a)$$

$$y_{AB} = \frac{Fax}{6EIl}(l^2 - x^2)$$

$$y_{BC} = \frac{F(x - l)}{6EI}[(x - l)^2 - a(3x - l)]$$

$$y_c = -\frac{Fa^2}{3EI}(l + a)$$

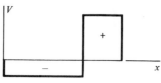

TABLE A-9

Shear, Moment, and ·
Deflection of Beams
(*Continued*)

11 One fixed and one simple support—
   center load

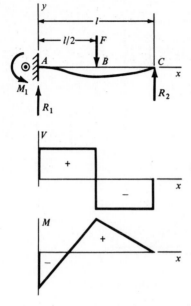

$$R_1 = \frac{11F}{16} \qquad R_2 = \frac{5F}{16} \qquad M_1 = -\frac{3Fl}{16}$$

$$V_{AB} = R_1 \qquad V_{BC} = -R_2$$

$$M_{AB} = \frac{F}{16}(11x - 3l) \qquad M_{BC} = \frac{5F}{16}(l - x)$$

$$y_{AB} = \frac{Fx^2}{96EI}(11x - 9l)$$

$$y_{BC} = \frac{F(l - x)}{96EI}(5x^2 + 2l^2 - 10lx)$$

12 One fixed and one simple support—
   intermediate load

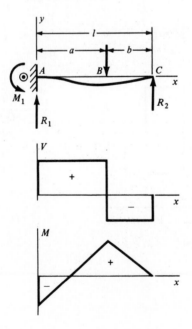

$$R_1 = \frac{Fb}{2l^3}(3l^2 - b^2) \qquad R_2 = \frac{Fa^2}{2l^3}(3l - a)$$

$$M_1 = \frac{Fb}{2l^2}(b^2 - l^2) \qquad V_{AB} = R_1$$

$$V_{AB} = R_1 \qquad V_{BC} = -R_2$$

$$M_{AB} = \frac{Fb}{2l^3}[b^2l - l^3 + x(3l^2 - b^2)]$$

$$M_{BC} = \frac{Fa^2}{2l^3}(3l^2 - 3lx - al + ax)$$

$$y_{AB} = \frac{Fbx^2}{12EIl^3}[3l(b^2 - l^2) + x(3l^2 - b^2)]$$

$$y_{BC} = y_{AB} - \frac{F(x - a)^3}{6EI}$$

TABLE A-9

**Shear, Moment, and
Deflection of Beams**
(Continued)

**13 One fixed and one simple support
—uniform load**

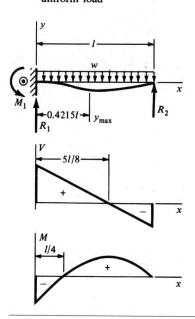

$$R_1 = \frac{5wl}{8} \qquad R_2 = \frac{3wl}{8} \qquad M_1 = -\frac{wl^2}{8}$$

$$V = \frac{5wl}{8} - wx$$

$$M = \frac{w}{8}(4x^2 + 5lx - l^2)$$

$$y = \frac{wx^2}{48EI}(l - x)(2x - 3l)$$

$$y_{max} = -\frac{wl^4}{185EI}$$

14 Fixed supports—center load

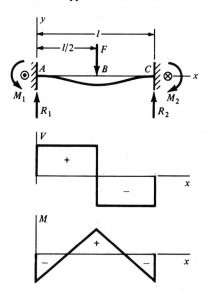

$$R_1 = R_2 = \frac{F}{2} \qquad M_1 = M_2 = -\frac{Fl}{8}$$

$$V_{AB} = -V_{BC} = \frac{F}{2}$$

$$M_{AB} = \frac{F}{8}(4x - l) \qquad M_{BC} = \frac{F}{8}(3l - 4x)$$

$$y_{AB} = \frac{Fx^2}{48EI}(4x - 3l)$$

$$y_{max} = -\frac{Fl^3}{192EI}$$

15 Fixed supports—intermediate load

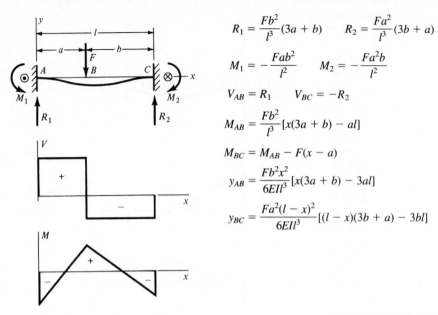

$$R_1 = \frac{Fb^2}{l^3}(3a + b) \qquad R_2 = \frac{Fa^2}{l^3}(3b + a)$$

$$M_1 = -\frac{Fab^2}{l^2} \qquad M_2 = -\frac{Fa^2b}{l^2}$$

$$V_{AB} = R_1 \qquad V_{BC} = -R_2$$

$$M_{AB} = \frac{Fb^2}{l^3}[x(3a + b) - al]$$

$$M_{BC} = M_{AB} - F(x - a)$$

$$y_{AB} = \frac{Fb^2x^2}{6EIl^3}[x(3a + b) - 3al]$$

$$y_{BC} = \frac{Fa^2(l - x)^2}{6EIl^3}[(l - x)(3b + a) - 3bl]$$

16 Fixed supports—uniform load

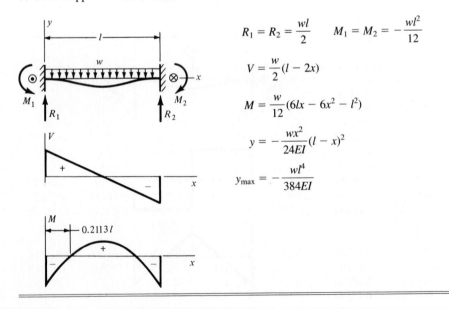

$$R_1 = R_2 = \frac{wl}{2} \qquad M_1 = M_2 = -\frac{wl^2}{12}$$

$$V = \frac{w}{2}(l - 2x)$$

$$M = \frac{w}{12}(6lx - 6x^2 - l^2)$$

$$y = -\frac{wx^2}{24EI}(l - x)^2$$

$$y_{max} = -\frac{wl^4}{384EI}$$

TABLE A-10

Cumulative Density Function of Normal (Gaussian) Distribution

$$\Phi(z_\alpha) = \int_{-\infty}^{z_\alpha} \frac{1}{\sqrt{2\pi}} \exp\left(-\frac{u^2}{2}\right) du$$

$$= \begin{cases} \alpha & z_\alpha \le 0 \\ 1-\alpha & z_\alpha > 0 \end{cases}$$

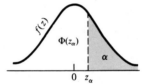

z_α	0.00	0.01	0.02	0.03	0.04	0.05	0.06	0.07	0.08	0.09
0.0	0.5000	0.4960	0.4920	0.4880	0.4840	0.4801	0.4761	0.4721	0.4681	0.4641
0.1	0.4602	0.4562	0.4522	0.4483	0.4443	0.4404	0.4364	0.4325	0.4286	0.4247
0.2	0.4207	0.4168	0.4129	0.4090	0.4052	0.4013	0.3974	0.3936	0.3897	0.3859
0.3	0.3821	0.3783	0.3745	0.3707	0.3669	0.3632	0.3594	0.3557	0.3520	0.3483
0.4	0.3446	0.3409	0.3372	0.3336	0.3300	0.3264	0.3238	0.3192	0.3156	0.3121
0.5	0.3085	0.3050	0.3015	0.2981	0.2946	0.2912	0.2877	0.2843	0.2810	0.2776
0.6	0.2743	0.2709	0.2676	0.2643	0.2611	0.2578	0.2546	0.2514	0.2483	0.2451
0.7	0.2420	0.2389	0.2358	0.2327	0.2296	0.2266	0.2236	0.2206	0.2177	0.2148
0.8	0.2119	0.2090	0.2061	0.2033	0.2005	0.1977	0.1949	0.1922	0.1894	0.1867
0.9	0.1841	0.1814	0.1788	0.1762	0.1736	0.1711	0.1685	0.1660	0.1635	0.1611
1.0	0.1587	0.1562	0.1539	0.1515	0.1492	0.1469	0.1446	0.1423	0.1401	0.1379
1.1	0.1357	0.1335	0.1314	0.1292	0.1271	0.1251	0.1230	0.1210	0.1190	0.1170
1.2	0.1151	0.1131	0.1112	0.1093	0.1075	0.1056	0.1038	0.1020	0.1003	0.0985
1.3	0.0968	0.0951	0.0934	0.0918	0.0901	0.0885	0.0869	0.0853	0.0838	0.0823
1.4	0.0808	0.0793	0.0778	0.0764	0.0749	0.0735	0.0721	0.0708	0.0694	0.0681
1.5	0.0668	0.0655	0.0643	0.0630	0.0618	0.0606	0.0594	0.0582	0.0571	0.0559
1.6	0.0548	0.0537	0.0526	0.0516	0.0505	0.0495	0.0485	0.0475	0.0465	0.0455
1.7	0.0446	0.0436	0.0427	0.0418	0.0409	0.0401	0.0392	0.0384	0.0375	0.0367
1.8	0.0359	0.0351	0.0344	0.0336	0.0329	0.0322	0.0314	0.0307	0.0301	0.0294
1.9	0.0287	0.0281	0.0274	0.0268	0.0262	0.0256	0.0250	0.0244	0.0239	0.0233
2.0	0.0228	0.0222	0.0217	0.0212	0.0207	0.0202	0.0197	0.0192	0.0188	0.0183
2.1	0.0179	0.0174	0.0170	0.0166	0.0162	0.0158	0.0154	0.0150	0.0146	0.0143
2.2	0.0139	0.0136	0.0132	0.0129	0.0125	0.0122	0.0119	0.0116	0.0113	0.0110
2.3	0.0107	0.0104	0.0102	0.00990	0.00964	0.00939	0.00914	0.00889	0.00866	0.00842
2.4	0.00820	0.00798	0.00776	0.00755	0.00734	0.00714	0.00695	0.00676	0.00657	0.00639
2.5	0.00621	0.00604	0.00587	0.00570	0.00554	0.00539	0.00523	0.00508	0.00494	0.00480
2.6	0.00466	0.00453	0.00440	0.00427	0.00415	0.00402	0.00391	0.00379	0.00368	0.00357
2.7	0.00347	0.00336	0.00326	0.00317	0.00307	0.00298	0.00289	0.00280	0.00272	0.00264
2.8	0.00256	0.00248	0.00240	0.00233	0.00226	0.00219	0.00212	0.00205	0.00199	0.00193
2.9	0.00187	0.00181	0.00175	0.00169	0.00164	0.00159	0.00154	0.00149	0.00144	0.00139

z_α	0.0	0.1	0.2	0.3	0.4	0.5	0.6	0.7	0.8	0.9
3	0.00135	0.0^3968	0.0^3687	0.0^3483	0.0^3337	0.0^3233	0.0^3159	0.0^3108	0.0^4723	0.0^4481
4	0.0^4317	0.0^4207	0.0^4133	0.0^5854	0.0^5541	0.0^5340	0.0^5211	0.0^5130	0.0^6793	0.0^6479
5	0.0^6287	0.0^6170	0.0^7996	0.0^7579	0.0^7333	0.0^7190	0.0^7107	0.0^8599	0.0^8332	0.0^8182
6	0.0^9987	0.0^9530	0.0^9282	0.0^9149	$0.0^{10}777$	$0.0^{10}402$	$0.0^{10}206$	$0.0^{10}104$	$0.0^{11}523$	$0.0^{11}260$

TABLE A-11

A Selection of International Tolerance Grades— Metric Series (Size Ranges Are for *Over* the Lower Limit and *Including* the Upper Limit. All Values Are in Millimeters)

BASIC SIZES	TOLERANCE GRADES					
	IT6	IT7	IT8	IT9	IT10	IT11
0–3	0.006	0.010	0.014	0.025	0.040	0.060
3–6	0.008	0.012	0.018	0.030	0.048	0.075
6–10	0.009	0.015	0.022	0.036	0.058	0.090
10–18	0.011	0.018	0.027	0.043	0.070	0.110
18–30	0.013	0.021	0.033	0.052	0.084	0.130
30–50	0.016	0.025	0.039	0.062	0.100	0.160
50–80	0.019	0.030	0.046	0.074	0.120	0.190
80–120	0.022	0.035	0.054	0.087	0.140	0.220
120–180	0.025	0.040	0.063	0.100	0.160	0.250
180–250	0.029	0.046	0.072	0.115	0.185	0.290
250–315	0.032	0.052	0.081	0.130	0.210	0.320
315–400	0.036	0.057	0.089	0.140	0.230	0.360

Souce: Preferred Metric Limits and Fits, ANSI B4.2-1978. See also BSI 4500.

TABLE A-12

Fundamental Deviations for Shafts—Metric Series
(Size Ranges Are for *Over* the Lower Limit and *Including* the Upper Limit. All Values Are in Millimeters)

BASIC SIZES	UPPER-DEVIATION LETTER					LOWER-DEVIATION LETTER				
	c	d	f	g	h	k	n	p	s	u
0–3	−0.060	−0.020	−0.006	−0.002	0	0	+0.004	+0.006	+0.014	+0.018
3–6	−0.070	−0.030	−0.010	−0.004	0	+0.001	+0.008	+0.012	+0.019	+0.023
6–10	−0.080	−0.040	−0.013	−0.005	0	+0.001	+0.010	+0.015	+0.023	+0.028
10–14	−0.095	−0.050	−0.016	−0.006	0	+0.001	+0.012	+0.018	+0.028	+0.033
14–18	−0.095	−0.050	−0.016	−0.006	0	+0.001	+0.012	+0.018	+0.028	+0.033
18–24	−0.110	−0.065	−0.020	−0.007	0	+0.002	+0.015	+0.022	+0.035	+0.041
24–30	−0.110	−0.065	−0.020	−0.007	0	+0.002	+0.015	+0.022	+0.035	+0.048
30–40	−0.120	−0.080	−0.025	−0.009	0	+0.002	+0.017	+0.026	+0.043	+0.060
40–50	−0.130	−0.080	−0.025	−0.009	0	+0.002	+0.017	+0.026	+0.043	+0.070
50–65	−0.140	−0.100	−0.030	−0.010	0	+0.002	+0.020	+0.032	+0.053	+0.087
65–80	−0.150	−0.100	−0.030	−0.010	0	+0.002	+0.020	+0.032	+0.059	+0.102
80–100	−0.170	−0.120	−0.036	−0.012	0	+0.003	+0.023	+0.037	+0.071	+0.124
100–120	−0.180	−0.120	−0.036	−0.012	0	+0.003	+0.023	+0.037	+0.079	+0.144
120–140	−0.200	−0.145	−0.043	−0.014	0	+0.003	+0.027	+0.043	+0.092	+0.170
140–160	−0.210	−0.145	−0.043	−0.014	0	+0.003	+0.027	+0.043	+0.100	+0.190
160–180	−0.230	−0.145	−0.043	−0.014	0	+0.003	+0.027	+0.043	+0.108	+0.210
180–200	−0.240	−0.170	−0.050	−0.015	0	+0.004	+0.031	+0.050	+0.122	+0.236
200–225	−0.260	−0.170	−0.050	−0.015	0	+0.004	+0.031	+0.050	+0.130	+0.258
225–250	−0.280	−0.170	−0.050	−0.015	0	+0.004	+0.031	+0.050	+0.140	+0.284
250–280	−0.300	−0.190	−0.056	−0.017	0	+0.004	+0.034	+0.056	+0.158	+0.315
280–315	−0.330	−0.190	−0.056	−0.017	0	+0.004	+0.034	+0.056	+0.170	+0.350
315–355	−0.360	−0.210	−0.062	−0.018	0	+0.004	+0.037	+0.062	+0.190	+0.390
355–400	−0.400	−0.210	−0.062	−0.018	0	+0.004	+0.037	+0.062	+0.208	+0.435

Source: Preferred Metric Limits and Fits, ANSI B4.2-1978. See also BSI 4500.

TABLE A-13

A Selection of International Tolerance Grades— Inch Series (Size Ranges Are for *Over* the Lower Limit and *Including* the Upper Limit. All Values Are in Inches, Converted from Table A-11)

BASIC SIZES	TOLERANCE GRADES					
	IT6	IT7	IT8	IT9	IT10	IT11
0–0.12	0.0002	0.0004	0.0006	0.0010	0.0016	0.0024
0.12–0.24	0.0003	0.0005	0.0007	0.0012	0.0019	0.0030
0.24–0.40	0.0004	0.0006	0.0009	0.0014	0.0023	0.0035
0.40–0.72	0.0004	0.0007	0.0011	0.0017	0.0028	0.0043
0.72–1.20	0.0005	0.0008	0.0013	0.0020	0.0033	0.0051
1.20–2.00	0.0006	0.0010	0.0015	0.0024	0.0039	0.0063
2.00–3.20	0.0007	0.0012	0.0018	0.0029	0.0047	0.0075
3.20–4.80	0.0009	0.0014	0.0021	0.0034	0.0055	0.0087
4.80–7.20	0.0010	0.0016	0.0025	0.0039	0.0063	0.0098
7.20–10.00	0.0011	0.0018	0.0028	0.0045	0.0073	0.0114
10.00–12.60	0.0013	0.0020	0.0032	0.0051	0.0083	0.0126
12.60–16.00	0.0014	0.0022	0.0035	0.0055	0.0091	0.0142

TABLE A-14

Fundamental Deviations for Shafts—Inch Series
(Size Ranges Are for *Over* the Lower Limit and *Including* the Upper Limit. All Values Are in Inches, Converted from Table A-12)

BASIC SIZES	UPPER-DEVIATION LETTER					LOWER-DEVIATION LETTER				
	c	d	f	g	h	k	n	p	s	u
0–0.12	−0.0024	−0.0008	−0.0002	−0.0001	0	0	+0.0002	+0.0002	+0.0006	+0.0007
0.12–0.24	−0.0028	−0.0012	−0.0004	−0.0002	0	0	+0.0003	+0.0005	+0.0007	+0.0009
0.24–0.40	−0.0031	−0.0016	−0.0005	−0.0002	0	0	+0.0004	+0.0006	+0.0009	+0.0011
0.40–0.72	−0.0037	−0.0020	−0.0006	−0.0002	0	0	+0.0005	+0.0007	+0.0011	+0.0013
0.72–0.96	−0.0043	−0.0026	−0.0008	−0.0003	0	+0.0001	+0.0006	+0.0009	+0.0014	+0.0016
0.96–1.20	−0.0043	−0.0026	−0.0008	−0.0003	0	+0.0001	+0.0006	+0.0009	+0.0014	+0.0019
1.20–1.60	−0.0047	−0.0031	−0.0010	−0.0004	0	+0.0001	+0.0007	+0.0010	+0.0017	+0.0024
1.60–2.00	−0.0051	−0.0031	−0.0010	−0.0004	0	+0.0001	+0.0007	+0.0010	+0.0017	+0.0028
2.00–2.60	−0.0055	−0.0039	−0.0012	−0.0004	0	+0.0001	+0.0008	+0.0013	+0.0021	+0.0034
2.60–3.20	−0.0059	−0.0039	−0.0012	−0.0004	0	+0.0001	+0.0008	+0.0013	+0.0023	+0.0040
3.20–4.00	−0.0067	−0.0047	−0.0014	−0.0005	0	+0.0001	+0.0009	+0.0015	+0.0028	+0.0049
4.00–4.80	−0.0071	−0.0047	−0.0014	−0.0005	0	+0.0001	+0.0009	+0.0015	+0.0031	+0.0057
4.80–5.60	−0.0079	−0.0057	−0.0017	−0.0006	0	+0.0001	+0.0011	+0.0017	+0.0036	+0.0067
5.60–6.40	−0.0083	−0.0057	−0.0017	−0.0006	0	+0.0001	+0.0011	+0.0017	+0.0039	+0.0075
6.40–7.20	−0.0091	−0.0057	−0.0017	−0.0006	0	+0.0001	+0.0011	+0.0017	+0.0043	+0.0083
7.20–8.00	−0.0094	−0.0067	−0.0020	−0.0006	0	+0.0002	+0.0012	+0.0020	+0.0048	+0.0093
8.00–9.00	−0.0102	−0.0067	−0.0020	−0.0006	0	+0.0002	+0.0012	+0.0020	+0.0051	+0.0102
9.00–10.00	−0.0110	−0.0067	−0.0020	−0.0006	0	+0.0002	+0.0012	+0.0020	+0.0055	+0.0112
10.00–11.20	−0.0118	−0.0075	−0.0022	−0.0007	0	+0.0002	+0.0013	+0.0022	+0.0062	+0.0124
11.20–12.60	−0.0130	−0.0075	−0.0022	−0.0007	0	+0.0002	+0.0013	+0.0022	+0.0067	+0.0130
12.60–14.20	−0.0142	−0.0083	−0.0024	−0.0007	0	+0.0002	+0.0015	+0.0024	+0.0075	+0.0154
14.20–16.00	−0.0157	−0.0083	−0.0024	−0.0007	0	+0.0002	+0.0015	+0.0024	+0.0082	+0.0171

TABLE A-15

Charts of Theoretical Stress-Concentration Factors K_t*

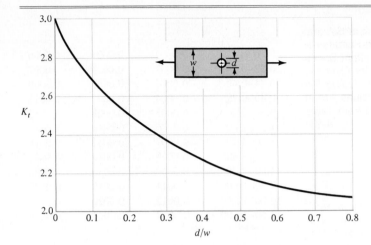

FIGURE A-15-1

Bar in tension or simple compression with a transverse hole. $\sigma_0 = F/A$, where $A = (w - d)t$ and t is the thickness.

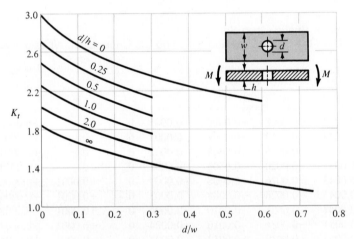

FIGURE A-15-2

Rectangular bar with a transverse hole in bending. $\sigma_0 = Mc/I$, where $I = (w - d)h^3/12$.

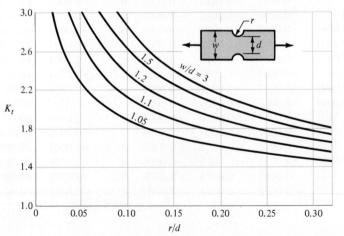

FIGURE A-15-3

Notched rectangular bar in tension or simple compression. $\sigma_0 = F/A$, where $A = dt$ and t is the thickness.

*Unless otherwise stated, these factors are from R. E. Peterson, "Design Factors for Stress Concentration," *Machine Design*, vol. 23, no. 2, February 1951, p. 169; no. 3, March 1951, p. 161; no. 5, May 1951, p. 159; no. 6, June 1951, p. 173; no. 7, July 1951, p. 155; reproduced with the permission of the author and publisher.

TABLE A-15

Charts of Theoretical Stress-Concentration Factors K_t (Continued)

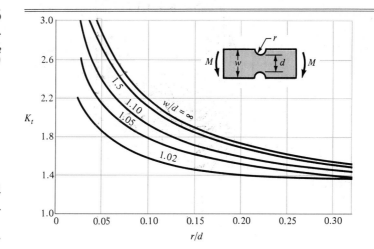

FIGURE A-15-4

Notched rectangular bar in bending. $\sigma_0 = Mc/I$, where $c = d/2$, $I = td^3/12$, and t is the thickness.

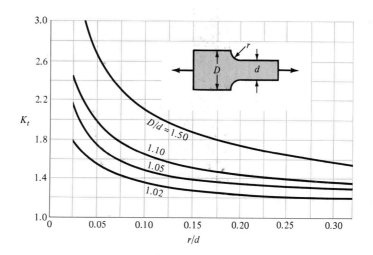

FIGURE A-15-5

Rectangular filleted bar in tension or simple compression. $\sigma_0 = F/A$, where $A = dt$ and t is the thickness.

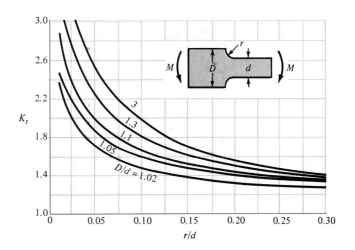

FIGURE A-15-6

Rectangular filleted bar in bending. $\sigma_0 = Mc/I$, where $c = d/2$, $I = td^3/12$, and t is the thickness.

TABLE A-15

Charts of Theoretical Stress-Concentration Factors K_t *(Continued)*

FIGURE A-15-7

Round shaft with shoulder fillet in tension. $\sigma_0 = F/A$, where $A = \pi d^2/4$.

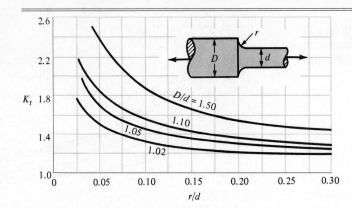

FIGURE A-15-8

Round shaft with shoulder fillet in torsion. $\tau_0 = Tc/J$, where $c = d/2$ and $J = \pi d^4/32$.

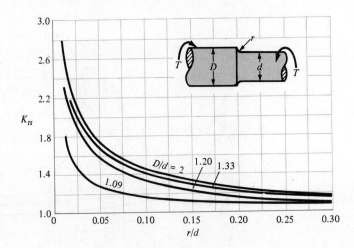

FIGURE A-15-9

Round shaft with shoulder fillet in bending. $\sigma_0 = Mc/I$, where $c = d/2$ and $I = \pi d^4/64$.

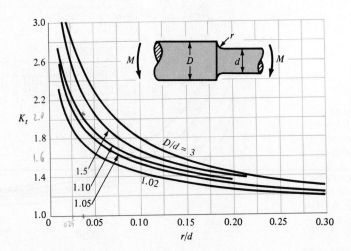

TABLE A-15

Charts of Theoretical Stress-Concentration Factors K_t
(Continued)

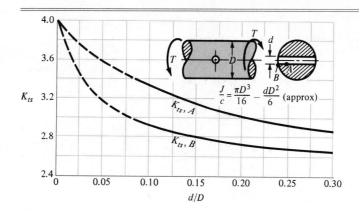

FIGURE A-15-10

Round shaft in torsion with transverse hole.

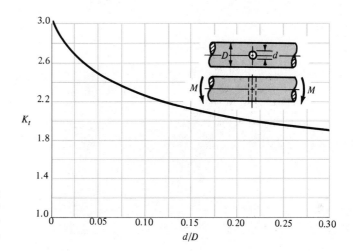

FIGURE A-15-11

Round shaft in bending with a transverse hole. $\sigma_0 = M/[(\pi D^3/32) - (dD^2/6)]$, approximately.

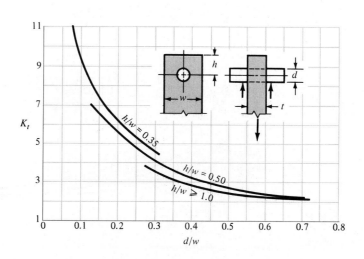

FIGURE A-15-12

Plate loaded in tension by a pin through a hole. $\sigma_0 = F/A$, where $A = (w - d)t$. When clearance exists, increase K_t 35 to 50 percent. (M. M. Frocht and H. N. Hill, "Stress Concentration Factors around a Central Circular Hole in a Plate Loaded through a Pin in Hole," *J. Appl. Mechanics*, vol. 7, no. 1, March 1940, p. A-5.)

TABLE A-15

Charts of Theoretical Stress-Concentration Factors K_t (Continued)

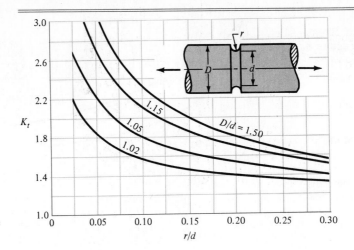

FIGURE A-15-13

Grooved round bar in tension. $\sigma_0 = F/A$, where $A = \pi d^2/4$.

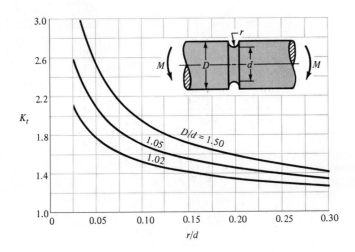

FIGURE A-15-14

Grooved round bar in bending. $\sigma_0 = Mc/I$, where $c = d/2$ and $I = \pi d^4/64$.

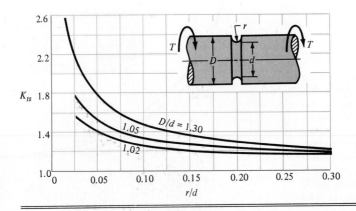

FIGURE A-15-15

Grooved round bar in torsion. $\tau_0 = Tc/J$, where $c = d/2$ and $J = \pi d^4/32$.

TABLE A-16

Approximate Stress-Concentration Factors K_t for Bending of
a Round Bar or Tube with a Transverse Round Hole
[The Nominal Bending Stress Is $\sigma_0 = M/Z_{net}$, where Z_{net} is a
Reduced Value of the Section Modulus and Is Defined by

$$Z_{net} = \frac{\pi A}{32D}(D^4 - d^4)$$

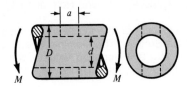

Values of A Are Listed in the Table. Use $d = 0$ for a Solid Bar]

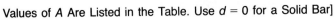

	d/D					
	0.9		0.6		0	
a/D	A	K_t	A	K_t	A	K_t
0.050	0.92	2.63	0.91	2.55	0.88	2.42
0.075	0.89	2.55	0.88	2.43	0.86	2.35
0.10	0.86	2.49	0.85	2.36	0.83	2.27
0.125	0.82	2.41	0.82	2.32	0.80	2.20
0.15	0.79	2.39	0.79	2.29	0.76	2.15
0.175	0.76	2.38	0.75	2.26	0.72	2.10
0.20	0.73	2.39	0.72	2.23	0.68	2.07
0.225	0.69	2.40	0.68	2.21	0.65	2.04
0.25	0.67	2.42	0.64	2.18	0.61	2.00
0.275	0.66	2.48	0.61	2.16	0.58	1.97
0.30	0.64	2.52	0.58	2.14	0.54	1.94

Source: R. E. Peterson, *Stress Concentration Factors*, Wiley, New York, 1974, pp. 146, 235.

TABLE A-16 *(Continued)*

Approximate Stress-Concentration Factors K_{ts} for a Round Bar or Tube Having a Transverse-Round Hole and Loaded in Torsion
[The Maximum Stress Occurs on the Inside of the Hole, Slightly Below the Shaft Surface. The Nominal Shear Stress Is $\tau_0 = TD/2J_{net}$, where J_{net} Is a Reduced Value of the Second Polar Moment of Area and Is Defined by

$$J_{net} = \frac{\pi A(D^4 - d^4)}{32}$$

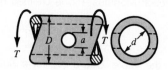

Values of A are Listed in the Table. Use $d = 0$ for a Solid Bar]

	d/D									
	0.9		0.8		0.6		0.4		0	
a/D	A	K_{ts}	A	K_{ts}	A	K_{ts}	A	K_{ts}	A	K_{ts}
0.05	0.96	1.78							0.95	1.77
0.075	0.95	1.82							0.93	1.71
0.10	0.94	1.76	0.93	1.74	0.92	1.72	0.92	1.70	0.92	1.68
0.125	0.91	1.76	0.91	1.74	0.90	1.70	0.90	1.67	0.89	1.64
0.15	0.90	1.77	0.89	1.75	0.87	1.69	0.87	1.65	0.87	1.62
0.175	0.89	1.81	0.88	1.76	0.87	1.69	0.86	1.64	0.85	1.60
0.20	0.88	1.96	0.86	1.79	0.85	1.70	0.84	1.63	0.82	1.58
0.25	0.87	2.00	0.82	1.86	0.81	1.72	0.80	1.63	0.79	1.54
0.30	0.80	2.18	0.78	1.97	0.77	1.76	0.75	1.63	0.74	1.51
0.35	0.77	2.41	0.75	2.09	0.72	1.81	0.69	1.63	0.68	1.47
0.40	0.72	2.67	0.71	2.25	0.68	1.89	0.64	1.63	0.63	1.44

Source: R. E. Peterson, *Stress Concentration Factors*, Wiley, New York, 1974, pp. 148, 244.

TABLE A-17

Preferred Sizes and Renard (R-Series) Numbers (When a Choice Can Be Made, Use One of these Sizes. However, Not All Parts or Items Are Available in All the Sizes Shown in the Table)

FRACTION OF INCHES

$\frac{1}{64}$, $\frac{1}{32}$, $\frac{1}{16}$, $\frac{3}{32}$, $\frac{1}{8}$, $\frac{5}{32}$, $\frac{3}{16}$, $\frac{1}{4}$, $\frac{5}{16}$, $\frac{3}{8}$, $\frac{7}{16}$, $\frac{1}{2}$, $\frac{9}{16}$, $\frac{5}{8}$, $\frac{11}{16}$, $\frac{3}{4}$, $\frac{7}{8}$, 1, $1\frac{1}{4}$, $1\frac{1}{2}$, $1\frac{3}{4}$, 2, $2\frac{1}{4}$, $2\frac{1}{2}$, $2\frac{3}{4}$, 3, $3\frac{1}{4}$, $3\frac{1}{2}$, $3\frac{3}{4}$, 4, $4\frac{1}{4}$, $4\frac{1}{2}$, $4\frac{3}{4}$, 5, $5\frac{1}{4}$, $5\frac{1}{2}$, $5\frac{3}{4}$, 6, $6\frac{1}{2}$, 7, $7\frac{1}{2}$, 8, $8\frac{1}{2}$, 9, $9\frac{1}{2}$, 10, $10\frac{1}{2}$, 11, $11\frac{1}{2}$, 12, $12\frac{1}{2}$, 13, $13\frac{1}{2}$, 14, $14\frac{1}{2}$, 15, $15\frac{1}{2}$, 16, $16\frac{1}{2}$, 17, $17\frac{1}{2}$, 18, $18\frac{1}{2}$, 19, $19\frac{1}{2}$, 20

DECIMAL INCHES

0.010, 0.012, 0.016, 0.020, 0.025, 0.032, 0.040, 0.05, 0.06, 0.08, 0.10, 0.12, 0.16, 0.20, 0.24, 0.30, 0.40, 0.50, 0.60, 0.80, 1.00, 1.20, 1.40, 1.60, 1.80, 2.0, 2.4, 2.6, 2.8, 3.0, 3.2, 3.4, 3.6, 3.8, 4.0, 4.2, 4.4, 4.6, 4.8, 5.0, 5.2, 5.4, 5.6, 5.8, 6.0, 7.0, 7.5, 8.0, 8.5, 9.0, 9.5, 10.0, 10.5, 11.0, 11.5, 12.0, 12.5, 13.0, 13.5, 14.0, 14.5, 15.0, 15.5, 16.0, 16.5, 17.0, 17.5, 18.0, 18.5, 19.0, 19.5, 20

MILLIMETERS

0.05, 0.06, 0.08, 0.10, 0.12, 0.16, 0.20, 0.25, 0.30, 0.40, 0.50, 0.60, 0.70, 0.80, 0.90, 1.0, 1.1, 1.2, 1.4, 1.5, 1.6, 1.8, 2.0, 2.2, 2.5, 2.8, 3.0, 3.5, 4.0, 4.5, 5.0, 5.5, 6.0, 6.5, 7.0, 8.0, 9.0, 10, 11, 12, 14, 16, 18, 20, 22, 25, 28, 30, 32, 35, 40, 45, 50, 60, 80, 100, 120, 140, 160, 180, 200, 250, 300

RENARD NUMBERS*

1st choice, R5:   1, 1.6, 2.5, 4, 6.3, 10

2d choice, R10:   1.25, 2, 3.15, 5, 8

3d choice, R20:   1.12, 1.4, 1.8, 2.24, 2.8, 3.55, 4.5, 5.6, 7.1, 9

4th choice, R40:   1.06, 1.18, 1.32, 1.5, 1.7, 1.9, 2.12, 2.36, 2.65, 3, 3.35, 3.75, 4.25, 4.75, 5.3, 6, 6.7, 7.5, 8.5, 9.5

*May be multiplied or divided by powers of 10.

TABLE A-18

Geometric Properties

PART 1   PROPERTIES OF SECTIONS

A = area

G = location of centroid

$I_x = \displaystyle\int x^2 \, dA$ = second moment of area about x axis

$I_{xy} = \displaystyle\int xy \, dA$ = mixed moment of area about x and y axes

$J_G = \displaystyle\int r^2 \, dA = \int (x^2 + y^2) \, dA = I_x + I_y$ = second polar moment of area about axis through G

$k_x^2 = I_x/A$ = squared radius of gyration about x axis

Rectangle

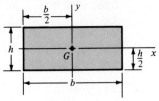

$A = bh \qquad I_x = \dfrac{bh^3}{12} \qquad I_y = \dfrac{b^3h}{12} \qquad I_{xy} = 0$

Circle

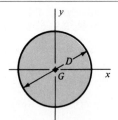

$A = \dfrac{\pi D^2}{4} \qquad I_x = I_y = \dfrac{\pi D^4}{64} \qquad I_{xy} = 0$

Hollow circle

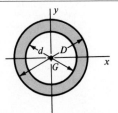

$A = \dfrac{\pi}{4}(D^2 - d^2) \qquad I_x = I_y = \dfrac{\pi}{64}(D^4 - d^4) \qquad I_{xy} = 0$

TABLE A-18
Geometric Properties
(Continued)

Right triangles

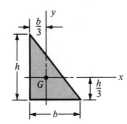

$$A = \frac{bh}{2} \qquad I_x = \frac{bh^3}{36} \qquad I_y = \frac{b^3h}{36} \qquad I_{xy} = -\frac{b^2h^2}{72}$$

Right triangles

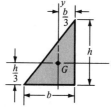

$$A = \frac{bh}{2} \qquad I_x = \frac{bh^3}{36} \qquad I_y = \frac{b^3h}{36} \qquad I_{xy} = \frac{b^2h^2}{72}$$

Quarter-circles

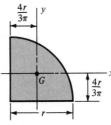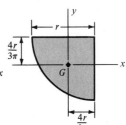

$$A = \frac{\pi r^2}{4} \qquad I_x = I_y = r^4\left(\frac{\pi}{16} - \frac{4}{9\pi}\right) \qquad I_{xy} = r^4\left(\frac{1}{8} - \frac{4}{9\pi}\right)$$

Quarter-circles

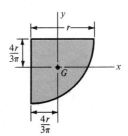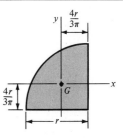

$$A = \frac{\pi r^2}{4} \qquad I_x = I_y = r^4\left(\frac{\pi}{16} - \frac{4}{9\pi}\right) \qquad I_{xy} = r^4\left(\frac{4}{9\pi} - \frac{1}{8}\right)$$

PART 2   PROPERTIES OF SOLIDS (ρ = DENSITY, WEIGHT/UNIT VOLUME)

Rods

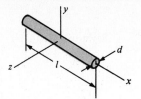

$$m = \frac{\pi d^2 \ell \rho}{4g} \qquad I_y = I_z = \frac{m\ell^2}{12}$$

Round disks

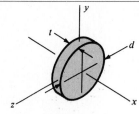

$$m = \frac{\pi d^2 t \rho}{4g} \qquad I_x = \frac{md^2}{8} \qquad I_y = I_z = \frac{md^2}{16}$$

Rectangular prisms

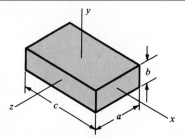

$$m = \frac{abc\rho}{g} \qquad I_x = \frac{m}{12}(a^2 + b^2) \qquad I_y = \frac{m}{12}(a^2 + c^2) \qquad I_z = \frac{m}{12}(b^2 + c^2)$$

Cylinders

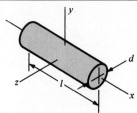

$$m = \frac{\pi d^2 \ell \rho}{4g} \qquad I_x = \frac{md^2}{8} \qquad I_y = I_z = \frac{m}{48}(3d^2 + 4\ell^2)$$

Hollow cylinders

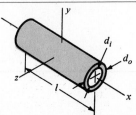

$$m = \frac{\pi \ell \rho}{4g} \qquad I_x = \frac{m}{8}(d_o^2 + d_i^2) \qquad I_y = I_z = \frac{m}{48}(3d_o^2 + 3d_i^2 + 4\ell^2)$$

TABLE A-19

American Standard Pipe

NOMINAL SIZE, in	OUTSIDE DIAMETER, in	THREADS PER INCH	WALL THICKNESS, in		
			STANDARD, NO. 40	EXTRA STRONG, NO. 80	DOUBLE EXTRA STRONG
$\frac{1}{8}$	0.405	27	0.070	0.098	
$\frac{1}{4}$	0.540	18	0.090	0.122	
$\frac{3}{8}$	0.675	18	0.093	0.129	
$\frac{1}{2}$	0.840	14	0.111	0.151	0.307
$\frac{3}{4}$	1.050	14	0.115	0.157	0.318
1	1.315	$11\frac{1}{2}$	0.136	0.183	0.369
$1\frac{1}{4}$	1.660	$11\frac{1}{2}$	0.143	0.195	0.393
$1\frac{1}{2}$	1.900	$11\frac{1}{2}$	0.148	0.204	0.411
2	2.375	$11\frac{1}{2}$	0.158	0.223	0.447
$2\frac{1}{2}$	2.875	8	0.208	0.282	0.565
3	3.500	8	0.221	0.306	0.615
$3\frac{1}{2}$	4.000	8	0.231	0.325	
4	4.500	8	0.242	0.344	0.690
5	5.563	8	0.263	0.383	0.768
6	6.625	8	0.286	0.441	0.884
8	8.625	8	0.329	0.510	0.895

TABLE A-20

Mechanical Properties of Some Hot-Rolled (HR) and Cold-Drawn (CD) Steels
[The Strengths Listed Are Estimated ASTM Minimum Values in the Size Range 18 to 32 mm ($\frac{3}{4}$ to $1\frac{1}{4}$ in). These Strengths Are Suitable for Use With the Design Factor Defined in Sec. 1-9, Provided the Materials Conform to ASTM A6 or A568 Requirements or Are Required in the Purchase Specifications. It is Well to Remember that a Numbering System Is Not a Specification. See Table 1-1 for Certain ASTM Steels]

1	2	3	4	5	6	7	8
UNS NO.	SAE AND/OR AISI NO.	PROCES-SING	TENSILE STRENGTH, MPa (kpsi)	YIELD STRENGTH, MPa (kpsi)	ELONGATION IN 2 in, %	REDUCTION IN AREA, %	BRINELL HARDNESS
G10060	1006	HR	300 (43)	170 (24)	30	55	86
		CD	330 (48)	280 (41)	20	45	95
G10100	1010	HR	320 (47)	180 (26)	28	50	95
		CD	370 (53)	300 (44)	20	40	105
G10150	1015	HR	340 (50)	190 (27.5)	28	50	101
		CD	390 (56)	320 (47)	18	40	111
G10180	1018	HR	400 (58)	220 (32)	25	50	116
		CD	440 (64)	370 (54)	15	40	126
G10200	1020	HR	380 (55)	210 (30)	25	50	111
		CD	470 (68)	390 (57)	15	40	131
G10300	1030	HR	470 (68)	260 (37.5)	20	42	137
		CD	520 (76)	440 (64)	12	35	149
G10350	1035	HR	500 (72)	270 (39.5)	18	40	143
		CD	550 (80)	460 (67)	12	35	163
G10400	1040	HR	520 (76)	290 (42)	18	40	149
		CD	590 (85)	490 (71)	12	35	170
G10450	1045	HR	570 (82)	310 (45)	16	40	163
		CD	630 (91)	530 (77)	12	35	179
G10500	1050	HR	620 (90)	340 (49.5)	15	35	179
		CD	690 (100)	580 (84)	10	30	197
G10600	1060	HR	680 (98)	370 (54)	12	30	201
G10800	1080	HR	770 (112)	420 (61.5)	10	25	229
G10950	1095	HR	830 (120)	460 (66)	10	25	248

Source: 1986 SAE Handbook, p. 2.15.

TABLE A-21

Mechanical Properties of Some Heat-Treated Steels
[These Are Typical Properties for Materials Normalized and Annealed. The Properties for Quenched and Tempered (Q&T) Steels Are from a Single Heat. Because of the Many Variables, the Properties Listed Might Be Considered Attainable but Should Not Be Treated as Average or as Minimum. In All Cases, Data Were Obtained from Specimens of Diameter 0.505 in, Machined from 1-in Rounds, and of Gauge Length 2 in. Unless Noted, All Specimens Were Oil-Quenched]

1	2	3	4	5	6	7	8
AISI NO.	TREATMENT	TEMPERATURE, °C (°F)	TENSILE STRENGTH, MPa (kpsi)	YIELD STRENGTH, MPa (kpsi)	ELONGATION, %	REDUCTION IN AREA, %	BRINELL HARDNESS
1030	Q&T*	205 (400)	848 (123)	648 (94)	17	47	495
	Q&T*	315 (600)	800 (116)	621 (90)	19	53	401
	Q&T*	425 (800)	731 (106)	579 (84)	23	60	302
	Q&T*	540 (1000)	669 (97)	517 (75)	28	65	255
	Q&T*	650 (1200)	586 (85)	441 (64)	32	70	207
	Normalized	925 (1700)	521 (75)	345 (50)	32	61	149
	Annealed	870 (1600)	430 (62)	317 (46)	35	64	137
1040	Q&T	205 (400)	779 (113)	593 (86)	19	48	262
	Q&T	425 (800)	758 (110)	552 (80)	21	54	241
	Q&T	650 (1200)	634 (92)	434 (63)	29	65	192
	Normalized	900 (1650)	590 (86)	374 (54)	28	55	170
	Annealed	790 (1450)	519 (75)	353 (51)	30	57	149
1050	Q&T*	205 (400)	1120 (163)	807 (117)	9	27	514
	Q&T*	425 (800)	1090 (158)	793 (115)	13	36	444
	Q&T*	650 (1200)	717 (104)	538 (78)	28	65	235
	Normalized	900 (1650)	748 (108)	427 (62)	20	39	217
	Annealed	790 (1450)	636 (92)	365 (53)	24	40	187
1060	Q&T	425 (800)	1080 (156)	765 (111)	14	41	311
	Q&T	540 (1000)	965 (140)	669 (97)	17	45	277
	Q&T	650 (1200)	800 (116)	524 (76)	23	54	229
	Normalized	900 (1650)	776 (112)	421 (61)	18	37	229
	Annealed	790 (1450)	626 (91)	372 (54)	22	38	179
1095	Q&T	315 (600)	1260 (183)	813 (118)	10	30	375
	Q&T	425 (800)	1210 (176)	772 (112)	12	32	363
	Q&T	540 (1000)	1090 (158)	676 (98)	15	37	321
	Q&T	650 (1200)	896 (130)	552 (80)	21	47	269
	Normalized	900 (1650)	1010 (147)	500 (72)	9	13	293
	Annealed	790 (1450)	658 (95)	380 (55)	13	21	192
1141	Q&T	315 (600)	1460 (212)	1280 (186)	9	32	415
	Q&T	540 (1000)	896 (130)	765 (111)	18	57	262
4130	Q&T*	205 (400)	1630 (236)	1460 (212)	10	41	467
	Q&T*	315 (600)	1500 (217)	1380 (200)	11	43	435
	Q&T*	425 (800)	1280 (186)	1190 (173)	13	49	380
	Q&T*	540 (1000)	1030 (150)	910 (132)	17	57	315
	Q&T*	650 (1200)	814 (118)	703 (102)	22	64	245
	Normalized	870 (1600)	670 (97)	436 (63)	25	59	197
	Annealed	865 (1585)	560 (81)	361 (52)	28	56	156
4140	Q&T	205 (400)	1770 (257)	1640 (238)	8	38	510
	Q&T	315 (600)	1550 (225)	1430 (208)	9	43	445

TABLE A-21 *(Continued)*

Mechanical Properties of Some Heat-Treated Steels
[These Are Typical Properties for Materials Normalized and Annealed. The Properties for Quenched and Tempered (Q&T) Steels Are from a Single Heat. Because of the Many Variables, the Properties Listed Might Be Considered Attainable but Should Not Be Treated as Average or as Minimum. In All Cases, Data Were Obtained from Specimens of Diameter 0.505 in, Machined from 1-in Rounds, and of Gauge Length 2 in. Unless Noted, All Specimens Were Oil-Quenched]

1 AISI NO.	2 TREATMENT	3 TEMPERATURE, °C (°F)	4 TENSILE STRENGTH, MPa (kpsi)	5 YIELD STRENGTH, MPa (kpsi)	6 ELONGATION, %	7 REDUCTION IN AREA, %	8 BRINELL HARDNESS
4140	Q&T	425 (800)	1250 (181)	1140 (165)	13	49	370
	Q&T	540 (1000)	951 (138)	834 (121)	18	58	285
	Q&T	650 (1200)	758 (110)	655 (95)	22	63	230
	Normalized	870 (1600)	1020 (148)	655 (95)	18	47	302
	Annealed	815 (1500)	655 (95)	417 (61)	26	57	197
4340	Q&T	315 (600)	1720 (250)	1590 (230)	10	40	486
	Q&T	425 (800)	1470 (213)	1360 (198)	10	44	430
	Q&T	540 (1000)	1170 (170)	1080 (156)	13	51	360
	Q&T	650 (1200)	965 (140)	855 (124)	19	60	280

*Water-quenched.

Source: ASM Metals Reference Book, 2d ed., American Society for Metals, Metals Park, Ohio, 1983.

TABLE A-22

Results of Tensile Tests of Some Metals*

NUMBER	MATERIAL	CONDITION	STRENGTH (TENSILE) YIELD, S_y, MPa (kpsi)	ULTIMATE, S_u, MPa (kpsi)	FRACTURE, σ_f, MPa (kpsi)	COEFFICIENT σ_0, MPa (kpsi)	STRAIN STRENGTH. EXPONENT m	FRACTURE STRAIN ε_f
1018	Steel	Annealed	220 (32.0)	341 (49.5)	628 (91.1)†	620 (90.0)	0.25	1.05
1144	Steel	Annealed	358 (52.0)	646 (93.7)	898 (130)†	992 (144)	0.14	0.49
1212	Steel	HR	193 (28.0)	424 (61.5)	729 (106)†	758 (110)	0.24	0.85
1045	Steel	Q&T 600°F	1520 (220)	1580 (230)	2380 (345)	1880 (273)†	0.041	0.81
4142	Steel	Q&T 600°F	1720 (250)	1930 (210)	2340 (340)	1760 (255)†	0.048	0.43
303	Stainless steel	Annealed	241 (35.0)	601 (87.3)	1520 (221)†	1410 (205)	0.51	1.16
304	Stainless steel	Annealed	276 (40.0)	568 (82.4)	1600 (233)†	1270 (185)	0.45	1.67
2011	Aluminum alloy	T6	169 (24.5)	324 (47.0)	325 (47.2)†	620 (90)	0.28	0.10
2024	Aluminum alloy	T4	296 (43.0)	446 (64.8)	533 (77.3)†	689 (100)	0.15	0.18
7075	Aluminum alloy	T6	542 (78.6)	593 (86.0)	706 (102)†	882 (128)	0.13	0.18

*Values from one or two heats and believed to be attainable using proper purchase specifications. The fracture strain may vary as much as 100 percent.
†Derived value.

Source: J. Datsko, "Solid Materials," chap. 7 in Joseph E. Shigley and Charles R. Mischke (eds.), *Standard Handbook of Machine Design,* McGraw-Hill, New York, 1986, pp. 7.47–7.50.

TABLE A-23

Mechanical Properties of Some Aluminum Alloys
[These Are *Typical* Properties for Sizes of About $\frac{1}{2}$ in; Similar Properties Can Be Obtained Using Proper Purchase Specifications. The Values Given for Fatigue Strength Correspond to $50(10^7)$ Cycles of Completely Reversed Stress. Aluminum Alloys Do Not Have an Endurance Limit. Yield Strengths Were Obtained by the 0.2 Percent Offset Method]

ALUMINUM ASSOCIATION NUMBER	TEMPER	STRENGTH			ELONGATION IN 2 in, %	BRINELL HARDNESS H_B
		YIELD, S_y, MPa (kpsi)	TENSILE, S_u, MPa (kpsi)	FATIGUE, S_f, MPa (kpsi)		
Wrought:						
2017	O	70 (10)	179 (26)	90 (13)	22	45
2024	O	76 (11)	186 (27)	90 (13)	22	47
	T3	345 (50)	482 (70)	138 (20)	16	120
3003	H12	117 (17)	131 (19)	55 (8)	20	35
	H16	165 (24)	179 (26)	65 (9.5)	14	47
3004	H34	186 (27)	234 (34)	103 (15)	12	63
	H38	234 (34)	276 (40)	110 (16)	6	77
5052	H32	186 (27)	234 (34)	117 (17)	18	62
	H36	234 (34)	269 (39)	124 (18)	10	74
Cast:						
319.0*	T6	165 (24)	248 (36)	69 (10)	2.0	80
333.0†	T5	172 (25)	234 (34)	83 (12)	1.0	100
	T6	207 (30)	289 (42)	103 (15)	1.5	105
355.0*	T6	172 (25)	241 (35)	62 (9)	3.0	80
	T7	248 (36)	262 (38)	62 (9)	0.5	85

*Sand casting.
†Permanent-mold casting.

TABLE A-24

Typical Properties of Gray Cast Iron
[The American Society for Testing Materials (ASTM) Numbering System for Gray Cast Iron Is Such that the Numbers Correspond to the *Minimum Tensile Strength* in kpsi. Thus an ASTM No. 20 Cast Iron Has a Minimum Tensile Strength of 20 kpsi. Note Particularly that the Tabulations Are *Typical Values*]

ASTM NUMBER	TENSILE STRENGTH S_{ut}, kpsi	COMPRESSIVE STRENGTH S_{uc}, kpsi	SHEAR MODULUS OF RUPTURE S_{su}, kpsi	MODULUS OF ELASTICITY, Mpsi		ENDURANCE LIMIT* S_e, kpsi	BRINELL HARDNESS H_B	FATIGUE STRESS-CONCENTRATION FACTOR K_f
				TENSION†	TORSION			
20	22	83	26	9.6–14	3.9–5.6	10	156	1.00
25	26	97	32	11.5–14.8	4.6–6.0	11.5	174	1.05
30	31	109	40	13–16.4	5.2–6.6	14	201	1.10
35	36.5	124	48.5	14.5–17.2	5.8–6.9	16	212	1.15
40	42.5	140	57	16–20	6.4–7.8	18.5	235	1.25
50	52.5	164	73	18.8–22.8	7.2–8.0	21.5	262	1.35
60	62.5	187.5	88.5	20.4–23.5	7.8–8.5	24.5	302	1.50

*Polished or machined specimens.
†The modulus of elasticity of cast iron in compression corresponds closely to the upper value in the range given for tension and is a more constant value than that for tension.

TABLE A-25

Decimal Equivalents of Wire and Sheet-Metal Gauges*
(All Sizes Are Given in Inches)

NAME OF GAUGE:	AMERICAN OR BROWN & SHARPE	BIRMINGHAM OR STUBS IRON WIRE	UNITED STATES STANDARD†	MANU-FACTURERS STANDARD	STEEL WIRE OR WASHBURN & MOEN	MUSIC WIRE	STUBS STEEL WIRE	TWIST DRILL
PRIN-CIPAL USE:	NONFERROUS SHEET, WIRE, AND ROD	TUBING, FERROUS STRIP, FLAT WIRE, AND SPRING STEEL	FERROUS SHEET AND PLATE, 480 lb/ft^3	FERROUS SHEET	FERROUS WIRE EXCEPT MUSIC WIRE	MUSIC WIRE	STEEL DRILL ROD	TWIST DRILLS AND DRILL STEEL
7/0	—	—	0.500	—	0.490 0			
6/0	0.580 0	—	0.468 75	—	0.461 5	0.004		
5/0	0.516 5	—	0.437 5	—	0.430 5	0.005		
4/0	0.460 0	0.454	0.406 25	—	0.393 8	0.006		
3/0	0.409 6	0.425	0.375	—	0.362 5	0.007		
2/0	0.364 8	0.380	0.343 75	—	0.331 0	0.008		
0	0.324 9	0.340	0.312 5	—	0.306 5	0.009		
1	0.289 3	0.300	0.281 25	—	0.283 0	0.010	0.227	0.228 0
2	0.257 6	0.284	0.265 625	—	0.262 5	0.011	0.219	0.221 0
3	0.229 4	0.259	0.25	0.239 1	0.243 7	0.012	0.212	0.213 0
4	0.204 3	0.238	0.234 375	0.224 2	0.225 3	0.013	0.207	0.209 0
5	0.181 9	0.220	0.218 75	0.209 2	0.207 0	0.014	0.204	0.205 5
6	0.162 0	0.203	0.203 125	0.194 3	0.192 0	0.016	0.201	0.204 0
7	0.144 3	0.180	0.187 5	0.179 3	0.177 0	0.018	0.199	0.201 0
8	0.128 5	0.165	0.171 875	0.164 4	0.162 0	0.020	0.197	0.199 0
9	0.114 4	0.148	0.156 25	0.149 5	0.148 3	0.022	0.194	0.196 0
10	0.101 9	0.134	0.140 625	0.134 5	0.135 0	0.024	0.191	0.193 5
11	0.090 74	0.120	0.125	0.119 6	0.120 5	0.026	0.188	0.191 0
12	0.080 81	0.109	0.109 357	0.104 6	0.105 5	0.029	0.185	0.189 0
13	0.071 96	0.095	0.093 75	0.089 7	0.091 5	0.031	0.182	0.185 0
14	0.064 08	0.083	0.078 125	0.074 7	0.080 0	0.033	0.180	0.182 0
15	0.057 07	0.072	0.070 312 5	0.067 3	0.072 0	0.035	0.178	0.180 0
16	0.050 82	0.065	0.062 5	0.059 8	0.062 5	0.037	0.175	0.177 0
17	0.045 26	0.058	0.056 25	0.053 8	0.054 0	0.039	0.172	0.173 0
18	0.040 30	0.049	0.05	0.047 8	0.047 5	0.041	0.168	0.169 5
19	0.035 89	0.042	0.043 75	0.041 8	0.041 0	0.043	0.164	0.166 0
20	0.031 96	0.035	0.037 5	0.035 9	0.034 8	0.045	0.161	0.161 0
21	0.028 46	0.032	0.034 375	0.032 9	0.031 7	0.047	0.157	0.159 0
22	0.025 35	0.028	0.031 25	0.029 9	0.028 6	0.049	0.155	0.157 0
23	0.022 57	0.025	0.028 125	0.026 9	0.025 8	0.051	0.153	0.154 0
24	0.020 10	0.022	0.025	0.023 9	0.023 0	0.055	0.151	0.152 0
25	0.017 90	0.020	0.021 875	0.020 9	0.020 4	0.059	0.148	0.149 5
26	0.015 94	0.018	0.018 75	0.017 9	0.018 1	0.063	0.146	0.147 0
27	0.014 20	0.016	0.017 187 5	0.016 4	0.017 3	0.067	0.143	0.144 0
28	0.012 64	0.014	0.015 625	0.014 9	0.016 2	0.071	0.139	0.140 5
29	0.011 26	0.013	0.014 062 5	0.013 5	0.015 0	0.075	0.134	0.136 0
30	0.010 03	0.012	0.012 5	0.012 0	0.014 0	0.080	0.127	0.128 5

TABLE A-25

Decimal Equivalents of Wire and Sheet-Metal Gauges
(All Sizes Are Given in Inches) *(Continued)*

NAME OF GAUGE:	AMERICAN OR BROWN & SHARPE	BIRMINGHAM OR STUBS IRON WIRE	UNITED STATES STANDARD†	MANU-FACTURERS STANDARD	STEEL WIRE OR WASHBURN & MOEN	MUSIC WIRE	STUBS STEEL WIRE	TWIST DRILL
PRIN-CIPAL USE:	NONFERROUS SHEET, WIRE, AND ROD	TUBING, FERROUS STRIP, FLAT WIRE, AND SPRING STEEL	FERROUS SHEET AND PLATE, 480 lb/ft³	FERROUS SHEET	FERROUS WIRE EXCEPT MUSIC WIRE	MUSIC WIRE	STEEL DRILL ROD	TWIST DRILLS AND DRILL STEEL
31	0.008 928	0.010	0.010 937 5	0.010 5	0.013 2	0.085	0.120	0.120 0
32	0.007 950	0.009	0.010 156 25	0.009 7	0.012 8	0.090	0.115	0.116 0
33	0.007 080	0.008	0.009 375	0.009 0	0.011 8	0.095	0.112	0.113 0
34	0.006 305	0.007	0.008 593 75	0.008 2	0.010 4	—	0.110	0.111 0
35	0.005 615	0.005	0.007 812 5	0.007 5	0.009 5	—	0.108	0.110 0
36	0.005 000	0.004	0.007 031 25	0.006 7	0.009 0	—	0.106	0.106 5
37	0.004 453	—	0.006 640 625	0.006 4	0.008 5	—	0.103	0.104 0
38	0.003 965	—	0.006 25	0.006 0	0.008 0	—	0.101	0.101 5
39	0.003 531	—	—	—	0.007 5	—	0.099	0.099 5
40	0.003 145	—	—	—	0.007 0	—	0.097	0.098 0

*Reproduced by courtesy of the Reynolds Metal Company. Specify sheet, wire, and plate by stating the gauge number, the gauge name, and the decimal equivalent in parentheses.

†Reflects present average unit weights of sheet steel.

TABLE A-26

Dimensions of Square and Hexagonal Bolts

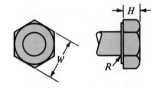

NOMINAL SIZE, in	SQUARE		REGULAR HEXAGONAL			HEAVY HEXAGONAL			STRUCTURAL HEXAGONAL		
	W	H	W	H	R_{min}	W	H	R_{min}	W	H	R_{min}
$\frac{1}{4}$	$\frac{3}{8}$	$\frac{11}{64}$	$\frac{7}{16}$	$\frac{11}{64}$	0.01						
$\frac{5}{16}$	$\frac{1}{2}$	$\frac{13}{64}$	$\frac{1}{2}$	$\frac{7}{32}$	0.01						
$\frac{3}{8}$	$\frac{9}{16}$	$\frac{1}{4}$	$\frac{9}{16}$	$\frac{1}{4}$	0.01						
$\frac{7}{16}$	$\frac{5}{8}$	$\frac{19}{64}$	$\frac{5}{8}$	$\frac{19}{64}$	0.01						
$\frac{1}{2}$	$\frac{3}{4}$	$\frac{21}{64}$	$\frac{3}{4}$	$\frac{11}{32}$	0.01	$\frac{7}{8}$	$\frac{11}{32}$	0.01	$\frac{7}{8}$	$\frac{5}{16}$	0.009
$\frac{5}{8}$	$\frac{15}{16}$	$\frac{27}{64}$	$\frac{15}{16}$	$\frac{27}{64}$	0.02	$1\frac{1}{16}$	$\frac{27}{64}$	0.02	$1\frac{1}{16}$	$\frac{25}{64}$	0.021
$\frac{3}{4}$	$1\frac{1}{8}$	$\frac{1}{2}$	$1\frac{1}{8}$	$\frac{1}{2}$	0.02	$1\frac{1}{4}$	$\frac{1}{2}$	0.02	$1\frac{1}{4}$	$\frac{15}{32}$	0.021
1	$1\frac{1}{2}$	$\frac{21}{32}$	$1\frac{1}{2}$	$\frac{43}{64}$	0.03	$1\frac{5}{8}$	$\frac{43}{64}$	0.03	$1\frac{5}{8}$	$\frac{39}{64}$	0.062
$1\frac{1}{8}$	$1\frac{11}{16}$	$\frac{3}{4}$	$1\frac{11}{16}$	$\frac{3}{4}$	0.03	$1\frac{13}{16}$	$\frac{3}{4}$	0.03	$1\frac{13}{16}$	$\frac{11}{16}$	0.062
$1\frac{1}{4}$	$1\frac{7}{8}$	$\frac{27}{32}$	$1\frac{7}{8}$	$\frac{27}{32}$	0.03	2	$\frac{27}{32}$	0.03	2	$\frac{25}{32}$	0.062
$1\frac{3}{8}$	$2\frac{1}{16}$	$\frac{29}{32}$	$2\frac{1}{16}$	$\frac{29}{32}$	0.03	$2\frac{3}{16}$	$\frac{29}{32}$	0.03	$2\frac{3}{16}$	$\frac{27}{32}$	0.062
$1\frac{1}{2}$	$2\frac{1}{4}$	1	$2\frac{1}{4}$	1	0.03	$2\frac{3}{8}$	1	0.03	$2\frac{3}{8}$	$\frac{15}{16}$	0.062
NOMINAL SIZE, mm											
M5	8	3.58	8	3.58	0.2						
M6			10	4.38	0.3						
M8			13	5.68	0.4						
M10			16	6.85	0.4						
M12			18	7.95	0.6	21	7.95	0.6			
M14			21	9.25	0.6	24	9.25	0.6			
M16			24	10.75	0.6	27	10.75	0.6	27	10.75	0.6
M20			30	13.40	0.8	34	13.40	0.8	34	13.40	0.8
M24			36	15.90	0.8	41	15.90	0.8	41	15.90	1.0
M30			46	19.75	1.0	50	19.75	1.0	50	19.75	1.2
M36			55	23.55	1.0	60	23.55	1.0	60	23.55	1.5

TABLE A-27

Dimensions of Hexagonal Cap Screws and Heavy Hexagonal Screws
(W = Width Across Flats; H = Height of Head; See Figure in Table A-26)

NOMINAL SIZE, in	MINIMUM FILLET RADIUS	TYPE OF SCREW CAP W	HEAVY W	HEIGHT H
$\frac{1}{4}$	0.015	$\frac{7}{16}$		$\frac{5}{32}$
$\frac{5}{16}$	0.015	$\frac{1}{2}$		$\frac{13}{64}$
$\frac{3}{8}$	0.015	$\frac{9}{16}$		$\frac{15}{64}$
$\frac{7}{16}$	0.015	$\frac{5}{8}$		$\frac{9}{32}$
$\frac{1}{2}$	0.015	$\frac{3}{4}$	$\frac{7}{8}$	$\frac{5}{16}$
$\frac{5}{8}$	0.020	$\frac{15}{16}$	$1\frac{1}{16}$	$\frac{25}{64}$
$\frac{3}{4}$	0.020	$1\frac{1}{8}$	$1\frac{1}{4}$	$\frac{15}{32}$
$\frac{7}{8}$	0.040	$1\frac{5}{16}$	$1\frac{7}{16}$	$\frac{35}{64}$
1	0.060	$1\frac{1}{2}$	$1\frac{5}{8}$	$\frac{39}{64}$
$1\frac{1}{4}$	0.060	$1\frac{7}{8}$	2	$\frac{25}{32}$
$1\frac{3}{8}$	0.060	$2\frac{1}{16}$	$2\frac{3}{16}$	$\frac{27}{32}$
$1\frac{1}{2}$	0.060	$2\frac{1}{4}$	$2\frac{3}{8}$	$\frac{15}{16}$
NOMINAL SIZE, mm				
M5	0.2	8		3.65
M6	0.3	10		4.15
M8	0.4	13		5.50
M10	0.4	16		6.63
M12	0.6	18	21	7.76
M14	0.6	21	24	9.09
M16	0.6	24	27	10.32
M20	0.8	30	34	12.88
M24	0.8	36	41	15.44
M30	1.0	46	50	19.48
M36	1.0	55	60	23.38

TABLE A-28

Dimensions of Hexagonal Nuts

NOMINAL SIZE, in	WIDTH W	HEIGHT H		
		REGULAR HEXAGONAL	THICK OR SLOTTED	JAM
$\frac{1}{4}$	$\frac{7}{16}$	$\frac{7}{32}$	$\frac{9}{32}$	$\frac{5}{32}$
$\frac{5}{16}$	$\frac{1}{2}$	$\frac{17}{64}$	$\frac{21}{64}$	$\frac{3}{16}$
$\frac{3}{8}$	$\frac{9}{16}$	$\frac{21}{64}$	$\frac{13}{32}$	$\frac{7}{32}$
$\frac{7}{16}$	$\frac{11}{16}$	$\frac{3}{8}$	$\frac{29}{64}$	$\frac{1}{4}$
$\frac{1}{2}$	$\frac{3}{4}$	$\frac{7}{16}$	$\frac{9}{16}$	$\frac{5}{16}$
$\frac{9}{16}$	$\frac{7}{8}$	$\frac{31}{64}$	$\frac{39}{64}$	$\frac{5}{16}$
$\frac{5}{8}$	$\frac{15}{16}$	$\frac{35}{64}$	$\frac{23}{32}$	$\frac{3}{8}$
$\frac{3}{4}$	$1\frac{1}{8}$	$\frac{41}{64}$	$\frac{13}{16}$	$\frac{27}{64}$
$\frac{7}{8}$	$1\frac{5}{16}$	$\frac{3}{4}$	$\frac{29}{32}$	$\frac{31}{64}$
1	$1\frac{1}{2}$	$\frac{55}{64}$	1	$\frac{35}{64}$
$1\frac{1}{8}$	$1\frac{11}{16}$	$\frac{31}{32}$	$1\frac{5}{32}$	$\frac{39}{64}$
$1\frac{1}{4}$	$1\frac{7}{8}$	$1\frac{1}{16}$	$1\frac{1}{4}$	$\frac{23}{32}$
$1\frac{3}{8}$	$2\frac{1}{16}$	$1\frac{11}{64}$	$1\frac{3}{8}$	$\frac{25}{32}$
$1\frac{1}{2}$	$2\frac{1}{4}$	$1\frac{9}{32}$	$1\frac{1}{2}$	$\frac{27}{32}$

NOMINAL SIZE, mm				
M5	8	4.7	5.1	2.7
M6	10	5.2	5.7	3.2
M8	13	6.8	7.5	4.0
M10	16	8.4	9.3	5.0
M12	18	10.8	12.0	6.0
M14	21	12.8	14.1	7.0
M16	24	14.8	16.4	8.0
M20	30	18.0	20.3	10.0
M24	36	21.5	23.9	12.0
M30	46	25.6	28.6	15.0
M36	55	31.0	34.7	18.0

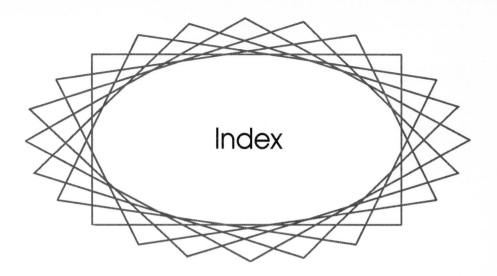

Index